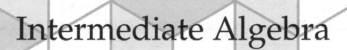

Intermediate Algebra

Richard Rusczyk
Mathew Crawford
Art of Problem Solving

Published by: AoPS Incorporated
                 P.O. Box 2185
                 Alpine, CA 91903-2185
                 (619) 659-1612
                 books@artofproblemsolving.com

ISBN-13: 978-1-934124-04-8

Visit the Art of Problem Solving website at http://www.artofproblemsolving.com

Cover image designed by Vanessa Rusczyk using KaleidoTile software.

Printed in the United States of America.

Second Printing 2010.

How to Use This Book

Organization

You can think of this book as divided into three sections:

- *Chapters 1 through 4.* If you completed Art of Problem Solving's *Introduction to Algebra*, then much of the first four chapters will be review. Don't just skip these four chapters; there is some new material, and some of the review material is extended to more challenging problems. If you have not completed *Introduction to Algebra*, then you should have experience factoring quadratic expressions before starting this book. If you have completed a typical Algebra 1 course in the United States, you may be ready for this text, but the first four chapters might not be review. Study them closely, and if they give you great difficulty, you may wish to study those topics in greater depth in *Introduction to Algebra*.

- *Chapters 5 through 16.* These chapters form the core of the book. Chapter 5 is the least important of the first 16 chapters of the book; much of the material can be read for a qualitative understanding, rather than for detailed study. Section 5.3 is the most important section of this chapter; the remaining sections can be read lightly. All students who use this book should study chapters 6 through 16 closely.

- *Chapters 17 through 20.* These chapters are primarily intended for those students preparing for mathematics competitions, or for those who want to pursue more advanced study in the subjects of these chapters.

Learn by Solving Problems

We believe that the best way to learn mathematics is by solving problems. Lots and lots of problems. In fact, we believe that the best way to learn mathematics is to try to solve problems that you don't know how to do. When you discover something on your own, you'll understand it much better than if someone had just told you.

Most of the sections of this book begin with several problems. The solutions to these problems will be covered in the text, but try to solve the problems *before* reading the section. If you can't solve some of the problems, that's OK, because they will all be fully solved as you read the section. Even if you can solve all of the problems, it's still important to read the section, both to make sure that your solutions are correct, and also because you may find that the book's solutions are simpler or easier to understand than your own.

Explanation of Icons

Throughout the book, you will see various shaded boxes and icons.

> **Concept:** This will be a general problem-solving technique or strategy. These are the "keys" to becoming a better problem solver!

> **Important:** This will be something important that you should learn. It might be a formula, a solution technique, or a caution.

> **WARNING!!** Beware if you see this box! This will point out a common mistake or pitfall.

> **Sidenote:** This box will contain material that, although interesting, is not part of the main material of the text. It's OK to skip over these boxes, but if you read them, you might learn something interesting!

> **Bogus Solution:** Just like the impossible cube shown to the left, there's something wrong with any "solution" that appears in this box.

> **Proof:** Some proofs in the text will be presented in boxes like this one. These are intended to be examples that the students can use as guides for how to write their own proofs.

Exercises, Review Problems, and Challenge Problems

Most sections end with several **Exercises**. These will test your understanding of the material that was covered in the section. You should try to solve *all* of the exercises. Exercises marked with a ★ are more difficult.

Each chapter concludes with many **Review Problems** and **Challenge Problems**. The Review Problems test your basic understanding of the material covered in the chapter. Your goal should be to solve most or all of the Review Problems for every chapter—if you're unable to do this, it means that you haven't yet mastered the material, and you should probably go back and read the chapter again.

The Challenge Problems are generally more difficult than the other problems in the book, and will really test your mastery of the material. Some of them are very hard—the hardest ones are marked with a ★. Don't expect to be able to solve all of the Challenge Problems on your first try—these are difficult problems even for experienced problem solvers. If you are able to solve a large number of Challenge Problems, then congratulations, you are on your way to becoming an expert problem solver!

Hints

Many problems come with one or more hints. You can look up the hints in the Hints section in the back of the book. The hints are numbered in random order, so that when you're looking up a hint to a problem you don't accidentally glance at the hint to the next problem at the same time.

It is very important that you first try to solve each problem without resorting to the hints. Only after you've seriously thought about a problem and are stuck should you seek a hint. Also, for problems which have multiple hints, use the hints one at a time; don't go to the second hint until you've thought about the first one.

Solutions

The solutions to all of the Exercises, Review Problems, and Challenge Problems are in the separate solutions book. If you are using this textbook in a regular school class, then your teacher may decide not to make this solutions book available to you, and instead present the solutions him/herself. However, if you are using this book to learn on your own, then you probably have a copy of the solutions book, in which case there are some very important things to keep in mind:

1. Make a serious attempt to solve each problem before looking at the solution. Don't use the solutions book as a crutch to avoid really thinking about the problem first. You should think *hard* about a problem before deciding to look at the solution. On the other hard, after you've thought hard about a problem, don't feel bad about looking at the solution if you're really stuck.

2. After you solve a problem, it's usually a good idea to read the solution, even if you think you know how to solve the problem. The solutions book might show you a quicker or more concise way to solve the problem, or it might have a completely different solution method than yours.

3. If you have to look at the solution in order to solve a problem, make a note of that problem. Come back to it in a week or two to make sure that you are able to solve it on your own, without resorting to the solution.

Resources

Here are some other good resources for you to further pursue your study of mathematics:

- The Art of Problem Solving Intermediate Series of texts. This series of texts will form a full curriculum for high-performing students in grades 8-12. In addition to this *Intermediate Algebra* text, the text *Intermediate Counting & Probability* is also available. More titles will be added to this series in 2009-2011.

- *The Art of Problem Solving* books, by Sandor Lehoczky and Richard Rusczyk. Whereas the book that you're reading right now will go into great detail of one specific subject area—algebra—*the Art of Problem Solving* books cover a wide range of problem solving topics across many different areas of mathematics.

- The www.artofproblemsolving.com website. The publishers of this book are also the webmasters of the Art of Problem Solving website, which contains many resources for students:

  - a discussion forum
  - online classes
  - resource lists of books, contests, and other websites
  - a LaTeX tutorial
  - a math and problem solving Wiki
  - and much more!

- You can hone your problem solving skills (and perhaps win prizes!) by participating in various math contests. For middle school students in the United States, the major contests are MOEMS, MATHCOUNTS, and the AMC 8. For U.S. high school students, some of the best-known contests are the AMC/AIME/USAMO series of contests (which are used to choose the U.S. team for the International Mathematical Olympiad), the American Regions Math League (ARML), the Mandel-brot Competition, the Harvard-MIT Mathematics Tournament, and the USA Mathematical Talent Search. More details about these contests are on page vii, and links to these and many other contests are available on the Art of Problem Solving website.

A Note to Teachers

We believe students learn best when they are challenged with hard problems that at first they may not know how to do. This is the motivating philosophy behind this book.

Rather than first introducing new material and then giving students exercises, we present problems at the start of each section that students should try to solve *before* new material is presented. The goal is to get students to discover the new material on their own. Often, complicated problems are broken into smaller parts, so that students can discover new techniques one piece at a time. After the problems, new material is formally presented in the text, and full solutions to each problem are explained, along with problem-solving strategies.

We hope that teachers will find that their stronger students will discover most of the material in this book on their own by working through the problems. Other students may learn better from a more traditional approach of first seeing the new material, then working the problems. Teachers have the flexibility to use either approach when teaching from this book.

Sections marked with a ⋆ contain supplementary material that may be safely skipped. In general, chapters are not equal in length, so different chapters may take different amounts of classroom time.

Links

The Art of Problem Solving website has a page containing links to websites with content relating to material in this book, as well as an errata list for the book. This page can be found at:

http://www.artofproblemsolving.com/BookLinks/IntermAlgebra/links.php

---

Extra! Occasionally, you'll see a box like this at the bottom of a page. This is an "Extra!" and
➠➠➠➠ might be a quote, some biographical or historical background, or perhaps an interesting
idea to think about.

Acknowledgements

Contests

We would like to thank the following contests for allowing us to use a wide selection of their problems in this book:

- The **American Mathematics Competitions**, a series of contests for U.S. middle and high school students. The **AMC 8**, **AMC 10**, and **AMC 12** contests are multiple-choice tests that are taken by over 400,000 students every year. Top scorers on the AMC 10 and AMC 12 are invited to take the **American Invitational Mathematics Examination (AIME)**, which is a more difficult, short-answer contest. Approximately 10,000 students every year participate in the AIME. Then, based on the results of the AMC and AIME contests, about 500 students are invited to participate in the **USA Mathematical Olympiad (USAMO)**, a 2-day, 9-hour examination in which each student must show all of his or her work. Results from the USAMO are used to invite a number of students to the Math Olympiad Summer Program, at which the U.S. team for the **International Mathematical Olympiad (IMO)** is chosen. More information about the AMC contests can be found on the AMC website at www.unl.edu/amc. (Note: Problems marked **AHSME** are from the American High School Math Exam, the predecessor of the AMC 12.)

- The **Mandelbrot Competition**, which was founded in 1990 by Sandor Lehoczky, Richard Rusczyk, and Sam Vandervelde. The aim of the Mandelbrot Competition is to provide a challenging, engaging mathematical experience that is both competitive and educational. Students compete both as individuals and in teams. The Mandelbrot Competition is offered at the national level for more advanced students and at the regional level for less experienced problem solvers. More information can be found at www.mandelbrot.org.

- The **USA Mathematical Talent Search (USAMTS)**, which was founded in 1989 by Professor George Berzsenyi. The USAMTS is a free mathematics competition open to all United States middle and high school students. As opposed to most mathematics competitions, the USAMTS allows students a full month to work out their solutions. Carefully written justifications are required for each problem. More information is available at www.usamts.org.

- The **American Regions Math League (ARML)**, which was founded in 1976. The annual ARML competition brings together nearly 2,000 of the nation's finest students. They meet, compete against, and socialize with one another, forming friendships and sharpening their mathematical skills. The contest is written for high school students, although some exceptional junior high students attend each year. The competition consists of several events, which include a team round, a power question (in which a team solves proof-oriented questions), an individual round, and two relay rounds. More information is available at www.arml.com. Some questions from the **New York State Math League (NYSML)** also appear in this text. NYSML is a New York statewide competition that inspired the start of ARML.

- The **Harvard-MIT Mathematics Tournament**, which is an annual math tournament for high school students, held at MIT and at Harvard in alternating years. It is run exclusively by MIT and Harvard students, most of whom themselves participated in math contests in high school. More information is available at web.mit.edu/hmmt/.

Some other sources for problems in the book are listed below.

MATHCOUNTS, the premier national math contest for U.S. middle school students.

The William Lowell Putnam Competition, an annual math competition for undergraduate students in North America.

Mathematical journals *Mathematics & Informatics Quarterly* (M&IQ), *Mathematical Mayhem*, and *Crux Mathematicorum*.

Various Canadian national competitions, such as the Canadian Invitational Mathematics Contest (CIMC), the Canadian Open Mathematics Challenge (COMC), the Canadian Mathematics Olympiad (CMO), and contests run by the Centre for Education in Mathematics and Computing (CEMC).

The Duke Math Meet and the Michigan Mathematics Prize Competition (MMPC).

Problems from various countries' national math contests. These problems are cited by country name in the text.

How We Wrote This Book

This book was written using the LaTeX document processing system. We thank the authors of the various LaTeX packages that we used while preparing this book, and also the authors of *The LaTeX Companion* for writing a reference book that is not only thorough but also very readable. The diagrams were prepared using METAPOST and Asymptote.

About Us

Even more than our other texts, this book is a collaborative effort of the staff of the Art of Problem Solving. Richard Rusczyk was the lead author of the text and primary editor of the solutions manual. Mathew Crawford prepared a draft of some portions of the text, and thorough notes on other portions. He also selected and organized many of the problems in the text, and wrote a draft of many of the solutions. Naoki Sato was the lead author of the solutions manual, and selected and organized many of the problems in the text. David Patrick read several drafts of the text and solutions and made many, many helpful suggestions. The following people also made contributions through proofreading, creating diagrams, suggesting extra material for the textbook, or writing problems for the text: Tom Belulovich, Yakov Berchenko-Kogan, Tim Black, Greg Brockman, Chris Chang, Lisa Davis, Kenan Diab, Larry Evans, Mike Kling, Joseph Laurendi, Daniel Li, Linda Liu, Sean Markan, Maria Monks, Jeff Nanney, Dimitar Popov, Tianren Qi, Adrian Sanborn, Nathan Savir, Beth Schaffer, Arnav Tripathy, Philip Tynan, Valentin Vornicu, Samson Zhou, and Olga Zverovich. Vanessa Rusczyk designed the cover.

Contents

4 Quadratics 79

5 Conics 106

6 Polynomial Division 162

7 Polynomial Roots Part I 193

13   Exponents and Logarithms   413

14   Radicals   456

15   Special Classes of Functions   476

16   Piecewise Defined Functions   508

CONTENTS

For Vanessa's casita. –RR

Extra!
➡➡➡➡

In this text, we use as the chapter headings images generated from a specific **cellular automaton** that is often referred to as **Rule 110**. A cellular automaton is a grid of cells in which some cells are initially assigned a value. The cellular automaton also has a set of rules that are used to assign values to subsequent cells. As an example, we'll explain Rule 110 in more detail.

In Rule 110, each **generation** of cells consists of a row of cells, and the value of each cell is either 1 or 0. We color a cell white if its value is 0 and black if its value is 1. Suppose our initial generation consists of a single black cell in an otherwise infinite row of white cells. Below, we show a small snippet of this generation that contains the black cell:

The next generation of the cellular automaton appears in the next row. The value of each cell in that row is determined by the values of the three cells directly above it. There are $2^3 = 8$ possible combinations of values for three consecutive cells, which we label A through H as follows:

We define a rule for the cellular automaton by determining what each combination of three cells produces. Rule 110 is defined as follows:

So, for example, combination A tells us that if there are three black cells in a row in one generation, then the cell directly below the center cell of these three black cells is white (that is, has value 0). Recalling that black cells have value 1 and white cells have value 0, we can read the results of the 8 possible combinations above as 01101110. Evaluating this as a binary number, we find $01101110_2 = 110$, which is where Rule 110 gets its name.

Now, let's see what happens when we apply Rule 110 to our first generation:

Most of the cells in the second generation are below three white cells in the first generation. Rule H tells us that such cells must be white. There are only three cells in the second generation that are not below three white cells; these are marked (1), (2) and (3) above. Reading left to right, cell (1) is below cells that read white-white-black. So, we apply Rule G to see that cell (1) is black. Similarly, the cells above cell (2) are white-black-white, so Rule F tells us to color it black. Finally, the cells above cell (3) are black-white-white, so Rule D tells us cell (3) is white.

All the rest of the cells in the second generation are white, and therefore our first two generations are:

(Again, the generations continue infinitely to the left and right, but all those cells are white, so they're not too interesting.)

Continuing in this way, we produce the next 10 generations:

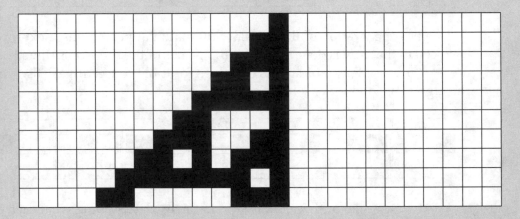

As chapter headings in this book, we use various snippets of the subsequent generations of Rule 110. Rule 110 has been closely studied by mathematicians and computer scientists. **Stephen Wolfram** conjectured that Rule 110 is **Turing complete**, which loosely means that it can be adapted to replicate the logic of any possible computer. This conjecture was proved by **Matthew Cook** while working with Wolfram, and the proof was published in 2004 in the journal *Complex Systems*.

On the website cited on page vi, you'll find links to more webpages with information about cellular automata in general, and Rule 110 in specific, including many images that display the intricate structure of Rule 110.

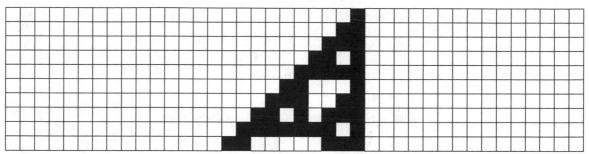

It all began in the remote past, with a simple turn of thought from the declarative to the interrogative, from "this plus this equals this" to "this plus what equals this?" – John Derbyshire

CHAPTER **1**

Basic Techniques for Solving Equations

We'll start by reviewing some basic techniques for solving linear equations. As we review these tactics, we'll show how these same strategies can also be used to solve inequalities and more complex equations.

1.1 Isolation

One of the most straightforward ways to solve a one-variable equation or inequality is simply to manipulate it until the variable is alone on one side and a constant is on the other side. In other words, we **isolate** the variable.

Problems

Problem 1.1: Solve the following equations:

(a) $3r - 2 + 2r = 7 - r + 3$

(b) $\dfrac{2t}{3} + 7 = 3(t - 9)$

Problem 1.2: Solve each of the following inequalities:

(a) $2 - 4x \geq 10$

(b) $7 - 3t < 5 - 2t$

(c) $3 - 3x < 4x + 7 \leq 8 - x$

Problem 1.3: Find all x such that $2x^2 + 5 = x^2 + 18$.

Problem 1.4: Find all z such that $\dfrac{1}{\sqrt{z-3}-1} + 2 = \dfrac{3}{\sqrt{z-3}-1}$.

Problem 1.5: In this problem, we find all values of c such that the equation

$$\frac{3}{2 - \frac{1}{x}} = c$$

has no solutions for x.

(a) Solve the equation for x in terms of c.

(b) Find all values of c for which there is no solution for x to the equation you found in part (a).

(c) What values of x cause division by 0 in the original equation? To what values of c do these values of x correspond in the equation you found in part (a)?

(d) List all values of c for which there is no solution for x to the original equation.

Problem 1.1: Solve the following equations:

(a) $3r - 2 + 2r = 7 - r + 3$ (b) $\frac{2t}{3} + 7 = 3(t - 9)$

Solution for Problem 1.1:

(a) Simplifying both sides gives $5r - 2 = 10 - r$. We add r to both sides to give $6r - 2 = 10$, then add 2 to both sides to give $6r = 12$. Finally, we divide by 6 to find $r = 2$.

We can test our answer by substituting $r = 2$ back into the equation. On the left we have $3r - 2 + 2r = 3 \cdot 2 - 2 + 2 \cdot 2 = 8$, and on the right we have $7 - r + 3 = 7 - 2 + 3 = 8$. So, the equation is satisfied when $r = 2$.

> **Important:** Whenever you solve an equation, you can check an answer by substituting it back into the original equation.

We won't do this on every problem in the text to save space, but it's a very useful habit to develop to prevent errors.

(b) Expanding the right side gives $\frac{2t}{3} + 7 = 3t - 27$. Subtracting $3t$ and subtracting 7 from both sides gives $-\frac{7t}{3} = -34$. Multiplying both sides by $-\frac{3}{7}$ gives us $t = \frac{102}{7}$.

□

Just as we isolate the variable to solve a linear equation, we isolate the variable to solve a linear inequality.

Problem 1.2: Solve each of the following inequalities:

(a) $2 - 4x \geq 10$ (b) $7 - 3t < 5 - 2t$ (c) $3 - 3x < 4x + 7 \leq 8 - x$

Solution for Problem 1.2:

(a) What's wrong with the following solution:

> **Bogus Solution:** Subtracting 2 from both sides gives $-4x \geq 8$. Dividing both sides by -4 gives $x \geq -2$ as our solution.

We see that we've made a mistake when we test our answer. If $x \geq -2$ is the solution, then $x = 0$ should satisfy the original inequality (because 0 is greater than -2). However, if $x = 0$, then it is not true that $2 - 4x \geq 10$. So, we've made a mistake.

Our mistake is that we forgot to reverse the direction of the inequality sign when we divided by -4. When we divide $-4x \geq 8$ by -4, we have $x \leq -2$.

> **WARNING!!** If we multiply or divide an inequality by a negative number, we must reverse the direction of the inequality sign.

(b) Adding $3t$ to both sides and subtracting 5 from both sides gives us $2 < t$. It's common to write the variable first when writing the solution to an equation or inequality, so we might also write our answer to this problem as $t > 2$.

> **WARNING!!** The inequality $2 < t$ is *not the same* as $t < 2$. If we rewrite an inequality by swapping the sides of the inequality, we must reverse the direction of the inequality sign. So, $2 < t$ is the same as $t > 2$.

(c) We are given three expressions linked together in what we call an **inequality chain**. Any value of x that satisfies the inequality chain must satisfy both $3 - 3x < 4x + 7$ and $4x + 7 \leq 8 - x$. Isolating x in $3 - 3x < 4x + 7$ gives us $-\frac{4}{7} < x$, while isolating x in $4x + 7 \leq 8 - x$ gives $x \leq \frac{1}{5}$. Therefore, we must have both $-\frac{4}{7} < x$ and $x \leq \frac{1}{5}$, which we can write as $-\frac{4}{7} < x \leq \frac{1}{5}$.

□

We say that an inequality is **strict** if the two sides cannot be equal, and that it is **nonstrict** if they can be equal. So, $t > 2$ is a strict inequality, while $x \geq -2$ is a nonstrict inequality.

Solutions to inequality problems are sometimes written using **interval notation**. For example, we can denote "all real numbers greater than 3 and less than 5" by the interval $(3, 5)$. We use "(" and ")" to indicate that the values 3 and 5 are not included. Notice that we don't write $(5, 3)$; the lesser end of the interval always comes first. To use this notation to indicate the solutions to the inequality $3 < r < 5$, we write $r \in (3, 5)$, where "$r \in$" means "r is in," and $(3, 5)$ indicates the real numbers greater than 3 and less than 5, as we just described.

To include a boundary value in the interval, we use "[" for the lower bound and "]" for the upper bound. For example, the statement $x \in (-3, 5]$ means $-3 < x \leq 5$ and $y \in [-12.2, 0]$ means $-12.2 \leq y \leq 0$.

Finally, to indicate an interval that has either no upper bound or no lower bound (or neither), we use the ∞ symbol. For example, we write $t > -3$ as $t \in (-3, +\infty)$. The "-3" part indicates that no numbers equal to -3 or lower than -3 are in the interval. The "$+\infty$)" part indicates that the interval continues forever in the positive direction. That is, there is no upper bound. So, the interval $(-3, +\infty)$ is all real numbers greater than -3. Similarly, the statement $w \in (-\infty, -2]$ is the same as $w \leq -2$. Notice that we always use "(" with $-\infty$ and ")" with $+\infty$, instead of "[" and "]". In both cases, there is not a boundary value to include in the interval, so we must use "(" or ")" instead of "[" or "]".

To write all real numbers in interval notation, we can write $(-\infty, +\infty)$. We often use the symbol $\mathbb{R}$ to denote "all real numbers." Therefore, we typically write the interval $(-\infty, \infty)$ as "$\mathbb{R}$", so $x \in \mathbb{R}$ means "x is a real number."

Problem 1.3: Find all x such that $2x^2 + 5 = x^2 + 18$.

Solution for Problem 1.3: Just as we isolate the variable to solve a linear equation, we isolate x^2 in this equation by subtracting x^2 from both sides, then subtracting 5 from both sides. This gives us $x^2 = 13$. We then take the square root of both sides to find $x = \pm\sqrt{13}$. $\square$

WARNING!! When taking the square root of both sides of an equation, we must not forget negative values. Specifically, if we have $x^2 = a^2$, then we have $x = \pm a$, where the $\pm$ sign indicates that x can equal either a or $-a$.

Here's another example:

Problem 1.4: Find all z such that $\dfrac{1}{\sqrt{z-3}-1} + 2 = \dfrac{3}{\sqrt{z-3}-1}$.

Solution for Problem 1.4: First, we notice that the denominators of the fractions are the same. So, we can combine the fractions by subtracting $1/(\sqrt{z-3}-1)$ from both sides, which gives us

$$2 = \frac{2}{\sqrt{z-3}-1}.$$

We get rid of the fraction by multiplying both sides by $\sqrt{z-3}-1$, which gives us $2\sqrt{z-3}-2 = 2$. Adding 2 to both sides, then dividing by 2 to isolate $\sqrt{z-3}$ gives us $\sqrt{z-3} = 2$. We get rid of the square root sign by squaring both sides, which yields $z - 3 = 4$, so $z = 7$.

WARNING!! Whenever we square an equation as a step in solving it, we have to check that our solutions are valid. If the solution to the squared equation does not satisfy our original equation, then the solution is called **extraneous**. An extraneous solution is not a valid solution to the original equation.

Here, if $z = 7$, we have $\sqrt{z-3}-1 = 2-1 = 1$, so our equation reads $1 + 2 = 3$, which is true. So, $z = 7$ is the only solution to the equation. $\square$

To see how extraneous solutions might occur, consider the equation $\sqrt{x} + 5 = 2$. Subtracting 5 gives $\sqrt{x} = -3$, which clearly has no solution. However, if we square $\sqrt{x} = -3$, we produce $x = 9$, which doesn't satisfy $\sqrt{x} + 5 = 2$. So, $x = 9$ is extraneous and there is no solution to $\sqrt{x} + 5 = 2$. The problem arises when we take the square root of $x = 9$ to evaluate $\sqrt{x}$ for our original equation—this gives us $\sqrt{x} = 3$, not $\sqrt{x} = -3$, because we define $\sqrt{x}$ for positive numbers x to be the *positive* number whose square is x.

One key step in each of the last two problems was isolating an expression. In Problem 1.3, we isolated x^2, and in Problem 1.4, we isolated $\sqrt{z-3}$.

Concept: Just as we can solve a linear equation by isolating the variable, we can often solve more complex equations by isolating an expression within the equation.

Problem 1.5: Find all values of c such that the equation

$$\frac{3}{2-\frac{1}{x}} = c.$$

has no solutions for x.

Solution for Problem 1.5: We want to know when there are no solutions for x, so we solve the equation for x in terms of c. Then, we can use the resulting expression to determine what values of c fail to give us a value for x.

Multiplying both sides of the equation by $2-\frac{1}{x}$ gives us $3 = 2c - \frac{c}{x}$. We isolate the term with x by adding $\frac{c}{x} - 3$ to both sides, which gives $\frac{c}{x} = 2c - 3$. Taking the reciprocal of both sides gives $\frac{x}{c} = \frac{1}{2c-3}$, and multiplying both sides by c gives

$$x = \frac{c}{2c-3}.$$

We cannot have $c = \frac{3}{2}$, because we cannot divide by 0. It appears that for any other value of c, we can use $x = \frac{c}{2c-3}$ to find the solution for x that results for that value of c. But we have to be careful—we cannot have a value of x that results in dividing by 0 in our original equation. So, we cannot have $x = 0$, and we cannot have $2 - \frac{1}{x} = 0$. The latter only occurs if $x = \frac{1}{2}$. Therefore, we must check for values of c in $x = \frac{c}{2c-3}$ that give $x = 0$ or $x = \frac{1}{2}$.

If $x = 0$ in $x = \frac{c}{2c-3}$, then we must have $c = 0$, so it appears that we cannot have $c = 0$ in our original equation. Indeed, letting $c = 0$ gives us

$$\frac{3}{2-\frac{1}{x}} = 0,$$

which clearly has no solution, because the numerator is not 0 for any x. If $x = \frac{1}{2}$ in $x = \frac{c}{2c-3}$, then we have $\frac{1}{2} = \frac{c}{2c-3}$. Cross-multiplying gives $2c - 3 = 2c$, which has no solutions. Therefore, there are no values of c for which $x = \frac{1}{2}$.

We conclude that the only values of c for which there is no solution for x in the original equation are $c = 0$ and $c = \frac{3}{2}$. For all other values of c, we can use $x = \frac{c}{2c-3}$ to find the x that satisfies the original equation for that value of c. $\square$

Exercises

1.1.1 Solve the following equations:

(a) $9 - 4x + 4 = -5x + 3 - x$

(b) $5(y+2)/2 = 7y - 4$

1.1.2 Solve the following inequalities. Write your solutions using interval notation.

(a) $8 - 2t \geq 11 + t$

(b) $7 - 2x \leq 6x - 1 < 23 - 2x$

1.1.3 Find all x such that $5x^2 + 2x - 1 = 4x^2 + 2x + 7$.

1.1.4 Find all values of a such that there is no value of b that satisfies the equation $\frac{2-b}{3-b} = 5a$.

1.1.5★ Find all values of t such that $t - 1$, $t + 1$, and 4 could be the lengths of the sides of a right triangle.

1.1.6★ Find all a such that

$$\frac{3}{1 - \sqrt{a - 2}} + \frac{3}{1 + \sqrt{a - 2}} = 6.$$

Hints: 115

1.2 Substitution

A group of equations for which we seek values that satisfy all of the equations at the same time is called a **system of equations**. In the next two sections, we explore the two most common strategies for solving systems of equations: substitution and elimination. We'll start each section by reviewing these methods by solving a system of two-variable linear equations. Then, we'll extend the strategies to more challenging systems.

Both strategies involve trying to create an equation that has a single variable. Then, we solve that equation for that variable. The key step in our first strategy, **substitution**, is to solve one of the equations for one of the variables, and then substitute the result into other equations in the system.

Problems

Problem 1.6: Consider the system of equations

$$x - y = 4,$$
$$2x + y = 29.$$

(a) Find x in terms of y using the first equation.

(b) Substitute your answer from (a) for x into the second equation to create a new equation. Solve this equation for y.

(c) Use your answers from (a) and (b) to find the values of x and y that satisfy the system.

Problem 1.7: Consider the following system of equations:

$$x - y = 3,$$
$$\frac{1}{x} + \frac{1}{y} = \frac{1}{2}.$$

(a) Solve the first equation for x in terms of y.

(b) Create a new equation by replacing x in the second equation with your expression for x from part (a).

(c) Find all solutions to the original system of equations.

Problem 1.8: Jackie and Rachel both worked during last summer and made $960 each. Rachel worked 16 hours more than Jackie, but Rachel earned $2 less per hour. How many hours did Jackie work?

Problem 1.9: Suppose that a, b, c, and d are constants such that $ax + b = cx + d$ for all values of x. In this problem, we show that we must have $a = c$ and $b = d$.

(a) The equation must hold for all values of x. Choose an appropriate value for x that allows you to deduce that $b = d$.

(b) Choose another value of x that, along with the fact that $b = d$, allows you to deduce that $a = c$.

We begin our study of substitution by solving a simple system of linear equations.

Problem 1.6: Solve the system of equations $x - y = 4$, $2x + y = 29$.

Solution for Problem 1.6: The first equation allows us to easily express x in terms of y, though we could isolate either variable in either linear equation. Adding y to both sides of the first equation gives us $x = y + 4$. This rearrangement of the first equation allows us to replace x in the second equation with the expression $y + 4$ to give

$$2(y + 4) + y = 29.$$

Solving for y gives $y = 7$, so $x = y + 4 = 7 + 4 = 11$. The solution to the system of equations is $x = 11$ and $y = 7$. We often write solutions to two-variable systems of equations as **ordered pairs**. For example, we write $(x, y) = (11, 7)$ to indicate $x = 11$ and $y = 7$. $\square$

Let's try a slightly more complicated system of equations.

Problem 1.7: Find x and y if $x - y = 3$ and $\frac{1}{x} + \frac{1}{y} = \frac{1}{2}$.

Solution for Problem 1.7: From the first equation, we have $x = y + 3$. Substituting for x in the second equation, we have

$$\frac{1}{y + 3} + \frac{1}{y} = \frac{1}{2}.$$

Multiplying this equation by $2y(y + 3)$ to get rid of the denominators gives us

$$2y + 2(y + 3) = y(y + 3).$$

Rearranging this equation gives us $y^2 - y - 6 = 0$. Factoring the left side gives us $(y - 3)(y + 2) = 0$, so we have $y = 3$ or $y = -2$. Since $x = y + 3$, the two solutions to our system of equations are $(x, y) = (6, 3)$ and $(x, y) = (1, -2)$. $\square$

As we saw in the previous problem, and will see in the next one as well, when one or more equations look complicated, substitution can quickly provide us with an equation that's not as intimidating.

Problem 1.8: Jackie and Rachel both worked during last summer and made $960 each. Rachel worked 16 hours more than Jackie, but Rachel earned $2 less per hour. How many hours did Jackie work?

Solution for Problem 1.8: First, we translate this word problem into algebra. We let h_J and h_R be the numbers of hours Jackie and Rachel worked, respectively, and we let w_J and w_R be their corresponding hourly wages. Notice that each variable is clearly related to what it stands for. If we had used a, b, c, and d instead, we would have had a harder time remembering what each variable represents.

> **Concept:** If you have to assign multiple variables in a problem, use variables that are clearly related to what they represent.

Here, we use h for hours and w for wages, and attach the subscripts J and R to stand for Jackie and Rachel. So, for example, it's easy to remember that h_R means the number of hours Rachel worked.

We can now turn the information in the problem into equations:

$$h_J w_J = 960, \tag{1.1}$$
$$h_R w_R = 960, \tag{1.2}$$
$$h_R = h_J + 16, \tag{1.3}$$
$$w_R = w_J - 2. \tag{1.4}$$

The last two equations are already set up for substitution into Equation (1.2). We replace h_R and w_R in Equation (1.2) with the right sides of Equations (1.3) and (1.4), and we have

$$(h_J + 16)(w_J - 2) = 960.$$

Expanding the left side gives $h_J w_J - 2h_J + 16w_J - 32 = 960$. Equation (1.1) tells us that $h_J w_J = 960$, so our equation now is $960 - 2h_J + 16w_J - 32 = 960$. Simplifying this equation gives us $h_J = 8w_J - 16$. Now we can substitute $h_J = 8w_J - 16$ into $h_J w_J = 960$ to create an equation with one variable:

$$(8w_J - 16)w_J = 960.$$

Simplifying this equation gives us the quadratic equation $w_J^2 - 2w_J - 120 = 0$, and factoring the quadratic gives $(w_J - 12)(w_J + 10) = 0$. Since $w_J > 0$, we know $w_J = 12$.

Finally, $h_J = 8w_J - 16 = 80$ is the number of hours Jackie worked.

> **WARNING!!** Everyone makes mistakes! After performing a great deal of problem solving and computation to find a solution, it's a good idea to check and see if your answer really works.

Jackie worked 80 hours at a wage of $12 per hour, for a total of

$$80 \text{ hours} \cdot \frac{\$12}{\text{hour}} = \$960.$$

Rachel worked 96 hours at a wage of $10 per hour, for a total of

$$96 \text{ hours} \cdot \frac{\$10}{\text{hour}} = \$960.$$

We see that our solution satisfies the conditions of the problem. $\square$

Problem 1.9: Suppose that a, b, c, and d are constants such that $ax + b = cx + d$ for all values of x. Show that $a = c$ and $b = d$.

Solution for Problem 1.9: Because the equation must be true for all values of x, we can assign any value to x in the equation. We start by letting $x = 0$, since this will eliminate the ax and cx terms in the equation, leaving us with $b = d$. Since $b = d$, our equation now is $ax + b = cx + b$, so $ax = cx$. Again, this must hold for all values of x, so we let $x = 1$ to find $a = c$. $\square$

Problem 1.9 gives us another use of substitution:

Concept: If you have an equation that must be true for all values of a variable, then substituting values for that variable can give you information about the equation. Specifically, setting the variable equal to 0 can eliminate some terms from the equation. Setting the variable equal to 1 is also often a useful simplifying substitution.

Exercises

1.2.1 Find all ordered pairs (x, y) such that $-x + 2y = 8$ and $x - y = -5$.

1.2.2 Find b if a is a constant such that $3ax - 8b = 4x + 6a$ for all values of x.

1.2.3 Find all ordered pairs (x, y) such that $x + y = 6$ and $x^2 + xy + 2y^2 = 32$.

1.2.4★ Solve the system of equations for x:

$$x + 2y - z = 5,$$
$$3x + 2y + z = 11,$$
$$(x + 2y)^2 - z^2 = 15.$$

(Source: COMC) **Hints:** 154

1.3 Elimination

Elimination is the process of solving or simplifying a system of equations by combining the equations to produce new equations with fewer variables.

Here's an example. Suppose that we know both $x + 2y = 7$ and $x - y = 4$. When we add twice the second equation to the first, we get

$$(x + 2y) + 2(x - y) = 7 + 2 \cdot 4.$$

This equation simplifies to $3x = 15$, which has only one variable instead of two. We say that we have "eliminated y."

Problems

Problem 1.10: In this problem we solve the following system of linear equations:

$$2x + 3y = 35,$$
$$13x - 6y = -2.$$

(a) Multiply the first equation by a constant to produce an equation whose y coefficient is the opposite of the y coefficient in the second equation.

(b) Add your result from (a) to the second equation to produce a new equation.

(c) Solve the original system of equations.

Problem 1.11: The sum of three numbers is 12. The third number is 2 less than the sum of the second number and three times the first number. Five times the first number is equal to the sum of the second and third number. Find the numbers.

Problem 1.12: Find all ordered pairs (x, y) that satisfy the following system of equations:

$$x^2 + xy = 126,$$
$$x^2 - xy = 36.$$

Problem 1.13: Consider the following system of equations:

$$xy^2 = 10^8, \quad \frac{x^3}{y} = 10^{10}.$$

(a) Find positive integers A and B such that the simplification of $(xy^2)^A \left(\frac{x^3}{y}\right)^B$ has only one variable.

(b) Solve the given system of equations. *(Source: COMC)*

Problem 1.14: Find all ordered pairs (x, y) such that $5x^2 - 3xy = 280$ and $3y^2 - 5xy = 56$.

Problem 1.10: Solve the following system of equations using elimination:

$$2x + 3y = 35,$$
$$13x - 6y = -2.$$

Solution for Problem 1.10: If the coefficient of y in one equation were the opposite of the coefficient of y in the other equation, then we could add the equations to eliminate y. So, we multiply the first equation by 2 to give us the system

$$4x + 6y = 70,$$
$$13x - 6y = -2.$$

Adding these equations eliminates y and leaves $17x = 68$. This equation is true whenever both the original equations are true. Solving $17x = 68$ gives us $x = 4$.

Now that we know $x = 4$, we can replace x with 4 in either of the original equations to solve for the value of y. For instance, from the first equation in the original system, we have $2(4) + 3y = 35$, so $y = 9$. Therefore, the solution to the system of equations is $(x, y) = (4, 9)$. $\square$

Notice that it is much simpler to eliminate y than x in the system in Problem 1.10. Had we wanted to eliminate x, we would have had to multiply the first equation by 13 and the second by 2, which is more complicated than simply multiplying just the first equation by 2.

> **Concept:** When using elimination to solve a system of equations, look for the easiest variables to eliminate first.

> **Problem 1.11:** The sum of three numbers is 12. The third number is 2 less than the sum of the second number and three times the first number. Five times the first number is equal to the sum of the second and third number. Find the numbers.

Solution for Problem 1.11: We begin by rewriting the given information as equations. Let the three numbers be a, b, and c, in that order. We have

$$a + b + c = 12,$$
$$c = 3a + b - 2,$$
$$5a = b + c.$$

We could go ahead and begin eliminating variables from the system, but first we get more organized by rewriting the equations with all the variables on the left side of the equations:

$$a + b + c = 12,$$
$$-3a - b + c = -2,$$
$$5a - b - c = 0.$$

Organizing our equations like this makes it easier to see ways to eliminate variables. There are a few quick ways to eliminate a variable by adding two equations, but our organization also helps us see that adding the first and third equation eliminates *two variables* and leaves us with just $6a = 12$, so $a = 2$. Letting $a = 2$ in our equations gives us

$$b + c = 10,$$
$$-b + c = 4,$$
$$-b - c = -10.$$

Note that the last equation is just -1 times the first equation, so the last of these three equations is redundant. Adding the first and second equations eliminates b and leaves $2c = 14$, so $c = 7$. Finally, we have $b = 10 - c = 3$, so our solution is $(a, b, c) = (2, 3, 7)$. $\square$

> **Concept:** Don't just blindly eliminate variables. Keep your eyes open for quick ways to eliminate variables.

Problem 1.12: Find all ordered pairs (x, y) that satisfy the following system of equations:

$$x^2 + xy = 126,$$
$$x^2 - xy = 36.$$

Solution for Problem 1.12: Elimination is not just a method for removing linear terms from a system of equations. Elimination can also help us to remove other expressions that complicate a system. Here, we add the given equations to get rid of the xy term. This gives us a single equation in one variable, $2x^2 = 162$, that we can easily solve to find $x = \pm 9$. When $x = 9$ in either of the original equations, we find $y = 5$. When $x = -9$, we have $y = -5$. So, our solutions are $(x, y) = (9, 5)$ and $(x, y) = (-9, -5)$. $\square$

Concept: Elimination can be an effective tool when working with systems of non-linear equations.

Problem 1.13: Find all ordered pairs (x, y) that satisfy both $xy^2 = 10^8$ and $\dfrac{x^3}{y} = 10^{10}$. *(Source: COMC)*

Solution for Problem 1.13: We can't eliminate x or y by simply adding these two equations, nor can we eliminate a variable by multiplying the two equations by nonzero constants and adding the results. However, we can use multiplication and division to eliminate whichever variable we like.

Solution 1: Squaring the second equation gives us $x^6/y^2 = 10^{20}$. Multiplying this by the first equation gives us

$$(xy^2)\left(\frac{x^6}{y^2}\right) = \left(10^8\right)\left(10^{20}\right),$$

so $x^7 = 10^{28}$, which means $x = 10^4 = 10000$. Substituting this value of x into $x^3/y = 10^{10}$, we find that $y = 10^2 = 100$, so our only solution is $(x, y) = (10000, 100)$.

Solution 2: Dividing the cube of the first equation by the second equation eliminates x:

$$\frac{\left(xy^2\right)^3}{\frac{x^3}{y}} = \frac{10^{24}}{10^{10}},$$

so $y^7 = 10^{14}$. Therefore, we have $y = 10^2 = 100$, and we find $x = 10^4$ by substituting $y = 100$ into either of the original equations. $\square$

Concept: We can use multiplication and division to eliminate variables in some systems of equations.

But, sometimes we have to do some work to be able to do so.

Problem 1.14: Find all ordered pairs (x, y) such that $5x^2 - 3xy = 280$ and $3y^2 - 5xy = 56$.

Solution for Problem 1.14: We might try eliminating the xy terms, but we'd still have x^2 and y^2 terms to deal with. Instead, we note that we can easily factor the left sides of both equations:

$$x(5x - 3y) = 280,$$
$$y(3y - 5x) = 56.$$

> **Concept:** Factoring is your friend. When stuck on an algebra problem, it's very often useful to factor expressions that appear in the problem.

The equations share a similar factor. In fact, we have $5x - 3y = -(3y - 5x)$, so $(5x - 3y)/(3y - 5x) = -1$. This means we can eliminate these factors by dividing the first equation by the second, which leaves us with $-x/y = 280/56 = 5$. We have to be careful when we do this division, though. What if $3y = 5x$? Then we will be dividing by 0. Fortunately, we can rule out the possibility that $3y = 5x$ by noting that this makes the left sides of our equations both equal to 0, so we can't have $3y = 5x$.

After dividing the first equation by the second, we have $-x/y = 5$. Solving for x in terms of y gives $x = -5y$. Substituting $x = -5y$ into the first equation in the original system gives $5(-5y)^2 - 3(-5y)y = 280$. Simplifying this equation gives $140y^2 = 280$, which has solutions $y = \pm\sqrt{2}$. Using $x = -5y$ again, we see that if $y = \sqrt{2}$, we have $x = -5\sqrt{2}$, and if $y = \sqrt{2}$, then $x = 5\sqrt{2}$. So, the solutions to the system of equations are $(x, y) = (5\sqrt{2}, -\sqrt{2})$ and $(x, y) = (-5\sqrt{2}, \sqrt{2})$. $\square$

Notice that a key step in our solution to Problem 1.14 was eliminating a whole expression, not just a variable.

> **Concept:** While we usually substitute for a variable or eliminate a variable, we can also sometimes substitute for or eliminate complicated expressions.

Exercises

1.3.1 Find all ordered pairs (a, b) such that $3a - 2b = -8$ and $5a + 4b = 5$.

1.3.2 Let x and y be real numbers satisfying $\frac{2}{x} = \frac{y}{3} = \frac{x}{y}$. Determine x^3. *(Source: Mandelbrot)*

1.3.3 Find all ordered pairs (x, y) such that $x^3y^2 = 2^{34}$ and $x^4y = 2^{32}$.

1.3.4 Find the solution to the system of equations,

$$x + 2z = 1,$$
$$-xy + 2z = 13,$$
$$xy + z^2 = -14.$$

1.3.5 Find all r and s such that $r^2 - 2rs = 27$ and $rs - 2s^2 = 9$.

Extra! *Basic research is what I am doing when I don't know what I am doing.*
⟫⟫⟫⟫
– Wernher von Braun

1.4 Larger Systems of Linear Equations

In this section, we use our strategies from the last two sections to discuss larger systems of linear equations. We also introduce some terminology used to describe systems of equations.

Problems

Problem 1.15: Consider the system of equations

$$x + y + z = 1,$$
$$2x + 2y + 3z = -2,$$
$$3x + y - z = 13.$$

(a) Use the first equation to eliminate the variable z in each of the other two equations.

(b) Solve the resulting system of two equations, and solve the original system above.

Problem 1.16: Find all solutions to the system of equations

$$7x + 5y - z = 28,$$
$$x - 4y - z = 5,$$
$$4x - 5y - 2z = 19.$$

Problem 1.17: Find all solutions to the system of equations

$$a + b + c = 6,$$
$$a + 2b + 3c = 14,$$
$$2a + 5b + 8c = 36.$$

Problem 1.18: Find $a - b + c$ if $3a - b + 5c = 44$ and $a + 2c = 12$.

Elimination is also effective with systems of equations involving many variables, as we see in the next problem:

Problem 1.15: Use elimination to solve the system

$$x + y + z = 1,$$
$$2x + 2y + 3z = -2,$$
$$3x + y - z = 13.$$

Solution for Problem 1.15: Why is this not the best start we can make:

Extra! *There are two kinds of people, those who finish what they start and so on.*

– Robert Byrne

> **Bogus Solution:** Adding the first equation to the third eliminates z and gives us $4x + 2y = 14$, and dividing this by 2 gives $2x + y = 7$. Subtracting the first equation from the third eliminates y and gives $2x - 2z = 12$, so $x - z = 6$. So, now we just have 2 two-variable equations.

Indeed, we have 2 two-variable equations, but there are different variables in the two equations. (We could use these two equations to find y and z in terms of x, and then use substitution to finish. See if you can figure out how.)

We need to eliminate the same variable in our first two eliminations so that we'll then have a system of 2 two-variable equations with the same variables. So, after eliminating z by adding the first and third equations, we then eliminate z from the first and second equations. We do so by multiplying the first equation by 3 to get $3x + 3y + 3z = 3$, then subtracting this from the second equation in the given system to find $-x - y = -5$.

We have reduced the system of three linear equations in three variables to a system of two linear equations in two variables:

$$2x + y = 7,$$
$$-x - y = -5.$$

Adding these two equations gives $x = 2$. Substituting $x = 2$ into $2x + y = 7$ gives us $y = 3$. Finally, we let $x = 2$ and $y = 3$ in the first equation of the original system, $x + y + z = 1$, to find $z = -4$. The single solution to the system is $(x, y, z) = (2, 3, -4)$. $\square$

In our solution to Problem 1.15, we first eliminate a variable from a system of three linear equations to get a system of two linear equations with two variables. We then eliminate a variable from that simpler system to solve the system.

> **Concept:** If a system of linear equations has solutions, they can be found by repeatedly eliminating variables from the system.

Following this process, we can find any solution(s) to any system of linear equations that has solutions, no matter how many variables and equations are in the system.

Note that in Problem 1.15, it doesn't matter which variable we choose to eliminate first. If we wanted to, we could have eliminated x or y first (instead of z).

Problem 1.16: Find all solutions to the system of equations

$$7x + 5y - z = 28,$$
$$x - 4y - z = 5,$$
$$4x - 5y - 2z = 19.$$

Solution for Problem 1.16: Subtracting the second equation from the first gives us $6x + 9y = 23$. Subtracting twice the second equation from the third gives us $2x + 3y = 9$.

Now we have a simpler system of equations. However, a problem occurs when we try to eliminate one of the remaining variables. If we subtract three times $2x + 3y = 9$ from $6x + 9y = 23$, we get $0 = -4$. There is no triple (x, y, z) that makes this statement true. Since the statement $0 = -4$ is the result of combining the equations in our system, any values of (x, y, z) that satisfy all three equations must also satisfy this equation. However, the equation $0 = -4$ is never true, so there are no solutions to the system of equations. $\square$

Definitions: When there is at least one solution to a system of equations, we say that the system is **consistent**. Otherwise, we say that the system is **inconsistent**.

The system in Problem 1.15 is a consistent system, and the system in Problem 1.16 is inconsistent.

So far, all the consistent systems we've solved had a finite number of solutions. This isn't always the case.

Problem 1.17: Find all solutions (a, b, c) to the system of equations

$$a + b + c = 6,$$
$$a + 2b + 3c = 14,$$
$$2a + 5b + 8c = 36.$$

Solution for Problem 1.17: We eliminate a by subtracting the first equation from the second, then subtracting twice the second equation from the third. This gives us a simpler system in two variables:

$$b + 2c = 8,$$
$$b + 2c = 8.$$

These two new equations are the same! There aren't any more equations we can use to eliminate variables or simplify the system—we've used all three of the original equations to find these two. So, every solution to the equation $b + 2c = 8$ can be used to give a solution to the system of equations. The equation $b + 2c = 8$ has infinitely many solutions, so there are infinitely many solutions to the system of equations.

To express all possible solutions to our system of equations, we note that $b + 2c = 8$ means that $b = 8 - 2c$. From $a + b + c = 6$, we have $a = 6 - b - c = 6 - (8 - 2c) - c = -2 + c$. For any value of c, we can use $a = -2 + c$ and $b = 8 - 2c$ to generate a solution to our system of equations. So, for all values of c, the triple $(a, b, c) = (-2 + c, 8 - 2c, c)$ gives us a solution to the system. We can check this by letting $(a, b, c) = (-2 + c, 8 - 2c, c)$ in each of our equations:

$$(-2 + c) + (8 - 2c) + c = 6,$$
$$(-2 + c) + 2(8 - 2c) + 3c = 14,$$
$$2(-2 + c) + 5(8 - 2c) + 8c = 36.$$

Indeed, all three equations are satisfied if $(a, b, c) = (-2 + c, 8 - 2c, c)$ for any value of c, so these are the solutions to our system.

We could instead write our solutions in terms of a different variable. For instance, since $a = -2 + c$, we have $c = a + 2$, which we can substitute into $(a, b, c) = (-2 + c, 8 - 2c, c)$ to see that $(a, b, c) = (a, -2a + 4, a + 2)$ is a solution for all values of a.

Don't worry about the many different ways we can represent the solutions to a system of equations that has infinitely many solutions. The most important point to understand now is how to determine that a given linear system has infinitely many solutions. □

As mentioned before, we can solve systems of linear equations by continually simplifying them, removing one variable at a time from the system. In general, we can find any solution(s) to a system of linear equations by simplifying the system one step at a time until we have a single linear equation with as few variables as possible (such as $x = 2$ in Problem 1.15, or $b + 2c = 8$ in Problem 1.17). That last remaining equation can then be used to describe all solutions to the system. If this remaining equation has no solutions, then the system has no solutions.

When we multiply two equations by different constants and add the results, the resulting equation is called a **linear combination** of the first two equations. For example, consider the two equations

$$2x + y = 3,$$
$$x + 7y = -5.$$

If we multiply the first equation by 2 and the second by -3 and add the results, we get $x - 19y = 21$. So, the equation $x - 19y = 21$ is a linear combination of our original two equations.

In Problem 1.17, each equation in the system is a linear combination of the other two. For example, when we add 3 times $a + 2b + 3c = 14$ to -1 times $2a + 5b + 8c = 36$, we get $a + b + c = 6$.

We call such a system of equations a **dependent** system.

> **Definitions:**
>
> - When some equation in a system of linear equations is a linear combination of the others, we call that system **dependent**. Dependent systems are also sometimes called **redundant** since at least one of the equations can be removed without changing the solutions to the system.
>
> - When no equation in a system of linear equations is a linear combination of the others, the system of equations is **independent**.

While the system in Problem 1.17 is an example of a dependent system, the systems in Problems 1.15 and 1.16 are examples of independent systems of linear equations.

Don't worry too much about all these terms right now. The main point to take from this discussion of larger systems of linear equations is that systems of linear equations can have no solution, a single unique solution, or infinitely many solutions. Using the tools we have discussed so far, you should now be able to determine if a linear system with a relatively small number of variables has a solution or not, and, if the solution is unique, you should be able to find it. When you study a field of mathematics called **linear algebra**, you will learn much more effective general strategies for working with systems of linear equations.

Problem 1.18: Find $a - b + c$ if $3a - b + 5c = 44$ and $a + 2c = 12$.

Solution for Problem 1.18: We might try to solve the system with substitution. We solve the second equation for a to get $a = 12 - 2c$. Substituting this into our first equation gives us $3(12 - 2c) - b + 5c = 44$.

Simplifying this equation gives us $b + c = -8$. Now what? We don't have any more equations left to work with, so it appears that we can't find a, b, and c.

However, we're not asked to find a, b, and c.

> **Concept:** Keep your eye on the ball. For some problems, you don't have to find the values of all the variables in the problem to solve the problem.

We're asked to find $a - b + c$. We know that $a = 12 - 2c$ and that $b + c = -8$. Solving the second equation for b gives $b = -8 - c$. Substituting our expressions for a and b into $a - b + c$ gives us

$$a - b + c = (12 - 2c) - (-8 - c) + c = 20.$$

Without even finding a, b, and c, we have found $a - b + c$!

We could also have tackled this problem by trying to combine $3a - b + 5c$ and $a + 2c$ to give $a - b + c$. We notice that subtracting twice the second expression from the first gives us the correct a and b terms. Let's see what happens with the c terms:

$$(3a - b + 5c) - 2(a + 2c) = a - b + c.$$

So, subtracting twice $a + 2c = 12$ from $3a - b + 5c = 44$ gives us $a - b + c = 20$. $\square$

You might find Problem 1.18 somewhat gimmicky (and you're more than a little right!), but it's an example of how we can sometimes evaluate an expression without knowing the values of the variables in the expression.

Exercises

1.4.1 Solve each of the following systems of equations:

(a) $2a + b + c = 2,$
$5a - b + 3c = 5,$
$7a + 4c = 8.$

(b) $2a + b + c = 3,$
$5a - b + 3c = 5,$
$7a + 4c = 8.$

(c) $x + y + z = 13,$
$x + y - z = -1,$
$x - y + z = 9.$

1.4.2 Consider the system of equations in Problem 1.17. Find constants k_1 and k_2 such that k_1 times the first equation in the system plus k_2 times the second equation in the system gives us the third equation in the system.

1.4.3 Find the value of $a^3b^7c^{14}$ given that $a^3b^2c = 108$ and $a^2b^3c^5 = 240$. **Hints:** 122

1.4.4 Solve each of the following systems of equations:

(a) $a + 3b - 2c = 18,$
$2a - 4b + 2c - d = -16,$
$-a + 2b - 5c + d = 23,$
$3b - 7c + d = 35.$

(b) $x_1 + x_2 + x_3 + x_4 = 1,$
$x_1 + x_2 - x_3 = 2,$
$x_2 + x_3 = 0.$

1.4.5★ Suppose we have a system of three linear equations in three variables such that a times the first equation plus b times the second equation equals the third equation, where a and b are nonzero constants.

(a) Must there be constants c and d such that c times the second equation plus d times the third equation equals the first equation?

(b) If the answer to part (a) is "yes," is it possible to find c and d in terms of a and b?

1.5 Summary

We often solve linear equations and inequalities with one variable by isolating the variable. We can also sometimes solve more complicated equations by isolating non-linear expressions such as x^2 or $\sqrt{z-3}$.

Two general strategies for solving systems of equations are:

- *Substitution.* Solve one of the equations for one variable in terms of the others, and then substitute the result into the other equations.

- *Elimination.* Add, subtract, multiply, or divide two equations so that one variable is eliminated, and the resulting equation has one fewer variable than the original equations.

As we saw with larger systems of equations, sometimes we have to use substitution or elimination several times to solve a problem.

Once we have determined how many solutions a system of equations has, there is some special terminology we can use to describe the system:

> **Definitions:**
>
> - When there is at least one solution to a system of equations, we say that the system is **consistent**. Otherwise, we say that the system is **inconsistent**.
>
> - When some equation in a system of linear equations is a linear combination of the others, we call that system **dependent**. Dependent systems are also sometimes called **redundant** since at least one of the equations can be removed without changing the solutions to the system.
>
> - When no equation in a system of linear equations is a linear combination of the others, the system of equations is **independent**.

Things To Watch Out For!

- If we multiply or divide an inequality by a negative number, we must reverse the direction of the inequality sign.

- When taking the square root of both sides of an equation, we must not forget negative values. Specifically, if we have $x^2 = a^2$, then we have $x = \pm a$, where the $\pm$ sign indicates that x can equal either a or $-a$.

- Whenever we square an equation as a step in solving it, we have to check that our solutions are valid. If a solution to the squared equation does not satisfy our original equation, then that solution is called **extraneous**. An extraneous solution is not a valid solution to the original equation.

- Everyone makes mistakes! After performing a great deal of problem solving and computation to find a solution, it's a good idea to check and see if your answer really works.

Problem Solving Strategies

Concepts:

- Just as we can solve a linear equation by isolating the variable, we can often solve more complex equations by isolating an expression within the equation.

- If you have to assign multiple variables in a problem, use variables that are clearly related to what they represent.

- If you have an equation that must be true for all values of a variable, then substituting values for that variable can give you information about the equation. Specifically, setting the variable equal to 0 can eliminate terms from the equation. Setting the variable equal to 1 is also often a useful simplifying substitution.

- When using elimination to solve a system of equations, look for the easiest variables to eliminate first.

- Elimination can be an effective tool when working with systems of nonlinear equations.

- We can use multiplication and division to eliminate variables in some systems of equations.

- Factoring is your friend. When stuck on an algebra problem, it's very often useful to try to factor expressions that appear in the problem.

- While we usually substitute for a variable or eliminate a variable, we can also sometimes substitute for or eliminate more complex expressions.

- Keep your eye on the ball. For some problems, you don't have to find the values of all the variables in the problem to solve the problem.

REVIEW PROBLEMS

1.19 Solve the following systems:

(a) $4x + 5y = 43,$
$9x - 2y = 57.$
$\boxed{x = 7, \; y = 3}$

(b) $2x + 4y = 18,$
$3x - 3y = -12.$
$\boxed{x = \frac{1}{3}, \; y = \frac{13}{3}}$

(c) $3x - y = 1,$
$12x - 4y = 4.$
$\boxed{\infty \text{ solutions}}$

1.20 Jeff is 4 times older than his daughter. Five years ago he was 9 times older than his daughter. How old is his daughter? $\boxed{8 \text{ years}}$

1.21 Find all solutions to each of the following systems of equations:

(a) $8x + y - z = 46,$
$3x + 4y - 2z = 27,$
$4x - y + z = 14.$
$\boxed{x = 5, \; y = 0, \; z = -6}$

(b) $2a - 3b + 5c = 17,$
$3a + b - 6c = -4,$
$a + 4b - 11c = -15.$
$\boxed{\text{no solutions}}$

(c) $x + y - z = 11,$
$3x - 2y + 2z = 3,$
$5x + 3y - 3z = 43.$
$\boxed{x = 5, \; y = z + 6, \; z = z}$

1.22 Solve the following system of equations for $a, b, c, d,$ and e:

$$a + b = 2,$$
$$b + c = 13,$$
$$c + 3d = 37,$$
$$3d + 4e = -23,$$
$$8e + 9a = -43.$$

$\boxed{a = 5, \; b = -3, \; c = 16, \; d = 7, \; e = -11}$

1.23 Solve each of the following inequalities. Write your answers in interval notation.

(a) $\frac{2}{9}(x - 1) \leq \frac{x}{18} + \frac{2}{9}$
$\boxed{x \leq \frac{8}{3}}$ $\boxed{(-\infty, \frac{8}{3}]}$

(b) $x - 1 \leq 3x + 2 \leq 2x + 6$
$\boxed{[-\frac{3}{2}, 4]}$

(c) $14 - (x - 1) > -(5 - 3x)$
$\boxed{(-\infty, 5)}$

1.24 Find all ordered pairs (x, y) that satisfy both $\sqrt{x} + \sqrt{y} = 7$ and $3\sqrt{x} - 4\sqrt{y} = -14.$ $\boxed{(4, 25)}$

1.25 Solve the following system of equations:

$$x_1 + x_2 - x_3 - x_4 = 1,$$
$$x_1 + 2x_2 + 3x_3 - x_4 = 2, \quad \boxed{\text{no solutions}}$$
$$3x_1 + 5x_2 + 5x_3 - 3x_4 = 6.$$

1.26 Find all values of r and s such that $\dfrac{1}{s - r} = \dfrac{2}{3r}$ and $\dfrac{1}{r} - \dfrac{1}{s} = \dfrac{1}{6}.$ $\boxed{(\frac{18}{5}, 9)}$

1.27 Find the value of $a + b + c$ given that

$$2a - b + 5c = 13,$$
$$2a + 3b + c = 75.$$

$\boxed{\frac{119}{4}}$

1.28 Find all ordered pairs (a, b) such that $a^2 9^b = 4$ and $a/3^b = 18$. $(6, -1)$

1.29 Find all ordered pairs (x, y) that satisfy both $x^2 + xy = 28$ and $y^2 + xy = -12$. $(-7, 3)$ and $(7, -3)$

1.30 Find all solutions to the system of equations $a - 2b = -4$, $a^2 - 2b^2 = -14$.

Challenge Problems

1.31 A tennis player computes her "win ratio" by dividing the number of matches she has won by the total number of matches she has played. At the start of a weekend, her win ratio is exactly .500. During the weekend she plays four matches, winning three and losing one. At the end of the weekend her win ratio is greater than .503. What is the largest number of matches that she could have won before the weekend began? *(Source: AIME)*

1.32 Find all x such that $-4 < \dfrac{1}{x} < 3$. **Hints:** 278

1.33 Suppose that a, b, c, d, e, and f are constants such that $ax^2 + bx + c = dx^2 + ex + f$ for all values of x. Prove that $a = d$, $b = e$, and $c = f$.

1.34 A right triangle has both a perimeter and an area of 30. Find the side lengths of the triangle.

1.35 Find all pairs of real numbers (a, b) such that $(x - a)^2 + (2x - b)^2 = (x - 3)^2 + (2x)^2$ for all x.

1.36 If $\dfrac{x^2 y}{z} = 24$ and $\dfrac{y^4 z}{x} = 30$, find the value of $\dfrac{x^8}{(yz)^5}$.

1.37 Ten people form a circle. Each picks a number and tells it to the two neighbors adjacent to him in the circle. Then each person computes and announces the average of the numbers of his two neighbors. The figure shows the average announced by each person (*not* the original number the person picked).

"10" "1" "2"
"9"   "3"
"8"   "4"
"7" "6" "5"

What number was picked by the person who announced the average 6? *(Source: AHSME)* **Hints:** 198

1.38 Find all values (a, b, c, d) that satisfy $ac = 6$, $ad = 4$, $bc = 9$, and $bd = 6$.

1.39 Find all ordered pairs (x, y) such that $9x + 3y - 2 = 0$ and $9x^2 + 3y^2 - 7x = 0$. *(Source: CEMC)*

1.40 Find all ordered pairs (x, y) of solutions to the system of equations,

$$\frac{2y - 3x}{xy} = -\frac{7}{12},$$
$$\frac{5x + 3y}{xy} = \frac{25}{4}.$$

Hints: 294

1.41 When we place four numbers in a 2×2 grid, we form a **matrix**. We find the **determinant** of such a 2×2 matrix as follows:

$$\begin{vmatrix} a & b \\ c & d \end{vmatrix} = ad - bc.$$

For example,

$$\begin{vmatrix} 3 & -2 \\ 5 & 7 \end{vmatrix} = (3)(7) - (-2)(5) = 31.$$

Suppose a, b, c, d, e, and f are constants in the system of linear equations

$$ax + by = e,$$
$$cx + dy = f.$$

Cramer's Rule states that if $\begin{vmatrix} a & b \\ c & d \end{vmatrix}$ is nonzero, then the solution to this system is

$$x = \frac{\begin{vmatrix} e & b \\ f & d \end{vmatrix}}{\begin{vmatrix} a & b \\ c & d \end{vmatrix}} \quad \text{and} \quad y = \frac{\begin{vmatrix} a & e \\ c & f \end{vmatrix}}{\begin{vmatrix} a & b \\ c & d \end{vmatrix}}.$$

Prove Cramer's Rule. What happens if the determinant $\begin{vmatrix} a & b \\ c & d \end{vmatrix}$ equals 0?

1.42★ For what values of k does the linear system of equations

$$kx + y + z = k,$$
$$x + ky + z = k,$$
$$x + y + kz = k,$$

(a) have no solution?

(b) have an infinite number of solutions?

(c) have precisely one solution?

(Source: CEMC) **Hints:** 104

1.43★ Let a, b, c be nonzero constants. Solve the system

$$ay + bx = c,$$
$$az + cx = b,$$
$$bz + cy = a,$$

for (x, y, z) in terms of a, b, and c.

1.44★ The binomial coefficients can be arranged in rows to form Pascal's Triangle (where row n is $\binom{n}{0}, \binom{n}{1}, \dots, \binom{n}{n}$). In which row of Pascal's Triangle do three consecutive entries occur that are in the ratio $3 : 4 : 5$? *(Source: AIME)* **Hints:** 325

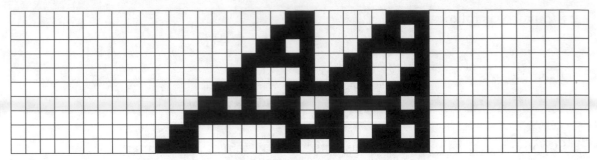

Any impatient student of mathematics or science or engineering who is irked by having algebraic symbolism thrust upon him should try to get along without it for a week. – Eric Temple Bell

CHAPTER 2

Functions Review

This chapter is a review of essential facts about functions that are covered in greater detail in *Introduction to Algebra*. If any of this is entirely new to you, we recommend working through the appropriate chapters in that text.

2.1 Function Basics

Suppose we have a machine that accepts any number, multiplies it by two, adds three to this product, and then outputs the result. Mathematically speaking, this machine is a **function** because there is only one possible output from the machine for each input to the machine. We can give this function a label, f, and write the function as

$$f(x) = 2x + 3.$$

This simple equation describes the machine. The "(x)" after f on the left side indicates that we are putting x into the function f. When speaking, $f(x)$ is read "f of x." The x in the equation $f(x) = 2x + 3$ is a **dummy variable**, which means that it is essentially a placeholder. When we put a specific number in our machine, we replace x with that number in the equation $f(x) = 2x + 3$ to determine what the machine outputs. For example,

$$f(5) = 2 \cdot 5 + 3 = 13,$$

so the machine outputs 13 when we put 5 into it.

Functions are really that simple. We define the function, then whenever we input a number to the function, we follow the definition to get an output. Usually, we use an equation, such as $f(x) = 2x + 3$, to define a function. For the obvious reason, f is the most commonly used label for functions.

In this chapter, we will only discuss functions that take real number inputs and give real number outputs. The **domain** of a function consists of all the values we are able to input to the function and get an output, and the **range** of the function consists of all the values that can possibly come out of the function.

For example, consider the function

$$f(x) = \frac{1}{x-3}.$$

The value $x = 3$ is not part of the domain of this function, because $\frac{1}{x-3}$ is not defined when $x = 3$. We can safely put any other value of x into this function, so the domain of f is "all real numbers except 3." Similarly, there is no value of x for which it is possible to make the function output 0. However, we can make the function output any other real number, so the range of f is "all real numbers except 0."

We call a function that can only output real numbers a **real-valued function**, or, even more simply, a **real function**. Unless a problem states otherwise, you can assume that all functions in this chapter are real functions. Moreover, you can assume that they are only defined for real number inputs. Later in this book we will explore a variety of functions that can take nonreal inputs and give nonreal outputs.

Problems

Problem 2.1: Let $f(x) = 2x^2 - 3$.

(a) Find $f(2)$.

(b) Find all values of t such that $f(t) = 47$.

(c) Find $f(3x + 1)$.

Problem 2.2: A function can be defined to accept multiple inputs. For example, let

$$g(a, b, c) = 3a - 2b + 7c^2.$$

(a) Evaluate $g(3, -5, -1)$. (b) Find b if $g(b, 2, -1) = 21$.

Problem 2.3: Suppose $t(x) = ax^4 + bx^2 + x + 5$, where a and b are constants. In this problem, we find $t(4)$ if $t(-4) = 3$.

(a) Use $t(-4) = 3$ to write an equation in terms of a and b.

(b) Express $t(4)$ in terms of a and b.

(c) Find $t(4)$.

Problem 2.4: Find the domain and range of each of the following functions:

(a) $f(x) = 2x - 3$ (c) $g(t) = \dfrac{2t}{t-1}$

(b) $f(x) = \sqrt{-2x+7}$ (d) $h(x) = 9x^2 + 4$

Problem 2.5: Suppose $p(x) = 2\sqrt{x} + 3$, but that $p(x)$ is only defined for $4 \le x \le 9$.

(a) Is there a value of x for which $p(x) = 5$?

(b) What is the range of p?

Problem 2.6: Find the domain of each of the following functions:

(a) $\quad f(x) = \dfrac{\sqrt{2x-5}}{x-3}$

(b) $\quad g(t) = \dfrac{2t-4}{\frac{1}{t} - \frac{1}{3t-4}}$

Problem 2.7: Notice that $\dfrac{x^2-x}{x-1} = \dfrac{(x-1)(x)}{x-1} = x$. Are $f(x) = \dfrac{x^2-x}{x-1}$ and $g(x) = x$ the same function?

Problem 2.1: Let $f(x) = 2x^2 - 3$.

(a) Find $f(2)$.

(b) Find all values of t such that $f(t) = 47$.

(c) Find $f(3x+1)$.

Solution for Problem 2.1:

(a) We simply replace x with 2 in the function definition, which gives us $f(2) = 2(2)^2 - 3 = 8 - 3 = 5$.

(b) If $f(t) = 47$, we must have $2t^2 - 3 = 47$. Adding 3 to both sides gives $2t^2 = 50$, and dividing by 2 gives $t^2 = 25$. The two values of t that satisfy this equation are $t = 5$ and $t = -5$. Checking our answer, we find $f(5) = 2(5)^2 - 3 = 47$ and $f(-5) = 2(-5)^2 - 3 = 47$.

> **Concept:** We can often check our answers in algebra problems by plugging the solutions we find back into the original question.

(c) We can input an entire expression into a function by replacing the dummy variable in the function definition with the expression. Moreover, we can replace the x in $f(x) = 2x^2 - 3$ with an expression that contains x. For example, we have

$$f(3x+1) = 2(3x+1)^2 - 3 = 2(9x^2 + 6x + 1) - 3 = 18x^2 + 12x - 1.$$

□

Problem 2.2: A function can be defined to accept multiple inputs. For example, let

$$g(a,b,c) = 3a - 2b + 7c^2.$$

(a) Evaluate $g(3,-5,-1)$.

(b) Find b if $g(b,2,-1) = 21$.

Solution for Problem 2.2:

(a) We replace the dummy variables a, b, and c in the function definition with the values 3, -5, and -1, in that order. This gives us $g(3,-5,-1) = 3(3) - 2(-5) + 7(-1)^2 = 26$.

(b) What's wrong with this solution:

Bogus Solution: We have

$$g(b, 2, -1) = 3(2) - 2b + 7(-1)^2 = -2b + 13,$$

so we seek the value of b such that $-2b + 13 = 21$, which is $b = -4$.

We see our mistake immediately if we check our answer. We have

$$g(-4, 2, -1) = 3(-4) - 2(2) + 7(-1)^2 = -9,$$

but we want $g(b, 2, -1) = 21$. Looking at our Bogus Solution above, we see that in finding an expression for $g(b, 2, -1)$, we set $a = 2$ and $c = -1$, and left b as b. But this isn't what we must do to evaluate $g(b, 2, -1)$.

WARNING!! When evaluating a function that accepts multiple inputs, make sure you assign the values to the dummy variables in the correct order.

To evaluate $g(b, 2, -1)$, we replace a in the function definition with b, we replace b with 2, and replace c with -1. This gives us

$$g(b, 2, -1) = 3b - 2(2) + 7(-1)^2 = 3b + 3.$$

So, we seek the value of b such that $3b + 3 = 21$, which gives us $b = 6$. Checking our answer, we find that $g(6, 2, -1) = 21$, as desired.

□

Problem 2.3: Suppose $t(x) = ax^4 + bx^2 + x + 5$, where a and b are constants. Find $t(4)$ if $t(-4) = 3$.

Solution for Problem 2.3: We have

$$t(4) = 256a + 16b + 4 + 5 = 256a + 16b + 9.$$

If we can find a and b, then we can evaluate $t(4)$. We turn to the only other piece of information we have, which is $t(-4) = 3$. We have

$$t(-4) = 256a + 16b - 4 + 5 = 256a + 16b + 1,$$

so $t(-4) = 3$ gives us $256a + 16b + 1 = 3$, which means $256a + 16b = 2$. Unfortunately, this doesn't tell us a or b. However, looking back at our expression for $t(4)$, we see that we don't need a and b. We need $256a + 16b$, which we know equals 2. So, we have $t(4) = 256a + 16b + 9 = 2 + 9 = 11$. □

Concept: Keep your eye on the ball. Sometimes we don't need to evaluate every variable in a problem in order to solve the problem.

Now that we have a little practice evaluating functions, let's try finding the domain and range of specific functions.

Problem 2.4: Find the domain and range of each of the following functions:

(a) $f(x) = 2x - 3$

(b) $f(x) = \sqrt{-2x + 7}$

(c) $g(t) = \dfrac{2t}{t - 1}$

(d) $h(x) = 9x^2 + 4$

Solution for Problem 2.4:

(a) We can input any real number to f, so the domain of f is all real numbers. It appears that any real number can be output from f. To show that the range of f is all real numbers, we show that for any real number y, we can find an x such that $f(x) = y$. Since $f(x) = 2x - 3$, we seek the value of x such that $2x - 3 = y$. Solving for x gives $x = (y + 3)/2$. So, for any real number y, if we let $x = (y + 3)/2$, then we have $f(x) = y$. This tells us that every real number is in the range of f. Therefore, both the domain and range are all real numbers. Remember, we can use the symbol $\mathbb{R}$ to indicate the real numbers, so we can write, "The domain and range of f are both $\mathbb{R}$."

(b) The square root of a negative number is not a real number, so we must have $-2x + 7 \geq 0$. This gives us $x \leq 7/2$, so the domain is all real numbers less than or equal to 7/2. In interval notation, the domain is $(-\infty, 7/2]$.

Intuitively, we might guess that because $-2x + 7$ can equal any nonnegative number, the range of f is all nonnegative real numbers. We can explicitly show that this is the range of f as we did in part (a). We let

$$y = f(x) = \sqrt{-2x + 7}.$$

We then solve this equation for x in terms of y. Squaring both sides of $y = \sqrt{-2x + 7}$ gives $y^2 = -2x + 7$. We solve for x by subtracting 7 from both sides, then dividing by -2, to find $x = (7 - y^2)/2$. So, for any nonnegative value of y, if we let $x = (7 - y^2)/2$, then we have

$$f(x) = f\left(\frac{7 - y^2}{2}\right) = \sqrt{-2\left(\frac{7 - y^2}{2}\right) + 7} = \sqrt{-7 + y^2 + 7} = \sqrt{y^2} = y.$$

This final step, $\sqrt{y^2} = y$, is only valid *because y is nonnegative*. We therefore see that all nonnegative real numbers are in the range of f, because for every nonnegative value of y, we can find an x such that $f(x) = y$. Since $\sqrt{-2x + 7}$ cannot be negative, no negative numbers are in the range of f. So, the range of f is all nonnegative real numbers, which we can write as $[0, +\infty)$.

(c) We cannot have $t = 1$ in $g(t) = 2t/(t - 1)$, since this would make the denominator equal to 0. There are no other restrictions on the input to g, so the domain is all real numbers except 1.

Finding the range of g is a little trickier. We let $y = g(t)$, so we have

$$y = \frac{2t}{t - 1}.$$

Then we solve for t in terms of y. Multiplying both sides by $t - 1$ gives $yt - y = 2t$. Solving this equation for t in terms of y gives

$$t = \frac{y}{y - 2}.$$

For any desired output y except $y = 2$, we can use this equation to find the input, t, to $g(t)$ that will produce the desired output. To see that 2 is definitely not in the range of g, note what happens when we try to solve $g(t) = 2$. We have $2t/(t - 1) = 2$, and multiplying both sides by $t - 1$ gives $2t = 2t - 2$, which clearly has no solution. Therefore, the range is all real numbers except 2. We can denote "all real numbers except 2" with interval notation as $(-\infty, 2) \cup (2, +\infty)$.

The "$\cup$" in $(-\infty, 2) \cup (2, +\infty)$ means "or," so, "$y \in (-\infty, 2) \cup (2, +\infty)$" means

y is in the interval $(-\infty, 2)$ or the interval $(2, +\infty)$.

> **Sidenote:** The **union** of two intervals consists of all numbers that are in either one or both of the intervals. As we just saw, we use the symbol $\cup$ to denote a union of two intervals, so $(-\infty, 2) \cup (2, +\infty)$ means "all numbers in the interval $(-\infty, 2)$ or the interval $(2, +\infty)$."
>
> The **intersection** of two intervals consists of all numbers that are in both of the intervals. We use the symbol $\cap$ to refer to the intersection of two intervals. For example, the numbers that are in both the interval $[3,7]$ and $[5,11]$ form the interval $[5,7]$, so we can write
>
> $$[3,7] \cap [5,11] = [5,7].$$
>
> On the other hand, the numbers that are either in the interval $[3,7]$ or $[5,11]$ or both form the interval $[3,11]$. Therefore, we have
>
> $$[3,7] \cup [5,11] = [3,11].$$

We also have a special notation for "all real numbers except a few specific values." We can write "all real numbers except 2" as $\mathbb{R}\backslash\{2\}$. If we wish to exclude several specific values, we just list them inside the curly braces. So, we write "all real numbers except 5, 6, and 7" as $\mathbb{R}\backslash\{5,6,7\}$.

(d) We can input any real number to the function $h(x) = 9x^2 + 4$, so the domain of h is $\mathbb{R}$. Because the square of a real number is always nonnegative, the expression $9x^2$ is always nonnegative. Therefore, the expression $9x^2 + 4$ must be greater than or equal to 4. Since $9x^2 + 4$ can equal any number that is greater than or equal to 4, the range of h is $[4, +\infty)$.

☐

Note that in parts (b) and (c), we find the domain by considering operations that we are not able to perform. Specifically, we cannot take the square root of a negative number in a real function, and we cannot divide by zero. This is typically how we find the domain when it is not obvious, as it is in parts (a) and (d), that the domain is all real numbers. Finding the range is also often a matter of considering the possible outputs of special expressions like square roots, perfect squares, or absolute values. But, as shown in part (c), the restrictions on the range can also be more subtle. There, we found a strategy that is often useful for determining the range of complicated functions:

> **Concept:** We can often find the range of a function by setting the function definition equal to a new variable, then solving for the dummy variable in terms of the new variable.

Sometimes functions are defined with explicit constraints on the domain of the function. The range of such a function consists only of those outputs that can be obtained from the permitted inputs. Here's an example:

Problem 2.5: Suppose $p(x) = 2\sqrt{x} + 3$, but that $p(x)$ is only defined for $4 \le x \le 9$. Find the range of p.

Solution for Problem 2.5: What's wrong with this solution:

> **Bogus Solution:** The output of $\sqrt{x}$ can be any nonnegative real number, so the result of $2\sqrt{x}+3$ can be any nonnegative real number greater than or equal to 3.

Our Bogus Solution completely ignores the restriction on the domain of p.

> **WARNING!!** We must take into account any restrictions on the domain of a function when finding the range of the function.

Since p is only defined for inputs x such that $4 \le x \le 9$, the range consists of all values of p that can result from inputting a value of x such that $4 \le x \le 9$. Because $4 \le x \le 9$, we have $2 \le \sqrt{x} \le 3$. Multiplying all parts of this inequality chain by 2 gives $4 \le 2\sqrt{x} \le 6$ and adding 3 to all parts of this chain gives $7 \le 2\sqrt{x}+3 \le 9$. Therefore, for all x such that $4 \le x \le 9$, we have $7 \le p(x) \le 9$.

We're not finished yet! We've only shown that all values in the range of p are between 7 and 9, inclusive (which means 7 and 9 are included). We haven't shown that all real numbers from 7 to 9 are in the range. This is somewhat "obvious," since $p(x)$ goes "smoothly" from 7 to 9 as x goes from 4 to 9. However, to explicitly prove that all possible values from 7 to 9 are in the range, we let $y = p(x) = 2\sqrt{x}+3$ and solve for x in terms of y. This gives us

$$x = \left(\frac{y-3}{2}\right)^2.$$

For any value of y from 7 to 9, we can use this equation to find the value of x for which $p(x)$ equals this y. (Make sure you see why all the resulting values of x are in the domain of p!) □

Finding the domain and range of more complicated functions can be pretty tricky. We'll try finding the domain of a couple of such functions.

> **Problem 2.6:** Find the domain of each of the following functions:
>
> (a) $\quad f(x) = \dfrac{\sqrt{2x-5}}{x-3}$ $\qquad\qquad$ (b) $\quad g(t) = \dfrac{2t-4}{\frac{1}{t}-\frac{1}{3t-4}}$

Solution for Problem 2.6:

(a) The input to a square root must be nonnegative, so we must have $2x-5 \ge 0$. This gives us $x \ge 5/2$. However, we cannot have $x = 3$, since this will make the denominator equal to 0.

> **WARNING!!** Sometimes there is more than one constraint on the domain of a function.

Combining these two constraints tells us that the domain is all real numbers greater than or equal to 5/2 except 3. We can write this in interval notation as $[5/2, 3) \cup (3, +\infty)$.

(b) First, we notice that we cannot have $t = 0$ or $3t - 4 = 0$, because we cannot divide by 0. This tells us that we must exclude $t = 0$ and $t = 4/3$ from the domain. However, we're not finished, because we cannot have the entire denominator of $g(t)$ equal to 0. Therefore, we must exclude from our domain those values of t for which

$$\frac{1}{t} - \frac{1}{3t-4} = 0.$$

Multiplying both sides by $t(3t-4)$ to get rid of the fractions gives us $3t - 4 - t = 0$, from which we find $t = 2$. So, the domain of g is all real numbers except 0, 4/3, and 2. We could write this as intervals with the unwieldy $(-\infty, 0) \cup (0, \frac{4}{3}) \cup (\frac{4}{3}, 2) \cup (2, +\infty)$. We can also write it as $\mathbb{R}\backslash\{0, \frac{4}{3}, 2\}$, which reads "all real numbers except 0, $\frac{4}{3}$, and 2."

□

Problem 2.7: Notice that $\dfrac{x^2 - x}{x - 1} = \dfrac{(x-1)(x)}{x-1} = x$. Are $f(x) = \dfrac{x^2 - x}{x - 1}$ and $g(x) = x$ the same function?

Solution for Problem 2.7: No! The function f is not defined for $x = 1$, even though we can simplify our expression for $f(x)$ to just x. The simplification

$$\frac{x^2 - x}{x - 1} = \frac{x(x - 1)}{x - 1} = x$$

is only valid if $x \neq 1$. If $x = 1$, then $\frac{x^2-x}{x-1}$ is $\frac{0}{0}$, which is not defined. In other words, 1 is not in the domain of f, but it is in the domain of g. So, f and g are not the same function. □

We finish this section with the relatively obvious observation that the sum, product, and quotient of two functions are all themselves functions. For example, if f and g are functions, we can define a new function h as $h = f + g$, where this means that $h(x) = f(x) + g(x)$. Note that $h(x)$ is only defined if both $f(x)$ and $g(x)$ are defined. Therefore, a number must be in the domains of both f and g in order to be in the domain of $f + g$.

In much the same way, we can define a new function as a linear combination of any two functions f and g. Specifically, if a and b are constants, then we can define a new function t as $t = a \cdot f + b \cdot g$, where this means that $t(x) = a \cdot f(x) + b \cdot g(x)$.

We can also define the function $p = f \cdot g$ such that $p(x) = f(x) \cdot g(x)$ and the function $q = f/g$ such that $q(x) = f(x)/g(x)$. Note that if $q = f/g$, then we must exclude values of x for which $g(x) = 0$ from the domain of q.

Exercises

2.1.1 Let $f(x) = x^2 - x - 6$.

(a) Compute $f(2)$.

(b) Find all x such that $f(x) = 0$.

(c) Find $f(x - 1)$

(d) Is there a real value of x such that $f(x) = 6$?

(e) ★ Find the range of f. **Hints:** 134

2.1.2 Find the domain and range of each of the following real-valued functions:

(a) $f(x) = |1 - x|$
(b) $f(t) = \sqrt{2 - t}$
(c) $h(x) = 1 - x^2$
(d) $g(u) = \frac{1}{1 + \frac{1}{u}}$

2.1.3 Let $f(x) = \sqrt{4 - x}$ and $g(x) = \sqrt{2x - 6}$. Find the domain of $f \cdot g$ and the domain of f/g. Explain why these domains are different.

2.1.4 Do the functions $f(x) = \sqrt{\dfrac{2x-5}{x-8}}$ and $g(x) = \dfrac{\sqrt{2x-5}}{\sqrt{x-8}}$ have the same domain? Why or why not?

2.1.5 Let $f(x) = \dfrac{x+1}{x-1}$. Suppose $|x| \neq 1$. Simplify $f(x) \cdot f(-x)$.

2.1.6 A function $g(x)$ is defined for $-3 \leq x \leq 4$ such that $g(x) = (2+x)^2$. Find the range of g.

2.1.7 Let $T(a, b, c) = 3a^b - c$.

(a) Find $T(2, 3, -5)$.

(b) Find all values of x such that $T(x, 2, 6) = 21$.

2.2 Graphing Functions

When we **graph an equation** in which the only variables are x and/or y, we plot all points (x, y) on the Cartesian plane that satisfy the equation. When we **graph a function** f, we graph the equation $y = f(x)$. For example, the graph of the function $f(x) = x^2$ is shown at right.

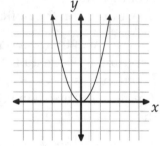

The **x-intercepts** of a graph are the points where the graph crosses the x-axis, while the **y-intercepts** are the points where the graph crosses the y-axis. For example, the point $(0, 0)$ is both an x-intercept and a y-intercept of the graph of $f(x) = x^2$. (Note: Some sources use "x-intercept" to refer to the x-coordinate of a point where a graph crosses the x-axis, and likewise for "y-intercept." So, these sources would describe both the x-intercept and the y-intercept of the graph of $f(x) = x^2$ as 0 instead of $(0, 0)$.)

In this section, we review the basics of graphing functions, getting information from the graph of a function, and transforming the graph of a function.

Problems

Problem 2.8: For each of the functions below, graph the function and find the x-intercepts and the y-intercepts of the graph.

(a) $f(x) = (x-6)^2 - 4$

(b) $f(x) = 4$

(c) $f(x) = \sqrt{x+5} - 3$

(d) $f(x) = |x+2| + 1$

Problem 2.9: The graph of $y = f(x)$ is pictured at right.

(a) According to the graph, what is $f(2)$?

(b) According to the graph, for what values of a is $f(a) = 2$?

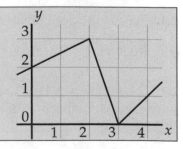

Problem 2.10:

    (a) Is it possible for a vertical line to intersect the graph of a function at two different points? Why or why not?

    (b) If there is no vertical line that passes through more than one point of a curve on the Cartesian plane, then is the curve the graph of a function?

Problem 2.11: At right is the graph of $y = f(x)$.

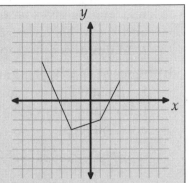

    (a) Name three points on the graph of $y = f(x) + 2$.

    (b) Explain why the graph of $y = f(x) + 2$ is a 2-unit vertical upward shift of the graph of $y = f(x)$.

    (c) Name three points on the graph of $y = f(x + 2)$.

    (d) Explain why the graph of $y = f(x + 2)$ is a 2-unit leftward shift of the graph of $y = f(x)$.

    (e) Find the domain and range of h if $h(x) = f(x + 3) - 4$.

Problem 2.12: Let $f(x)$ be the function graphed in the previous problem.

    (a) Draw the graph of $y = -f(x)$.

    (b) Draw the graph of $y = f(-x)$.

    (c) Draw the graph of $y = 2f(x)$.

    (d) Draw the graph of $y = f(2x)$.

Problem 2.8: For each of the functions below, graph the function and find the x-intercepts and the y-intercepts of the graph.

    (a) $f(x) = (x - 6)^2 - 4$ (c) $f(x) = \sqrt{x + 5} - 3$

    (b) $f(x) = 4$ (d) $f(x) = |x + 2| + 1$

Solution for Problem 2.8:

    (a) The graph of $y = (x - 6)^2 - 4$ is an upward-opening parabola with vertex $(6, -4)$. (We will discuss parabolas in much more detail in Chapter 5.) We find several points on the graph by finding y for various values of x. A table with these points is at left below, and the graph is at right below.

x	$y = f(x)$
3	5
4	0
5	-3
6	-4
7	-3
8	0
9	5

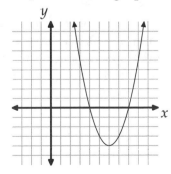

The y-intercept of the graph is the point where the graph intersects the y-axis. All points on the y-axis have an x-coordinate of 0, so evaluating $f(0)$ gives us the y-coordinate of the y-intercept of the graph of $y = f(x)$. Since $f(0) = (0 - 6)^2 - 4 = 32$, the y-intercept of the graph is $(0, 32)$.

The x-intercepts of the graph are those points where the graph intersects the x-axis. The y-coordinate of such a point is 0, so the x-coordinates of the x-intercepts are the solutions to the equation $f(x) = 0$. Therefore, we seek the values of x such that $(x - 6)^2 - 4 = 0$. Expanding and rearranging gives us $x^2 - 12x + 32 = 0$. Factoring gives $(x - 4)(x - 8) = 0$, so our solutions are $x = 4$ and $x = 8$. Therefore, the x-intercepts of our graph are $(4, 0)$ and $(8, 0)$. (Notice that we could also read these off our graph.)

(b) Since $f(x) = 4$, the graph of $y = f(x)$ is the graph of $y = 4$, which is a horizontal line as shown at right. This line meets the y-axis at $(0, 4)$, so this is the y-intercept. The graph never hits the x-axis, so it does not have an x-intercept.

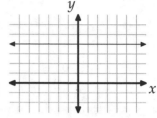

(c) As with the first part, we choose values of x, and then find y such that $y = f(x)$. Notice that we start with $x = -5$ and then substitute greater values of x, because the domain of f is $[-5, +\infty)$.

x	$y = f(x)$
-5	-3
-4	-2
-1	-1
4	0
11	1

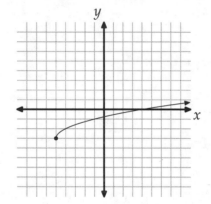

When $x = 0$, we have $y = f(0) = -3 + \sqrt{5}$, so the y-intercept is $(0, -3 + \sqrt{5})$. When $y = 0$, we have $0 = \sqrt{x + 5} - 3$. Adding 3 to both sides gives $3 = \sqrt{x + 5}$, and squaring both sides of this gives $9 = x + 5$. Solving this gives $x = 4$, so $(4, 0)$ is the x-intercept of the graph.

(d) If $x \geq -2$, we have $|x + 2| = x + 2$, so if $x \geq -2$, we have $f(x) = x + 2 + 1 = x + 3$. So, when graphing $y = f(x)$, we graph $y = x + 3$ for $x \geq -2$. Similarly, if $x < -2$, we have $|x + 2| = -x - 2$, so graphing $y = f(x)$ for $x < -2$ is the same as graphing $y = -x - 2 + 1 = -x - 1$. The result is shown at right.

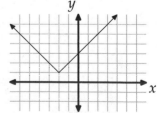

When $x = 0$, we have $y = f(0) = 3$, so $(0, 3)$ is the y-intercept of the graph. When $y = 0$, we have $0 = |x + 2| + 1$. Subtracting 1 gives $|x + 2| = -1$, which has no solutions because absolute value is always nonnegative. So, the graph has no x-intercept. (We can also see this from our graph.)

$\square$

Now that we've used the definition of a function to get information about the graph of the function, let's try going the other direction—using a graph to get information about the function.

Problem 2.9: The graph of $y = f(x)$ is pictured at right.

(a) According to the graph, what is $f(2)$?

(b) According to the graph, for what values of a is $f(a) = 2$?

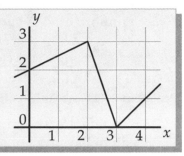

Solution for Problem 2.9:

(a) The graph of $y = f(x)$ passes through $(2, 3)$, so we have $f(2) = 3$.

> **Important:** If the point (a, b) is on the graph of $y = f(x)$, then $f(a) = b$. Conversely, if $f(a) = b$, then (a, b) is on the graph of $y = f(x)$.

(b) To determine the values of a such that $f(a) = 2$, we locate the points on the graph of $y = f(x)$ such that $y = 2$. The x-coordinates of these points then are our desired values of a. Since $(0, 2)$ is on the graph of $y = f(x)$, we have $f(0) = 2$.

There is another point on the graph with y-coordinate equal to 2, but it's not immediately obvious what the x-coordinate of this point is. However, this point is on the line segment connecting $(2, 3)$ to $(3, 0)$, so we can use the equation of the line through these two points to find the desired x-coordinate. The slope of this line is $(3 - 0)/(2 - 3) = -3$, so a point-slope equation of the line is $y - 0 = -3(x - 3)$, or $y = -3x + 9$. When $y = 2$ in this equation, we have $x = 7/3$, so the line passes through $(7/3, 2)$. Therefore, we have $f(7/3) = 2$, so the two values of a for which $f(a) = 2$ are $a = 0$ and $a = 7/3$.

□

Problem 2.10:

(a) Is it possible for a vertical line to intersect the graph of a function at two different points? Why or why not?

(b) If there is no vertical line that passes through more than one point of a curve on the Cartesian plane, then is the curve the graph of a function?

Solution for Problem 2.10:

(a) For each value of x in the domain of f, there is exactly one value of $f(x)$. So, for each such value of x, there is only one point (x, y) on the graph of $y = f(x)$. Therefore, it is impossible for the graph of a function to pass through two points with the same x-coordinate. Because the equation of any vertical line has the form $x = a$ for some constant a, this means that it is impossible for a vertical line to pass through more than one point on the graph of a function.

(b) If there is not a vertical line that passes through more than one point of a graph, then for every value of x, there is at most one value of y such that (x, y) is on the graph. In other words, for every x for which there is a point (x, y) on the graph, there is exactly one corresponding value of y, so the graph does indeed represent a function.

□

> **Important:** A curve on the Cartesian plane is the graph of some function if and only
> ⚠ if every vertical line passes through no more than one point on the curve.
> We call this the **vertical line test**.

Problem 2.11: At right is the graph of $y = f(x)$.

(a) Explain why the graph of $y = f(x) + 2$ is a 2-unit vertical upward shift of the graph of $y = f(x)$.

(b) Explain why the graph of $y = f(x + 2)$ is a 2-unit leftward shift of the graph of $y = f(x)$.

(c) Find the domain and range of h if $h(x) = f(x + 3) - 4$.

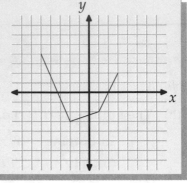

Solution for Problem 2.11:

(a) Suppose $x = a$, where a is in the domain of f. Then, the point $(a, f(a))$ is on the graph of $y = f(x)$ and the point $(a, f(a) + 2)$ is on the graph of $y = f(x) + 2$. Therefore, every point of the graph of $y = f(x) + 2$ is 2 units above the corresponding point on the graph of $y = f(x)$. So, the graph of $y = f(x) + 2$ is a 2-unit upward shift of the graph of $y = f(x)$. The graph of $y = f(x) + 2$ is solid in the graph at right, and the graph of $y = f(x)$ is dashed.

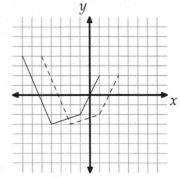

Similarly, if k is positive, then the graph of $y = f(x) + k$ is a k-unit upward shift of the graph of $y = f(x)$, and the graph of $y = f(x) - k$ is a k-unit downward shift of the graph of $y = f(x)$.

(b) Suppose $x = a$, where a is in the domain of f. Then, the point $(a, f(a))$ is on the graph of $y = f(x)$ and the point $(a, f(a + 2))$ is on the graph of $y = f(x + 2)$. Unfortunately, it isn't so clear how $f(a)$ and $f(a + 2)$ are related. To get a point on the graph of $y = f(x + 2)$ with y-coordinate equal to $f(a)$, we have to let $x = a - 2$. When $x = a - 2$ in $y = f(x + 2)$, we have $y = f(a)$, so $(a - 2, f(a))$ is on the graph of $y = f(x + 2)$.

We've shown that for any a in the domain of f, the point $(a, f(a))$ is on the graph of $y = f(x)$ and the point $(a - 2, f(a))$ is on the graph of $y = f(x + 2)$. Therefore, every point on the graph of $y = f(x + 2)$ is 2 units to the left of the corresponding point on the graph of $y = f(x)$. So, the graph of $y = f(x + 2)$ is a 2-unit leftward shift of the graph of $y = f(x)$. The graph of $y = f(x + 2)$ is solid in the graph at right above, and the graph of $y = f(x)$ is dashed.

Similarly, if $k > 0$, then the graph of $y = f(x + k)$ is a k-unit *leftward* shift of the graph of $y = f(x)$ and the graph of $y = f(x - k)$ is a k-unit *rightward* shift of the graph of $y = f(x)$.

> **WARNING!!** A common mistake is to think that the graph of $y = f(x + k)$ is a
> ☢ *rightward* shift of the graph of $y = f(x)$ when k is positive. This is
> incorrect; make sure you see why.

(c) The graph of $y = f(x+3) - 4$ is the result of shifting the graph of $y = f(x)$ downward 4 units and leftward 3 units. The leftward shift means the domain of h is 3 units "to the left" of the domain of f. The domain of f is $[-5, 3]$, so the domain of h is $[-8, 0]$. The downward shift means the range of h is 4 units "below" the range of f. The range of f is $[-3, 4]$, so the range of h is $[-7, 0]$.

In Problem 2.11, we learned the effect of adding a constant to the input or to the output of a function:

> **Important:** When $k > 0$, the graph of $y = f(x) + k$ results from shifting the graph of $y = f(x)$ *vertically* upward by k units, while the graph of $y = f(x + k)$ results from shifting the graph of $y = f(x)$ *horizontally* to the left by k units. Similarly, the graph of $y = f(x) - k$ is a k-unit downward shift of the graph of $y = f(x)$ and the graph of $y = f(x - k)$ is a k-unit rightward shift.

Problem 2.12: At right is the graph of $y = f(x)$.

(a) Find the graph of $y = -f(x)$.

(b) Find the graph of $y = f(-x)$.

(c) Find the graph of $y = 2f(x)$.

(d) Find the graph of $y = f(2x)$.

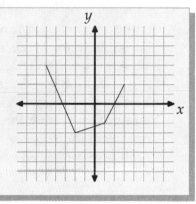

Solution for Problem 2.12:

(a) As before, we let $x = a$, where a is in the domain of f. Then, the point $(a, f(a))$ is on the graph of $y = f(x)$ and the point $(a, -f(a))$ is on the graph of $y = -f(x)$. So, the y-coordinate of each point on the graph of $y = -f(x)$ is the opposite of the y-coordinate of the corresponding point on the graph of $y = f(x)$. Therefore, we form the graph of $y = -f(x)$ by reflecting the graph of $y = f(x)$ over the x-axis. At right, the graph of $y = -f(x)$ is solid and the graph of $y = f(x)$ is dashed.

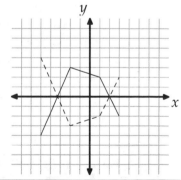

> **Extra!** The word **algebra** is derived from the works of the ninth century Arabic mathematician Abu Ja'far Muhammad ibn Musa al-Khwarizmi. His book *al-Kitab al-mukhtasar fi hisab al-jabr wa'l-muqabala* classified quadratic equations and discussed approaches to solve them. However, like many early mathematicians, all of al-Khwarizmi's "algebra" was rhetorical and geometric, rather than symbolic. For example, he described the equation $x^2 + 10x = 39$ as "*One square and ten roots are equal to 39 units.*" Imagine doing all the mathematics in this book with words rather than symbols!

(b) We let $x = a$, where a is in the domain of f. Then, the point $(a, f(a))$ is on the graph of $y = f(x)$ and the point $(a, f(-a))$ is on the graph of $y = f(-x)$. We can find a point on the graph of $y = f(-x)$ with $f(a)$ as the y-coordinate by letting $x = -a$, which tells us that $(-a, f(a))$ is on the graph of $y = f(-x)$. Therefore, the graph of $y = f(-x)$ is the result of reflecting the graph of $y = f(x)$ over the y-axis. At right, the graph of $y = f(-x)$ is solid and the graph of $y = f(x)$ is dashed.

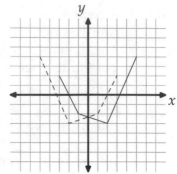

Important: The graph of $y = -f(x)$ is the result of reflecting the graph of $y = f(x)$ over the x-axis.
The graph of $y = f(-x)$ is the result of reflecting the graph of $y = f(x)$ over the y-axis.

(c) For each point $(a, f(a))$ on the graph of $y = f(x)$, the point $(a, 2f(a))$ is on the graph of $y = 2f(x)$. The latter point is twice as far from the x-axis as the former. Similarly, each point on the graph of $y = 2f(x)$ can be found by doubling the distance from the x-axis of the corresponding point on the graph of $y = f(x)$. Therefore, the graph of $y = 2f(x)$ can be found by scaling the graph of $y = f(x)$ vertically by a factor of 2 away from the x-axis. At right, the graph of $y = 2f(x)$ is solid and the graph of $y = f(x)$ is dashed.

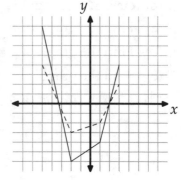

(d) For each point $(2a, f(2a))$ on the graph of $y = f(x)$, the point $(a, f(2a))$ is on the graph of $y = f(2x)$. The latter point is *half* as far from the y-axis as the former. Similarly, each point on the graph of $y = f(2x)$ can be found by halving the distance from the y-axis of the corresponding point on the graph of $y = f(x)$. Therefore, the graph of $y = f(2x)$ can be found by scaling the graph of $y = f(x)$ horizontally by a factor of $1/2$ relative to the y-axis. At right, the graph of $y = f(2x)$ is solid and the graph of $y = f(x)$ is dashed.

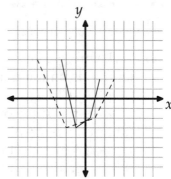

□

Important: When $k > 0$, the graph of $y = kf(x)$ results from scaling the graph of $y = f(x)$ *vertically* relative to the x-axis by a factor of k, while the graph of $y = f(kx)$ results from scaling the graph of $y = f(x)$ *horizontally* relative to the y-axis by a factor of $1/k$.

WARNING!! The graph of $y = f(kx)$ is a horizontal scaling of the graph of $y = f(x)$ by a factor of $1/k$, not by k. So, the graph of $y = f(3x)$ is the result of compressing the graph of $y = f(x)$ towards the y-axis, not stretching it away from the y-axis.

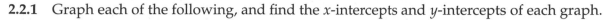

Exercises

2.2.1 Graph each of the following, and find the x-intercepts and y-intercepts of each graph.

(a) $f(x) = 3x - 7$ (b) $f(x) = 2 - |x|$ (c) $f(x) = x^2 - 5x + 6$

2.2.2 Determine if each of the following graphs represents a function:

(a) 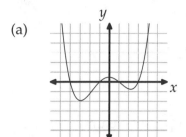   (b) 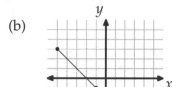   (c)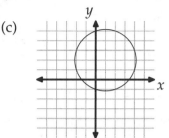

2.2.3 Suppose $f(9) = 2$. For each of the parts below, find a point that must be on the graph of the given equation.

(a) $y = f(x - 3) + 5$ (b) $y = 2f(x/4)$ (c) $y = 2f(3x - 1) + 7$

2.2.4★ Suppose $f(4) = 8$. Note that the point $(2, 8)$ is on the graph of $y = f(2x)$ and that the point $(5, 8)$ is on the graph of $y = f(2x - 6)$. So, the point we have found on the graph of $y = f(2x - 6)$ is 3 units to the right of the point we found on the graph of $y = f(2x)$. Which of the following is true: (a) we made a mistake; (b) it's a coincidence—not every point on $y - f(2x - 6)$ is 3 to the right of a point on $y - f(2x)$; (c) the graph of $y = f(2x - 6)$ is a 3-unit rightward shift of the graph of $y = f(2x)$. Explain your answer. **Hints:** 285

2.3 Composition

Sometimes we want to hook two (or more) machines together, taking the output from one machine and putting it into another. When we connect functions together like this, we are performing a **composition** of the functions. For example, the expression $f(g(x))$ means we put our input x into function g, and then take the output, $g(x)$, and put that into function f. A composition of functions is also sometimes indicated with the symbol ∘. For example, when we write $h = f \circ g$, we define the function h such that $h(x) = f(g(x))$.

Problems

Problem 2.13: Suppose $f(x) = 3x + 7$ and $g(x) = 2\sqrt{x - 3}$.

(a) Find $f(g(3))$ and $g(f(3))$.

(b) Find a if $f(g(a)) = 25$.

(c) Is $g(f(-7))$ defined? Why or why not?

(d) Suppose that $h = g \circ f$. What is the domain of h?

Problem 2.14:

(a) Let $h(x) = f(g(x))$, where f and g are functions, and suppose $h(x)$ is defined for all values of x in the domain of g. Explain why the range of g must be part of the domain of f.

(b) Again, let $h(x) = f(g(x))$, where f and g are functions. Suppose every value in the range of g is in the domain of f. Must the domain of h be the same as the domain of g?

Problem 2.15: Let $f(a) = a - 2$ and $F(a, b) = b^2 + a$.

(a) Compute $f(4)$.

(b) Find $F(3, f(4))$. *(Source: AHSME)*

Problem 2.16:

(a) Let $f(x) = 3x - 1$. Find $f(f(x))$ and $f(f(f(x)))$.

(b) Suppose $g(x) = ax + b$, where a and b are real constants. Find all possible pairs (a, b) such that $g(g(x)) = 9x + 28$.

Problem 2.17: Let $f(x) = \sqrt{x}$ and $g(x) = x^2$. Is it true that $f(g(x))$ and $g(f(x))$ are the same function? Why or why not?

Problem 2.13: Suppose $f(x) = 3x + 7$ and $g(x) = 2\sqrt{x - 3}$.

(a) Find $f(g(3))$ and $g(f(3))$.

(b) Find a if $f(g(a)) = 25$.

(c) Is $g(f(-7))$ defined? Why or why not?

(d) Suppose that $h = g \circ f$. What is the domain of h?

Solution for Problem 2.13:

(a) We have $g(3) = 2\sqrt{3 - 3} = 0$, so $f(g(3)) = f(0) = 3(0) + 7 = 7$. Similarly, we have $f(3) = 3(3) + 7 = 16$, so $g(f(3)) = g(16) = 2\sqrt{16 - 3} = 2\sqrt{13}$. Notice that $f(g(3)) \neq g(f(3))$.

> **WARNING!!** If $f(x)$ and $g(x)$ are functions, then the functions $f(g(x))$ and $g(f(x))$ are *not necessarily the same function*.

(b) We have
$$f(g(a)) = f(2\sqrt{a - 3}) = 3(2\sqrt{a - 3}) + 7 = 6\sqrt{a - 3} + 7,$$
so $f(g(a)) = 25$ means
$$6\sqrt{a - 3} + 7 = 25.$$
Subtracting 7 from both sides and then dividing both sides by 6 gives $\sqrt{a - 3} = 3$. Squaring both sides gives us $a - 3 = 9$, so $a = 12$. Checking our answer, we find $f(g(12)) = f(6) = 25$, as expected.

(c) We have $f(-7) = 3(-7) + 7 = -14$, so $g(f(-7)) = g(-14)$. However, the value -14 is not in the domain of g. Since $g(-14)$ is not defined, $g(f(-7))$ is not defined.

(d) Since $h(x) = g(f(x))$, in order for x to be in the domain of h, both $f(x)$ and $g(f(x))$ must be defined. The domain of f is all real numbers; however, as we saw in the previous part, $g(f(x))$ is not defined for all real values of x.

In order for $g(f(x))$ to be defined, the value of $f(x)$ must be in the domain of g. The domain of g is $[3, +\infty)$. Therefore, in order for $g(f(x))$ to be defined, we must have $f(x) \geq 3$. So, we must have $3x + 7 \geq 3$, which gives us $x \geq -4/3$. For all values of x such that $x \geq -4/3$, the expression $g(f(x))$ is defined. For any value of x less than $-4/3$, the expression $g(f(x))$ is not defined. Therefore, the domain of h is $[-4/3, +\infty)$.

☐

Problem 2.14:

(a) Let $h(x) = f(g(x))$, where f and g are functions, and suppose $h(x)$ is defined for all values of x in the domain of g. Explain why the range of g must be part of the domain of f.

(b) Again, let $h(x) = f(g(x))$, where f and g are functions. Suppose every value in the range of g is in the domain of f. Must the domain of h be the same as the domain of g?

Solution for Problem 2.14:

(a) Since $h(x) = f(g(x))$ and $h(x)$ is defined for all values of x in the domain of g, the expression $f(g(x))$ must be defined for all values of x in the domain of g. Because $f(g(x))$ is defined, the value of $g(x)$ is in the domain of f. So, every number in the range of g must be in the domain of f.

(b) The expression $f(g(x))$ is not defined if $g(x)$ is not defined. Therefore, all values in the domain of h must be in the domain of g. Conversely, if x is in the domain of g, then $g(x)$ is defined. Furthermore, we are given that every value in the range of g is in the domain of f, so $f(g(x))$ is defined for all x in the domain of g. Therefore, every value in the domain of g is in the domain of h. Since every value in the domain of g is in the domain of h, and vice versa, we conclude that the domain of h is the same as the domain of g.

☐

Problem 2.15: If $f(a) = a - 2$ and $F(a, b) = b^2 + a$, find the value of $F(3, f(4))$. *(Source: AHSME)*

Solution for Problem 2.15: We note that $f(4) = 4 - 2 = 2$, so
$$F(3, f(4)) = F(3, 2) = 2^2 + 3 = 7.$$

☐

Although the expression $F(3, f(4))$ looks a little intimidating, the process of evaluating it is just simple plug-and-chug arithmetic.

> **WARNING!!** Don't be intimidated by notation. Often the notation is a fancy way of representing very simple ideas.

We'll see the value of notation as we move to more complicated problems throughout this book. However, we'll usually introduce notation with simple problems like Problem 2.15 so you can get used to the notation before using it on harder problems.

Problem 2.16:

(a) Let $f(x) = 3x - 1$. Find $f(f(x))$ and $f(f(f(x)))$.

(b) Suppose $g(x) = ax + b$, where a and b are real constants. Find all possible pairs (a, b) such that $g(g(x)) = 9x + 28$.

Solution for Problem 2.16:

(a) We have

$$f(f(x)) = f(3x - 1) = 3(3x - 1) - 1 = 9x - 4.$$

We can now use this result to find $f(f(f(x)))$:

$$f(f(f(x))) = f([f(f(x))]) = f(9x - 4) = 3(9x - 4) - 1 = 27x - 13.$$

(b) In order to find a and b, we first need an expression for $g(g(x))$. We have

$$g(g(x)) = g(ax + b) = a(ax + b) + b = a^2x + ab + b.$$

So, our equation $g(g(x)) = 9x + 28$ is now $a^2x + ab + b = 9x + 28$. In order for this equation to be true for all values of x, the coefficient of x on both sides must be the same and the constant terms on both sides must be the same. Therefore, we have the system of equations

$$a^2 = 9,$$
$$ab + b = 28.$$

From $a^2 = 9$, we have $a = \pm 3$. Letting $a = 3$ in the second equation gives $b = 7$ and letting $a = -3$ in the second equation gives $b = -14$. So, our two possible pairs are $(a, b) = (3, 7)$ and $(a, b) = (-3, -14)$.

$\square$

We have a special notation for repeatedly feeding the output of a function back into the function itself. In this text, we write $f^2(x)$ to mean $f(f(x))$. Likewise, $f^n(x)$ refers to applying the function $f(x)$ exactly n times; for example, $f^5(x) = f(f(f(f(f(x)))))$.

> **WARNING!!** Some sources use $f^2(x)$ to mean $f(x) \cdot f(x)$ instead of $f(f(x))$. Likewise, $f^n(x)$ is sometimes used to denote $[f(x)]^n$. You'll often have to determine by context whether $f^n(x)$ refers to a function applied n times, or to a function raised to the nth power.

Problem 2.17: Let $f(x) = \sqrt{x}$ and $g(x) = x^2$. Is it true that $f(g(x))$ and $g(f(x))$ are the same function? Why or why not?

Solution for Problem 2.17: At first glance, we might be tempted to reason as follows:

> **Bogus Solution:** We have $f(g(x)) = \sqrt{x^2} = x$ and $g(f(x)) = (\sqrt{x})^2 = x$, so $f(g(x))$ and $g(f(x))$ are the same.

This Bogus Solution contains two significant errors. First, the domain of f is all nonnegative numbers, so the domain of $g(f(x))$ is all nonnegative numbers. However, because the domain of g is all reals while the range of g is all nonnegative numbers, the expression $f(g(x))$ is defined for all real x. Therefore, the domain of $f(g(x))$ is all real numbers. So, the two functions $g(f(x))$ and $f(g(x))$ have different domains, which means that the functions are not the same. For example, $g(f(-1))$ is undefined, because $f(-1)$ is undefined, but $f(g(-1)) = f(1) = 1$.

The other error occurs when we write $\sqrt{x^2} = x$. This statement is only true if $x \geq 0$. If $x < 0$, then we have $\sqrt{x^2} = -x$. So, it is not true that $f(g(x)) = x$ for all x. If $x \geq 0$, then $f(g(x)) = x$, and if $x < 0$, then $f(g(x)) = -x$. We can capture both of these by writing $f(g(x)) = |x|$. $\square$

Exercises

2.3.1 Check the answer to part (b) of Problem 2.16 by evaluating $g(g(x))$ for $g(x) = 3x + 7$ and for $g(x) = -3x - 14$.

2.3.2 Let $f(x) = x^2 - 2x$ and $g(x) = \sqrt{1-x}$.

(a) Find $f(g(x))$.

(b) Find the domain of $f(g(x))$.

(c) Is $g(f(-2))$ defined? Why or why not?

2.3.3 Are both $f(f^2(x))$ and $f^2(f(x))$ the same as $f^3(x)$?

2.3.4 Let a be a constant, and let $f(x) = ax$. Find an expression for $f^n(x)$ in terms of a and n, where n is a positive integer.

2.3.5 Suppose h, f, and g are functions such that $h(x) = f(g(x))$. If $(3,0)$ is the only x-intercept of the graph of g, then can we determine the x-intercepts or y-intercepts of h?

2.4 Inverse Functions

You're probably very familiar with using Control-Z when typing on a computer: that's the "undo" command for many applications. Being able to "undo" the work of a function can be extremely useful. The machine that "undoes" the work of a function f is called the **inverse function** of f.

Definition: Function g is the inverse of function f if and only if

$$g(f(x)) = x \text{ for all values of } x \text{ in the domain of } f, \text{ and}$$
$$f(g(x)) = x \text{ for all values of } x \text{ in the domain of } g.$$

From our definition, we see that if g is the inverse function of f, then f is the inverse function of g.

We often write the inverse of f as f^{-1}.

WARNING!! When working with functions, f^{-1} does not mean $\frac{1}{f}$! The expression f^{-1} is a special notation that denotes the inverse of the function f.

Problems

Problem 2.18: In this problem we find the inverse of the function $f(x) = 3x - 5$.

(a) Let g be the inverse of f, so that we have $f(g(x)) = x$ and $g(f(x)) = x$. Which of these equations is easier to use to find $g(x)$?

(b) Use the definition of f to solve the equation $f(y) = x$ for y.

(c) What is $g(x)$?

(d) Check your answer by confirming that $g(f(x)) = x$.

Problem 2.19: Let $f(x) = \dfrac{-5 - x}{3 - 2x}$.

(a) Find $f^{-1}(7)$, *without finding* $f^{-1}(x)$.

(b) Find $f^{-1}(x)$.

Problem 2.20:

(a) Does the function $g(x) = 2x^2 - 9$ have an inverse? If it does, find it. If it doesn't, explain why it doesn't.

(b) Does the function $h(x) = 3\sqrt{2x + 1}$ have an inverse? If it does, find it. If it doesn't, explain why it doesn't. (Be careful on this part!)

Problem 2.21:

(a) Explain how to determine from the graph of a function whether or not the function has an inverse.

(b) How are the graph of a function and the graph of its inverse related?

Problem 2.22: Suppose functions f and g both have an inverse. Show that if $h = f \circ g$, then the inverse of h is $g^{-1} \circ f^{-1}$.

Problem 2.18: Find the inverse of the function $f(x) = 3x - 5$.

Solution for Problem 2.18: Suppose g is the inverse of f. Then, we must have

$$g(f(x)) = f(g(x)) = x.$$

The equation $f(g(x)) = x$ is more helpful, because we know $f(x)$. So, we can put $g(x)$ into $f(x)$, which gives us $f(g(x)) = 3g(x) - 5$. Therefore, we must have $3g(x) - 5 = x$.

Important: Just as we can sometimes solve an equation for a variable by isolating the variable, we can sometimes solve an equation for a function by isolating that function.

We isolate $g(x)$ by adding 5 to both sides, then dividing by 3. This gives us $g(x) = \frac{x+5}{3}$. We can check our work by confirming that $f(g(x)) = g(f(x)) = x$. Therefore, we have $f^{-1}(x) = \frac{x+5}{3}$. $\square$

Note that we found the inverse of f by solving the equation $f(g(x)) = x$ for $g(x)$. Make sure you see why this produces a function g that is the inverse of f.

Important: If f has an inverse g, then we can often find that inverse by solving the equation $f(g(x)) = x$ for $g(x)$.

If you find the idea of "solving for a function" confusing, you can instead let $y = g(x)$ and solve for y. For example, when $f(x) = 3x - 5$, we can solve for y in $f(y) = x$ to find the inverse of f. Since $f(y) = x$, we have $3y - 5 = x$. Solving for y gives $y = \frac{x+5}{3}$, so the inverse of f is $f^{-1}(x) = \frac{x+5}{3}$.

Problem 2.19: Let $f(x) = \dfrac{-5 - x}{3 - 2x}$.

(a) Find $f^{-1}(7)$, *without finding* $f^{-1}(x)$.

(b) Find $f^{-1}(x)$.

Solution for Problem 2.19:

(a) Suppose $y = f^{-1}(7)$. Because f^{-1} is the inverse of f, we have $f(f^{-1}(x)) = x$ for all values of x. Specifically, we have $f(f^{-1}(7)) = 7$, so because $y = f^{-1}(7)$, we have $f(y) = 7$. We also could have deduced $f(y) = 7$ by noting that $y = f^{-1}(7)$ means that the y is the input to f that produces 7 as an output.

Important: If function f has an inverse, then $f(a) = b$ means that $f^{-1}(b) = a$, and vice versa.

Since $f(y) = 7$, we have
$$\frac{-5 - y}{3 - 2y} = 7.$$

Multiplying both sides by $3 - 2y$ gives $-5 - y = 7(3 - 2y)$. Solving this equation gives us $y = 2$. Therefore, $f^{-1}(7) = 2$.

(b) As we just learned, we can find the inverse of f by solving the equation $f(y) = x$ for y in terms of x. Using our definition of f, the equation $f(y) = x$ becomes
$$\frac{-5 - y}{3 - 2y} = x.$$

Multiplying both sides by $3 - 2y$ gives $-5 - y = x(3 - 2y)$. Isolating the terms with y on the left side gives $2xy - y = 3x + 5$, so $y(2x - 1) = 3x + 5$. Dividing by $2x - 1$ gives
$$f^{-1}(x) = y = \frac{3x + 5}{2x - 1}.$$

With this, we can check our answer to the first part by directly evaluating $f^{-1}(7) = 2$.

$\square$

We say that a function is **invertible** if it has an inverse. As you might have guessed, not all functions are invertible.

> **Problem 2.20:**
> (a) Does the function $g(x) = 2x^2 - 9$ have an inverse? If it does, find it. If it doesn't, explain why it doesn't.
>
> (b) Does the function $h(x) = 3\sqrt{2x+1}$ have an inverse? If it does, find it. If it doesn't, explain why it doesn't.

Solution for Problem 2.20:

(a) Both $g(1)$ and $g(-1)$ equal -7. If g has an inverse, should this inverse return 1 or -1 when -7 is input? We can't tell! Therefore, g is not reversible; it does not have an inverse.

> **Important:** If a function gives the same output for two different inputs, then the function does not have an inverse.

On the other hand, if a function never gives the same output for two different inputs, then the function is reversible—we can tell from the output of the function what input produced that output.

> **Important:** If there are no two different inputs to a function that produce the same output, then the function has an inverse.

(b) We start by looking for two different inputs to h that give the same output. After inputting a few values to h, we notice that as we input higher and higher values of x, we get higher and higher values of $h(x)$. This makes us think that h might indeed have an inverse, since it seems impossible to find two different inputs that give the same output. So, we try to find the inverse of h by solving the equation $h(y) = x$ for y. This gives us the equation

$$3\sqrt{2y+1} = x.$$

Squaring both sides gives us $9(2y + 1) = x^2$. Solving for y gives

$$y = \frac{x^2}{18} - \frac{1}{2}.$$

This isn't the end of the story, however. Recall our definition of inverse functions. Functions f and h are inverse functions of each other if and only if

$$h(f(x)) = x \text{ for all values of } x \text{ in the domain of } f, \text{ and}$$
$$f(h(x)) = x \text{ for all values of } x \text{ in the domain of } h.$$

We let $f(x) = \frac{x^2}{18} - \frac{1}{2}$. Our work above suggests that this is the inverse of h. However, if f is the inverse of h, then h is the inverse of f. So, we should have $h(f(x)) = x$ for all x in the domain of f. Unfortunately, we have $f(-3) = 0$, but $h(f(-3)) = h(0) = 3$, so we don't have $h(f(x)) = x$ when $x = -3$. The problem here is that the function $f(x) = \frac{x^2}{18} - \frac{1}{2}$ *does not have an inverse*. We can see this quickly by noting that $f(3) = f(-3) = 0$. So, where did we go wrong?

Let's evaluate $h(f(x))$ and see if it really does equal x. Maybe this will shed some light on where we went wrong. We have

$$h(f(x)) = 3\sqrt{2f(x) + 1} = 3\sqrt{\frac{x^2}{9} - 1 + 1} = 3\sqrt{\frac{x^2}{9}} = \sqrt{x^2}.$$

Here, we have to be careful.

> **WARNING!!** We only have $\sqrt{x^2} = x$ if x is nonnegative.

Now, we've found our problem. We only have $h(f(x)) = x$ if $\sqrt{x^2} = x$. This only occurs if x is nonnegative. So, if we restrict the domain of f to nonnegative numbers, then we have $h(f(x)) = x$ for all values of x in the domain of f. We can quickly confirm that $f(h(x)) = x$ for all x in the domain of h as well. So, the inverse of h is $h^{-1}(x) = \frac{x^2}{18} - \frac{1}{2}$, where the function h^{-1} is only defined for nonnegative values of x.

□

> **Problem 2.21:**
> (a) Explain how to determine from the graph of a function whether or not the function has an inverse.
>
> (b) How are the graph of a function and the graph of its inverse related?

Solution for Problem 2.21:

(a) A function has an inverse if and only if it is reversible. That is, a function has an inverse if we can tell from the output of the function what the input was. If f has an inverse, then when we graph $y = f(x)$, there can be no two points on the graph with the same y-coordinate. If there were, then we would have two different inputs (the x-coordinates of points on the graph) that provide the same output (the common y-coordinate) for the function, so it couldn't have an inverse. "No two points on the graph have the same y-coordinate" means the same thing as "No horizontal line intersects the graph in more than one point." Therefore, we have a **horizontal line test** for whether or not a function has an inverse:

> **Important:** A function has an inverse if and only if there does not exist a horizontal line that passes through more than one point on the graph of the function.

(b) Suppose the point (a, b) is on the graph of $y = f(x)$. Then, we have $f(a) = b$, so $f^{-1}(b) = a$. Because $f^{-1}(b) = a$, the point (b, a) is on the graph of $y = f^{-1}(x)$. Therefore, if (a, b) is on the graph of a function, then (b, a) is on the graph of its inverse.

> **Extra!** *I've missed more than 9000 shots in my career. I've lost almost 300 games. 26 times, I've been trusted to take the game winning shot and missed. I've failed over and over and over again in my life. And that is why I succeed.*
>
> – Michael Jordan

The fact that the points on the graph of $y = f^{-1}(x)$ can be found by reversing the coordinates of the points on the graph of $y = f(x)$ has a geometric interpretation. We just learned that if (a, b) is on the graph of $y = f(x)$, then (b, a) is on the graph of $y = f^{-1}(x)$. In the graph at right, we have plotted the points $(5, 6)$, $(6, -3)$, and $(-7, 1)$. We have also plotted the points that result when the coordinates of these three points are reversed. We see that the new points are the result of reflecting the old points over the line $y = x$. The graph of $y = x$ is the dashed line in the diagram at right.

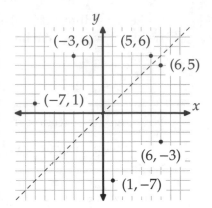

In general, reversing the coordinates of a point has the same effect as reflecting the point over the graph of $y = x$. To prove this, we let point P be (a, b) and Q be (b, a). The slope of $\overline{PQ}$ is -1, so $\overline{PQ}$ is perpendicular to the graph of $y = x$. Furthermore, the midpoint of $\overline{PQ}$ is $(\frac{a+b}{2}, \frac{a+b}{2})$, which is on the graph of $y = x$. So, the graph of $y = x$ is the perpendicular bisector of $\overline{PQ}$, which means that P and Q are images of each other upon reflection over the graph of $y = x$. So, each point on the graph of $y = f^{-1}(x)$ is the mirror image of a point on the graph of $y = f(x)$ when reflected over the line $y = x$ (and vice versa). Therefore, the entire graph of $y = f^{-1}(x)$ is the mirror image of the graph of $y = f(x)$ when reflected over the line $x = y$.

> **Important:** If the function f has an inverse f^{-1}, then the graph of $y = f^{-1}(x)$ is the reflection of the graph of $y = f(x)$ over the line $x = y$.

□

> **Problem 2.22:** Suppose functions f and g both have an inverse. Show that if $h = f \circ g$, then the inverse of h is $g^{-1} \circ f^{-1}$.

Solution for Problem 2.22: Let $t = g^{-1} \circ f^{-1}$. To show that t is the inverse of h, we must show that $h(t(x)) = x$ and $t(h(x)) = x$. We'll start with $h(t(x))$. Using our definitions of h and t, we have

$$h(t(x)) = f(g(g^{-1}(f^{-1}(x)))).$$

That looks pretty scary! But seeing both g and g^{-1} right next to each other reminds us that $g(g^{-1}(x)) = x$ for all values of x. In other words, the output of $g \circ g^{-1}$ is always the same as the input. So, when we put $f^{-1}(x)$ into $g \circ g^{-1}$, we get $f^{-1}(x)$ out, which means

$$g(g^{-1}(f^{-1}(x))) = f^{-1}(x).$$

So, using this in our expression for $h(t(x))$ above, we have

$$h(t(x)) = f(g(g^{-1}(f^{-1}(x)))) = f(f^{-1}(x)).$$

Because f^{-1} is the inverse of f, we have $f(f^{-1}(x)) = x$, so $h(t(x)) = x$.

We can do essentially the same thing for $t(h(x))$. We start with

$$t(h(x)) = g^{-1}(f^{-1}(f(g(x)))).$$

Because f is the inverse of f^{-1}, we have $f^{-1}(f(g(x))) = g(x)$, so $t(h(x)) = g^{-1}(f^{-1}(f(g(x)))) = g^{-1}(g(x)) = x$, as desired. Because $h(t(x)) = t(h(x)) = x$ when $h = f \circ g$ and $t = g^{-1} \circ f^{-1}$, the function $g^{-1} \circ f^{-1}$ is the inverse of $f \circ g$. $\square$

Exercises

2.4.1 If $f(x) = \frac{x-1}{x-2}$, what is the value of $f^{-1}(3)$?

2.4.2 Determine whether or not each of the following functions has an inverse. If the function has an inverse, find it.

(a) $f(x) = 2x - 7$

(b) $f(x) = \dfrac{1}{2x + 3}$

(c) $f(x) = \sqrt{2 - x}$

(d) $f(x) = |5 - x|$

(e) $f(x) = \sqrt{4 - x} + \sqrt{x - 2}$

(f)★ $f(x) = x^2 - 6x + 3$, where $x \geq 4$

2.4.3 Describe all ordered pairs (a, b) such that the inverse of $f(x) = ax + b$ satisfies $f(x) = f^{-1}(x)$ for all x.

2.4.4 Let $g(x) = \frac{ax+b}{cx+d}$, where a, b, c, and d are positive and $ad \neq bc$. Which of $\frac{a}{b}$, $\frac{a}{c}$, and $\frac{a}{d}$ *cannot* be in the domain of g^{-1}? **Hints:** 336

2.5 Summary

A **function** gives a single output for each input. The **domain** of a function consists of all the values we are able to input to the function and get an output. Meanwhile, the **range** of the function consists of all the values that can possibly come out of the function.

When we graph the function f, we graph the equation $y = f(x)$ on the Cartesian plane. Therefore, if the point (a, b) is on the graph of $y = f(x)$, then $f(a) = b$. Conversely, if $f(a) = b$, then (a, b) is on the graph of $y = f(x)$.

> **Important:** A curve on the Cartesian plane is the graph of some function if and only if every vertical line passes through no more than one point on the curve. We call this the **vertical line test**.

> **Important:** When $k > 0$, the graph of $y = f(x) + k$ results from shifting the graph of $y = f(x)$ *vertically* upward by k units, while the graph of $y = f(x + k)$ results from shifting the graph of $y = f(x)$ *horizontally* to the left by k units. Similarly, the graph of $y = f(x) - k$ is a k-unit downward shift of the graph of $y = f(x)$ and the graph of $y = f(x - k)$ is a k-unit rightward shift.

> **Important:** The graph of $y = -f(x)$ is the result of reflecting the graph of $y = f(x)$ over the x-axis. The graph of $y = f(-x)$ is the result of reflecting the graph of $y = f(x)$ over the y-axis.

> **Important:** When $k > 0$, the graph of $y = kf(x)$ results from scaling the graph of $y = f(x)$ *vertically* by a factor of k, while the graph of $y = f(kx)$ results from scaling the graph of $y = f(x)$ *horizontally* by a factor of $1/k$.

When we perform a **composition** of two functions, we place the output from the first function into the second. For example, both $f(g(x))$ and $(f \circ g)(x)$ indicate that we take the output $g(x)$ and input this result to f. We have a special notation for when we feed the output of a function back into the function itself. In this text, we write $f^2(x)$ to mean $f(f(x))$. Likewise, $f^n(x)$ refers to applying the function $f(x)$ exactly n times; for example, $f^5(x) = f(f(f(f(f(x)))))$.

Finally, we have a special name for a pair of functions such that each function "undoes" the action of the other:

> **Definition:** Functions f and g are **inverse functions** of each other if and only if
>
> $$g(f(x)) = x \text{ for all values of } x \text{ in the domain of } f, \text{ and}$$
> $$f(g(x)) = x \text{ for all values of } x \text{ in the domain of } g.$$
>
> If f is the inverse function of g, then g is the inverse function of f.

Not every function has an inverse. If a function gives the same output for two different inputs, then the function does not have an inverse. If there are no two different inputs to a function that produce the same output, then the function has an inverse. If f has an inverse g, then we can often find that inverse by solving the equation $f(g(x)) = x$ for $g(x)$.

We usually denote the inverse of the function f as f^{-1}. If f has an inverse, then the graph of f^{-1} is the reflection of the graph of f over the line $y = x$.

> **Important:** A function has an inverse if and only if there does not exist a horizontal line that passes through more than one point on the graph of the function.

Things To Watch Out For!

- When evaluating a function that accepts multiple inputs, make sure you assign the values to the dummy variables in the correct order.

- We must take into account any restrictions on the domain of a function when finding the range of the function.

- Sometimes there is more than one constraint on the domain of a function.

- A common mistake is to think that the graph of $y = f(x + k)$ is a *rightward* shift of the graph of $y = f(x)$ when k is positive. This is incorrect; make sure you see why.

- The graph of $y = f(kx)$ is a horizontal scaling of the graph of $y = f(x)$ by a factor of $1/k$, not by k. So, the graph of $y = f(3x)$ is the result of compressing the graph of $y = f(x)$ towards the y-axis, not stretching it away from the y-axis.

- Some sources use $f^2(x)$ to mean $f(x) \cdot f(x)$ instead of $f(f(x))$. Likewise, $f^n(x)$ is sometimes used to denote $[f(x)]^n$. You'll often have to determine by context whether $f^n(x)$ refers to a function applied n times, or to a function raised to the nth power.

- Don't be intimidated by notation. Often the notation is a fancy way of representing very simple ideas.

- When working with functions, f^{-1} does not mean $\frac{1}{f}$! The expression f^{-1} is a special notation that denotes the inverse of the function f.

Problem Solving Strategies

Concepts:
- We can often check our answers in algebra problems by plugging the solutions we find back into the original question.

- Keep your eye on the ball. Sometimes we don't need to evaluate every variable in a problem in order to solve the problem.

- We can often find the range of a function by setting the function definition equal to a new variable, then solving for the dummy variable in terms of the new variable.

REVIEW PROBLEMS

$y\left(3+\sqrt{x+1}\right)=1$
$x = \frac{8y^2 - 6y + 1}{y^2}$

2.23 Find the domain and range of each of the following functions:

(a) $A(x) = 4x^2 + 1$ $D \to (-\infty, \infty)$
$R \to [1, \infty)$

(b) $o(x) = 3 + \sqrt{16 - (x-3)^2}$ $D \to [-1, 7]$
$R \to [3, 4]$

(c) $P(x) = \dfrac{1}{3 + \sqrt{x+1}}$ $D \to [-1, \infty)$
$R \to (-\infty, \infty)$

(d) $S(x) = \dfrac{12x - 9}{6 - 9x}$ $D \to (-\infty, \frac{2}{3}) \cup (\frac{2}{3}, \infty)$
$R \to (-\infty, -\frac{3}{4}) \cup (-\frac{3}{4}, \infty)$

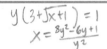

$y(6-9x) = 12x - 9$
$6y + 9 = 12x + 9xy$
$x = \dfrac{2y+3}{4+3y}$

2.24 Find the domain of each of the following functions:

(a) $f(x) = \dfrac{1}{\sqrt{2x-5}} + \sqrt{9 - 3x}$ $D \to [\frac{5}{2}, 3]$
$R \to$

(b) $f(x) = |\sqrt{x} - 2| + |\sqrt{x-2}|$

(c) $g(x) = \sqrt{|x| - 2} + \sqrt{|x-2|}$

2.25 Let $f(x) = \dfrac{4x}{x+2}$ and $g(x) = \dfrac{2x}{x+4}$. Find $f(g(x))$.

2.26 Let $f(x) = ax^2 + bx + c$ and $g(x) = ax^2 - bx + c$. If $f(1) = g(1) + 2$ and $f(2) = 2$, find $g(2)$.

2.27 At right is the graph of $y = f(x)$. Graph each of the following:

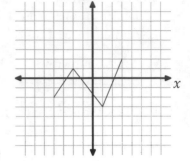

(a) $y = f(x - 3)$

(b) $y = 2f(x) + 1$

(c) $y = f(-x)$

(d) $y = f(2 - x)$

(e)★ $y = \frac{1}{2}f(2x - 1) + 3$ **Hints: 24**

2.28 Suppose that $f(f(x)) = g(f(x))$ for all real x. Must f and g be the same function?

2.29 For each of the following three graphs, determine if the graph could represent a function. If the graph can represent a function, then determine if the function has an inverse. If it does have an inverse, then graph the inverse.

(a) 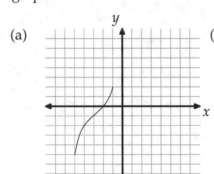 (b) 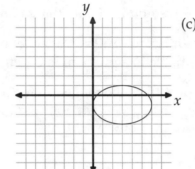 (c)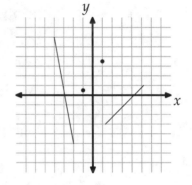

2.30 If $f(f(x)) = f(x^2)$, then must there be some constant c such that $f(x) = c$ for all values of x in the domain of f?

2.31 Let $f(x, y, z) = \dfrac{x - z}{y - z}$. If $f(a, b, c) = 1$, what is the value of $f(a, c, b)$?

2.32 Suppose that $f(x) = -x^2 + bx + c$ and $g(x) = dx + e$, where b, c, d, and e are real constants. Is it possible that $f(g(x)) = x^2$? Why or why not?

2.33 Let $f(x) = ax^2 + bx + c$ and $g(x) = (5a + 2b + c)x$. If $f(1) = 0$ and $f(3) = 2$, compute $g(2)$.

2.34 Must the inverse of a linear function $f(x) = ax + b$, where $a \neq 0$, be a linear function?

2.35 Suppose that a function $f(x)$ is defined for all real x. How can you use the graph of $f(x)$ to produce the graph of $y = |f(x)|$?

2.36 Let $f(t) = \dfrac{t}{1 - t}$, where $t \neq 1$. If $y = f(x)$, then x can be expressed as which of the following:

$$f(1/y), \ -f(y), \ -f(-y), \ f(-y), \ f(y)?$$

(Source: AHSME)

2.37 The function f defined by $f(x) = \dfrac{cx}{2x + 3}$ satisfies $f(f(x)) = x$ for all real numbers x except $-3/2$. Find c. *(Source: AHSME)*

2.38 Suppose that $f(x)$ has an inverse, $f^{-1}(x)$. Must $f^{-1}(2x)$ be the inverse of $f(2x)$? If so, prove it. If not, find an expression that is the inverse of $f(2x)$.

2.39 The function $f(x) = x^2$ is not invertible. However, if we let $g(x) = \sqrt{x}$, then $f(g(x)) = (\sqrt{x})^2 = x$. So, why isn't f invertible?

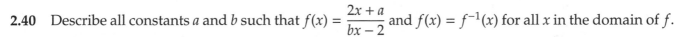

Challenge Problems

2.40 Describe all constants a and b such that $f(x) = \dfrac{2x+a}{bx-2}$ and $f(x) = f^{-1}(x)$ for all x in the domain of f.

2.41 Suppose that a function $f(x)$ has domain $(-1, 1)$. Find the domains of the following functions:

   (a) $f(x+1)$ (b) $f(1/x)$ (c) $f(\sqrt{x})$ (d)⋆ $f\left(\frac{x+1}{x-1}\right)$

2.42 Find the domain and range of the function $f(x) = \sqrt{2-x-x^2}$. **Hints:** 165

2.43 Let $f(x^2+1) = x^4 + 5x^2 + 3$. What is $f(x^2-1)$? *(Source: AMC 12)* **Hints:** 277

2.44 Suppose that a function $f(x)$ is defined for all real x. How can we obtain the graph of $f(|x|)$ from the graph of $f(x)$? **Hints:** 35

2.45 If $f(f(x)) = x$ for all x for which $f(x)$ is defined and $f(0) = 2003$, find all roots of the equation $f(x) = 0$.

2.46 Find $g^{-1}(3)$ given that $g(x) = \dfrac{3x+1}{2x+g(x)}$.

2.47 For which constants a, b, c, and d does the function $f(x) = \dfrac{ax+b}{cx+d}$ have an inverse?

2.48 Suppose that $g(x) = f(x)f(|x|)$, where $f(x)$ is a function with $\mathbb{R}$ as its domain. Must all x-intercepts of the graph of f be x-intercepts of the graph of g? Must all x-intercepts of the graph of g be x-intercepts of the graph of f?

2.49 At right is the graph of $y = f(x)$. Draw the graph of $y = \dfrac{3}{2}f(2x-2)-1$.

2.50⋆ The function f defined by

$$f(x) = \frac{ax+b}{cx+d},$$

where a, b, c, and d are nonzero real numbers, has the properties $f(19) = 19$, $f(97) = 97$, and $f(f(x)) = x$ for all values of x except $-d/c$. Find the unique number that is not in the range of f. *(Source: AIME)* **Hints:** 308

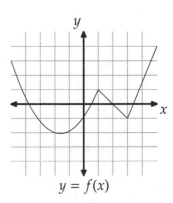

$y = f(x)$

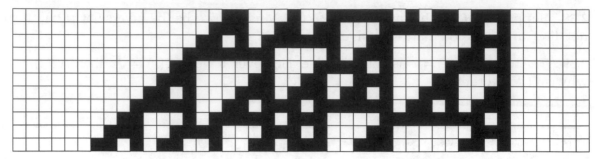

Don't let us make imaginary evils, when you know we have so many real ones to encounter. – Oliver Goldsmith

CHAPTER **3**

Complex Numbers

Many great discoveries, inventions, and mathematical advances have been the result of people assuming the impossible is not so impossible, after all. One such step in mathematics came when some mathematicians confronted equations like $x^2 = -1$.

Your first instinct might be, "No solutions. A positive number squared is positive. A negative number squared is positive. 0 squared is 0. So, there's no way the square of a number can be negative!" And for centuries, your first instinct was common knowledge among mathematicians, until eventually some mathematicians started wondering, "What if there were a number that, when squared, gave -1 as a result? What would such a number be like?"

Some mathematicians were not impressed by such musings. The great mathematician and philosopher René Descartes even scoffed at such an idea. No real number has a negative square; such a number is just **imaginary**, he believed. The name stuck. The great mathematician Leonhard Euler would later give us i, the symbol we use to represent a number whose square is -1:

$$i^2 = -1.$$

An **imaginary number** is a number whose square is a real number that is not positive. The "real numbers" Descartes was used to are those numbers we have worked with so far in this book. This name stuck, too. The numbers we can plot on the number line are called **real numbers**.

A **complex number** is a number of the form $a + bi$, where a and b are real numbers and $i^2 = -1$. We call a the **real part** of $a + bi$ and we call b the **imaginary part**. If $b \neq 0$, then $a + bi$ is **nonreal**. If $a = 0$, then the complex number $a + bi$ is called an **imaginary number**. (Some sources refer to imaginary numbers as "pure imaginary.") All real numbers and all imaginary numbers are complex numbers. Just as we use the symbol $\mathbb{R}$ to refer to "all real numbers," we use the symbol $\mathbb{C}$ to refer to "all complex numbers."

3.1 Arithmetic of Complex Numbers

Problems

Problem 3.1: Find all values of z such that $z^2 + 81 = 0$.

Problem 3.2: Let $w = 5 - 3i$, $x = -1 + 4i$, $y = 2 - i$, and $z = 1 + 2i$. Compute each of the following:

(a) $x + y + z$ (c) $wz - 2w - z$ (e) $2w^2z + 3wz^2$

(b) xyz (d) $w^2 + 2wz + z^2$

Problem 3.3: Evaluate i^{2007}.

Problem 3.4: Let $f(x) = (-2 + 3i)x^2 - (7 + 2i)x + 37 - 13i$, where the domain of f is all complex numbers. Determine each of the following:

(a) $f(1)$ (b) $f(i)$ (c) $f(2 + i)$

Problem 3.5:

(a) Find all real values of k such that the product $(3 + 2i)(3 + ki)$ is a real number.

(b) Let $z = a + bi$, where a and b are real numbers, and let $\bar{z} = a - bi$. Prove that $z \cdot \bar{z}$ is real.

Problem 3.6: Write $\dfrac{1}{3 + 2i}$ as a complex number. In other words, find real numbers a and b such that $\dfrac{1}{3 + 2i} = a + bi$.

Problem 3.7: Let $w = 2 - 5i$ and $z = -3 + i$. Express each of the following in the form $a + bi$, where a and b are real numbers.

(a) $\dfrac{1}{w}$ (c) $\dfrac{z}{w}$ (e) $\dfrac{z}{1 + i} + \dfrac{\bar{z}}{1 - i}$

(b) $\dfrac{1}{\bar{w}}$ (d) $\dfrac{\bar{z}}{\bar{w}}$

Problem 3.8: Find all complex numbers z such that $\dfrac{2z - 3i}{z + 4} = -5 + i$.

Problem 3.1: Find all values of z such that $z^2 + 81 = 0$.

Solution for Problem 3.1: Isolating z gives $z^2 = -81$. Since the square of z is negative, z must be imaginary. We might immediately see that our solutions are $z = \pm 9i$, because $i^2 = -1$ and both 9^2 and $(-9)^2$ equal 81. We could also "take the square root" of both sides of $z^2 = -81$ and write $z = \pm \sqrt{-81}$, where $\sqrt{-81}$ is a shorthand for $i\sqrt{81}$. So, we have $z = \pm i\sqrt{81} = \pm 9i$. $\square$

> **WARNING!!** We have to be careful about using notation like $\sqrt{-81}$. The real-valued
> function $f(x) = \sqrt{x}$ is only defined for $x \geq 0$. When we write $\sqrt{-81}$ as
> an intermediate step to solving an equation like $z^2 = -81$, we should
> keep in mind that we mean $i\sqrt{81}$.

Problem 3.2: Let $w = 5 - 3i$, $x = -1 + 4i$, $y = 2 - i$, and $z = 1 + 2i$. Compute each of the following:

(a) $x + y + z$ (c) $wz - 2w - z$ (e) $2w^2z + 3wz^2$

(b) xyz (d) $w^2 + 2wz + z^2$

Solution for Problem 3.2:

(a) We add complex numbers by adding the real parts and the imaginary parts separately:

$$x + y = (-1 + 4i) + (2 - i) = (-1 + 2) + (4 - 1)i = 1 + 3i,$$

so $x + y + z = (x + y) + z = (1 + 3i) + (1 + 2i) = (1 + 1) + (3 + 2)i = 2 + 5i$.

(b) We use the distributive property to multiply complex numbers:

$$\begin{aligned}
xy = (-1 + 4i)(2 - i) &= -1(2 - i) + 4i(2 - i) \\
&= -2 + i + 8i - 4i^2 \\
&= -2 + 9i - 4(-1) \\
&= 2 + 9i.
\end{aligned}$$

> **WARNING!!** Be careful with signs when working with complex numbers. One
> common error is to incorrectly simplify an expression like $1 - 4i^2$ as
> $1 - 4$, rather than $1 + 4$. Make sure you see why $1 - 4i^2 = 1 + 4$.

Similarly, for any two complex numbers $a + bi$ and $c + di$, we have

$$\begin{aligned}
(a + bi)(c + di) &= a(c + di) + bi(c + di) \\
&= ac + adi + bci + bdi^2 \\
&= ac + adi + bci - bd \\
&= (ac - bd) + (ad + bc)i.
\end{aligned}$$

You shouldn't memorize this as a formula; this is just an example of the expansion of the product of two binomials using the distributive property.

Since $xy = 2 + 9i$, we have $xyz = (xy)z = (2 + 9i)(1 + 2i) = 2(1 + 2i) + 9i(1 + 2i) = -16 + 13i$.

(c) Since $wz = (5 - 3i)(1 + 2i) = 5 + 10i - 3i - 6i^2 = 11 + 7i$, we have

$$wz - 2w - z = (5 - 3i)(1 + 2i) - 2(5 - 3i) - (1 + 2i) = 11 + 7i - 10 + 6i - 1 - 2i = 11i.$$

(d) We note that $w^2 + 2wz + z^2 = (w + z)^2$, so

$$w^2 + 2wz + z^2 = (w + z)^2 = [(5 - 3i) + (1 + 2i)]^2 = (6 - i)^2 = 6^2 + 2(6)(-i) + (-i)^2.$$

In simplifying the last expression on the right, we note that $(-i)^2 = (-1)^2(i)^2 = -1$, so we have $w^2 + 2wz + z^2 = 36 - 12i - 1 = 35 - 12i$.

> **Concept:**   Factoring and other manipulations can make some calculations easier.

(e) The given expression factors as $wz(2w + 3z)$, so

$$2w^2z + 3wz^2 = wz(2w + 3z) = (5 - 3i)(1 + 2i)[2(5 - 3i) + 3(1 + 2i)]$$
$$= (5 - 3i)(1 + 2i)(13) = (11 + 7i)(13) = 143 + 91i.$$

Notice how factoring saved a lot of work. Without factoring we would have had to perform several multiplications of binomials, but factoring allowed us to get the answer with just one.

□

Problem 3.3: Evaluate i^{2007}.

Solution for Problem 3.3:   At first, it's not obvious how to compute i^{2007}, so we experiment with smaller powers of i, hoping to find a pattern. We have $i^1 = i$, $i^2 = -1$, $i^3 = i^2 \cdot i = -i$, and $i^4 = i^2 \cdot i^2 = (-1)^2 = 1$. Then, the powers repeat: $i^5 = i^4 \cdot i = i$, $i^6 = i^4 \cdot i^2 = i^2$, $i^7 = i^4 \cdot i^3 = i^3$, and $i^8 = i^4 \cdot i^4$. Therefore, we see that:

> **Important:**   The powers of i repeat in blocks of 4:
>
> $$
> \begin{aligned}
> i^1 &= i^5 = i^9 = \cdots = i, \\
> i^2 &= i^6 = i^{10} = \cdots = -1, \\
> i^3 &= i^7 = i^{11} = \cdots = -i, \\
> i^4 &= i^8 = i^{12} = \cdots = 1.
> \end{aligned}
> $$

Because 2007 leaves a remainder of 3 when divided by 4, we have $i^{2007} = i^3 = -i$. □

In Chapter 2, we reviewed functions with real domains and real ranges. We'll continue practicing complex number arithmetic by considering a function that accepts nonreal inputs and has nonreal numbers in its range.

Problem 3.4: Let $f(x) = (-2 + 3i)x^2 - (7 + 2i)x + 37 - 13i$, where the domain of f is all complex numbers. Determine each of the following:

(a) $f(1)$ (b) $f(i)$ (c) $f(2 + i)$

Solution for Problem 3.4:   Just as with the real functions we have studied, we substitute each input for x to determine the value of the function f for that input.

(a) $f(1) = (-2 + 3i) \cdot 1^2 - (7 + 2i) \cdot 1 + 37 - 13i = 28 - 12i$.

(b) $f(i) = (-2 + 3i) \cdot i^2 - (7 + 2i) \cdot i + 37 - 13i = (-2 + 3i)(-1) - (7i - 2) + 37 - 13i = 41 - 23i$.

(c) $f(2 + i) = (-2 + 3i) \cdot (2 + i)^2 - (7 + 2i) \cdot (2 + i) + 37 - 13i = (-2 + 3i)(3 + 4i) - (12 + 11i) + 37 - 13i = (-18 + i) - (12 + 11i) + 37 - 13i = 7 - 23i.$

□

As we've seen in our examples so far in this chapter, the product of two complex numbers is itself a complex number. Let's take a look at when the product of two nonreal complex numbers turns out to be a real number.

Problem 3.5:

(a) Find all values of k such that the product $(3 + 2i)(3 + ki)$ is a real number.

(b) Let $z = a + bi$, where a and b are real numbers, and let $\bar{z} = a - bi$. Prove that $z \cdot \bar{z}$ is real.

Solution for Problem 3.5:

(a) Expanding the given product, we have $(3 + 2i)(3 + ki) = 9 - 2k + (6 + 3k)i$. The result is a real number when $6 + 3k = 0$, so $k = -2$.

(b) In part (a) we saw that the product of the nonreal numbers $3 + 2i$ and $3 - 2i$ is real. Notice that the only difference between these two numbers is the sign of the imaginary part. Let's see if we can generalize our observation by checking whether or not the product $(a + bi)(a - bi)$ is a real number. We find
$$(a + bi)(a - bi) = a(a - bi) + bi(a - bi) = a^2 - abi + abi - b^2i^2 = a^2 + b^2,$$
which is real because a and b are real.

□

Definition: If $z = a + bi$, then $a - bi$ is called the **conjugate** of z. We denote the conjugate of z as $\bar{z}$. So, we have $\overline{a + bi} = a - bi$. Together, a number and its conjugate are referred to as a **conjugate pair**.

Important: The product of a complex number and its conjugate is a real number.

Note that in order to prove this fact in Problem 3.5, we wrote a general complex number in terms of its real and imaginary parts by writing $z = a + bi$.

Concept: In general, if you are given a problem involving a general complex number z, it may help to write z in the form $a + bi$, and then express the problem in terms of a and b.

We've covered addition, subtraction, and multiplication of complex numbers. Let's put conjugates to work by turning to division.

Problem 3.6: Write $\dfrac{1}{3 + 2i}$ as a complex number. In other words, find real numbers a and b such that $\dfrac{1}{3 + 2i} = a + bi.$

Solution for Problem 3.6: Knowing that $(3 + 2i)(3 - 2i)$ is real, we multiply the given fraction by $\frac{3-2i}{3-2i}$, which equals 1, to produce an equivalent fraction with a real denominator:

$$\frac{1}{3 + 2i} = \frac{1}{3 + 2i} \cdot \frac{3 - 2i}{3 - 2i} = \frac{3 - 2i}{(3 + 2i)(3 - 2i)} = \frac{3 - 2i}{3^2 + 2^2} = \frac{3 - 2i}{13}.$$

Therefore, we have $\dfrac{1}{3 + 2i} = \dfrac{3}{13} - \dfrac{2}{13}i.$ $\square$

The fact that $z\bar{z}$ is real for all complex numbers z gives us a way to divide by a complex number.

> **Important:** If w and z are complex numbers and $z \neq 0$, then we express the quotient w/z as a complex number by multiplying both the numerator and denominator by the conjugate of the denominator, $\bar{z}$:
>
> $$\frac{w}{z} = \frac{w}{z} \cdot \frac{\bar{z}}{\bar{z}} = \frac{w\bar{z}}{z\bar{z}}.$$
>
> Since $z\bar{z}$ is always real, we can easily express $w\bar{z}/(z\bar{z})$ as a complex number.

Let's give this technique some practice.

> **Problem 3.7:** Let $w = 2 - 5i$ and $z = -3 + i$. Express each of the following in the form $a + bi$, where a and b are real numbers.
>
> (a) $\dfrac{1}{w}$ (c) $\dfrac{z}{w}$ (e) $\dfrac{z}{1 + i} + \dfrac{\bar{z}}{1 - i}$
>
> (b) $\dfrac{1}{\bar{w}}$ (d) $\dfrac{\bar{z}}{\bar{w}}$

Solution for Problem 3.7:

(a) $\dfrac{1}{2 - 5i} \cdot \dfrac{2 + 5i}{2 + 5i} = \dfrac{2 + 5i}{(2 - 5i)(2 + 5i)} = \dfrac{2 + 5i}{2^2 + 5^2} = \dfrac{2}{29} + \dfrac{5i}{29}.$

(b) $\dfrac{1}{2 + 5i} \cdot \dfrac{2 - 5i}{2 - 5i} = \dfrac{2 - 5i}{(2 + 5i)(2 - 5i)} = \dfrac{2 - 5i}{2^2 + 5^2} = \dfrac{2}{29} - \dfrac{5i}{29}.$

(c) $\dfrac{-3 + i}{2 - 5i} \cdot \dfrac{2 + 5i}{2 + 5i} = \dfrac{(-3 + i)(2 + 5i)}{(2 - 5i)(2 + 5i)} = -\dfrac{11}{29} - \dfrac{13i}{29}.$

(d) $\dfrac{-3 - i}{2 + 5i} \cdot \dfrac{2 - 5i}{2 - 5i} = \dfrac{(-3 - i)(2 - 5i)}{(2 + 5i)(2 - 5i)} = -\dfrac{11}{29} + \dfrac{13i}{29}.$

(e) We could simplify each fraction, and then add. Instead, we recognize that the denominators in the problem are conjugates, so their product is real. Therefore, the common denominator of the two fractions is a real number:

$$\frac{-3 + i}{1 + i} + \frac{-3 - i}{1 - i} = \frac{(-3 + i)(1 - i) + (-3 - i)(1 + i)}{(1 + i)(1 - i)} = \frac{-2 + 4i + -2 - 4i}{1^2 + 1^2} = \frac{-4}{2} = -2.$$

$\square$

Notice that the answer to part (b) is the conjugate of the answer to part (a), and the answer to part (d) is the conjugate of the answer to part (c). This isn't a coincidence. We'll explore why in Section 3.3. See if you can figure out why on your own before getting there.

Problem 3.8: Find z such that $\dfrac{2z - 3i}{z + 4} = -5 + i$.

Solution for Problem 3.8: We begin by multiplying both sides by the denominator of the left side, which gives us $2z - 3i = (-5 + i)(z + 4)$. Expanding the right side then gives $2z - 3i = -5z - 20 + iz + 4i$. Putting all terms with z on one side and terms without z on the other gives us

$$7z - iz = -20 + 7i.$$

Factoring z out of the left side gives $z(7 - i) = -20 + 7i$. Dividing both sides by $7 - i$ then gives us

$$z = \frac{-20 + 7i}{7 - i} = \frac{-20 + 7i}{7 - i} \cdot \frac{7 + i}{7 + i} = \frac{-147 + 29i}{50} = -\frac{147}{50} + \frac{29}{50}i.$$

□

Sidenote: Mathematicians frequently use w and z as variables to represent complex numbers, and use x and y as variables to represent real numbers. Of course, this is not a hard-and-fast rule! As you'll see in this book, the variable x is so universally common that we will have many equations in terms of x to which some solutions turn out to be nonreal.

Exercises

3.1.1 Find all z such that $4z^2 + 12 = 0$.

3.1.2 Let $a = 3 + 4i$ and $b = 12 - 5i$. Compute each of the following:

(a) $a - b$

(b) ab

(c) $a^2 + 3a + 2$

(d) $\dfrac{a}{\overline{b}} + \dfrac{\overline{a}}{b}$

3.1.3 Write each of the following as a complex number:

(a) $\overline{2i} + \overline{7 - 2i}$

(b) $\dfrac{1}{2 + 3i + \overline{1 + 2i}}$

3.1.4 $f(x) = \dfrac{x^6 + x^4}{x + 1}$. Find each of the following:

(a) $f(i)$

(b) $f(-i)$

(c) $f(i - 1)$

3.1.5 What is $(i - i^{-1})^{-1}$? *(Source: AHSME)*

3.1.6 Solve each of the following equations for z:

(a) $\dfrac{z + 3i}{z - 3} = 2$

(b) $\dfrac{1 + 2i}{3z} = 4 + 5i$

(c) $3z + \dfrac{z}{1 + i} = 10 - 4i$

3.1.7 Let $S = i^n + i^{-n}$, where n is an integer. Find the total number of possible distinct values of S. *(Source: AHSME)*

3.2 The Complex Plane

A complex number can be represented as a point in the **complex plane**. Like the coordinate plane, the complex plane has two axes, a horizontal **real axis** for the real part, and a vertical **imaginary axis** for the imaginary part. So, plotting the complex number $x + yi$ on the complex plane is the same as plotting the point (x, y) in the Cartesian coordinate plane. The figure at right shows the numbers $1 + 2i$, $-5 - 3i$, and $2 - 4i$ plotted in the complex plane.

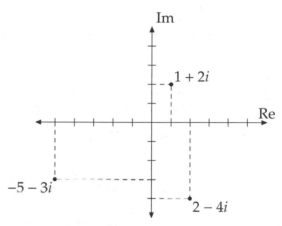

Just as the point $(0, 0)$ is called the origin on the Cartesian plane, the point that represents 0 on the complex plane is also called the **origin** of the complex plane.

The real axis of the complex plane is labeled Re and the imaginary axis is labeled Im. Similarly, we can use Re(z) to refer to the real part of the complex number z, and use Im(z) to refer to the imaginary part of z. So, for example Re($2 - 4i$) = 2 and Im($2 - 4i$) = -4. Some sources use $\Re(z)$ and $\Im(z)$ in place of Re(z) and Im(z).

| Problems |

Problem 3.9:

(a) Let $w = 2 + 3i$. Plot w, $\overline{w}$, and $-w$ on the complex plane.

(b) For any complex number z, how is z related to $\overline{z}$ on the complex plane? How is z related to $-z$ on the complex plane?

Problem 3.10:

(a) Let $w = 2 + 4i$ and $z = 3 - 2i$. Plot w, z, and $w + z$ on the complex plane. Connect the origin to w, then connect w to $w + z$, then connect $w + z$ to z, then connect z to the origin. Notice anything interesting?

(b) Suppose w and z are nonzero complex numbers such that w/z is not a real constant. (In other words, there is no real number a such that $w = az$.) Connect the origin to w, then connect w to $w + z$, then connect $w + z$ to z, then connect z to the origin. What type of quadrilateral must result? Why?

Problem 3.11:
(a) What is the distance between $3 + 4i$ and the origin in the complex plane?

(b) What is the distance between $a + bi$ and the origin in the complex plane?

Problem 3.12: Let $|a + bi| = \sqrt{a^2 + b^2}$.
(a) Evaluate $|5 - 12i|$ and $|3 - 3i|$.

(b) Suppose $w = 3 - 5i$ and $z = -2 + 7i$. Find $|w - z|$. Find the distance between w and z on the complex plane. Notice anything interesting?

(c) For any complex numbers w and z, how is $|w - z|$ related to the distance between w and z on the complex plane?

Problem 3.9:   For any complex number z, how is z related to $\bar{z}$ on the complex plane? How is z related to $-z$ on the complex plane?

Solution for Problem 3.9: We start by experimenting with a specific value of z.

> **Concept:** Trying specific examples can be a good way to discover general relation-
> ships.

Suppose $z = 2 + 3i$. Then, we have $\bar{z} = 2 - 3i$ and $-z = -2 - 3i$. We see that $\bar{z}$ is the reflection of z over the real axis and $-z$ is the reflection of $\bar{z}$ over the imaginary axis. Is this just a coincidence?

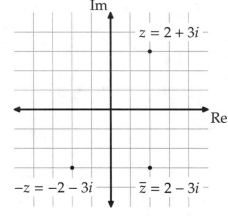

To see if this is true in general, we let $z = a + bi$. Then, we have $\bar{z} = a - bi$ and $-z = -a - bi$. So, z and $\bar{z}$ have the same real part and opposite imaginary parts. Therefore, z and $\bar{z}$ are reflections of each other over the real axis. Similarly, $\bar{z}$ and $-z$ have the same imaginary part, but opposite real parts, so $\bar{z}$ and $-z$ are reflections of each other over the imaginary axis.

But how are z and $-z$ related? Because $z = a + bi$ and $-z = -a - bi$, z and $-z$ are in exactly opposite directions from the origin. We also see that z and $-z$ are the same distance from the origin. Therefore, $-z$ is a 180° rotation of z around the origin on the complex plane. (Note: We sometimes refer to a 180° rotation of a point A around a point P as a "reflection of point A through point P.") □

> **Important:** For every complex number z, we have
>
>   • The point $-z$ is a 180° rotation of point z about the origin.
>
>   • The points z and $\bar{z}$ are reflections of each other over the real axis.
>
>   • The points $\bar{z}$ and $-z$ are reflections of each other over the imaginary axis.

Don't memorize these relationships! If you understand why they are true, they should become obvious to you.

Problem 3.10: Suppose w and z are nonzero complex numbers such that w/z is not a real constant. (In other words, there is no real number a such that $w = az$.) Connect the origin to w, then connect w to $w + z$, then connect $w + z$ to z, then connect z to the origin. What type of quadrilateral must result? Why?

Solution for Problem 3.10: Again, we start with an example. Suppose $w = 2 + 4i$ and $z = 3 - 2i$, so $w + z = 5 + 2i$. In the diagram at right, we connect the points as suggested. We appear to have a parallelogram.

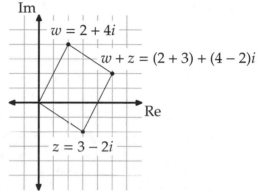

To see if this is always the case, we let $w = a + bi$ and $z = c + di$. So, we have $w + z = (a + c) + (b + d)i$. If $a \neq 0$, then the slope between w and the origin is b/a, as is the slope between $w + z$ and z. If $a = 0$, then both of these lines are vertical lines. In either case, these two sides of our quadrilateral are parallel. Similarly, the slopes of the other two sides are either both equal to d/c or they are both undefined (when both sides are vertical). So, these two sides are parallel as well. Therefore, both pairs of opposite sides of the quadrilateral are parallel, which means the quadrilateral is a parallelogram. $\square$

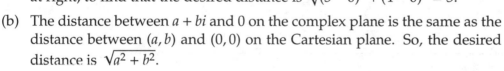

Problem 3.11:

(a) What is the distance between $3 + 4i$ and the origin in the complex plane?

(b) What is the distance between $a + bi$ and the origin in the complex plane?

Solution for Problem 3.11:

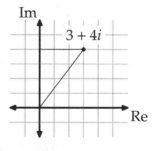

(a) Finding the distance between $3 + 4i$ and the origin on the complex plane is the same as finding the distance between $(3, 4)$ and $(0, 0)$ on the Cartesian plane. We can use the distance formula, or build a right triangle as shown at right, to find that the desired distance is $\sqrt{(3 - 0)^2 + (4 - 0)^2} = 5$.

(b) The distance between $a + bi$ and 0 on the complex plane is the same as the distance between (a, b) and $(0, 0)$ on the Cartesian plane. So, the desired distance is $\sqrt{a^2 + b^2}$.

$\square$

Just as we use $|x|$ to denote the distance between a real number x and 0 on the number line, we use $|z|$ to represent the distance between z and the origin on the complex plane. We call $|z|$ the **magnitude** of z.

> **Important:** The **magnitude** of z, denoted by $|z|$, equals the distance from z to the origin on the complex plane. If $z = a + bi$, we have
>
> $$|z| = \sqrt{a^2 + b^2}.$$

Problem 3.12:

(a) Evaluate $|5 - 12i|$ and $|3 - 3i|$.

(b) Suppose $w = 3 - 5i$ and $z = -2 + 7i$. Find $|w - z|$. Find the distance between w and z on the complex plane. Notice anything interesting?

(c) For any complex numbers w and z, how is $|w - z|$ related to the distance between w and z on the complex plane?

Solution for Problem 3.12:

(a) We have $|5 - 12i| = \sqrt{5^2 + (-12)^2} = 13$ and $|3 - 3i| = \sqrt{3^2 + (-3)^2} = 3\sqrt{2}$.

(b) We have $w - z = 5 - 12i$, and in the previous part we found that $|5 - 12i| = 13$. We plot the two points in the diagram at right. We see that the horizontal distance between them is 5 and the vertical distance between them is 12, so the segment connecting the two points is the hypotenuse of a right triangle with legs of lengths 5 and 12. Therefore, the Pythagorean Theorem tells us that the distance between w and z on the complex plane is $\sqrt{5^2 + 12^2} = 13$.

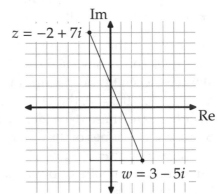

Notice that the distance between w and z equals $|w - z|$. Is this a coincidence?

(c) No, it's not a coincidence. Suppose $w = a + bi$ and $z = c + di$. Finding the distance between w and z in the complex plane is the same as finding the distance between (a, b) and (c, d) on the Cartesian plane. So, the distance between w and z is $\sqrt{(a - c)^2 + (b - d)^2}$.

Because $w - z = (a + bi) - (c + di) = (a - c) + (b - d)i$, we have

$$|w - z| = \sqrt{(a - c)^2 + (b - d)^2}.$$

This tells us that:

> **Important:** The distance between complex numbers w and z on the complex plane is $|w - z|$.

□

Exercises

3.2.1 Plot each of the following in the complex plane:

(a) $4 + 7i$

(b) $-6 - 2i$

(c) $(3 + i)(-2 + 5i)$

3.2.2 Find the magnitude of each of the following complex numbers:

(a) $24 - 7i$

(b) $2 + 2\sqrt{3}i$

(c) $(1 + 2i)(2 + i)$

3.2.3 Let $w = 3 + 5i$ and $z = 12 + 2i$. Find the area of the convex quadrilateral in the complex plane that has vertices $w, z, \overline{w},$ and $\overline{z}$.

3.2.4 Find the distance between the points $4 + 7i$ and $-3 - 17i$ in the complex plane.

3.2.5 Show that the midpoint of the segment connecting z_1 and z_2 on the complex plane is $(z_1 + z_2)/2$.

3.2.6

(a) Find the magnitude of $\dfrac{1 + 2i}{2 + i}$.

(b) Find the magnitude of $\dfrac{6 + 11i}{11 + 6i}$. (You can use a calculator for this part.)

(c) Notice anything interesting? Can you generalize your observations from the first two parts?

3.2.7★ Four complex numbers lie at the vertices of a square in the complex plane. Three of the numbers are $1 + 2i$, $-2 + i$ and $-1 - 2i$. What is the fourth number? *(Source: AMC 12)*

3.3   Real and Imaginary Parts

Problems

Problem 3.13: Let z and w be complex numbers.

(a) Let $z = a + bi$ and $w = c + di$, where $a, b, c,$ and d are real. Show that $\overline{z + w} = \overline{z} + \overline{w}$.

(b) Show that $\overline{zw} = \overline{z} \cdot \overline{w}$.

Problem 3.14:

(a) Show that $\overline{\overline{z}} = z$ for all complex numbers z.

(b) Show that $\overline{z} = z$ if and only if z is real.

(c) Show that $\overline{z} = -z$ if and only if z is imaginary.

Problem 3.15:

(a) Prove that $z\overline{z} = |z|^2$ for all complex numbers z.

(b) Prove that $|zw| = |z||w|$ for all complex numbers z and w.

Problem 3.16: Solve the equation $z + 2\overline{z} = 6 - 4i$ for z.

Problem 3.17: In this problem, we find all complex numbers z such that $z^2 = 21 - 20i$.

(a) Let $z = a + bi$ in the given equation. Find a system of equations involving a and b.

(b) Solve for b in terms of a in one of the equations, and substitute the expression you found for b into the other equation.

(c) Solve the equation you formed in part (b) for all possible values of a. **Hints:** 90

(d) Find all complex numbers z such that $z^2 = 21 - 20i$.

Problem 3.18: In this problem, we find $|z|$ if the complex number z satisfies $z + |z| = 2 + 8i$.

(a) Let $r = |z|$. Express z in terms of r.

(b) What is the real part of z? What is the imaginary part of z? Find $|z|$.

(c) Solve the problem again by letting $z = a + bi$.

(Source: AHSME)

Problem 3.13: Let z and w be complex numbers.

(a) Show that $\overline{z + w} = \bar{z} + \bar{w}$.

(b) Show that $\overline{zw} = \bar{z} \cdot \bar{w}$.

Solution for Problem 3.13: Let $z = a + bi$ and $w = c + di$, where $a, b, c,$ and d are real numbers.

(a) Since $z + w = a + bi + c + di = (a + c) + (b + d)i$, the conjugate of $z + w$ is

$$\overline{z + w} = \overline{(a + c) + (b + d)i} = (a + c) - (b + d)i.$$

Similarly, the conjugate of z is $\bar{z} = a - bi$, and the conjugate of w is $\bar{w} = c - di$, and their sum is $a - bi + c - di = (a + c) - (b + d)i$. So, we have $\overline{z + w} = \bar{z} + \bar{w}$.

(b) We have

$$zw = (a + bi)(c + di) = ac + adi + bci + bdi^2 = (ac - bd) + (ad + bc)i,$$

so the conjugate of zw is $\overline{zw} = (ac - bd) - (ad + bc)i$. Also, we have

$$\bar{z} \cdot \bar{w} = (a - bi)(c - di) = ac - adi - bci + bdi^2 = (ac - bd) - (ad + bc)i,$$

so $\overline{zw} = \bar{z} \cdot \bar{w}$.

$\square$

> **Important:** For any complex numbers w and z, we have $\overline{w + z} = \bar{w} + \bar{z}$ and $\overline{wz} = \bar{w} \cdot \bar{z}$.

We'll be seeing these two very important relationships again when we study nonreal roots of polynomials in Section 8.2.

Problem 3.14:

(a) Show that $\bar{\bar{z}} = z$ for all complex numbers z.

(b) Show that $\bar{z} = z$ if and only if z is real.

(c) Show that $\bar{z} = -z$ if and only if z is imaginary.

Solution for Problem 3.14: Let $z = a + bi$, where a and b are real numbers.

(a) Since $\bar{z} = a - bi$, we have $\bar{\bar{z}} = \overline{a - bi} = a + bi = z$.

(b) Since $\bar{z} = a - bi$, the equation $\bar{z} = z$ becomes $a - bi = a + bi$. Subtracting a from both sides gives $-bi = bi$. Therefore, we have $2bi = 0$, so $b = 0$. This gives us $z = a + bi = a + 0i = a$, so z is real. This shows that $\bar{z} = z$ only if z is real.

We're not finished with our proof! We have only shown that when $\bar{z} = z$, then z is real. (In other words, we have $\bar{z} = z$ *only if* z is real.) We have not shown that the equation $\bar{z} = z$ is true for all real numbers. (That is, that $\bar{z} = z$ *if* z is real.)

To show that $\bar{z} = z$ whenever z is real, we let $z = a$, where a is a real number. We then have $\bar{z} = \bar{a} = a$, so $\bar{z} = z$ if z is real.

> **Important:** Suppose A and B are two mathematical statements. The statement "A if and only if B" means both of the following:
>
> - If A is true, then B is also true.
>
> - If B is true, then A is also true.

Therefore, as we just saw, proving an "if and only if" statement requires proving two statements, not just one.

(c) Since $\bar{z} = a - bi$, the equation $\bar{z} = -z$ becomes $a - bi = -a - bi$. Adding bi to both sides gives $a = -a$, so $a = 0$. Therefore, we have $z = a + bi = bi$, so z is imaginary. This shows that $\bar{z} = -z$ only if z is imaginary.

As with part (b), this is an "if and only if" statement. We must also show that $\bar{z} = -z$ if z is imaginary. Letting $z = bi$, we have $\bar{z} = \overline{bi} = -bi = -z$, as desired.

□

> **Problem 3.15:**
> (a) Prove that $z\bar{z} = |z|^2$ for all complex numbers z.
> (b) Prove that $|zw| = |z||w|$ for all complex numbers z and w.

Solution for Problem 3.15: Let $z = a + bi$ and $w = c + di$ for real numbers a, b, c, and d.

(a) Since $\bar{z} = a - bi$, we have $z\bar{z} = (a + bi)(a - bi) = a^2 + b^2 = \left(\sqrt{a^2 + b^2}\right)^2 = |a + bi|^2 = |z|^2$.

> **Extra!** In 1806, bookkeeper Jean-Robert Argand published his concept of a geometric interpretation of complex numbers. Since he wasn't a professional mathematician, he used his knowledge of the bookselling industry to get his work out. A copy of his work reached the mathematician Legendre, who sent a copy to François Français, yet neither could figure out who the author was—apparently Argand had left out his name!
>
> François Français passed away, and his brother Jacques Français discovered Argand's book among his deceased brother's papers. Jacques Français then published a work in 1813 featuring a geometric representation of complex numbers, but instead of claiming that concept as his own, he mentioned in his paper that the idea was from some anonymous mathematician's work, and so Argand came forward and his work was recognized. For this reason, some now call the complex plane the **Argand plane**, and call graphs of complex number relationships on this plane **Argand diagrams**.

(b) First, we note that $zw = (a + bi)(c + di) = ac - bd + (ad + bc)i$. So, we have

$$|zw| = |ac - bd + (ad + bc)i|.$$

Next, we use the fact that $|r + si| = \sqrt{r^2 + s^2}$ for any real numbers r and s. Here, we have $r = ac - bd$ and $s = ad + bc$. So, we have:

$$
\begin{aligned}
|zw| = |ac - bd + (ad + bc)i| &= \sqrt{(ac - bd)^2 + (ad + bc)^2} \\
&= \sqrt{a^2c^2 - 2abcd + b^2d^2 + a^2d^2 + 2abcd + b^2c^2} \\
&= \sqrt{a^2c^2 + a^2d^2 + b^2c^2 + b^2d^2} \\
&= \sqrt{a^2(c^2 + d^2) + b^2(c^2 + d^2)} \\
&= \sqrt{(a^2 + b^2)(c^2 + d^2)} \\
&= \sqrt{a^2 + b^2}\sqrt{c^2 + d^2} \\
&= |a + bi||c + di| = |z||w|.
\end{aligned}
$$

We could also have used part (a) together with a little cleverness to solve part (b):

$$|zw|^2 = (zw)(\overline{zw}) = zw\overline{z} \cdot \overline{w} = (z\overline{z})(w\overline{w}) = |z|^2|w|^2.$$

Since all the magnitudes are nonnegative numbers, we have $|zw| = |z||w|$.

$\square$

> **Important:** For any complex numbers z and w, we have $z\overline{z} = |z|^2$ and $|zw| = |z||w|$.

Problem 3.16: Solve $z + 2\overline{z} = 6 - 4i$ for z.

Solution for Problem 3.16: Letting $z = a + bi$, for some real numbers a and b, the given equation becomes $(a + bi) + 2(a - bi) = 6 - 4i$. Simplifying the left side gives $3a - bi = 6 - 4i$. Since two complex numbers are equal if and only if their real parts are equal and their imaginary parts are equal, we have $3a = 6$ and $-b = -4$. Solving these two equations gives $a = 2$ and $b = 4$, so $z = a + bi = 2 + 4i$. $\square$

Problem 3.17: Find all complex numbers z such that $z^2 = 21 - 20i$.

Solution for Problem 3.17: We might start by taking the square root of both sides, so that $z = \pm\sqrt{21 - 20i}$. But what does $\sqrt{21 - 20i}$ really mean? The real-valued function $f(x) = \sqrt{x}$ is only defined for $x \geq 0$; if $x \geq 0$, then $\sqrt{x}$ is the nonnegative number whose square is x. Unfortunately, this doesn't help us define the square root of a nonreal number. Instead of trying to define $\sqrt{21 - 20i}$, let's look for complex numbers whose square is $21 - 20i$.

Let $z = a + bi$, so we have $(a + bi)^2 = 21 - 20i$. Expanding the left side gives

$$a^2 - b^2 + 2abi = 21 - 20i.$$

Since a and b are real, we can equate the real parts of both sides to get $a^2 - b^2 = 21$, and equate the imaginary parts to find $2ab = -20$. Solving $2ab = -20$ for b gives $b = -10/a$. Substituting this into $a^2 - b^2 = 21$ gives

$$a^2 - \frac{100}{a^2} = 21.$$

Multiplying this by a^2 gives us $a^4 - 21a^2 - 100 = 0$. Letting $c = a^2$ in this equation yields the quadratic equation $c^2 - 21c - 100 = 0$, which has roots 25 and -4. However, a is real and $a^2 = c$, so c cannot be -4.

Therefore, we have $c = a^2 = 25$, so $a = \pm 5$. Since $b = -10/a$ we have $b = -2$ when $a = 5$, and $b = 2$ when $a = -5$. So, the two possible values of z are $\pm(5 - 2i)$. $\square$

Having found the two complex numbers whose squares equal $21 - 20i$, we can see some of the difficulty that arises from defining $\sqrt{21 - 20i}$. Which of the values $5 - 2i$ or $-5 + 2i$ should we take to be the square root? We can't simply call one "positive" and one "negative."

Due to these difficulties, we try to avoid using expressions like $\sqrt{21 - 20i}$, and instead think of the "numbers whose square is $21 - 20i$." We may sometimes refer to these two numbers as "the square roots" of $21 - 20i$, and may even refer to one or the other as "a square root" of $21 - 20i$. But we generally won't choose one or the other to equal $\sqrt{21 - 20i}$ the way we choose 5 to equal $\sqrt{25}$.

We have these same difficulties when trying to define $\sqrt{-81}$. For example, which of $9i$ or $-9i$ is "the" square root of -81? However, as we saw in Problem 3.1, we may occasionally stumble on expressions like $\sqrt{-81}$ when solving equations like $z^2 = -81$. When this happens, the square root should always be preceded by $\pm$, as in $z = \pm\sqrt{-81}$. In other words, we are referring to both "square roots" of -81; that is, we are considering both numbers whose squares are -81.

Problem 3.18: The complex number z satisfies $z + |z| = 2 + 8i$. What is $|z|$? (*Source: AHSME*)

Solution for Problem 3.18: Solution 1: Keep your eye on the ball. We want $|z|$, and our equation involves $|z|$, so we let $|z| = r$. Then, we have $z = 2 + 8i - |z| = (2 - r) + 8i$. Written this way, we can see that the complex number z has real part $2 - r$ and imaginary part 8. Therefore, we have

$$|z|^2 = (2 - r)^2 + 8^2 = r^2 - 4r + 68.$$

Concept: Often, dealing with $|z|^2$ is nicer than dealing with $|z|$, since squaring $|z|$ gets rid of the square root sign in the definition of $|z|$.

We also have $|z|^2 = r^2$, and combining this with our equation above gives $r^2 = r^2 - 4r + 68$. Solving this equation gives us $r = 17$, so $|z| = 17$.

Solution 2: Let $z = a + bi$. Then, the given equation is

$$a + bi + \sqrt{a^2 + b^2} = 2 + 8i.$$

The only imaginary term on the left is bi, so we have $b = 8$. Equating the real terms of both sides then gives $a + \sqrt{a^2 + 64} = 2$. Isolating the radical gives $\sqrt{a^2 + 64} = 2 - a$ and squaring both sides gives $a^2 + 64 = 4 - 4a + a^2$. Solving this equation gives us $a = -15$. Because we squared the equation $\sqrt{a^2 + 64} = 2 - a$, we must check if the solution is extraneous. It isn't, so $|z| = |-15 + 8i| = 17$. $\square$

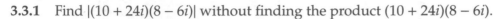

Exercises

3.3.1 Find $|(10 + 24i)(8 - 6i)|$ without finding the product $(10 + 24i)(8 - 6i)$.

3.3.2 Show that $|z| = |\bar{z}|$.

3.3.3 Solve $3z + 4\bar{z} = 12 - 5i$ for z.

3.3.4 Solve each of the following equations:

    (a) $z^2 = 2i$ (b) $z^2 = -5 + 12i$ (c) $z^2 = 24 - 10i$

3.3.5 Let $z = 3 + 4i$ and $w = 5 - 12i$. Evaluate the following expressions:

    (a) $\left|\dfrac{1}{z}\right|$ (b) $\dfrac{1}{|z|}$ (c) $\left|\dfrac{z}{w}\right|$ (d) $\dfrac{|z|}{|w|}$

Compare your answers in parts (a) and (b), and compare your answers in parts (c) and (d). Notice anything interesting? (You should!) Prove that your observations will hold for any nonzero complex numbers.

3.3.6 Find all complex numbers z such that $z/\bar{z}$ is

    (a)   a real number.   (b)   an imaginary number.

3.3.7★ Two solutions of $x^4 - 3x^3 + 5x^2 - 27x - 36 = 0$ are pure imaginary numbers. Find these two solutions. *(Source: ARML)* **Hints:** 213

3.4 Graphing in the Complex Plane

We graph an equation in terms of x and y in the Cartesian plane by plotting all the points (x, y) that satisfy the equation. Similarly, we graph an equation in terms of a complex number z in the complex plane by plotting all the values of z that satisfy the equation.

Problems

Problem 3.19: Show that $\text{Re}(z) = \dfrac{z + \bar{z}}{2}$ and $\text{Im}(z) = \dfrac{z - \bar{z}}{2i}$.

Problem 3.20: Graph in the complex plane all complex numbers z that satisfy each of the following:

    (a) $\dfrac{z + \bar{z}}{2} = 6$ (c) $(3 + 2i)z + (3 - 2i)\bar{z} = 36$

    (b) $z - \bar{z} = -2\sqrt{7}i$

Problem 3.21: Let w be a constant complex number and c be a nonnegative real constant. Describe the graphs of each of the following:

    (a) $|z| = 3$ (c) $|z + 5 - 4i| = 2\sqrt{2}$

    (b) $|z - 4| = 3$ (d) $|z - w| = c$

Problem 3.22: Find and graph all z such that $|z - 3| = |z + 2i|$.

Problem 3.19: Show that $\text{Re}(z) = \dfrac{z + \bar{z}}{2}$ and $\text{Im}(z) = \dfrac{z - \bar{z}}{2i}$.

Solution for Problem 3.19: Let $z = a + bi$ so that $\bar{z} = a - bi$. Then

$$\frac{z + \bar{z}}{2} = \frac{(a + bi) + (a - bi)}{2} = \frac{2a}{2} = a,$$

$$\frac{z - \bar{z}}{2i} = \frac{(a + bi) - (a - bi)}{2i} = \frac{2bi}{2i} = b.$$

So, $\text{Re}(z) = a = \dfrac{z + \bar{z}}{2}$ and $\text{Im}(z) = b = \dfrac{z - \bar{z}}{2i}$. $\square$

 Important: For any complex number z, we have

$$\text{Re}(z) = \frac{z + \bar{z}}{2} \quad \text{and} \quad \text{Im}(z) = \frac{z - \bar{z}}{2i}.$$

Problem 3.20: Graph in the complex plane all complex numbers z that satisfy each of the following:

(a) $\dfrac{z + \bar{z}}{2} = 6$ (b) $z - \bar{z} = -2\sqrt{7}i$ (c) $(3 + 2i)z + (3 - 2i)\bar{z} = 36$

Solution for Problem 3.20:

(a) In Problem 3.19, we found that $\text{Re}(z) = \dfrac{z + \bar{z}}{2}$. So, we seek all complex numbers z such that $\text{Re}(z) = 6$. These form a vertical line in the complex plane. This line is shown in the graph on the left below.

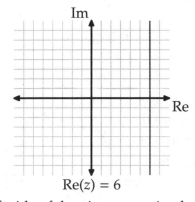

Re(z) = 6

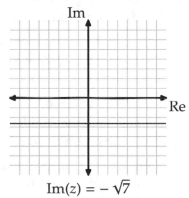

Im(z) = $-\sqrt{7}$

(b) The left side of the given equation looks a lot like our expression for $\text{Im}(z)$ from Problem 3.19. We only have to divide both sides by $2i$ to match this form on the left side. Doing so gives us

$$\frac{z - \bar{z}}{2i} = -\sqrt{7},$$

so we have $\text{Im}(z) = -\sqrt{7}$. All z with imaginary part $-\sqrt{7}$ satisfy this equation, so the graph of the given equation is the horizontal line $\sqrt{7}$ units below the real axis, which is shown at right above.

(c) We can't easily relate the given equation to either of the forms we found in Problem 3.19. So, we start by writing z as $a + bi$, where a and b are real numbers. The given equation becomes

$$(3 + 2i)(a + bi) + (3 - 2i)(a - bi) = 36.$$

Expanding and rearranging this equation gives us $3a - 2b = 18$. This is simply a linear equation. To graph it on the complex plane, we note that $a = \text{Re}(z)$ and $b = \text{Im}(z)$, so we have

$$3\text{Re}(z) - 2\text{Im}(z) = 18.$$

$3\text{Re}(z) - 2\text{Im}(z) = 18$

The graph of this line is shown at right. (If you don't see why this is the graph of the equation, try graphing the line $3x - 2y = 18$ on the Cartesian plane. In the complex plane, $\text{Re}(z)$ takes the place of x and $\text{Im}(z)$ takes the place of y.)

□

In the next problem, we explore how distance can help us graph in the complex plane.

Problem 3.21: Let w be a constant complex number and c be a nonnegative real constant. Describe the graphs of each of the following:

(a) $|z| = 3$

(b) $|z - 4| = 3$

(c) $|z + 5 - 4i| = 2\sqrt{2}$

(d) $|z - w| = c$

Solution for Problem 3.21:

(a) If $|z| = 3$, then the distance from z to the origin of the complex plane is 3. Therefore, z must be on the circle centered at the origin with radius 3. Moreover, every point z on this circle satisfies the equation $|z| = 3$, because every point on this circle is 3 units from the origin. Therefore, the graph $|z| = 3$ describes a circle of radius 3 centered the origin.

(b) The equation $|z - 4| = 3$ tells us that the distance between z and 4 in the complex plane is 3, so the graph is a circle of radius 3 centered at the point that represents 4. Since $4 = 4 + 0i$, the center of the circle is on the real axis, 4 units to the right of the origin.

(c) Writing $z + 5 - 4i$ as $z - (-5 + 4i)$, we see that the graph of the equation $|z - (-5 + 4i)| = 2\sqrt{2}$ consists of all points that are $2\sqrt{2}$ from the point $-5 + 4i$. Therefore, the graph is a circle of radius $2\sqrt{2}$ centered at $-5 + 4i$.

(d) Generalizing our work from parts (a)-(c), we note that the graph of $|z - w| = c$ consists of all points that are a distance c from the complex number w. In other words, the graph is a circle of radius c centered at w (shown at right for $w = 2 + i$ and $c = 6$).

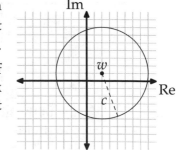

□

Problem 3.22: Find and graph all z such that $|z - 3| = |z + 2i|$.

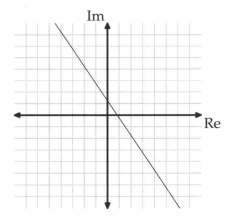

Solution for Problem 3.22: Solution 1: Algebra. Letting $z = a + bi$, for real numbers a and b, the equation becomes $|a + bi - 3| = |a + bi + 2i|$, or

$$|(a - 3) + bi| = |a + (b + 2)i|.$$

Applying the definition of the magnitude of a complex number to both sides of this equation, we have

$$\sqrt{(a - 3)^2 + b^2} = \sqrt{a^2 + (b + 2)^2}.$$

Squaring both sides of this equation and expanding the squares of binomials gives

$$a^2 - 6a + 9 + b^2 = a^2 + b^2 + 4b + 4.$$

Simplifying this equation gives $6a + 4b = 5$, so our equation is $6\text{Re}(z) + 4\text{Im}(z) = 5$. The graph of this equation is a line, shown at right above. We can write the equation $6\text{Re}(z) + 4\text{Im}(z) = 5$ in terms of z and $\bar{z}$ by noting that $\text{Re}(z) = (z + \bar{z})/2$ and $\text{Im}(z) = (z - \bar{z})/(2i)$. Substituting these into $6\text{Re}(z) + 4\text{Im}(z) = 5$ gives us $(3 - 2i)z + (3 + 2i)\bar{z} = 5$.

Solution 2: Geometric intuition. We rewrite the equation so that both magnitudes are of differences between complex numbers, $|z - 3| = |z - (-2i)|$. The quantity on the left side of the equation is the distance from z to 3 in the complex plane. Similarly, the right side is the distance from z to $-2i$ in the complex plane. Therefore, our equation tells us that z is equidistant from 3 and $-2i$ in the complex plane. So, z must be on the perpendicular bisector of the segment that connects 3 and $-2i$ in the complex plane.

The slope of the line connecting 3 and $-2i$ is $2/3$, so the slope of its perpendicular bisector is $-\frac{3}{2}$. The perpendicular bisector passes through the midpoint of 3 and $-2i$, which is $\frac{3-2i}{2} = \frac{3}{2} - i$. Using this point and the slope of the line, we can write a point-slope form of the line:

$$\text{Im}(z) - (-1) = -\frac{3}{2}\left(\text{Re}(z) - \frac{3}{2}\right).$$

Rearranging this equation gives us $6\text{Re}(z) + 4\text{Im}(z) = 5$, as before. $\square$

The key to our second solution was using the geometric interpretations of $|z - 3|$ and $|z + 2i|$.

> **Concept:** Many complex number concepts have a strong geometric significance as well as an algebraic one, so if you have trouble understanding an algebraic relationship regarding complex numbers, try considering it in the complex plane.

Exercises

3.4.1 Graph each of the following equations on the complex plane.

(a) $z + \bar{z} = 4$

(b) $(1 - 2i)z + (-1 - 2i)\bar{z} = -6i$

(c) $|z - 2 + i| = 6$

(d) $|z - i| = |z + i|$

3.4.2   In each of the following parts, find an equation whose graph in the complex plane is the following:

(a)   A circle centered at $-2 + 5i$ with radius $3\sqrt{2}$.

(b)   The perpendicular bisector of the segment connecting $4 + i$ and $7 - 2i$.

(c)   The line that passes through $3 + i$ and has slope 2.

3.4.3   Let S be the set of points z in the complex plane such that $(3 + 4i)z$ is a real number. Describe the graph of S. *(Source: AMC 12)*

3.4.4★   Let O be the origin of the complex plane, and let w and z be two nonzero complex numbers. Draw the line joining O to w, and the line joining O to z. Show that these two lines are perpendicular if and only if w/z is imaginary.

3.4.5★   Let u be a complex number, and let v be the reflection of u over the line $\text{Re}(z) = \text{Im}(z)$. Express v in terms of u and $\bar{u}$. **Hints:** 301

3.5   Summary

A **complex number** is a number of the form $a + bi$, where a and b are real numbers and $i^2 = -1$.

Definitions: For the complex number $a+bi$, where a and b are real, we have the following definitions:

- We call a the **real part** and b the **imaginary part** of $a + bi$. We use the notation $\text{Re}(z)$ and $\text{Im}(z)$ to denote the real and imaginary parts of a complex number z. So, if $z = a + bi$, then we have $\text{Re}(z) = a$ and $\text{Im}(z) = b$.

- The **conjugate** of $a + bi$ is $a - bi$, and is denoted by $\overline{a + bi}$.

- The **magnitude** of $a + bi$ is defined as $\sqrt{a^2 + b^2}$, and is denoted by $|a + bi|$.

We add two complex numbers by adding their real parts and adding their imaginary parts:

$$(a + bi) + (c + di) = a + c + bi + di = (a + c) + (b + d)i.$$

We multiply two complex numbers by using the distributive property and the fact that $i^2 = -1$:

$$(a + bi)(c + di) = (ac - bd) + (ad + bc)i.$$

For every complex number $z = a + bi$, the number $z\bar{z} = a^2 + b^2$ is real. We can use this to write the ratio of two complex numbers as a single complex number:

$$\frac{w}{z} = \frac{w\bar{z}}{z\bar{z}} = \frac{w\bar{z}}{|z|^2}.$$

If $a + bi = c + di$ where $a, b, c,$ and d are real numbers, then $a = c$ and $b = d$. In other words, complex numbers are equal if and only if both their real and imaginary parts are equal.

> **Concept:** In general, if you are given a problem involving a complex number z, it may help to write z in the form $a + bi$, and then express the problem in terms of a and b.

For any complex numbers w and z, we have

(i) $\overline{z + w} = \overline{z} + \overline{w}$.

(ii) $\overline{zw} = \overline{z} \cdot \overline{w}$.

(iii) $\overline{\overline{z}} = z$.

(iv) $z\overline{z} = |z|^2$.

(v) $|zw| = |z||w|$.

(vi) $\text{Re}(z) = \dfrac{z + \overline{z}}{2}$ and $\text{Im}(z) = \dfrac{z - \overline{z}}{2i}$.

A complex number can be represented as a point in the **complex plane**. Like the coordinate plane, the complex plane has two axes, the **real axis** for the real part, and the **imaginary axis** for the imaginary part. So, the complex number $x + yi$ corresponds to the point (x, y) in the Cartesian coordinate plane. The figure at right shows the numbers $1 + 2i$, $5 - 3i$, and $2 - 4i$ plotted in the complex plane. Just as we can graph equations in terms of two real variables on the Cartesian plane, we can graph equations in terms of a single complex variable on the complex plane.

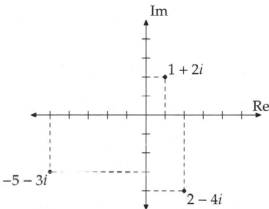

The magnitude of a complex number has a geometric significance, as well. For any complex numbers w and z, the expression $|z|$ equals the distance from z to the origin in the complex plane, and the expression $|w - z|$ equals the distance between w and z in the complex plane.

Problem Solving Strategies

> **Concepts:**
> - Factoring and other manipulations can make some calculations easier.
>
> - Trying specific examples can be a good way to discover general relationships.
>
> - Often, dealing with $|z|^2$ is nicer than dealing with $|z|$, since squaring $|z|$ gets rid of the square root sign in the definition of $|z|$.
>
> - Many complex number concepts have a strong geometric significance as well as an algebraic one.

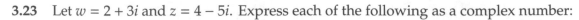

REVIEW PROBLEMS

3.23 Let $w = 2 + 3i$ and $z = 4 - 5i$. Express each of the following as a complex number:

(a) $2w - 3z$

(b) $\dfrac{1}{w}$

(c) $\dfrac{2}{\overline{w} + z}$

(d) $\dfrac{w^3 + 2w^2 z + wz^2}{w^2 z + wz^2}$

(e) $\dfrac{-y + xi}{x + yi}$

(f) $\dfrac{(1-i)^4}{(1+i)^3}$

3.24 Solve the equation $-3 - x^2 = 9 + x^2$.

3.25 Simplify $\dfrac{1}{1 + \dfrac{1}{1 - \dfrac{1}{1+i}}}$.

3.26 Find the complex number z such that $\dfrac{z}{1+z} = -1 + i$.

3.27 Four complex numbers are plotted in the plane, as shown. One of them is z. The other three are $-z$, $\overline{z}$, and $-\overline{z}$. Which is which?

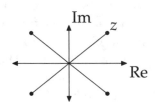

3.28 Let w and z be complex numbers.

(a) Show that $w\overline{z} + \overline{w}z$ is real.

(b) Show that $w\overline{z} - \overline{w}z$ is imaginary.

3.29 Find two complex numbers whose squares equal $5 - 12i$.

3.30 Find all real numbers c such that $7 + i$ is 5 units from $10 + ci$ on the complex plane.

3.31 Find the area of the region enclosed by the graph of $|z - 4 + 5i| = 2\sqrt{3}$.

3.32 The diagram to the right shows several numbers in the complex plane. The circle has radius 1 and is centered at the origin. One of these numbers is the reciprocal of F. Which one? *(Source: AHSME)*

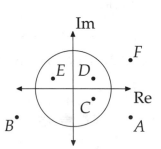

3.33 Find all complex numbers z such that $|z + 1 - i| = |z - 2|$.

3.34 The product of the complex numbers $a + bi$ and $c + di$ is real, where a, b, c, and d are real numbers. Prove that the product of the complex numbers $b + ai$ and $d + ci$ is also real.

3.35 Describe the graph of each of the following:

(a) $z - \overline{z} = -8i$

(b) $(4-i)z - (4+i)\overline{z} = 16i$

(c) $|7 + i - 2z| = 4$

3.36 Graph $|z - 5 + 2i| \le 4$ on the complex plane.

3.37 Simplify $(i+1)^{3200} - (i-1)^{3200}$.

3.38 Show that $|z - 1|^2 + |z + 1|^2 = 4$ for all z such that $|z| = 1$.

Challenge Problems

3.39 Evaluate $i + 2i^2 + 3i^3 + 4i^4 + \cdots + 64i^{64}$.

3.40 Find all complex numbers z such that $|z - 1| = |z + 3| = |z - i|$. *(Source: ARML)* **Hints:** 163

3.41 Find the area of the region of points z in the complex plane such that

$$|z + 4 - 4i| \leq 4\sqrt{2} \qquad \text{and} \qquad |z - 4 - 4i| \geq 4\sqrt{2}.$$

3.42 Let w and z be complex numbers and $z \neq 0$. Prove that $\dfrac{w}{z} + \dfrac{\overline{w}}{z}$ is a real number.

3.43 Show that the points w, x, y, and z are the vertices of a parallelogram in the complex plane if and only if the sum of some two of them is equal to the sum of the other two. **Hints:** 56

3.44★ Define a sequence of complex numbers by $z_1 = 0$, $z_{n+1} = z_n^2 + i$ for $n \geq 1$. How far away from the origin is z_{111}? *(Source: AHSME)* **Hints:** 60

3.45★ A function f is defined on the complex numbers by $f(z) = (a + bi)z$, where a and b are positive numbers. This function has the property that the image of each point in the complex plane is equidistant from that point and the origin. Given that $|a + bi| = 8$, find the value of b^2. *(Source: AIME)* **Hints:** 330, 146

3.46 For a nonzero complex number z, let $f(z) = 1/z$.

  (a) Show that $f(f(z)) = z$ for all $z \neq 0$.

  (b)★ Let $w = f(z)$. As z varies along the line $(1 + 2i)z - (1 - 2i)\bar{z} = i$, what curve does w trace?
     Hints: 59

3.47★ Find all ordered pairs (a, b) such that $a + \dfrac{a + 8b}{a^2 + b^2} = 2$ and $b + \dfrac{8a - b}{a^2 + b^2} = 0$. **Hints:** 143

3.48★ Let u and v be complex numbers such that $|u| = |v| = 1$, $u \neq v$, and $u \neq -v$. Show that

$$\frac{4uv}{(u + v)^2}$$

is real, and that

$$\frac{u + v}{u - v}$$

is imaginary. **Hints:** 334

Extra!
➠➠➠➠ Suppose we defined a "special number" as an ordered pair of real numbers, so that $(3, -2)$ and $(-2, 7)$ are "special numbers." Next, let's define addition and multiplication of these "special numbers" as follows:

$$(a, b) + (c, d) = (a + c, b + d),$$
$$(a, b) \cdot (c, d) = (ac - bd, ad + bc).$$

It turns out that this definition of "special numbers" can be used as an alternative definition of complex numbers. To get a feel for why, let's compare a few sums and products of "special numbers" with their i-notation counterparts:

$$(3, 7) + (-3, 4) = (0, 11) \qquad\qquad (3, 4) \cdot (-2, 1) = (-10, -5)$$
$$(3 + 7i) + (-3 + 4i) = 11i \qquad\qquad (3 + 4i) \cdot (-2 + i) = -10 - 5i$$

$$(0, 1) \cdot (0, 1) = (-1, 0) \qquad\qquad (4, -1) \cdot (4, 1) = (17, 0)$$
$$i \cdot i = -1 \qquad\qquad (4 - i)(4 + i) = 17$$

So, we see that the first number in the ordered pair (a, b) of a "special number" corresponds to what we have been calling the "real part" of a complex number and the second number in the ordered pair corresponds to what we have been calling the "imaginary part." (If you don't see why, note that $(1, 0) \cdot (1, 0) = (1, 0)$, but $(0, 1) \cdot (0, 1) = (-1, 0)$.)

We can also define the magnitude of (a, b) the same way we have defined $|a + bi|$ by letting $(a, b) = \sqrt{a^2 + b^2}$. Moreover, we can define the conjugate of (a, b) as $\overline{(a, b)} = (a, -b)$. Just to make sure, let's check the product of (a, b) and its "conjugate:"

$$(a, b) \cdot (a, -b) = (a^2 + b^2, -ab + ab) = (a^2 + b^2, 0).$$

As expected, the second component of the product is 0.

This ordered pair notation very obviously lends itself to the geometric depiction of complex numbers on the complex plane. Moreover, it was this formal definition of complex numbers that helped mathematicians accept the concept of nonreal numbers.

The first use of imaginary numbers is often attributed to Girolamo Cardano in the 16th century, who referred to square roots of negative numbers. For around two centuries thereafter, mathematicians used the notation $\sqrt{-1}$ until the great mathematician Leonhard Euler introduced i in 1777.

Despite all the interesting results mathematicians found regarding imaginary numbers, many were very uncomfortable with them, since they appeared to have no "physical" meaning. Mathematicians would add graphical representations of complex numbers in the 17th and 18th centuries, but it wasn't until 1833 that Lord William Rowan Hamilton gave complex numbers the definition described above. With this description of complex numbers, mathematicians could use complex numbers without having to ever encounter such an intuitively "unreal" concept as $\sqrt{-1}$.

(Source: MacTutor History of Mathematics archive)

I'm very well acquainted, too, with matters mathematical,
I understand equations, both the simple and quadratical. – Gilbert & Sullivan

CHAPTER **4**

_____Quadratics

An expression of the form

$$ax^2 + bx + c,$$

where a, b, and c are constants and $a \neq 0$, is called a **quadratic**. The solutions to the equation $ax^2 + bx + c = 0$ are called the **roots** of the quadratic. The roots are also sometimes called **zeros** of the quadratic. The ax^2 term of $ax^2 + bx + c$ is called the **quadratic term**, the bx term is the **linear term**, and c is the **constant term**. In this chapter, we review techniques for manipulating and solving quadratic equations.

4.1 Factoring Quadratics

When we expand the product of two linear expressions, such as

$$(x - 2)(x - 3) = x^2 - 5x + 6,$$

the result is a quadratic. When we reverse this process, writing a quadratic as the product of two linear terms, we say we **factor** the quadratic.

Problems

Problem 4.1: Find the solutions to each of the following equations.

(a) $x^2 - 5x + 6 = 0$

(b) $x^2 + x - 20 = 0$

(c) $r^4 - 13r^2 + 36 = 0$

Problem 4.2: The quadratic $x^2 - 12x - 540$ has two integer roots. In this problem, we find those roots.

(a) How can we tell from the coefficients of the quadratic that both roots are divisible by 6?

(b) Find the roots of the quadratic.

Problem 4.3: Find the solutions to each of the following equations:

(a) $3t^2 + 11t + 10 = 0$

(b) $2x^2 - 7x = 15$

(c) $36y^2 - 25 = 0$

(d) $9t^2 - 2t + 5 = 25 - 90t$

Problem 4.4:

(a) Factor the expression $6x^2 + x - 15$.

(b) Find all possible values of the ratio of x to y if x and y are nonzero numbers that satisfy the equation $6x^2 + xy = 15y^2$.

Problem 4.5: How many integers x satisfy the equation $(x^2 - x - 1)^{x+2} = 1$? *(Source: AHSME)*

We start by factoring quadratics in which the coefficient of the quadratic term is 1.

Problem 4.1: Find the solutions to each of the following equations.

(a) $x^2 - 5x + 6 = 0$

(b) $x^2 + x - 20 = 0$

(c) $r^4 - 13r^2 + 36 = 0$

Solution for Problem 4.1: We know that the product $(x + p)(x + q)$ will give us an x^2 term when expanded, so we guess that the quadratics in parts (a) and (b) can be factored in this form. Expanding this product gives

$$(x + p)(x + q) = x^2 + px + qx + pq = x^2 + (p + q)x + pq.$$

To factor each quadratic, we must find the appropriate p and q.

(a) From the coefficient of x, we have $p + q = -5$. The constant term tells us that $pq = 6$. Because pq is positive, the numbers p and q have the same sign. Because $p + q$ is negative, we know that they are both negative. Now we just try different pairs of p and q that multiply to 6, looking for a pair that add to -5. We find that if p and q are -2 and -3, then $p + q = -5$ and $pq = 6$, so

$$x^2 - 5x + 6 = (x - 2)(x - 3).$$

Therefore, our equation is $(x - 2)(x - 3) = 0$, so our solutions are $x = 2$ and $x = 3$.

(b) Here, we have $p + q = 1$ and $pq = -20$. Because pq is negative, p and q have opposite signs. Trying different pairs of numbers that multiply to -20, we find that 5 and -4 satisfy both equations. (We might have tried this first, since $p + q = 1$ tells us that p and q are not far apart in magnitude.) So, we have $(x + 5)(x - 4) = 0$, which gives us $x = -5$ and $x = 4$ as solutions.

(c) This equation isn't a quadratic, but we can still factor it as one. If we don't see how at first, we can let $z = r^2$, which makes our equation $z^2 - 13z + 36 = 0$.

> **Concept:** Some equations can be turned into quadratic equations with a substitution.

Factoring the left side gives $(z - 4)(z - 9) = 0$, from which we have $z = 4$ or $z = 9$. Since $z = r^2$, we have $r^2 = 4$ or $r^2 = 9$. From $r^2 = 4$, we have $r = 2$ and $r = -2$, and $r^2 = 9$ gives us $r = 3$ and $r = -3$ as solutions.

□

Let's try a slightly harder quadratic.

> **Problem 4.2:** Find the roots of $f(x) = x^2 - 12x - 540$.

Solution for Problem 4.2: Suppose $f(x) = (x + p)(x + q)$. As before, we have $pq = -540$ and $p + q = -12$. We could start searching through all the pairs of numbers that multiply to -540, but that might take quite a while.

Instead, we first notice that 12 and 540 are both even. If x is odd, then x^2 is odd while both $-12x$ and -540 are even, so the expression $x^2 - 12x - 540$ is odd. This tells us that the quadratic cannot equal 0 if x is odd. So, if the quadratic has integer roots, they must be even.

Next, we notice that 12 and 540 are both divisible by 3. Therefore, the last two terms of $x^2 - 12x - 540$ are always divisible by 3 if x is an integer. So, $f(x)$ can only equal 0 for an integer value of x if x is divisible by 3, since otherwise $x^2 - 12x - 540$ will not be divisible by 3. This tells us that both roots are divisible by 3.

Now, because an integer that is divisible by 2 and by 3 must also be divisible by 6, we know that both roots are divisible by 6 if they are integers. If $f(x) = (x + p)(x + q)$, then the roots are $-p$ and $-q$. So, if the roots are integers, then both p and q must be divisible by 6. Therefore, we let $p = 6p_1$ and $q = 6q_1$, and we have $(6p_1)(6q_1) = -540$ and $6p_1 + 6q_1 = -12$. Simplifying these gives us $p_1 q_1 = -15$ and $p_1 + q_1 = -2$. These are much easier to handle than our original equations! We quickly find that p_1 and q_1 are -5 and 3, so p and q are $6 \cdot (-5) = -30$ and $6 \cdot 3 = 18$. This gives us

$$x^2 - 12x - 540 = (x - 30)(x + 18),$$

so the roots of $f(x)$ are $x = 30$ and $x = -18$. □

> **Concept:** Using a little number theory can help factor quadratics.

Let's try some quadratics in which the coefficient of the quadratic term is not 1.

> **Problem 4.3:** Find the solutions to each of the following equations:
>
> (a) $3t^2 + 11t + 10 = 0$
>
> (b) $2x^2 - 7x = 15$
>
> (c) $36y^2 - 25 = 0$
>
> (d) $9t^2 - 2t + 5 = 25 - 90t$

Solution for Problem 4.3:

(a) Because the quadratic term is $3t^2$, we know that $(t + p)(t + q)$ won't do, since this will produce a t^2 term instead of a $3t^2$ term. Instead, we try $(3t + p)(t + q)$, since this will produce a $3t^2$ term. Expanding this product gives

$$(3t + p)(t + q) = 3t^2 + (p + 3q)t + pq.$$

So, we seek p and q such that $p + 3q = 11$ and $pq = 10$. Therefore, p and q are both positive. Trying different pairs of numbers that multiply to 10, we find $p = 5$ and $q = 2$. (Notice that it *does* matter which is p and which is q.) We therefore have $(3t + 5)(t + 2) = 0$, so we must have $3t + 5 = 0$ or $t + 2 = 0$. Solving these two equations gives the solutions $t = -5/3$ and $t = -2$.

(b) First, we move all terms to the left, which gives $2x^2 - 7x - 15 = 0$. We guess that the quadratic factors as $(2x + p)(x + q)$. Expanding this product gives $2x^2 + (p + 2q)x + pq$, so we seek p and q such that $p + 2q = -7$ and $pq = -15$. We know that p and q have opposite signs, and we find that $p = 3$ and $q = -5$ satisfy both equations. So, we have $2x^2 - 7x - 15 = (2x + 3)(x - 5)$. This gives us $x = -3/2$ and $x = 5$ as solutions to the original equation.

(c) We can solve this equation without factoring. Adding 25 to both sides gives $36y^2 = 25$, then taking the square root of both sides gives $6y = \pm 5$. Therefore, our solutions are $y = \pm 5/6$.

We also can use the **difference of squares factorization**, $a^2 - b^2 = (a - b)(a + b)$, to note that $36y^2 - 25 = (6y - 5)(6y + 5)$.

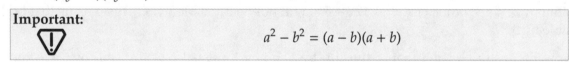

Important:

$$a^2 - b^2 = (a - b)(a + b)$$

We'll use the difference of squares factorization often throughout this book.

(d) Putting all terms on the same side gives us $9t^2 + 88t - 20 = 0$. We would like to factor this as $(At + B)(Ct + D)$. Expanding this product and comparing the result to our quadratic gives us

$$AC = 9,$$
$$AD + BC = 88,$$
$$BD = -20.$$

From $AC = 9$, we see that A and C are factors of 9, and from $BD = -20$, we know that B and D are factors of 20. Next, we consider the middle equation. Because 88 is much larger than 9 and 20, we know that we must choose large factors of 9 to pair with large factors of 20 when assigning values of A, B, C, and D. In other words, $AD + BC$ is pretty large, so we need both A and D (or both B and C) to be pretty large.

For example, if we let $A = 3$, $C = 3$, $B = -4$ and $D = 5$, then $AD + BC = 3$, which is way too small. If we instead let $A = 9$ and $D = 10$, we have $C = 1$ and $B = -2$, so $AD + BC = 88$. This gives

$$9t^2 + 88t - 20 = (9t - 2)(t + 10),$$

which means the solutions are $t = 2/9$ and $t = -10$.

$\square$

> **Concept:** Factoring quadratics is just a number game in which we use the signs, the divisors, and the magnitudes of the coefficients of the quadratic to find the quadratic's factors.

> **Problem 4.4:** Find all possible values of the ratio of x to y if x and y are nonzero numbers such that $6x^2 + xy = 15y^2$.

Solution for Problem 4.4: What's wrong with this solution:

> **Bogus Solution:** Subtracting $10xy$ from both sides of the equation gives us $6x^2 - 9xy = -10xy + 15y^2$. We can then factor both sides as
>
> $$3x(2x - 3y) = -5y(2x - 3y).$$
>
> Dividing both sides by $2x - 3y$, we get $3x = -5y$, so $x/y = -5/3$.

While this solution begins with a nice, clever step to allow us to factor, dividing that factor out might involve division by 0, so we lose a possible solution. Here are two valid solutions:

Solution 1: Keep your eye on the ball. We seek the ratio of x to y, so we let $r = x/y$, which means $x = ry$. The given equation then is

$$6(ry)^2 + (ry)y = 15y^2.$$

Since y is nonzero, we can divide by y^2, which gives $6r^2 + r = 15$. Rearranging and factoring this equation gives us $(3r + 5)(2r - 3) = 0$, so $r = -5/3$ or $r = 3/2$.

Solution 2: It's a quadratic; factor it. Subtracting $15y^2$ from both sides gives $6x^2 + xy - 15y^2 = 0$. Even though there are two variables, the expression $6x^2 + xy - 15y^2$ is still a quadratic expression. We can view this as a quadratic with x as the variable and y as a constant. However, if we don't see at first how to factor it, we can look at a simpler, related problem.

> **Concept:** When facing a problem you don't know how to do, try relating it to a problem you know how to do.

Letting $y = 1$ gives $6x^2 + x - 15$, which we can factor as $(2x - 3)(3x + 5)$.

Returning to $6x^2 + xy - 15y^2$, we see that $6x^2 + x - 15$ is the same as $6x^2 + xy - 15y^2$, but with the y's removed. So maybe the factorization of $6x^2 + xy - 15y^2$ is the factorization of $6x^2 + x - 15$, but with the y's "put back in." Let's try it:

$$(2x - 3y)(3x + 5y) = 2x(3x + 5y) - 3y(3x + 5y) = 6x^2 + xy - 15y^2.$$

Success! We have $6x^2 + xy - 15y^2 = (2x - 3y)(3x + 5y)$. Solving $(2x - 3y)(3x + 5y) = 0$ gives us $x/y = 3/2$ and $x/y = -5/3$, as before. $\square$

Here is an example of how factoring can help solve more complicated equations.

Problem 4.5: How many integers x satisfy the equation $(x^2 - x - 1)^{x+2} = 1$? *(Source: AHSME)*

Solution for Problem 4.5: If a and b are integers such that $a^b = 1$, then at least one of the following must be true:

- $b = 0$ *and* $a \neq 0$. Any nonzero number raised to the 0th power equals 1. So, if $x + 2 = 0$ and $x^2 - x - 1$ is nonzero, then x is a solution to the equation. Solving $x + 2 = 0$ gives us $x = -2$, for which $x^2 - x - 1 = 5$. So, $x = -2$ is a solution.

- $a = 1$. We note that 1 raised to any power is 1. If $x^2 - x - 1 = 1$, then rearranging and factoring gives $(x - 2)(x + 1) = 0$, so $x = 2$ and $x = -1$ are solutions since any power of 1 is equal to 1.

- $a = -1$ *and* b *is even*. If $x^2 - x - 1 = -1$, rearranging and factoring gives $x(x - 1) = 0$, so $x = 0$ and $x = 1$ need to be considered. When $x = 0$, we have $(x^2 - x - 1)^{x+2} = (-1)^2 = 1$, so $x = 0$ is a solution. When $x = 1$, we have $(x^2 - x - 1)^{x+2} = (-1)^3 = -1$, so $x = 1$ is not a solution.

In total, there are 4 integer solutions: $-2, -1, 0$, and 2. $\square$

Exercises

4.1.1 Factor and find the roots of each of the following quadratics:

(a) $x^2 + 19x + 90$ (c) $6a^2 + 13a - 8$

(b) $2y^2 - 15y + 18$ (d) $7x^2 + 296x - 215$

4.1.2 Find the solutions to each of the following equations:

(a) $3x^4 + 26 = 19x^2$ (b) $x + \dfrac{679}{x} = 104$ (c) $\sqrt{x^2 + x - 7} = \sqrt{x - 3}$

4.1.3 Find all possible values of a/b if $a^2 + 4b^2 = 4ab$.

4.1.4 Fix the Bogus Solution to Problem 4.4 by rearranging the equation $3x(2x - 3y) = -5y(2x - 3y)$ into an equation that's easy to factor.

4.1.5 Find the solutions to $(x^4 - 11x^3 + 24x^2) - (4x^2 - 44x + 96) = 0$.

4.2 Relating Roots and Coefficients

While factoring quadratics in which the coefficient of the quadratic term is 1, you might have noticed an interesting relationship between the roots of a quadratic and its coefficients. If not, find the roots of $x^2 - 7x + 10$ and see if you can find the relationship.

In this section, we explore this relationship in more detail.

Problem 4.6: Suppose the roots of the quadratic $ax^2 + bx + c$ are r and s.

(a) Why must we have $ax^2 + bx + c = a(x - r)(x - s)$ for all values of x?

(b) Express $r + s$ in terms of only a and b.

(c) Express rs in terms of only a and c.

Problem 4.7: Find the sum and the product of the roots of each of the following quadratic equations:

(a) $x^2 - 13x + 36 = 0$ (b) $-2006x^2 = 2007x + 1977$

Problem 4.8: Let r and s be the roots of $x^2 - px + q = 0$. In this problem, we find $r^2 + s^2$ in terms of p and q.

(a) Find equations relating r and s to p and q.

(b) Express $r^2 + s^2$ in terms of $r + s$ and rs.

(c) Find $r^2 + s^2$ in terms of p and q. *(Source: AHSME)*

Problem 4.9: Find real numbers a and b if $a + 2i$ and $b - 5i$ are the roots of $x^2 + (-4 + 3i)x + (13 + i) = 0$.

Problem 4.10: The sum of the zeros, the product of the zeros, and the sum of the coefficients of the function $f(x) = ax^2 + bx + c$ are equal. Show that all three of these quantities must equal a. *(Source: AMC 12)*

Problem 4.6: Express the sum and product of the roots of the quadratic $ax^2 + bx + c$ in terms of a, b, and c.

Solution for Problem 4.6: In order to look for a relationship between the roots and the coefficients of a general quadratic, we write the quadratic using its roots. Suppose the roots are r and s. Then, the product $(x - r)(x - s)$ will give us a quadratic with roots r and s. However, to produce the quadratic $ax^2 + bx + c$, we must have a as the coefficient of the quadratic term. So, we multiply $(x - r)(x - s)$ by a, and we have

$$ax^2 + bx + c = a(x - r)(x - s). \tag{4.1}$$

Expanding the right side of Equation (4.1), we have

$$ax^2 + bx + c = ax^2 - a(r + s)x + ars. \tag{4.2}$$

These two quadratics are the same for all values of x, so the corresponding coefficients of the two quadratics must be the same. Equating the coefficients of each power of x, we get

$$a = a,$$
$$b = -a(r + s),$$
$$c = ars.$$

While the first of these equations is true by design (we chose to multiply $(x - r)(x - s)$ by a in order to match the coefficients of the x^2 term in Equation (4.2)), the other two equations tell an interesting story. Isolating the sum and product of r and s in these equations, we have

$$r + s = -\frac{b}{a} \quad \text{and} \quad rs = \frac{c}{a}.$$

In other words, the sum of the roots of $ax^2 + bx + c$ is $-b/a$, and the product of those roots is c/a. □

These relationships between the roots and coefficients of a quadratic are called **Vieta's Formulas**.

> **Important:** If r and s are the roots of the quadratic $ax^2 + bx + c$, then
>
> $$r + s = -\frac{b}{a} \quad \text{and} \quad rs = \frac{c}{a}.$$

When using these formulas, we will often simply write, "By Vieta...." For example, by Vieta, we know that the sum of the roots of $x^2 - 4x + 3$ is $-(-4)/1 = 4$.

Problem 4.7: Find the sum and the product of the roots of each of the following quadratic equations:

(a) $x^2 - 13x + 36 = 0$

(b) $-2006x^2 = 2007x + 1977$

Solution for Problem 4.7:

(a) We can factor this quadratic as $(x - 4)(x - 9) = 0$, so the roots are 4 and 9. Or, we could have used Vieta to find that the sum of the roots is $-(-13)/1 = 13$ and the product of the roots is $36/1 = 36$.

(b) Factoring doesn't look like much fun on this problem, so let's try Vieta. What's wrong with this solution:

> **Bogus Solution:** We have $a = -2006$, $b = 2007$, and $c = 1977$, so the sum of the roots is $-b/a = 2007/2006$ and the product of the roots is $c/a = -1977/2006$.

The problem here is that Vieta's Formulas are for quadratics in the form $ax^2 + bx + c$. So, we have to move all terms to one side of the equation before proceeding with Vieta. Reorganizing the terms on one side of the equation gives us $2006x^2 + 2007x + 1977 = 0$. So, the sum of the roots is $-2007/2006$ and their product is $1977/2006$.

□

We can also use Vieta to find other relationships between the roots and the coefficients of a quadratic.

Problem 4.8: If r and s are the roots of $x^2 - px + q = 0$, then find $r^2 + s^2$ (in terms of p and q). *(Source: AHSME)*

Solution for Problem 4.8: While it's not immediately obvious how to express $r^2 + s^2$ in terms of p and q, we can use Vieta to relate r and s to p and q:

$$r + s = p,$$
$$rs = q.$$

Our goal is to find an expression for $r^2 + s^2$. But Vieta's Formulas don't directly provide us with any expressions involving squares of the roots. However, we can get an equation involving $r^2 + s^2$ by squaring both sides of $r + s = p$. This gives us $r^2 + 2rs + s^2 = p^2$. Now, subtracting $2rs$ from both sides gives us what we want:

$$r^2 + s^2 = p^2 - 2rs = p^2 - 2q.$$

□

We call a two-variable expression **symmetric** if it remains the same when the variables are swapped. For example, $r^2 + s^2$ is symmetric because swapping the variables gives $s^2 + r^2$, which is the same as $r^2 + s^2$. But $r + 2s$ is not symmetric because swapping the variables gives $s + 2r$, which is different from $r + 2s$. As shown in Problem 4.8, we can use $r + s$ and rs to evaluate more complicated symmetric expressions in r and s.

> **Concept:** Symmetric expressions involving the roots of a quadratic are a clue to try using Vieta's Formulas.

Problem 4.9: Find real numbers a and b if $a + 2i$ and $b - 5i$ are the roots of $x^2 + (-4 + 3i)x + (13 + i) = 0$.

Solution for Problem 4.9: Nothing in our proof of Vieta's Formulas prevents the roots or the coefficients of the quadratic from being nonreal. So, we can use Vieta for quadratics in which the roots and/or the coefficients are not real. By Vieta, the sum of the roots of the given quadratic is $-(-4 + 3i)/1 = 4 - 3i$, so

$$4 - 3i = a + 2i + b - 5i = (a + b) - 3i,$$

and the product of the roots is $(13 + i)/1 = 13 + i$, so

$$13 + i = (a + 2i)(b - 5i) = ab - 5ai + 2bi - 10i^2 = (ab + 10) + (-5a + 2b)i.$$

Equating real parts of $(a+b) - 3i = 4 - 3i$ gives $a + b = 4$. Equating the real parts of $(ab + 10) + (-5a + 2b)i = 13 + i$ gives $ab + 10 = 13$, and equating the imaginary parts gives $-5a + 2b = 1$.

Solving the system of linear equations $a + b = 4$ and $-5a + 2b = 1$ gives us $a = 1$ and $b = 3$, which is consistent with the equation $ab + 10 = 13$. Therefore, we have $a = 1$ and $b = 3$. □

Problem 4.10: The sum of the zeros, the product of the zeros, and the sum of the coefficients of the function $f(x) = ax^2 + bx + c$ are equal. Show that all three of these quantities must equal a. (*Source: AMC 12*)

Solution for Problem 4.10: By Vieta, the sum of the zeros is $-b/a$ and their product is c/a. So, we can write the given information as

$$-\frac{b}{a} = \frac{c}{a} = a + b + c.$$

From the first two parts of this equation chain, we see that $-b = c$. Therefore, we have $a + b + c = a + b - b = a$, so the sum of the coefficients, and therefore the sum and product of the zeros, equals a. □

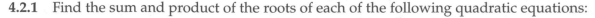

Exercises

4.2.1   Find the sum and product of the roots of each of the following quadratic equations:

(a) $x^2 - 5x + 7 = 0$

(b) $2x^2 = 7x + 120$

(c) $12x^2 + 13x + 14 = 15x^2 + 16x + 17$

4.2.2   Find the product of the roots of the equation $(x - 1)(x - 2) + (x - 3)(x - 4) + (x - 5)(x - 6) = 0$.

4.2.3   One root of the equation $a(b - c)x^2 + b(c - a)x + c(a - b) = 0$ is $x = 1$. Find the other root in terms of a, b, and c. *(Source: AHSME)*

4.2.4   Let $p(x) = 2x^2 + 7x + c$, where $c \neq 0$. If r_1 and r_2 are the roots of $p(x)$, then find the value of $\frac{1}{r_1} + \frac{1}{r_2}$ in terms of c.

4.2.5   Let m and n be the roots of $ax^2 + bx + c = 0$. Prove that if $m^2 + n^2 = 1$, then $2ac = b^2 - a^2$.

4.2.6   Back on page 85, we made the claim that if two quadratics are the same for all values of the variable, then the corresponding coefficients of the two quadratics are the same. In other words, if $ax^2 + bx + c = dx^2 + ex + f$ for all x, then we must have $a = d$, $b = e$, and $c = f$. However, we didn't prove this fact! Remedy that oversight now by proving it yourself.

4.3   Completing the Square

Not all quadratics can be factored easily. For example, try factoring $x^2 + 6x - 1$. However, we can still find the roots of such quadratics using a process called **completing the square**. When we complete the square of a quadratic $ax^2 + bx + c$, we write it in the form

$$a(x - h)^2 + k,$$

for some constants a, h, and k. In this section, we review how to complete the square, and how to use completing the square to solve quadratic equations.

Problems

Problem 4.11: Expand each of the following squares of binomials:

(a) $(x + 4)^2$

(b) $(4t - 5)^2$

Problem 4.12: Suppose k is a constant. In terms of k, what constant must be added to $x^2 + kx$ to produce a quadratic that is the square of a binomial?

Problem 4.13: Complete the square in each of the following quadratics by writing each in the form $a(x + p)^2 + q$, where a, p, and q are constants.

(a) $x^2 + 8x + 2$

(b) $-6x^2 - 18x - 7$

Problem 4.14: For each of the equations below, add a constant to both sides such that the left side becomes the square of a binomial. Then, solve the equation.

(a) $t^2 - 4t = 18$

(b) $4y^2 + 24y = -37$

Problem 4.15: Find a formula for the roots of the quadratic equation $ax^2 + bx + c$, where a, b, and c are constants and $a \neq 0$. Test your formula on the quadratic $x^2 - 5x + 6 = 0$. Does your formula give the correct roots?

Problem 4.16: Find the solutions of each of the following equations:

(a) $x^2 + 4x + 7 = 0$

(b) $3x^2 = 10x + 24$

(c)$\star$ $x^2 - (11 - 2i)x + 23 - 11i = 0$

Problem 4.17: Two numbers sum to 6 and have product -19. Find the two numbers.

Problem 4.18: Suppose $f(x) = x + \sqrt{x}$ and $g(x) = x + 1/4$. In this problem, we evaluate $g(f(g(f(g(f(7))))))$.

(a) Express $g(f(x))$ as the square of a binomial.

(b) Use your answer to part (a) to find $g(f(g(f(g(f(7))))))$ quickly.

Problem 4.11: Expand each of the following squares of binomials:

(a) $(x + 4)^2$

(b) $(4t - 5)^2$

Solution for Problem 4.11:

(a) We have $(x + 4)(x + 4) = x^2 + 4x + 4x + 4^2 = x^2 + 8x + 16$.

(b) We have $(4t - 5)(4t - 5) = (4t)^2 - 5(4t) - 5(4t) + (-5)^2 = 16t^2 - 40t + 25$.

$\square$

Problem 4.11 gives us two examples of the expansion of the square of a binomial. In general, we have
$$(a + b)^2 = a^2 + ab + ab + b^2 = a^2 + 2ab + b^2.$$

Important:
$$(a + b)^2 = a^2 + 2ab + b^2$$

Problem 4.12: Suppose k is a constant. In terms of k, what constant must be added to $x^2 + kx$ to produce a quadratic that is the square of a binomial?

Solution for Problem 4.12: We have $x^2 + 2xb + b^2 = (x + b)^2$. We can turn the $2xb$ into kx by letting $b = k/2$, which gives us

$$x^2 + kx + \left(\frac{k}{2}\right)^2 = \left(x + \frac{k}{2}\right)^2.$$

So, if we add $(k/2)^2$ to $x^2 + kx$, we produce a quadratic that is the square of a binomial. □

> **Important:** Adding the square of half the coefficient of the linear term to a quadratic of the form $x^2 + kx$ produces a quadratic that is the square of a binomial.

Let's see how we use this to complete the square.

Problem 4.13: Complete the square in each of the following quadratics by writing each in the form $a(x + p)^2 + q$, where a, p, and q are constants.

(a) $x^2 + 8x + 2$

(b) $-6x^2 - 18x - 7$

Solution for Problem 4.13:

(a) We must add $(8/2)^2 = 16$ to $x^2 + 8x$ in order to produce a quadratic that is the square of a binomial. However, when we add 16 to an expression, we must also subtract 16, so that our new expression is still equivalent to the original $x^2 + 8x + 2$. This gives us

$$x^2 + 8x + 2 = x^2 + 8x + 16 - 16 + 2 = (x + 4)^2 - 14.$$

(b) Let $f(x) = -6x^2 - 18x - 7$. We know what to do when the coefficient of x^2 is 1, so we factor -6 out of the first two terms, to give

$$f(x) = -6x^2 - 18x - 7 = -6(x^2 + 3x) - 7.$$

To complete the square in $x^2 + 3x$, we must add $(3/2)^2 = 9/4$ to the quadratic. However, if we just add 9/4 to $f(x)$, we have

$$f(x) + \frac{9}{4} = -6(x^2 + 3x) - 7 + \frac{9}{4}.$$

This doesn't give us a square of a binomial inside the parentheses. In order to get the 9/4 inside the parentheses, we instead add $-6(9/4)$ to $f(x)$:

$$f(x) + (-6)\left(\frac{9}{4}\right) = -6(x^2 + 3x) - 7 + (-6)\left(\frac{9}{4}\right) = -6\left(x^2 + 3x + \frac{9}{4}\right) - 7.$$

Subtracting $(-6)(9/4)$ from both sides of $f(x) + (-6)\left(\frac{9}{4}\right) = -6\left(x^2 + 3x + \frac{9}{4}\right) - 7$ gives

$$f(x) = -6\left(x^2 + 3x + \frac{9}{4}\right) - 7 - (-6)\left(\frac{9}{4}\right) = -6\left(x + \frac{3}{2}\right)^2 + \frac{13}{2}.$$

Note that this is the same as starting with $f(x) = -6(x^2 + 3x) - 7$, and then both adding and subtracting $(-6)(9/4)$ to the right side:

$$f(x) = -6(x^2 + 3x) - 7 = -6(x^2 + 3x) + (-6)\left(\frac{9}{4}\right) - (-6)\left(\frac{9}{4}\right) - 7$$

$$= -6\left(x^2 + 3x + \frac{9}{4}\right) + \frac{27}{2} - 7$$

$$= -6\left(x + \frac{3}{2}\right)^2 + \frac{13}{2}.$$

We can check our answer by expanding our final result. This gives us

$$-6\left(x + \frac{3}{2}\right)^2 + \frac{13}{2} = -6\left(x^2 + 3x + \frac{9}{4}\right) + \frac{13}{2} = -6x^2 - 18x - 7,$$

which is indeed our original quadratic.

□

Now that we know how to complete the square, let's use our new tool to solve a couple equations.

Problem 4.14: For each of the equations below, add a constant to both sides such that the left side becomes the square of a binomial. Then, solve the equation.

(a) $t^2 - 4t = 18$ (b) $4y^2 + 24y = -37$

Solution for Problem 4.14:

(a) We add $(-4/2)^2 = 4$ to both sides of our original equation to produce the square of a binomial on the left side:

$$t^2 - 4t + 4 = 18 + 4.$$

As planned, the left side is now the square of a binomial, so we have $(t - 2)^2 = 22$. Taking the square root of both sides, we have $t - 2 = \pm \sqrt{22}$, so our solutions are $t = 2 \pm \sqrt{22}$.

(b) We factor 4 out of the left side, which gives us $4(y^2 + 6y) = -37$. Now, we want the quadratic on the left side to be the square of a binomial. We recognize that $(y + 3)^2 = y^2 + 6y + 9$, so we need to add 9 inside the parentheses on the left. To do so, we add $4(9)$ to both sides, which gives us

$$4(y^2 + 6y) + 4(9) = 4(y^2 + 6y + 9)$$

on the left and $-37 + 4(9) = -1$ on the right. Factoring the quadratic then gives us

$$4(y + 3)^2 = -1.$$

Taking the square root of both sides of this equation gives us $2(y + 3) = \pm i$. Dividing by 2 gives us $y + 3 = \pm \frac{i}{2}$, so our solutions are $y = -3 \pm \frac{i}{2}$.

We could also have divided by 4 in our first step, to get $y^2 + 6y = -\frac{37}{4}$. See if you can finish the problem from here.

□

Since we can complete the square for any quadratic, we can complete the square of a general quadratic to produce a formula that will give us the roots of any quadratic.

Problem 4.15: Find a formula for the roots of the quadratic equation $ax^2 + bx + c$, where a, b, and c are constants and $a \neq 0$.

Solution for Problem 4.15: Since $a \neq 0$, we can divide the whole equation by a to make the coefficient of x^2 equal to 1:

$$x^2 + \frac{b}{a}x + \frac{c}{a} = 0.$$

This lets us use the result of Problem 4.12. We complete the square by adding $\left(\frac{b/a}{2}\right)^2 = \left(\frac{b}{2a}\right)^2$ to both sides:

$$x^2 + \frac{b}{a}x + \left(\frac{b}{2a}\right)^2 + \frac{c}{a} = \left(\frac{b}{2a}\right)^2.$$

Because $x^2 + \frac{b}{a}x + \left(\frac{b}{2a}\right)^2 = \left(x + \frac{b}{2a}\right)^2$, we now have $\left(x + \frac{b}{2a}\right)^2 + \frac{c}{a} = \left(\frac{b}{2a}\right)^2$. We then isolate the square of the binomial:

$$\left(x + \frac{b}{2a}\right)^2 = \left(\frac{b}{2a}\right)^2 - \frac{c}{a}.$$

Writing the terms on the right with a common denominator gives us

$$\left(x + \frac{b}{2a}\right)^2 = \frac{b^2}{4a^2} - \frac{c}{a} = \frac{b^2}{4a^2} - \frac{4ac}{4a^2} = \frac{b^2 - 4ac}{4a^2},$$

so we have

$$\left(x + \frac{b}{2a}\right)^2 = \frac{b^2 - 4ac}{4a^2}.$$

Then we take the square root of both sides, and remember the $\pm$:

$$x + \frac{b}{2a} = \pm\sqrt{\frac{b^2 - 4ac}{4a^2}} = \pm\frac{\sqrt{b^2 - 4ac}}{\sqrt{4a^2}} = \pm\frac{\sqrt{b^2 - 4ac}}{2a}.$$

Subtracting $b/2a$ from both sides gives us our desired formula:

$$x = \frac{-b \pm \sqrt{b^2 - 4ac}}{2a}.$$

$\square$

WARNING!! In one of our steps above, we replace $\sqrt{4a^2}$ with $2a$. However, it is not true that $\sqrt{4a^2} = 2a$ if $a < 0$, so we can't write $\sqrt{4a^2} = 2a$ for all a. Why is it OK for us to do so above?

While we can't write $\sqrt{4a^2} = 2a$ for all a, we *can* write $\pm\sqrt{4a^2} = \pm 2a$, because $\pm\sqrt{4a^2}$ means "take both numbers that, when squared, give us $4a^2$." These two numbers can be depicted as $\pm 2a$ *whether or not a is positive or negative.*

> **Important:** The **quadratic formula** states that the solutions to the quadratic equation
> $$ax^2 + bx + c = 0$$
> are
> $$x = \frac{-b \pm \sqrt{b^2 - 4ac}}{2a}.$$

Before we use the quadratic formula to solve a bunch of problems, we should test it on a quadratic we know how to solve.

> **Concept:** Using sample cases to test a formula you derive is a good way to check your formula.

We'll try $x^2 - 5x + 6 = 0$. We know we can factor the quadratic to get $(x-2)(x-3) = 0$, so the solutions are $x = 2$ and $x = 3$. Let's see what the quadratic formula gives us. In this case, $a = 1$, $b = -5$, and $c = 6$, so

$$x = \frac{-(-5) \pm \sqrt{(-5)^2 - 4(1)(6)}}{2(1)} = \frac{5 \pm \sqrt{25 - 24}}{2} = \frac{5 \pm 1}{2}.$$

Therefore, our solutions are $x = (5 + 1)/2 = 3$ and $x = (5 - 1)/2 = 2$, as expected.

Let's try it on quadratics we can't factor easily.

> **Problem 4.16:** Find the roots of each of the following:
>
> (a) $x^2 + 4x + 7 = 0$ (b) $3x^2 = 10x + 24$ (c)★ $x^2 - (11 - 2i)x + 23 - 11i = 0$

Solution for Problem 4.16:

(a) Applying the quadratic formula, the roots of $x^2 + 4x + 7$ are

$$\frac{-4 \pm \sqrt{4^2 - 4(1)(7)}}{2} = \frac{-4 \pm \sqrt{-12}}{2} = \frac{-4 \pm \sqrt{12}\sqrt{-1}}{2} = \frac{-4 \pm (2\sqrt{3})(i)}{2} = -2 \pm i\sqrt{3}.$$

(b) We first move all the terms to one side so we can apply the quadratic formula correctly. This gives us $3x^2 - 10x - 24 = 0$. So, applying the quadratic formula gives

$$x = \frac{-(-10) \pm \sqrt{(-10)^2 - 4(3)(-24)}}{2(3)} = \frac{10 \pm \sqrt{388}}{6} = \frac{10 \pm 2\sqrt{97}}{6} = \frac{5 \pm \sqrt{97}}{3}.$$

We can check this quickly by noticing that Vieta tells us that the sum of the roots of $3x^2 - 10x - 24$ is $10/3$ and the product of the roots is $-24/3 = -8$. Adding and multiplying the roots we found, we have

$$\frac{5 + \sqrt{97}}{3} + \frac{5 - \sqrt{97}}{3} = \frac{10}{3},$$
$$\left(\frac{5 + \sqrt{97}}{3}\right)\left(\frac{5 - \sqrt{97}}{3}\right) = \frac{25 - 97}{9} = -8.$$

(c) Nothing in our proof of the quadratic formula assumes that the coefficients are real, so we can use the quadratic formula when the coefficients are nonreal. Applying the quadratic formula gives us

$$\frac{(11 - 2i) \pm \sqrt{(11 - 2i)^2 - 4(1)(23 - 11i)}}{2} = \frac{11 - 2i \pm \sqrt{117 - 44i - 92 + 44i}}{2} = \frac{11 - 2i \pm 5}{2}.$$

Therefore, the roots are

$$\frac{11 - 2i + 5}{2} = 8 - i \qquad \text{and} \qquad \frac{11 - 2i - 5}{2} = 3 - i.$$

We can check our work by confirming that the sum of the roots is $11 - 2i$ and their product is $23 - 11i$, as required by Vieta.

$\square$

Problem 4.17: Two numbers sum to 6 and have product -19. Find the two numbers.

Solution for Problem 4.17: Solution 1: Make a system of equations. Let r and s be the numbers. We then have $r + s = 6$ and $rs = -19$. Solving the first equation for r gives $r = 6 - s$. Substituting this into the second equation gives $(6 - s)s = -19$. Rearranging this equation gives $s^2 - 6s - 19 = 0$. The quadratic formula then gives us

$$s = \frac{-(-6) \pm \sqrt{(-6)^2 - 4(1)(-19)}}{2(1)} = \frac{6 \pm \sqrt{36 + 76}}{2} = \frac{6 \pm 4\sqrt{7}}{2} = 3 \pm 2\sqrt{7}.$$

If $s = 3 + 2\sqrt{7}$, then $r = 6 - s = 3 - 2\sqrt{7}$. If $s = 3 - 2\sqrt{7}$, then $r = 6 - s = 3 + 2\sqrt{7}$. In both cases, the two numbers are $3 - 2\sqrt{7}$ and $3 + 2\sqrt{7}$.

Solution 2: Use Vieta. We know the sum and the product of two numbers. This reminds us of Vieta's Formulas. Suppose the numbers are the roots of the quadratic $x^2 + bx + c = 0$. Then, $-b$ is the sum of the numbers and c is their product. Therefore, we must have $b = -6$ and $c = -19$, so the numbers are the roots of the quadratic $x^2 - 6x - 19 = 0$. Applying the quadratic formula gives us the same answers as in our first solution. $\square$

Our second solution shows us a useful way to apply Vieta's Formulas:

> **Concept:** When we have information about the sum and the product of two numbers, we can let the numbers be the roots of a quadratic and use Vieta to construct a quadratic whose roots are the numbers.

Problem 4.18: Suppose $f(x) = x + \sqrt{x}$ and $g(x) = x + 1/4$. Evaluate $g(f(g(f(g(f(7))))))$.

Solution for Problem 4.18: We might try just plugging 7 into $f(x)$ and grinding out the answer. Our first step gives us $f(7) = 7 + \sqrt{7}$. Next, we have $g(f(7)) = 7 + \sqrt{7} + \frac{1}{4} = \frac{29}{4} + \sqrt{7}$. Sticking that into $f(x)$ doesn't look like fun. Let's take a step back and look for another approach.

We're repeatedly sticking the output of f into g, so let's look at $g(f(x))$:

$$g(f(x)) = x + \sqrt{x} + \frac{1}{4}.$$

If we don't see how to simplify this right away, we can try a few values of x to see if anything interesting happens. We choose perfect squares to avoid radicals in our results:

$$g(f(1)) = \frac{9}{4},$$
$$g(f(4)) = \frac{25}{4},$$
$$g(f(9)) = \frac{49}{4}.$$

Why do we keep getting perfect squares divided by 4 as our results? This makes us wonder if $g(f(x))$ is itself the square of an expression.

> **Important:** Not only must you know how to create a perfect square of a binomial, you must know how to recognize one when you see one.

We have $\left(\sqrt{x}\right)^2 = x$, $(1/2)^2 = 1/4$, and $2(\sqrt{x})(1/2) = \sqrt{x}$, so

$$g(f(x)) = x + \sqrt{x} + \frac{1}{4} = \left(\sqrt{x} + \frac{1}{2}\right)^2.$$

Aha! Suppose we let $h(x) = g(f(x))$. Our problem is to evaluate

$$g(f(g(f(g(f(7)))))) = g(f(g(f(h(7))))) = g(f(h(h(7)))) = h(h(h(7))).$$

We have $h(7) = \left(\sqrt{7} + \frac{1}{2}\right)^2$. We don't expand the square, though, because our next step is to evaluate $h(h(7))$:

$$h(h(7)) = \left(\sqrt{h(7)} + \frac{1}{2}\right)^2.$$

Using our expression for $h(7)$ from above, we have $\sqrt{h(7)} = \sqrt{7} + \frac{1}{2}$, so

$$h(h(7)) = \left(\sqrt{h(7)} + \frac{1}{2}\right)^2 = \left(\sqrt{7} + 1\right)^2.$$

Therefore, we have $h(h(h(7))) = \left(\sqrt{h(h(7))} + \frac{1}{2}\right)^2 = \left(\sqrt{7} + \frac{3}{2}\right)^2 = \frac{37}{4} + 3\sqrt{7}.$ $\square$

Exercises

4.3.1 Complete the square for each of the following quadratics:

(a) $x^2 + 8x$

(b) $y^2 - 6y + 1$

(c) $2z^2 + 20z + 3$

(d) $9x^2 + \frac{1}{2}x + \frac{4}{3}$

4.3.2 Find all real solutions to each of the following quadratic equations:

(a) $x^2 + 8x = 12$

(b) $y^2 - 6y + 1 = -4$

(c) $2z^2 + 20z + 3 = -30$

(d) $9x^2 + \frac{1}{2}x + \frac{4}{3} = 2$

4.3.3 What positive number is 8 more than its reciprocal?

4.3.4 Let $g(x) = -7ax^2 + 3a^2x + 13$ for some constant a. If $g(3) = -59$, find all possible values of a.

4.3.5 Find the roots of each of the following quadratic equations:

(a) $2x^2 + 17x = -21$

(b) $y^2 - 16y + 51 = 0$

(c) $z^2 + 4z + 5 = 0$

(d)★ $\sqrt{5}x^2 - (8\sqrt{5} + 10)x + (20\sqrt{5} + 40) = 0$

4.3.6 Let $g(t) = 2t^2 - 8\sqrt{5}t + 25$. Find all values of t such that $g(t) = -13$.

4.3.7 Find all z such that $z^2 - (3 - 8i)z - (14 + 12i) = 0$.

4.4 The Discriminant

| Problems |

> **Problem 4.19:** *Without finding the roots,* determine whether the roots of each of the following quadratics are nonreal, irrational, or rational. Also, determine the number of distinct roots each quadratic has.
>
> (a) $x^2 - 4x + 20 = 0$
>
> (b) $2x^2 - 3x - 7 = 0$
>
> (c) $r^2 - r - 462 = 0$
>
> (d) $-\frac{t^2}{2} + 4t - 8 = 0$

The quadratic formula gives us a quick way to determine the nature of the roots of a quadratic with real coefficients *without even determining the roots of the quadratic.* To see how, we note that the roots of $ax^2 + bx + c$ are

$$x = \frac{-b}{2a} \pm \frac{\sqrt{b^2 - 4ac}}{2a}.$$

So, if a, b, and c are real, these roots are nonreal if and only if $b^2 - 4ac < 0$. Moreover, the $\pm$ tells us that these roots form a conjugate pair, since $-b/2a$ is the real part of each root and the $\pm$ tells us that the imaginary parts of the roots are opposites.

If the coefficients are real and $b^2 - 4ac = 0$, then both roots equal $-\frac{b}{2a}$. We say that such a quadratic has a **double root** at $x = -\frac{b}{2a}$, and it has only one distinct root. Finally, if $b^2 - 4ac > 0$, then the quadratic has two different real roots.

If the coefficients are rational, we can go a step further. If $b^2 - 4ac$ is the square of a rational number, then both roots are rational, since each equals the sum or difference of two rational numbers divided by another rational number. Otherwise, the roots are both irrational.

Important: The **discriminant** of the quadratic $ax^2 + bx + c$ is $b^2 - 4ac$. When a, b, and c are real, we can use the discriminant to determine the nature of the roots of the quadratic:

- If $b^2 - 4ac \geq 0$, the roots of the quadratic are real.

- If $b^2 - 4ac < 0$, the roots of the quadratic are a nonreal conjugate pair.

Furthermore, if $b^2 - 4ac = 0$, then the quadratic has a double root.
   If the coefficients of the quadratic are rational, then the roots of the quadratic are rational if and only if $b^2 - 4ac$ is the square of a rational number.

Problem 4.19:   *Without finding the roots*, determine whether the roots of each of the following quadratics are nonreal, irrational, or rational. Also, determine the number of distinct roots the quadratic has.

(a) $x^2 - 4x + 20 = 0$

(b) $2x^2 - 3x - 7 = 0$

(c) $r^2 - r - 462 = 0$

(d) $-\frac{t^2}{2} + 4t - 8 = 0$

Solution for Problem 4.19:

(a)   The discriminant is $(-4)^2 - 4(1)(20) = -64$, so the two distinct roots are not real.

(b)   The discriminant is $(-3)^2 - 4(2)(-7) = 65$, which is positive but not the square of a rational number. So, the roots are two distinct irrational numbers.

(c)   The discriminant is $(-1)^2 - 4(1)(-462) = 1849 = 43^2$, which is the square of a rational number. The coefficients of the quadratic are rational, so the roots are two different rational numbers.

(d)   The discriminant is $(4)^2 - 4\left(-\frac{1}{2}\right)(-8) = 0$, so this quadratic has one distinct root. The coefficients of the quadratic are rational, so the double root is rational.

□

 Problems

Problem 4.20:
   (a)   Find the quadratic with real coefficients and quadratic term x^2 that has $3 + 2i$ as a root.

   (b)   Find the quadratic with rational coefficients and quadratic term x^2 that has $-3 + 2\sqrt{7}$ as a root.

Problem 4.21: In this problem, we find all pairs of real numbers (x, y) such that there is at least one real number t for which $y = 2tx - t^2$.

   (a)   Treat x and y as constants, and write the equation as a quadratic with t as the variable. Find the discriminant of the quadratic in terms of x and y.

   (b)   Use the discriminant from part (a) to describe all pairs of real numbers (x, y) such that the quadratic has real roots. For all of these pairs, is it true that the equation $y = 2tx - t^2$ has at least one real solution t?

Problem 4.20:

(a) Find the quadratic with real coefficients and quadratic term x^2 that has $3 + 2i$ as a root.

(b) Find the quadratic with rational coefficients and quadratic term x^2 that has $-3 + 2\sqrt{7}$ as a root.

Solution for Problem 4.20:

(a) Since the roots are nonreal but the coefficients are real, the roots form a conjugate pair. Therefore, the other root is $\overline{3 + 2i} = 3 - 2i$. To find the quadratic, we note that the sum of the roots is $3 + 2i + 3 - 2i = 6$ and the product of the roots is $(3 + 2i)(3 - 2i) = 13$. Using Vieta's Formulas, we now see that the quadratic is $x^2 - 6x + 13$.

(b) The roots of $ax^2 + bx + c$ are

$$x = \frac{-b}{2a} \pm \frac{\sqrt{b^2 - 4ac}}{2a}.$$

If a, b, and c are rational, the only way a square root can be part of a root is from the second term. The $\pm$ before that term tells us that if $-3 + 2\sqrt{7}$ is one root of a quadratic with rational coefficients, then $-3 - 2\sqrt{7}$ is the other root. To build the quadratic, we again use Vieta. The sum of the roots is -6 and their product is $(-3 + 2\sqrt{7})(-3 - 2\sqrt{7}) = -19$, so the quadratic is $x^2 + 6x - 19$.

□

Problem 4.21: Find all pairs of real numbers (x, y) such that there is at least one real number t for which $y = 2tx - t^2$.

Solution for Problem 4.21: We wish to know when we can find real values of t that satisfy the equation, so we view the given equation as a quadratic equation in t,

$$t^2 - 2xt + y = 0.$$

The coefficient of t is $-2x$ and the "constant term" of the quadratic (i.e., the term without t) is y. If t is real, the discriminant of the quadratic must be nonnegative, so we must have

$$(-2x)^2 - 4(1)(y) \geq 0.$$

Simplifying this inequality gives us $x^2 \geq y$. Therefore, if the ordered pair (x, y) satisfies $x^2 \geq y$, then there is at least one real solution t to the equation $y = 2tx - t^2$. □

The hardest part of Problem 4.21 was figuring out where to start. The subsequent algebraic manipulation in the problem was pretty straightforward. What helped us get started was focusing on just one of the three variables in the given equation. Our goal was to find the circumstances under which real values of t exist, so we rewrote the equation as a quadratic in the variable t. This allowed us to use the discriminant to answer the question.

> **Concept:** If an equation with multiple variables is quadratic in one of the variables, then viewing the equation as a quadratic in that variable allows us to use all our quadratic equation tools on the problem.

> ▌ **Exercises** ▶

4.4.1 One of the roots of a quadratic with real coefficients is $6 + \sqrt{5}i$. Find the other root.

4.4.2 Find all a such that $ax^2 - 5x + 9 = 0$ has only one distinct root.

4.4.3 Is it possible for a quadratic with real coefficients to have a double root that is nonreal? Why or why not?

4.4.4 In part (c) of Problem 4.16, we found the roots of the quadratic to be $3 - i$ and $8 - i$. Why is it possible for the quadratic to have roots that are not a conjugate pair?

4.4.5 It is given that one root of $2x^2 + rx + s = 0$, with r and s real numbers, is $3 + 2i$. Find s. *(Source: AHSME)*

4.4.6 Find the real values of K for which the equation $x = K^2(x - 1)(x - 2)$ has real roots.

4.5 Quadratic Inequalities

> ▌ **Problems** ▶

Problem 4.22:
  (a) Find all x such that $(x - 2)(x - 3) \geq 0$.
  (b) Find all t such that $2t^2 + 9t + 4 < 0$.
  (c) Find all r such that $r^2 - 4r + 2 > 0$.

Problem 4.23: Let $p(x) = 2x^2 + 8x + 27$.
  (a) Complete the square to write $p(x)$ in the form $a(x - h)^2 + k$, where a, h, and k are constants.
  (b) Use part (a) to determine all values of x for which $p(x) > 0$.

Problem 4.24: Let $f(x) = \dfrac{5x^2 - 4x + 8}{x^2 + 1}$, where the domain of f is all real numbers. In this problem we find the range of f.
  (a) For what real values of k is there a real value of x such that $f(x) = k$?
  (b) Find the range of $f(x)$.

Problem 4.22:
  (a) Find all x such that $(x - 2)(x - 3) \geq 0$.
  (b) Find all t such that $2t^2 + 9t + 4 < 0$.
  (c) Find all r such that $r^2 - 4r + 2 > 0$.

Solution for Problem 4.22:

(a) Both $x - 2$ and $x - 3$ are nonnegative for $x \geq 3$, so $(x - 2)(x - 3) \geq 0$ when $x \geq 3$. Similarly, both $x - 2$ and $x - 3$ are nonpositive for $x \leq 2$, so $(x - 2)(x - 3) \geq 0$ when $x \leq 2$. Therefore, all x such that $x \leq 2$ or $x \geq 3$ satisfy the inequality. In interval notation, our solution is $x \in (-\infty, 2] \cup [3, +\infty)$. Recall from page 28 that the "$\cup$" in this statement means "or," so, "$x \in (-\infty, 2] \cup [3, +\infty)$" means

$$x \text{ is in the interval } (-\infty, 2] \text{ or the interval } [3, +\infty).$$

Note that we use "]" after the 2 and "[" before the 3 to indicate that 2 and 3 are valid solutions.

(b) Factoring the quadratic gives us $(2t + 1)(t + 4) < 0$. The product $(2t + 1)(t + 4)$ is only negative if $2t + 1$ and $t + 4$ have opposite signs. We have $2t + 1 > 0$ for $t > -1/2$ and $2t + 1 < 0$ for $t < -1/2$. Similarly, $t + 4 > 0$ for $t > -4$ and $t + 4 < 0$ for $t < -4$. We can capture this information in a table, as shown on

	$t + 4$	$2t + 1$	$(t + 4)(2t + 1)$
$t > -\frac{1}{2}$	$+$	$+$	$+$
$-4 < t < -\frac{1}{2}$	$+$	$-$	$-$
$t < -4$	$-$	$-$	$+$

the right. Our table quickly shows us that $(2t + 1)(t + 4) < 0$ for $-4 < t < -1/2$. Notice that neither $t = -1/2$ nor $t = -4$ satisfy the inequality because the inequality is strict. In interval notation, our solution is $t \in (-4, -\frac{1}{2})$.

(c) We can't easily factor this quadratic. However, we can use the quadratic formula to find that the roots of $r^2 - 4r + 2$ are $2 \pm \sqrt{2}$. So, we can write the inequality as

$$\left(r - (2 + \sqrt{2})\right)\left(r - (2 - \sqrt{2})\right) > 0.$$

Both of the factors on the left are positive if $r > 2 + \sqrt{2}$ and they are both negative if $r < 2 - \sqrt{2}$, so the inequality is satisfied if $r \in (-\infty, 2 - \sqrt{2}) \cup (2 + \sqrt{2}, +\infty)$. Notice that we use ")" and "(" in the interval notation because $2 - \sqrt{2}$ and $2 + \sqrt{2}$ do not satisfy the strict inequality.

We also could have solved the problem by completing the square. We must add $(-4/2)^2 = 4$ to $r^2 - 4r$ to complete the square. Adding 4 to both sides of our original inequality gives us $r^2 - 4r + 4 + 2 > 4$, so we have $(r - 2)^2 > 2$. From this inequality, we see that we must have either $r - 2 > \sqrt{2}$ or $r - 2 < -\sqrt{2}$, which leads to the same solution as before.

Concept: Solving the same problem in two different ways is a great way to check your answer.

Problem 4.23: Let $p(x) = 2x^2 + 8x + 27$. Find all real values of x such that $p(x) > 0$.

Solution for Problem 4.23: We start by trying to find the roots of $p(x)$. Factoring doesn't work, and when we find the roots of $p(x)$, we see that the discriminant is $8^2 - 4(2)(27) = 8(8) - 8(27) = 8(-19)$, which is negative. So, the roots of $p(x)$ are not real. Now what?

Completing the square worked when the roots were irrational. Let's try it on this quadratic, too.

Concept: Completing the square isn't just for finding roots of quadratics.

We have $p(x) = 2(x^2 + 4x) + 27$. Completing the square gives us

$$p(x) = 2(x^2 + 4x) + 2(4) - 2(4) + 27 = 2(x + 2)^2 + 19.$$

If x is real, then $2(x + 2)^2$ is nonnegative. Therefore, $p(x)$ equals 19 plus some nonnegative number. In other words, $p(x)$ is positive for all real x. $\square$

As an Exercise, you'll extend Problem 4.23 to show that:

> **Important:** Suppose $f(x)$ is a quadratic with nonreal roots. If the coefficient of x^2 is
> positive, then $f(x) > 0$ for all x. If the coefficient of x^2 is negative, then
> $f(x) < 0$ for all x.

Problem 4.24: Let $f(x) = \dfrac{5x^2 - 4x + 8}{x^2 + 1}$, where the domain of f is all real numbers. Find the range of f.

Solution for Problem 4.24: If k is in the range of f, then there must be some real value of x for which $f(x) = k$. So, the equation

$$\frac{5x^2 - 4x + 8}{x^2 + 1} = k.$$

must have a real solution for x. Multiplying this equation by $x^2 + 1$, and then rearranging, gives us

$$(k - 5)x^2 + 4x + (k - 8) = 0.$$

There is at least one real solution x when the discriminant is nonnegative. Therefore, we must have $4^2 - 4(k - 5)(k - 8) \geq 0$. Dividing both sides by 4 gives $4 - (k - 5)(k - 8) \geq 0$. Expanding and rearranging gives $-k^2 + 13k - 36 \geq 0$, so multiplying by -1 (and reversing the inequality sign) gives $k^2 - 13k + 36 \leq 0$. Factoring the quadratic gives $(k - 4)(k - 9) \leq 0$, and solving this inequality gives us $k \in [4, 9]$. So, only for values of k in this interval is there a value of x such that $f(x) = k$. This means $[4, 9]$ is the range of f. $\square$

Exercises

4.5.1 Find all values of t such that $t^2 - 13t \leq -40$.

4.5.2 Find all real values of x such that $-x^2 + 7x - 13 > 0$.

4.5.3 Let $f(x) = \frac{7x^2 - 4x + 4}{x^2 + 1}$, where the domain of f is all real numbers. Find the range of f.

4.5.4

(a) Let $f(x)$ be a quadratic with nonreal roots. Show that if the coefficient of x^2 in $f(x)$ is positive, then $f(x) > 0$ for all x.

(b) Let $f(x)$ be a quadratic with nonreal roots. Show that if the coefficient of x^2 in $f(x)$ is negative, then $f(x) < 0$ for all x.

(c) What if a quadratic $f(x)$ has a real double root? What then can we say about the possible values of $f(x)$?

4.5.5 If x is real and $4y^2 + 4xy + x + 6 = 0$, then find the complete set of values of x for which y is real. *(Source: AHSME)* **Hints:** 269

4.5.6 Find all real values of x such that $x^3 - 6x^2 + 9x > 0$.

4.6 Summary

An expression of the form $ax^2 + bx + c$, where a, b, and c are constants and $a \neq 0$, is called a **quadratic**. The solutions to the equation $ax^2 + bx + c = 0$ are called the **roots** or **zeros** of the quadratic.

There are two common methods for finding the roots of quadratics. First, we can sometimes easily **factor** the quadratic by writing it as the product of two linear expressions. Setting each of the linear expressions equal to zero gives us the roots. Second, the **quadratic formula** tells us that the roots of $ax^2 + bx + c = 0$ are

$$x = \frac{-b \pm \sqrt{b^2 - 4ac}}{2a}.$$

We prove the quadratic formula by **completing the square**, which means writing a quadratic expression $ax^2 + bx + c$ in the form $a(x + p)^2 + q$ for some constants a, p, and q.

Important: The **discriminant** of the quadratic $ax^2 + bx + c$ is $b^2 - 4ac$. When a, b, and c are real, we can use the discriminant to determine the nature of the roots of the quadratic:

- If $b^2 - 4ac \geq 0$, the roots of the quadratic are real.

- If $b^2 - 4ac < 0$, the roots of the quadratic are a nonreal conjugate pair.

Furthermore, if $b^2 - 4ac = 0$, then the quadratic has a double root.

If the coefficients of the quadratic are rational, then the roots of the quadratic are rational if and only if $b^2 - 4ac$ is the square of a rational number.

Important: **Vieta's Formulas** for a quadratic tell us that if r and s are the roots of the quadratic $ax^2 + bx + c$, then

$$r + s = -\frac{b}{a} \quad \text{and} \quad rs = \frac{c}{a}.$$

We can solve many quadratic inequalities by factoring the quadratic, then determining for what values of the variable each factor is positive or negative. If $f(x)$ is a quadratic that does not have real roots, then either $f(x) > 0$ for all x or $f(x) < 0$ for all x.

Problem Solving Strategies

Concepts:
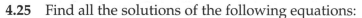

- Some equations can be turned into quadratic equations with a substitution.

- Using a little number theory can greatly help factor quadratics.

- When facing a problem you don't know how to do, try relating it to a problem you know how to do.

- We call a two-variable expression **symmetric** if it remains the same when the variables are swapped. Symmetric expressions involving the roots of a quadratic are a clue to try using Vieta's Formulas.

- Using sample cases to test a formula you derive is a good way to check your formula.

- When we have information about the sum and the product of two numbers, we can let the numbers be the roots of a quadratic and use Vieta to construct a quadratic whose roots are the numbers.

- If an equation with multiple variables is quadratic in one of the variables, then viewing the equation as a quadratic in that variable allows us to use all our quadratic equation tools on the problem.

- Solving the same problem in two different ways is a great way to check your answer.

REVIEW PROBLEMS

4.25 Find all the solutions of the following equations:

(a) $x^2 + 12x + 27 = 0$ (b) $x^2 - 6x + 1 = 0$ (c) $2y^2 - 5y + 2 = 0$ (d) $3x^2 + 15x = 7$

4.26 Let $f(x) = a^2x^2 + \frac{5}{2}ax + 3$ and $f(2) = 2$. Find all possible values of the constant a.

4.27 Find the value of x if x is positive and $x - 1$ is the reciprocal of $x + \frac{1}{2}$.

4.28 The sum of the squares of the roots of the equation $x^2 + 2hx = 3$ is 10. Find $|h|$. *(Source: AHSME)*

4.29 Solve each of the following equations:

(a) $\sqrt{6a^2 + 5a + 21} = 14$ (b) $\sqrt{4x^2 + 20x + 25} = 2x + 5$

4.30 Let f be a function for which $f(x/3) = x^2 + x + 1$. Find the sum of all values of z for which $f(3z) = 7$. *(Source: AMC 12)*

4.31 What is the domain of the function $f(x) = \dfrac{x+7}{\frac{1}{x}+\frac{1}{x^2-6}}$?

4.32 Let a and b be the roots of the equation $x^2 - mx + 2 = 0$. Suppose that $a + \frac{1}{b}$ and $b + \frac{1}{a}$ are the roots of the equation $x^2 - px + q = 0$. What is q? *(Source: AMC 10)*

4.33 One of the roots of $x^2 - ax + 2a + 3 = 0$ is 3 times the other. Find all possible values of a.

4.34 Find all x such that $6x^2 + 5x < 4$. *(Source: AHSME)*

4.35 Find all values of k such that $x^2 + kx + 27 = 0$ has two distinct real solutions for x.

4.36 Find all possible values of x/y if x and y are nonzero real numbers such that $x^2 = xy + 12y^2$.

4.37 Find x if $y = \frac{8}{x^2+4}$ and $x + y = 2$. *(Source: AHSME)*

4.38 The Bolts charge \$50 for each ticket, and they sell 20,000 tickets per game. For each dollar they raise their prices, they will sell 100 fewer tickets. For what values of k can the Bolts increase ticket prices by \$$k$ and have more than \$1,000,000 in revenue per game?

4.39 Find all x such that $x^2 + 50x - 2079 \geq 0$.

4.40 Let $P(x)$ be a quadratic polynomial such that $P(0) = -1$, $P(1) = 9$, and $P(2) = 25$. Find $P(-1)$.

4.41 Find all real solutions to $(x^2 - 5x + 5)^{x^2-9x+20} = 1$.

4.42 Find all solutions to the system of equations,

$$x^2 + yz = 39,$$
$$x - yz = -33,$$
$$y + z = 13.$$

4.43 Find the roots of $x^2 + \left(a - \dfrac{1}{a}\right)x - 1 = 0$ in terms of a.

Challenge Problems

4.44 If $\frac{x^2-bx}{ax-c} = \frac{m-1}{m+1}$ has solutions for x such that each solution is the negative of the other, then find m in terms of a and b. *(Source: AHSME)*

4.45 Find constants a and b such that $b - a$ is as small as possible, and the entire graph of the equation $y = (1 - x^2)/(1 + x^2)$ lies within $a < y \leq b$.

4.46 Let t be a real number such that $|t| \leq 2$. Show that both roots of the quadratic $z^2 + tz + 1 = 0$ have magnitude 1.

4.47 Prove that if $\dfrac{a+b}{a} = \dfrac{b}{a+b}$ then a and b can't *both* be real numbers. **Hints:** 191

4.48 Find the positive difference between the roots of $x^2 - px + \dfrac{p^2-1}{4} = 0$. **Hints:** 133

4.49 Find all ordered pairs (x, y) that satisfy both $x^2 - xy = 6$ and $xy + y^2 = 4$.

4.50 A quadratic expression is **monic** if the coefficient of its quadratic term is 1. Let $f(x)$, $g(x)$, and $h(x)$ be monic quadratics such that the sum of the roots of $f(x)$ is a, the sum of the roots of $g(x)$ is b, and the sum of the roots of $h(x)$ is c. Find the sum of the solutions to $f(x) + g(x) + h(x) = 0$ in terms of a, b, and c.

4.51 Let a and b be nonzero real numbers such that $a^2 + b^2 + 2a + 1 = 2ab + 2b$. Find the value of $\dfrac{b-1}{a}$.
Hints: 34

4.52 Let a be a constant such that the roots of $ax^2 - 6x + 12 = 0$ are r and s. Find the value of $\dfrac{1}{r^2} + \dfrac{1}{s^2}$ in terms of a.

4.53 Find the sum of the real roots of $(2 + (2 + (2 + (2 + x)^2)^2)^2)^2 = 15129$.

4.54 Prove that if a quadratic has real coefficients such that the sum of the coefficients is 0, then the quadratic has real roots. **Hints:** 208

4.55 Find all ordered pairs (r, s) that satisfy both $s^2 - 1 = 4r^2 + 4r$ and $4r^2 + s^2 = 1 + 3rs$.

4.56 Prove that if $(r + 1)(s + 1) = 13$, then there is some constant c such that r and s are roots of the quadratic equation $x^2 - cx + (12 - c) = 0$. **Hints:** 303

4.57 Find all real solutions (x, y) of the system $x^2 + y = 12 = y^2 + x$. *(Source: HMMT)* **Hints:** 160

4.58 Let a and b be real numbers such that $0 < a < b$ and $a^2 + b^2 = 6ab$. Find $\dfrac{a+b}{a-b}$. **Hints:** 264

4.59 Find all ordered pairs (x, y) that satisfy $x + y = \dfrac{9}{2}$ and $2x^2y^2 - 13xy + 18 = 0$. *(Source: CEMC)*
Hints: 292

4.60 Find all the values of m for which the zeros of $2x^2 - mx - 8$ differ by $m - 1$. *(Source: HMMT)*

4.61★ Find all x such that $\dfrac{x-a}{b} + \dfrac{x-b}{a} = \dfrac{b}{x-a} + \dfrac{a}{x-b}$, where a and b are constants.

4.62★ Solve the following system of equations in real numbers:

$$2x_1 = x_5^2 - 23,$$
$$4x_2 = x_1^2 + 7,$$
$$6x_3 = x_2^2 + 14,$$
$$8x_4 = x_3^2 + 23,$$
$$10x_5 = x_4^2 + 34.$$

(Source: Austria) **Hints:** 177

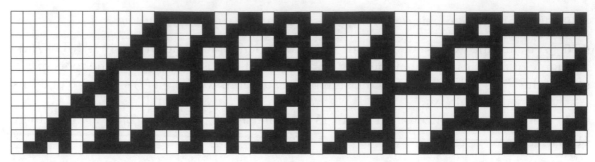

Along a parabola life like a rocket flies,
Mainly in darkness, now and then on a rainbow. – Andrei Voznesensky

CHAPTER **5**

Conics

A **locus** is a collection of points that satisfy specific criteria. For example, the locus of points on the Cartesian plane that satisfy the equation $x = y$ is a line, and the locus of points in a plane that are 2 units from a given point P is a circle with radius 2. In this chapter, we explore some special loci ("loci" is the plural of "locus"), and we discuss equations whose graphs on the Cartesian plane produce these loci.

The loci we study in this chapter are collectively called **conic sections**, or **conics** for short. The diagrams below illustrate why. In each diagram, a plane intersects a pair of congruent cones that share a vertex and axis, but open in opposite directions.

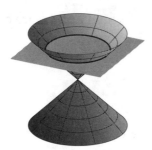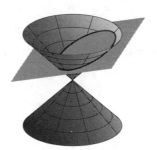

If the plane is parallel to a line on the surface of the cones such that the line is not in the plane, then the intersection of the plane and the cones forms a shape called a **parabola**. If the plane is perpendicular to the axis, then we get a **circle**. If it intersects one cone in a closed curve, we have an **ellipse**. Finally, we produce a **hyperbola** when the plane intersects both cones as in the final diagram.

Sidenote: What happens when the plane passes through the vertex of the cones?

In this chapter, we'll learn definitions of the conic sections, and the important points, lines, and segments associated with each type of conic section. We'll use these definitions to create general formulas that can be used to solve specific problems, and we'll also learn how to use our definitions to tackle some problems without resorting to formulas.

The many formulas we will prove are not very important, so you shouldn't feel the need to memorize them all. You can usually just look them up when you need them.

As stated in the introduction to this text on page iii, the details of this chapter are not as important as those presented in most of the other chapters of this book. The two main items you should learn from this chapter are the importance of using completing the square to work with quadratic expressions, and a qualitative understanding of equations whose graphs are conic sections.

5.1 Parabolas

Geometrically speaking, a **parabola** is the locus of points in a plane that are equidistant from a given line and a point not on the line. The line is called the **directrix** of the parabola and the point is the parabola's **focus**.

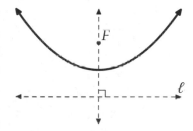

The solid curve at right is a parabola. The dashed horizontal line ℓ below the parabola is the parabola's directrix and point F "inside" the parabola is its focus. The line through the focus perpendicular to the directrix is the parabola's **axis of symmetry**. As the name indicates (and we'll prove in the first problem below), a parabola is symmetric about its axis of symmetry. The point where the axis intersects the parabola is the **vertex** of the parabola. As shown at right, the vertex of a parabola with a horizontal directrix is the bottommost point of the parabola.

You may already know that the graph of a quadratic produces a parabola. In this section, we learn how our geometric definition of a parabola produces a quadratic equation.

Problems

Problem 5.1: In the diagram, F is the focus of the parabola, P is on the parabola, ℓ is the directrix, and Q is the foot of the altitude from P to ℓ. Line k passes through the focus and is perpendicular to the directrix. In this problem, we prove that the parabola is symmetric about line k.

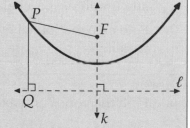

Let P' be the image of P upon reflection over the line k. Our goal is to show that P' is also on the parabola.

(a) Explain why $PQ = PF$.

(b) Explain why $P'F = PF$.

(c) Explain why the distance from P' to ℓ equals PQ.

(d) Explain why P' is on the parabola. Why does this mean that the parabola is symmetric about line k?

Problem 5.2: Show that the vertex of a parabola is the midpoint of the segment that connects the focus of the parabola to the point where the parabola's axis of symmetry intersects the parabola's directrix.

Problem 5.3: Suppose a parabola on the Cartesian plane has focus $(0, t)$ and vertex $(0, 0)$, where $t > 0$.

(a) First, we show that the equation whose graph is this parabola can be written in the form $y = ax^2$, where $a = 1/(4t)$.

    (i) Find the equation whose graph is the directrix of the parabola.

    (ii) Suppose that (x, y) is a point on the parabola. In terms of x, y, and t, how far is the point (x, y) from the focus? How far is the point from the directrix?

    (iii) Because (x, y) is on the parabola, what must be true about the two distances you found in part (ii)?

    (iv) Find an equation that must be satisfied by x and y. Show that your equation can be written in the form $y = ax^2$, where $a = 1/(4t)$. Does the same equation result if $t < 0$?

(b) Let a be a nonzero constant. Find the focus and directrix of the graph of $y = ax^2$ in terms of a.

(c) Graph the equation $y = ax^2$ for several values of a. How does the value of a affect the graph?

Problem 5.4:

(a) Find an equation whose graph is a parabola with vertex $(0, 0)$ and focus $(0, 2)$.

(b) Suppose we wish to find an equation whose graph is a parabola with vertex $(-6, 3)$ and focus $(-6, 5)$. How can we use our answer to part (a) to quickly find the desired equation?

(c) Use the strategy from part (b) and your final equation from Problem 5.3 to find an equation whose graph is a parabola with vertex (h, k) and focus $(h, k + t)$.

(d) Find the vertex, focus, axis of symmetry, and directrix of the graph of $y = a(x - h)^2 + k$, where a, h, and k are constants, and $a \neq 0$.

Problem 5.5: Graph the equations $y = (x + 2)^2 + 3$ and $x = (y + 2)^2 + 3$.

Problem 5.6:

(a) A parabola has vertex $(-2, 5)$ and directrix $y = \frac{7}{2}$. Find an equation whose graph is this parabola, and graph the parabola.

(b) Find an equation whose graph is a parabola with vertex $(4, -1)$ and focus $(0, -1)$, and graph the parabola.

Problem 5.7:

(a) Find the vertex and axis of symmetry of the graph of $y^2 - 8y + x = 2$.

(b) Find the vertex and axis of symmetry of the graph of $2x^2 + 14x - y = 1$.

Problem 5.1: Prove that a parabola is symmetric about the line through its focus that is perpendicular to its directrix. In other words, show why this line earns the name "axis of symmetry."

Solution for Problem 5.1: We let ℓ be the directrix, F be the focus, P be a point on the parabola, and k be the line through F perpendicular to ℓ. To show that k is a line of symmetry of the parabola, we consider the reflection of P over k, which we call point P'. We wish to show that P' is on the parabola.

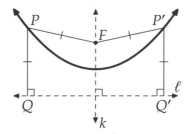

To show that P' is on the parabola, we must show that it is just as far from ℓ as it is from F. Since P is on the parabola, we already know that $PF = PQ$, where Q is the foot of the altitude from P to ℓ (so PQ is the distance from P to ℓ). Because P' is the image of P upon reflection over k, and F is its own image upon this reflection, we have $P'F = PF$. Meanwhile, because $\overline{PQ}$ is parallel to k, the image of $\overline{PQ}$, which is $\overline{P'Q'}$, is parallel to both k and $\overline{PQ}$. Since $\overline{PQ} \perp \ell$ and $\overline{PQ} \parallel \overline{P'Q'}$, we know that $\overline{P'Q'} \perp \ell$ as well. This means that $P'Q'$ is the distance from P' to ℓ. Since $\overline{P'Q'}$ is the image of $\overline{PQ}$ upon reflection over k, we have $P'Q' = PQ$. Finally, because $PF = PQ$, $PF = P'F$, and $PQ = P'Q'$, we have $P'F = P'Q'$. So, P' is equidistant from F and from ℓ, which means that P' is on the parabola. Therefore, the image of each point on the parabola upon reflection over k is also on the parabola, which means that k is a line of symmetry of the parabola.

This is why the line through the focus of a parabola that is perpendicular to the directrix is called the **axis of symmetry**. $\square$

Problem 5.2: Show that the vertex of a parabola is the midpoint of the segment that connects the focus of the parabola to the point where the parabola's axis of symmetry intersects the parabola's directrix.

Solution for Problem 5.2: Let V be the point where the axis intersects the parabola, and let Z be the point where the axis meet the directrix. Because the axis is perpendicular to the directrix, $\overline{VZ}$ is perpendicular to the directrix. Therefore, VZ is the distance between the vertex and the directrix. Since V is on the parabola, it is equidistant from the focus and the directrix. Letting the focus be F, we therefore have $VF = VZ$. Because $F, V,$ and Z are different points on the same line such that $VF = VZ$, point V is the midpoint of $\overline{FZ}$, as desired. $\square$

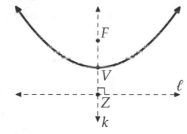

Problem 5.3: Suppose a parabola on the Cartesian plane has focus $(0, t)$ and vertex $(0, 0)$, where $t \neq 0$.

(a) Let (x, y) be a point on the parabola. Find an equation that must be satisfied by x and y. Show that your equation can be written in the form $y = ax^2$, where $a = 1/(4t)$.

(b) Let a be a nonzero constant. Find the vertex, focus, and directrix of the graph of $y = ax^2$ in terms of a.

(c) Graph the equation $y = ax^2$ for several values of a. How does the value of a affect the graph?

Solution for Problem 5.3:

(a) We'll first tackle the case in which $t > 0$. The axis of symmetry of a parabola passes through the vertex and the focus. Since the vertex and focus are both on the y-axis, the axis of symmetry of this parabola is vertical. The axis of symmetry and directrix of a parabola are perpendicular, so the directrix, which we'll call ℓ, is horizontal. Moreover, the vertex of a parabola is equidistant from the focus and the directrix. So, because the vertex is t below the focus, the directrix is t below the vertex. This means that the directrix is the graph of $y = -t$.

Let point P be on the parabola, where $P = (x, y)$. Let the focus be point F and the directrix be line ℓ. By the definition of a parabola, the length of $\overline{PF}$ equals the distance from P to ℓ. We let Q be the foot of the altitude from P to ℓ, so PQ is the distance from P to ℓ.

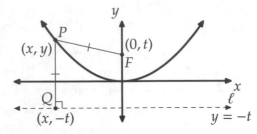

The distance formula gives us $PF = \sqrt{(x-0)^2 + (y-t)^2}$. P and F cannot be on opposite sides of ℓ, since this would force PF to be greater than the distance from P to ℓ. Therefore, P is above ℓ, as shown, so $y > -t$ and the distance PQ is $y - (-t) = y + t$. This distance must equal PF, so we have

$$\sqrt{(x-0)^2 + (y-t)^2} = y + t.$$

We square both sides to get rid of the radical, and we have $x^2 + (y-t)^2 = (y+t)^2$. Subtracting $(y-t)^2$ from both sides gives

$$x^2 = (y+t)^2 - (y-t)^2 = 4ty.$$

Solving this equation for y gives $y = \frac{1}{4t}x^2$.

If $t < 0$, then the focus is below the directrix, as shown at right. The distance from $P = (x, y)$ to ℓ is $-t - y$, but PF equals the same expression as before, so we have

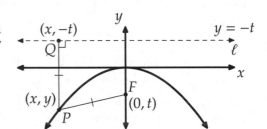

$$\sqrt{(x-0)^2 + (y-t)^2} = -t - y.$$

Because $(-t - y)^2 = (y + t)^2$, squaring this equation leads to the same equation as the one we found for $t > 0$.

We could have knocked off both cases at once by noting that in either case, the distance from P to ℓ is $|y - (-t)|$. This expression equals $|y + t|$, and it must equal PF, so we have

$$\sqrt{(x-0)^2 + (y-t)^2} = |y + t|.$$

Squaring both sides gives $(x-0)^2 + (y-t)^2 = (|y+t|)^2$. We have $(|y+t|)^2 = (y+t)^2$, because $(|z|)^2 = z^2$ for all z, so $(x - 0)^2 + (y - t)^2 = (y + t)^2$, as before.

We can reverse our steps in Problem 5.3 to show that the graph of any equation of the form $y = \frac{1}{4t}x^2$, where t is a nonzero constant, is a parabola with vertex $(0,0)$, focus $(0,t)$, and directrix $y = -t$.

(b) We just found that a parabola with vertex $(0,0)$, focus $(0,t)$, and directrix $y = -t$ is the graph of the equation $y = \frac{1}{4t}x^2$. If we let $a = \frac{1}{4t}$, then this equation is $y = ax^2$. Solving $a = \frac{1}{4t}$ for t in terms of a gives $t = \frac{1}{4a}$. The vertex is still just $(0,0)$, but using this expression for t allows us to write the focus as $(0, \frac{1}{4a})$ and the directrix as $y = -\frac{1}{4a}$.

(c) Below are graphs of $y = ax^2$ for $a = 5$, $a = 1/5$, and $a = -1/5$.

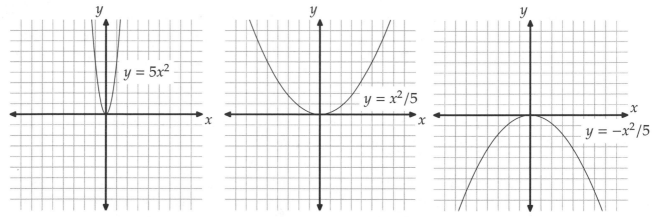

These three graphs give us a good idea how the coefficient of x^2 affects the graph of the quadratic. The most striking difference is that the graph of $y = -x^2/5$ opens downward, while our other two graphs open upward. This suggests that whenever the coefficient of x^2 is negative, the graph opens downward, and when this coefficient is positive, the graph opens upward.

We can see why this is true by considering very large values of x. When x gets very large and $y = x^2/5$, then y becomes a large positive number. When x gets very large and $y = -x^2/5$, then y becomes much less than 0.

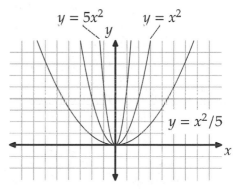

The coefficient of x^2 also affects the width of the parabola. The graph of $y = 5x^2$ is much narrower than that of $y = x^2$, which is narrower than that of $y = x^2/5$. To see this effect more clearly, consider the graphs of $y = 5x^2$, $y = x^2$, and $y = x^2/5$ all on the same Cartesian plane at right.

Another feature we note from these graphs is that we have an algebraic explanation for why a parabola is symmetric about its axis. Focusing on the graph of $y = x^2$, we see that for every point (x, x^2) on the graph with $x \neq 0$, there is a point $(-x, x^2)$ also on the graph. The point $(-x, x^2)$ is the reflection of the point (x, x^2) over the y-axis, so the parabola is symmetric about the y-axis, which is its axis of symmetry.

□

Problem 5.4:

(a) Find an equation whose graph is a parabola with vertex $(0, 0)$ and focus $(0, 2)$.

(b) Suppose we wish to find an equation whose graph is a parabola with vertex $(-6, 3)$ and focus $(-6, 5)$. How can we use our answer to part (a) to quickly find the desired equation?

(c) Use the strategy from part (b) and your final equation from Problem 5.3 to find an equation whose graph is a parabola with vertex (h, k) and focus $(h, k + t)$.

(d) Find the vertex, focus, axis of symmetry, and directrix of the graph of $y = a(x - h)^2 + k$, where a, h, and k are constants.

Solution for Problem 5.4:

(a) A parabola with vertex $(0,0)$ and focus $(0, \frac{1}{4a})$ is the graph of the equation $y = ax^2$. Therefore, to find an equation whose graph is a parabola with vertex $(0,0)$ and focus $(0,2)$, we must find the value of a such that $\frac{1}{4a} = 2$. Solving this equation gives us $a = \frac{1}{8}$, so the parabola is the graph of $y = \frac{1}{8}x^2$.

(b) Let $\mathcal{P}$ be the parabola with vertex $(-6,3)$ and focus $(-6,5)$. The vertex of $\mathcal{P}$ is 2 units below the focus, just as in part (a). But the vertex of $\mathcal{P}$ is not the origin, so it appears at first that we cannot use what we know about the graphs of equations of the form $y = ax^2$.

However, if we shift the parabola with vertex $(0,0)$ and focus $(0,2)$ left 6 units and up 3 units, we get a parabola with vertex $(-6,3)$ and focus $(-6,5)$. Mathematically speaking, this "shift" is a **translation**. But is the result of this translation the only parabola with vertex $(-6,3)$ and focus $(-6,5)$?

To answer this question, we note that once we know the vertex and the focus of a parabola, we can find the directrix of the parabola. Moreover, there is only one possible parabola with a given focus and directrix. Combining these observations tells us that there is only one possible parabola with a given focus and vertex. Therefore, there is only one parabola with vertex $(-6,3)$ and focus $(-6,5)$, so the result of our translation is the parabola $\mathcal{P}$ we seek.

Now, we're ready to find the equation that has $\mathcal{P}$ as its graph. $\mathcal{P}$ is the result of translating the parabola in part (a) left 6 units and up 3 units. Therefore, for every point (x, y) on the graph of $y = \frac{1}{8}x^2$, there is a point (x', y') on $\mathcal{P}$ where $(x', y') = (x - 6, y + 3)$. For example, the vertex of $\mathcal{P}$ is $(0 - 6, 0 + 3) = (-6, 3)$.

Parabola $\mathcal{P}$ is graphed as a solid curve at right, and the graph of $y = \frac{1}{8}x^2$ is dashed. Notice that every point on $\mathcal{P}$ is 6 units to the left and 3 units above a corresponding point on the dashed parabola.

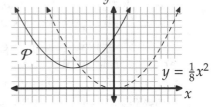

We can now use the relationship $(x', y') = (x - 6, y + 3)$ to find an equation whose graph is $\mathcal{P}$. For any point (x', y') on $\mathcal{P}$, we must have $x' = x - 6$ and $y' = y + 3$, where (x, y) is a point on the graph of $y = \frac{1}{8}x^2$. We solve $x' = x - 6$ and $y' = y + 3$ for x and y, respectively, to find $x = x' + 6$ and $y = y' - 3$. Since (x, y) must satisfy $y = \frac{1}{8}x^2$, we substitute our expressions for x and y to get an equation relating x' and y':

$$y' - 3 = \frac{1}{8}(x' + 6)^2.$$

Therefore, $\mathcal{P}$ consists of every point (x', y') that satisfies $y' - 3 = \frac{1}{8}(x' + 6)^2$. In other words, $\mathcal{P}$ is the graph of the equation $y - 3 = \frac{1}{8}(x + 6)^2$.

We can test our equation by confirming that the vertex satisfies the equation. Indeed, when $x = -6$ and $y = 3$, both sides of the equation equal 0. This check doesn't mean our equation is definitely correct; there are infinitely many equations whose graphs pass through $(-6, 3)$. However, quick checks like this help us catch errors.

You may have noticed that our using a translation to produce the equation $y - 3 = \frac{1}{8}(x + 6)^2$ from the equation $y = \frac{1}{8}x^2$ is a lot like our discussion of transforming functions in Section 2.2. It's not only "a lot like" transforming functions; it's exactly the same! If we let $f(x) = \frac{1}{8}x^2$, then the parabola in part (a) is the graph of f. In part (b), we must shift this graph to the left 6 units and up 3. Our work in Section 2.2 tells us that the result is the graph of $y = f(x + 6) + 3$, which is $y = \frac{1}{8}(x + 6)^2 + 3$, or $y - 3 = \frac{1}{8}(x + 6)^2$.

(c) Let $\mathcal{P}$ be the parabola with vertex (h, k) and focus $(h, k + t)$. We follow the same steps as in part (b) to find an equation whose graph is $\mathcal{P}$.

The graph of the equation $y = \frac{1}{4t}x^2$ is a parabola with vertex $(0, 0)$ and focus $(0, t)$. So, the vertex and focus of $\mathcal{P}$ are h to right and k above the vertex and focus, respectively, of the graph of $y = \frac{1}{4t}x^2$. Therefore, $\mathcal{P}$ is the result of shifting the graph of $y = \frac{1}{4t}x^2$ right h units and up k units. In other words, for every point (x, y) on the graph of $y = \frac{1}{4t}x^2$, there is a point (x', y') on $\mathcal{P}$ where $(x', y') = (x + h, y + k)$. This means $\mathcal{P}$ consists of all points (x', y') such that $x' = x + h$ and $y' = y + k$. Solving these equations for x and y, respectively, gives $x = x' - h$ and $y = y' - k$. Because (x, y) must satisfy the equation $y = \frac{1}{4t}x^2$, we can substitute our expressions for x and y to find

$$y' - k = \frac{1}{4t}(x' - h)^2.$$

As before, we can get rid of the prime symbols and write that $\mathcal{P}$ is the graph of the equation $y - k = \frac{1}{4t}(x - h)^2$. We can also write this equation as $y = \frac{1}{4t}(x - h)^2 + k$.

(d) We just learned that the graph of the equation $y = \frac{1}{4t}(x - h)^2 + k$ is a parabola with vertex (h, k) and focus $(h, k + t)$. If we let $a = \frac{1}{4t}$, then the equation becomes $y = a(x - h)^2 + k$. The vertex is still (h, k), and we can solve $a = \frac{1}{4t}$ for t to find that the focus is $(h, k + \frac{1}{4a})$.

Because the focus has the same x-coordinate as the vertex, the axis of symmetry is the vertical line through the vertex, which is $x = h$. Since the axis of symmetry is vertical, the directrix is horizontal. The y-coordinate of the vertex is k, the y-coordinate of the focus is $k + \frac{1}{4a}$, and the vertex is equidistant from the focus and directrix, so the directrix must be the line $y = k - \frac{1}{4a}$.

$\square$

> **Important:** The **standard form** of an equation whose graph is a parabola that opens vertically (upward or downward) is
>
> $$y = a(x - h)^2 + k,$$
>
> where a, h, and k are constants with $a \neq 0$. If the $a > 0$, then the parabola opens upward, and if $a < 0$, then the parabola opens downward. The vertex of the parabola is (h, k) and the axis of symmetry is $x = h$.

The power of this standard form is that *every equation whose graph is a parabola that opens upward or downward can be written in this form.* Once we have the equation in this form, we can easily determine the vertex and axis of symmetry of the parabola as described above. We can also note that the focus of such a parabola is $\left(h, k + \frac{1}{4a}\right)$ and the directrix is $y = k - \frac{1}{4a}$.

Now, let's graph a couple particular parabolas.

Problem 5.5: Graph the equations $y = (x + 2)^2 + 3$ and $x = (y + 2)^2 + 3$.

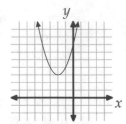

Solution for Problem 5.5: From our work in Problem 5.4, we recognize $y = (x + 2)^2 + 3$ as an equation that is in the standard form of an equation whose graph is an upward-opening parabola. Since $(x + 2)^2$ is nonnegative for all real x, the smallest possible value of $(x + 2)^2 + 3$ occurs when $(x + 2)^2 = 0$, which is when $x = -2$. When $x = -2$, we have $y = 3$, so the bottommost point of the parabola, which is the vertex of the parabola, is $(-2, 3)$.

We find more points on the graph by choosing various values of x and find the corresponding values of y. We thereby produce the graph at right above.

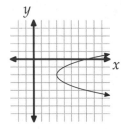

At first glance, the equation $x = (y + 2)^2 + 3$ resembles the standard form we discussed in Problem 5.4. However, in this equation, the y term is squared, rather than the x term. We can get a sense for the graph the same way we found the graph of $y = (x + 2)^2 + 3$. First, we note that $(y + 2)^2$ is nonnegative for all real y, and is zero for $y = -2$. Therefore, $(y + 2)^2 + 3$ is minimized when $y = -2$, which makes $x = 3$. For all other values of y, the value of x is greater than 3, which means that $(x, y) = (3, -2)$ is the leftmost point of the graph.

Choosing more values of y and computing the corresponding values of x, we produce the graph at left above. As expected, the graph is a parabola with vertex $(3, -2)$, but its axis of symmetry is horizontal rather than vertical. $\square$

Using similar methods to those we used in Problem 5.4, we can show that the graph of any equation of the form $x = a(y - k)^2 + h$ is a horizontally-opening parabola:

> **Important:** The **standard form** of an equation whose graph is a parabola that opens horizontally (rightward or leftward) is
>
> $$x = a(y - k)^2 + h,$$
>
> where a, h, and k are constants with $a \neq 0$. If $a > 0$, then the parabola opens rightward and if $a < 0$, it opens leftward. The vertex of the parabola is (h, k) and the axis of symmetry is $y = k$.

We can also show that the graph of $x = a(y - k)^2 + h$ has focus $\left(h + \frac{1}{4a}, k\right)$ and directrix $y = h - \frac{1}{4a}$.

Combining our observations about the graphs of equations of the form $y = a(x - h)^2 + k$ and $x = a(y - k)^2 + h$, where a, h, and k are constants with $a \neq 0$, we have the following:

> **Important:** Suppose we have an equation of the form $Ax^2 + By^2 + Cx + Dy + E = 0$, where A, B, C, D, E are constants such that one of A or B is zero and the other is nonzero. The graph of the equation is a parabola. If $A \neq 0$, then the parabola opens upward or downward. If $B \neq 0$, then the parabola opens leftward or rightward.

An equation does not have to be in the form described above in order to have a graph that is a parabola. Specifically, if the axis of symmetry of a parabola on the Cartesian plane is not horizontal or vertical, then the equation whose graph is this parabola does not fit either of the standard forms we discuss in this section. We need more advanced mathematical tools to study these parabolas thoroughly, so we'll focus on vertically-opening and horizontally-opening parabolas in this text.

Problem 5.6:

(a) A parabola has vertex $(-2, 5)$ and directrix $y = \frac{7}{2}$. Find an equation whose graph is this parabola, and graph the parabola.

(b) Find an equation whose graph is a parabola with vertex $(4, -1)$ and focus $(0, -1)$, and graph the parabola.

Solution for Problem 5.6:

(a) Because the vertex is $\frac{3}{2}$ above the directrix, the focus is $\frac{3}{2}$ above the vertex, at $(-2, \frac{13}{2})$. Therefore, any point (x, y) on the parabola is equidistant from $(-2, \frac{13}{2})$ (the focus) and $y = \frac{7}{2}$ (the directrix), so we have

$$\sqrt{(x+2)^2 + \left(y - \frac{13}{2}\right)^2} = y - \frac{7}{2}.$$

Squaring both sides and rearranging gives

$$(x+2)^2 = \left(y - \frac{7}{2}\right)^2 - \left(y - \frac{13}{2}\right)^2 = \left(y - \frac{7}{2} + y - \frac{13}{2}\right)\left(y - \frac{7}{2} - \left(y - \frac{13}{2}\right)\right) = 6y - 30.$$

Solving for y in terms of x gives $y = \frac{1}{6}(x+2)^2 + 5$.

We could also have used the standard form of the equation of a parabola to find the desired equation. Specifically, we know that since the directrix is a horizontal line, the equation must be of the form

$$y = a(x - h)^2 + k.$$

Furthermore, since the vertex is $(-2, 5)$ and the vertex must be (h, k), we have $h = -2$ and $k = 5$. Now all we have to do is find a. For that, we use the directrix. The directrix must have the form $y = k - \frac{1}{4a}$. Since we are given that the directrix is $y = \frac{7}{2}$ and we found above that $k = 5$, we must have $5 - \frac{1}{4a} = \frac{7}{2}$. Therefore, $a = \frac{1}{6}$, so our equation is

$$y = \frac{1}{6}(x+2)^2 + 5.$$

To graph the parabola, we start with the vertex $(-2, 5)$. Since x is squared and a is positive, the parabola opens upward. We then choose values of x greater than -2 and find the corresponding values of y to graph the right side of the parabola. The left side is the reflection of the right side over the axis of symmetry $x = -2$, and we have the graph shown at right.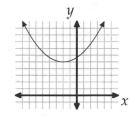

(b) The focus and vertex are on the same horizontal line, so the axis of symmetry is horizontal. Therefore, the directrix is vertical. Since the focus is 4 to the left of the vertex, the directrix must be 4 to the right of the vertex, which means the directrix is the graph of $x = 8$.

Because the parabola has focus $(0, -1)$ and directrix $x = 8$, every point (x, y) on the parabola must satisfy

$$\sqrt{(x - 0)^2 + (y + 1)^2} = 8 - x.$$

We have $8 - x$ on the right rather than $x - 8$ because the fact that the focus is to the left of the directrix tells us that every point on the parabola is to the left of the directrix. Therefore, if (x, y) is on the parabola, we have $x < 8$, and the distance from this point to the directrix is $8 - x$.

Squaring both sides of this equation gives us $x^2 + (y + 1)^2 = x^2 - 16x + 64$, and solving for x gives us $x = -\frac{1}{16}(y + 1)^2 + 4$.

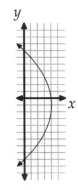

As in part (a), we could also have used our understanding of the standard form of a parabola. The vertex $(4, -1)$ is 4 units to the right of the focus $(0, -1)$, so the parabola opens leftward, and its equation has the form $x = a(y - k)^2 + h$. From the vertex, we have $h = 4$ and $k = -1$. From the x-coordinate of the focus we have $h + \frac{1}{4a} = 0$, from which we find $\frac{1}{4a} = -h = -4$, so $a = -\frac{1}{16}$. Therefore, our equation is

$$x = -\frac{1}{16}(y + 1)^2 + 4.$$

To graph this equation, we again start with the vertex $(4, -1)$. We then choose values of y (because we have x in terms of y in our equation above) and find corresponding values of x to produce the graph shown at right.

$\square$

In Problem 5.6, we showed two methods to get the equation for each parabola to highlight what's important in this chapter. In each case, using our standard form looks easier. This shows the power of understanding how to use standard forms of equations, and why we bother developing them. However, our other approach, starting from the geometric definition of a parabola, shows that even when we don't know or remember our formulas, all is not lost. We can start from the fundamentals and follow the footsteps of the work we did when deriving the standard form.

Problem 5.7:

(a) Find the vertex and axis of symmetry of the graph of $y^2 - 8y + x = 2$.

(b) Find the vertex and axis of symmetry of the graph of $2x^2 + 14x - y = 1$.

Solution for Problem 5.7:

(a) First, we note that y is squared, not x. So, the parabola opens left or right. If we write the equation in the form $x = a(y - k)^2 + h$, we can easily identify the vertex and axis.

We start by solving for x to find $x = -y^2 + 8y + 2$. We then complete the square on the right side. To do so, we first factor -1 out of the first two terms, to get $x = -1(y^2 - 8y) + 2$. We then complete the square by adding $(-1)(16) - (-1)(16)$ (which equals 0) to the right side to give

$$x = -(y^2 - 8y) + (-1)(16) - (-1)(16) + 2 = -1(y^2 - 8y + 16) + 18 = -1(y - 4)^2 + 18.$$

What's wrong with this next step:

Bogus Solution: We have $a = -1$, $h = 4$, and $k = 18$, so the vertex is $(4, 18)$.

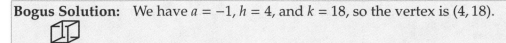

We see our mistake immediately if we check to see if the point $(4, 18)$ satisfies the original equation. The vertex is on the parabola, so this point should satisfy the equation. However, when we substitute $(x, y) = (4, 18)$ in the left side of the equation $y^2 - 8y + x = 2$, we have $y^2 - 8y + x = 18^2 - 8(18) + 4 = 184$, so clearly something is wrong.

We overlooked which variable is squared. Our equation in standard form is $x = -1(y - 4)^2 + 18$. Since y is squared, the parabola opens horizontally. We can read off the correct answer by comparing $x = -1(y - 4)^2 + 18$ to the standard form we found earlier, $x = -1(y - k)^2 + h$. We then have $a = -1$, $h = 18$, and $k = 4$, so the vertex is $(18, 4)$ and the axis is $y = 4$.

We can also simply think about the equation to determine the axis and vertex. We can see that the parabola opens leftward by noting that as y gets much larger than 0 or much smaller than 0, the corresponding value of x is much less than 0. Since $x = -1(y - 4)^2 + 18$ and $-1(y - 4)^2 \leq 0$ for all real y, the rightmost point on the parabola occurs when $y = 4$. This makes our squared term 0, and gives $x = 18$, so our vertex is $(18, 4)$. The axis of symmetry is the horizontal line through the vertex (because the parabola opens horizontally), so the axis is $y = 4$.

(b) We again start by isolating the non-squared variable, to get $y = 2x^2 + 14x - 1$, then we complete the square to get

$$y = 2\left(x + \frac{7}{2}\right)^2 - \frac{51}{2}.$$

This fits our standard form, $y = a(x - h)^2 + k$, of a parabola that opens vertically. What's wrong with this next step:

Bogus Solution: We therefore have $a = 2$, $h = \frac{7}{2}$, and $k = -\frac{51}{2}$, so the vertex is $\left(\frac{7}{2}, -\frac{51}{2}\right)$.

We see our error by putting the supposed vertex into our equation in standard form. We quickly see it doesn't work, because the expression $2(x + \frac{7}{2})^2$ isn't 0 when $x = \frac{7}{2}$ (so $y \neq -\frac{51}{2}$ when $x = \frac{7}{2}$).

This time, we made a sign error. Notice that in our standard form, we have $(x - h)^2$, not $(x + h)^2$. So, for our equation $y = 2\left(x + \frac{7}{2}\right)^2 - \frac{51}{2}$, we have $h = -\frac{7}{2}$, not $h = \frac{7}{2}$. We also have $k = -\frac{51}{2}$, so the vertex is $\left(-\frac{7}{2}, -\frac{51}{2}\right)$ and the axis is $x = -\frac{7}{2}$.

> **WARNING!!** Be very wary of sign errors when working with parabolas (or other conic sections). One quick way to check for mistakes when working with parabolas is to plug the point that you think is the vertex into your equation. If the resulting equation isn't true, then you've made a mistake somewhere.

We could have avoided the sign error by thinking about the equation rather than reaching for our formulas right away. Since $y = 2\left(x + \frac{7}{2}\right)^2 - \frac{51}{2}$ and $2\left(x + \frac{7}{2}\right)^2$ is always nonnegative, the bottommost point on the graph will occur when $2\left(x + \frac{7}{2}\right)^2 = 0$, which gives us $x = -\frac{7}{2}$, so $y = -\frac{51}{2}$. Therefore, the vertex is $\left(-\frac{7}{2}, -\frac{51}{2}\right)$. The axis of symmetry is the vertical line through the vertex, which is $x = -\frac{7}{2}$. $\square$

> **Exercises**

5.1.1 Find an equation whose graph is a parabola with directrix $y = -1$ and vertex $(1,3)$.

5.1.2

(a) What is the directrix of a parabola with focus $(3,5)$ and vertex $(2,5)$?

(b) What is the directrix of a parabola with focus $(3,7)$ and vertex $(2,5)$? **Hints:** 250

5.1.3 Find the vertex and axis of symmetry of the parabolas that result from graphing the following equations. Draw the graph of each.

(a) $y = \frac{1}{4}(x - 3)^2 + 2$

(b) $x = 5y^2 - 20y + 23$

5.1.4 For each of the following parts, find an equation whose graph is a parabola with the given properties.

(a) The focus is $(2,2)$ and the directrix is $x = 4$.

(b) The vertex is $(4,6)$, the axis of symmetry is horizontal, and the parabola passes through $(-2,2)$.

5.1.5★ The **latus rectum** of a parabola is a line segment passing through the focus of the parabola such that the segment is parallel to the directrix and has both endpoints on the parabola, as shown in the diagram. Show that the latus rectum of the graph of the equation $y = a(x - h)^2 + k$ has length $1/a$. **Hints:** 327

Latus rectum

Focus

5.2 Problem Solving With Parabolas

In Section 5.1, we learned the standard form for parabolas, but this is not the only useful tool for working with parabolas. In fact, you've already learned some other tools for dealing with parabolas.

> **Problems**

Problem 5.8: A parabola has vertex $(4,3)$ and its axis of symmetry parallel to the y-axis. If one x-intercept is $(1,0)$, find the other x-intercept. *(Source: CMO)*

Problem 5.9: A parabolic arch has a height of 20 feet and a span of 30 feet. How high is the arch 6 feet from its center?

20

6

30

Problem 5.10: Parabola $\mathcal{P}$ has a vertical axis of symmetry and passes through the points $(4, 5)$, $(-2, 11)$, and $(-4, 21)$. Find an equation whose graph is $\mathcal{P}$.

Problem 5.8: A parabola has vertex $(4, 3)$ and its axis of symmetry parallel to the y-axis. If one x-intercept is $(1, 0)$, find the other x-intercept. *(Source: CMO)*

Solution for Problem 5.8: Solution 1: Find the equation. The axis of symmetry is vertical, so the parabola is the graph of an equation of the form $y = a(x - h)^2 + k$. From the vertex, we have $h = 4$ and $k = 3$, so our equation is $y = a(x - 4)^2 + 3$. The parabola passes through the point $(1, 0)$, so we can find a by plugging this point into our equation. This gives $0 = a(1 - 4)^2 + 3$, so $a = -\frac{1}{3}$ and our equation is $y = -\frac{1}{3}(x - 4)^2 + 3$.

To find the x-intercepts of the graph, we let $y = 0$, which gives $0 = -\frac{1}{3}(x - 4)^2 + 3$. Subtracting 3, and then multiplying by -3, gives $(x - 4)^2 = 9$, so $x - 4 = \pm 3$. This gives us $x = 7$ and $x = 1$ as solutions, so $(7, 0)$ is the other x-intercept.

Hmmm.... The x-intercepts are $(1, 0)$ and $(7, 0)$, which are 3 to the left and 3 to the right, respectively, of the axis of symmetry $x = 4$. Aha! There's a much faster solution!

Solution 2: Forget the formulas, think about the problem. The axis of symmetry of the parabola is the vertical line through the vertex. The vertex is $(4, 3)$, so the axis is $x = 4$. The point $(1, 0)$ is on the parabola, so its reflection over $x = 4$, which is $(7, 0)$, is also on the parabola. $\square$

Concept: Don't depend too heavily on your formulas. A little thinking about a problem can go a long way.

Concept: Symmetry can be a very powerful problem-solving tool.

Problem 5.9: A parabolic arch has a height of 20 feet and a span (width at its base) of 30 feet. How high is the arch 6 feet from its center?

Solution for Problem 5.9: We know how to describe parabolas on the Cartesian plane with equations, so we create a model of the arch on the Cartesian plane. We have our choice of what point to choose as the origin when placing the arch on the Cartesian plane. We want to do so in a way that will make our algebra as simple as possible. Specifically, we want many of the points we know something about to have 0 as one or both of its coordinates. In order to make calculations simplest, we let the parabola be a downward opening parabola whose line of symmetry is the y-axis. We place its vertex at $(0, 20)$ so that its span is the distance between its x-intercepts, which are symmetric about the y-axis.

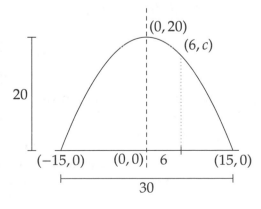

119

Since the span is 30, these intercepts are $(15,0)$ and $(-15,0)$. We therefore seek the value of c such that $(6,c)$ is on the parabola.

Here are a couple ways we can find the equation for our parabola:

Method 1: Use standard form. The parabola opens downward and has vertex $(0,20)$, so its equation is $y = a(x-0)^2 + 20$ for some negative value of a. Since the parabola passes through $(15,0)$, we let $x = 15$ and $y = 0$ in this equation to find $a = -\frac{20}{225} = -\frac{4}{45}$.

Method 2: Use the roots. Since the x-intercepts of our graph are $(15,0)$ and $(-15,0)$, the roots of our desired quadratic are 15 and -15. Therefore, our quadratic is $y = a(x+15)(x-15)$ for some constant a. Since $(x,y) = (0,20)$ must also satisfy this equation, we have $20 = a(0-15)(0+15)$, which gives $a = -\frac{4}{45}$, as before.

Our equation for the parabola is therefore $y = -\frac{4}{45}x^2 + 20$. If $(6,c)$ is on the graph of this equation, we have
$$c = -\frac{4}{45} \cdot 6^2 + 20 = -\frac{16}{5} + 20 = \frac{84}{5}.$$

So, the height of the arch 6 feet from its center is $\frac{84}{5} = 16.8$ feet. $\square$

Problem 5.10: Parabola $\mathcal{P}$ has a vertical axis of symmetry and passes through the points $(4,5)$, $(-2,11)$, and $(-4,21)$. Find an equation whose graph is $\mathcal{P}$.

Solution for Problem 5.10: We might start by noting that because $\mathcal{P}$ has a vertical axis of symmetry, the desired equation has the form $y = a(x-h)^2 + k$. We can then use the three given points to produce the system of equations
$$5 = a(4-h)^2 + k,$$
$$11 = a(-2-h)^2 + k,$$
$$21 = a(-4-h)^2 + k.$$

At first glance this system of equations looks pretty scary. (But it's not impossible to solve; see if you can do so.) Fortunately, there's an easier way to approach the problem. Standard form is not the only form we know for a quadratic. We can also write the equation in the form $y = ax^2 + bx + c$. Plugging the three given points into this form gives a much simpler system:
$$16a + 4b + c = 5,$$
$$4a - 2b + c = 11,$$
$$16a - 4b + c = 21.$$

Subtracting the last equation from the first gives $b = -2$. Subtracting the second from the first gives $12a + 6b = -6$. Letting $b = -2$ in this equation gives $a = \frac{1}{2}$. Finally, $c = 5 - 16a - 4b = 5$, so our desired equation is $y = \frac{1}{2}x^2 - 2x + 5$. $\square$

A common thread in the three problems in this section is the importance of being flexible in your approach to problems. We used symmetry, roots of quadratics, standard form, and general form.

> **Concept:** Don't get locked into one way of thinking about quadratics and parabolas.

Exercises

5.2.1 Find an equation whose graph is a horizontally-opening parabola that passes through $(20, -3)$, $(12, 1)$, and $(33, 2)$.

5.2.2 A parabolic arch is 30 meters wide at its base. A fly sits on the arch 25 meters above a point on the ground that is 5 meters from one point where the arch touches the ground. How high is the center of the arch?

5.2.3 The graph of $y = x^2 + 2x - 2$ is reflected over the line $y = x$. How many points of intersections does the original graph have with its reflection?

5.2.4 If the graph of the equation $y = ax^2 + 6$ is tangent to the graph of $y = x$, then what is a? *(Source: Mandelbrot)* (A line is **tangent** to a parabola if it touches the parabola at one point, but is otherwise always "outside" the parabola.)

5.3 Maxima and Minima of Quadratics

Problems

Problem 5.11: Let $f(x) = x^2 + 6x + 5$.
 (a) Graph f.
 (b) Find the smallest possible value of $f(x)$ if x is a real number.
 (c) How is the value of x for which $f(x)$ is minimized related to the roots of $f(x)$?

Problem 5.12: Let $f(x) = ax^2 + bx + c$, where a, b, and c are real numbers and $a > 0$. In terms of a, b, and/or c, find the real value of x that makes $f(x)$ as small as possible. What happens if a is negative?

Problem 5.13: Let k be a real number. Find the minimum possible distance between $k(1 + 3i)$ and $2i$ in the complex plane.

Problem 5.14: Greg can bicycle at 20 miles per hour on pavement and 10 miles per hour on gravel. Greg is building a rectangular track with one side paved and three sides gravel. Greg would like to build the track so that it takes him 3 hours to ride all the way around the track. In this problem, we find how long the paved side should be if he wishes his track to enclose as large an area as possible. (We assume the track itself has negligible width.)
 (a) Let the dimensions of the track be x and y, where the paved side has length x. Use Greg's speed on pavement and gravel to find an equation for x and y, and find an expression in terms of x that we wish to maximize.
 (b) Determine how long the paved side should be if he wishes his track to enclose as large an area as possible.

Problem 5.15: Let $f(x) = \sqrt{ax^2 + bx}$. For how many real values of a is there at least one positive value of b for which the domain of f and the range f are the same? *(Source: AMC 12)*

Problem 5.11: Let $f(x) = x^2 + 6x + 5$.

 (a) Graph f.

 (b) Find the smallest possible value of $f(x)$ if x is a real number.

 (c) How is the value of x for which $f(x)$ is minimized related to the roots of $f(x)$?

Solution for Problem 5.11:

 (a) Since $f(x)$ is a quadratic, its graph is a parabola. We complete the square to give $f(x) = (x + 3)^2 - 4$, and we see that the graph of f is an upward-opening parabola with vertex $(-3, -4)$, as shown at right.

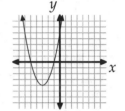

 (b) Our graph gives the answer right away. The smallest value of $f(x)$ corresponds to the y-coordinate of the lowest point on the graph of f. This lowest point is the vertex, $(-3, -4)$, so the smallest possible value of $f(x)$ is -4, which occurs when $x = -3$.

   Algebraically speaking, we can see that -4 is the smallest possible value of $f(x)$ after we have completed the square to find $f(x) = (x + 3)^2 - 4$. Since $(x + 3)^2 \geq 0$ for all real x, we have $(x + 3)^2 - 4 \geq -4$, which means $f(x) \geq -4$. We also have $f(-3) = -4$, so -4 is indeed the minimum of $f(x)$.

 (c) Factoring gives us $f(x) = (x + 1)(x + 5)$, so the roots of $f(x)$ are -1 and -5. The value of x that minimizes $f(x)$ is the average of these roots. Our graphical approach to finding the minimum explains why. The vertex of a parabola is on the axis of symmetry. The parabola's x-intercepts, whose x-coordinates are the roots of $f(x)$, are symmetric about the axis of symmetry. So, the x-coordinate of the vertex is the average of the roots of $f(x)$.

□

   By thinking about the graph of a parabola, we have now learned how to find the minimum or maximum of a quadratic expression:

> **Concept:** To find the minimum or maximum value of a quadratic, complete the square in the quadratic.

   We can use this tactic to generalize our solution to Problem 5.11.

Problem 5.12: Let $f(x) = ax^2 + bx + c$, where a, b, and c are real numbers and $a > 0$. In terms of a, b, and/or c, find the real value of x that makes $f(x)$ as small as possible. What happens if a is negative?

Solution for Problem 5.12:   As we often do with generalizations, we use the solution to a specific case as our guide. In Problem 5.11, we completed the square, so we do the same here. Rather than walk through all the steps again, we recall that we completed the square for $ax^2 + bx + c$ when proving the quadratic formula back on page 92, and we found

$$f(x) = a\left(x + \frac{b}{2a}\right)^2 + c - \frac{b^2}{4a}.$$

The graph of f is an upward-opening parabola with vertex $\left(-\frac{b}{2a}, c - \frac{b^2}{4a}\right)$. Therefore, its lowest point has y-coordinate $c - \frac{b^2}{4a}$, which occurs when $x = -\frac{b}{2a}$, which means that the minimum value of $f(x)$ is $f\left(-\frac{b}{2a}\right) = c - \frac{b^2}{4a}$.

If a is negative, the parabola opens downwards, so $\left(-\frac{b}{2a}, c - \frac{b^2}{4a}\right)$ is the highest point on the parabola. Therefore, $f\left(-\frac{b}{2a}\right) = c - \frac{b^2}{4a}$ is the *maximum* value of $f(x)$. $\square$

Important: Let $f(x) = ax^2 + bx + c$, where $a, b,$ and c are constants.

- If $a > 0$, then the *minimum* value of $f(x)$ is $c - \frac{b^2}{4a}$, which occurs when $x = -\frac{b}{2a}$.

- If $a < 0$, then the *maximum* value of $f(x)$ is $c - \frac{b^2}{4a}$, which occurs when $x = -\frac{b}{2a}$.

You don't have to memorize these formulas. As we have seen, we simply complete the square in a quadratic to find its minimum or maximum.

Notice that this result also verifies our observation in Problem 5.11 that the value of x that minimizes $f(x) = ax^2 + bx + c$ is the average of the roots of $f(x)$. Vieta's Formulas tell us that the sum of the roots of $f(x)$ is $-\frac{b}{a}$, so the average of the roots is $-\frac{b}{2a}$. Furthermore, this relationship holds even when the roots are nonreal.

Problem 5.13: Let k be a real number. Find the minimum possible distance between $k(1 + 3i)$ and $2i$ in the complex plane.

Solution for Problem 5.13: As discussed in Section 3.2, the distance between $2i$ and the point $k(1 + 3i)$ in the complex plane is

$$|k(1 + 3i) - 2i| = |k + (3k - 2)i| = \sqrt{k^2 + (3k - 2)^2}.$$

To minimize $\sqrt{k^2 + (3k - 2)^2}$, we must minimize $k^2 + (3k - 2)^2$. This is just a quadratic, which we know how to minimize. Expanding the square and combining like terms gives

$$k^2 + (3k - 2)^2 = k^2 + 9k^2 - 12k + 4 = 10k^2 - 12k + 4.$$

We could complete the square, but instead we use the result of Problem 5.12 to note that this quadratic is minimized when $k = -(-12)/(2 \cdot 10) = 3/5$. Therefore, the minimum value of the quadratic is $10\left(\frac{3}{5}\right)^2 - 12\left(\frac{3}{5}\right) + 4 = \frac{2}{5}$, which means the minimum distance from $k(1 + 3i)$ to $2i$ is $\sqrt{2/5} = \sqrt{10}/5$. $\square$

Problem 5.14: Greg can bicycle at 20 miles per hour on pavement and 10 miles per hour on gravel. Greg is building a rectangular track with one side paved and three sides gravel. Greg would like to build the track so that it takes him 3 hours to ride all the way around the track. How long should the paved side be if he wishes his track to enclose as large an area as possible? (You can assume the track itself has negligible width.)

Solution for Problem 5.14: We start as we usually do with word problems—we turn the words into math. Let the dimensions of the rectangular track be x and y miles, where one of the x-mile sides is paved. Therefore, we wish to maximize xy. Of course, we need more information about x and y to maximize xy.

We turn to the only other information we have, the fact that it should take Greg 3 hours to ride all the way around the track. Because Greg rides 20 miles per hour on paved road, he can complete the paved side in $\frac{x}{20}$ hours. Similarly, he can cover the other three sides in $\frac{x+2y}{10}$ hours. So, we must have $\frac{x}{20} + \frac{x+2y}{10} = 3$. Multiplying both sides by 20 to get rid of fractions, then rearranging, gives $3x + 4y = 60$.

We now wish to maximize xy given that $3x + 4y = 60$. Specifically, we want to find the value of x that gives us this maximum, so we solve $3x + 4y = 60$ for y in terms of x, which gives $y = 15 - \frac{3}{4}x$. Substituting this into our area expression gives

$$xy = x\left(15 - \frac{3}{4}x\right) = -\frac{3}{4}x^2 + 15x.$$

We complete the square to find

$$-\frac{3}{4}x^2 + 15x = -\frac{3}{4}\left(x^2 - 20x\right) = -\frac{3}{4}\left(x^2 - 20x + 100\right) + \frac{3}{4}(100) = -\frac{3}{4}(x - 10)^2 + 75.$$

The expression $-\frac{3}{4}(x - 10)^2 + 75$ is maximized when $x = 10$. (Alternatively, we could have simply noted that the maximum of $-\frac{3}{4}x^2 + 15x$ occurs when $x = -\frac{b}{2a} = -15/(2(-3/4)) = 10$.)

Therefore, Greg should make the paved side 10 miles long. $\square$

Problem 5.15: Let $f(x) = \sqrt{ax^2 + bx}$. For how many real values of a is there at least one positive value of b for which the domain of f and the range f are the same? *(Source: AMC 12)*

Solution for Problem 5.15: First, we take a little time to understand exactly what the question is asking.

Concept: If a problem is difficult to understand, take extra time reading and re-reading the problem until you are sure you understand what the problem is. You don't want to spend a lot of time working on the wrong problem!

Our problem requires us to count the values of a for which we can find a b that makes the domain and range of f the same. We start our search for such a and b by finding the domain and range of f in terms of a and b.

There are no negative numbers in the range of f because $\sqrt{ax^2 + bx}$ cannot be negative. For the domain of f, we must have $ax^2 + bx \geq 0$ in order for f to be defined. We solved some quadratic inequalities in Section 4.5 by factoring and considering the signs of the factors. Trying the same here, we get $x(ax + b) \geq 0$. Solutions to this inequality are in the domain of f. In the problem, we are told that b must be positive. However, we don't know if a is positive, negative, or 0. So, we have 3 cases to consider:

- *Case 1: $a > 0$.* We know that there cannot be negative numbers in the range of f, so we check if there are negative numbers in the domain of f when $a > 0$. If a and b are positive, then the solution to the inequality $x(ax + b) \geq 0$ is $x \in (-\infty, -\frac{b}{a}] \cup [0, +\infty)$. If you don't see this right away, note that if $a > 0$, then the graph of $g(x) = x(ax + b)$ is an upward-opening parabola. Since $g(x)$ has roots $x = 0$ and $x = -\frac{b}{a}$, the graph of g is below the y-axis for $0 < x < -\frac{b}{a}$, and above or on the y-axis otherwise. Therefore, there are negative numbers in the domain of f.

  Since there are no negative numbers in the range of f, but there are negative numbers in the domain if $a > 0$, there are no positive values of a that satisfy the problem.

- *Case 2: $a = 0$.* In this case, our inequality becomes $bx \geq 0$, which is true for all $x \geq 0$. Therefore, if $a = 0$, the domain of f is all nonnegative numbers. Because $f(x) = \sqrt{bx}$ when $a = 0$, the range of f is also all nonnegative numbers when $b > 0$. We see this by noting that for any nonnegative y, setting $x = y^2/b$ gives $\sqrt{bx} = y$, so y is in the range of f. Therefore, $a = 0$ satisfies the problem.

- *Case 3: $a < 0$.* When $a < 0$ and $b > 0$, then the solution to the inequality $x(ax + b) \geq 0$ is $x \in [0, -\frac{b}{a}]$. (Note that $a < 0$ and $b > 0$, so $-\frac{b}{a}$ is positive.) We can visualize the solution to this inequality with graphing. If $a < 0$, then the graph of $g(x) = x(ax + b)$ is a downward-opening parabola. Since $g(x)$ has roots $x = 0$ and $x = -\frac{b}{a}$, the graph of g is above or on the y-axis for $0 \leq x \leq -\frac{b}{a}$, and below the y-axis otherwise.

  The domain of f when $a < 0$ is therefore $[0, -\frac{b}{a}]$. Next, we focus on the range of f. We first note that the smallest possible value of $f(x)$ is 0, which we attain at both $x = 0$ and $x = -b/a$. But what is the largest possible value of f? To find the maximum value of f, we must first maximize $ax^2 + bx$. Now, we're in familiar territory. The maximum of this expression occurs when $x = -\frac{b}{2a}$. So, if $a < 0$, the maximum of $f(x)$ is

$$f\left(-\frac{b}{2a}\right) = \sqrt{a\left(-\frac{b}{2a}\right)^2 + b\left(-\frac{b}{2a}\right)} = \sqrt{-\frac{b^2}{4a}} = \frac{b}{2\sqrt{-a}}.$$

This must match the maximum value of our domain, $-\frac{b}{a}$, so we have $\frac{b}{2\sqrt{-a}} = -\frac{b}{a}$. Rearranging this equation gives $a = -2\sqrt{-a}$. Squaring both sides gives $a^2 = -4a$, from which we find $a = 0$ or $a = -4$. We have therefore found another value of a, namely $a = -4$, for which the domain and range of f are the same for some positive value of b.

Combining all our cases, we have 2 values of a for which there exists a positive value of b such that the domain and range of $f(x) = \sqrt{ax^2 + bx}$ are the same. □

Exercises

5.3.1 Find the largest possible value of $f(x) = -2x^2 + 6x + 9$.

5.3.2   A farmer is going to fence his field by first forming a rectangle with his fence, then cutting this rectangle into three congruent rectangles with two additional fences, where these additional fences must be parallel to each other. He has 80 yards of fence. What is the maximum area, in square yards, of field that he can enclose in this manner?

5.3.3   Suppose that p and q are constants such that the smallest possible value of $x^2 + px + q$ is 0. Express q in terms of p.

5.3.4   Suppose the temperature of each point (x, y) in the plane is given by the function $x^2 + y^2 - 4x + 6y$. Find the coldest point in the plane and determine its temperature.

5.3.5★   A table is to be constructed by gluing together 68 $1 \times 1 \times 1$ wooden cubes. All four legs and the rectangular top will be formed by the cubes. The four legs must be the same length and need to be one cube thick, and the top is one cube thick, as well. What is the maximum possible volume of the space between the table's top and the floor, excluding the space taken up by the table's legs? **Hints:** 3

5.4   Circles

The locus of points in a plane that are a fixed distance from a given point in that plane is called a **circle**. The fixed distance is the **radius** of the circle, and the given point is the **center** of the circle.

You're probably already familiar with representing circles in the Cartesian plane with equations. In this chapter, we review this material and apply it to a couple more challenging problems.

Problems

Problem 5.16: Show that the circle with center (h, k) and radius r, where $r > 0$, is the graph of the equation $(x - h)^2 + (y - k)^2 = r^2$.

Problem 5.17: Find the distance from $(5, 7)$ to the nearest point on the graph of $2x^2 + 4x + 2y^2 - 8y = 6$.

Problem 5.18: Let x and y be real numbers satisfying the equation $x^2 - 4x + y^2 + 3 = 0$. Find the maximum and minimum values of $x^2 + y^2$. *(Source: NYSML)*

Problem 5.19: A car travels due east at $\frac{2}{3}$ mile per minute on a long, straight road. At the same time, a circular storm, whose radius is 51 miles, moves southeast at $\frac{\sqrt{2}}{2}$ mile per minute. At time $t = 0$, the center of the storm is 110 miles due north of the car. At time $t = t_1$ minutes, the car enters the storm circle, and at time $t = t_2$ minutes, the car leaves the storm circle. Find $(t_1 + t_2)/2$. *(Source: AIME)* **Hints:** 346

Problem 5.16:   Show that the circle with center (h, k) and radius r, where $r > 0$, is the graph of the equation $(x - h)^2 + (y - k)^2 = r^2$.

Solution for Problem 5.16:   Suppose that the point (x, y) is on the circle with center (h, k) and radius r.

Then, the distance between (x, y) and (h, k) is r, so the distance formula gives us

$$\sqrt{(x - h)^2 + (y - k)^2} = r.$$

Conversely, any point (x, y) that satisfies this equation is r away from (h, k), which means that (x, y) is on the circle with center (h, k) and radius r. Because $(x - h)^2 + (y - k)^2$ is nonnegative, the equation above is equivalent to

$$(x - h)^2 + (y - k)^2 = r^2.$$

The graph of this equation is a circle with center (h, k) and radius r. $\square$

> **Important:** The standard form of an equation whose graph is a circle with center (h, k) and radius r is
> $$(x - h)^2 + (y - k)^2 = r^2.$$

Problem 5.17: Find the distance from $(5, 7)$ to the nearest point on the graph of $2x^2 + 4x + 2y^2 - 8y = 6$.

Solution for Problem 5.17: We begin by finding the graph of the given equation. We first divide by 2 to make the coefficients of x^2 and y^2 equal to 1. This gives $x^2 + 2x + y^2 - 4y = 3$. We complete the square in x and in y by adding both 1 and 4 to both sides. This gives us $x^2 + 2x + 1 + y^2 - 4y + 4 = 3 + 1 + 4$, so

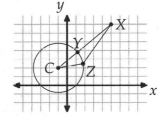

$$(x + 1)^2 + (y - 2)^2 = 8.$$

The graph of this equation is a circle with center $(-1, 2)$ and radius $\sqrt{8} = 2\sqrt{2}$. The circle is shown in the graph at right. The center of the circle is labeled C, and the point $(5, 7)$ is labeled X.

The nearest point on the circle to X is the point where $\overline{CX}$ intersects the circle. To see why, we let $\overline{CX}$ intersect the circle at Y, and let Z be some other point on the circle. The Triangle Inequality tells us that $CZ + XZ > CX$. We also have $CX = XY + CY$ and $CY = CZ$ (because both are radii of the circle), so $CZ + XZ > CX$ becomes $CY + XZ > XY + CY$. This means that $XZ > XY$, so Y is closer to X than any other point on the circle is.

Because $\overline{CY}$ is a radius of the circle, we have $CY = 2\sqrt{2}$. The distance formula gives us

$$CX = \sqrt{(5 + 1)^2 + (7 - 2)^2} = \sqrt{61},$$

so $XY = CX - CY = \sqrt{61} - 2\sqrt{2}$. $\square$

Problem 5.18: Let x and y be real numbers satisfying the equation $x^2 - 4x + y^2 + 3 = 0$. Find the maximum and minimum values of $x^2 + y^2$. *(Source: NYSML)*

Solution for Problem 5.18: Let $k = x^2 + y^2$. We now seek an expression in terms of k that we can maximize and minimize. We might at this point substitute k for $x^2 + y^2$ in the given equation, but before we do so, we recognize the form of the two equations $x^2 + y^2 = k$ and $x^2 - 4x + y^2 + 3 = 0$. The graphs of these two equations are circles; maybe we can solve this problem with geometry instead of algebra.

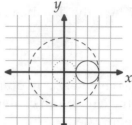

The graph of $x^2 + y^2 = k$ is a circle with center $(0, 0)$ and radius $\sqrt{k}$. Completing the square in x and rearranging the equation $x^2 - 4x + y^2 + 3 = 0$ gives $(x-2)^2 + y^2 = 1$. The graph of this equation is a circle with center $(2, 0)$ and radius 1. We'll call this circle C, and its graph is the solid circle shown at right. Every point (x, y) that satisfies $x^2 - 4x + y^2 + 3 = 0$ must be on this circle. If we also have $x^2 + y^2 = k$, then the point (x, y) must be on the circle with center $(0, 0)$ and radius $\sqrt{k}$.

So, to find the largest and smallest possible values of k, we must simply find the largest and smallest circles centered at the origin that intersect circle C. The largest is shown as a dashed circle, and has radius 3. The smallest is shown as a dotted circle, and has radius 1. Since the radius in each case equals $\sqrt{k}$, the largest possible value of k is 9 and the smallest possible value is 1. $\square$

> **Concept:** Be alert for equations you know how to graph easily. Sometimes geometry can be used to solve algebra problems.

Problem 5.19: A car travels due east at $\frac{2}{3}$ mile per minute on a long, straight road. At the same time, a circular storm, whose radius is 51 miles, moves southeast at $\frac{\sqrt{2}}{2}$ mile per minute. At time $t = 0$, the center of the storm is 110 miles due north of the car. At time $t = t_1$ minutes, the car enters the storm circle, and at time $t = t_2$ minutes, the car leaves the storm circle. Find $(t_1 + t_2)/2$. *(Source: AIME)*

Solution for Problem 5.19: We have a long, wordy problem, so we first try to turn all these words into equations and expressions we can manipulate. We set the problem up on the Cartesian plane, since this will give us a way to model the location of the car and the storm at any time t. We let the car start at the origin. Because it travels due east at $\frac{2}{3}$ mile per minute, its location is $(\frac{2}{3}t, 0)$ after t minutes.

The storm is a little harder to work with. At time $t = 0$, the center of the storm is 110 miles north of the car, at $(0, 110)$. Because the storm moves southeast at $\frac{1}{2}\sqrt{2}$ mile per minute, its center moves south $\frac{1}{2}$ mile and east $\frac{1}{2}$ mile each minute. Therefore, the center of the storm at time t is $(\frac{t}{2}, 110 - \frac{t}{2})$. The radius of the storm is always 51, so the equation whose graph represents the storm's boundary on the Cartesian plane is

$$\left(x - \frac{t}{2}\right)^2 + \left(y - \left(110 - \frac{t}{2}\right)\right)^2 = 51^2.$$

The values of t at which the car enters and leaves the storm correspond to the values of t at which the point representing the car is on the circle representing the storm. So, we seek the values of t such that $(\frac{2}{3}t, 0)$ satisfies the equation for our storm. Setting $x = \frac{2}{3}t$ and $y = 0$ in our storm equation gives us

$$\left(\frac{2t}{3} - \frac{t}{2}\right)^2 + \left(0 - \left(110 - \frac{t}{2}\right)\right)^2 = 51^2.$$

Simplifying the left side gives us

$$\left(\frac{2t}{3} - \frac{t}{2}\right)^2 + \left(-110 + \frac{t}{2}\right)^2 = \frac{t^2}{36} + \frac{t^2}{4} - 110t + 110^2 = \frac{5t^2}{18} - 110t + 110^2,$$

so our equation is $\frac{5t^2}{18} - 110t + 110^2 = 51^2$. Solving that doesn't look like much fun. Before we dive into the quadratic formula, we reread the problem and notice that all we want is the average of the solutions to this equation, not the solutions themselves.

> **Concept:** Keep your eye on the ball! Don't forget what you're looking for in a problem—you don't always have to solve for every variable in a problem to solve the problem itself.

The average of the solutions to this quadratic is half the sum of the quadratic's roots. Vieta's Formulas give us this sum quickly: $-(-110)/(5/18) = 396$. Half this sum is our desired average, 198. □

___

5.4.1 Find the center and the radius of the graph of $4x^2 + 4y^2 + 12x + 16y + 9 = 0$.

5.4.2 Find an equation whose graph is a circle that has a diameter with endpoints $(3, 2)$ and $(7, -4)$.

5.4.3 A **chord** of a circle is a line segment whose endpoints are on the circle. Find the length of the common chord of the two circles whose equations are $x^2 + y^2 = 4$ and $x^2 + y^2 - 6x + 2 = 0$. *(Source: CEMC)*

5.4.4 Square $ABCD$ has sides of length 4, and M is the midpoint of $\overline{CD}$. A circle with radius 2 and center M intersects a circle with radius 4 and center A at points P and D. What is the distance from P to $\overline{AD}$? *(Source: AMC)*

5.5 Ellipses

Suppose we have points F_1 and F_2 in a plane. For any positive constant k greater than the distance F_1F_2, the locus of all points P in the plane such that $PF_1 + PF_2$ equals k is called an **ellipse**, an example of which is shown at right. The points F_1 and F_2 are the **foci** of the ellipse. ("Foci" is the plural of **focus**.) The midpoint of the segment connecting the foci is the **center** of the ellipse.

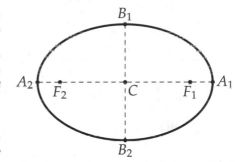

A line segment with its endpoints on the ellipse is called a **chord** of the ellipse. The chord that passes through both foci is the **major axis** of the ellipse, while the chord through the center that is perpendicular to the major axis is called the **minor axis**. So, in the given diagram, $\overline{A_1A_2}$ is the major axis and $\overline{B_1B_2}$ is the minor axis of the ellipse above. These axes meet at the center, point C.

> **Sidenote:** ♪ You can make your own ellipse with two pins, a string, and a pencil. Push your pins into a piece of paper on a hard surface. These are the foci of your ellipse. Make sure that your string is at least twice as long as the distance between the pins. Tie the ends of the string together so that you now have a loop. Place your loop around both pins. Finally, make your ellipse: hook your pencil inside the loop, pull the loop taut against the pins and pencil, then, keeping the string taut and pencil perpendicular to the paper throughout, trace the entire boundary you can reach with the pencil. Why does this process produce an ellipse?

Problems

Problem 5.20: Suppose the ellipse $\mathcal{E}$ has foci $F_1 = (c, 0)$ and $F_2 = (-c, 0)$ for some positive c. Moreover, suppose that for every point P on $\mathcal{E}$, we have $PF_1 + PF_2 = 2a$ for some nonzero constant a such that $a > c$.

(a) Why must we have $a > c$?

(b) In terms of a and/or c, find the coordinates of the points on the x-axis that are on $\mathcal{E}$, and prove that $PF_1 + PF_2$ equals the length of the major axis of $\mathcal{E}$ for any point P on $\mathcal{E}$.

(c) Suppose $\mathcal{E}$ passes through the point $(0, b)$, where $b > 0$. Find b in terms of a and c.

(d) Suppose $P = (x, y)$ is on $\mathcal{E}$. Use the equation $PF_1 + PF_2 = 2a$ to show that $\mathcal{E}$ is the graph of the equation

$$\frac{x^2}{a^2} + \frac{y^2}{b^2} = 1.$$

(This part requires a considerable amount of algebraic manipulation to get rid of square root symbols; don't give up too quickly!)

Problem 5.21: Consider the graph of the equation $\dfrac{x^2}{a^2} + \dfrac{y^2}{b^2} = 1$.

(a) What happens to the graph if $a = b$?

(b) In which of the three graphs below is a/b largest? The smallest?

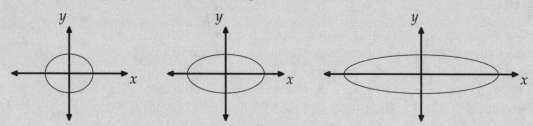

(c) What happens if $a < b$?

Problem 5.22: Suppose that ellipse $\mathcal{E}$ has (h, k) as its center, $(h + c, k)$ as a focus, and a major axis of length $2a$. Show that $\mathcal{E}$ is the graph of the equation

$$\frac{(x - h)^2}{a^2} + \frac{(y - k)^2}{b^2} = 1,$$

where $b^2 = a^2 - c^2$.

Problem 5.23: Let C be the center of an ellipse, A_1 and A_2 be the endpoints of the ellipse's major axis, B_1 and B_2 be the endpoints of its minor axis, and F_1 and F_2 be its foci.

(a) Why must $B_1F_1 = B_1F_2$? Why must $B_1F_1 + B_1F_2 = A_1A_2$?

(b) Show that B_1F_1 equals half the length of the major axis of the ellipse.

(c) Find an equation that relates the distance between the foci to the lengths of the axes of the ellipse.

Problem 5.24: Graph each of the ellipses described below, and find an equation that produces the given graph. Then, find the center, foci, and the lengths of both axes of each ellipse.

   (a) The center is $(4, -2)$, the major axis is horizontal with length 6, and $(4, -3)$ is one endpoint of the minor axis.

   (b) The foci are $(-3, -1)$ and $(-3, 3)$, and the ellipse passes through $(-5, 1)$.

Problem 5.25: Find the center, foci, and the lengths of both axes of the graph of each of the equations below, and then graph each equation.

   (a) $\dfrac{(x-1)^2}{16} + \dfrac{(y+7)^2}{25} = 1$
                (b) $x^2 + 4y^2 + 6x + 16y + 24 = 0$

Problem 5.26: What is the graph of each of the following equations:

   (a) $\dfrac{x^2}{4} + \dfrac{(y-2)^2}{9} = 0$
                (b) $9x^2 + 4y^2 - 16y + 52 = 0$

Problem 5.27: The graphics program we used to write this book doesn't have a command to graph an ellipse. It does, however, have a command to graph a circle given its center and radius. It also allows the user to scale a graph horizontally or vertically. (When we "scale a graph horizontally," we multiply the x-coordinate of every point on the graph by some fixed constant.) In this problem, we figure out how to use this program to produce the ellipse at right, and explain why this process does indeed produce the correct ellipse.

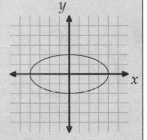

   (a) Draw a circle on the Cartesian plane that you can scale horizontally by a factor of 2 to produce the shown ellipse. What is the equation whose graph is your circle?

   (b) Suppose that (x, y) is on the circle you produced in part (a). What point must be on the graph produced by scaling your part (a) graph horizontally by a factor of 2?

   (c) Find the equation whose graph results from scaling the circle in part (a) horizontally by a factor of 2.

   (d) Does this process of scaling a circle horizontally produce an ellipse?

   (e) Take a guess at a formula for the area of an ellipse with axes of lengths $2a$ and $2b$.

Problem 5.20: Suppose the ellipse $\mathcal{E}$ has foci $F_1 = (c, 0)$ and $F_2 = (-c, 0)$ for some positive c. Moreover, suppose that for every point P on $\mathcal{E}$, we have $PF_1 + PF_2 = 2a$ for some nonzero constant a such that $a > c$. Find the equation whose graph is this ellipse.

Solution for Problem 5.20: We start by wondering why we must have $a > c$. The Triangle Inequality tells us that $PF_1 + PF_2 \geq F_1F_2$, where equality holds if and only if P is on $\overline{F_1F_2}$. Since $PF_1 + PF_2 = 2a$ and $F_1F_2 = 2c$, we must have $a \geq c$, with equality if and only if P is on $\overline{F_1F_2}$. Therefore, if $a < c$, then there are no points P that satisfy $PF_1 + PF_2 = 2a$; and if $a = c$, then only points on $\overline{F_1F_2}$ satisfy $PF_1 + PF_2 = 2a$ (and our locus of points such that $PF_1 + PF_2 = 2a$ is a line segment). So, we must have $a > c$ to produce an ellipse as our locus.

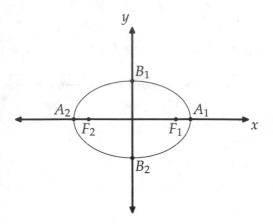

To find the equation whose graph is this ellipse, we first find a few points on the ellipse. We aim first at the easy points: the points where the ellipse intersects the axes. Let $A_1 = (x_1, 0)$ be a point where $\mathcal{E}$ intersects the x-axis. Because A_1 is on the ellipse, we have $A_1F_1 + A_1F_2 = 2a$. We know that A_1 is not on $\overline{F_1F_2}$, since the sum of the distances from any point on $\overline{F_1F_2}$ to the endpoints of this segment is $2c$, not $2a$. So, suppose $x_1 > c$, so A_1 is to the right of F_1. Then, we have $A_1F_1 = x_1 - c$ and $A_1F_2 = x_1 + c$, so $A_1F_1 + A_1F_2 = 2x_1$. Since this must equal $2a$, we have $x_1 = a$, which means $A_1 = (a, 0)$. Similarly, the other point where $\mathcal{E}$ intersects the x-axis is $A_2 = (-a, 0)$. Therefore, the length of the major axis is $2a$, and this length equals the constant sum $PF_1 + PF_2$ for points P on the ellipse.

> **Important:** If P is on an ellipse with foci F_1 and F_2, then $PF_1 + PF_2$ equals the length of the major axis of the ellipse.

Next, suppose that $B_1 = (0, b)$, with $b > 0$, is a point where $\mathcal{E}$ intersects the y-axis. Then, since $F_1 = (c, 0)$ and $F_2 = (-c, 0)$, we have

$$B_1F_1 + B_1F_2 = \sqrt{b^2 + c^2} + \sqrt{b^2 + c^2} = 2\sqrt{b^2 + c^2}.$$

Because B_1 is on the ellipse, this sum must equal $2a$, which gives us $\sqrt{b^2 + c^2} = a$. Squaring both sides of this and rearranging the result gives us $b^2 = a^2 - c^2$. Similarly, the point $B_2 = (0, -b)$, where $b^2 = a^2 - c^2$ and $b > 0$, is on $\mathcal{E}$. So, the minor axis has length $2b$.

We now have the endpoints of both axes. Let's try to find an equation whose graph is the whole ellipse. Let $P = (x, y)$ be a point on the ellipse. From $PF_1 + PF_2 = 2a$, we have

$$\sqrt{(x - c)^2 + y^2} + \sqrt{(x + c)^2 + y^2} = 2a.$$

Unfortunately, we have no choice but to square this equation to get rid of the radicals. Instead of squaring right away, we first move one of the radicals to the right to get

$$\sqrt{(x - c)^2 + y^2} = 2a - \sqrt{(x + c)^2 + y^2}.$$

This allows us to square the equation without having to multiply the two radical expressions. Squaring this equation gives $(x - c)^2 + y^2 = 4a^2 - 4a\sqrt{(x + c)^2 + y^2} + (x + c)^2 + y^2$. The y^2 terms cancel, and we can isolate the remaining square root to find

$$4a\sqrt{(x + c)^2 + y^2} = 4a^2 + (x + c)^2 - (x - c)^2 = 4a^2 + 4xc.$$

Dividing by 4 gives us $a\sqrt{(x+c)^2+y^2} = a^2 + xc$. Squaring again gives $a^2(x+c)^2 + a^2y^2 = a^4 + 2a^2xc + x^2c^2$. Expanding the square on the left gives

$$a^2x^2 + 2a^2xc + a^2c^2 + a^2y^2 = a^4 + 2a^2xc + x^2c^2.$$

The $2a^2xc$ terms cancel. We then move the x^2 and y^2 terms to the left and the other terms to the right to give

$$(a^2 - c^2)x^2 + a^2y^2 = a^4 - a^2c^2.$$

Aha! We recognize $a^2 - c^2 = b^2$ on the left. We find b^2 again on the right when we factor a^2 out: $a^4 - a^2c^2 = a^2(a^2 - c^2) = a^2b^2$. Now, our equation is $b^2x^2 + a^2y^2 = a^2b^2$. Dividing by a^2b^2 gives us the surprisingly simple

$$\frac{x^2}{a^2} + \frac{y^2}{b^2} = 1.$$

☐

We can reverse the steps above to show that:

> **Important:** The graph of the equation $\frac{x^2}{a^2} + \frac{y^2}{b^2} = 1$, where $a > b > 0$, is an ellipse with major axis of length $2a$, minor axis of length $2b$, and foci $(c, 0)$ and $(-c, 0)$, where $c = \sqrt{a^2 - b^2}$.

The endpoints of the horizontal axis of the graph of $\frac{x^2}{a^2} + \frac{y^2}{b^2} = 1$ are $(a, 0)$ and $(-a, 0)$. To get a sense for why, we note that in the equation $\frac{x^2}{a^2} + \frac{y^2}{b^2} = 1$, both $\frac{x^2}{a^2}$ and $\frac{y^2}{b^2}$ are squares, so they are both nonnegative. Since their sum is 1 and both are nonnegative, each must be no greater than 1. We can only have $\frac{x^2}{a^2} = 1$ if $\frac{y^2}{b^2} = 0$, so we only have $x = a$ or $x = -a$ if $y = 0$. Moreover, we cannot have $x^2 > a^2$, so x can be no farther from 0 than a is. Therefore, $(a, 0)$ and $(-a, 0)$ are the most extreme points horizontally from the center. Similarly, $(0, b)$ and $(0, -b)$ are the most extreme points vertically from the center.

With this in mind, we can now get a feel for this equation by thinking about what happens as we vary a and b.

Problem 5.21: Consider the graph of the equation $\dfrac{x^2}{a^2} + \dfrac{y^2}{b^2} = 1$.

(a) What happens to the graph if $a = b$?

(b) In which of the three graphs below is a/b largest? The smallest?

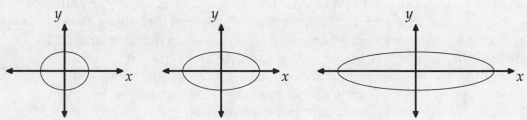

(c) What happens if $a < b$?

Solution for Problem 5.21:

(a) If $a = b$, then we can rewrite the equation as $x^2 + y^2 = a^2$. The graph of this equation is a circle with center $(0,0)$ and radius a. But how can this be? We're studying ellipses, not circles.

> **Important:**   Circles are ellipses.
>
> ⚠

But if circles are ellipses, what are the foci of a circle? We know that the foci of an ellipse that is the graph of the equation $\frac{x^2}{a^2} + \frac{y^2}{b^2} = 1$ are the points $(c,0)$ and $(-c,0)$, where $c = \sqrt{a^2 - b^2}$. But when $a = b$, we have $c = 0$, which puts both of our foci at the center of the circle, $(0,0)$.

Now, we see why a circle can be considered an ellipse. If we let the foci F_1 and F_2 of an ellipse be the same point, which we'll call O, then our condition that "$PF_1 + PF_2$ is constant" becomes "$2PO$ is constant." In other words, every point on the ellipse is the same distance from O, which means the ellipse is a circle.

(b) We know we have a circle when $a = b$. To analyze the shape of the ellipses that result as we choose values of a that are larger than b, we consider the endpoints of the axes of the ellipses.

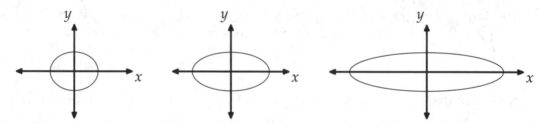

The endpoints of the major axis of the ellipse are $(a,0)$ and $(-a,0)$, while the endpoints of the minor axis of the ellipse are $(0,b)$ and $(0,-b)$. As we saw in part (a), when $a = b$, we have a circle. If a/b is slightly greater than 1, then the ratio of the length of the major axis to the length of the minor axis is slightly greater than 1. This means the ellipse is almost a circle, as in the ellipse at left above. Increasing the ratio a/b increases the ratio of the major axis to the minor axis. The larger this ratio is, the more "stretched" the resulting ellipse is. So, the ellipse at right above has the largest ratio a/b and the one at left above has the smallest ratio a/b.

> **Sidenote:**   **Eccentricity** is one measure of how "stretched" an ellipse is. The eccentricity of an ellipse is defined as the ratio of the distance between the ellipse's foci to the length of its major axis. So, the eccentricity of a circle is 0, and the greater the eccentricity of an ellipse, the more "stretched" the ellipse is.

(c) Now that we've tackled what happens if $a = b$ and as a gets greater and greater than b, we wonder what happens if $a < b$. We start by finding a few points on the resulting graph when $a < b$. The easiest, as usual, are those on the coordinate axes. As before, $(a,0)$ and $(-a,0)$ are on the graph, as are $(0,b)$ and $(0,-b)$. However, this time, we have $a < b$. In other words, the major axis is vertical, not horizontal! So, the graph is still an ellipse, but the foci and major axis are on the y-axis, not the x-axis. At right, we have the graph of an ellipse in which the major axis is vertical.

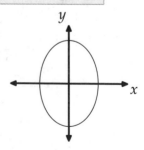

□

We have just discovered that the major axis of the graph of an equation of the form $\frac{x^2}{a^2} + \frac{y^2}{b^2} = 1$ is horizontal when $a > b$ and vertical when $b > a$. We don't need to memorize this. As described in Problem 5.21, we can simply think about where the endpoints of the axes are to see when the major axis is vertical and when it is horizontal.

We now have a pretty good handle on ellipses centered at the origin. Of course, not all ellipses are centered at the origin.

Problem 5.22: Suppose that ellipse $\mathcal{E}$ has (h, k) as its center, $(h + c, k)$ as a focus, and a major axis of length $2a$. Show that $\mathcal{E}$ is the graph of the equation

$$\frac{(x - h)^2}{a^2} + \frac{(y - k)^2}{b^2} = 1,$$

where $b^2 = a^2 - c^2$.

Solution for Problem 5.22: Back in Section 5.1, we started with a parabola with its vertex at the origin, and then translated the parabola so that its vertex was at (h, k). We use a similar strategy here.

In Problem 5.20, we found that an ellipse with foci $(c, 0)$ and $(-c, 0)$ and major axis of length $2a$ is the graph of the equation

$$\frac{x^2}{a^2} + \frac{y^2}{b^2} = 1,$$

where $a > b > 0$ and $b^2 = a^2 - c^2$. Let this ellipse be $\mathcal{D}$. The origin is the center of this ellipse, since it is the midpoint of the major and minor axes.

Because (h, k) is the center of $\mathcal{E}$ and $(h + c, k)$ is one focus of $\mathcal{E}$, the other focus is c to the left of the center, at $(h - c, k)$. So, the center and the foci of $\mathcal{E}$ result from translating the center and foci of $\mathcal{D}$ right h units and up k units. Furthermore, the major axes of the two ellipses are equal in length, so we suspect that $\mathcal{E}$ is the result of translating $\mathcal{D}$ right h units and up k units.

To show this is the case, we let F_1 and F_2 be the foci of $\mathcal{D}$ and F_1' and F_2' be the foci of $\mathcal{E}$. We also let $P = (x, y)$ be on $\mathcal{D}$ and let $P' = (x + h, y + k)$, so P' is the result of translating P right h units and up k units. We wish to show that P' is on $\mathcal{E}$.

Since P is on $\mathcal{D}$, we have $PF_1 + PF_2 = 2a$. Because $\overline{P'F_1'}$ is a translation of $\overline{PF_1}$, we have $P'F_1' = PF_1$. Similarly, we have $P'F_2' = PF_2$, so $P'F_1' + P'F_2' = PF_1 + PF_2 = 2a$, which means that P' is on ellipse $\mathcal{E}$. Conversely, every point on $\mathcal{E}$ is the result of translating some point on $\mathcal{D}$ right h units and up k units.

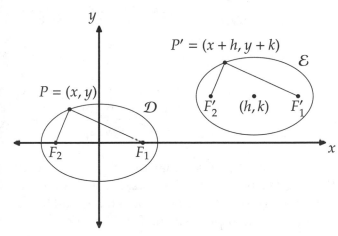

So, for every point (x, y) on the graph of $\frac{x^2}{a^2} + \frac{y^2}{b^2} = 1$, there is a point (x', y') on the graph of $\mathcal{E}$, where $(x', y') = (x + h, y + k)$. We therefore have $x = x' - h$ and $y = y' - k$, where (x, y) is on $\mathcal{D}$ and (x', y') is on

$\mathcal{E}$. Substituting these into our equation for $\mathcal{D}$ tells us that $\mathcal{E}$ consists of all points (x', y') such that

$$\frac{(x' - h)^2}{a^2} + \frac{(y' - k)^2}{b^2} = 1.$$

Therefore, we have:

> **Important:** The **standard form** of an equation whose graph is an ellipse with a horizontal major axis is
>
> $$\frac{(x - h)^2}{a^2} + \frac{(y - k)^2}{b^2} = 1,$$
>
> where $a > b > 0$. The center of the ellipse is (h, k), its major axis has length $2a$, its minor axis has length $2b$, and its foci are $(h \pm c, k)$, where $c = \sqrt{a^2 - b^2}$.

Note that we don't have to memorize the expressions for the lengths of the axes of the graph of

$$\frac{(x - h)^2}{a^2} + \frac{(y - k)^2}{b^2} = 1.$$

Since the major axis is horizontal, its endpoints have $y = k$, just like the center. This leaves us $\frac{(x-h)^2}{a^2} = 1$, so $(x - h)^2 = a^2$, which means $x - h = \pm a$. This tells us that the endpoints of the major axis are a to the left and right of the center, so the major axis has length $2a$. In other words, thinking about the values of x and y that make each fraction equal 1 tells us the lengths of the axes. $\square$

In much the same way, we can show that for an ellipse with a vertical major axis, we have:

> **Important:** The **standard form** of an equation whose graph is an ellipse with a vertical major axis is
>
> $$\frac{(x - h)^2}{a^2} + \frac{(y - k)^2}{b^2} = 1,$$
>
> where $b > a > 0$. The center of the ellipse is (h, k), its major axis has length $2b$, its minor axis has length $2a$, and its foci are $(h, k \pm c)$, where $c = \sqrt{b^2 - a^2}$.

Make sure you see that there's nothing really new here. This is basically the same information that we learned in Problem 5.22. You shouldn't memorize all this information. Even the relationship between c (the distance from each focus and the center) and the lengths of the axes has a nice a geometric relationship that prevents us from having to memorize a formula for a, b, and c.

Problem 5.23: Show that in an ellipse, the distance between an endpoint of the minor axis to either focus equals half the length of the major axis. Use this to find an equation that relates the distance between the foci to the lengths of the axes.

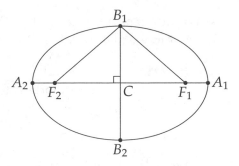

Solution for Problem 5.23: We start with a diagram at right. Let C be the center of the ellipse, A_1 and A_2 be the endpoints of the major axis, B_1 and B_2 be the endpoints of the minor axis, and F_1 and F_2 be the foci. We wish to relate B_1F_1 to A_1A_2, the length of the major axis, so we note that $B_1F_1 + B_1F_2 = A_1A_2$ because B_1 is on the ellipse. All we have to do now is take care of B_1F_2. Our diagram suggests that $B_1F_1 = B_1F_2$. To see that this is the case, we note that $CF_1 = CF_2$ because the center of the ellipse is the midpoint of the segment connecting the foci. Since we also have $\overline{B_1C} \perp \overline{F_1F_2}$, we can apply the Pythagorean Theorem to triangles $\triangle B_1CF_1$ and $\triangle B_1CF_2$ to see that

$$B_1F_1 = \sqrt{(CF_1)^2 + (CB_1)^2} = \sqrt{(CF_2)^2 + (CB_1)^2} = B_1F_2.$$

We also could have determined this by noting that the two right triangles are congruent.

Combining $B_1F_1 = B_1F_2$ and $B_1F_1 + B_1F_2 = A_1A_2$ gives us $2B_1F_1 = A_1A_2$, so $B_1F_1 = A_1A_2/2$. The Pythagorean Theorem applied to $\triangle B_1CF_1$ gives us $(B_1F_1)^2 = (CB_1)^2 + (CF_1)^2$. Substituting $B_1F_1 = A_1A_2/2$ into this gives $(A_1A_2/2)^2 = (CB_1)^2 + (CF_1)^2$. Multiplying by 4 gives us

$$(A_1A_2)^2 = 4(CB_1)^2 + 4(CF_1)^2 = (2CB_1)^2 + (2CF_1)^2.$$

Since C is the midpoint of $\overline{B_1B_2}$, we have $2CB_1 = B_1B_2$. Similarly, we have $2CF_1 = F_1F_2$, so the equation $(A_1A_2)^2 = (2CB_1)^2 + (2CF_1)^2$ gives us $(A_1A_2)^2 = (B_1B_2)^2 + (F_1F_2)^2$. This means we have

$$(\text{major axis length})^2 = (\text{minor axis length})^2 + (\text{distance between foci})^2.$$

$\square$

This problem revealed a couple useful facts about ellipses:

> **Important:** Let B be an endpoint of the minor axis of an ellipse, F be a focus of the ellipse, and C be the center of an ellipse. Then, $\triangle BFC$ is a right triangle, BC equals half the length of the minor axis, and BF equals half the length of the major axis. Therefore, the length of the major axis, the length of the minor axis, and the distance between the foci can be related by the Pythagorean Theorem:
>
> $(\text{major axis length})^2 = (\text{minor axis length})^2 + (\text{distance between foci})^2.$

As with our other formulas, this one doesn't need to be memorized. Once you understand it, you can produce it whenever you need it.

To recap, we're now able to relate many notable parts of the graph of an ellipse with horizontal and vertical axes to the equation that produced the graph. We first put the equation in the form

$$\frac{(x-h)^2}{a^2} + \frac{(y-k)^2}{b^2} = 1.$$

The center is (h, k), since this point makes both fractions 0. The endpoints of the horizontal axis are those points that make the first fraction 1 and the second 0, and the endpoints of the vertical axis make the first fraction 0 and the second 1. Finally, we can find the distance from the foci to the center with the relationship we just discovered with the Pythagorean Theorem.

If the axes of an ellipse are not horizontal and vertical, then the equation that produces the ellipse cannot be written in the standard form $\frac{(x-h)^2}{a^2} + \frac{(y-k)^2}{b^2} = 1$. We will learn how to handle such ellipses in a later text in the *Art of Problem Solving* series.

We'll now get a little practice working with specific ellipses.

Problem 5.24: Graph each of the ellipses described below, and find an equation that produces the given graph. Then, find the center, foci, and the lengths of both axes of each ellipse.

(a) The center is $(4, -2)$, the major axis is horizontal with length 6, and $(4, -3)$ is one endpoint of the minor axis.

(b) The foci are $(-3, -1)$ and $(-3, 3)$, and the ellipse passes through $(-5, 1)$.

Solution for Problem 5.24:

(a) We start with a sketch of the graph. Because the major axis is horizontal with length 6, its endpoints are 3 to the left and to the right of the center, at $(1, -2)$ and $(7, -2)$, as shown. Since the given endpoint of the minor axis is 1 unit below the center, the other endpoint is 1 unit above the center, at $(4, -1)$ as shown.

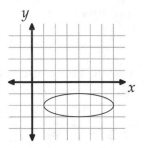

To find the equation whose graph is this ellipse, we start with the standard form

$$\frac{(x - h)^2}{a^2} + \frac{(y - k)^2}{b^2} = 1.$$

From the center $(4, -2)$, we know that our numerators on the left are $(x - 4)^2$ and $(y + 2)^2$. From the lengths of the axes, the denominator of the x term is $(6/2)^2 = 9$ and the denominator of the y term is $(2/2)^2 = 1$. (The x term has the larger denominator because the major axis is horizontal.)

Finally, we can find the distance between the foci with the relationship

$$(\text{major axis length})^2 = (\text{minor axis length})^2 + (\text{distance between foci})^2,$$

from which we find that the foci are $2\sqrt{2}$ apart. The major axis is horizontal, so the foci are $\sqrt{2}$ to the left and to the right of the center, at $(4 + \sqrt{2}, -2)$ and $(4 - \sqrt{2}, -2)$.

(b) We start with a sketch of the ellipse. At first, we only know one point on the ellipse, so we'll have to see what information about the ellipse we can get from the foci. The foci of this ellipse are on the same vertical line, so the major axis is vertical. The center is the midpoint of the segment connecting the foci, which is $(-3, 1)$. Aha! This point is 2 units to the right of the given point on the ellipse $(-5, 1)$. From this, we know that the minor axis has length $2 \cdot 2 = 4$. The minor axis is horizontal and the center is $(-3, 1)$, so the equation whose graph is the ellipse is

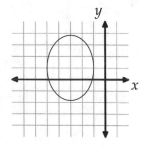

$$\frac{(x + 3)^2}{2^2} + \frac{(y - 1)^2}{b^2} = 1$$

for some value of b. (We know the denominator of the first fraction is $(4/2)^2$ because the minor axis is horizontal with length 4. We can even check our work at this point by noting that $(-5, 1)$ satisfies this equation.) To find the denominator of the second fraction, we have to find the length of the major axis. Since

$$(\text{major axis length})^2 = (\text{minor axis length})^2 + (\text{distance between foci})^2,$$

the length of the major axis is $4\sqrt{2}$. Therefore, the denominator of the $(y-1)^2$ term in the equation is $(4\sqrt{2}/2)^2 = 8$, and our equation is

$$\frac{(x+3)^2}{4} + \frac{(y-1)^2}{8} = 1.$$

Now that we have the length of the major axis, we see that the endpoints of the major axis are $(-3, 1 + 2\sqrt{2})$ and $(-3, 1 - 2\sqrt{2})$. Using $2\sqrt{2} \approx 2.8$, we can locate these endpoints on the Cartesian plane and produce the graph shown.

□

Problem 5.25: Graph each of the equations below, then find the center, foci, and the lengths of both axes of the graph.

(a) $\dfrac{(x-1)^2}{16} + \dfrac{(y+7)^2}{25} = 1$ (b) $x^2 + 4y^2 + 6x + 16y + 24 = 0$

Solution for Problem 5.25:

(a) First, we note that the denominator of the y term is larger than that of the x term, so the major axis is vertical. The numerators are both zero for $(x, y) = (1, -7)$, so this point is the center of the ellipse. The endpoints of the major axis are directly above and below the center, so they occur when $x = 1$, which leaves us $\frac{(y+7)^2}{25} = 1$, so $y + 7 = \pm\sqrt{25}$. This tells us that the endpoints of the major axis are $\sqrt{25}$ above and below the center, so the major axis has length $2\sqrt{25} = 10$. Similarly, the minor axis has length $2\sqrt{16} = 8$.

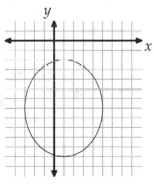

Using the relationship among the distance between the foci and the lengths of the axes, we find that the foci are $\sqrt{100 - 64} = 6$ apart, so they are 3 from the center. Because the major axis is vertical, the foci are 3 above and 3 below the center, at $(1, -10)$ and $(1, -4)$. Using the center and the lengths of the axes, we can produce the graph at right.

(b) The equation has both an x^2 and a y^2 term, but the coefficients of these two terms are different. So, we know the graph of the equation isn't a circle, since the coefficients of x^2 and y^2 are the same in the general form of an equation whose graph is a circle. Therefore, we suspect the graph is an ellipse, so we try to write the equation in the standard form of an equation whose graph is an ellipse.

We start by completing the square. We first group the x terms and the y terms, and move the constant to the other side. This gives us

$$x^2 + 6x + 4y^2 + 16y = -24.$$

Adding 9 to both sides completes the square in x. Factoring 4 out of each of the y terms gives $4(y^2 + 4y)$, so we must add $4 \cdot 4$ to both sides to complete the square in y:

$$x^2 + 6x + 9 + 4(y^2 + 4y + 4) = -24 + 9 + 4(4).$$

Therefore, our equation is $(x + 3)^2 + 4(y + 2)^2 = 1$. Uh-oh. Where are the denominators on the left side? We can view the denominator of the $(x + 3)^2$ term as 1, but what about the $4(y + 2)^2$ term; do we let that denominator be 1 also? But then what do we do with the 4?

We move the 4 to the denominator of the $4(y + 2)^2$ term by writing this term with a denominator of $\frac{1}{4}$. This makes our equation

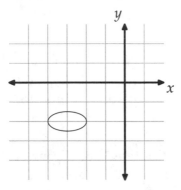

$$\frac{(x + 3)^2}{1} + \frac{(y + 2)^2}{1/4} = 1.$$

We know how to deal with this equation. The center is $(-3, -2)$. The major axis is horizontal because the denominator of the x term is greater than that of the y term. The length of the major axis is $2\sqrt{1} = 2$ and the length of the minor axis is $2\sqrt{\frac{1}{4}} = 1$. The foci are $\sqrt{1 - \frac{1}{4}} = \frac{\sqrt{3}}{2}$ units horizontally from the center, so the foci are $(-3 + \frac{\sqrt{3}}{2}, -2)$ and $(-3 - \frac{\sqrt{3}}{2}, -2)$. Finally, we can produce the graph shown at right.

□

Problem 5.26: What is the graph of each of the following equations:

(a) $\dfrac{x^2}{4} + \dfrac{(y - 2)^2}{9} = 0$ (b) $9x^2 + 4y^2 - 16y + 52 = 0$

Solution for Problem 5.26:

(a) A quick glance at the left side might make us suspect that the graph of this equation is an ellipse. However, the right side is 0, not 1. Because the numerators of the terms on the left are squares, both terms are nonnegative for all real x and y. Therefore, we can only have 0 on the left side if both terms equal 0, which happens only when $(x, y) = (0, 2)$. So, the graph of this equation consists of this single point.

(b) The $4x^2$ and $9y^2$ terms suggest that the graph is an ellipse, so we try to write the equation in the standard form for an ellipse. Completing the square in y and moving the constant to the right gives $9x^2 + 4(y - 2)^2 = -36$. Then, we divide by 36 to get

$$\frac{x^2}{4} + \frac{(y - 2)^2}{9} = -1.$$

As in part (a), the two terms on the left must be nonnegative, so their sum cannot be negative. Therefore, no ordered pair (x, y) of real numbers satisfies this equation. Hence, the graph of this equation is empty—there are no points on the Cartesian plane whose coordinates satisfy the equation.□

In part (b) of Problem 5.25, we rearranged an equation of the form $Ax^2 + By^2 + Cx + Dy + E = 0$, where A, B, C, D, and E are constants with $AB > 0$, into the standard form of an ellipse. However, in Problem 5.26, we see that not all such equations can be put in the standard form of an ellipse. Therefore, we see that:

> **WARNING!!** Not every equation of the form $Ax^2 + By^2 + Cx + Dy + E = 0$, where ☢ A, B, C, D, E are constants with $AB > 0$, has a graph that is an ellipse. The graph of the equation may be a point, or may be empty.

In order to tell if the graph of an equation of this form is an ellipse, we try to write it in the standard form of an ellipse by making one side $\frac{(x-h)^2}{a^2} + \frac{(y-k)^2}{b^2}$ and the other side constant. If the constant is positive, we can divide by it to produce our standard form, and the graph is an ellipse. If the constant is nonpositive, then the graph is not an ellipse.

> **Sidenote:** If the graph of $Ax^2 + By^2 + Cx + Dy + E = 0$, where A, B, C, D, E are constants ♪ with $AB > 0$, is a point, the graph is sometimes referred to as a **degenerate** ellipse.

Problem 5.27: The graphics program we used to write this book doesn't have a command to graph an ellipse. It does, however, have a command to graph a circle given its center and radius. It also allows the user to scale a graph horizontally or vertically. (When we "scale a graph horizontally," we multiply the x-coordinate of every point on the graph by a fixed constant.) How can we use this program to produce the ellipse at right?

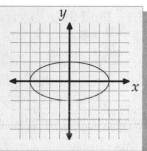

Solution for Problem 5.27: We've already seen that a circle is a specific type of ellipse. This wasn't a surprise, because an ellipse looks like a circle that has been "stretched." Our graphics program allows us to stretch a graph horizontally or vertically, so maybe we can simply stretch a circle into an ellipse.

Our desired ellipse has a horizontal major axis with length 8 and a vertical minor axis with length 4. So, we start with a circle with diameter 4 (shown at right), and we plan to scale it horizontally by a factor of 2 to make our ellipse.

We have to show that this scaling does produce an ellipse, and not just some curve that looks a lot like an ellipse but isn't quite an ellipse. To do this, we start with the equation whose graph is the circle, and consider the effect of the scaling on this equation. The equation of the circle is simply $x^2 + y^2 = 4$. But what does the scaling do?

The scaling doubles the x-coordinate of every point on the graph. So, for every point (x, y) on the graph of $x^2 + y^2 = 4$, there is a point (x', y') on the scaled graph such that $(x', y') = (2x, y)$. From $x' = 2x$, we have $x = x'/2$. Letting $x = x'/2$ and $y = y'$ in $x^2 + y^2 = 4$ gives us $(x'/2)^2 + (y')^2 = 4$, which looks a lot like an equation of an ellipse. Dividing by 4 then gives us

$$\frac{(x')^2}{16} + \frac{(y')^2}{4} = 1.$$

Graphing this equation does indeed give us the desired ellipse with horizontal major axis of length 8 and vertical minor axis of length 4. $\square$

There's nothing special about scaling horizontally. If we similarly scale a circle in any direction, we produce an ellipse. With some more advanced geometric tools, we can show that if we scale a closed figure in some direction by a factor k, then the area of the scaled figure is k times the area of the original figure. As an Exercise, you'll use this fact to explain why we have the following:

> **Important:** The area of an ellipse with major axis of length $2a$ and minor axis of length $2b$ is $ab\pi$.

Exercises

5.5.1 Find an equation whose graph is the ellipse shown at right.

5.5.2 Find the foci, center, and lengths of the axes of the graph of the equation $x^2 + 4y^2 + 8y = -3$. Graph the equation.

5.5.3 Find the foci, the lengths of the major and minor axes, and the center of the graph of the equation $x^2 + 2y^2 - 12y + 4x + 2 = 0$.

5.5.4 An ellipse has foci at $(0,0)$ and $(14,0)$ and passes through the vertex of the parabola with equation $y = x^2 - 10x + 37$. Find the length of the major axis of this ellipse.

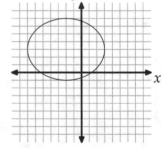

5.5.5 Scaling a closed figure in some direction by a factor k produces a figure with area k times the area of the original figure. Use this fact to explain why the area of an ellipse with major axis of length $2a$ and minor axis of length $2b$ is $ab\pi$.

5.6 Hyperbolas

Suppose we have points F_1 and F_2 in a plane. For any positive constant k, the locus of all points P in the plane such that $|PF_1 - PF_2| = k$ is called a **hyperbola**, an example of which is shown at right. As you can see, unlike the other conic sections we have studied, a hyperbola consists of two pieces, which are called **branches** of the hyperbola.

As with an ellipse, the points F_1 and F_2 are the **foci** of the hyperbola. The line connecting the foci meets the hyperbola at the **vertices** of the hyperbola (just as with a parabola, the vertex of each branch of a hyperbola is an "extreme" point of a branch of the hyperbola). The midpoint of the segment connecting a hyperbola's vertices is the hyperbola's **center**.

Problems

Problem 5.28:

(a) Let $\mathcal{H}$ be a hyperbola with foci $F_1 = (c, 0)$ and $F_2 = (-c, 0)$ and vertices $V_1 = (a, 0)$ and $V_2 = (-a, 0)$, where a and c are positive constants with $c > a$. Show that if P is on the hyperbola, then $|PF_1 - PF_2| = 2a$. Suppose that P is closer to F_2 than to F_1. Then, how can we write the equation $|PF_1 - PF_2| = 2a$ without the absolute value signs?

(b) Let P be (x, y). Show that if P is on $\mathcal{H}$, then $\frac{x^2}{a^2} - \frac{y^2}{b^2} = 1$, where $b^2 = c^2 - a^2$. Let $\mathcal{H}'$ be a hyperbola with foci $(h \pm c, k)$ and vertices $(h \pm a, k)$, where a, c, h, and k are positive constants and $c > a$. Find an equation whose graph is $\mathcal{H}'$.

Problem 5.29:

(a) Graph the equation $\frac{x^2}{9} - \frac{y^2}{4} = 1$ by finding several points on the graph.

(b) Graph the equations $\frac{x}{3} = \frac{y}{2}$ and $\frac{x}{3} = -\frac{y}{2}$ on the same Cartesian plane as your graph in part (a).

(c) Is it possible for your graphs in parts (a) and (b) to meet?

(d) What happens to these graphs as x gets very large? Why?

(e) How can you use the graphs of $\frac{x-h}{a} = \frac{y-k}{b}$ and $\frac{x-h}{a} = -\frac{y-k}{b}$ to help you graph the hyperbola $\frac{(x-h)^2}{a^2} - \frac{(y-k)^2}{b^2} = 1$?

Problem 5.30:

(a) Graph the hyperbola $\frac{x^2}{a^2} - \frac{y^2}{b^2} = 1$ for various values of a and b. How do the values of a and b affect the graph?

(b) What happens if the y term is positive and the x term is negative? For example, what is the graph of $\frac{y^2}{9} - \frac{x^2}{4} = 1$?

Problem 5.31: Find the center and vertices of the graph of each equation given below, and then graph the equation.

(a) $\dfrac{(y+1)^2}{9} - \dfrac{(x-2)^2}{16} = 1$

(b) $4y^2 - x^2 + 16y + 6x = -23$

Problem 5.32: Back in Problem 5.26, we looked at equations that produce "degenerate" ellipses. These equations almost exactly matched the standard form of an equation whose graph is an ellipse, except that the isolated constant term was 0 or negative rather than being 1. In this problem, we investigate corresponding equations that look like they have hyperbolas as their graphs. Find the graphs of each of the following:

(a) $\dfrac{x^2}{4} - \dfrac{(y-2)^2}{9} = 0$

(b) $\dfrac{x^2}{4} - \dfrac{(y-2)^2}{9} = -1$

Problem 5.28:

(a) Let $\mathcal{H}$ be a hyperbola with foci $F_1 = (c, 0)$ and $F_2 = (-c, 0)$ and vertices $V_1 = (a, 0)$ and $V_2 = (-a, 0)$, where a and c are positive constants with $c > a$. Find an equation whose graph is $\mathcal{H}$.

(b) Let $\mathcal{H}'$ be a hyperbola with foci $(h \pm c, k)$ and vertices $(h \pm a, k)$, where $a, c, h,$ and k are positive constants. Find an equation whose graph is $\mathcal{H}'$.

Solution for Problem 5.28:

(a) We start with the definition of a hyperbola. Suppose the point $P = (x, y)$ is on the hyperbola. Then, the difference $|PF_1 - PF_2|$ is constant for all possible P. But what is this constant?

We find this constant difference the same way we found the constant sum for an ellipse.

> **Concept:** When faced with a new problem that's very similar to a problem you know how to do, use your solution to the known problem as a guide for the new problem.

We consider a specific point on the hyperbola, namely, vertex V_1. Since $V_1F_1 = c - a$ and $V_1F_2 = c + a$, we have $|V_1F_1 - V_1F_2| = |(c - a) - (c + a)| = |-2a| = 2a$. Because V_1 is on the hyperbola and $|V_1F_1 - V_1F_2| = 2a$, we know that the constant difference $|PF_1 - PF_2|$ is $2a$.

To get rid of the absolute value, we'll first investigate points on the hyperbola that are closer to F_2 than to F_1. If P is closer to F_2 than to F_1, then we have $PF_1 > PF_2$. This means $PF_1 - PF_2 > 0$, so $|PF_1 - PF_2| = PF_1 - PF_2$. Therefore, the equation $|PF_1 - PF_2| = 2a$ is $PF_1 - PF_2 = 2a$. We can use the distance formula to find PF_1 and PF_2, so we now have an equation for our hyperbola:

$$\sqrt{(x - c)^2 + y^2} - \sqrt{(x + c)^2 + y^2} = 2a.$$

This is very close to the initial equation we had on page 132 for an ellipse, which was

$$\sqrt{(x - c)^2 + y^2} + \sqrt{(x + c)^2 + y^2} = 2a.$$

The only difference between these two equations is the sign before the second radical. We have a situation very much like our ellipse equation, so we use the same strategy we used to simplify the ellipse equation. We first isolate one radical,

$$\sqrt{(x - c)^2 + y^2} = 2a + \sqrt{(x + c)^2 + y^2},$$

then square both sides, to get

$$(x - c)^2 + y^2 = (2a)^2 + 2(2a)\left(\sqrt{(x + c)^2 + y^2}\right) + (x + c)^2 + y^2.$$

Rearranging, we have

$$4a\sqrt{(x + c)^2 + y^2} = (x - c)^2 + y^2 - (x + c)^2 - y^2 - (2a)^2$$
$$= x^2 - 2xc + c^2 + y^2 - x^2 - 2xc - c^2 - y^2 - 4a^2$$
$$= -4xc - 4a^2.$$

Dividing both sides by 4 and squaring again will get rid of the radical and give us

$$a^2((x+c)^2 + y^2) = x^2c^2 + 2xca^2 + a^4.$$

Expanding the left side gives us $a^2x^2 + 2xca^2 + a^2c^2 + a^2y^2 = x^2c^2 + 2xca^2 + a^4$. Bringing the x's and y's to one side and moving the constants to the other gives

$$x^2(c^2 - a^2) - a^2y^2 = a^2(c^2 - a^2).$$

Dividing both sides by $a^2(c^2 - a^2)$ gives us

$$\frac{x^2}{a^2} - \frac{y^2}{c^2 - a^2} = 1.$$

Finally, we let $b^2 = c^2 - a^2$, and we have $\frac{x^2}{a^2} - \frac{y^2}{b^2} = 1$.

We've now shown that if $P = (x, y)$ is on the hyperbola such that $PF_1 > PF_2$, then (x, y) satisfies the equation $\frac{x^2}{a^2} - \frac{y^2}{b^2} = 1$. If P is closer to F_1 than to F_2, we have $PF_1 < PF_2$, so $PF_1 - PF_2 < 0$. In this case, we have $|PF_1 - PF_2| = -(PF_1 - PF_2)$, so the equation $|PF_1 - PF_2| = 2a$ becomes $PF_1 - PF_2 = -2a$. We can then work through essentially the same steps as above to find that (x, y) satisfies $\frac{x^2}{a^2} - \frac{y^2}{b^2} = 1$ in this case, as well.

(b) The center of the hyperbola in part (a) is $(0, 0)$ and the center of the hyperbola in this part is (h, k). Otherwise, the two hyperbolas are the same. That is, hyperbola $\mathcal{H}'$ in this part is the result of translating hyperbola $\mathcal{H}$ rightward h units and upward k units. So, for each point (x, y) on $\mathcal{H}$, there is a point (x', y') on $\mathcal{H}'$ such that $(x', y') = (x + h, y + k)$. Therefore, we have $x = x' - h$ and $y = y' - k$. Substituting these into our equation in part (a), we have

$$\frac{(x' - h)^2}{a^2} - \frac{(y' - k)^2}{b^2} = 1,$$

where $b^2 = c^2 - a^2$. We can reverse all these steps to show that:

> **Important:** The graph of the equation
>
> $$\frac{(x - h)^2}{a^2} - \frac{(y - k)^2}{b^2} = 1$$
>
> is a hyperbola with center (h, k), vertices $(h \pm a, k)$, and foci $(h \pm c, k)$, where $c^2 = a^2 + b^2$. This is the standard form of an equation whose graph is a hyperbola that opens horizontally.

$\square$

You shouldn't need to memorize that the vertices of the graph of

$$\frac{(x - h)^2}{a^2} - \frac{(y - k)^2}{b^2} = 1$$

are $(h \pm a, k)$. The vertices are the points closest to the center of the hyperbola. These points occur when the negative term of the left side of the equation is 0. When this happens, we have $\frac{(x-h)^2}{a^2} = 1$, so $x - h = \pm a$, which tells us that the two resulting points on the graph are a to the left and right from the center. If the negative term of our equation is nonzero, then $y \neq k$, and the fraction $\frac{(x-h)^2}{a^2}$ is greater than 1, which means that x is more than a from h. Therefore, (x, y) is both horizontally and vertically farther from the center than the points we found by letting $y = k$, so the points that result from letting the negative term equal 0 are the vertices of the hyperbola.

Problem 5.29:

(a) Graph the equation $\frac{x^2}{9} - \frac{y^2}{4} = 1$ by finding several points on the graph.

(b) Graph the equations $\frac{x}{3} = \frac{y}{2}$ and $\frac{x}{3} = -\frac{y}{2}$.

(c) Is it possible for your graphs in parts (a) and (b) to meet?

(d) What happens to these graphs as x gets very large? Why?

(e) How can you use the graphs of $\frac{x-h}{a} = \frac{y-k}{b}$ and $\frac{x-h}{a} = -\frac{y-k}{b}$ to help you graph the hyperbola $\frac{(x-h)^2}{a^2} - \frac{(y-k)^2}{b^2} = 1$?

Solution for Problem 5.29:

(a) Below is our table and our graph. Note that because the equation only has x^2 and y^2 and no other variable expressions, if the point (x, y) is on the graph, then so are $(x, -y)$, $(-x, y)$, and $(-x, -y)$. So, aside from first row (the vertices of the hyperbola), each row in the table gives us four points on the hyperbola. (Several of the y-coordinates are approximations.)

x	y
±3	0
±4	±1.76
±5	±2.67
±6	±3.46
±7	±4.22
±8	±4.94
±9	±5.66

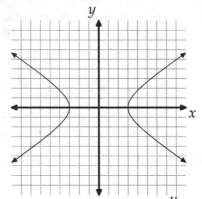

(b) We add the graphs of the lines to the graph of our hyperbola, and the result is shown at right.

The lines appear to "frame" the hyperbola. It looks like the hyperbola gets closer and closer to the lines, but doesn't intersect them.

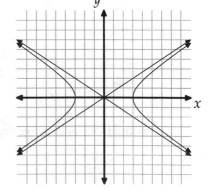

(c) If the hyperbola intersects the graph of $\frac{x}{3} = \frac{y}{2}$, then it must be at a point (x, y) that satisfies the system of equations

$$\frac{x^2}{9} - \frac{y^2}{4} = 1,$$
$$\frac{x}{3} = \frac{y}{2}.$$

But squaring the second equation gives us $\frac{x^2}{9} = \frac{y^2}{4}$, and rearranging this gives $\frac{x^2}{9} - \frac{y^2}{4} = 0$. This equation and the first equation in our system above, $\frac{x^2}{9} - \frac{y^2}{4} = 1$, cannot both be true, so the hyperbola cannot intersect the graph of $\frac{x}{3} = \frac{y}{2}$. Similarly, it cannot intersect the graph of $\frac{x}{3} = -\frac{y}{2}$.

(d) We investigate what happens as x gets larger by finding points on the upper portion of the right branch and on the graph of $\frac{x}{3} = \frac{y}{2}$ for larger and larger values of x. At right is a table showing the y-coordinates (rounded to the nearest hundredth) of points on the hyperbola and the line for various values of x. We see that the y-coordinates get closer and closer to each other.

x	Hyperbola's y	Line's y
6	3.46	4
10	6.36	6.67
20	13.18	13.33
50	33.27	33.33
100	66.64	66.67

Let's take a look at why the y-coordinates get closer and closer to each other. We'll focus on why the line $\frac{x}{3} = \frac{y}{2}$ approaches the upper half of the right branch of the hyperbola when x is large. Solving for y in $\frac{x}{3} = \frac{y}{2}$ gives $y = \frac{2x}{3}$. Solving for y^2 in $\frac{x^2}{9} - \frac{y^2}{4} = 1$ gives $y^2 = 4(\frac{x^2}{9} - 1)$. We take the positive square root because we are interested in the upper half of the right branch, and we have

$$y = 2\sqrt{\frac{x^2}{9} - 1}.$$

Intuitively, we can see that as x gets very large, the value of $\frac{x^2}{9} - 1$ is (relatively speaking) very close to $\frac{x^2}{9}$, which makes $y = 2\sqrt{\frac{x^2}{9} - 1}$ very close to $y = 2\sqrt{\frac{x^2}{9}} = \frac{2x}{3}$. We can be a little more precise with our explanation by noting that if $x > 0$, we have

$$y = 2\sqrt{\frac{x^2}{9} - 1} = 2\sqrt{\left(\frac{x^2}{9}\right)\left(1 - \frac{9}{x^2}\right)} = \frac{2x}{3}\sqrt{1 - \frac{9}{x^2}}.$$

As x gets large, the value of $\frac{9}{x^2}$ gets closer and closer to 0, so $1 - \frac{9}{x^2}$ gets closer and closer to 1. This means that y gets closer and closer to $2x/3$, as we saw in our table.

We call a line an **asymptote** of a graph if the graph gets closer and closer to, but does not intersect, the line as x and/or y gets far from 0. As we have just seen, the lines $\frac{x}{3} = \frac{y}{2}$ and $\frac{x}{3} = -\frac{y}{2}$ are asymptotes of the hyperbola $\frac{x^2}{9} - \frac{y^2}{4} = 1$.

Similarly, we can show that:

> **Important:** The lines $\dfrac{x-h}{a} = \pm\dfrac{y-k}{b}$ are asymptotes of the graph of
>
> $$\frac{(x-h)^2}{a^2} - \frac{(y-k)^2}{b^2} = 1.$$

(e) Suppose we wish to graph the equation $\frac{(x-h)^2}{a^2} - \frac{(y-k)^2}{b^2} = 1$. We know that the graph is a hyperbola with center (h, k) and vertices $(h \pm a, k)$, but this just gives us two points on the hyperbola. To get a quick sketch of the hyperbola, we can draw the asymptotes of the hyperbola first, then use the asymptotes as a guide to fill out the hyperbola.

The asymptotes of the hyperbola are $\frac{x-h}{a} = \pm\frac{y-k}{b}$. Multiplying both sides by b gives the equation $y - k = \pm\frac{b}{a}(x - h)$. Therefore, the asymptotes are the lines through (h, k) with slopes b/a and $-b/a$.

One quick way to graph these asymptotes is to make a rectangle centered at (h, k) with vertical sides of length $2b$ and horizontal sides of length $2a$. The midpoints of the vertical sides then are the vertices of the hyperbola, and the asymptotes of the hyperbola contain the diagonals of the rectangle. The result is shown below, with the sides of the rectangle dashed and asymptotes drawn with thin lines.

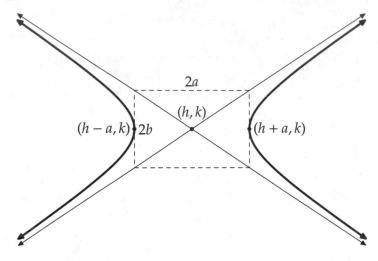

☐

If we add the graph of $\frac{(x-h)^2}{a^2} + \frac{(y-k)^2}{b^2} = 1$ to the graph we just made of $\frac{(x-h)^2}{a^2} - \frac{(y-k)^2}{b^2} = 1$ (including the rectangle we made to plot the asymptotes), we have a visual sense for how these equations are related:

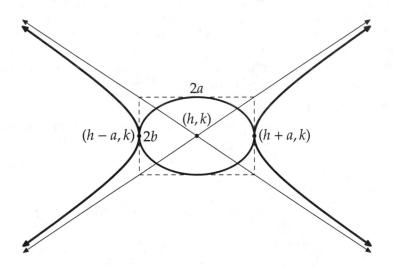

Problem 5.30:

(a) Graph the hyperbola $\frac{x^2}{a^2} - \frac{y^2}{b^2} = 1$ for various values of a and b. How do the values of a and b affect the graph?

(b) What happens if the y term is positive and the x term negative? For example, what is the graph of $\frac{y^2}{9} - \frac{x^2}{4} = 1$?

Solution for Problem 5.30:

(a) In our first three hyperbolas below, we keep b fixed and vary a. In the second three hyperbolas, we keep a fixed and vary b. We use the asymptotes of each hyperbola as described in the last part of the previous problem to quickly generate the graphs.

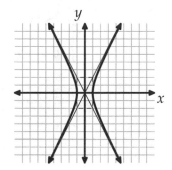

Figure 5.1: $\dfrac{x^2}{1} - \dfrac{y^2}{4} = 1$

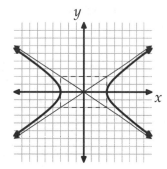

Figure 5.2: $\dfrac{x^2}{9} - \dfrac{y^2}{4} = 1$

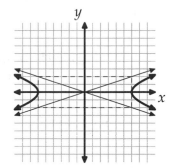

Figure 5.3: $\dfrac{x^2}{36} - \dfrac{y^2}{4} = 1$

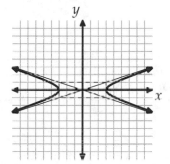

Figure 5.4: $\dfrac{x^2}{9} - \dfrac{y^2}{1} = 1$

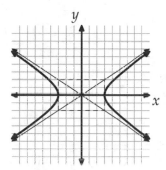

Figure 5.5: $\dfrac{x^2}{9} - \dfrac{y^2}{4} = 1$

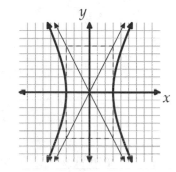

Figure 5.6: $\dfrac{x^2}{9} - \dfrac{y^2}{36} = 1$

First, we notice that increasing a moves the branches farther apart. This makes sense because the vertices of a hyperbola are $2a$ apart. We also notice that decreasing a or increasing b makes the hyperbola "open more widely." Examining the asymptotes explains why. Decreasing a or increasing b makes the slopes of the asymptotes, $\frac{b}{a}$ and $-\frac{b}{a}$, farther from 0. In other words, the asymptotes are "steeper," thereby making the hyperbola open wider. So, as $\frac{b}{a}$ increases, the hyperbola opens wider, and as $\frac{b}{a}$ decreases, the hyperbola opens more narrowly.

> **Sidenote:** Just as with ellipses, we define the **eccentricity** of a hyperbola as the ratio of the distance between the foci to the distance between the vertices. Because $c^2 = a^2 + b^2$, we have
>
> $$\frac{c}{a} = \frac{\sqrt{a^2 + b^2}}{a} = \sqrt{1 + \frac{b^2}{a^2}} = \sqrt{1 + \left(\frac{b}{a}\right)^2}.$$
>
> So, the "wider" a hyperbola is (meaning $\frac{b}{a}$ is large), the more eccentric it is.

(b) We've seen that having y squared instead of x squared in a parabola means the parabola opens horizontally rather than vertically. Similarly, when the y term has the larger denominator in the standard form equation of an ellipse, then the foci are on a vertical line, rather than on a horizontal line. So, we're not surprised to find that the graph of $\frac{y^2}{9} - \frac{x^2}{4} = 1$ is a hyperbola that opens up and down rather than left and right. The graph is shown at right.

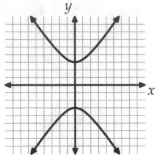

We can also find the graph of $\frac{y^2}{9} - \frac{x^2}{4} = 1$ by noting that for each point (a, b) on this graph, there is a point (b, a) on the graph of $\frac{x^2}{9} - \frac{y^2}{4} = 1$, and vice versa. We've seen something like this before! Back on page 48, we noted that the points (a, b) and (b, a) are reflections of each other about the line $y = x$. Therefore, the graph of $\frac{y^2}{9} - \frac{x^2}{4} = 1$ is the image of the graph of $\frac{x^2}{9} - \frac{y^2}{4} = 1$ when reflected over the line $y = x$.

□

> **Important:** The standard form of an equation whose graph is a hyperbola that opens vertically is
> $$\frac{(y-k)^2}{a^2} - \frac{(x-h)^2}{b^2} = 1.$$
> The center of the hyperbola is (h, k), its vertices are $(h, k \pm a)$, and its foci are $(h, k \pm c)$, where $c^2 = a^2 + b^2$. The asymptotes of the hyperbola are $y - k = \pm\frac{a}{b}(x - h)$.

> **Important:** Don't worry too much about memorizing all these various forms and formulas. It's more important to understand how we derive them, and how we apply them once we have found them. You can usually look them up when you need them.

Problem 5.31: Find the center and vertices of the graph of each equation given below, then graph the equation.

(a) $\dfrac{(y+1)^2}{9} - \dfrac{(x-2)^2}{16} = 1$

(b) $4y^2 - x^2 + 16y + 6x = -23$

Solution for Problem 5.31:

(a) What's wrong with this start:

> **Bogus Solution:** We have $h = 2$, $k = -1$, $a = 4$, and $b = 3$. So, the center is $(2, -1)$. The y term is positive, so the branches of the hyperbola open vertically. So, the vertices are $(2, -1 \pm 4)$, which are $(2, 3)$ and $(2, -5)$.

The problem with this start is that the standard form of a hyperbola that opens up and down is $\frac{(y-k)^2}{a^2} - \frac{(x-h)^2}{b^2} = 1$. So, we have $a = 3$, not $a = 4$. We could have caught our mistake by checking our work:

Important: You can easily check if you have the vertices of a hyperbola correct. They must satisfy the equation of the hyperbola.

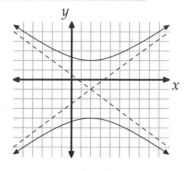

This shows yet another reason you shouldn't simply memorize formulas without thinking about them. If you don't have a formula quite right, and don't understand it, you won't notice you've made a mistake.

Returning to the problem, we know the hyperbola opens upward and downward, and the center is $(2, -1)$. Rather than relying on our formulas, we note that the negative term is 0 for $x = 2$, which gives us $(y + 1)^2/9 = 1$. From this, we find that our vertices are 3 above and below the center, at $(2, 2)$ and $(2, -4)$.

We draw our asymptotes, $y + 1 = \pm\frac{3}{4}(x - 2)$, and then we use these and our vertices to graph the hyperbola, as shown at right above.

(b) We first complete the square in x and y. Factoring 4 out of the y terms and -1 out of the x terms, we have $-1(x^2 - 6x) + 4(y^2 + 4y) = -23$. We complete the square by adding $(-1)(9)$ for x and $4(4)$ for y, and we have

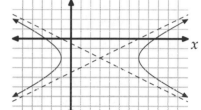

$$-1(x^2 - 6x + 9) + 4(y^2 + 4y + 4) = -23 + (-1)(9) + (4)(4) = -16.$$

Finally, we divide by -16 (make sure you see why we must divide by -16, not just 16) to give

$$\frac{(x - 3)^2}{16} - \frac{(y + 2)^2}{4} = 1.$$

The center of the hyperbola is $(3, -2)$. The negative term is 0 for $y = -2$, from which we find that the vertices are $(7, -2)$ and $(-1, -2)$.

As before, we graph the asymptotes, $y + 2 = \pm\frac{2}{4}(x - 3)$, and use these along with the vertices to graph the hyperbola, shown above.

$\square$

Problem 5.32: Find the graphs of each of the following:

(a) $\dfrac{x^2}{4} - \dfrac{(y - 2)^2}{9} = 0$ (b) $\dfrac{x^2}{4} - \dfrac{(y - 2)^2}{9} = -1$

Solution for Problem 5.32:

(a) From the left side, it looks like the graph of the equation is a hyperbola. However, the right side is 0, not 1. We can add $\frac{(y-2)^2}{9}$ to both sides to get $\frac{x^2}{4} = \frac{(y-2)^2}{9}$. Taking the square root of both sides then gives us $\frac{x}{2} = \pm\frac{y-2}{3}$. These two possibilities give us the equations $3x = 2(y - 2)$ and $3x = -2(y - 2)$. The graphs of these are the two lines shown at right.

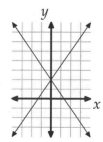

The original equation is very close to an equation that we know has a hyperbola as its graph. So, we look for how this equation is related to that hyperbola. In our graph at right, we have graphed with dashed lines the hyperbola that is the graph of $\frac{x^2}{4} - \frac{(y-2)^2}{9} = 1$. From the graph, we quickly see the relationship—the lines that form the graph of $\frac{x^2}{4} - \frac{(y-2)^2}{9} = 0$ are the asymptotes of the graph of $\frac{x^2}{4} - \frac{(y-2)^2}{9} = 1$.

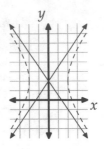

The graph of an equation such as $\frac{x^2}{4} - \frac{(y-2)^2}{9} = 0$ is considered a **degenerate** hyperbola.

(b) Our first instinct might be to guess that the graph of this equation is not a hyperbola, just like the graph of the equation $\frac{x^2}{4} + \frac{(y-2)^2}{9} = -1$ is not an ellipse. However, multiplying the equation given in this part by -1 produces an equation that is in the standard form of a hyperbola:

$$\frac{(y-2)^2}{9} - \frac{x^2}{4} = 1.$$

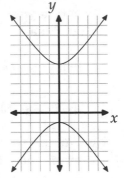

Graphing this equation produces the hyperbola at right. Algebraically speaking, while there are no real solutions to the equation $\frac{x^2}{4} + \frac{(y-2)^2}{9} = -1$, there are solutions to the equation $\frac{x^2}{4} - \frac{(y-2)^2}{9} = -1$.

☐

We now summarize what we've learned about the possible graphs that can be produced from an equation which is quadratic in x and/or y:

> **Important:** Graphs of equations of the form $Ax^2 + By^2 + Cx + Dy + E = 0$, where A, B, C, D, and E are constants such that not both A and B are zero, can be grouped into three categories:
>
> - *One of A or B is 0.* The graph is a parabola.
>
> - $AB > 0$. In other words, A and B have the same sign. The graph is an ellipse (a circle if $A = B$), a point, or empty.
>
> - $AB < 0$. The graph is either a hyperbola or a pair of lines.

These are not the only equations that produce parabolas, ellipses, or hyperbolas as their graphs.

 Problems

Problem 5.33: In this problem, we examine the graph of the equation $xy = 1$.
  (a) Graph the equation. Does it look like a hyperbola?
  (b) What lines appear to be the asymptotes of the graph in part (a)?
  (c) Assuming the graph is a hyperbola, what are its vertices?

Problem 5.34: Find the graph of the equation $(x - 3)(y + 1) = -4$.

Problem 5.33: In this problem, we examine the graph of the equation $xy = 1$.

(a) Graph the equation. Does it look like a hyperbola?

(b) What lines appear to be the asymptotes of the graph in part (a)?

(c) Assuming the graph is a hyperbola, what are its vertices?

Solution for Problem 5.33:

(a) We solve the equation for y to find $y = \frac{1}{x}$. Choosing various values of x, we generate the graph at right.

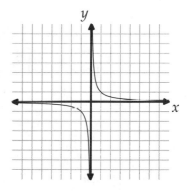

   The graph sure looks like a hyperbola, but this hyperbola doesn't open horizontally or vertically. Instead, it appears to be the result of rotating a horizontally-opening hyperbola 45 degrees counterclockwise. In a later text in the *Art of Problem Solving* series we'll study trigonometry, which we can use to prove that this graph is the result of rotating the graph of $\frac{x^2}{2} - \frac{y^2}{2} = 1$ forty-five degrees counterclockwise about the origin.

> **Important:** If k is a nonzero constant, then the graph of the equation $xy = k$ is a hyperbola.

(b) The graph appears to approach, but not intersect, either coordinate axis. Clearly, we can't have $x = 0$ or $y = 0$ in the equation $xy = 1$, since this will give us $0 = 1$, which can never be true. So, the graph of $xy = 1$ does not intersect either axis. However, because we have $y = \frac{1}{x}$, as x gets large, y gets very close to 0. The same happens if x is much less than 0. Therefore, the x-axis, which is the graph of $y = 0$, is an asymptote of the graph.

   Similarly, when y is very large, x is close to 0. The graph of $xy = 1$ never intersects the y-axis, but approaches the y-axis as y gets large. So, the y-axis is also an asymptote of the graph.

   Note that the asymptotes of the graph are perpendicular. We call a hyperbola with perpendicular asymptotes a **rectangular hyperbola**.

(c) For each point $(a, \frac{1}{a})$ on the graph of $xy = 1$, we see that the point $(\frac{1}{a}, a)$ is also on the graph. In other words, the graph $xy = 1$ is symmetric about the line $x = y$. The line that divides both branches into two symmetric pieces must pass through the vertices of the hyperbola, so the vertices of the hyperbola occur where $x = y$. Setting $x = y$ in $xy = 1$ gives us $y^2 = 1$, so $y = \pm 1$. Therefore, the vertices are $(1, 1)$ and $(-1, -1)$.

□

Problem 5.34: Find the graph of the equation $(x - 3)(y + 1) = -4$.

Extra! For any set of five points in a plane such that no three of the points lie on the same line, there is a single conic section that passes through all five points.

Solution for Problem 5.34: We suspect that the graph is a hyperbola, both because the equation is similar to the equation in the previous problem, and because the problem is in the "Hyperbola" section. To graph the equation, we solve for y in terms of x. We know that we can't have $x = 3$, since this would give us $0 = -4$. So, we can divide both sides by $x - 3$, and then subtract 1, to find $y = -\frac{4}{x-3} - 1$. Choosing several values of x and computing the corresponding y gives us the graph at right, which sure looks like a hyperbola.

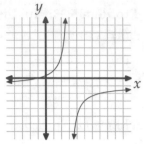

Let's take a look at how we can reason our way to the graph from the equation $(x-3)(y+1) = -4$ once we know the graph is a hyperbola. First, we could have seen that the lines $x = 3$ and $y = -1$ are asymptotes. If $x = 3$ or $y = -1$ in $(x-3)(y+1) = -4$, we will have $0 = -4$, which is impossible. Therefore, the graph doesn't intersect either $x = 3$ or $y = -1$. However, from $y = -\frac{4}{x-3} - 1$, we see that if x is just above 3, but very close to 3, then y is the negative of a large number. So, $x = 3$ is an asymptote of the graph. Since $\frac{4}{x-3}$ is very close to 0 when x is far from 0, we see that $y = -\frac{4}{x-3} - 1$ approaches, but never equals, -1 when x is far from 0. So, the line $y = -1$ is an asymptote as well. The asymptotes meet at $(3, -1)$, so this point is the center of the hyperbola.

The asymptotes divide the Cartesian plane into four quarters. The branches of the hyperbola are in the upper left and lower right quarters, which we also could have reasoned from the equation $(x-3)(y+1) = -4$. If $x > 3$ and $y > -1$, then $(x-3)(y+1)$ is positive, so no points with $x > 3$ and $y > -1$ are on the graph of $(x-3)(y+1) = -4$. Similarly, we cannot have $x < 3$ and $y < -1$. Therefore, the hyperbola is in the upper left and lower right quarters formed by the asymptotes.

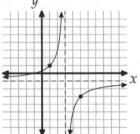

The vertices of the hyperbola are the intersections of the hyperbola and the line through $(3, -1)$ with slope -1. This line has equation $y + 1 = -(x - 3)$. Substituting this into $(x-3)(y+1) = -4$ gives us $-(x-3)^2 = -4$, from which we have $x - 3 = \pm 2$. Therefore, x equals 5 or 1, and our vertices are $(5, -3)$ and $(1, 1)$. We have our vertices and our asymptotes, and now we can use these as a guide to graph the hyperbola.

The resemblance of the equation $(x-3)(y+1) = -4$ and its graph to the equation $xy = 1$ and its graph suggests we might have used our function transformation tools from Section 2.2 to find the graph of $(x-3)(y+1) = -4$ starting with the graph of $xy = 1$. To see how, we solve both equations for y, so that we have $y = \frac{1}{x}$ and $y = -\frac{4}{x-3} - 1$. Letting $f(x) = \frac{1}{x}$ and $g(x) = -\frac{4}{x-3} - 1$, we seek to relate $f(x)$ and $g(x)$. We do so in a series of transformations, as shown on the following page.

Extra! When a ray reflects off a curve, it behaves as if it is being
▶▶▶▶ reflected by the line tangent to the curve at the point where the ray meets the curve. In the diagram, the incoming ray is parallel to the axis of symmetry of the parabola, and the reflection of the ray passes through the focus. This is not a coincidence! As shown in the various dotted rays, the reflection of every ray parallel to the axis of symmetry passes through the focus. This is why satellite dishes are parabolic. When we point our satellite dish so that its axis is pointed directly at the satellite, the rays coming from the satellite are all essentially parallel to the axis of the dish, so if we place the receiver for the satellite signals at the focus of the dish, all the rays will be focused on the receiver.

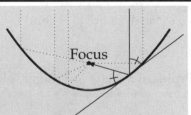

Transformation	Effect on Graph
$-f(x) = -\frac{1}{x}$	Reflect graph of $f(x)$ over the y-axis
$-4f(x) = -\frac{4}{x}$	Scale the graph of $-f(x)$ vertically by 4
$-4f(x-3) = -\frac{4}{x-3}$	Shift the graph of $-4f(x)$ right 3 units
$-4f(x-3) - 1 = -\frac{4}{x-3} - 1$	Shift the graph of $-4f(x-3)$ down 1 unit

Our final expression for $-4f(x-3) - 1$ matches $g(x)$, so our final graph is the graph of g. The series of transformations is shown below:

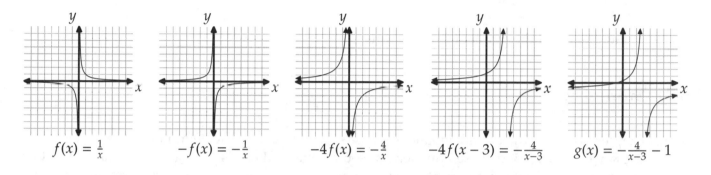

$f(x) = \frac{1}{x}$ $-f(x) = -\frac{1}{x}$ $-4f(x) = -\frac{4}{x}$ $-4f(x-3) = -\frac{4}{x-3}$ $g(x) = -\frac{4}{x-3} - 1$

□

Similarly, we can show that if t is a nonzero constant, then the graph of $(x-h)(y-k) = t$ is a hyperbola with center (h, k).

We have not yet seen all the forms of equations whose graphs are parabolas, ellipses, or hyperbolas. In a later text in the *Art of Problem Solving* series, we'll consider what happens when we include an xy term along with x^2 and y^2 terms.

Exercises

5.6.1 A hyperbola centered at the origin has one focus at $(-13, 0)$ and one vertex at $(5, 0)$. Find an equation whose graph is this hyperbola.

5.6.2 For each of the graphs below, find an equation whose graph is shown. Each graph is a hyperbola.

(a)

(b)

Note: The graph in (a) passes through $(6, 7)$ and the graph in (b) passes through $(0, 6)$.

5.6.3 Find the center, asymptotes, and vertices of the graph of each equation below, and then graph each equation.

(a) $\dfrac{(x-3)^2}{9} - y^2 = 1$

(b) $x^2 + 12x - 3y^2 + 18y + 21 = 0$

5.6.4 If a and b are positive constants such that the graph of $\dfrac{(x-h)^2}{a^2} - \dfrac{(y-k)^2}{b^2} = 1$ is a rectangular hyperbola, then what is a/b?

5.6.5 Hyperbola $\mathcal{H}$ is the graph of the equation $\dfrac{(x-3)^2}{a^2} - \dfrac{y^2}{b^2} = 1$. One of the points of intersection of $\mathcal{H}$ and the line $y = ax$ has x-coordinate 0. Find the x-coordinates of the points of intersection of the line $y = bx$ and $\mathcal{H}$.

5.6.6★ Find the foci of the graph of $xy = 1$. **Hints:** 315

5.7 Summary

In this chapter, we saw how to find equations whose graphs fit geometric definitions of certain curves. The process we used to find these equations and understand them is far more important than the equations themselves, or the many related facts we discovered. Many areas of study require building mathematical models. The process for doing so is often very similar to what we did in this chapter: build equations that depend on certain constants (like our h, k, and a for parabolas, for example), then try different values of those constants to see how they affect your model.

A **locus** is a set of points that satisfy certain given conditions. We covered four special loci in this chapter:

- **Parabola:** The locus of points in a plane equidistant from a given line (the **directrix**) and a point (the **focus**) not on the line. The line through the focus that is perpendicular to the directrix is the **axis of symmetry**, which divides the parabola into two pieces that are mirror images of each other. The axis intersects the parabola at the **vertex** of the parabola.

  The standard form of a vertically-opening parabola is $y = a(x - h)^2 + k$, where $a \neq 0$. The graph of this equation has vertex (h, k) and axis $x = h$.

  The standard form of a horizontally-opening parabola is $x = a(y - k)^2 + h$, where $a \neq 0$. The graph of this equation has vertex (h, k) and axis $y = k$.

- **Circle:** The locus of points in a plane a specified distance (the **radius**) from a given point (the **center**).

  The standard form of a circle is $(x - h)^2 + (y - k)^2 = r^2$, where $r > 0$. The graph of this equation has center (h, k) and radius r.

- **Ellipse:** The locus of points P in a plane such that $PF_1 + PF_2$ is a given constant, where F_1 and F_2 are two points in the plane. The points F_1 and F_2 are the **foci** (plural of **focus**) of the ellipse, and the midpoint of $\overline{F_1F_2}$ is the **center** of the ellipse. The chord (a line segment with both endpoints on the ellipse) that passes through both foci is the **major axis** of the ellipse, and the chord through the center that is perpendicular to the major axis is the **minor axis**.

The standard form for an ellipse is $\frac{(x-h)^2}{a^2} + \frac{(y-k)^2}{b^2} = 1$. The center is (h,k). The axes have lengths $2a$ and $2b$. If $a > b$, then the major axis is horizontal; if $a < b$, then the major axis is vertical. (If $a = b$, then the graph is a circle.) The distance between the foci can be found with the relationship:

$$(\text{major axis length})^2 = (\text{minor axis length})^2 + (\text{distance between foci})^2.$$

- **Hyperbola:** The locus of points P in a plane such that $|PF_1 - PF_2|$ is a given constant, where F_1 and F_2 are two points in the plane called the **foci** of the hyperbola. The line connecting the foci meets the hyperbola at two points called the **vertices** of the hyperbola.

The standard form for a horizontally-opening hyperbola is $\frac{(x-h)^2}{a^2} - \frac{(y-k)^2}{b^2} = 1$. The center is (h,k), the vertices are $(h \pm a, k)$, and the foci are $(h \pm c, k)$, where $c^2 = a^2 + b^2$. The **asymptotes** (lines the graph approaches, but never intersects) of the hyperbola are $\frac{x-h}{a} = \pm \frac{y-k}{b}$.

The standard form for a vertically-opening hyperbola is $\frac{(y-k)^2}{a^2} - \frac{(x-h)^2}{b^2} = 1$. The center is (h,k), the vertices are $(h, k \pm a)$, and the foci are $(h, k \pm c)$, where $c^2 = a^2 + b^2$. The asymptotes of the hyperbola are $\frac{y-k}{a} = \pm \frac{x-h}{b}$.

Collectively, these curves are known as **conic sections**.

>
> **Important:** Don't waste time memorizing all the formulas we discussed in this chapter. The formulas themselves are not terribly important—you can usually look them up when you need them. It's far more important to understand how we derived the formulas and how to use them.

> **Important:** Graphs of equations of the form $Ax^2 + By^2 + Cx + Dy + E = 0$, where A, B, C, D, and E are constants such that not both A and B are zero, can be grouped into three categories:
>
> - *One of A or B is 0.* The graph is a parabola.
> - $AB > 0$. In other words, A and B have the same sign. The graph is an ellipse (a circle if $A = B$), a point, or empty.
> - $AB < 0$. The graph is either a hyperbola or a pair of lines.

We used our understanding of parabolas and completing the square to learn how to find the maximum or minimum of a quadratic expression.

>
> **Important:** Let $f(x) = ax^2 + bx + c$, where a, b, and c are constants.
>
> - If $a > 0$, then the *minimum* value of $f(x)$ is $c - \frac{b^2}{4a}$, which occurs when $x = -\frac{b}{2a}$.
> - If $a < 0$, then the *maximum* value of $f(x)$ is $c - \frac{b^2}{4a}$, which occurs when $x = -\frac{b}{2a}$.

Problem Solving Strategies

Concepts:

- Don't depend too heavily on your formulas. A little thinking about a problem can go a long way.

- Symmetry can be a very powerful problem-solving tool.

- Don't get locked into one way of thinking about quadratics and parabolas.

- To find the minimum or maximum value of a quadratic, complete the square in the quadratic.

- If a problem is difficult to understand, take extra time reading and re-reading the problem until you are sure you understand what the problem is. You don't want to spend a lot of time working on the wrong problem!

- Be alert for equations you know how to graph easily. Sometimes geometry can be used to solve algebra problems.

- Keep your eye on the ball! Don't forget what you're looking for in a problem—you don't always have to solve for every variable in a problem to solve the problem itself.

- When faced with a new problem that's very similar to a problem you know how to do, use your solution to the known problem as a guide for the new problem.

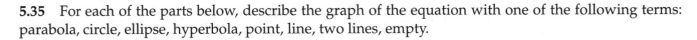

REVIEW PROBLEMS

5.35 For each of the parts below, describe the graph of the equation with one of the following terms: parabola, circle, ellipse, hyperbola, point, line, two lines, empty.

(a) $x^2 = -y + 7x - 3$

(b) $\dfrac{(x-3)^2}{16} + \dfrac{y^2}{4} = 9$

(c) $x^2 + y^2 + 18y = -81$

(d) $x^2 + 4y^2 + 2x = 0$

(e) $6x^2 + 12x + 6y^2 = 18y$

(f) $x = \dfrac{4}{y-2}$

(g) $x^2 + 2x = y^2 + 4y + 3$

(h) $x = \dfrac{y}{x+1}$

(i) $x^2 - 4y^2 + 8y = 0$

5.36 Find the vertex and axis of symmetry of the graph of $y = 4x^2 - 12x + 8$.

5.37 Find the vertex and axis of symmetry of the graph of $x = \dfrac{1}{6}(y-4)^2 - 1$. Graph the equation.

5.38 The vertex of the parabola $x = 4y^2 + 6y + c$ lies on the y-axis. Find the value of c.

5.39 Find an equation whose graph is a parabola with vertex $(1, 5)$ and focus $(1, 3)$.

5.40 For each of the following, find an equation whose graph is shown in the diagram.

(a)
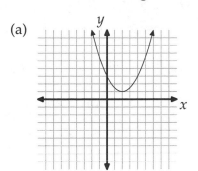

(c)
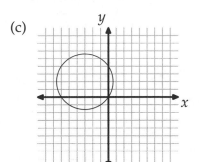

(b)
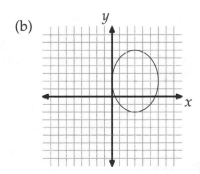

(d)
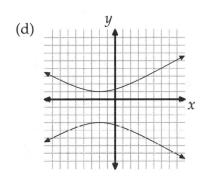

The hyperbola in part (d) passes through $(-2, 1)$, $(-2, -3)$, and $(4, 3)$.

5.41 For how many values of a is it true that the line $y = x + a$ passes through the vertex of the parabola $y = x^2 + a^2$? *(Source: AMC)*

5.42

(a) Find the minimum value of $p(y) = 2y^2 - 4y + 19$.

(b) Find the maximum value of $37 - 16r - r^2$.

(c) Find the minimum value of the function $f(x) = x^4 - 3x^2 - 13$.

(d) Find the minimum value of $x^2 + 4y^2 - 6x + 4y + 5$.

5.43 GetThere Airlines currently charges \$200 per ticket, and sells 40,000 tickets. For every \$10 they increase the ticket price, they sell 1000 fewer tickets. How much should they charge to maximize their revenue?

5.44 Find the smallest possible value of the length of a diagonal of a rectangle with perimeter 36. (Note: You cannot simply assume the rectangle is a square; you must show that the shortest possible diagonal occurs when the rectangle is a square.)

5.45 Find the area enclosed in the graph of $x^2 + y^2 = 16x + 32y$.

5.46 Find r if r is positive and the line whose equation is $x + y = r$ is tangent to the circle whose equation is $x^2 + y^2 = r$. *(Source: AHSME)*

5.47 The circumcircle of a triangle is the circle that passes through all three vertices of the triangle. Find an equation whose graph is the circumcircle of a triangle with vertices $(-2, 5)$, $(-4, -3)$, and $(0, -3)$.

5.48 Find an equation whose graph is an ellipse with axes parallel to the coordinate axes such that the endpoints of the axes of the ellipse are $(0, 2)$, $(3, 6)$, $(3, -2)$, and $(6, 2)$.

5.49 Graph each of the equations below, then find the foci, center, and lengths of the axes of each graph.

(a) $\dfrac{(x-3)^2}{9} + y^2 = 1$

(b) $4x^2 + y^2 + 16x - 6y = 11$

5.50 Ellipse $\mathcal{E}$ has a horizontal major axis, one focus at $(4, 1)$, and one axis with an endpoint at $(6, 5)$. Find an equation whose graph is this ellipse.

5.51 Find the center, asymptotes, and vertices of the graphs of each equation below, then graph each equation.

(a) $x^2 - 4(y+1)^2 = 1$

(b) $2x^2 - y^2 + 8x + 6y = 9$

5.52 Al begins on a vertex of the hyperbola $\dfrac{(y-k)^2}{a^2} - \dfrac{(x-h)^2}{b^2} = 1$. He walks in a straight line to the closest focus of this hyperbola, and then he moves to the farthest vertex. Find an expression, in terms of a, b, k, and h, for the total distance Al walked.

5.53 One of the graphs of the following equations comes within 1 unit of the point $(265, 346)$. Which one?

(a) $y^2 + x + 4y + 7 = 0$

(b) $16x^2 - 96x - 9y^2 - 54y = 81$

(c) $x^2 + y^2 + 4y - 17 = 0$

(d) $25x^2 + 9y^2 - 36y = 9$

Challenge Problems

5.54 Points A and B are on the parabola $y = 4x^2 + 7x - 1$, and the origin of the Cartesian plane is the midpoint of $\overline{AB}$. What is the length of $\overline{AB}$? *(Source: AMC)*

5.55 A **latus rectum** of an ellipse is a line segment with both endpoints on the ellipse that is parallel to the minor axis and passes through a focus of the ellipse. Find the length of a latus rectum of the graph of $\dfrac{x^2}{a^2} + \dfrac{y^2}{b^2} = 1$ in terms of a and b.

5.56 A parabola with equation $y = ax^2 + bx + c$ is reflected about the x-axis. The parabola and its reflection are translated horizontally five units in opposite directions to become the graphs of $y = f(x)$ and $y = g(x)$, respectively. Which of the following describes the graph of $y = (f + g)(x)$?

(A) a parabola tangent to the x-axis

(B) a parabola not tangent to the x-axis

(C) a horizontal line

(D) a non-horizontal line

(E) the graph of a cubic function

(Source: AMC) **Hints:** 119

5.57 If x is real, then compute the maximum value of $\dfrac{3x^2 + 9x + 17}{3x^2 + 9x + 7}$. *(Source: ARML)*

5.58 Let $f(x) = x^2 + 6x + 1$, and let R denote the set of points (x, y) in the coordinate plane such that $f(x) + f(y) \leq 0$ and $f(x) - f(y) \leq 0$. Find the area of R. *(Source: AMC)* **Hints:** 75

5.59 A **chord** of a parabola is a segment with both endpoints on the parabola. Two parabolas have the same focus, namely the point $(3, -28)$. Their directrixes are the x-axis and y-axis, respectively. Compute the slope of their common chord. *(Source: NYSML)* **Hints:** 209

5.60 Find the largest value of x for which $x^2 + y^2 = x + y$ has a solution, if x and y are real. *(Source: ARML)*

5.61 P is a fixed point on the diameter $\overline{AB}$ of a circle. Prove that for any chord $\overline{CD}$ of the circle that is parallel to $\overline{AB}$, we have $PC^2 + PD^2 = PA^2 + PB^2$. **Hints:** 58

5.62 Let $(2, 3)$ be the vertex of a parabola whose equation is of the form $y = ax^2 + bx + c$. If the graph of the parabola has two x-intercepts with positive x-coordinates, compute the greatest possible integral value of k such that $a = \frac{k}{2000}$. *(Source: ARML)* **Hints:** 202

5.63 An equilateral triangle is inscribed in the ellipse whose equation is $x^2 + 4y^2 = 4$. One vertex of the triangle is $(0, 1)$, and one altitude is contained in the y-axis. Find the length of each side of the triangle. *(Source: AIME)*

5.64 Given that $x^2 + y^2 = 14x + 6y + 6$, what is the largest possible value that $3x + 4y$ can have? *(Source: AHSME)*

5.65 Three circles with radius 2 are drawn in a plane such that each circle is tangent to the other two. Let the centers of the circles be A, B, and C. Point X is on the circle with center C such that $AX + XB = AC + CB$. Find the area of triangle AXB. *(Source: USAMTS)* **Hints:** 66

5.66★ Let $f(x) = (x + 3)^2 + \frac{9}{4}$ for $x \geq -3$. Compute the shortest possible distance between a point on the graph of f and a point on the graph of f^{-1}. *(Source: ARML)* **Hints:** 39, 36

5.67 Determine the unique pair of real numbers (x, y) that satisfy the equation

$$(4x^2 + 6x + 4)(4y^2 - 12y + 25) = 28.$$

(Source: USAMTS)

5.68★ Find an equation whose graph is a parabola with vertex $(1, 1)$ and focus $(2, 2)$.

5.69★ Consider the two functions

$$f(x) = x^2 + 2bx + 1 \quad \text{and} \quad g(x) = 2a(x + b),$$

where the constants a and b may be considered as a point (a, b) in an ab-plane. Let S be the set of such points (a, b) for which the graphs of $y = f(x)$ and $y = g(x)$ do NOT intersect (in the xy-plane). What is the area of S? *(Source: AHSME)* **Hints:** 357

5.70★ The graph of $2x^2 + xy + 3y^2 - 11x - 20y + 40 = 0$ is an ellipse in the first quadrant of the xy-plane. (This means that all points on the ellipse have nonnegative coordinates.) Let a and b be the maximum and minimum values of $\frac{y}{x}$ over all points (x, y) on the ellipse. What is the value of $a + b$? *(Source: AMC)* **Hints:** 270

Life is too short for long division – Unknown

CHAPTER **6**

Polynomial Division

A **polynomial** in one variable is an expression of the form

$$a_n x^n + a_{n-1} x^{n-1} + a_{n-2} x^{n-2} + \cdots + a_1 x + a_0,$$

where x is a variable and each a_i is a constant. Each individual summand $a_i x^i$ is called a **term**, and each number a_i is the **coefficient** of the corresponding x^i. The **degree** of a polynomial is the greatest integer n such that the coefficient a_n is not equal to 0. So for example, the polynomial $2x^4 - x^3 + 3x^2 - x - 7$ has degree 4, and the coefficient of x^3 in this polynomial is -1. The degree of a nonzero constant polynomial is zero, and the degree of the polynomial 0 is undefined. We call the polynomial 0 the **zero polynomial**, and any other polynomial is a **nonzero polynomial**.

We say the function f is a **polynomial function** if $f(x)$ is a polynomial, and we denote the degree of $f(x)$ as $\deg f$. We will often refer to a polynomial function simply as a polynomial.

There are some special names for polynomials of certain degrees. For example, a polynomial of degree 1 is called **linear**, and a polynomial of degree 2 is called **quadratic**. The following table lists some names used to describe polynomials of specific degrees.

Degree	Name
0	constant
1	linear
2	quadratic
3	cubic
4	quartic

Polynomials are also sometimes described according to the number of terms they contain. A polynomial with one term is a **monomial**, with two terms is a **binomial**, and with three terms is a **trinomial**. For instance, $3x^4$ is a quartic monomial, $2x^2 + 1$ is a quadratic binomial, and $8x^5 - 3x^4 + 2x^3$ is a trinomial of degree 5.

6.1 Polynomial Review

In this section, we review polynomial addition, multiplication, and evaluation.

Problems

Problem 6.1: Let $f(x) = 2x^2 + 5x + 13$ and $g(x) = x^3 - 3x - 4$.
  (a) Find $f(g(3))$.
  (b) Find $f(x) + g(x)$ and determine its degree.
  (c) For what value of the constant c does the degree of $x \cdot f(x) + c \cdot g(x)$ equal 2?
  (d) Find $f(x)g(x)$ and determine its degree.

Problem 6.2: Prove that if f and g are nonzero polynomials, then $\deg(f \cdot g) = (\deg f) + (\deg g)$.

Problem 6.3:
  (a) Let $f(x) = 3x^3 - 4x^2 + ax - 11$ be a polynomial such that $f(1) = 2$. Find the value of a.
  (b) Suppose $(3x - 1)^7 = a_7 x^7 + a_6 x^6 + \cdots + a_0$ for constants $a_0, a_1, \ldots, a_7$. Find $a_7 + a_6 + \cdots + a_1 + a_0$.
  (Source: AHSME)

Problem 6.1: Let $f(x) = 2x^2 + 5x + 13$ and $g(x) = x^3 - 3x - 4$.
  (a) Find $f(g(3))$.
  (b) Find $f(x) + g(x)$ and determine its degree.
  (c) For what value of the constant c does the degree of $x \cdot f(x) + c \cdot g(x)$ equal 2?
  (d) Find $f(x)g(x)$ and determine its degree.

Solution for Problem 6.1:

  (a) We have $g(3) = 3^3 - 3(3) - 4 = 14$, so $f(g(3)) = f(14) = 2(14)^2 + 5(14) + 13 = 475$.

  (b) We add the two polynomials the way we add any two expressions—we combine like terms:

$$f(x) + g(x) = 2x^2 + 5x + 13 + x^3 - 3x - 4$$
$$= x^3 + 2x^2 + (5x - 3x) + (13 - 4)$$
$$= x^3 + 2x^2 + 2x + 9.$$

Since x^3 is the highest power of x in $f(x) + g(x)$, the degree of $f(x) + g(x)$ is 3.

  (c) We have

$$x \cdot f(x) + c \cdot g(x) = 2x^3 + 5x^2 + 13x + cx^3 - 3cx - 4c$$
$$= (2 + c)x^3 + 5x^2 + (13 - 3c)x - 4c.$$

If the degree of this polynomial is 2, then there must be no x^3 term. So, we must have $2 + c = 0$, which gives us $c = -2$. When $c = -2$, we have $x \cdot f(x) + c \cdot g(x) = 5x^2 + 19x + 8$, which does indeed have degree 2.

(d) We can use the distributive property to multiply any two polynomials:

$$(2x^2 + 5x + 13)(x^3 - 3x - 4) = 2x^2(x^3 - 3x - 4) + 5x(x^3 - 3x - 4) + 13(x^3 - 3x - 4)$$
$$= (2x^5 - 6x^3 - 8x^2) + (5x^4 - 15x^2 - 20x) + (13x^3 - 39x - 52) \qquad (\diamond)$$
$$= 2x^5 + 5x^4 + 7x^3 - 23x^2 - 59x - 52.$$

So, the degree of the product is 5.

$\square$

We can also multiply two polynomials by "stacking" them, just as we do for long multiplication of numbers. Here's an example:

$$
\begin{array}{rrrrrrl}
 & & x^3 & & -3x & -4 & \\
\times & & & 2x^2 & +5x & +13 & \\
\hline
 & & +13x^3 & & -39x & -52 & (\clubsuit) \\
 & +5x^4 & & -15x^2 & -20x & & (\clubsuit) \\
+ \;\; 2x^5 & & -6x^3 & -8x^2 & & & (\clubsuit) \\
\hline
2x^5 & +5x^4 & +7x^3 & -23x^2 & -59x & -52 &
\end{array}
$$

The lines marked ($\clubsuit$) in our "stacked" multiplication result from multiplying $x^3 - 3x - 4$ by each term, in turn, of $2x^2 + 5x + 13$. In other words, we're just doing the distributive property. We can see this clearly by noticing that the polynomials in these lines are the same as the polynomials that appear in the line marked ($\diamond$) in our distributive expansion above.

Notice in our final part above that $\deg f = 2$, $\deg g = 3$, and $\deg(f \cdot g) = 2 + 3 = 5$. Is this just a coincidence?

Problem 6.2: Prove that if f and g are nonzero polynomials, then $\deg(f \cdot g) = (\deg f) + (\deg g)$.

Solution for Problem 6.2: Let $\deg f = n$ and $\deg g = m$. To relate the degrees of $f(x)$ and $g(x)$ to $f(x) \cdot g(x)$, we start by writing $f(x)$ and $g(x)$ in general form:

$$f(x) = a_n x^n + a_{n-1} x^{n-1} + a_{n-2} x^{n-2} + \cdots + a_1 x + a_0,$$
$$g(x) = b_m x^m + b_{m-1} x^{m-1} + b_{m-2} x^{m-2} + \cdots + b_1 x + b_0.$$

Because the degree of $f(x)$ is n, we know that a_n is nonzero. (Otherwise, $f(x)$ would have no x^n term.) Similarly, we know that b_m is nonzero because the degree of $g(x)$ is m. When we take the product $f(x) \cdot g(x)$, the highest power of x will occur when we multiply the highest powers of x in $f(x)$ and $g(x)$ to get $(a_n x^n)(b_m x^m) = a_n b_m x^{n+m}$. Because a_n and b_m are nonzero, we know that $a_n b_m \neq 0$, so the product $f(x) \cdot g(x)$ has a term with x^{n+m}. All other terms in $f(x) \cdot g(x)$ have a lower power of x. Since x^{n+m} is the highest power of x in $h(x)$, the degree of h is $n + m$. Since the degree of f is n and the degree of g is m, we see that

$$(\deg f) + (\deg g) = \deg h.$$

This relationship holds even if $f(x)$ or $g(x)$ is just a nonzero constant, because the degree of a nonzero constant is 0. However, our rule above breaks if we try to define the degree of the polynomial 0. To see why, suppose $f(x) = 0$, $g(x) = x^2$, and $h(x) = f(x) \cdot g(x)$. Then, we have $h(x) = 0$. Our relationship above gives us

$$\deg f + 2 = \deg h,$$

so we have $(\deg 0) + 2 = (\deg 0)$. There's no value we can assign to $(\deg 0)$ to make this equation true, so we can't define the degree of 0.

> **WARNING!!** The degree of the polynomial 0 is undefined.
> ☢

□

> **Problem 6.3:**
>
> (a) Let $f(x) = 3x^3 - 4x^2 + ax - 11$ be a polynomial such that $f(1) = 2$. Find the value of a.
>
> (b) Suppose $(3x - 1)^7 = a_7 x^7 + a_6 x^6 + \cdots + a_1 + a_0$ for constants $a_0, a_1, \ldots, a_7$. Find $a_7 + a_6 + \cdots + a_1 + a_0$. (Source: AHSME)

Solution for Problem 6.3:

(a) Substituting $x = 1$ into the polynomial, we have $f(1) = 3 - 4 + a - 11 = a - 12$. Since $f(1) = 2$, we have $a - 12 = 2$, so $a = 14$.

(b) Multiplying out $(3x - 1)^7$ looks a little scary. However, a quick glance at part (a) gives us an idea. In part (a), we see that $f(1) = 3 - 4 + a - 11$. These numbers look familiar—they're the coefficients of $f(x)$! This is no coincidence; when we let $x = 1$, each term is equal to the coefficient of that term. Returning to the problem in this part, we see that letting $x = 1$ gives us the expression $a_7 + a_6 + \cdots + a_1 + a_0$ on the right, and our equation then is

$$(3 \cdot 1 - 1)^7 = a_7 + a_6 + \cdots + a_1 + a_0.$$

Evaluating the left side gives us $a_7 + a_6 + \cdots + a_1 + a_0 = 2^7 = 128$.

□

> **Concept:** Substituting a value for the variable in a polynomial can give us useful
> information about the coefficients of the polynomial. In particular, letting the variable equal 1 gives the sum of the coefficients of the polynomial.

Speaking of coefficients, we have a little more vocabulary before we continue. The term of a polynomial with the highest power of the variable is called the **leading term** and the coefficient of this term is the **leading coefficient**. If the leading coefficient of the polynomial is 1, then the polynomial is called a **monic polynomial**.

| Exercises |

6.1.1 Let $g(x) = 2x^2 + 5x + 7$. If $h(x)$ is a polynomial such that $h(x) = [g(x) + 3][g(x) + 9]$, then what is the degree of $h(x)$?

6.1.2 Let $f(x) = x^4 + x^2 - 2$ and $g(x) = 2x^3 + 3x^2 + 4x + 5$.

   (a) What is the degree of $f(x)$?

   (b) What is the degree of $g(x)$?

   (c) Find the degree of $f(g(x))$.

   (d) Find the degree of $f^4(x)$.

   (e) Find the sum of the coefficients in the expansion of $g(f(x))$.

6.1.3 If $ax^3 + bx^2 + cx + d = (x^2 + 2x - 8)(x - 3) - (x - 2)(x^2 + 5x + 4)$ for all values of x, then what is the value of $a + b + c + d$?

6.1.4 Find the degree of $f(g(x))$ given that f and g are polynomials such that the degree of f is m and the degree of g is n, where m and n are nonnegative integers.

6.2 Introduction to Polynomial Division

Problems

Problem 6.4: Suppose that a, b, and c are constants such that

$$(2x^2 + 3x + 7)(ax^2 + bx + c) = 2x^4 + 11x^3 + 9x^2 + 13x - 35.$$

   (a) Use the coefficient of x^4 to find a.

   (b) Use the coefficient of x^3 to find b.

   (c) Find c. Check your answer by expanding the product on the left side using your values of a, b, and c.

Problem 6.5: In this problem we divide $x^3 + 7x^2 - x + 11$ by $x - 5$. Let $g_1(x) = \dfrac{x^3 + 7x^2 - x + 11}{x - 5}$, so $(x - 5)g_1(x) = x^3 + 7x^2 - x + 11$.

   (a) Why must the highest degree term of $g_1(x)$ be x^2?

   (b) Let $g_1(x) = x^2 + g_2(x)$, so that we have

$$\frac{x^3 + 7x^2 - x + 11}{x - 5} = x^2 + g_2(x).$$

      Find $(x - 5)g_2(x)$ by multiplying this equation by $x - 5$. Find the highest degree term of $g_2(x)$.

   (c) Let $g_2(x) = ax + g_3(x)$, where ax is the highest degree term from part (b), so we have

$$\frac{x^3 + 7x^2 - x + 11}{x - 5} = x^2 + ax + g_3(x).$$

      Find $g_3(x)$.

   (d) Combine the previous parts to write $\dfrac{x^3 + 7x^2 - x + 11}{x - 5} = q(x) + \dfrac{r}{x - 5}$, where $q(x)$ is a polynomial and r is a constant.

Problem 6.4: Find the constants a, b, c such that

$$(2x^2 + 3x + 7)(ax^2 + bx + c) = 2x^4 + 11x^3 + 9x^2 + 13x - 35.$$

Solution for Problem 6.4: The only way to produce an x^4 term when expanding the product on the left side is when multiplying $(2x^2)(ax^2) = 2ax^4$. This must equal the $2x^4$ on the right, so we must have $2a = 2$, which gives us $a = 1$. Now we seek the constants b and c such that

$$(2x^2 + 3x + 7)(x^2 + bx + c) = 2x^4 + 11x^3 + 9x^2 + 13x - 35.$$

Looking at the x^4 term helped us find a, so let's take a look at the x^3 term. On the right, we have $11x^3$, and the only x^3 terms that will appear in the expansion on the left side are $(2x^2)(bx)$ and $(3x)(x^2)$. So, we must have

$$2bx^3 + 3x^3 = 11x^3.$$

Therefore, we must have $2b + 3 = 11$, so $b = 4$. We could move on to x^2 to find c, or we can use a bit of a shortcut by noticing that the only constant term in the expansion of $(2x^2 + 3x + 7)(ax^2 + bx + c)$ is $7c$, so we must have $7c = -35$ to match the constant term of the given quartic polynomial. This gives us $c = -5$. Checking our answer, we find that

$$(2x^2 + 3x + 7)(x^2 + 4x - 5) = 2x^4 + 11x^3 + 9x^2 + 13x - 35.$$

$\square$

This section is titled "Introduction to Polynomial Division," but the last problem is a multiplication problem. What does it have to do with division?

One way to think of division is as the "reverse" of multiplication. In other words, when trying to evaluate the division $35/7$, we can ask ourselves, "By what must we multiply 7 to get 35?" Because $7 \times 5 = 35$, we know that $35/7 = 5$. In much the same way, because our work in Problem 6.4 tells us that

$$(2x^2 + 3x + 7)(x^2 + 4x - 5) = 2x^4 + 11x^3 + 9x^2 + 13x - 35,$$

we know that

$$\frac{2x^4 + 11x^3 + 9x^2 + 13x - 35}{2x^2 + 3x + 7} = x^2 + 4x - 5.$$

Let's try using this insight to perform a polynomial division.

Problem 6.5: Let $f(x) = x^3 + 7x^2 - x + 11$ and $d(x) = x - 5$. Find a polynomial $q(x)$ and a constant r such that

$$\frac{f(x)}{d(x)} = q(x) + \frac{r}{d(x)}.$$

Solution for Problem 6.5: Let

$$g_1(x) = \frac{x^3 + 7x^2 - x + 11}{x - 5}.$$

We hope to write $g_1(x)$ in the form $q(x) + r/(x - 5)$, where $q(x)$ is a polynomial and r is a constant. We multiply the equation by $x - 5$ to get rid of the denominator, and we have

$$(x - 5)g_1(x) = x^3 + 7x^2 - x + 11.$$

This means we seek the expression that we must multiply $x-5$ by to get x^3+7x^2-x+11. The leading term on the right is x^3, so we know the leading term of $(x-5)g_1(x)$ must also be x^3. So, we let $g_1(x) = x^2 + g_2(x)$ for some expression $g_2(x)$, since this makes the leading term of $(x-5)(x^2 + g_2(x))$ be x^3. So, now we have

$$(x-5)(x^2 + g_2(x)) = x^3 + 7x^2 - x + 11.$$

Expanding the product on the left gives

$$(x-5)x^2 + (x-5)g_2(x) = x^3 + 7x^2 - x + 11.$$

Subtracting $(x-5)x^2$ from both sides gives us

$$(x-5)g_2(x) = x^3 + 7x^2 - x + 11 - (x-5)x^2 = 12x^2 - x + 11.$$

We need a $12x^2$ term in the expansion of $(x-5)g_2(x)$ to match the $12x^2$ term in $12x^2 - x + 11$, so we let $g_2(x) = 12x + g_3(x)$, which gives us

$$(x-5)(12x + g_3(x)) = 12x^2 - x + 11.$$

Expanding the product on the left gives

$$(x-5)(12x) + (x-5)g_3(x) = 12x^2 - x + 11,$$

so we have

$$(x-5)g_3(x) = 12x^2 - x + 11 - (x-5)(12x) = 59x + 11.$$

We need a $59x$ term in the expansion of $(x-5)g_3(x)$ to match the $59x$ on the right, so we let $g_3(x) = 59 + g_4(x)$, which gives us

$$(x-5)(59 + g_4(x)) = 59x + 11.$$

Expanding the left side gives us

$$59(x-5) + (x-5)g_4(x) = 59x + 11.$$

Simplifying this equation, we have

$$(x-5)g_4(x) = 306.$$

Uh-oh. Clearly $g_4(x)$ isn't a polynomial. Did we make a mistake?

No! Dividing this equation by $x - 5$ gives us

$$g_4(x) = \frac{306}{x-5}.$$

If we look back through our steps, we have

$$\begin{aligned}
g_1(x) &= x^2 + g_2(x) \\
&= x^2 + 12x + g_3(x) \\
&= x^2 + 12x + 59 + g_4(x) \\
&= x^2 + 12x + 59 + \frac{306}{x-5}.
\end{aligned}$$

So, we have

$$\frac{x^3 + 7x^2 - x + 11}{x - 5} = x^2 + 12x + 59 + \frac{306}{x - 5},$$

which means our desired $q(x)$ and r are $q(x) = x^2 + 12x + 59$ and $r = 306$. $\square$

When we divide one integer into another, we have a quotient and a remainder. For example, 233 divided by 5 gives a quotient of 46 and a remainder of 3:

$$\frac{233}{5} = 46 + \frac{3}{5}.$$

Multiplying this equation by the divisor, we see that we can also express our division in terms of multiplication and addition:

$$233 = 5 \times 46 + 3.$$

The remainder of a division is always less than the divisor (the number we are dividing by).

In general, we can express the fact that n divided by d gives a quotient q with remainder r (where $0 \le r < d$) as

$$\frac{n}{d} = q + \frac{r}{d}.$$

Multiplying both sides by d gives us

$$n = qd + r.$$

So, when dividing an integer n by another integer d, we seek the integers q and r such that $n = qd + r$ and $0 \le r < d$.

Dividing polynomials is essentially the same game. When we divide the polynomial $f(x)$ by the polynomial $d(x)$, we seek polynomials $q(x)$ and $r(x)$ such that

$$\frac{f(x)}{d(x)} = q(x) + \frac{r(x)}{d(x)}.$$

Multiplying both sides by $d(x)$ gives us

$$f(x) = q(x) \cdot d(x) + r(x).$$

When dividing one integer into another, the remainder must be less than the divisor. For polynomials, our measure of the "size" of a polynomial is its degree. When dividing the polynomial $d(x)$ into $f(x)$, the remainder must either be 0, or the degree of the remainder must be less than the degree of the divisor.

> **Important:** When the polynomial $f(x)$ is divided by the polynomial $d(x)$, the quotient $q(x)$ and remainder $r(x)$ are the polynomials such that
>
> $$f(x) = q(x) \cdot d(x) + r(x)$$
>
> and either $r(x) = 0$ or $\deg r < \deg d$.

If the remainder is 0 when the polynomial $f(x)$ is divided by the polynomial $d(x)$, then we say that $f(x)$ is **divisible** by $d(x)$. Again, this is just like division of integers: if the remainder is 0 when the integer n is divided by an integer d, then n is divisible by d.

We've already seen a couple examples of polynomial division. In Problem 6.4, we found that

$$\frac{2x^4 + 11x^3 + 9x^2 + 13x - 35}{2x^2 + 3x + 7} = x^2 + 4x - 5,$$

so the quotient when $2x^4 + 11x^3 + 9x^2 + 13x - 35$ is divided by $2x^2 + 3x + 7$ is $x^2 + 4x - 5$, and the remainder is 0. Therefore, $2x^4 + 11x^3 + 9x^2 + 13x - 35$ is divisible by $2x^2 + 3x + 7$.

In Problem 6.5, we found that

$$\frac{x^3 + 7x^2 - x + 11}{x - 5} = x^2 + 12x + 59 + \frac{306}{x - 5}. \qquad (\spadesuit)$$

The quotient is $x^2 + 12x + 59$ and the remainder is 306.

> **WARNING!!** The remainder is 306, not $306/(x - 5)$, just as the remainder when 33
> ☢ is divided by 5 is 3, not 3/5.

Multiplying Equation ($\spadesuit$) by $x - 5$ gives us

$$x^3 + 7x^2 - x + 11 = (x - 5)(x^2 + 12x + 59) + 306.$$

We can check our work by expanding the product on the right side:

$$
\begin{aligned}
(x - 5)(x^2 + 12x + 59) + 306 &= x(x^2 + 12x + 59) - 5(x^2 + 12x + 59) + 306 \\
&= x^3 + 12x^2 + 59x - 5x^2 - 60x - 295 + 306 \\
&= x^3 + 7x^2 - x + 11.
\end{aligned}
$$

This matches our original $f(x)$, so our quotient and remainder are correct.

> **Concept:** Just as you can check integer division with multiplication and addition,
> 🔑 you can check polynomial division. If $f(x)$ divided by $d(x)$ gives a quotient
> of $q(x)$ and remainder of $r(x)$, we must have
> $$d(x) \cdot q(x) + r(x) = f(x).$$

You may have found a somewhat faster solution to Problem 6.5 on your own. However, we can use our solution above to build a method of long division for polynomials that is very much like the long division we perform to divide one integer into another. We'll now walk through our solution to Problem 6.5 again, and develop our long division method as we go.

Our initial set-up is just like integer long division; this is shown at right. When we are finished, line (6.1) will contain the quotient. On line (6.2), we place the polynomial we are dividing inside the division bracket and we place the divisor on the left.

$$
\begin{array}{r}
 \hspace{4cm} (6.1) \\
x - 5 \,\overline{\smash{)}\, x^3 + 7x^2 - x + 11} \quad (6.2)
\end{array}
$$

Our first step in our solution to Problem 6.5 was to find the highest degree term of $g_1(x)$ such that

$$(x - 5)g_1(x) = x^3 + 7x^2 - x + 11.$$

Since x times the highest degree term of $g_1(x)$ must equal x^3, the highest degree term of $g_1(x)$ is $x^3/x = x^2$. This is the leading term of our quotient.

In our long division, we find the leading term of our quotient by dividing the leading term of the divisor $(x - 5)$ into the leading term of $x^3 + 7x^2 - x + 11$. We place the resulting x^2 in line (6.1) as the leading term of our quotient.

$$
\begin{array}{r}
x^2 \qquad\qquad\qquad\quad (6.1) \\
x - 5 \overline{\smash{\big)}\, x^3 + 7x^2 - x + 11}\ (6.2)
\end{array}
$$

Next in our solution to Problem 6.5, we write $g_1(x) = x^2 + g_2(x)$, so $(x-5)(x^2 + g_2(x)) = x^3 + 7x^2 - x + 11$. Expanding the left side gives us $(x - 5)x^2 + (x - 5)g_2(x) = x^3 + 7x^2 - x + 11$. Rearranging this gives us

$$(x - 5)g_2(x) = x^3 + 7x^2 - x + 11 - (x - 5)x^2 = 12x^2 - x + 11. \quad (\spadesuit)$$

So, $(x - 5)g_2(x)$ gives us what's left when $(x - 5)x^2$ is subtracted from $x^3 + 7x^2 - x + 11$.

In our long division, we multiply the quotient term we just found (x^2) by the divisor $(x-5)$, and subtract this product from x^3+7x^2-x+11 on line (6.3). The subtraction is exactly the same as the subtraction on line ($\spadesuit$). Notice that the resulting difference, $12x^2 - x + 11$, on line (6.4) is the same as the final right side on line ($\spadesuit$) above.

$$
\begin{array}{r}
x^2 \qquad\qquad\qquad\qquad (6.1) \\
x - 5 \overline{\smash{\big)}\, x^3\ +\ 7x^2\ -\ x\ +\ 11}\ (6.2) \\
\underline{x^3\ -\ 5x^2 \qquad\qquad}\ (6.3) \\
12x^2\ -\ x\ +\ 11\ (6.4)
\end{array}
$$

Now, we simply repeat the process, finding the highest degree term of $g_2(x)$ using our equation

$$(x - 5)g_2(x) = 12x^2 - x + 11.$$

Notice that we are essentially just dividing $x - 5$ into $12x^2 - x + 11$ in this step. The highest degree term of $g_2(x)$ times x gives us $12x^2$, so the highest degree term of $g_2(x)$ is $12x^2/x = 12x$.

In our long division, we are also now dividing $x - 5$ into the polynomial $12x^2 - x + 11$ on line (6.4). We find the next term of the quotient by dividing the leading term of the divisor $x - 5$ into the leading term of our difference $12x^2 - x + 11$ on line (6.4). This gives us $12x^2/x = 12x$, which we place as our next term of the quotient in line (6.1).

$$
\begin{array}{r}
x^2\ +\ 12x \qquad\qquad\quad (6.1) \\
x - 5 \overline{\smash{\big)}\, x^3\ +\ 7x^2\ -\ x\ +\ 11}\ (6.2) \\
\underline{x^3\ -\ 5x^2 \qquad\qquad}\ (6.3) \\
12x^2\ -\ x\ +\ 11\ (6.4)
\end{array}
$$

Next up in our solution to Problem 6.5, we write $g_2(x) = 12x + g_3(x)$, so we have

$$(x-5)(12x + g_3(x)) = 12x^2 - x + 11.$$

As before, we expand the left side to get $(x-5)(12x) + (x-5)g_3(x) = 12x^2 - x + 11$. Rearranging then gives

$$(x-5)g_3(x) = 12x^2 - x + 11 - (x-5)(12x) = 59x + 11. \quad (\clubsuit)$$

So, $(x-5)g_3(x)$ gives us what's left when $(x-5)(12x)$ is subtracted from $12x^2 - x + 11$.

Now back to our long division. We multiply the quotient term we just found ($12x$) by the divisor ($x-5$), and subtract this product from $12x^2 - x + 11$ on line (6.4). The subtraction is exactly the same as the subtraction on line ($\clubsuit$). Notice that the resulting difference on line (6.6), $59x + 11$, is the same as the final right side on line ($\clubsuit$) above. This step should look very familiar. It's essentially the same process that produced line (6.4).

$$
\begin{array}{r}
x^2 + 12x \qquad\qquad\quad (6.1) \\[2pt]
\hline
x-5\,\big)\,x^3 + 7x^2 - x + 11 \quad (6.2) \\[2pt]
\underline{x^3 - 5x^2} \qquad\qquad\qquad (6.3) \\[2pt]
12x^2 - x + 11 \quad (6.4) \\[2pt]
\underline{12x^2 - 60x} \qquad\quad (6.5) \\[2pt]
59x + 11 \quad (6.6)
\end{array}
$$

You know what's coming next. Our equation $(x-5)g_3(x) = 59x + 11$ tells us that the highest degree term of $g_3(x)$ is $59x/x = 59$. So, we have $g_3(x) = 59 + g_4(x)$, and $(x-5)(59 + g_4(x)) = 59x + 11$. Expanding the left side and rearranging gives

$$(x-5)g_4 = 59x + 11 - (x-5)(59) = 306.$$

We can't divide anymore, so 306 is our remainder and our quotient is $x^2 + 12x + 59$.

We continue with our long division by dividing $x - 5$ into $59x + 11$ on line (6.6). This tells us the next term of our quotient is $59x/x = 59$, which we place on line (6.1). We then subtract the product of this term and the divisor ($59(x-5)$) on line (6.7), which gives us a difference of 306 on line (6.8). We can't divide any further, because the leading term of the divisor doesn't divide into the leading term of line (6.8). So, line (6.1) gives us our quotient and line (6.8) gives us our remainder, and we have

$$
\begin{array}{r}
x^2 + 12x + 59 \qquad\quad (6.1) \\[2pt]
\hline
x-5\,\big)\,x^3 + 7x^2 - x + 11 \quad (6.2) \\[2pt]
\underline{x^3 - 5x^2} \qquad\qquad\qquad\quad (6.3) \\[2pt]
12x^2 - x + 11 \quad (6.4) \\[2pt]
\underline{12x^2 - 60x} \qquad\qquad\quad (6.5) \\[2pt]
59x + 11 \quad (6.6) \\[2pt]
\underline{59x - 295} \quad (6.7) \\[2pt]
306 \quad (6.8)
\end{array}
$$

$$\frac{x^3 + 7x - x + 11}{x - 5} = x^2 + 12x + 59 + \frac{306}{x - 5}.$$

Now, it's your turn to try long division. Don't be intimidated by more complicated divisors such as $2x - 1$ or $x^2 - 3x + 5$. The long division process for these works exactly the same way it did above for $x - 5$. At each step, we are dividing the divisor into a polynomial. We divide the leading term of the polynomial by the leading term of our divisor to get the next term in the quotient. Then, we subtract the product of this next term and the divisor from the polynomial. This gives us a new polynomial to divide by the divisor in our next step.

Problems

Problem 6.6: Find the quotient and remainder when $x + 6$ is divided into $3x^2 + 3x - 7$.

Problem 6.7: Steve divided $2x - 1$ into $6x^4 - 7x^3 + 22x^2 - 24x - 13$ and came up with a quotient of $6x^3 - x^2 + 21x - 3$ and a remainder of -16. Looking at his answer, he immediately knew he had made a mistake. How could he quickly tell he made a mistake, and what are the correct quotient and remainder?

Problem 6.8: Steve's at it again. He divides $x^2 - 3x + 5$ into $x^4 - 3x^3 + 7x - 1$, and gets a quotient of x with a remainder of $2x - 1$. Again, he immediately realizes he has made a mistake. How could he quickly tell he made a mistake, and what are the correct quotient and remainder?

Problem 6.9: In this problem we explore why the quotient and remainder are unique when one polynomial is divided into another. Suppose we are dividing the polynomial $d(x)$ into $f(x)$. Let $q_1(x)$ and $r_1(x)$ be polynomials such that either $r_1(x) = 0$ or $\deg r_1 < \deg d$, and

$$f(x) = d(x) \cdot q_1(x) + r_1(x).$$

Let $q_2(x)$ and $r_2(x)$ also be polynomials such that either $r_2(x) = 0$ or $\deg r_2 < \deg d$, and

$$f(x) = d(x) \cdot q_2(x) + r_2(x).$$

We will show that $q_2(x)$ and $q_1(x)$ must be the same, and that $r_2(x)$ and $r_1(x)$ must be the same.

(a) Eliminate $f(x)$ from our two equations by subtracting the second from the first.

(b) Rearrange the equation in part (a) by putting the remainders on the right side and the other terms on the left side.

(c) Compare the degrees of both sides of the equation from part (b) to explain why $q_1(x)$ and $q_2(x)$ must be the same, and $r_1(x)$ and $r_2(x)$ must be the same.

Problem 6.6: Find the quotient and remainder when $x + 6$ is divided into $3x^2 + 3x - 7$.

Solution for Problem 6.6: We divide the leading term of $x + 6$ into the leading term of $3x^2+3x-7$ to get the first term of our quotient, $3x^2/x = 3x$. We place this term on line (6.6.1), and then place the product of this term and the divisor on line (6.6.3). We then subtract this product on line (6.6.3) from $3x^2 + 3x - 7$ to get line (6.6.4). Then, we divide the leading term of $x + 6$ into the leading term on line (6.6.4), which gives us the next term of our quotient, -15. We place this term on line (6.6.1), then subtract the product of this term and $x + 6$ from line (6.6.4). The degree of the difference (line (6.6.6)) is less than the degree of $x + 6$, so we can't divide anymore.

$$
\begin{array}{r}
3x \quad - \quad 15 \qquad (6.6.1)\\
x+6\overline{\smash{\big)}\,3x^2 + 3x - 7} \quad (6.6.2)\\
3x^2 + 18x \qquad (6.6.3)\\
\hline
-15x - 7 \quad (6.6.4)\\
-15x - 90 \quad (6.6.5)\\
\hline
83 \quad (6.6.6)
\end{array}
$$

The quotient is $3x - 15$ and the remainder is 83. Checking our answer, we have

$$(3x - 15)(x + 6) + 83 = 3x^2 + 3x - 90 + 83 = 3x^2 + 3x - 7,$$

as expected. $\square$

> **Problem 6.7:** Steve divided $2x - 1$ into $6x^4 - 7x^3 + 22x^2 - 24x - 13$ and came up with a quotient of $6x^3 - x^2 + 21x - 3$ and a remainder of -16. Looking at his answer, he immediately knew he had made a mistake. How could he quickly tell he made a mistake, and what are the correct quotient and remainder?

Solution for Problem 6.7: If Steve's answer is correct, then we must have

$$6x^4 - 7x^3 + 22x^2 - 24x - 13 = (6x^3 - x^2 + 21x - 3)(2x - 1) - 16.$$

We don't have to multiply out the entire right side to see that the two sides are not the same. The leading term of the right side is clearly $12x^4$, which doesn't match the leading term on the left. So, Steve made a mistake. Let's figure out what the answer should have been. (As an extra challenge, see if you can figure out how Steve made his mistake.)

As before, we form our quotient by dividing the leading term of the divisor into the leading terms of each of the polynomials on lines (6.7.2), (6.7.4), (6.7.6), and (6.7.8). On lines (6.7.3), (6.7.5), (6.7.7), and (6.7.9), we subtract the product of the divisor and each successive term of the quotient, thereby creating the next polynomial into which we divide $2x - 1$. The degree of the polynomial we produce on line (6.7.10) is less than the degree of the divisor, so we can't divide any further (and line (6.7.10) is our remainder).

$$
\begin{array}{r}
3x^3 - 2x^2 + 10x - 7 \quad (6.7.1) \\
2x - 1 \overline{\smash{\big)}\ 6x^4 - 7x^3 + 22x^2 - 24x - 13} \quad (6.7.2) \\
\underline{6x^4 - 3x^3} \qquad\qquad\qquad\qquad (6.7.3) \\
-4x^3 + 22x^2 - 24x - 13 \quad (6.7.4) \\
\underline{-4x^3 + 2x^2} \qquad\qquad\qquad (6.7.5) \\
20x^2 - 24x - 13 \quad (6.7.6) \\
\underline{20x^2 - 10x} \qquad\qquad (6.7.7) \\
-14x - 13 \quad (6.7.8) \\
\underline{-14x + 7} \quad (6.7.9) \\
-20 \quad (6.7.10)
\end{array}
$$

So, the quotient is $3x^3 - 2x^2 + 10x - 7$ and the remainder is -20. We can check this in the usual way:

$$(2x - 1)(3x^3 - 2x^2 + 10x - 7) - 20 = 2x(3x^3 - 2x^2 + 10x - 7) - 1(3x^3 - 2x^2 + 10x - 7) - 20$$
$$= 6x^4 - 4x^3 + 20x^2 - 14x - 3x^3 + 2x^2 - 10x + 7 - 20$$
$$= 6x^4 - 7x^3 + 22x^2 - 24x - 13.$$

This matches the original polynomial we are dividing into on line (6.7.2), so our quotient and remainder are correct. $\square$

> **Problem 6.8:** Steve's at it again. He divides $x^2 - 3x + 5$ into $x^4 - 3x^3 + 7x - 1$, and gets a quotient of x with a remainder of $2x - 1$. Again, he immediately realizes he has made a mistake. How could he quickly tell he made a mistake, and what are the correct quotient and remainder?

Solution for Problem 6.8: If Steve is correct, then we must have $x^4 - 3x^3 + 7x - 1 = (x^2 - 3x + 5)(x) + 2x - 1$, but the degree of the right side is not the same as the degree of the left side. Therefore, the two sides cannot be the same.

We'll have to fix Steve's mistake. Before we get started with our division, we notice that the polynomial we are dividing has no x^2 term. (Hmmm.... Maybe that's what Steve overlooked.)

> **WARNING!!** When dividing one polynomial by another, be careful not to overlook "missing" terms. These terms aren't really missing—their coefficients are 0. Include such terms in your long division by writing them with a coefficient of 0.

Even though our divisor is not linear, we can use long division just as we have done before. To find the first term of our quotient, we divide the leading term of the divisor into the leading term of the polynomial on line (6.8.2). This gives us $x^4/x^2 = x^2$ as the first term of the quotient, and we place this term on line (6.8.1). We then subtract the product $x^2(x^2 - 3x + 5)$ on line (6.8.3). Notice that both the x^4 and x^3 terms

$$
\begin{array}{r}
x^2 \qquad\quad - 5 \qquad\qquad (6.8.1) \\
x^2 - 3x + 5 \,\overline{\big)\, x^4 - 3x^3 + 0x^2 + 7x - 1} \quad (6.8.2) \\
\underline{x^4 - 3x^3 + 5x^2} \qquad\qquad\qquad (6.8.3) \\
-5x^2 + 7x - 1 \quad (6.8.4) \\
\underline{-5x^2 + 15x - 25} \quad (6.8.5) \\
-8x + 24 \quad (6.8.6)
\end{array}
$$

are eliminated. So, when we divide the leading term of the divisor into the leading term of the difference on line (6.8.4), the next term of our quotient is a constant term, $-5x^2/x^2 = -5$, not a linear term. We then subtract $(-5)(x^2 - 3x + 5)$ on line (6.8.5). The resulting difference on line (6.8.6) has a smaller degree than the divisor, so we can't divide anymore.

The quotient is $x^2 - 5$ and the remainder is $-8x + 24$. As before, we check our work:

$$
\begin{aligned}
(x^2 - 3x + 5)(x^2 - 5) - 8x + 24 &= (x^2 - 3x + 5)(x^2) + (x^2 - 3x + 5)(-5) - 8x + 24 \\
&= x^4 - 3x^3 + 5x^2 - 5x^2 + 15x - 25 - 8x + 24 \\
&= x^4 - 3x^3 + 7x - 1.
\end{aligned}
$$

□

Throughout this section, we have assumed that it is always possible to find a quotient, $q(x)$, and remainder, $r(x)$, such that $f(x) = q(x) \cdot d(x) + r(x)$ and either $\deg r < \deg d$ or $r(x) = 0$. Our long division process makes it intuitively obvious why this is true, but we won't go through all the details proving this fact. We will, however, prove that there's not more than one possible quotient and remainder for each polynomial division.

> **Problem 6.9:** Show that for any two polynomials $f(x)$ and $d(x)$, where $d(x) \neq 0$, there is not more than one pair of polynomials $q(x)$ and $r(x)$ such that $f(x) = q(x) \cdot d(x) + r(x)$, and either $r(x) = 0$ or $\deg r < \deg d$.

Solution for Problem 6.9: Suppose we have

$$ f(x) = q_1(x) \cdot d(x) + r_1(x) $$

and

$$ f(x) = q_2(x) \cdot d(x) + r_2(x), $$

where $\deg r_1 < \deg d$ (or $r_1(x) = 0$) and $\deg r_2 < \deg d$ (or $r_2(x) = 0$).

Our two equations have a lot of different polynomials in them. We can at least eliminate $f(x)$ by subtracting the second equation from the first. This gives us

$$ 0 = q_1(x) \cdot d(x) - q_2(x) \cdot d(x) + r_1(x) - r_2(x). $$

We bring the two terms with $d(x)$ to the left and factor $d(x)$ out of both terms to give

$$d(x)(q_2(x) - q_1(x)) = r_1(x) - r_2(x).$$

The left side of this equation either is 0 or has degree at least $\deg d$. However, because the degree of a remainder cannot equal or exceed that of the divisor, the degrees of $r_1(x)$ and $r_2(x)$ cannot equal or exceed $\deg d$, which means that $\deg(r_1 - r_2) \leq \deg d$ or $r_1(x) - r_2(x) = 0$. Therefore, the right side of the equation either is 0 or has degree less than $\deg d$. Combining this with our observation about the degree of the left side, we see that the only way the two sides are equal for all x is if both sides are 0.

Since $r_1(x) - r_2(x) = 0$, we have $r_1(x) = r_2(x)$. Since $d(x)(q_2(x) - q_1(x)) = 0$ and $d(x)$ is a nonzero polynomial, we must have $q_2(x) - q_1(x) = 0$, so $q_2(x) = q_1(x)$ for all x. (This glosses over the possibility that $q_2(x) = q_1(x)$ for all x *except where* $d(x) = 0$. That this is impossible is a consequence of the Identity Theorem, which we study in Section 7.6.)

We have therefore shown that for any two polynomials $f(x)$ and $d(x)$, there cannot be more than one pair of polynomials $q(x)$ and $r(x)$ such that either $r(x) = 0$ or $\deg r < \deg d$, and

$$f(x) = q(x) \cdot d(x) + r(x).$$

Note that we've only proved that there is no more than one pair of polynomials $q(x)$ and $r(x)$ that satisfy these conditions. We haven't proved that these polynomials $q(x)$ and $r(x)$ must always exist, though our process for polynomial division described earlier in this section gives us some intuition for why they must exist. $\square$

> **Important:** The quotient and remainder in polynomial division are unique. In other words, for any two polynomials $f(x)$ and $d(x)$, there's exactly one pair of polynomials $q(x)$ and $r(x)$ such that either $r(x) = 0$ or $\deg r < \deg d$, and
>
> $$f(x) = q(x) \cdot d(x) + r(x).$$

Exercises

6.2.1 Find the quotient and remainder for each of the following polynomial divisions:

(a) $x^2 - 19x + 17$ divided by $x + 7$

(b) $3x^4 + 13x^2 - 9x + 122$ divided by $x^2 + x + 4$

(c) $4x^3 + 2x - 1$ divided by $2x + 3$

(d) $19x^5$ divided by $x - 1$

6.2.2 Find a constant c such that there is no remainder when $x^3 + cx^2 + 4x - 21$ is divided by $x - 3$.

6.2.3 Find a if the remainder is a constant when $x^3 + 3x^2 + ax + 13$ is divided by $x^2 + 3x - 2$.

6.2.4 Teresa divides $3x^4 + 2x^3 - 7x^2 + 4x - 1$ by $x + 2$ and gets a quotient of $3x^3 - 4x^2 + x + 2$ and a remainder of 5. How can Teresa quickly realize that she made a mistake without performing the division again, and without multiplying $x + 2$ by her quotient?

6.3 Synthetic Division

Dividing polynomials by binomials of the form $x - a$ will prove to be very useful. But long division is, well, pretty long. Fortunately, there's a more compact way to divide a linear binomial into a polynomial. We call this method **synthetic division**.

We'll walk through an example of synthetic division, then let you try.

$$x - 5 \overline{\smash{\big)}\ x^3 + 7x^2 - x + 11} \quad \text{(6.2)}$$
(6.1)

$$
\begin{array}{c|cccc}
5 & 1 & 7 & -1 & 11 & (\heartsuit) \\
  &   &   &   &   & (\diamond) \\
\hline
  &   &   &   &   & (\spadesuit)
\end{array}
$$

Above, we have the initial set-up for long division on the left and for synthetic division on the right. When we divide a polynomial by $x - a$ using synthetic division, we start by placing a to the left of our division bracket and the coefficients of the polynomial we are dividing (including coefficients of 0 for any missing terms) inside the bracket.

> **WARNING!!** When we divide by $x - a$ using synthetic division, we put a to the left of the division bracket, not $-a$. So, for example, when we divide by $x - 5$, we put 5 to the left of the division bracket.

We have one column inside the bracket for each power of the variable in the polynomial, just as we have one column in our long division for each power of the variable. The row ($\spadesuit$) is used for our quotient and remainder.

$$1x^2$$
$$x - 5 \overline{\smash{\big)}\ x^3 + 7x^2 - x + 11} \quad \text{(6.2)}$$
(6.1)

$$
\begin{array}{c|cccc}
5 & 1 & 7 & -1 & 11 & (\heartsuit) \\
  &   &   &   &   & (\diamond) \\
\hline
  & 1 &   &   &   & (\spadesuit)
\end{array}
$$

Because we are dividing by $x - 5$, the leading term of the quotient has the same coefficient as the leading term of the polynomial we are dividing. So, we copy the leading coefficient of the polynomial we are dividing to the quotient row ($\spadesuit$).

> **WARNING!!** Synthetic division only works when the coefficient of the linear term in the divisor is 1.

Later in this section, we'll learn how to handle divisors in which the coefficient of the linear term is not 1.

> **Extra!** Synthetic division is also occasionally referred to as **Ruffini's rule**, after Italian mathematician **Paolo Ruffini**, who was supposedly one of the first to use this method of polynomial division. Though Ruffini was appointed professor of the foundations of analysis (a branch of mathematics) at the University of Modena in 1788 when he was only 23, he also held degrees in philosophy, medicine, and surgery.

$$\begin{array}{r} 1x^2 \hspace{3.2cm} (6.1) \\ x-5\overline{\smash{\big)}\,x^3 + 7x^2 - x + 11} \quad (6.2) \\ \underline{x^3 - 5x^2} \hspace{2.2cm} (6.3) \end{array}$$

$$\begin{array}{r|rrrr} 5 & 1 & 7 & -1 & 11 \quad (\heartsuit) \\ & & 5 & & \quad (\diamond) \\ \hline & 1 & & & \quad (\spadesuit) \end{array}$$

In the long division above, after finding the first term of the quotient, we will subtract $x^2(x-5)$ from the polynomial on line (6.3) above. We know that the x^3 terms will cancel in this subtraction, so we focus on the x^2 terms. Notice how the $-5x^2$ term is formed: it is the product of the term we just put in the quotient (x^2) and the constant term of the divisor (-5). Therefore, the coefficient equals the product of the coefficient of the term we just put in the quotient and the constant term of the divisor.

Turning to our synthetic division, the number to the left of the line is the negative of the constant term of the divisor, and the number we just placed in the quotient row ($\spadesuit$) is the coefficient of the term we just found in the quotient. So, we place the product of these two numbers, $5 \cdot 1 = 5$, in the next column on row ($\diamond$). The 5 we place in this row corresponds to the $-5x^2$ in line (6.3) of our long division. The 5 in row ($\diamond$) is the *opposite of* the coefficient of x^2 in line (6.3) because the sign of the number to the left of the line in the synthetic division is the *opposite of* the constant term of the divisor.

$$\begin{array}{r} 1x^2 + 12x \hspace{2.7cm} (6.1) \\ x-5\overline{\smash{\big)}\,x^3 + 7x^2 - x + 11} \quad (6.2) \\ \underline{x^3 - 5x^2} \hspace{2.6cm} (6.3) \\ 12x^2 - x + 11 \quad (6.4) \end{array}$$

$$\begin{array}{r|rrrr} 5 & 1 & 7 & -1 & 11 \quad (\heartsuit) \\ & & 5 & & \quad (\diamond) \\ \hline & 1 & 12 & & \quad (\spadesuit) \end{array}$$

In the long division, we subtract $-5x^2$ on line (6.3) from the $7x^2$ on line (6.2) to give us $12x^2$ on line (6.4). This is the leading term of the next polynomial into which we divide $x - 5$. So, the next term of our quotient is $12x^2/x = 12x$.

Turning to our synthetic division, the 5 in row ($\diamond$) is the opposite of the corresponding coefficient of x^2 in line (6.3). Therefore, instead of subtracting the 5 in row ($\diamond$) from the 7 in row ($\heartsuit$), we *add* the 5 in row ($\diamond$) and the 7 in row ($\heartsuit$) to produce the 12 in row ($\spadesuit$). This represents adding $7x^2 + 5x^2$ (which is the same as the subtraction $7x^2 - (-5x^2)$) to get $12x^2$, and it indicates that the coefficient of this sum is the next term in our quotient.

Sidenote: In addition to preferring synthetic division because it's more compact and faster than long division, we also prefer it because it involves addition at each step rather than subtraction. We're less likely to make sign errors or other arithmetic errors with addition than subtraction.

That the coefficient of the leading term of our polynomial on line (6.4) is also the coefficient of the next term in the quotient is a direct result of the fact that the coefficient of the linear term of the divisor $x - 5$ is 1. This is why we must have this coefficient equal to 1 in order to use synthetic division.

$$
\begin{array}{r}
1x^2 + 12x \qquad\qquad\quad (6.1)\\
x-5\,\big|\,\overline{x^3 + 7x^2 - \ x + 11}\ \ (6.2)\\
\underline{x^3 - 5x^2}\qquad\qquad (6.3)\\
12x^2 - \ x + 11\ \ (6.4)\\
\underline{12x^2 - 60x}\qquad (6.5)
\end{array}
$$

$$
\begin{array}{r|rrrr}
5 & 1 & 7 & -1 & 11 & (\heartsuit)\\
  &   & 5 & 60 &   & (\diamond)\\ \hline
  & 1 & 12&   &   & (\spadesuit)
\end{array}
$$

We repeat our steps to produce the next term of the quotient. In the long division, we subtract $12x(x-5) = 12x^2 - (5)(12)x$ on line (6.5), which is the same as adding $-12x^2 + 60x$. In our synthetic division, we multiply the 5 to the left of the line by the 12 in line ($\spadesuit$) and place the product in the next column of line ($\diamond$).

$$
\begin{array}{r}
1x^2 + 12x + 59 \qquad\qquad (6.1)\\
x-5\,\big|\,\overline{x^3 + 7x^2 - \ x + 11}\ \ (6.2)\\
\underline{x^3 - 5x^2}\qquad\qquad\quad (6.3)\\
12x^2 - \ x + 11\ \ (6.4)\\
\underline{12x^2 - 60x}\qquad\quad (6.5)\\
59x + 11\ \ (6.6)
\end{array}
$$

$$
\begin{array}{r|rrrr}
5 & 1 & 7 & -1 & 11 & (\heartsuit)\\
  &   & 5 & 60 &   & (\diamond)\\ \hline
  & 1 & 12& 59 &   & (\spadesuit)
\end{array}
$$

Our next step is subtracting line (6.5) from line (6.4). The x^2 terms cancel, and subtracting $-1x-(-60x)$ gives us $59x$ as the leading term of our polynomial on line (6.6). Turning to our synthetic division, we *add* the -1 in row ($\heartsuit$) to the 60 below it in row ($\diamond$) to produce the 59 in row ($\spadesuit$). This represents our adding $-1x + 60x$ (which is the same as $-1x - (-60x)$) to get $59x$, and it indicates that the coefficient of the next term in our quotient is 59.

$$
\begin{array}{r}
1x^2 + 12x + 59 \qquad\qquad\quad (6.1)\\
x-5\,\big|\,\overline{x^3 + 7x^2 - \ x + \ 11}\ \ (6.2)\\
\underline{x^3 - 5x^2}\qquad\qquad\qquad (6.3)\\
12x^2 - \ x + \ 11\ \ (6.4)\\
\underline{12x^2 - 60x}\qquad\qquad (6.5)\\
59x + \ 11\ \ (6.6)\\
\underline{59x - 295}\ \ (6.7)\\
306\ \ (6.8)
\end{array}
$$

$$
\begin{array}{r|rrr|r}
5 & 1 & 7 & -1 & 11 & (\heartsuit)\\
  &   & 5 & 60 & 295 & (\diamond)\\ \hline
  & 1 & 12& 59 & 306 & (\spadesuit)
\end{array}
$$

We repeat the process one more time to find the remainder. Subtracting $59(x-5)$ on line (6.7) is the same as adding $-59x + 295$. The x terms cancel, and we have $11 + 295 = 306$ as our remainder. Over in the synthetic division, multiplying 5 by the 59 in row ($\spadesuit$) gives us the 295 in row ($\diamond$) that corresponds to the -295 we subtract in row (6.7) in our long division. Adding the 11 in row ($\heartsuit$) to the 295 in row ($\diamond$) to get 306 in row ($\spadesuit$) corresponds to our subtraction $11 - (-295)$ to find the remainder in our long division.

We typically separate the remainder from the quotient with a vertical bar in row ($\spadesuit$) of the synthetic division. This helps us avoid making a mistake when reading the quotient and remainder from row (♠). The number after the bar is our remainder, and the other numbers are the coefficients of the quotient. The number just before the bar is the constant term. The next number to the left is the coefficient of the linear term, the next number to the left is the quadratic term's coefficient, and so on. So, our synthetic division gives us a quotient of $x^2 + 12x + 59$ with a remainder of 306. □

Below is one more example, in which we divide $x - 2$ into $3x^4 - 9x^3 + 2x^2 + 17x - 24$.

$$
\begin{array}{r}
3x^3 - 3x^2 - 4x + 9 \\
x-2\overline{\smash{\big)}\,3x^4 - 9x^3 + 2x^2 + 17x - 24} \\
\underline{3x^4 - 6x^3} \\
-3x^3 + 2x^2 + 17x - 24 \\
\underline{-3x^3 + 6x^2} \\
-4x^2 + 17x - 24 \\
\underline{-4x^2 + 8x} \\
+9x - 24 \\
\underline{+9x - 18} \\
-6
\end{array}
$$

$$
\begin{array}{r|rrrrr}
2 & 3 & -9 & 2 & 17 & -24 & (\heartsuit) \\
 & & 6 & -6 & -8 & 18 & (\diamond) \\
\hline
 & 3 & -3 & -4 & 9 & -6 & (\spadesuit)
\end{array}
$$

Notice that all the numbers in our synthetic division, or their opposites, appear as coefficients (or constants) in the long division.

> **Concept:** Synthetic division is nothing new. It's just a more compact way to represent long division of a polynomial $f(x)$ by an expression of the form $x - a$, where a is a constant.

Problems

Problem 6.10: Divide $x + 5$ into $x^3 - 3x^2 + 5x - 2$ using long division, then using synthetic division.

Problem 6.11: In this problem we use synthetic division to divide $2x - 7$ into $f(x) = 4x^2 - 6x + 9$.
(a) Write $2x - 7$ in the form $c(x - a)$, where c and a are constants.
(b) Use synthetic division to divide your $x - a$ from part (a) into $f(x)$.
(c) Use your result from (b) to find the quotient and remainder when $f(x)$ is divided by $2x - 7$.

Problem 6.12: Perform the following divisions with synthetic division:
(a) $(x^3 - x + 7)/(x + 3)$
(b) $(3y^4 - 4y^3 + 5y^2 - 11y + 2)/(2 + 3y)$

Problem 6.10: Divide $x + 5$ into $x^3 - 3x^2 + 5x - 2$ using long division, then using synthetic division.

Solution for Problem 6.10: What's wrong with the synthetic division below:

Bogus Solution:

$$
\begin{array}{r|rrrr}
5 & 1 & -3 & 5 & -2 \\
  &   & 5 & 10 & 75 \\
\hline
  & 1 & 2 & 15 & 73 \\
\end{array}
$$

When we divide a polynomial by $x - a$ with synthetic division, we put a on the left of the division bracket. Here, we are dividing by $x + 5$, which is $x - (-5)$. So, we should put -5, not 5, to the left of the division bracket.

> **WARNING!!** When we divide a polynomial by $x - a$ with synthetic division, we put
> a on the left of the division bracket. So, when we divide by $x + c$, we
> put $-c$ to the left of the division bracket, not c.

Below are the correct long and synthetic divisions.

$$
\begin{array}{r}
x^2 - 8x + 45 \\
x + 5 \overline{\smash{\big)}\, x^3 - 3x^2 + 5x - 2} \\
\underline{x^3 + 5x^2} \\
-8x^2 + 5x - 2 \\
\underline{-8x^2 - 40x} \\
45x - 2 \\
\underline{45x + 225} \\
-227
\end{array}
$$

$$
\begin{array}{r|rrrr}
-5 & 1 & -3 & 5 & -2 \\
   &   & -5 & 40 & -225 \\
\hline
   & 1 & -8 & 45 & -227 \\
\end{array}
$$

The quotient is $x^2 - 8x + 45$ and the remainder is -227. □

Problem 6.11: Divide $2x - 7$ into $4x^2 - 6x + 9$.

Solution for Problem 6.11: Why is the following solution wrong:

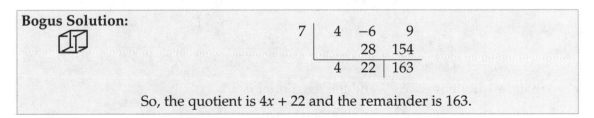

Bogus Solution:

$$
\begin{array}{r|rrr}
7 & 4 & -6 & 9 \\
  &   & 28 & 154 \\
\hline
  & 4 & 22 & 163 \\
\end{array}
$$

So, the quotient is $4x + 22$ and the remainder is 163.

We see that we've made a mistake when we check our answer, because we have

$$(2x - 7)(4x + 22) + 163 = 8x^2 + 16x + 9.$$

Uh-oh. If we had divided correctly, we should have gotten $4x^2 - 6x + 9$ when we multiplied $2x - 7$ by the quotient and added the remainder. What went wrong?

We forgot that synthetic division only works if the leading coefficient of the divisor is 1. So, are we stuck using long division on this problem?

No! We are performing the division

$$\frac{4x^2 - 6x + 9}{2x - 7}.$$

We would like to convert the division into an equivalent problem such that the leading coefficient of the divisor is 1. We can make the coefficient of x equal to 1 in the divisor by factoring 2 out of $2x - 7$ to give us $2(x - \frac{7}{2})$:

$$\frac{4x^2 - 6x + 9}{2x - 7} = \frac{4x^2 - 6x + 9}{2(x - \frac{7}{2})} = \frac{1}{2} \cdot \frac{4x^2 - 6x + 9}{x - \frac{7}{2}}.$$

Aha! We can divide $x - \frac{7}{2}$ into $4x^2 - 6x + 9$ using synthetic division.

Our synthetic division is shown at right. We appear to have a quotient of $4x + 8$ and a remainder of 37. Let's check our answer:

$$(2x - 7)(4x + 8) + 37 = 8x^2 - 12x - 19.$$

$$\begin{array}{r|rrr} \frac{7}{2} & 4 & -6 & 9 \\ & & 14 & 28 \\ \hline & 4 & 8 & 37 \end{array}$$

Uh-oh. Wrong again. But why?

Our synthetic division tells us that

$$\frac{4x^2 - 6x + 9}{x - \frac{7}{2}} = 4x + 8 + \frac{37}{x - \frac{7}{2}}.$$

However, we want to perform the division $(4x^2 - 6x + 9)/(2x - 7)$. Looking back at how we came up with $(4x^2 - 6x + 9)/(x - \frac{7}{2})$ tells us how to use the result of this division to find the desired quotient and remainder:

$$\begin{aligned} \frac{4x^2 - 6x + 9}{2x - 7} &= \frac{1}{2} \cdot \frac{4x^2 - 6x + 9}{x - \frac{7}{2}} \\ &= \frac{1}{2}\left(4x + 8 + \frac{37}{x - \frac{7}{2}}\right) \\ &= 2x + 4 + \frac{37}{2\left(x - \frac{7}{2}\right)} \\ &= 2x + 4 + \frac{37}{2x - 7}. \end{aligned}$$

Now, we see that the quotient is $2x + 4$ and the remainder is 37.

We've been wrong twice, so we better check our answer:

$$(2x - 7)(2x + 4) + 37 = 4x^2 - 6x + 9.$$

Phew. It works. □

Problem 6.11 shows us that we can use synthetic division to divide by a linear expression with a leading coefficient besides 1. However, we must first turn the division into an equivalent problem in which the divisor has leading coefficient 1. As we saw in our solution to Problem 6.11, we must be careful about the remainder. Here are two ways to make sure you have your division correct when dividing $f(x)$ by $d(x)$:

1. Write your answer in the form

$$\frac{f(x)}{d(x)} = q(x) + \frac{r(x)}{d(x)},$$

where $d(x)$ is the original divisor. In Problem 6.11, doing this enabled us to get the correct remainder.

2. Check your answer! Once you have found $q(x)$ and $r(x)$, expand $d(x) \cdot q(x) + r(x)$. If the result is not $f(x)$, then you made a mistake.

For one more example, we divide $9x^3 - 24x^2 + 12x - 14$ by $-3x - 1$. First, we convert the problem into a division by an expression with leading coefficient 1, by factoring -3 out of the divisor:

$$\frac{9x^3 - 24x^2 + 12x - 14}{-3x - 1} = \frac{9x^3 - 24x^2 + 12x - 14}{-3(x + \frac{1}{3})}$$

$$= -\frac{1}{3} \cdot \frac{9x^3 - 24x^2 + 12x - 14}{x + \frac{1}{3}}$$

We divide $x + \frac{1}{3}$ into $9x^3 - 24x^2 + 12x - 14$ at right. Note that we put $-\frac{1}{3}$ to the left of the division bracket, not $\frac{1}{3}$. Our synthetic division gives us

$$\begin{array}{r|rrrr} -\frac{1}{3} & 9 & -24 & 12 & -14 \\ & & -3 & 9 & -7 \\ \hline & 9 & -27 & 21 & -21 \end{array}$$

$$\frac{9x^3 - 24x^2 + 12x - 14}{x + \frac{1}{3}} = 9x^2 - 27x + 21 + \frac{-21}{x + \frac{1}{3}}.$$

So, we have

$$\frac{9x^3 - 24x^2 + 12x - 14}{-3x - 1} = -\frac{1}{3} \cdot \frac{9x^3 - 24x^2 + 12x - 14}{x + \frac{1}{3}}$$

$$= -\frac{1}{3}\left(9x^2 - 27x + 21 + \frac{-21}{x + \frac{1}{3}}\right)$$

$$= -3x^2 + 9x - 7 + \frac{1}{-3} \cdot \frac{-21}{x + \frac{1}{3}}$$

$$= -3x^2 + 9x - 7 + \frac{-21}{-3x - 1}.$$

Notice that the denominator at the end is our original divisor. So, when we divide $9x^3 - 24x^2 + 12x - 14$ by $-3x - 1$, the quotient is $-3x^2 + 9x - 7$ and the remainder is -21. We check our answer and find

$$(-3x - 1)(-3x^2 + 9x - 7) - 21 = 9x^3 - 24x^2 + 12x - 14,$$

so our quotient and remainder are correct.

We'll do a couple more for practice.

Problem 6.12: Perform the following divisions with synthetic division:

(a) $(x^3 - x + 7)/(x + 3)$

(b) $(3y^4 - 4y^3 + 5y^2 - 11y + 2)/(2 + 3y)$

Solution for Problem 6.12:

(a) What's wrong with this solution:

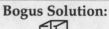

Bogus Solution:

$$
\begin{array}{r|rrr}
-3 & 1 & -1 & 7 \\
 & & -3 & 12 \\
\hline
 & 1 & -4 & \big|\ 19
\end{array}
$$

Our first clue that something is wrong is that our quotient is $x - 4$. When we divide a cubic by a linear expression, we should get a quadratic as a quotient. So, we made a mistake, but where?

Our mistake was leaving out the $0x^2$ term in $x^3 - x + 7$ in our division.

WARNING!! Be on the lookout for "missing terms" when doing synthetic division. They aren't really missing—the coefficients of these terms are 0. So, we include 0s for these coefficients in synthetic division.

The correct synthetic division is shown at right. We find

$$x^3 - x + 7 = (x + 3)(x^2 - 3x + 8) - 17.$$

$$
\begin{array}{r|rrrr}
-3 & 1 & 0 & -1 & 7 \\
 & & -3 & 9 & -24 \\
\hline
 & 1 & -3 & 8 & \big|\ -17
\end{array}
$$

So, the quotient is $x^2 - 3x + 8$ and the remainder is -17.

(b) First, we have to adjust for the fact that the coefficient of y in the divisor is not 1. We have

$$\frac{3y^4 - 4y^3 + 5y^2 - 11y + 2}{2 + 3y} = \frac{1}{3} \cdot \frac{3y^4 - 4y^3 + 5y^2 - 11y + 2}{y + \frac{2}{3}}.$$

From our synthetic division at right, we have

$$\frac{3y^4 - 4y^3 + 5y^2 - 11y + 2}{2 + 3y} = \frac{1}{3} \cdot \frac{3y^4 - 4y^3 + 5y^2 - 11y + 2}{y + \frac{2}{3}}$$

$$
\begin{array}{r|rrrrr}
-\frac{2}{3} & 3 & -4 & 5 & -11 & 2 \\
 & & -2 & 4 & -6 & \frac{34}{3} \\
\hline
 & 3 & -6 & 9 & -17 & \big|\ \frac{40}{3}
\end{array}
$$

$$= \frac{1}{3}\left(3y^3 - 6y^2 + 9y - 17 + \frac{40/3}{y + \frac{2}{3}}\right)$$

$$= y^3 - 2y^2 + 3y - \frac{17}{3} + \frac{1}{3} \cdot \frac{40/3}{y + \frac{2}{3}}$$

$$= y^3 - 2y^2 + 3y - \frac{17}{3} + \frac{40/3}{3y + 2}.$$

So, the quotient is $y^3 - 2y^2 + 3y - \frac{17}{3}$ and the remainder is $\frac{40}{3}$.

$\square$

Exercises

6.3.1 Find the quotient and remainder when $x^3 + 4x^2 - 5x + 19$ is divided by $x - 3$.

6.3.2   Check the two divisions we performed in Problem 6.12 by multiplying the quotient by the divisor, then adding the remainder.

6.3.3   Find the quotient and remainder when $x^5 - 23x^3 + 11x^2 - 14x + 20$ is divided by $x + 5$.

6.3.4   Find the quotient and remainder when $4x^4 - 10x^3 + 14x^2 + 7x - 19$ is divided by $2x - 1$.

6.3.5   Find the quotient and remainder when $3z^3 - 4z^2 - 14z + 3$ is divided by $3z + 5$.

6.3.6   Find the quotient and remainder when $x^4 + 3x^3 - x^2 + 7x - 1$ is divided by $2 - x$.

6.3.7   Stan divided $x^3 + 3x^2 - 7x + 5$ by $2x - 3$ using synthetic division. He found a quotient of $x^2 + 6x + 11$ and a remainder of 38. By looking at his answer, he immediately realized that he had made a mistake. How could he tell? What mistake did he probably make?

6.4   The Remainder Theorem

Problems

Problem 6.13: Let $f(x) = x^4 - 3x^3 + 7x^2 - x + 5$.
  (a) What is the remainder when $f(x)$ is divided by $x - 3$?
  (b) What is $f(3)$?

Problem 6.14: In this problem we prove the **Remainder Theorem** for polynomials. Suppose $q(x)$ and $r(x)$ are the quotient and remainder, respectively, when the polynomial $f(x)$ is divided by $x - a$, where a is a constant.
  (a) What is the degree of $r(x)$?
  (b) Write an equation expressing $f(x)$ in terms of $q(x)$, $r(x)$, and $x - a$.
  (c) Use part (b) to show that $f(a) = r(a)$.
  (d) Why do (a) and (c) together tell us that the remainder when the polynomial $f(x)$ is divided by $x - a$ is $f(a)$?

Problem 6.15: Let $f(x) = x^4 - 6x^3 - 9x^2 + 20x + 7$. Use synthetic division to find $f(7)$.

Problem 6.16: Find the remainder when $2x^{10} - 3ix^8 + (1 + i)x^2 - (3 + 2i)x + 1$ is divided by $x - i$.

Problem 6.17: Let $f(x) = x^{10} - 2x^5 + 3$.
  (a) Find the remainder when $f(x)$ is divided by $x - 2$.
  (b) Find the remainder when $f(x)$ is divided by $2x - 4$.

Extra!   *Imagination is the beginning of creation. You imagine what you desire, you will what you imagine*
➤➤➤➤   *and at last you create what you will.*

– George Bernard Shaw

Problem 6.18: Let $P(x)$ be a polynomial such that when $P(x)$ is divided by $x - 19$, the remainder is 99, and when $P(x)$ is divided by $x - 99$, the remainder is 19. In this problem we find the remainder when $P(x)$ is divided by $(x - 19)(x - 99)$. *(Source: AHSME)*

(a) Find $P(19)$ and $P(99)$.

(b) Let $r(x)$ be the desired remainder. Write an equation representing the division of $P(x)$ by $(x - 19)(x - 99)$.

(c) What is the degree of $r(x)$?

(d) Choose two convenient values of x and substitute these into your equation from (b). Use the results to find $r(x)$.

Problem 6.19: Find the remainder when the polynomial $x^{81} + x^{49} + x^{25} + x^9 + x$ is divided by $x^3 - x$. *(Source: Great Britain)*

Problem 6.13: Let $f(x) = x^4 - 3x^3 + 7x^2 - x + 5$.

(a) What is the remainder when $f(x)$ is divided by $x - 3$?

(b) What is $f(3)$?

Solution for Problem 6.13:

(a) We find the remainder using the synthetic division shown at right. The remainder is 65.

(b) We have $f(3) = 3^4 - 3 \cdot 3^3 + 7 \cdot 3^2 - 3 + 5 = 65$.

$$\begin{array}{c|ccccc} 3 & 1 & -3 & 7 & -1 & 5 \\ & & 3 & 0 & 21 & 60 \\ \hline & 1 & 0 & 7 & 20 & 65 \end{array}$$

□

Hmm.... We see that $f(3)$ equals the remainder when $f(x)$ is divided by $x - 3$. Is this a coincidence?

Problem 6.14: Suppose $q(x)$ and $r(x)$ are the quotient and remainder, respectively, when the polynomial $f(x)$ is divided by $x - a$, where a is a constant. Is it always true that the remainder equals $f(a)$?

Solution for Problem 6.14: Because $q(x)$ and $r(x)$ are the quotient and remainder, respectively, when the polynomial $f(x)$ is divided by $x - a$, we have

$$f(x) = (x - a)q(x) + r(x).$$

Because either the remainder is 0 or its degree is less than the degree of the divisor $x - a$, we see that $r(x)$ must be a constant. So, we let $r(x) = c$ for some constant c, and we have

$$f(x) = (x - a)q(x) + c.$$

This equation must hold for all x, and we're interested in $f(a)$, so we let $x = a$ in the equation. This gives

$$f(a) = (a - a)q(a) + c = 0 \cdot q(a) + c = c.$$

Therefore, when a polynomial $f(x)$ is divided by $x - a$, the remainder is $f(a)$. □

> **Important:** The **Remainder Theorem** states that when a polynomial $f(x)$ is divided
> by $x - a$, the remainder is $f(a)$.

Problem 6.15: Let $f(x) = x^4 - 6x^3 - 9x^2 + 20x + 7$. Use synthetic division to find $f(7)$.

Solution for Problem 6.15: We can put $x = 7$ into $f(x)$ to evaluate $f(7)$, but the Remainder Theorem allows us to evaluate $f(7)$ with less complicated arithmetic. The synthetic division at right tells us that the remainder when $f(x)$ is divided by $x - 7$ is 49. Therefore, we have $f(7) = 49$. □

$$
\begin{array}{r|rrrrr}
7 & 1 & -6 & -9 & 20 & 7 \\
  &   & 7 & 7 & -14 & 42 \\
\hline
  & 1 & 1 & -2 & 6 & 49
\end{array}
$$

The Remainder Theorem can also be used to find remainders without division.

Problem 6.16: Find the remainder when $2x^{10} - 3ix^8 + (1 + i)x^2 - (3 + 2i)x + 1$ is divided by $x - i$.

Solution for Problem 6.16: We could use synthetic division, even though some of our coefficients are nonreal. However, because we're dividing a 10th degree polynomial, this synthetic division would be pretty lengthy. Instead, we use the Remainder Theorem. We let $f(x) = 2x^{10} - 3ix^8 + (1 + i)x^2 - (3 + 2i)x + 1$. By the Remainder Theorem, the remainder when we divide $f(x)$ by $x - i$ is

$$
\begin{aligned}
f(i) &= 2 \cdot i^{10} - 3i \cdot i^8 + (1 + i) \cdot i^2 - (3 + 2i) \cdot i + 1 \\
&= 2(-1) - 3i(1) + (1 + i)(-1) - 3i - 2i^2 + 1 \\
&= -7i.
\end{aligned}
$$

□

Problem 6.17: Let $f(x) = x^{10} - 2x^5 + 3$. Find the remainder when $f(x)$ is divided by $2x - 4$.

Solution for Problem 6.17: What's wrong with this solution:

> **Bogus Solution:** The remainder is $f(4) = 4^{10} - 2 \cdot 4^5 + 3 = 1046531$.

The Remainder Theorem tells us that when $f(x)$ is divided by $x - a$, the remainder is $f(a)$. The value of $f(4)$ equals the remainder when $f(x)$ is divided by $x - 4$, not $2x - 4$.

> **WARNING!!** The Remainder Theorem applies to division by linear expressions with leading coefficient 1.

While we can't directly apply the Remainder Theorem, we can use our proof of the Remainder Theorem as guidance. Because we are dividing by a linear expression, the remainder must be a constant. So, we seek the constant r such that

$$
f(x) = (2x - 4)q(x) + r
$$

for all values of x. When proving the Remainder Theorem, we eliminated $q(x)$ by setting x equal to a value that made the divisor 0. Here, we let $x = 2$ to find $f(2) = (2 \cdot 2 - 4)q(2) + r = r$. So, our remainder is $f(2) = 2^{10} - 2 \cdot 2^5 + 3 = 963$. $\square$

Notice that the key step in this solution is not to apply the Remainder Theorem directly, but rather to use the same general technique we used to prove the Remainder Theorem.

> **Concept:** Don't just memorize important theorems. Learn how to prove them. The methods used to prove important theorems can also be used to solve other problems to which it is not easy to directly apply these theorems.

Here, the technique we use is:

> **Concept:** Suppose $q(x)$ and $r(x)$ are the quotient and remainder, respectively, when the polynomial $f(x)$ is divided by the polynomial $d(x)$. One powerful way to learn more about $r(x)$ is to express the division as
>
> $$f(x) = q(x) \cdot d(x) + r(x),$$
>
> and then eliminate the $q(x) \cdot d(x)$ term by choosing values of x such that $d(x) = 0$. Combining this with the fact that $\deg r$ is no greater than $\deg d$ will often allow us to determine $r(x)$.

This strategy works even when $d(x)$ is not linear.

Problem 6.18: Let $P(x)$ be a polynomial such that when $P(x)$ is divided by $x - 19$, the remainder is 99, and when $P(x)$ is divided by $x - 99$, the remainder is 19. What is the remainder when $P(x)$ is divided by $(x - 19)(x - 99)$? *(Source: AHSME)*

Solution for Problem 6.18: Because we are dividing by a quadratic, the degree of the remainder is no greater than 1. So, the remainder is $ax + b$, for some constants a and b. Therefore, we have

$$P(x) = (x - 19)(x - 99)Q(x) + ax + b,$$

where $Q(x)$ is the quotient when $P(x)$ is divided by $(x - 19)(x - 99)$. We can eliminate the $Q(x)$ term by letting $x = 19$ or by letting $x = 99$. Doing each in turn gives us the system of equations

$$P(19) = 19a + b = 99,$$
$$P(99) = 99a + b = 19.$$

Solving this system of equations gives us $a = -1$ and $b = 118$. So, the remainder is $-x + 118$. $\square$

Problem 6.19: Find the remainder when the polynomial $x^{81} + x^{49} + x^{25} + x^9 + x$ is divided by $x^3 - x$. *(Source: Great Britain)*

Solution for Problem 6.19: Let $P(x) = x^{81} + x^{49} + x^{25} + x^9 + x$ and let $Q(x)$ be the quotient when $P(x)$ is divided by $x^3 - x$. We are dividing by a cubic, so the degree of the remainder is no more than 2.

Therefore, the remainder is $ax^2 + bx + c$ for some constants $a, b,$ and c. So, we have

$$P(x) = (x^3 - x)Q(x) + ax^2 + bx + c.$$

Since $x^3 - x = x(x+1)(x-1)$, we can let $x = -1, 0,$ or 1 to make $x^3 - x$ equal to 0. Doing each in turn gives us the system of equations

$$P(-1) = a - b + c,$$
$$P(0) = c,$$
$$P(1) = a + b + c.$$

Since $P(0) = 0$, we have $c = 0$. Adding the first and third equations then gives $2a = P(-1) + P(1) = -5 + 5 = 0$, so $a = 0$. Finally, from $P(1) = a + b + c$, we have $b = P(1) = 5$, so the desired remainder is $5x$. □

Exercises

6.4.1 Let $g(x) = 2x^6 - x^4 + 4x^2 - 8$. Find the remainder when $g(x)$ is divided by each of the following:

(a) $x + 1$ (c) $x + 3$ (e) $x^2 + 4x + 3$

(b) $x - 1$ (d) $x - 3$ (f) $x^2 - 4x + 3$

6.4.2 When $y^2 + my + 2$ is divided by $y - 1$, the quotient is $f(y)$ and the remainder is R_1. When $y^2 + my + 2$ is divided by $y + 1$, the quotient is $g(y)$ and the remainder is R_2. If $R_1 = R_2$, then find m. *(Source: AHSME)*

6.4.3 Find the remainder when x^{100} is divided by $(x - 1)(x - 2)$. *(Source: AHSME)* (You can leave constants in the form a^b in your answer.)

6.4.4 Suppose $q(x)$ and $r(x)$ are the quotient and remainder, respectively, when the polynomial $f(x)$ is divided by the polynomial $d(x)$. Show that if $x = a$ is a root of $d(x)$, then $r(a) = f(a)$.

6.4.5★ Find the remainder when $x^{100} - 4x^{98} + 5x + 6$ is divided by $x^3 - 2x^2 - x + 2$. **Hints:** 91

6.5 Summary

A **polynomial** in one variable is an expression or function of the form

$$a_n x^n + a_{n-1} x^{n-1} + a_{n-2} x^{n-2} + \cdots + a_1 x + a_0,$$

where x is a variable and each a_i is a constant. Each individual summand $a_i x^i$ is called a **term**, and each number a_i is the **coefficient** of the corresponding x^i. The **degree** of a polynomial is the greatest integer n such that the coefficient a_n is not equal to 0. (The degree of the polynomial 0 is undefined.)

> **Important:** When the polynomial $f(x)$ is divided by the polynomial $d(x)$, the quotient $q(x)$ and remainder $r(x)$ are the polynomials such that
>
> $$f(x) = q(x) \cdot d(x) + r(x)$$
>
> and either $r(x) = 0$ or $\deg r < \deg d$.

We covered a method for performing polynomial division that is a lot like "long division" of integers. We then also covered a shorthand, called **synthetic division**, for dividing a polynomial by a linear polynomial in which the linear term has coefficient 1. We can also use synthetic division even if the coefficient of the linear term of the divisor is not 1, by converting the division into a division in which this coefficient is 1.

> **Important:** The **Remainder Theorem** states that when a polynomial $f(x)$ is divided by $x - a$, the remainder is $f(a)$.

Things To Watch Out For!

- When dividing one polynomial by another, be careful not to overlook "missing" terms. These terms aren't really missing—their coefficients are 0. Include such terms in your long division by writing them with a coefficient of 0.

- When we divide by $x - a$ using synthetic division, we put a to the left of the division bracket, not $-a$. So, when we divide by $x + c$, we put $-c$ to the left of the division bracket, not c.

- Synthetic division only works when the coefficient of the linear term in the divisor is 1.

Problem Solving Strategies

>
> **Concepts:**
> - Substituting a value for the variable in a polynomial can give us useful information about the coefficients of the polynomial. In particular, letting the variable equal 1 gives the sum of the coefficients of the polynomial.
>
> - Just as you can check integer division with multiplication and addition, you can check polynomial division. If $f(x)$ divided by $d(x)$ gives a quotient of $q(x)$ and remainder of $r(x)$, we must have
>
> $$d(x) \cdot q(x) + r(x) = f(x).$$
>
> - Don't just memorize important theorems. Learn how to prove them. The methods used to prove important theorems can also be used to solve other problems to which it is not easy to directly apply these theorems.
>
> *Continued on the next page. . .*

Concepts: ... *continued from the previous page*

- Suppose $q(x)$ and $r(x)$ are the quotient and remainder, respectively, when the polynomial $f(x)$ is divided by the polynomial $d(x)$. One powerful way to learn more about $r(x)$ is to express the division as

$$f(x) = q(x) \cdot d(x) + r(x),$$

and then eliminate the $q(x) \cdot d(x)$ term by choosing values of x such that $d(x) = 0$. Combining this with the fact that $\deg r$ is no greater than $\deg d$ will often allow us to determine $r(x)$.

REVIEW PROBLEMS

6.20 Let $f(x) = x^2 + 3x + 3$ and $g(x) = 2x^2 - x + 2$. Find all constants c such that $h(x) = f(x) + c \cdot g(x)$ is a binomial.

6.21 Find the quotient and remainder in each of the following divisions:

(a) $9x^4 + x^3 - 12x + 21$ divided by $x + 4$.

(b) $x^3 - 3x^2 - 9x + 27$ divided by $x - 3$.

(c) $2y^6 + 3y^4 + 4y^2 + 5$ divided by $y^2 - 1$.

(d) $8x^3 + 16x^2 - 7x + 4$ divided by $5 + 2x$.

(e) $3r^4 + 16r^3 - 5r + 19$ divided by $3r - 2$.

(f) $t^4 - t^2 + 3t - 7$ divided by $t^2 - 3t + 8$.

6.22 If $P(x)$ is a polynomial in x, and $x^{23} + 23x^{17} - 18x^{16} - 24x^{15} + 108x^{14} = (x^4 - 3x^2 - 2x + 9) \cdot P(x)$ for all values of x, compute the sum of the coefficients of $P(x)$. *(Source: ARML)*

6.23 In Exercise 6.4.1, you should have found that $g(x) = 2x^6 - x^4 + 4x^2 - 8$ leaves the same remainder when divided by $x + 1$ and by $x - 1$, and that $g(x)$ leaves the same remainder when divided by $x + 3$ and by $x - 3$. Is it true that for every polynomial $f(x)$, the remainder when $f(x)$ is divided by $x - a$ is the same as when it is divided by $x + a$, for any constant a?

6.24 Beth divided $2x^3 - x + 5$ by $x + 3$ and came up with a quotient of $2x + 7$ and a remainder of 26. She immediately knew from her answer that she had made a mistake. How did she know so quickly that she had erred? Can you figure out what mistake she made?

6.25 Find the remainder when $x^{13} + 1$ is divided by $x - 1$. *(Source: AHSME)*

6.26 Suppose the polynomial $f(x)$ is divided by the polynomial $d(x)$ to give a quotient $q(x)$ and remainder $r(x)$. If $\deg f = 12$ and $\deg r = 2$, then what are the possible values of $\deg q$?

6.27 If $x^4 + 4x^3 + 6px^2 + 4qx + r$ is exactly divisible by $x^3 + 3x^2 + 9x + 3$, then find $(p + q)r$. *(Source: AHSME)*

6.28 Find the remainder when $13x^6 + 3x^4 + 9x^3 + 2x^2 + 17$ is divided by $x^2 - 1$.

6.29 $f(x)$ is a polynomial of degree greater than 3. If $f(1) = 2$, $f(2) = 3$ and $f(3) = 5$, find the remainder when $f(x)$ is divided by $(x - 1)(x - 2)(x - 3)$. *(Source: ARML)*

6.30 Show that if the remainder when $f(x)$ is divided by $x - a$ is r, then the remainder when $f(x)$ is divided by $c(x - a)$ is also r for any nonzero constant c.

Challenge Problems

6.31 Let $f(x) = x^2 - 3x + 3$ and $g(x) = 3x^3 - 5x^2 - 3x + 12$. The solutions of the equation $f(x) = 0$ are also solutions of the equation $g(x) = 0$. The equation $g(x) = 0$ has a third solution besides the two solutions of $f(x) = 0$; what is this third solution?

6.32 Show that if a polynomial $f(x)$ leaves a remainder of the form $px + q$ when it is divided by $(x - a)(x - b)(x - c)$, where a, b, and c are all distinct, then $(b - c)f(a) + (c - a)f(b) + (a - b)f(c) = 0$. **Hints:** 20

6.33 How can we use synthetic division to divide $x^4 - 3x^3 + 4x^2 + 11x - 9$ by $x^2 - 3x + 2$? **Hints:** 162

6.34★ When $P(x) = x^{81} + Lx^{57} + Gx^{41} + Hx^{19} + 2x + 1$ is divided by $x - 1$, the remainder is 5, and when $P(x)$ is divided by $x - 2$, the remainder is -4. However, $x^{81} + Lx^{57} + Gx^{41} + Hx^{19} + Kx + R$ is exactly divisible by $(x - 1)(x - 2)$. If L, G, H, K, and R are real, compute the ordered pair (K, R). *(Source: NYSML)*

6.35★ Let $P(x) = (x - 1)(x - 2)(x - 3)$. For how many polynomials $Q(x)$ does there exist a polynomial $R(x)$ of degree 3 such that $P(Q(x)) = P(x) \cdot R(x)$? *(Source: AMC 12)* **Hints:** 231, 356, 224

6.36★ Find the remainder when the polynomial $x^{81} + x^{49} + x^{25} + x^9 + x$ is divided by $x^3 + x$. *Note: This is not the exact same problem as Problem 6.19!* **Hints:** 317

6.37★ Find a polynomial $f(x)$ of degree 5 such that $f(x) - 1$ is divisible by $(x - 1)^3$ and $f(x)$ is itself divisible by x^3. **Hints:** 286, 267

Extra! The ancient Greeks made many geometric discoveries, but their algebraic accomplishments were much more limited. One of the main reasons they didn't make much progress with algebra is that they had not developed sophisticated notation like the notation we now use. Indeed, much of their algebraic reasoning was done in geometric terms, such as the following translation of Euclid's *Elements*:

> *To a given straight line to apply a parallelogram equal to a given rectilineal figure and deficient by a parallelogrammic figure similar to the given one: thus the given rectilineal figure must not be greater than the parallelogram described on the half of the straight line and similar to the defect.*

In his book *Unknown Quantity* about the history of algebra, John Derbyshire referred to this passage when he wrote, "This, say Bashmakova and Smirnova, is tantamount to solving the quadratic equation $x(a - x) = S$. I am willing to take their word for it." Me too.

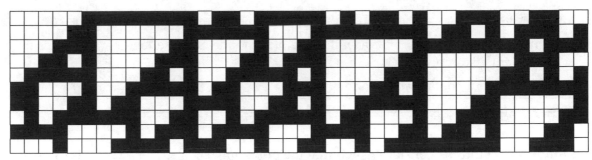

When solving problems, dig at the roots instead of just hacking at the leaves. – Anthony J. D'Angelo

CHAPTER 7

Polynomial Roots Part I

The number r is a **root** of the polynomial $f(x)$ if $f(r) = 0$. Roots of polynomials are also sometimes called **zeros** of the polynomial. In this chapter, we investigate tactics for finding rational roots of polynomials.

7.1 The Factor Theorem

Recall that a polynomial $d(x)$ evenly divides a polynomial $f(x)$ if the remainder is 0 when we divide $f(x)$ by $d(x)$. We then call $d(x)$ a **factor** of $f(x)$. In this section, we use the Remainder Theorem from Section 6.4 to find linear factors of polynomials.

Problems

Problem 7.1: Let $f(x) = 6x^3 - tx^2 + 7x - 12$, where t is a constant such that $f(x)$ is divisible by $x - 2$.

(a) Find $f(2)$ in terms of t.   (b) Find t.

Problem 7.2: Let $p(x)$ be a polynomial.

(a) Show that if $x - a$ is a factor of $p(x)$, then $p(a) = 0$.

(b) Show that if $p(a) = 0$, then $x - a$ is a factor of $p(x)$.

Problem 7.3: The polynomial $f(x) = x^4 + ax^3 + bx^2 + cx + d$ has roots 1, 3, 5, and 7. Determine all the coefficients of $f(x)$.

Problem 7.4: Let $g(x)$ and $f(x)$ be nonzero polynomials.

(a) If $g(x)$ is a factor of $f(x)$, then must every root of $g(x)$ also be a root of $f(x)$?

(b) If every root of $g(x)$ is a root of $f(x)$, then must $g(x)$ divide $f(x)$ evenly?

Problem 7.5: Let $p(x) = 3x^4 + 5x^3 - 49x^2 + 11x + 30$ and note that $p(-2/3) = p(3) = 0$.

(a) Find two linear factors of $p(x)$. Divide $p(x)$ by one of these factors. Why are the roots of the quotient also roots of $p(x)$?

(b) Find all four roots of $p(x)$.

Problem 7.1: Find the constant t such that the polynomial $6x^3 - tx^2 + 7x - 12$ is divisible by $x - 2$.

Solution for Problem 7.1: We let $f(x) = 6x^3 - tx^2 + 7x - 12$. If $f(x)$ is divisible by $x - 2$, then the remainder is 0 when $f(x)$ is divided by $x - 2$. By the Remainder Theorem, if the remainder when $f(x)$ is divided by $x - 2$ equals 0, then we must have $f(2) = 0$.

Since $f(2) = 0$, we let $x = 2$ in our definition of $f(x)$ and we have $6 \cdot 2^3 - t \cdot 2^2 + 7 \cdot 2 - 12 = 0$. Solving for t gives us $t = 12.5$.

At right, we confirm that $x - 2$ divides $f(x)$ evenly when we have $t = 12.5$. We find that

$$6x^3 - 12.5x^2 + 7x - 12 = (x - 2)(6x^2 - 0.5x + 6),$$

so $x - 2$ is a factor of $6x^3 - 12.5x^2 + 7x - 12$.

$$
\begin{array}{r|rrrr}
2 & 6 & -12.5 & 7 & -12 \\
  &   & 12 & -1 & 12 \\
\hline
  & 6 & -0.5 & 6 & 0
\end{array}
$$

Not only can we use synthetic division to test our solution, but we could also have used it to find the answer. At right, we synthetically divide $x - 2$ into $f(x) = 6x^3 - tx^2 + 7x - 12$. In order for $x - 2$ to divide $f(x)$ evenly, the remainder must be 0. Solving $50 - 4t = 0$ gives us $t = 12.5$, as before. $\square$

$$
\begin{array}{r|rrrr}
2 & 6 & -t & 7 & -12 \\
  &   & 12 & 24 - 2t & 62 - 4t \\
\hline
  & 6 & 12 - t & 31 - 2t & 50 - 4t
\end{array}
$$

Problem 7.2: Let $p(x)$ be a polynomial. Prove that $x - a$ is a factor of $p(x)$ if and only if $p(a) = 0$.

Solution for Problem 7.2: We begin by proving that if $x - a$ is a factor of $p(x)$, then $p(a) = 0$.

If $x - a$ is a factor of $p(x)$, then $p(x) = (x - a)q(x)$ for some polynomial $q(x)$. Letting $x = a$ gives us $p(a) = (a - a)q(a) = 0 \cdot q(a) = 0$, so a is a root of $p(x)$.

Next, we prove that if a is a root of $p(x)$, then $x - a$ is a factor of $p(x)$.

If $p(a)$ is a root of $p(x)$, then $p(a) = 0$. By the Remainder Theorem, the remainder when $p(x)$ is divided by $x - a$ is $p(a)$. Since $p(a) = 0$, this remainder is 0. This means that $x - a$ is a factor of $p(x)$. $\square$

Our solution to Problem 7.2 gives us a proof of the **Factor Theorem**:

> **Important:** The **Factor Theorem** states that the expression $x - a$, where a is a constant, is a factor of the polynomial $p(x)$ if and only if $p(a) = 0$.

This theorem tells us that factors and roots of polynomials go hand in hand. Since the polynomial $p(x)$ has the factor $x - a$ *if and only if* $p(a) = 0$, we know each of the following:

- If $x - a$ is a factor of $p(x)$, then $p(a) = 0$.

- If $p(a) = 0$, then $x - a$ is a factor of $p(x)$.

In other words, if a is a constant and $p(x)$ is a polynomial, then the statements "$x - a$ is a factor of $p(x)$" and "$p(a) = 0$" mean the same thing.

As you might have suspected from its name, the Factor Theorem can be used to help us factor polynomials if we know their roots.

Problem 7.3: The polynomial $f(x) = x^4 + ax^3 + bx^2 + cx + d$ has roots $1, 3, 5$, and 7. Determine all the coefficients of $f(x)$.

Solution for Problem 7.3: From the roots of $f(x)$, we note that $x - 1$, $x - 3$, $x - 5$, and $x - 7$ are all factors of $f(x)$. But are these the only ones? We see that they are the only factors by working through the roots one at a time.

Because 1 is a root of $f(x)$, we have $f(x) = (x - 1)g_1(x)$, for some polynomial $g_1(x)$. Since the leading term of $f(x)$ is x^4, the leading term of $g_1(x)$ must be x^3. Next, because 3 is a root of $f(x)$, we have $f(3) = 0$. This means that $(3 - 1)g_1(3) = 0$, so $g_1(3) = 0$, which means 3 is a root of $g_1(x)$. This means that $g_1(x) = (x - 3)g_2(x)$, where $g_2(x)$ is a monic quadratic polynomial.

We now have $f(x) = (x-1)(x-3)g_2(x)$. Since 5 is a root of $f(x)$, we have $(5-1)(5-3)g_2(5) = 0$, so $g_2(5) = 0$ and $g_2(x) = (x - 5)g_3(x)$ for some linear polynomial $g_3(x)$. Finally, because $f(x) = (x-1)(x-3)(x-5)g_3(x)$ and 7 is a root of $f(x)$, we find that $g_3(7) = 0$, which means that $g_3(x)$ is the linear polynomial $x - 7$. So, we can now deduce that

$$\begin{aligned}
f(x) &= (x - 1)(x - 3)(x - 5)(x - 7) \\
&= (x^2 - 4x + 3)(x^2 - 12x + 35) \\
&= x^2(x^2 - 12x + 35) - 4x(x^2 - 12x + 35) + 3(x^2 - 12x + 35) \\
&= x^4 - 16x^3 + 86x^2 - 176x + 105.
\end{aligned}$$

Notice that the coefficient of x^3 is $-(1 + 3 + 5 + 7)$ and the constant term is $1 \cdot 3 \cdot 5 \cdot 7$. We'll be exploring this "coincidence" in Section 8.3. Maybe you can figure out before then why this happens. □

In general, if we know all the roots of a polynomial, we don't have to go through the long-winded process of our solution to Problem 7.3 to write the polynomial as a product of the corresponding factors. We can follow essentially the same steps to show the following:

Important: Let the polynomial $f(x) = a_n x^n + a_{n-1}x^{n-1} + a_{n-2}x^{n-2} + \cdots + a_1 x + a_0$ have roots $r_1, r_2, r_3, \ldots, r_n$. Then we can write $f(x)$ as

$$f(x) = a_n(x - r_1)(x - r_2) \cdots (x - r_n).$$

We can use this to show that a polynomial with degree n cannot have more than n roots. If a polynomial $f(x)$ has $n + 1$ roots, then it can be written as

$$f(x) = c(x - r_1)(x - r_2) \cdots (x - r_{n+1}),$$

where c is a constant and $r_1, r_2, r_3, \ldots, r_{n+1}$ are the roots. There are $n + 1$ roots, so there are $n + 1$ linear factors. When this product is expanded, the highest degree term is formed by taking x from each of the $n + 1$ linear factors, which gives us the term cx^{n+1}. Therefore, the degree of f is $n + 1$, not n. Similarly, any polynomial with more than n roots has degree greater than n, so no polynomial with degree n can have more than n roots.

With tools more advanced than we will cover in this text, we can show that every one-variable polynomial with degree n has *exactly* n roots. This fact is so fundamental to the study of algebra that it is called the **Fundamental Theorem of Algebra**.

> **Important:** The **Fundamental Theorem of Algebra** states that every one-variable polynomial of degree n has exactly n complex roots.

You can get a sense for why this theorem is true by noting that multiplying n one-variable linear terms will produce a polynomial with degree n. The Fundamental Theorem of Algebra tells us that this process runs in reverse as well—if $f(x)$ is a polynomial with degree n, then we can write $f(x)$ as

$$f(x) = a_n(x - r_1)(x - r_2) \cdots (x - r_n),$$

where a_n is a constant and the constants $r_1, r_2, r_3, \ldots, r_n$ are the roots of the polynomial. These roots need not be distinct. For example, the polynomial $(x - 1)^3$ has roots 1, 1, and 1. That is, 1 is a triple root of $(x - 1)^3$. We therefore say that the root 1 of $(x - 1)^3$ has **multiplicity** 3.

We have already seen that the Factor Theorem tells us that if $g(x)$ is a linear polynomial, then $g(x)$ is a factor of the polynomial $f(x)$ if and only if the root of $g(x)$ is a root of $f(x)$. But what if $g(x)$ is not linear?

Problem 7.4: Let $g(x)$ and $f(x)$ be nonzero polynomials.

(a) If $g(x)$ is a factor of $f(x)$, then must every root of $g(x)$ also be a root of $f(x)$?

(b) If every root of $g(x)$ is a root of $f(x)$, then must $g(x)$ divide $f(x)$ evenly?

Solution for Problem 7.4:

(a) If $g(x)$ is a factor of $f(x)$, then we have $f(x) = g(x)q(x)$ for some polynomial $q(x)$. Therefore, if r is a root of $g(x)$, then we have $f(r) = g(r)q(r) = 0 \cdot q(r) = 0$, so r is also a root of $f(x)$. This means that if $g(x)$ is a factor of $f(x)$, then every root of $g(x)$ is a root of $f(x)$.

(b) Suppose $g(x) = (x - 1)^2$ and $f(x) = x(x - 1)$. Then, every root of $g(x)$ is also a root of $f(x)$, but $g(x)$ does not evenly divide $f(x)$. Therefore, the converse of part (a) is not true in general.

□

> **Important:** If $g(x)$ is a factor of $f(x)$, then every root of $g(x)$ is a root of $f(x)$.

Let's see an example of why this fact is useful.

> **Problem 7.5:** Let $p(x) = 3x^4 + 5x^3 - 49x^2 + 11x + 30$ and note that $p(-2/3) = p(3) = 0$. Find all four roots of $p(x)$.

Solution for Problem 7.5: Since $p(-2/3) = p(3) = 0$, the Factor Theorem tells us that $x + 2/3$ and $x - 3$ are factors of $p(x)$. Now, we can use synthetic division to help us factor $p(x)$. We first divide $p(x)$ by $x - 3$ (at right) to get $3x^3 + 14x^2 - 7x - 10$. Therefore, we have

$$\begin{array}{r|rrrrr} 3 & 3 & 5 & -49 & 11 & 30 \\ & & 9 & 42 & -21 & -30 \\ \hline & 3 & 14 & -7 & -10 & 0 \end{array}$$

$$p(x) = (x - 3)(3x^3 + 14x^2 - 7x - 10).$$

Since $3x^3 + 14x^2 - 7x - 10$ is a factor of $p(x)$, the roots of $3x^3 + 14x^2 - 7x - 10$ are also roots of $p(x)$. Therefore, we can continue our search for roots of $p(x)$ by finding roots of $3x^3 + 14x^2 - 7x - 10$.

We don't have to search far because we know that $x + 2/3$ is a factor of $p(x)$. So, we divide $3x^3 + 14x^2 - 7x - 10$ by $x + 2/3$ (at right) to get $3x^2 + 12x - 15$. We now have

$$\begin{array}{r|rrrr} -2/3 & 3 & 14 & -7 & -10 \\ & & -2 & -8 & 10 \\ \hline & 3 & 12 & -15 & 0 \end{array}$$

$$p(x) = (x - 3)(x + 2/3)(3x^2 + 12x - 15).$$

All three coefficients of the quadratic are divisible by 3, so we factor out a 3 from the quadratic to find

$$p(x) = 3(x - 3)(x + 2/3)(x^2 + 4x - 5).$$

> **Concept:** If all the coefficients of a polynomial you must factor are divisible by the same integer (other than 1), then factor that integer out so you can work with a simpler polynomial.

We finish factoring $p(x)$ by factoring the quadratic $x^2 + 4x - 5$ as $(x + 5)(x - 1)$, so we have

$$p(x) = 3(x - 3)(x + 2/3)(x + 5)(x - 1).$$

Therefore, the roots of $p(x)$ are -5, $-2/3$, 1, and 3. Note that because $3(x + 2/3) = 3x + 2$, we can also write the factorization as $p(x) = (x - 3)(3x + 2)(x + 5)(x - 1)$. $\square$

> **Important:** If we find that a is a root of the polynomial $p(x)$, we can simplify our search for more roots of $p(x)$ by dividing $p(x)$ by $x - a$. Since a is a root of $p(x)$, this division will leave no remainder, and we will have
>
> $$p(x) = (x - a)q(x).$$
>
> We can then continue our search for roots of $p(x)$ by searching for roots of $q(x)$.

Exercises

7.1.1 For what values of a is $x + 2$ a factor of $x^4 - 2x^3 + ax^2 - ax + 7$?

7.1.2 How can we quickly tell that $x - 1$ is a factor of $x^5 + 6x^4 - 7x^3 + 2x^2 - 2$ *without performing the long division*?

7.1.3 Let $f(x) = 6x^4 + 19x^3 - 36x^2 - 49x + 60$. One root of $f(x)$ is $-5/3$. Find all four roots of $f(x)$.

7.1.4 Suppose $f(x)$ and $q(x)$ are polynomials such that $f(x) = (x-a)q(x)$. Is it true that $q(x) = 0$ whenever $f(x) = 0$?

7.1.5 The polynomial $p(x) = 3x^3 - 20x^2 + kx + 12$ is divisible by $x - 3$ for some constant k. Factor $p(x)$ completely.

7.2   Integer Roots

Finding roots of polynomials can be extremely difficult. However, if we are looking for integer roots of a polynomial with integer coefficients, we can use the coefficients of the polynomial to considerably narrow our search.

In the next three sections, we'll explore methods for finding rational roots of polynomials. In Chapter 8, we'll study polynomials that have irrational or nonreal roots.

Problems

Problem 7.6: In this problem we find the roots of $f(x) = x^3 - 5x^2 - 77x + 441$.

(a) Consider $f(2)$:
$$f(2) = 2^3 - 5(2^2) - 77(2) + 441.$$

How can we determine that $f(2) \neq 0$ without evaluating $f(2)$?

(b) Consider $f(5)$:
$$f(5) = 5^3 - 5(5^2) - 77(5) + 441.$$

Which of the four terms on the right are divisible by 5? Which are not? Is 5 a root of $f(x)$?

(c) Suppose that n is an integer that is a solution to the equation $f(n) = 0$. Show that n is a divisor (positive or negative) of 441.

(d) Find the roots of $f(x)$.

Problem 7.7: Suppose k is a root of $f(x) = a_n x^n + a_{n-1} x^{n-1} + a_{n-2} x^{n-2} + \cdots + a_1 x + a_0$, where k is a nonzero integer, and all the coefficients of $f(x)$ are integers. Why must k divide a_0?

Extra! The polynomial $f(n) = n^2 + n + 41$ cannot be factored as the product of two polynomials with integer coefficients. Evaluate $f(1)$, $f(2)$, $f(3)$, and $f(4)$. Do you notice anything interesting about the integers that result? If so, explore further. *(Continued on page 209.)*

Problem 7.8: Let $f(x) = x^3 - 3x^2 - 10x + 24$.

(a) Graph $f(x)$.

(b) Notice that $f(0) = 24$ and $f(3) = -6$. Explain why there must be a root between 0 and 3.

(c) Notice that $f(1) = 12$ and $f(5) = 24$. Does this mean that there is not a root between 1 and 5?

(d) Find the roots of $f(x)$.

Problem 7.9: Let $f(x) = x^4 - 5x^3 - 4x^2 + 44x - 48$.

(a) Factor $f(x)$ and find its roots.

(b) Graph $f(x)$.

Problem 7.10: Let $g(x) = x^6 + 15x^5 + 81x^4 + 193x^3 + 198x^2 + 72x$.

(a) How can we quickly see that 0 is a root of $g(x)$?

(b) How can we quickly see that $g(x)$ has no positive roots?

(c) Find the roots of $g(x)$.

Problem 7.11: In this problem, we solve the inequality $x^3 + 5x^2 - 16x - 80 \leq 0$.

(a) Factor the left side.

(b) Find all x that satisfy the inequality.

Problem 7.6: Find all solutions to the equation $x^3 - 5x^2 - 77x + 441 = 0$.

Solution for Problem 7.6: Let $f(x) = x^3 - 5x^2 - 77x + 441$. We wish to find the roots of $f(x)$. We could just go hunting for roots by evaluating $f(a)$ for various values of a. We'll start with integer values of a, since they're easier to work with than fractions:

$$f(0) = 0^2 - 5(0^2) - 77(0) + 441 = 441,$$
$$f(1) = 1^3 - 5(1^2) - 77(1) + 441 = 360,$$
$$f(2) = 2^3 - 5(2^2) - 77(2) + 441 = 275,$$
$$f(3) = 3^3 - 5(3^2) - 77(3) + 441 = 192,$$
$$f(4) = 4^3 - 5(4^2) - 77(4) + 441 = 117,$$
$$f(5) = 5^3 - 5(5^2) - 77(5) + 441 = 56.$$

Hmmm.... This search might take a very long time. Let's see if we can find more clever method. First, we note that $f(0)$, $f(2)$, and $f(4)$ are all odd. Is this just a coincidence, or will we always get an odd output when we put an even number in for x?

Suppose x is an even integer. Then, the first three terms of $f(x)$ are even, but the last term, 441, is odd. Therefore, the sum of these four terms is odd. So, because 0 is even and $f(x)$ is odd for even x, we cannot have $f(x) = 0$ if x is an even integer. This tells us that $f(x)$ cannot have a root that is divisible by 2.

Considering values of x that are divisible by 2 narrowed our search for roots. Let's look at other possible divisors of x.

> **Concept:** If a strategy gives you some information about a problem, but doesn't completely solve it, then try that strategy again in a slightly different way.

If x is a multiple of 3, then the first three terms of $f(x)$ are divisible by 3. The constant term, 441, is also divisible by 3. Therefore, if x is divisible by 3, then all four terms of $f(x)$ are divisible by 3. So, it is possible to have $f(x) = 0$ if x is divisible by 3.

That didn't help too much. Let's try multiples of 5. (We skip multiples of 4 because we already know that the roots can't be even.) If x is divisible by 5, then all terms of $x^3 - 5x^2 - 77x + 441$ except the constant term are divisible by 5. Therefore, if x is divisible by 5, then $f(x)$ equals a multiple of 5 plus 441. This sum cannot be divisible by 5, so $f(x)$ cannot equal 0 if x is a multiple of 5.

We don't have to try multiples of 6 since there are no even roots. If x is divisible by 7, then all four terms of $f(x)$ are divisible by 7. So, it is possible to have $f(x) = 0$ if x is divisible by 7.

We have a pattern emerging. In each case, we let x be divisible by some positive integer d. If d does not divide 441 (such as when $d = 2$ or $d = 5$), then it seems that we cannot have $f(x) = 0$. If d does divide 441 (such as when $d = 3$ or $d = 7$), then it seems that we cannot eliminate the possibility that $f(x) = 0$. But does the pattern hold in general?

We suspect our hunt for integer solutions of $x^3 - 5x^2 - 77x + 441 = 0$ depends on whether or not $441/x$ is an integer. So, we isolate 441 in this equation, then divide by x (clearly, $x = 0$ is not a solution, so we can divide by it). This gives us

$$\frac{441}{x} = -x^2 + 5x + 77.$$

All three terms on the right are integers when x is an integer, so $441/x$ must be an integer if the integer x is a solution to the equation. Therefore, if $f(x) = 0$ for an integer x, then x must evenly divide 441.

This narrows our search for integer roots a great deal. Since 6 doesn't divide 441 but 7 does, we try dividing $f(x)$ by $x - 7$ next to see if 7 is a root of $f(x)$. We find that $x - 7$ divides $f(x)$ evenly, and we have $f(x) = (x - 7)(x^2 + 2x - 63)$. Factoring the quadratic gives us

$$
\begin{array}{r|rrrr}
7 & 1 & -5 & -77 & 441 \\
  &   & 7 & 14 & -441 \\
\hline
  & 1 & 2 & -63 & 0 \\
\end{array}
$$

$$f(x) = (x - 7)(x - 7)(x + 9).$$

So, there are two solutions to the equation $f(x) = 0$, namely $x = 7$ and $x = -9$. Because $f(x)$ has two factors equal to $x - 7$, we say that 7 is a **double root** of $f(x)$. So, we say that $f(x)$ has 3 roots, which are 7, 7, and -9, but only 2 *distinct* (different) roots, which are 7 and -9. $\square$

The number of times a number is a root of a polynomial is called the **multiplicity** of the root. So, the root 7 of $f(x) = (x - 7)(x - 7)(x + 9)$ has multiplicity 2, while the root -9 has multiplicity 1.

Our work in Problem 7.6 suggests that any nonzero integer root of a polynomial with integer coefficients must divide the constant term of the polynomial. Let's see if this is the case.

> **Problem 7.7:** Suppose $x = k$ is a root of $f(x) = a_n x^n + a_{n-1} x^{n-1} + a_{n-2} x^{n-2} + \cdots + a_1 x + a_0$, where k is a nonzero integer, and all the coefficients of $f(x)$ are integers. Why must k divide a_0?

Solution for Problem 7.7: We take essentially the same approach we took in Problem 7.6. If $x = k$ is a root of $f(x)$, then we must have $a_n k^n + a_{n-1} k^{n-1} + a_{n-2} k^{n-2} + \cdots + a_1 k + a_0 = 0$. Since k is a nonzero integer, we can divide by k to find

$$a_n k^{n-1} + a_{n-1} k^{n-2} + a_{n-2} k^{n-3} + \cdots + a_1 + \frac{a_0}{k} = 0.$$

Isolating the fraction then gives us

$$\frac{a_0}{k} = -a_n k^{n-1} - a_{n-1} k^{n-2} - a_{n-2} k^{n-3} - \cdots - a_1.$$

Since k and all the coefficients are integers, the right side is an integer. Therefore, a_0/k must be an integer, which tells us that k divides a_0. So, if k is a root of $f(x)$, then it must divide the constant term of $f(x)$.

We can also show that a_0 is a multiple of k by isolating a_0 in

$$a_n k^n + a_{n-1} k^{n-1} + a_{n-2} k^{n-2} + \cdots + a_1 k + a_0 = 0.$$

This gives us $a_0 = -(a_n k^n + a_{n-1} k^{n-1} + a_{n-2} k^{n-2} + \cdots + a_1 k) = -k(a_n k^{n-1} + a_{n-1} k^{n-2} + a_{n-2} k^{n-3} + \cdots + a_1)$. Because all the coefficients and k are integers, the expression $(a_n k^{n-1} + a_{n-1} k^{n-2} + a_{n-2} k^{n-3} + \cdots + a_1)$ is an integer. Therefore, a_0 equals k times an integer, which means a_0 is a multiple of k. $\square$

We indicate that a number d divides a number n by writing $d \mid n$.

> **Important:** Let $f(x) = a_n x^n + a_{n-1} x^{n-1} + a_{n-2} x^{n-2} + \cdots + a_0$, where the coefficients of $f(x)$ are integers. If r is a nonzero integer root of $f(x)$, then $r \mid a_0$.

This fact considerably reduces the number of integers we must try when searching for roots of a polynomial with integer coefficients. If $f(x)$ is a polynomial with integer coefficients and k is an integer that does not divide the constant term of $f(x)$, then we know that $x - k$ does not divide $f(x)$.

Now, we'll get some practice finding integer roots of polynomials, and find a few more strategies for finding roots quickly.

Problem 7.8: Let $f(x) = x^3 - 3x^2 - 10x + 24$. Graph $f(x)$ and find its roots.

Solution for Problem 7.8: If $f(x)$ has any integer roots, each of these roots must evenly divide 24, the constant term of $f(x)$. We start with 1 and -1 because it's so easy to evaluate $f(1)$ and $f(-1)$.

> **Concept:** We start our search for nonzero integer roots of any polynomial $f(x)$ with 1 and -1, since it's so easy to evaluate $f(1)$ and $f(-1)$.

Unfortunately, neither of these work, as we find $f(1) = 12$ and $f(-1) = 30$. So, we must continue our search. Since 3 divides 24 evenly, we check if 3 is a root. (We see that $x^3 - 3x^2 = 0$ for $x = 3$, so evaluating $f(3)$ is particularly easy.) We find that $f(3) = -6$, so 3 is not a root. However, we notice that $f(3) < 0$ while $f(1) > 0$. Because $f(3)$ is negative and $f(1)$ is positive, we guess that there is a value between 1 and 3 that is a root. Are we right?

The only integer between 1 and 3 is 2, and it divides 24 evenly, so it could be a root. The synthetic division at right shows that 2 is indeed a root, and we have $f(x) = (x-2)(x^2 - x - 12)$. Factoring the quadratic gives us $f(x) = (x-2)(x+3)(x-4)$, so the roots are 2, -3, and 4.

$$
\begin{array}{r|rrrr}
2 & 1 & -3 & -10 & 24 \\
  &   & 2 & -2 & -24 \\
\hline
  & 1 & -1 & -12 & 0
\end{array}
$$

We can use the graph of $f(x)$ to see why our guess that there is a root between 1 and 3 was correct. The roots of $f(x)$ tell us where the graph of $f(x)$ meets the x-axis, and we choose a few more values of x to fill out the graph. Also, $f(x)$ is much greater than 0 for large values of x, and is much less than 0 when x is much less than 0. This information gives us a good idea what the graph looks like.

x	$f(x)$
-4	-48
-3	0
-2	24
-1	30
0	24
1	12
2	0
3	-6
4	0
5	24

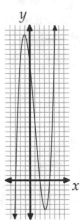

The graph passes through $(1, 12)$ and $(3, -6)$. So, as the graph goes from $x = 1$ to $x = 3$, it goes from above the x-axis to below the x-axis. Somewhere in between, it must cross the x-axis. Therefore, there must be a root between 1 and 3. This argument assumes that $f(x)$ is **continuous**, which means that its graph has no "jumps" in it. In other words, we can draw the portion of the graph between any two points without lifting our pencil. Proving that $f(x)$ is continuous requires too advanced tools to cover in this book. $\square$

WARNING!! Once we determined both that there is a root between 1 and 3, and that 2 is the only integer between 1 and 3 that divides 24, we still had to check if 2 is in fact a root.

To see why we can't immediately conclude that 2 is a root, consider the polynomial $g(x) = x^3 - 14x^2 - 16x + 15$. We have $g(0) = 15$ and $g(-3) = -90$, so there is a root between 0 and -3. Moreover, -1 is the only integer between 0 and -3 that divides 15 evenly. However, we have $g(-1) = 16$, so -1 is not a root of $g(x)$. The root of $g(x)$ that is between 0 and -3 is not an integer. (See if you can find it!)

Our method of determining that there is a root of $f(x)$ between 1 and 3 is a very handy tool for simplifying the search for roots of a polynomial.

Important: If $f(x)$ is a polynomial and a and b are real numbers such that $f(a)$ and $f(b)$ have different signs, then there is a root of $f(x)$ between a and b.

Notice that we cannot determine whether or not there is a root between a and b if $f(a)$ and $f(b)$ are the same sign. For example, we have $f(1) = 12$ and $f(5) = 24$ in Problem 7.8. Even though $f(1)$ and $f(5)$ are the same sign, we cannot deduce that there are no roots between 1 and 5. (Indeed, as we saw, there are two roots of $f(x)$ between 1 and 5.)

Problem 7.9: Let $f(x) = x^4 - 5x^3 - 4x^2 + 44x - 48$. Factor $f(x)$, find its roots, and graph it.

Solution for Problem 7.9: We start our hunt for roots by trying divisors of 48. We start with 1 and -1 because it's so easy to evaluate $f(1)$ and $f(-1)$.

> **Concept:** We start our search for nonzero integer roots of any polynomial $f(x)$ with 1 and -1, since it's so easy to evaluate $f(1)$ and $f(-1)$.

Unfortunately, we have $f(1) = -12$ and $f(-1) = -90$, so neither 1 nor -1 is a root. Since $f(1)$ is closer to 0, we try 2 next instead of -2. The synthetic division at right shows us that 2 is a root, and we have

$$\begin{array}{r|rrrrr} 2 & 1 & -5 & -4 & 44 & -48 \\ & & 2 & -6 & -20 & 48 \\ \hline & 1 & -3 & -10 & 24 & 0 \end{array}$$

$$f(x) = (x - 2)(x^3 - 3x^2 - 10x + 24).$$

We continue our search for roots by finding roots of $x^3 - 3x^2 - 10x + 24$. We find that $x = 2$ is a root of this polynomial as well.

> **WARNING!!** Don't forget about multiplicity. If we find that a is a root of $f(x)$, we can't forget that a may be a double root, or triple root, etc. So, we check if $x - a$ also divides the quotient $f(x)/(x - a)$.

Dividing $x - 2$ into $x^3 - 3x^2 - 10x + 24$ gives us a quotient of $x^2 - x - 12$. So, we have

$$f(x) = (x - 2)(x^3 - 3x^2 - 10x + 24) = (x - 2)^2(x^2 - x - 12).$$

Factoring the quadratic then gives us $f(x) = (x - 2)^2(x - 4)(x + 3)$, so the roots are -3, 2, 2, and 4.

When graphing f, the roots tell us where graph intersects the x-axis. We choose a few more values of x in the table at left below to find a few more points on the graph. The graph of f extends far off the bottom and top of the diagram shown below. To indicate this, we place arrows where the graph extends off the diagram. The graph is so steep near $x = -3$ that it almost appears to be a vertical line.

x	$f(x)$
-4	288
-3	0
-2	-96
-1	-90
0	-48
1	-12
2	0
3	-6
4	0
5	72

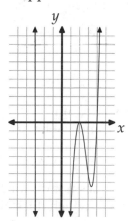

From our table, we see that $f(x)$ is so much less than 0 for most values of x between -3 and 2 that our graph of $y = f(x)$ extends well beyond the bottom of the shown graph. However, we do see interesting behavior of the graph near $x = 2$. It looks like the graph just touches the x-axis, but doesn't cross it. We know that the graph passes through $(2, 0)$. We can look at the factored form of $f(x)$ to see why the graph behaves the way it does near $x = 2$:

$$f(x) = (x - 2)^2(x - 4)(x + 3).$$

If $2 < x < 4$, then $x - 4$ is negative, $x + 3$ is positive, and $(x - 2)^2$ is positive, so $f(x)$ is negative for $2 < x < 4$. Similarly, if $-3 < x < 2$, then $(x - 2)^2$ is positive, $x - 4$ is negative, and $x + 3$ is positive, so $f(x)$ is negative for $-3 < x < 2$ as well. We capture all of this information in the table at right. Therefore, the graph of $f(x)$ is below the x-axis both just to the left and just to the right of $(2, 0)$. (Notice that we could have omitted the $(x - 2)^2$ column, since $(x - 2)^2$ is always nonnegative.)

	$x - 4$	$(x - 2)^2$	$x + 3$	$f(x)$
$x > 4$	$+$	$+$	$+$	$+$
$2 < x < 4$	$-$	$+$	$+$	$-$
$-3 < x < 2$	$-$	$+$	$+$	$-$
$x < -3$	$-$	$+$	$-$	$+$

When x is very far from 0, the value of $f(x)$ is primarily determined by the x^4 term. So, in both cases, $f(x)$ is a large positive number. This is depicted in our graph by the fact that the graph of $f(x)$ goes far above the x-axis both when it is far to the right of the y-axis and when it is far to the left of the y-axis. $\square$

Problem 7.10: Find the roots of $g(x) = x^6 + 15x^5 + 81x^4 + 193x^3 + 198x^2 + 72x$.

Solution for Problem 7.10: Our first step here is pretty easy. There's no constant term, so there is an x in every term. This means we can factor out an x, and that 0 is a root of $g(x)$:

$$g(x) = x(x^5 + 15x^4 + 81x^3 + 193x^2 + 198x + 72).$$

Important: If a polynomial has no constant term, then 0 is a root of the polynomial.

We continue our search for roots with $x^5 + 15x^4 + 81x^3 + 193x^2 + 198x + 72$. Let this polynomial be $q(x)$. Before we start trying factors of 72, we notice that the coefficients of $q(x)$ are all positive. Therefore, whenever x is positive, each term of $q(x)$ is positive, which means $q(x)$ is positive. So, there are no positive roots. This cuts our search in half.

We start with -1 because $q(-1)$ is easy to evaluate. We find that $q(-1) = 0$, so $x + 1$ is a factor of $q(x)$. Synthetic division gives us

$$g(x) = xq(x) = x(x + 1)(x^4 + 14x^3 + 67x^2 + 126x + 72).$$

That worked, so we try it again. We find that $x + 1$ divides $x^4 + 14x^3 + 67x^2 + 126x + 72$ because $1 - 14 + 67 - 126 + 72 = 0$, and we have

$$g(x) = x(x + 1)^2(x^3 + 13x^2 + 54x + 72).$$

We quickly see that $x + 1$ does not divide $x^3 + 13x^2 + 54x + 72$ (because $-1 + 13 - 54 + 72 \neq 0$). So we try $x + 2$, which fails, then $x + 3$, which does divide $x^3 + 13x^2 + 54x + 72$. Then, all we have left is a quadratic, and we find

$$g(x) = x(x+1)^2(x+3)(x^2+10x+24) = x(x+1)^2(x+3)(x+4)(x+6).$$

So, the roots of $g(x)$ are $-6, -4, -3, -1, -1,$ and 0. $\square$

Finding roots is not the only application of factoring.

Problem 7.11: Solve the inequality $x^3 + 5x^2 - 16x - 80 \le 0$.

Solution for Problem 7.11: We tackled quadratic inequalities by factoring them, so we try factoring the polynomial on the lesser side of the inequality. We find that

$$x^3 + 5x^2 - 16x - 80 = (x-4)(x^2+9x+20) = (x-4)(x+4)(x+5).$$

So, our inequality is $(x-4)(x+4)(x+5) \le 0$.

Let $f(x) = (x-4)(x+4)(x+5)$. We can find the solutions to $f(x) < 0$ by considering the signs of the factors of $f(x)$ for different values of x. All three factors of $f(x)$ are positive for $x > 4$, so $f(x) > 0$ if $x > 4$. The factor $x - 4$ is negative and the other two factors are positive if $-4 < x < 4$, so $f(x) < 0$ when $-4 < x < 4$. Similarly, $f(x)$ is positive if $-5 < x < -4$ and $f(x)$ is negative if $x < -5$. We can organize all this information in a table, as shown.

	$x-4$	$x+4$	$x+5$	$f(x)$
$x > 4$	$+$	$+$	$+$	$+$
$-4 < x < 4$	$-$	$+$	$+$	$-$
$-5 < x < -4$	$-$	$-$	$+$	$+$
$x < -5$	$-$	$-$	$-$	$-$

Because the inequality is nonstrict, the roots of $f(x)$ also satisfy it. So, the values of x that satisfy the inequality are $x \le -5$ and $-4 \le x \le 4$. In interval notation, our solution is $x \in (-\infty, -5] \cup [-4, 4]$.

We can also reason our way to the answer by thinking about the graph of $y = (x-4)(x+4)(x+5)$. When x is much less than 0, then $f(x)$ is also much less than 0. As x gets larger, the graph of $y = f(x)$ approaches the x-axis, crossing the x-axis at $(-5, 0)$ because -5 is a root of $f(x)$. There are no roots smaller than -5, so the graph of $y = f(x)$ is below the x-axis for all x less than -5. So, $f(x) < 0$ for $x < -5$. After passing through $(-5, 0)$, the graph goes above the x-axis, because as x goes from just below -5 to just above it, the sign of $x + 5$ changes from negative to positive. The signs of $x - 4$ and $x + 4$ stay negative, so $f(x) > 0$ just after the graph passes through $(-5, 0)$.

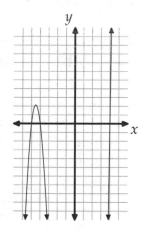

The graph again crosses the x-axis at $(-4, 0)$, because -4 is a root of $f(x)$. There are no roots of $f(x)$ between -5 and -4, so the graph of $y = f(x)$ is above the x-axis for all x such that $-5 < x < -4$. Therefore, $f(x) > 0$ for all x such that $-5 < x < -4$.

After passing through $(-4, 0)$, the graph goes below the x-axis, because the sign of $x + 4$ changes from negative to positive as x goes from just below -4 to just above it. The signs of $x + 5$ and $x - 4$ don't change, so $f(x) < 0$ just after the graph passes through $(-4, 0)$. The graph next crosses the x-axis at $(4, 0)$, so for x between -4 and 4, the graph of $y = f(x)$ is below the x-axis. (In fact, it is so far below the x-axis for most x between -4 and 4 that the graph extends well off the bottom of the diagram shown.) Therefore, we have $f(x) < 0$ for all x such that $-4 < x < 4$. As x goes from a little less than 4 to a little more than 4, the value of $f(x)$ increases so fast from being much less than 0 to being much greater than 0 that the graph is almost vertical. Finally, for all $x > 4$, we have $f(x) > 0$. Combining the intervals of x

for which our graph is below the x-axis with the points where the graph intersects the x-axis (because the inequality is nonstrict), we have the same answer as before, $x \in (-\infty, -5] \cup [-4, 4]$.

Our consideration of the graph of $y = f(x)$ is essentially the same approach to the problem as the table. In both, we determine the sign of $f(x)$ for different intervals of x. We use the roots of $f(x)$ to determine the intervals we must consider. □

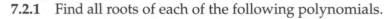

7.2.1 Find all roots of each of the following polynomials.

(a) $f(x) = x^3 - 4x^2 - 11x + 30$

(b) $g(t) = t^4 + 5t^3 - 19t^2 - 65t + 150$

(c) $f(x) = -x^5 - 12x^4 + 6x^3 + 64x^2 - 93x + 36$

(d) $h(y) = 6y^3 - 5y^2 - 22y + 24$

7.2.2 Find all solutions to the inequalities below.

(a) $t^3 + 10t^2 + 17t > 28$

(b) $r^2(6r + 12 - r^2) \le 5(14r + 15)$

(c) $r^4 + r^3 + 5r^2 - 13r - 18 \le 0$

7.2.3 There are four roots of $f(x) = x^4 - 8x^3 + 24x^2 - 32x + 16$. We can easily test to find that $f(2) = 0$. We can then check all the other divisors of 16, both positive and negative, and find that no divisors of 16 besides 2 are roots of the polynomial. Is it correct to deduce that the other three roots of $f(x)$ are not integers?

7.2.4★ Suppose that $f(x)$ is a polynomial with integer coefficients such that $f(2) = 3$ and $f(7) = -7$. Show that $f(x)$ has no integer roots. **Hints:** 1, 319

7.3 Rational Roots

A **rational number** is a number that can be expressed in the form p/q, where p and q are integers. A number that is not rational is called an **irrational number**. Just as we can use the coefficients of a polynomial with integer coefficients to narrow the search for integer roots, we can also use these coefficients to narrow the search for non-integer rational roots.

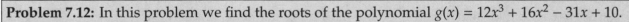

Problem 7.12: In this problem we find the roots of the polynomial $g(x) = 12x^3 + 16x^2 - 31x + 10$.

(a) Can you find any integers n such that $g(n) = 0$?

(b) Suppose $g(p/q) = 0$, where p and q are integers and p/q is in reduced form. Rewrite $g(p/q) = 0$ using the definition of g.

(c) Get rid of the fractions in your equation from (b) by multiplying by the appropriate power of q.

(d) What terms in your equation from (c) have p? Why must p divide 10?

(e) What terms in your equation from (c) have q? Why must q divide 12?

(f) Find all the roots of $g(x)$.

Problem 7.13: Let $f(x)$ be the polynomial $f(x) = a_n x^n + a_{n-1}x^{n-1} + a_{n-2}x^{n-2} + \cdots + a_1 x + a_0$, where all the coefficients are integers and both a_n and a_0 are nonzero. Let p and q be integers such that p/q is a fraction in simplest terms, and $f(p/q) = 0$. In this problem, we show that p divides a_0 and q divides a_n.

(a) Suppose $f(p/q) = 0$. Why must we have

$$a_n p^n + a_{n-1}p^{n-1}q + a_{n-2}p^{n-2}q^2 + a_{n-3}p^{n-3}q^3 + \cdots + a_1 pq^{n-1} + a_0 q^n = 0?$$

(b) Which terms in the equation in (a) must be divisible by p? Use this to show that a_0 must be divisible by p.

(c) Which terms in the equation in (a) must be divisible by q? Use this to show that a_n must be divisible by q.

Problem 7.14: Find all the roots of each of the following polynomials:

(a) $f(x) = 12x^3 - 107x^2 - 15x + 54$.

(b) $g(x) = 30x^4 - 133x^3 - 121x^2 + 189x - 45$.

Problem 7.15: Find all r such that $12r^4 - 16r^3 > 41r^2 - 69r + 18$.

We've found a way to narrow our search for integer roots. How about rational roots?

Problem 7.12: Find all x such that $12x^3 + 16x^2 - 31x + 10 = 0$.

Solution for Problem 7.12: We let $f(x) = 12x^3 + 16x^2 - 31x + 10$ and begin our search for roots of $f(x)$ by seeing if we can find any integer roots. An integer root of $f(x)$ must be a divisor of its constant term, which is $10 = 2^1 \cdot 5^1$. We try each divisor (positive and negative):

$$
\begin{aligned}
f(1) &= 12 \cdot 1^3 & +16 \cdot 1^2 & -31 \cdot 1 & +10 = 7, \\
f(-1) &= 12 \cdot (-1)^3 & +16 \cdot (-1)^2 & -31 \cdot (-1) & +10 = 45, \\
f(2) &= 12 \cdot 2^3 & +16 \cdot 2^2 & -31 \cdot 2 & +10 = 108, \\
f(-2) &= 12 \cdot (-2)^3 & +16 \cdot (-2)^2 & -31 \cdot (-2) & +10 = 40, \\
f(5) &= 12 \cdot 5^3 & +16 \cdot 5^2 & -31 \cdot 5 & +10 = 1755, \\
f(-5) &= 12 \cdot (-5)^3 & +16 \cdot (-1)^2 & -31 \cdot (-5) & +10 = -935, \\
f(10) &= 12 \cdot 10^3 & +16 \cdot 10^2 & -31 \cdot 10 & +10 = 13300, \\
f(-10) &= 12 \cdot (-10)^3 & +16 \cdot (-10)^2 & -31 \cdot (-10) & +10 = -10080.
\end{aligned}
$$

(We might also have checked some of these a little faster with synthetic division.) We know we don't have to check any other integers, since any integer root must divide the constant term of $f(x)$. Since all of these fail, we know there are no integer roots of $f(x)$. Back to the drawing board.

A little number theory helped us find integer roots, so we hope that a similar method might help us find rational roots. Letting p and q be relatively prime integers (meaning they have no common positive divisor besides 1), we examine the equation $f(p/q) = 0$:

$$f\left(\frac{p}{q}\right) = 12\left(\frac{p}{q}\right)^3 + 16\left(\frac{p}{q}\right)^2 - 31\left(\frac{p}{q}\right) + 10 = 0.$$

Multiplying by q^3 to get rid of all the denominators, we have the equation

$$12p^3 + 16p^2q - 31pq^2 + 10q^3 = 0. \tag{7.1}$$

Previously, we isolated the last term of such an expansion in order to find information about integer roots. Doing so here gives us

$$10q^3 = p(-12p^2 - 16pq + 31q^2).$$

Dividing this equation by p gives us

$$\frac{10q^3}{p} = -12p^2 - 16pq + 31q^2.$$

The right side of this equation must be an integer, so $10q^3/p$ must also be an integer. Therefore, p divides $10q^3$. But, since p and q are relatively prime, p must divide 10.

If we isolate the first term of the left side of Equation (7.1) instead of the last term, we have

$$12p^3 = q(-16p^2 + 31pq - 10q^2).$$

Dividing this equation by q gives us

$$\frac{12p^3}{q} = -16p^2 + 31pq - 10q^2.$$

As before, the right side is an integer, so the left side must be as well. Thus q divides $12p^3$. Because p and q are relatively prime, we know that q must divide 12.

Combining what we have learned about p and q, we see that if p/q is a root of $f(x)$, then p is a divisor of 10 and q is a divisor of 12.

Now, we make a list of possible rational roots p/q:

$$\pm 1, \pm 2, \pm 5, \pm 10, \pm\frac{1}{2}, \pm\frac{5}{2}, \pm\frac{1}{3}, \pm\frac{2}{3}, \pm\frac{5}{3}, \pm\frac{10}{3}, \pm\frac{1}{4}, \pm\frac{5}{4}, \pm\frac{1}{6}, \pm\frac{5}{6}, \pm\frac{1}{12}, \pm\frac{5}{12}.$$

While this is not a terribly small list, it is a *complete list of possible rational roots of a polynomial with leading coefficient 12 and constant term 10*. We already know that $f(x)$ has no integer roots, so we continue our search for roots with the fractions in the list above. We start with the fractions with small denominators, since these will be easiest to check.

After a little experimentation, we find that $f(1/2) = 0$ via the synthetic division at right. (You may have found one of the other roots first.) We therefore have

$$f(x) = \left(x - \frac{1}{2}\right)(12x^2 + 22x - 20).$$

$$\begin{array}{r|rrrr} \frac{1}{2} & 12 & 16 & -31 & 10 \\ & & 6 & 11 & -10 \\ \hline & 12 & 22 & -20 & 0 \end{array}$$

We continue our search by finding roots of $12x^2 + 22x - 20$. We find $12x^2 + 22x - 20 = 2(6x^2 + 11x - 10) = 2(3x - 2)(2x + 5)$, so we have

$$f(x) = 2\left(x - \frac{1}{2}\right)(3x - 2)(2x + 5).$$

When factoring polynomials, we often factor constants out of each linear factor so that the coefficient of x in each factor is 1. This makes identifying the roots particularly easy:

$$f(x) = 2\left(x - \frac{1}{2}\right)(3)\left(x - \frac{2}{3}\right)(2)\left(x + \frac{5}{2}\right) = 12\left(x - \frac{1}{2}\right)\left(x - \frac{2}{3}\right)\left(x + \frac{5}{2}\right).$$

Now we can easily see that the roots of $f(x)$ are $\frac{1}{2}$, $\frac{2}{3}$, and $-\frac{5}{2}$. $\square$

In our solution, we found a way to narrow our search for rational roots of $f(x)$ in much the same way we can narrow our search for integer roots. Let's see if we can apply this method to any polynomial with integer coefficients.

Problem 7.13: Let $f(x)$ be the polynomial $f(x) = a_n x^n + a_{n-1} x^{n-1} + a_{n-2} x^{n-2} + \cdots + a_1 x + a_0$, where all the coefficients are integers and both a_n and a_0 are nonzero. Let p and q be integers such that p/q is a fraction in simplest terms, and $f(p/q) = 0$. Show that p divides a_0 and q divides a_n.

Solution for Problem 7.13: We use our solution to Problem 7.12 as a guide. Since $f(p/q) = 0$, we have

$$a_n\left(\frac{p}{q}\right)^n + a_{n-1}\left(\frac{p}{q}\right)^{n-1} + a_{n-2}\left(\frac{p}{q}\right)^{n-2} + \cdots + a_1\left(\frac{p}{q}\right) + a_0 = 0.$$

We get rid of the fractions by multiplying both sides by q^n, which gives us

$$a_n p^n + a_{n-1} p^{n-1} q + a_{n-2} p^{n-2} q^2 + \cdots + a_1 p q^{n-1} + a_0 q^n = 0. \tag{7.2}$$

To see why q must divide a_n, we divide Equation (7.2) by q to produce a_n/q in the first term:

$$\frac{a_n p^n}{q} + a_{n-1} p^{n-1} + a_{n-2} p^{n-2} q + \cdots + a_1 p q^{n-2} + a_0 q^{n-1} = 0.$$

Isolating this first term gives us

$$\frac{a_n p^n}{q} = -a_{n-1} p^{n-1} - a_{n-2} p^{n-2} q - \cdots - a_1 p q^{n-2} - a_0 q^{n-1}.$$

Each term on the right side of this equation is an integer, so the entire right side is an integer. Therefore, $a_n p^n/q$ must be an integer. Since p/q is in lowest terms and $a_n p^n/q$ is an integer, we know that q divides a_n.

We take essentially the same approach to show that $p \mid a_0$. We divide Equation (7.2) by p and isolate the term with a_0 in it, which gives us

$$\frac{a_0 q^n}{p} = -a_n p^{n-1} - a_{n-1} p^{n-2} q - a_{n-2} p^{n-3} q^2 - \cdots - a_1 q^{n-1}.$$

Each term on the right is an integer, so the entire right side equals an integer. So, the expression $a_0 q^n/p$ must equal an integer, which means that p must divide a_0. $\square$

Extra! If $f(n) = n^2 + n + 41$, then $f(1)$, $f(2)$, $f(3)$, and $f(4)$ are all prime. Maybe $f(n)$ is prime for all integers n. Is it? *(Continued on page 220)*

In Problem 7.13, we have proved the **Rational Root Theorem**:

> **Important:** Let $f(x)$ be the polynomial
>
> $$f(x) = a_n x^n + a_{n-1} x^{n-1} + a_{n-2} x^{n-2} + \cdots + a_1 x + a_0,$$
> where all the a_i are integers, and both a_n and a_0 are nonzero. If p and q are relatively prime integers and $f(p/q) = 0$, then $p \mid a_0$ and $q \mid a_n$.

Problem 7.14: Find all the roots of each of the following polynomials:

(a) $f(x) = 12x^3 - 107x^2 - 15x + 54$

(b) $g(x) = 30x^4 - 133x^3 - 121x^2 + 189x - 45$

Solution for Problem 7.14:

(a) We first look for easy roots by evaluating $f(1)$ and $f(-1)$. We find that $f(1) = 12 - 107 - 15 + 54 = -56$ and $f(-1) = -12 - 107 + 15 + 54 = -50$. Since $f(0) = 54$, we see that there is a root between -1 and 0 (because $f(-1) < 0$ and $f(0) > 0$) and a root between 0 and 1 (because $f(0) > 0$ and $f(1) < 0$).

> **Concept:** We often start our hunt for rational roots by evaluating $f(-1)$, $f(0)$, and $f(1)$, because these are easy to compute and the results may tell us where to continue our search.

We continue our hunt for roots with $\frac{1}{2}$, which may be a root because 2 divides the leading coefficient of $f(x)$. Using synthetic division, we find $f(\frac{1}{2}) = 21\frac{1}{4}$, so there is a root between $\frac{1}{2}$ and 1. We try $\frac{2}{3}$ and it works, giving us

$$f(x) = \left(x - \frac{2}{3}\right)(12x^2 - 99x - 81) = 3\left(x - \frac{2}{3}\right)(4x^2 - 33x - 27).$$

Factoring the quadratic then gives us $f(x) = 3\left(x - \frac{2}{3}\right)(4x + 3)(x - 9)$. So, the roots of $f(x)$ are $-\frac{3}{4}, \frac{2}{3}$, and 9.

(b) We start our hunt for roots by finding $g(0) = -45$, $g(1) = -80$, and $g(-1) = -192$. Unfortunately, that doesn't help too much; we can't immediately tell if there are any roots between -1 and 0, or between 0 and -1. So, we test 3 and -3, knowing that 2 and -2 cannot be roots because the constant term of $g(x)$ is odd. By synthetic division, we find that $g(3) = -1728$. This has the same sign as $g(1)$, so we continue with $g(-3)$. We find that $g(-3) = 4320$, so because $g(-3) > 0$ and $g(-1) < 0$, we know there is a root between -3 and -1. (Note that we didn't need to find the actual value of $g(-3)$. Once it is clear that $g(-3)$ is a large positive number, we know that there is a root between -1 and -3.) With synthetic division, we find that $g(-2)$ is also a large positive number, so there is a root between -2 and -1.

We try $-3/2$. We know that $-3/2$ might be a root of $g(x)$ because -3 divides the constant term and 2 divides the leading coefficient. If $-3/2$ doesn't work, we will at least narrow our search to between -2 and

$-\frac{3}{2}$	30	-133	-121	189	-45
		-45	267	-219	45
	30	-178	146	-30	0

−3/2 or between −3/2 and −1. Fortunately, the synthetic division shows us that

$$g(x) = \left(x + \frac{3}{2}\right)(30x^3 - 178x^2 + 146x - 30) = 2\left(x + \frac{3}{2}\right)(15x^3 - 89x^2 + 73x - 15).$$

We now note that $15x^3 - 89x^2 + 73x - 15$ has no negative roots because if x is negative, each term of this polynomial is negative. So, we only have to search for positive roots. We already know that 1 and 3 don't work, and the only two other positive integers that divide 15 are 5 and 15. We find that 5 works, and we have

$$g(x) = 2\left(x + \frac{3}{2}\right)(x - 5)(15x^2 - 14x + 3).$$

Now, we can factor the quadratic or use the quadratic formula to find

$$g(x) = 2\left(x + \frac{3}{2}\right)(x - 5)(3x - 1)(5x - 3) = 30\left(x + \frac{3}{2}\right)(x - 5)\left(x - \frac{1}{3}\right)\left(x - \frac{3}{5}\right).$$

The roots of $g(x)$ are −3/2, 1/3, 3/5, and 5. Notice that there are roots between 0 and 1. Why doesn't testing $g(0)$ and $g(1)$ reveal the fact that these roots exist? (You'll have a chance to answer this question as an Exercise.)

□

Problem 7.15: Find all r such that $12r^4 - 16r^3 > 41r^2 - 69r + 18$.

Solution for Problem 7.15: First, we move all the terms to the left, which gives us

$$12r^4 - 16r^3 - 41r^2 + 69r - 18 > 0.$$

Now, we must factor the polynomial on the left. Let this polynomial be $f(r)$. Because $f(0) = -18$ and $f(1) = 6$, there is a root between 0 and 1. We find that $f(1/2) = 5$, so the root is between 0 and 1/2. Trying 1/3 gives us

$$f(r) = \left(r - \frac{1}{3}\right)(12r^3 - 12r^2 - 45r + 54) = 3\left(r - \frac{1}{3}\right)(4r^3 - 4r^2 - 15r + 18).$$

Continuing with $4r^3 - 4r^2 - 15r + 18$, we find that −1 and 2 are not roots, but −2 is, and we have

$$f(2) = 3\left(r - \frac{1}{3}\right)(r + 2)(4r^2 - 12r + 9) = 3\left(r - \frac{1}{3}\right)(r + 2)(2r - 3)^2.$$

So, our inequality is $3\left(r - \frac{1}{3}\right)(r + 2)(2r - 3)^2 > 0$.

The expression on the left side of the inequality equals 0 when $r = -2$, 1/3, or 3/2. We consider the intervals between (and beyond) these values to build the table at right. We see that $f(r) > 0$ for $r \in (-\infty, -2) \cup (\frac{1}{3}, \frac{3}{2}) \cup (\frac{3}{2}, +\infty)$. The inequality is strict, so the roots of $f(r)$ do not satisfy the inequality. □

	$r + 2$	$r - \frac{1}{3}$	$(2r - 3)^2$	$f(r)$
$r < -2$	−	−	+	+
$-2 < r < \frac{1}{3}$	+	−	+	−
$\frac{1}{3} < r < \frac{3}{2}$	+	+	+	+
$r > \frac{3}{2}$	+	+	+	+

 **Exercises**

7.3.1 Find all roots of the following polynomials:

(a) $g(y) = 12y^3 - 28y^2 - 9y + 10$ 　　　　(b) $f(x) = 45x^3 + 48x^2 + 17x + 2$

7.3.2 In part (b) of Problem 7.14, we saw that $g(0)$ and $g(1)$ are both negative, and yet we still found roots between 0 and 1. How can that be? Shouldn't $g(0)$ and $g(1)$ have different signs if $g(x) = 0$ for some x between 0 and 1?

7.3.3 Suppose $f(x) = x^3 + \frac{3}{4}x^2 - 4x - 3$. Notice that $f(2) = 0$, so 2 is a root of $f(x)$. But 2 doesn't evenly divide the constant term of $f(x)$. This seems to violate the Rational Root Theorem! Does it really?

7.3.4 In Problem 7.15, we built a table to find the values of r for which $3(r - \frac{1}{3})(r + 2)(2r - 3)^2 > 0$. How could we have solved this inequality by thinking about the graph of $y = 3(x - \frac{1}{3})(x + 2)(2x - 3)^2$?

7.3.5 Solve the two inequalities below:

(a) $24x^3 + 26x^2 \geq 21x + 9$ 　　　　(b)$\star$ $6s^4 + 13s^3 - 2s^2 + 35s - 12 < 0$

7.4 Bounds

In this section, we learn how to use synthetic division to determine bounds on the roots of a polynomial.

Problems

Problem 7.16:

(a) Let $f(x) = x^3 + 13x^2 + 39x + 27$. Without even finding the roots, how can we tell that there are no positive values of x for which $f(x) = 0$?

(b) Let $f(x) = 3x^3 - 28x^2 + 51x - 14$. Without finding the roots, how can we tell that there are no negative values of x for which $f(x) = 0$?

Problem 7.17: Let $f(x) = a_n x^n + a_{n-1} x^{n-1} + \cdots + a_1 x + a_0$.

(a) Show that if all the coefficients of $f(x)$ have the same sign (positive or negative), then $f(x)$ has no positive roots.

(b) Show that if $a_i > 0$ for all odd i and $a_i < 0$ for all even i, then $f(x)$ has no negative roots.

(c) Is it possible for $f(x)$ to have negative roots if a_i is negative for all odd i and positive for all even i?

Problem 7.18: Let $g(x) = x^4 - 3x^3 - 12x^2 + 52x - 48$.

(a) Use synthetic division to divide $g(x)$ by $x - 6$.

(b) What do you notice about the coefficients in the quotient and the remainder from part (a)?

(c) How can you use your answer to part (b) to deduce that there are no roots of $g(x)$ greater than 6? (In other words, how do you know you don't have to test 8, 12, 16, etc.?)

Problem 7.19: Let $f(x)$ be a polynomial and a be a positive constant.

(a) Let $f(x) = (x-a)q_1(x) + r_1$, so that $q_1(x)$ and r_1 are the quotient and remainder, respectively, when $f(x)$ is divided by $x - a$. If r_1 is positive and the coefficients of $q_1(x)$ are all nonnegative, explain why $f(x)$ has no roots that are greater than a.

(b) Why must a be positive to draw the conclusions we found in part (a)?

Problem 7.16:

(a) Let $f(x) = x^3 + 13x^2 + 39x + 27$. Without even finding the roots, how can we tell that there are no positive values of x for which $f(x) = 0$?

(b) Let $f(x) = 3x^3 - 28x^2 + 51x - 14$. Without finding the roots, how can we tell that there are no negative values of x for which $f(x) = 0$?

Solution for Problem 7.16:

(a) If $x > 0$, then each term of $f(x)$ is positive, so $f(x)$ is positive. Thus, there are no positive roots of $f(x)$.

(b) If $x < 0$, then all four terms of $f(x)$ are negative, so $f(x)$ is negative, which means there are no negative roots of $f(x)$.

□

Problem 7.16 suggests a general method we can sometimes use to quickly determine if a polynomial has no positive roots or has no negative roots.

Problem 7.17: Let $f(x) = a_nx^n + a_{n-1}x^{n-1} + \cdots + a_1x + a_0$.

(a) Show that if all the coefficients of $f(x)$ have the same sign (positive or negative), then $f(x)$ has no positive roots.

(b) Show that if $a_i > 0$ for all odd i and $a_i < 0$ for all even i, then $f(x)$ has no negative roots.

(c) Is it possible for $f(x)$ to have negative roots if a_i is negative for all odd i and positive for all even i?

Solution for Problem 7.17:

(a) Suppose that all the a_i are positive. Then, if $x > 0$, every term of $f(x)$ is positive, so $f(x) > 0$. Therefore, $f(x)$ has no positive roots. Similarly, if all the a_i are negative and $x > 0$, then every term of $f(x)$ is negative. So, $f(x) < 0$, which means $f(x)$ has no positive roots.

(b) If i is odd and $x < 0$, then $a_i > 0$ and $x^i < 0$, so $a_ix^i < 0$. Similarly, if i is even and $x < 0$, then $a_i < 0$ and $x^i > 0$, so $a_ix^i < 0$ in this case, as well. Therefore, every term a_ix^i of $f(x)$ is negative if $x < 0$, so $f(x) < 0$ for all negative values of x. This tells us that $f(x)$ has no negative roots.

(c) We follow essentially the same steps as in the last part. If i is odd and $x < 0$, then $a_i < 0$ and $x^i < 0$, so $a_ix^i > 0$. If i is even and $x < 0$, then $a_i > 0$ and $x^i > 0$, so $a_ix^i > 0$. Therefore, a_ix^i is positive for all i, which means $f(x) > 0$. So, $f(x)$ cannot have negative roots.

□

> **Important:** If the coefficients of a polynomial all have the same sign, then the polynomial has no positive roots. If the coefficients of a polynomial are nonzero and alternate in sign, then the polynomial has no negative roots.

As an Exercise, you'll explore what information we can deduce if some of the coefficients equal 0. But now, we'll apply the same reasoning we used on these first two problems to determine upper and lower bounds for roots of polynomials.

> **Problem 7.18:** Find the quotient and remainder when $g(x) = x^4 - 3x^3 - 12x^2 + 52x - 48$ is divided by $x - 6$. Explain how the quotient and remainder tell us that there are no roots of $g(x)$ greater than 6.

Solution for Problem 7.18: Our synthetic division at right shows us that

$$g(x) = (x - 6)(x^3 + 3x^2 + 6x + 88) + 480.$$

$$
\begin{array}{r|rrrrr}
6 & 1 & -3 & -12 & 52 & -48 \\
  &   & 6 & 18 & 36 & 528 \\
\hline
  & 1 & 3 & 6   & 88 & 480
\end{array}
$$

To explore whether or not $g(x)$ has any roots that are greater than 6, suppose $x > 6$ and consider each expression on the right side of our equation. The expression $x - 6$ is positive, and because the coefficients of $x^3 + 3x^2 + 6x + 88$ are all positive, this polynomial is positive, too. Finally, the remainder is positive for all values of x. So, if $x > 6$, the equation

$$g(x) = (x - 6)(x^3 + 3x^2 + 6x + 88) + 480$$

tells us that $g(x)$ is the product of two positive numbers, plus another positive number. Therefore, we know that $g(x)$ cannot equal 0. We've found that $g(x)$ has no roots greater than 6. So, we say that 6 is an **upper bound** for the roots of $g(x)$. $\square$

Let's generalize Problem 7.18.

> **Problem 7.19:** Let $f(x)$ be a polynomial and a be a positive constant.
>
> (a) Let $f(x) = (x - a)q_1(x) + r_1$, so that $q_1(x)$ and r_1 are the quotient and remainder, respectively, when $f(x)$ is divided by $x - a$. If r_1 is positive and the coefficients of $q_1(x)$ are all nonnegative, explain why $f(x)$ has no roots that are greater than a.
>
> (b) Why must a be positive to draw the conclusion we found in part (a)?

Solution for Problem 7.19:

(a) We follow essentially the same argument we used in Problem 7.18.

> **Concept:** To prove a general mathematical statement, it's often helpful to work out a specific example of that statement, then build your general proof by following your example.

If $x > a$, then $x - a$ is positive. If the coefficients of $q_1(x)$ are all nonnegative, then $q_1(x)$ is nonnegative for all positive values of x. The value of r_1 does not depend on x and we are given that r_1 is positive. Therefore, if $x > a$ in

$$f(x) = (x - a)q_1(x) + r_1,$$

then $f(x)$ is positive. This means that $f(x)$ cannot have any roots greater than a.

(b) In showing that $f(x) > 0$ if $x > a$ in part (a), we used the fact that x is positive because a is positive. However, if a is negative, then x is not positive for all $x > a$. If x is not necessarily positive, then we cannot deduce that $q_1(x)$ is nonnegative merely from the fact that the coefficients of $q_1(x)$ are nonnegative. So, we don't know the sign of $q_1(x)$, which means we cannot determine the sign of $(x - a)q_1(x) + r_1$. Therefore, $f(x)$ may have roots greater than a.

$\square$

> **Important:** Let $f(x) = (x - a)q_1(x) + r_1$, so that $q_1(x)$ and r_1 are the quotient and remainder, respectively, when $f(x)$ is divided by $x - a$, where a is a *positive* constant. If r_1 is positive and the coefficients of $q_1(x)$ are all nonnegative, then $f(x)$ has no roots that are greater than a.

> **WARNING!!** The rule above does not apply if a is negative.

 Exercises

7.4.1 Find the solutions to each of the following:

(a) $6r^3 + 31r^2 + 34r = 15$

(b) $4r^4 - 28r^3 + 61r^2 - 42r + 9 = 0$

7.4.2 Larry is finding the roots of a polynomial $f(x)$. He notes that $f(1) = 34$ and $f(0) = -3$, so he knows that there is a root between 0 and 1. He finds that $x = 1/3$ is a root. Can he then deduce that all the other roots are either greater than 1 or less than 0?

7.4.3 In the text, we noted that if the coefficients of a polynomial all have the same sign, then the polynomial has no positive roots. What if some of the coefficients equal 0, but all the others still have the same sign? Then is it possible for the polynomial to have a positive root?

7.4.4★ Consider the set of all equations $x^3 + a_2x^2 + a_1x + a_0 = 0$, where a_2, a_1, a_0 are real constants and $|a_i| \le 2$ for $i = 0, 1, 2$. Let r be the largest positive real number that satisfies at least one of these equations. Which of the inequalities below does r satisfy?

(A) $1 \le r < \dfrac{3}{2}$ (B) $\dfrac{3}{2} \le r < 2$ (C) $2 \le r < \dfrac{5}{2}$ (D) $\dfrac{5}{2} \le r < 3$ (E) $3 \le r < \dfrac{7}{2}$.

(Source: AHSME) **Hints:** 197

7.5 Graphing and the Fundamental Theorem of Algebra

In this section, we get a further feel for the Fundamental Theorem of Algebra by tackling a few questions about graphs of polynomials.

Problems

> **Problem 7.20:** Let $f(x)$ be a polynomial of degree n, where $n > 0$. Show that the graphs of $y = c$ and $y = f(x)$ cannot intersect at more than n points.

Problem 7.21: The graph of a polynomial intersects the x-axis at two distinct points. Does this mean that the polynomial is a quadratic?

Problem 7.22: At left below is the graph of $y = f(x)$, where $f(x)$ is a polynomial.

   (a)   Explain why the degree of the polynomial is at least 5.

   (b)   Explain why the polynomial must have an odd degree.

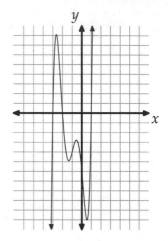

Figure 7.1: Diagram for Problem 7.22   Figure 7.2: Diagram for Problem 7.23

Problem 7.23: At right above is the graph of $y = f(x)$, where $f(x)$ is a polynomial. The only intercepts of the graph are $(-1,0)$, $(1,0)$, $(3,0)$, and $(0,3)$, and we have $\deg f < 5$. Find $f(x)$.

Problem 7.20: Let $f(x)$ be a polynomial of degree n, where $n > 0$. Show that the graphs of $y = c$ and $y = f(x)$ cannot intersect at more than n points.

Solution for Problem 7.20: For each point (x, y) at which the graphs of $y = c$ and $y = f(x)$ intersect, we must have $f(x) = c$. Subtracting c from both sides, we have $f(x) - c = 0$. The values of x for which $f(x) - c = 0$ are the roots of $f(x) - c$. Because c is a constant, the degree of $f(x) - c$ equals the degree of $f(x)$, which is n. So, by the Fundamental Theorem of Algebra, there are n roots of $f(x) - c$, and therefore n values of x such that $f(x) - c = 0$. This tells us that there are no more than n real values of x such that $f(x) - c = 0$, so there are no more than n points (x, y) at which the graphs of $y = c$ and $y = f(x)$ intersect. $\square$

Problem 7.21: The graph of a polynomial intersects the x-axis at two distinct points. Does this mean that the polynomial is a quadratic?

Solution for Problem 7.21: Let the polynomial be $f(x)$. Because the graph of $y = f(x)$ intersects the x-axis at two different points, there are two real values of x such that $f(x) = 0$. So, $f(x)$ has two distinct real roots. However, this doesn't say anything about nonreal roots of $f(x)$, or about double roots (or about triple roots, etc.). For example, at left below is the graph of $y = (x - 2)^2(x + 2)^2$ and at right is the graph of $y = x^4 - 16$.

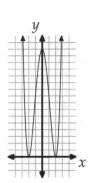

Figure 7.3: Graph of $y = (x - 2)^2(x + 2)^2$

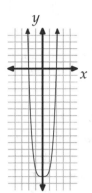

Figure 7.4: Graph of $y = x^4 - 16$

In both cases, the graph of the polynomial has only two x-intercepts, but the polynomial has degree 4. In the first case, the polynomial has two double roots, while in the second, the polynomial has two nonreal roots.

While we cannot say for sure that the polynomial is a quadratic, the fact that the polynomial has at least 2 roots tells us that its degree is at least 2, because the degree of a polynomial equals the number of roots it has. □

Problem 7.22: At right below is the graph of $y = f(x)$, where $f(x)$ is a polynomial.

(a) Explain why the degree of the polynomial is at least 5.

(b) Explain why the polynomial must have an odd degree.

Solution for Problem 7.22:

(a) The graph of $y = -4$ (dashed in the diagram) intersects the graph at 5 points. Therefore, there are 5 values of x for which $f(x) = -4$, so there are 5 values of x such that $f(x) + 4 = 0$. Since $f(x) + 4$ is a polynomial with 5 distinct roots, the degree of $f(x) + 4$ is at least 5. Since $\deg f > 0$, the degree of $f(x) + 4$ equals $\deg f$, so the degree of f is at least 5.

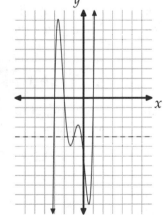

(b) When x is much more than 0 or much less than 0, the value of $f(x)$ is largely determined by its leading term. If a polynomial $f(x)$ has even degree, then its leading term is ax^{2n} for some constant a and some integer n. When x is much more than 0 or much less than 0, the term x^{2n} is a large positive number. So, the sign of ax^{2n} is same as the sign of a, *and it is the same both when x is much more than 0 and when x is much less than 0.* In our graph, for large values of x, the value of $f(x)$ is positive, but when x is much less than 0, the value of $f(x)$ is negative. Therefore, the degree of $f(x)$ cannot be even, so the degree of $f(x)$ must be odd.

□

Notice that we cannot be sure that the degree of the polynomial graphed in Problem 7.22 is exactly

5. We only know that the degree is at least 5. For example, in Problem 7.21, we graphed $f(x) = x^4 - 16$. No horizontal line hits more than 2 points on the graph of $f(x)$, but the degree of $f(x)$ is 4, not 2. So, the number of intersection points of a horizontal line with the graph of a polynomial only gives a lower bound for the degree of the polynomial.

Problem 7.23: At right below is the graph of $y = f(x)$, where $f(x)$ is a polynomial. The only intercepts of the graph are $(-1,0)$, $(1,0)$, $(3,0)$, and $(0,3)$, and we have deg $f < 5$. Find $f(x)$.

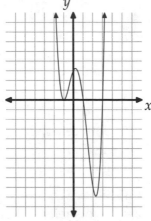

Solution for Problem 7.23: First, we note that the degree of $f(x)$ is at least 4, because there is a horizontal line ($y = 1$, for example) that passes through 4 points on the graph of $y = f(x)$. Since we know also that deg $f < 5$, we now know that deg $f = 4$. From our graph, we see that -1, 1, and 3 are all roots of $f(x)$. But what is the other root? Is it possible that somewhere off to the left or right, the graph intersects the x-axis again?

No; we know the graph cannot intersect the x-axis yet again, because this would require the graph to intersect the line $y = 1$ again, which would force the polynomial to have degree greater than 4. Therefore, we know that the three x-intercepts shown are the only points at which the graph intersects the x-axis. So, where is the other root?

One possibility is that $f(x)$ has a double root. Examining the three x-intercepts of the graph, we see that at one of them, $(-1,0)$, the graph does not cross the x-axis. Specifically, we see that $f(x)$ is positive both for x slightly larger than -1 and for x slightly smaller than -1. This is exactly the behavior we expect if $(x + 1)^2$ divides $f(x)$ evenly, since $(x + 1)^2$ is positive both for $x > -1$ and $x < -1$. We've found our double root! (If $x + 1$ were a single root, then $f(x)$ would change sign as x goes from slightly less than -1 to slightly greater than -1.) So, now that we know all four roots of $f(x)$, we know that

$$f(x) = c(x + 1)^2(x - 1)(x - 3)$$

for some constant c. The graph passes through $(0,3)$, so $f(0) = 3$. Combining $f(0) = 3$ with our expression for $f(x)$ above gives us $c = 1$, so $f(x) = (x + 1)^2(x - 1)(x - 3)$. $\square$

Exercises

7.5.1 Sketch the graphs of each of the following functions. Your graphs don't have to be precisely correct, but they should have the correct intercepts and be above or below the x-axis at the right places.

(a) $f(x) = 0.2(x - 3)(x + 2)(x + 3)$

(b) $f(x) = -(x - 2)^2(x - 1)^2(x + 1)^2$

(c) $f(x) = 0.1(x + 1)^3(x - 2)(x - 4)$

7.5.2

(a) What are the real roots of the polynomial $x^4 - 8x^2 + 16 = 0$?

(b) Solve the inequality $x^4 - 8x^2 + 16 < 0$ for real values of x.

(c) Graph the function $f(x) = x^4 - 8x^2 + 16$.

7.5.3 Let $f(x)$ be the polynomial that is graphed at right. The graph passes through $(-4, 0)$, $(-3, 0)$, $(2, 0)$, and the origin.

(a) Explain how we know that -4 is not a double root of $f(x)$.

(b) What is the minimum possible degree of f?

(c) Suppose f has the degree found in part (b) and passes through $(-2, 4)$. Find $f(x)$.

7.5.4 Suppose that $f(x)$ is a polynomial with degree 6. Is it possible for the graph of $y = x$ to intersect the graph of $y = f(x)$ at 7 different points? Why or why not?

7.6 Algebraic Applications of the Fundamental Theorem

Suppose that we are given two polynomials, $p_1(x)$ and $p_2(x)$. How can we tell if they are the same polynomial? At first, this appears to be an obvious question, because we can tell if two polynomials are the same simply by comparing their coefficients. However, what if we don't know the coefficients, but instead just know the values of $p_1(x)$ and $p_2(x)$ for a few values of x?

To investigate this question, let's play a game called "Guess the Polynomial." In this game, I think of a polynomial $f(x)$, and I give you $f(x)$ for certain values of x. Then, based on these values, you guess what polynomial I am thinking of. Ready?

OK, here are the values: $f(1) = 1$, $f(2) = 8$, $f(3) = 27$, and $f(4) = 64$. What do you think $f(x)$ is?

You're probably guessing that $f(x) = x^3$, right? Sorry, but I was thinking of the polynomial

$$f(x) = x^3 + (x - 1)(x - 2)(x - 3)(x - 4)\left(16{,}000{,}000x^{427} - 473.15x^{101} - \frac{\pi}{\sqrt{2}}x^{23} + x^5 + 99\right).$$

But you were close. Thanks for playing "Guess the Polynomial."

Well, obviously that wasn't a very fair game. But I'm not trying to be whimsical (not completely, anyway); I want to show that the values of a polynomial are not enough to deduce what the polynomial is. We need to know something else about the polynomial, something more fundamental. The key piece of additional information is the degree of the polynomial.

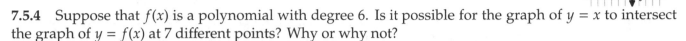

Problem 7.24: Suppose $f(x)$ is a polynomial such that

$$f(x) = a_n x^n + a_{n-1}x^{n-1} + a_{n-2}x^{n-2} + \cdots + a_1 x + a_0,$$

and there are at least $n + 1$ different values of x for which $f(x) = 0$.

(a) If $a_n \neq 0$, then what does the Fundamental Theorem of Algebra tell us about the number of different values of x for which $f(x) = 0$?

(b) Why must we have $f(x) = 0$ for all values of x?

Problem 7.25: Let $f(x)$ be a quadratic polynomial such that $f(2) = 4$, $f(3) = 9$, and $f(4) = 16$.

(a) Let $g(x) = f(x) - x^2$. What's the greatest possible degree of g?

(b) Find at least three solutions to $g(x) = 0$.

(c) Prove that $g(x) = 0$ for all values of x.

(d) What does this tell us about $f(x)$?

Problem 7.26: Let $p_1(x)$ and $p_2(x)$ be polynomials of degree at most n. In this problem, we prove that if $p_1(x) = p_2(x)$ for $n + 1$ distinct values of x, then $p_1(x) = p_2(x)$ for all values of x.

(a) Consider the polynomial $p(x) = p_1(x) - p_2(x)$. What is the greatest possible degree of $p(x)$?

(b) Based on your answer from (a), what is the greatest number of roots the polynomial $p(x)$ can have if $p(x)$ is not 0 for all values of x?

(c) If $p_1(x) = p_2(x)$ for $n + 1$ distinct values of x, then how at least how many roots does $p_1(x) - p_2(x)$ have?

(d) Combine your answers to parts (b) and (c) to prove that $p_1(x) = p_2(x)$ for all values of x.

Problem 7.27:

(a) Consider the system of equations

$$\begin{aligned}
a_1 + 8a_2 + 27a_3 + 64a_4 &= 1, \\
8a_1 + 27a_2 + 64a_3 + 125a_4 &= 27, \\
27a_1 + 64a_2 + 125a_3 + 216a_4 &= 125, \\
64a_1 + 125a_2 + 216a_3 + 343a_4 &= 343.
\end{aligned}$$

These four equations determine a_1, a_2, a_3, and a_4. Show that

$$a_1 m^3 + a_2(m + 1)^3 + a_3(m + 2)^3 + a_4(m + 3)^3 = (2m - 1)^3$$

for all values of m.

(b) Prove that $a_1 + a_2 + a_3 + a_4 = 8$ and $64a_1 + 27a_2 + 8a_3 + a_4 = 729$.
(Source: Mandelbrot)

Problem 7.28: There is a unique polynomial $P(x)$ of the form $P(x) = 7x^7 + c_1 x^6 + c_2 x^5 + \cdots + c_6 x + c_7$ such that $P(1) = 1$, $P(2) = 2, \ldots$, and $P(7) = 7$.

(a) Find the roots of $P(x) - x$.

(b) Find the constants $a, r_1, r_2, r_3, \ldots, r_7$ such that $P(x) - x = a(x - r_1)(x - r_2) \cdots (x - r_7)$ for all values of x.

(c) Determine $P(0)$. *(Source: Mandelbrot)*

Extra! The polynomial $f(n) = n^2 + n + 41$ is prime for all integers n such that $0 \le n < 40$, but
⟫⟫⟫⟫ $f(40) = 40^2 + 40 + 41 = 40(41) + 41 = 41^2$, which is not prime. This might make us wonder if there is any polynomial that is prime for all integers. Try to find one (but don't try too hard). *(Continued on page 226.)*

Problem 7.29: Let $g(x)$ and $f(x)$ be polynomials. Back in Problem 7.4 we showed that it is not necessarily true that $g(x)$ is a factor of $f(x)$ if every root of $g(x)$ is a root of $f(x)$. We showed this by considering $g(x) = (x-1)^2$ and $f(x) = x(x-1)$. Notice that our choice of $g(x)$ has a double root.

What if $g(x)$ does not have any multiple roots? Specifically, suppose that the roots of the polynomial $g(x)$ are distinct, and that each of these roots is a root of the polynomial $f(x)$. Must $g(x)$ be a factor of $f(x)$?

Problem 7.24: Suppose $f(x)$ is a polynomial such that

$$f(x) = a_n x^n + a_{n-1} x^{n-1} + a_{n-2} x^{n-2} + \cdots + a_1 x + a_0$$

and there are at least $n+1$ different values of x for which $f(x) = 0$. Show that $f(x) = 0$ for all x.

Solution for Problem 7.24: If any of the a_i are nonzero, then $\deg f \leq n$. Therefore, by the Fundamental Theorem of Algebra, if any of the a_i are nonzero, then $f(x)$ cannot have more than n roots. However, we are told that there are at least $n+1$ values of x for which $f(x) = 0$. So, it is impossible for any of the a_i to be nonzero (since then the Fundamental Theorem of Algebra would force $f(x)$ to have no more than n roots). Therefore, all the a_i must be 0, which means $f(x) = 0$ for all values of x. $\square$

Important: If $f(x)$ is a polynomial such that

$$f(x) = a_n x^n + a_{n-1} x^{n-1} + a_{n-2} x^{n-2} + \cdots + a_1 x + a_0,$$

and there are at least $n+1$ different values of x for which $f(x) = 0$, then $f(x) = 0$ for all values of x. That is, $a_n = a_{n-1} = a_{n-2} = \cdots = a_0 = 0$.

Problem 7.25: Suppose that $f(x)$ is a quadratic polynomial such that $f(2) = 4$, $f(3) = 9$, and $f(4) = 16$. Prove that $f(x) = x^2$.

Solution for Problem 7.25: By inspection, we note that $f(x) = x^2$ satisfies $f(2) = 4$, $f(3) = 9$, and $f(4) = 16$. However, maybe there are other quadratics $f(x)$ such that $f(2) = 4$, $f(3) = 9$, and $f(4) = 16$. How can we tell if $f(x) = x^2$ is the only possible one?

We let $g(x) = f(x) - x^2$ and focus on $g(x)$, because we know that 2, 3, and 4 are all roots of $g(x)$. Therefore, the Factor Theorem tells us that $x - 2$, $x - 3$, and $x - 4$ are all factors of $g(x)$. So, we have

$$g(x) = (x-2)(x-3)(x-4)q(x),$$

for some polynomial $q(x)$. This tells us that either the degree of $g(x)$ is at least 3, or $q(x) = g(x) = 0$ for all values of x. However, because $g(x) = f(x) - x^2$ and $f(x)$ is a quadratic, we know that the degree of $g(x)$ cannot be larger than 2. Therefore, we must have $q(x) = g(x) = 0$ for all x. Finally, since $f(x) = g(x) + x^2$, we know that $f(x) = x^2$ for all x.

Notice that we have used the Fundamental Theorem of Algebra in this problem. The degree of $f(x) - x^2$ is at most 2. If $f(x) - x^2$ is not 0 for all x, then by the Fundamental Theorem of Algebra, it must have no more than 2 roots. However, we know that 2, 3, and 4 are all roots of $f(x) - x^2$. So, we violate the Fundamental Theorem of Algebra if $f(x)$ is anything except x^2. (If $f(x) = x^2$, then $f(x) - x^2 = 0$ for all x, so we cannot apply the Fundamental Theorem of Algebra to $f(x) - x^2$ because its degree is undefined.) $\Box$

Let's see if we can generalize our work in Problem 7.25.

Problem 7.26: Let $p_1(x)$ and $p_2(x)$ be polynomials of degree at most n. Prove that if $p_1(x) = p_2(x)$ for $n + 1$ distinct values of x, then $p_1(x) = p_2(x)$ for all x.

Solution for Problem 7.26: In Problem 7.25, we saw that $f(x) = x^2$ for $x = 2, 3$, and 4, so we examined the polynomial $f(x) - x^2$ to prove that $f(x) = x^2$. Let's try the same general strategy here.

Let $x_1, x_2, \ldots, x_{n+1}$ be the $n + 1$ distinct values of x for which $p_1(x) = p_2(x)$. We let $p(x) = p_1(x) - p_2(x)$. Since $p_1(x_i) = p_2(x_i)$ for $1 \le i \le n + 1$, we have $p(x_1) = p(x_2) = \cdots = p(x_{n+1}) = 0$, which means that either $\deg p \ge n + 1$ or $p(x) = 0$ for all x. However, because $p_1(x)$ and $p_2(x)$ have degree at most n, there are no terms of $p(x)$ with degree greater than n. Therefore, we cannot have $\deg p \ge n + 1$, so we must have $p(x) = 0$ for all x. $\Box$

> **Important:** Let $p_1(x)$ and $p_2(x)$ be polynomials of degree at most n. If $p_1(x) = p_2(x)$ for $n + 1$ distinct values of x, then $p_1(x) = p_2(x)$ for all x. We call this the **Identity Theorem** for polynomials.

Problem 7.27: Consider the system of equations

$$
\begin{aligned}
a_1 + 8a_2 + 27a_3 + 64a_4 &= 1, \\
8a_1 + 27a_2 + 64a_3 + 125a_4 &= 27, \\
27a_1 + 64a_2 + 125a_3 + 216a_4 &= 125, \\
64a_1 + 125a_2 + 216a_3 + 343a_4 &= 343.
\end{aligned}
$$

These four equations determine a_1, a_2, a_3, and a_4. Find $a_1 + a_2 + a_3 + a_4$ and $64a_1 + 27a_2 + 8a_3 + a_4$. (Source: Mandelbrot)

Solution for Problem 7.27: In each equation in the system, the coefficients on the left side form a sequence of consecutive cubes. The constants on the right are the cubes of consecutive odd integers. We therefore see that all four equations have the form

$$m^3 a_1 + (m + 1)^3 a_2 + (m + 2)^3 a_3 + (m + 3)^3 a_4 = (2m - 1)^3,$$

where m is 1, 2, 3, or 4. The right side of this equation is a cubic polynomial in m. We can also view the left side as a cubic polynomial, where m is the variable and a_1, a_2, a_3, and a_4 are constants. Since the two cubic polynomials are equal for the four values $m = 1, 2, 3$, and 4, the Identity Theorem tells us that they must be the same cubic, so we have

$$m^3 a_1 + (m + 1)^3 a_2 + (m + 2)^3 a_3 + (m + 3)^3 a_4 = (2m - 1)^3$$

for *all values of m.*

We wish to evaluate $a_1 + a_2 + a_3 + a_4$. In the expansion of

$$m^3 a_1 + (m+1)^3 a_2 + (m+2)^3 a_3 + (m+3)^3 a_4,$$

the coefficient of m^3 is $a_1 + a_2 + a_3 + a_4$. (We don't have to multiply the whole thing out; we can see that the leading term of each of $(m+1)^3$, $(m+2)^3$ and $(m+3)^3$ is m^3 without expanding the cubes.) This must match the coefficient of m^3 in $(2m-1)^3$, so we have $a_1 + a_2 + a_3 + a_4 = 8$.

We can't easily find $64a_1 + 27a_2 + 8a_3 + a_4$ just by matching up coefficients. However, since

$$m^3 a_1 + (m+1)^3 a_2 + (m+2)^3 a_3 + (m+3)^3 a_4 = (2m-1)^3$$

for all m, we might be able to choose a value of m that enables us to evaluate $64a_1 + 27a_2 + 8a_3 + a_4$. We first try $m = 4$, since this makes the coefficient of a_1 equal to 64. Unfortunately, this makes the coefficient of a_2 equal to 125. So, we instead try $m = -4$, which makes the coefficient of a_1 equal to -64 and gives us

$$-64a_1 - 27a_2 - 8a_3 - a_4 = -729.$$

Aha! Multiplying by -1 gives us $64a_1 + 27a_2 + 8a_3 + a_4 = 729$. $\square$

Our key step in Problem 7.27 was noticing that all four equations in the given system had the same form. This enabled us to express all four equations using a single equation.

Problem 7.28: There is a unique polynomial $P(x)$ of the form

$$P(x) = 7x^7 + c_1 x^6 + c_2 x^5 + \cdots + c_6 x + c_7$$

such that $P(1) = 1$, $P(2) = 2, \ldots$, and $P(7) = 7$. Find $P(0)$. *(Source: Mandelbrot)*

Solution for Problem 7.28: The statements $P(1) = 1$, $P(2) = 2, \ldots$, and $P(7) = 7$ all have the same form, so we write them all at once by noting that $P(x) = x$ for $x = 1, 2, 3, \ldots, 7$. Now, we're in familiar territory. We rearrange this equation to give $P(x) - x = 0$ for $x = 1, 2, 3, \ldots, 7$, so these seven values are all roots of the polynomial $P(x) - x$. Because $P(x)$ is a 7th degree polynomial, we know from the Fundamental Theorem of Algebra that $P(x) - x$ has exactly 7 roots. We've already found the seven roots, so we know that

$$P(x) - x = a(x-1)(x-2)(x-3)(x-4)(x-5)(x-6)(x-7)$$

for some constant a. Because the leading term of $P(x) - x$ is $7x^7$, the leading term of the right side must also be $7x^7$. Therefore, we have $a = 7$, and

$$P(x) - x = 7(x-1)(x-2)(x-3)(x-4)(x-5)(x-6)(x-7).$$

So, we find $P(0) = 7(0-1)(0-2)(0-3)(0-4)(0-5)(0-6)(0-7) + 0 = -35280.$ $\square$

Our solution to Problem 7.28 revisits an important use of the Fundamental Theorem of Algebra that we'll be seeing many more times throughout this book:

Important: Let the polynomial $f(x) = a_n x^n + a_{n-1} x^{n-1} + a_{n-2} x^{n-2} + \cdots + a_1 x + a_0$ have roots $r_1, r_2, r_3, \ldots, r_n$. Then we can write $f(x)$ as

$$f(x) = a_n(x - r_1)(x - r_2) \cdots (x - r_n).$$

We'll often find it useful to express a polynomial in terms of its roots instead of its coefficients.

We started this chapter with Factor Theorem. We'll finish with one more look at relationships between factors and roots of polynomials.

Let $g(x)$ and $f(x)$ be polynomials. Back in Problem 7.4 we showed that it is not necessarily true that $g(x)$ is a factor of $f(x)$ if every root of $g(x)$ is a root of $f(x)$. We showed this by considering $g(x) = (x-1)^2$ and $f(x) = x(x-1)$. Notice that our choice of $g(x)$ has a double root.

What if $g(x)$ does not have any multiple roots?

Problem 7.29: Suppose that the roots of the polynomial $g(x)$ are distinct, and that each of these roots is a root of the polynomial $f(x)$. Must $g(x)$ be a factor of $f(x)$?

Solution for Problem 7.29: Let $q(x)$ and $r(x)$ be the quotient and remainder, respectively, when $f(x)$ is divided by $g(x)$. We then have $f(x) = q(x)g(x) + r(x)$, where $r(x) = 0$ or $\deg r < \deg g$. Let the degree of $g(x)$ be n, so that it has n roots. Let these roots be $s_1, s_2, \ldots, s_n$. Since $g(s_i) = 0$ for each i with $1 \le i \le n$, we have

$$f(s_i) = q(s_i)g(s_i) + r(s_i) = q(s_i) \cdot 0 + r(s_i) = r(s_i)$$

for $1 \le i \le n$. But these s_i are roots of $f(x)$, so $f(s_i) = 0$. Therefore, we have $r(s_i) = 0$ for $1 \le i \le n$. But this means that $r(x)$ has at least n roots, because the s_i are distinct. However, we must have $\deg r < \deg g$ if r is nonzero. We therefore conclude that $r(x)$ must be 0 for all x, which means that $g(x)$ is a factor of $f(x)$. $\square$

Exercises

7.6.1

(a) Show that if f is a polynomial of degree 4, such that $f(0) = f(1) = f(2) = f(3) = 1$ and $f(4) = 0$, then $f(5) = -4$.

(b) Show that if f is a polynomial of degree n, such that $f(0) = f(1) = \cdots = f(n-1) = 1$ and $f(n) = 0$, then $f(n+1) = -n$.

7.6.2 Find all possible polynomials $p(x)$ of degree 3 such that $f(1) = 0$, $f(2) = 7$, $f(3) = 26$, and $f(4) = 63$.

7.6.3 A certain polynomial $p(x)$ leaves a remainder of a when divided by $x - a$, a remainder of b when divided by $x - b$, and a remainder of c when divided by $x - c$, where a, b, and c are different. What is the remainder when $p(x)$ is divided by $(x-a)(x-b)(x-c)$?

7.6.4★ Solve the system

$$\begin{aligned} a + b + c + d &= 1, \\ 8a + 4b + 2c + d &= 16, \\ 27a + 9b + 3c + d &= 81, \\ 64a + 16b + 4c + d &= 256. \end{aligned}$$

Hints: 205

7.6.5★ Let $P(x)$ be a polynomial whose degree is 1996. If $P(n) = \frac{1}{n}$ for $n = 1, 2, 3, \ldots, 1997$, compute the value of $P(1998)$. *(Source: ARML)* **Hints:** 324

7.7 Summary

The value r is a **root** of the polynomial $f(x)$ if $f(r) = 0$. Roots of polynomials are also sometimes called **zeros** of the polynomial. We say that a polynomial $d(x)$ evenly divides a polynomial $f(x)$ if the remainder is 0 when we divide $f(x)$ by $d(x)$. We then call $d(x)$ a **factor** of $f(x)$.

> **Important:** The **Factor Theorem** states that the expression $x - a$, where a is a constant, is a factor of the polynomial $p(x)$ if and only if $p(a) = 0$.

> **Important:** The **Fundamental Theorem of Algebra** states that every one-variable polynomial of degree n has exactly n complex roots.

> **Important:** Let the polynomial $f(x) = a_n x^n + a_{n-1} x^{n-1} + a_{n-2} x^{n-2} + \cdots + a_1 x + a_0$ have roots $r_1, r_2, r_3, \ldots, r_n$. Then we can write $f(x)$ as
> $$f(x) = a_n(x - r_1)(x - r_2) \cdots (x - r_n).$$

> **Important:** If we find that a is a root of the polynomial $p(x)$, we can simplify our search for more roots of $p(x)$ by dividing $p(x)$ by $x - a$. Since a is a root of $p(x)$, this division will leave no remainder, and we will have
> $$p(x) = (x - a)q(x).$$
> We can then continue our search for roots of $p(x)$ by searching for roots of $q(x)$.

We discussed several methods for narrowing our search for rational roots of polynomials.

> **Important:** Let $f(x)$ be the polynomial
> $$f(x) = a_n x^n + a_{n-1} x^{n-1} + a_{n-2} x^{n-2} + \cdots + a_1 x + a_0,$$
> where all the a_i are integers and both a_n and a_0 are nonzero. If p and q are relatively prime integers and $f(p/q) = 0$, then $p \mid a_0$ and $q \mid a_n$. We call this the **Rational Root Theorem**.

> **Important:** If $f(x)$ is a polynomial and a and b are real numbers such that $f(a)$ and $f(b)$ have different signs, then there is a root of $f(x)$ between a and b.

> **Important:** If a polynomial has no constant term, then 0 is a root of the polynomial.

> **Important:** If the coefficients of a polynomial all have the same sign, then the polynomial has no positive roots. If the coefficients of a polynomial are nonzero and alternate in sign, then the polynomial has no negative roots.

We applied the Fundamental Theorem of Algebra to various questions about graphs of polynomials, and also used it to prove the **Identity Theorem** for polynomials:

> **Important:** Let $p_1(x)$ and $p_2(x)$ be polynomials of degree at most n. If $p_1(x) = p_2(x)$ for $n + 1$ distinct values of x, then $p_1(x) = p_2(x)$ for all x.

One common application of the Identity Theorem is:

> **Important:** If $f(x)$ is a polynomial such that
> $$f(x) = a_n x^n + a_{n-1}x^{n-1} + a_{n-2}x^{n-2} + \cdots + a_1 x + a_0,$$
> and there are at least $n + 1$ different values of x for which $f(x) = 0$, then $f(x) = 0$ for all values of x. That is, $a_n = a_{n-1} = a_{n-2} = \cdots = a_0 = 0$.

Problem Solving Strategies

> **Concepts:**
> - If all the coefficients of a polynomial you must factor are divisible by the same integer (other than 1), then factor that integer out so you can work with a simpler polynomial.
>
> - If a strategy gives you some information about a problem, but doesn't completely solve it, then try that strategy again in a slightly different way.
>
> - We often start our hunt for rational roots by evaluating $f(-1)$, $f(0)$, and $f(1)$, because these are easy to compute and the results may tell us where to continue our search.
>
> - To prove a general mathematical statement, it's often helpful to work out a specific example of that statement, then build your general proof by following your example.

> **Extra!** French mathematician **Adrien-Marie Legendre** proved that no polynomial (and no ratio of polynomials) gives a prime number as output for every integer input.

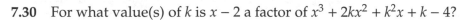

7.30 For what value(s) of k is $x - 2$ a factor of $x^3 + 2kx^2 + k^2x + k - 4$?

7.31 Let $f(x) = x^2 + 4x$. For what values of x is $f(f(x)) = f(x)$? *(Source: NYSML)*

7.32 Let $p(x) = 2x^3 - x^2 - 5x + 3$. Note that $p(0) = 3$ and $p(1) = -1$, so there is a root between 0 and 1. Applying the Rational Root Theorem, we next compute $p(1/2) = 1/2$. But there aren't any other fractions between 0 and 1 such that the numerator divides the constant term of $p(x)$ and the denominator divides the leading coefficient of $p(x)$. So, how can there possibly be a root between 0 and 1?

7.33 Compute the positive integer x such that $4x^3 - 41x^2 + 10x = 1989$. *(Source: NYSML)*

7.34 Let $f(x)$ be a quartic polynomial with integer coefficients and four integer roots.

(a) If the constant term of $f(x)$ is 6, is it possible for 3 to be a root of $f(x)$?

(b) If the constant term of $f(x)$ is 6, is it possible for 3 to be a double root of $f(x)$?

7.35 Why must the coefficients of a polynomial be integers in order to apply the Rational Root Theorem? It's pretty clear that the leading coefficient and the constant term of a polynomial have to be integers in order for the Rational Root Theorem to be applicable, but why can we not have non-integer fractions among the other coefficients?

7.36 Solve the following equations:

(a) $2x^3 + 35 = -19x^2 - 52x$

(b) $16y^2(y + 3) = 61y - 12$

7.37 Find all sets of four consecutive positive integers such that the sum of the cubes of three of them equals the cube of the fourth.

7.38 Find all the roots of each of the following:

(a) $f(r) = 6r^3 - 31r^2 - 31r + 6$

(c) $h(t) = 12t^3 - 40t^2 + 13t + 30$

(b) $g(t) = 6t^4 + 41t^3 + 88t^2 + 67t + 14$

7.39 Kai has programmed his calculator to accept as inputs two positive integers, a and b, that have no common divisors. The calculator then compares 30 times the cube of a/b plus 11 times the square of a/b to the number that is 12 more than $59a/b$. When these two quantities are equal, the calculator lets him play a game. Unfortunately, Kai has forgotten what two integers to input to be able to play his game. What integers should Kai input?

7.40 Solve the following inequalities:

(a) $6t^5 + 4t^4 \geq 2t^4$

(b) $2x^4 + 13x^3 + 9x^2 - 40x + 16 \geq 0$

7.41 In Section 7.4, we noted that "If the coefficients of a polynomial are nonzero and alternate in sign, then the polynomial has no negative roots."

(a) Suppose some of the coefficients of the polynomial $f(x)$ equal 0. Then, if the nonzero coefficients alternate in sign, can we still deduce that $f(x)$ has no negative roots?

(b) How can we amend our statement above to address a polynomial that has some coefficients equal to zero?

7.42 What is the maximum number of points of intersection of the graphs of two different fourth degree polynomial functions $y = p(x)$ and $y = q(x)$, each with leading coefficient 1? *(Source: AHSME)*

7.43 Sketch the graphs of each of the following functions. Your graphs don't have to be precisely correct, but should have the correct intercepts and be above or below the x-axis in the right places.

(a) $f(x) = -(x^2 - 1)(x + 4)$

(b) $f(x) = (x - 1)^2(x + 2)(x + 3)$

7.44 The graph at right has y-intercepts $(0, 3)$, $(0, -2)$, and $(0, -4)$, and it has x-intercept $(3, 0)$. Find an equation whose graph matches this graph.

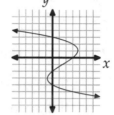

7.45 Let $f(x)$ be a polynomial with degree n, where $n > 1$. Prove that no line intersects the graph of $y = f(x)$ in more than n points. (Note: Your proof must address all possible lines, not just horizontal lines.)

7.46 If $f(x)$ is a monic quartic polynomial such that $f(-1) = -1$, $f(2) = -4$, $f(-3) = -9$, and $f(4) = -16$, find $f(1)$. *(Source: HMMT)*

Challenge Problems

7.47 How many real solutions are there to the equation $2(x - 1)^2(2x + 3)(x + 5)^3 = .001$?

7.48 Let $f(x) = x^3 - x^2 - 13x + 24$. Find three pairs (x, y) such that if $y = f(x)$, then $x = f(y)$.

7.49 The polynomial $g(x) = x^3 - x^2 - (m^2 + m)x + 2m^2 + 4m + 2$ is divisible by $x - 4$ and all of the roots of $g(x)$ are integers. Find m.

7.50 Show that if a polynomial $f(x)$ leaves the same remainder when divided by $x - a$ and by $x + a$ for all constants a, then the polynomial does not have any terms with odd degree. **Hints:** 49

7.51 Given an integer $n > 2$, for how many different rational numbers x does there exist a polynomial $P(x)$ of degree $n - 1$ with $P(0) = 0$, and with all integer coefficients, such that $54x^n + P(x) = 315$? **Hints:** 253

7.52 Let $P(x) = x^4 + ax^3 + bx^2 + cx + d$, where a, b, c, and d are constants. If $P(1) = 10$, $P(2) = 20$, and $P(3) = 30$, compute

$$\frac{P(12) + P(-8)}{10}.$$

(Source: ARML) **Hints:** 42

7.53★ Let $f(x) = x^2 + (n - a)x + a$, where a and n are integers such that $1 \le a \le n \le 49$. For how many values of n does $f(x)$ have irrational roots for all possible values of a?

7.54 Prove that if all coefficients of a quadratic equation are odd integers, then the roots of the equation cannot be rational.

7.55★ Shown is the complete graph of $y = f(x)$, a polynomial function of degree 10 whose domain is restricted to $[1, 5]$. Function f is symmetric about $x = 3$. Compute the number of solutions to the equation $f(x) = f(f(x))$. *(Source: ARML)* **Hints:** 30

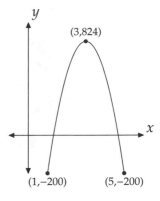

7.56★ In order to get to Mars, you must win a video game. The video game chooses 10 points (a, b), where a and b are single-digit integers, and places a disk with radius 1/3 on each of the points. You must find a polynomial f such that the graph of f hits all 10 discs. However, you must choose your polynomial *before seeing where the disks are*. Find a polynomial that guarantees you a trip to Mars. **Hints:** 151

7.57★ Let $f(x) = x^n + a_1 x^{n-1} + a_2 x^{n-2} + \cdots + a_{n-1}x + a_n$, where all the coefficients are integers. Suppose there are four distinct integers p, q, r, and s such that $f(p) = f(q) = f(r) = f(s) = 5$. Show that there is no integer k such that $f(k) = 8$. *(Source: Canada)* **Hints:** 217

7.58★ Find all pairs of nonzero integers (a, b) such that $(a^2 + b)(a + b^2) = (a - b)^3$. *(Source: USAMO)* **Hints:** 46

Extra! Having learned the quadratic formula, you might wonder if there is a similar formula
➡➡➡➡ for polynomials with higher degrees. Finding such formulas was a notable quest for centuries. In 1545, Girolamo Cardano published his method to solve any cubic and the method his pupil, Ludovico Ferrari, derived for solving quartic equations.

Both methods were much more complicated than the quadratic formula, and both were controversial, since Cardano's and Ferrari's work was based on a method showed to them by Nicolo Fontana Tartaglia for solving equations of the for $x^3 + bx = c$, where b and c are constants. Tartaglia had not published his results, and had allegedly sworn Cardano and Ferrari to secrecy.

Following Cardano's and Ferrari's successes, mathematicians spent over two centuries searching for a method to solve quintic equations. They searched in vain. Amateur mathematician Paolo Ruffini (about whom we read on page 177 with respect to synthetic division) was one of the first to suggest in print that it was impossible to find a general solution like those of Cardano and Ferrari for cubics and quartics. Being an amateur, Ruffini found his work largely overlooked, perhaps because mathematicians of his day (he published in 1799—over 250 years after Cardano and Ferrari) mostly still believed a solution for quintics was possible.

Later examination of Ruffini's work would expose some flaws in his arguments, but he was one of the first to be barking up the right tree. The great mathematician Niels Abel would finally put the problem to rest in 1824, proving that Ruffini was correct in his claim that there is no general solution for quintics as there is for quadratics, cubics, and quartics. See if you can figure out how to use the fact that there is no general solution for quintics to show that there is no general solution for higher degree polynomials, either.

The roots of education are bitter, but the fruit is sweet. – Aristotle

CHAPTER **8**

Polynomial Roots Part II

We continue our investigation of roots of polynomials by considering polynomials with roots that are not rational numbers. We'll consider both irrational roots and nonreal roots. We then will explore a useful relationship between the roots and the coefficients of a polynomial.

8.1 Irrational Roots

The **Pythagoreans**, an ancient Greek society of mathematicians and philosophers headed by the great mathematician **Pythagoras**, believed that "all is number." However, as we shall see, they were missing a few numbers.

The Pythagoreans started with the numbers we now know as the **positive integers**:

$$1, 2, 3, 4, 5, \ldots$$

They didn't accept zero as a number, nor did they use what we now call negative numbers. They did, however, use ratios of positive integers, such as

$$\frac{4}{3}, \frac{19}{4}, \frac{203}{2}, \frac{1}{17}.$$

Nowadays, we use the term **rational numbers** to describe ratios of integers, including 0 and negative ratios. Just as we use $\mathbb{R}$ to refer to the real numbers and $\mathbb{C}$ for the set of complex numbers, we use $\mathbb{Z}$ to refer to the set of all integers and $\mathbb{Q}$ for the set of rational numbers.

At first, the Pythagoreans believed that every number could be expressed as the ratio of two positive integers. Then, one of them made a startling discovery. The Greek **Hippasus** proved that the square root of 2 could not be written as a ratio of two positive integers.

We use the term **irrational number** to describe a real number that is not rational; that is, an irrational number cannot be written as the ratio of two integers. According to legend, Hippasus made his discovery while at sea and was rewarded by being thrown overboard. But his fate didn't change the fact that there are, indeed, irrational numbers. And the discovery of new numbers didn't stop there. Centuries later, mathematicians of various cultures would introduce 0, negative numbers, and nonreal numbers.

We'll start this section with a proof that $\sqrt{2}$ is irrational, but don't worry, we won't throw you overboard if you're able to prove it yourself!

 Problems

Problem 8.1: In this problem, we prove that $\sqrt{2}$ cannot be written as a ratio of two positive integers.

  (a) Suppose that $\sqrt{2} = p/q$, where p and q are positive integers that have no common factors. Square this equation and use the resulting equation to prove that p must be even.

  (b) Part (a) tells us that $p = 2r$ for some positive integer r. Substitute this into your equation in (a) and use the result to prove that q is even.

  (c) We started by assuming that $\sqrt{2} = p/q$, where p and q are positive integers that have no common factors. Why do your conclusions in parts (a) and (b) show that this is impossible?

Problem 8.2:

  (a) Prove that the ratio of two rational numbers must be rational.

  (b) Prove that if $\sqrt{d}$ is irrational, and b and c are rational numbers such that $b + c\sqrt{d} = 0$, then we must have $b = c = 0$.

Problem 8.3: If $b, c,$ and d are rational and $\sqrt{d}$ is irrational, then let $\underline{b + c\sqrt{d}} = b - c\sqrt{d}$. So, for example, we have $\underline{-2 + 3\sqrt{2}} = -2 - 3\sqrt{2}$ and $\underline{4 - 5\sqrt{3}} = 4 + 5\sqrt{3}$.

  For the following parts, let $t_i = b_i + c_i\sqrt{d}$, where $b_i, c_i,$ and d are rational and $\sqrt{d}$ is irrational.

  (a) Show that $\underline{t_1 + t_2} = \underline{t_1} + \underline{t_2}$.

  (b) Show that $\underline{t_1 t_2} = \underline{t_1} \cdot \underline{t_2}$.

  (c) Show that if $f(x)$ is a polynomial with rational coefficients such that t_1 is a root of $f(x)$, then $\underline{t_1}$ is also a root of $f(x)$.

Problem 8.1: Show that $\sqrt{2}$ is irrational.

Solution for Problem 8.1: We show that $\sqrt{2}$ cannot be written as the ratio of two positive numbers by using **proof by contradiction**. To do so, we start by assuming that $\sqrt{2}$ *is* rational, and then show that this assumption leads to an impossible conclusion. Assuming $\sqrt{2}$ is rational, it can be written as a fraction p/q, where p and q are integers and p/q is in lowest terms. So, we have

$$\sqrt{2} = \frac{p}{q}.$$

We square both sides to get $2 = p^2/q^2$, so $2q^2 = p^2$. Since p and q are integers, this means that p is even. So, we have $p = 2r$, for some integer r. Substituting this into $2q^2 = p^2$ gives $2q^2 = (2r)^2 = 4r^2$. Dividing both sides of $2q^2 = 4r^2$ by 2 gives us $q^2 = 2r^2$. So, now we see that q is even, too! However, we assumed that p/q is in lowest terms, so p and q can't both be divisible by 2.

Because our assumption (that $\sqrt{2}$ can be written as a ratio of positive integers in lowest terms) leads to the impossible conclusion that the ratio is also *not* in lowest terms, we must conclude that our original assumption is wrong. Specifically, we conclude that $\sqrt{2}$ cannot be written as a ratio of integers. $\square$

> **Important:** We can sometimes prove a statement by showing that if we assume the statement is false, then an impossible conclusion results. We call this strategy **proof by contradiction** or **indirect proof**.

Essentially the same argument that we used in Problem 8.1 can be used to show that if d is a positive integer that is not a perfect square, then $\sqrt{d}$ is irrational.

Problem 8.2:
(a) Prove that the ratio of two rational numbers must be rational.

(b) Prove that if $\sqrt{d}$ is irrational, and b, c, and d are rational numbers such that $b + c\sqrt{d} = 0$, then we must have $b = c = 0$.

Solution for Problem 8.2:

(a) Let p and q be two nonzero rational numbers. Because p and q are rational, we have $p = m/n$ and $q = r/s$ for some integers m, n, r, and s. Therefore, we have

$$\frac{p}{q} = \frac{m/n}{r/s} = \frac{ms}{rn}.$$

Since m, n, r, and s are integers, both ms and rn are integers. So, we can write the ratio p/q as the ratio of two integers, which means p/q is rational.

(b) If $b + c\sqrt{d} = 0$, then we have $c\sqrt{d} = -b$. From here, there are two possibilities. First, we could have $c = 0$, which makes $b = 0$, too. Second, if $c \neq 0$, then $b \neq 0$ also, and dividing our equation by c gives $\sqrt{d} = -b/c$. This means that we can express $\sqrt{d}$ as the ratio of two nonzero rational numbers, $-b$ and c. From part (a), we know that the ratio of these two rational numbers is also rational. But this means $\sqrt{d}$ is equal to a rational number, which we are given is not true. So, we conclude that we cannot have $c \neq 0$, which leaves $b = c = 0$ as the only possibility.

$\square$

> **Important:** If $\sqrt{d}$ is irrational, and b and c are rational numbers such that $b + c\sqrt{d} = 0$, then $b = c = 0$.

Back in Section 3.1, we introduced a special notation for complex conjugates. Specifically, we let $\overline{a + bi} = a - bi$. We'll now introduce a similar notation for numbers like $1 + 2\sqrt{3}$. If b, c, and d are rational and $\sqrt{d}$ is irrational, then we let $\underline{b + c\sqrt{d}} = b - c\sqrt{d}$. So, for example, we have $\underline{-2 + 3\sqrt{2}} = -2 - 3\sqrt{2}$ and $\underline{4 - 5\sqrt{3}} = 4 + 5\sqrt{3}$. We call $b - c\sqrt{d}$ the **radical conjugate** of $b + c\sqrt{d}$.

The notation $\underline{b + c\sqrt{d}}$ is not standard. There isn't a standard notation for radical conjugates the way there is for complex conjugates. However, this new terminology and notation will help us with our next proof.

Problem 8.3: Let $t_i = b_i + c_i\sqrt{d}$, where b_i, c_i, and d are rational and $\sqrt{d}$ is irrational.

(a) Show that $\underline{t_1 + t_2} = \underline{t_1} + \underline{t_2}$,

(b) Show that $\underline{t_1 t_2} = \underline{t_1} \cdot \underline{t_2}$.

(c) Show that if $f(x)$ is a polynomial with rational coefficients such that t_1 is a root of $f(x)$, then $\underline{t_1}$ is also a root of $f(x)$.

Solution for Problem 8.3:

(a) We have $t_1 + t_2 = (b_1 + c_1\sqrt{d}) + (b_2 + c_2\sqrt{d}) = (b_1 + b_2) + (c_1 + c_2)\sqrt{d}$, so $\underline{t_1 + t_2} = (b_1 + b_2) - (c_1 + c_2)\sqrt{d}$. Because $\underline{t_1} = b_1 - c_1\sqrt{d}$ and $\underline{t_2} = b_2 - c_2\sqrt{d}$, we have

$$\underline{t_1} + \underline{t_2} = b_1 - c_1\sqrt{d} + b_2 - c_2\sqrt{d} = (b_1 + b_2) - (c_1 + c_2)\sqrt{d} = \underline{t_1 + t_2}.$$

(b) We have

$$\underline{t_1 t_2} = \overline{(b_1 + c_1\sqrt{d})(b_2 + c_2\sqrt{d})} = \underline{(b_1 b_2 + c_1 c_2 d) + ((b_1 c_2 + b_2 c_1)\sqrt{d})} = (b_1 b_2 + c_1 c_2 d) - ((b_1 c_2 + b_2 c_1)\sqrt{d})$$

and $\underline{t_1} \cdot \underline{t_2} = (b_1 - c_1\sqrt{d})(b_2 - c_2\sqrt{d}) = (b_1 b_2 + c_1 c_2 d) - ((b_1 c_2 + b_2 c_1)\sqrt{d})$, so $\underline{t_1 t_2} = \underline{t_1} \cdot \underline{t_2}$.

(c) Let $f(x) = a_n x^n + a_{n-1}x^{n-1} + \cdots + a_1 x + a_0$. Applying parts (a) and (b) gives us

$$f(\underline{t_1}) = a_n(\underline{t_1})^n + a_{n-1}(\underline{t_1})^{n-1} + \cdots + a_1(\underline{t_1}) + a_0$$
$$= a_n(\underline{t_1^n}) + a_{n-1}(\underline{t_1^{n-1}}) + \cdots + a_1(\underline{t_1}) + a_0$$
$$= \underline{a_n t_1^n} + \underline{a_{n-1}t_1^{n-1}} + \cdots + \underline{a_1 t_1} + \underline{a_0}$$
$$= \underline{a_n t_1^n + a_{n-1}t_1^{n-1} + \cdots + a_1 t_1 + a_0}$$
$$= \underline{f(t_1)}.$$

If t_1 is a root of $f(x)$, we have $f(t_1) = 0$, so $f(\underline{t_1}) = \underline{f(t_1)} = \underline{0} = 0$, which means $\underline{t_1}$ is also a root of $f(x)$.

$\square$

Important: If $b + c\sqrt{d}$, where b, c, and d are rational and $\sqrt{d}$ is irrational, is the root of a polynomial with rational coefficients, then $b - c\sqrt{d}$ is also a root of the polynomial.

> **WARNING!!** ☢ The notation $b + c\sqrt{d} = b - c\sqrt{d}$ is not standard! We only introduced it here to be able to write our solution to Problem 8.3 more simply. We won't use this notation again in this book.

Problems

Problem 8.4: Let $f(x) = x^3 - x - 2$.
  (a) Show that $f(x)$ has no rational roots.
  (b) Show that $f(x)$ has at least one irrational root.

Problem 8.5: Find a polynomial with rational coefficients that has $2 - \sqrt{3}$ and $3 + 2\sqrt{5}$ as roots.

Problem 8.6: Find a polynomial with rational coefficients that has $\sqrt{3 + \sqrt{7}}$ as a root.

Problem 8.7: Let $g(x) = 4x^4 - 16x^2 + 15$.
  (a) Find all roots of $g(x)$.
  (b) Factor $g(x)$ completely into factors with rational coefficients.
  (c) Factor $g(x)$ completely into factors with real coefficients.

Problem 8.8: Let $r = \sqrt{2} + \sqrt{3}$.
  (a) Show that r is a root of $x^4 - 10x^2 + 1 = 0$.
  (b) Prove that r is irrational.

Problem 8.9: Prove that $\sqrt[3]{5} + \sqrt[3]{25}$ is irrational.

Problem 8.4: Show that $f(x) = x^3 - x - 2$ has at least one irrational root.

Solution for Problem 8.4: We have $f(0) = -2$ and $f(2) = 4$, so there is a root between 0 and 2. By the Rational Root Theorem, the only possible rational root of $f(x)$ between 0 and 2 is 1. However, we have $f(1) = -2$, so 1 is not a root. Therefore, there is an irrational root between 0 and 2. □

Problem 8.5: Find a polynomial with rational coefficients that has $2 - \sqrt{3}$ and $3 + 2\sqrt{5}$ as roots.

Solution for Problem 8.5: Let $f(x)$ be our desired polynomial. Because $2 - \sqrt{3}$ is a root and $f(x)$ has rational coefficients, we know that $2 + \sqrt{3}$ is also a root. Similarly, because $3 + 2\sqrt{5}$ is a root of $f(x)$, so is $3 - 2\sqrt{5}$. We can now guess that

$$f(x) = \left(x - (2 - \sqrt{3})\right)\left(x - (2 + \sqrt{3})\right)\left(x - (3 - 2\sqrt{5})\right)\left(x - (3 + 2\sqrt{5})\right)$$

will work. We'll look at a couple ways to multiply this out without getting lost in a forest of square roots. Both involve taking advantage of the radical conjugates.

First, we might note that because $2 - \sqrt{3}$ and $2 + \sqrt{3}$ are radical conjugates, their sum and their product are integers, namely $(2 - \sqrt{3}) + (2 + \sqrt{3}) = 4$ and $(2 - \sqrt{3})(2 + \sqrt{3}) = 1$. So, by Vieta's Formulas, the monic quadratic with these numbers as roots is $x^2 - 4x + 1$, which is nice because it has no radicals in its coefficients. Similarly, we have $(3 + 2\sqrt{5}) + (3 - 2\sqrt{5}) = 6$ and $(3 + 2\sqrt{5})(3 - 2\sqrt{5}) = 9 - 20 = -11$, so the product of the second two factors in our expression for $f(x)$ is $x^2 - 6x - 11$. Therefore, the polynomial

$$f(x) = (x^2 - 4x + 1)(x^2 - 6x - 11) = x^4 - 10x^3 + 14x^2 + 38x - 11$$

has rational coefficients and has $2 - \sqrt{3}$ and $3 + 2\sqrt{5}$ as roots.

Another way we can use the radical conjugates is to rewrite the product $(x - (2 - \sqrt{3}))(x - (2 + \sqrt{3}))$ as a factored difference of squares, and then expand the product:

$$\left(x - (2 - \sqrt{3})\right)\left(x - (2 + \sqrt{3})\right) = \left((x - 2) + \sqrt{3}\right)\left((x - 2) - \sqrt{3}\right) = (x - 2)^2 - \left(\sqrt{3}\right)^2 = x^2 - 4x + 1.$$

Similarly, we can multiply the other two factors of $f(x)$ to find

$$\left(x - (3 - 2\sqrt{5})\right)\left(x - (3 + 2\sqrt{5})\right) = \left((x - 3) + 2\sqrt{5}\right)\left((x - 3) - 2\sqrt{5}\right) = (x - 3)^2 - \left(2\sqrt{5}\right)^2 = x^2 - 6x - 11,$$

so $f(x) = (x^2 - 4x + 1)(x^2 - 6x - 11) = x^4 - 10x^3 + 14x^2 + 38x - 11$, as before. □

Problem 8.6: Find a polynomial with rational coefficients that has $\sqrt{3 + \sqrt{7}}$ as a root.

Solution for Problem 8.6: Clearly, this can't be the only root of the polynomial, since the polynomial must have rational coefficients. However, it's not immediately clear what other numbers must be roots in order to create a polynomial with rational coefficients. So, we take an educated guess.

Concept: There's nothing wrong with guessing! Guessing has a long and glorious tradition in math and science. Guessing, and then confirming that your guess is correct, is a perfectly valid way to solve problems.

We know that if $0 + b\sqrt{c}$, where b and c are rational but $\sqrt{c}$ is not, is a root of a polynomial with rational coefficients, then $0 - b\sqrt{c}$ is also a root. So, we guess that another root of our polynomial must be $-\sqrt{3 + \sqrt{7}}$. The equation $x^2 = 3 + \sqrt{7}$ clearly has $\pm\sqrt{3 + \sqrt{7}}$ as solutions, so a quadratic with these two ugly numbers as roots is

$$x^2 - (3 + \sqrt{7}).$$

That's not so bad. But we still have one irrational coefficient. Writing $x^2 - (3 + \sqrt{7})$ as $(x^2 - 3) - \sqrt{7}$, we see that can get rid of the square root by multiplying by $(x^2 - 3) + \sqrt{7}$, which gives us the polynomial

$$\left(x^2 - 3 - \sqrt{7}\right)\left(x^2 - 3 + \sqrt{7}\right) = (x^2 - 3)^2 - \left(\sqrt{7}\right)^2 = x^4 - 6x^2 + 2.$$

Success! Notice that the other two roots of this polynomial are $\pm\sqrt{3 - \sqrt{7}}$. □

CHAPTER 8. POLYNOMIAL ROOTS PART II

Problem 8.7: Let $g(x) = 4x^4 - 16x^2 + 15$.

(a) Find all roots of $g(x)$.

(b) Factor $g(x)$ completely into factors with rational coefficients.

(c) Factor $g(x)$ completely into factors with real coefficients.

Solution for Problem 8.7:

(a) We can factor $g(x)$ as a quadratic by letting $y = x^2$. This gives us

$$4x^4 - 16x^2 + 15 = 4y^2 - 16y + 15 = (2y - 3)(2y - 5).$$

Since $y = x^2$, we have

$$g(x) = (2x^2 - 3)(2x^2 - 5).$$

So, the roots of $2x^2 - 3$ and $2x^2 - 5$ are the roots of $g(x)$. Therefore, the roots of $g(x)$ are $\pm\sqrt{6}/2$ and $\pm\sqrt{10}/2$.

(b) If $g(x)$ has a factor of the form $ax + b$, where a and b are rational, then the rational number $x = -b/a$ is a root of $g(x)$. However, we have already found that $g(x)$ has no rational roots. Therefore, $g(x)$ does not have any linear factors with rational coefficients. So, our factorization above, $g(x) = (2x^2 - 3)(2x^2 - 5)$, is the most we can factor $g(x)$ such that all the factors have rational coefficients.

(c) The roots of $g(x)$ are $\pm\frac{\sqrt{6}}{2}$ and $\pm\frac{\sqrt{10}}{2}$, and the leading coefficient of $g(x)$ is 4, so

$$g(x) = 4\left(x + \frac{\sqrt{6}}{2}\right)\left(x - \frac{\sqrt{6}}{2}\right)\left(x + \frac{\sqrt{10}}{2}\right)\left(x - \frac{\sqrt{10}}{2}\right).$$

We also could have factored $2x^2 - 3$ and $2x^2 - 5$ in $g(x) = (2x^2 - 3)(2x^2 - 5)$ as differences of squares to find

$$g(x) = \left(\sqrt{2}x - \sqrt{3}\right)\left(\sqrt{2}x + \sqrt{3}\right)\left(\sqrt{2}x - \sqrt{5}\right)\left(\sqrt{2}x + \sqrt{5}\right).$$

The advantage of our first factorization is that it is much easier to read the roots of $g(x)$ than it is in the second factorization.

☐

The final two parts of Problem 8.7 show that we can sometimes factor a polynomial into more factors if we allow irrational coefficients among the factors than we can if we only allow rational coefficients.

When we factor a polynomial as completely as possible such that each factor has rational coefficients, we say we "factor the polynomial over the rational numbers." Similarly, if we factor the polynomial as completely as possible such that each factor has real coefficients, we "factor the polynomial over the reals." Likewise, we can "factor a polynomial over the integers" (meaning the factors have integer coefficients) or "factor a polynomial over the complex numbers" (meaning the factors have complex coefficients).

The Rational Root Theorem also gives us another method to prove that a number is irrational.

Problem 8.8: Let $r = \sqrt{2} + \sqrt{3}$. Prove that r is irrational.

Solution for Problem 8.8: In Problem 8.4, we deduced that a polynomial had an irrational root by first showing that it had a real root, then showing that it didn't have any rational roots. This gives us an idea for the present problem. If we can find a polynomial $f(x)$ with rational coefficients but no rational roots such that $f(r) = 0$, then we can deduce that r is irrational. Problem 8.6 gives us a guide for building a polynomial with $\sqrt{2} + \sqrt{3}$ as a root.

Let $f(x)$ be a polynomial with rational coefficients and r as a root. We start by guessing that $\sqrt{2} - \sqrt{3}$ is also a root of $f(x)$. Because $(\sqrt{2} + \sqrt{3}) + (\sqrt{2} - \sqrt{3}) = 2\sqrt{2}$ and $(\sqrt{2} + \sqrt{3})(\sqrt{2} - \sqrt{3}) = -1$, a quadratic with $\sqrt{2} + \sqrt{3}$ and $\sqrt{2} - \sqrt{3}$ as roots is

$$x^2 - 2\sqrt{2}x - 1.$$

We can use the same tactic we used in our solution to Problem 8.6 to get rid of the square root. We write $x^2 - 2\sqrt{2}x - 1$ as $x^2 - 1 - 2\sqrt{2}x$. Multiplying by $x^2 - 1 + 2\sqrt{2}x$ will get rid of the radicals:

$$\left(x^2 - 1 - 2\sqrt{2}x\right)\left(x^2 - 1 + 2\sqrt{2}x\right) = (x^2 - 1)^2 - \left(2\sqrt{2}x\right)^2 = x^4 - 10x^2 + 1.$$

So, $f(x) = x^4 - 10x^2 + 1$ has $\sqrt{2} + \sqrt{3}$ as a root. By the Rational Root Theorem, the only possible rational roots of $f(x)$ are 1 and -1. Since $f(1) = -8$ and $f(-1) = -8$, we know that $f(x)$ has no rational roots. Therefore, all the real roots of $f(x)$ are irrational. Since $\sqrt{2} + \sqrt{3}$ is one of these roots, we know that it is irrational. $\square$

Problem 8.9: Prove that $\sqrt[3]{5} + \sqrt[3]{25}$ is irrational.

Solution for Problem 8.9: We'd like to take the same approach as in the previous problem. We let $r = \sqrt[3]{5} + \sqrt[3]{25}$, and we try to find a polynomial $f(x)$ with integer coefficients and r as a root.

We start off by guessing that f must have degree 3, because we expect we'll have to cube r to get rid of the cube roots. We can simplify our steps when cubing r by letting $a = \sqrt[3]{5}$ and $b = \sqrt[3]{25}$, so $r = a + b$. This makes the algebraic manipulation of cubing r a little easier:

$$r^3 = (a + b)^3 = (a + b)(a + b)^2 = (a + b)(a^2 + 2ab + b^2) = a^3 + 3a^2b + 3ab^2 + b^3.$$

Because $a^3 = 5$ and $b^3 = 25$, we have

$$r^3 = 30 + 3a^2b + 3ab^2.$$

We can factor $3a^2b + 3ab^2$ on the right to give $r^3 = 30 + 3ab(a + b)$.

> **Concept:** When in doubt, factor. Factoring often simplifies expressions.

But what can we do with $3ab(a + b)$?

> **WARNING!!** When you make a substitution to help with algebra, don't forget what your variables stand for.

We have $r = a+b$ and $ab = \sqrt[3]{125} = 5$, so $3ab(a+b) = 15r$, and we have $r^3 = 30+15r$. Rearranging this gives $r^3 - 15r - 30 = 0$, and we have our polynomial! The number $r = \sqrt[3]{5} + \sqrt[3]{25}$ is a root of $f(x) = x^3 - 15x - 30$ because $r^3 - 15r - 30 = 0$. Clearly, r is positive, so we only have to test $f(x)$ for positive rational roots. We can quickly work through the possibilities given by the Rational Root Theorem (namely, the divisors of 30) and see that there are no positive rational roots of $f(x)$. Therefore, the root $\sqrt[3]{5} + \sqrt[3]{25}$ of $f(x)$ is irrational. $\square$

Sidenote: A real number that is not a root of any nonzero polynomial with rational coefficients is called a **transcendental number**. For example, the number $\sqrt{2}$ is not transcendental, because it is a root of $x^2 - 2$. All transcendental numbers are irrational, because any rational number r is a root of the polynomial $x - r$.

Naturally, we also have a name for a number that is a root of some nonzero polynomial with rational coefficients. Such a number is called an **algebraic number**. All rational numbers are algebraic, as are all square roots. (Of course, as we have seen in several problems in this section, rational numbers and square roots are not the only algebraic numbers!)

One interesting fact about transcendental numbers is that, given a segment of length 1, it is impossible to construct a segment of length k with straightedge and compass if k is transcendental. In 1882, mathematician Ferdinand von Lindemann used this fact to put to rest a problem that had baffled mathematicians for millennia. The problem, known as "squaring the circle," was to construct a circle with area equal to the area of a given square. Performing this construction essentially requires being able to construct a segment of length $\sqrt{\pi}$ given a segment of length 1. Ferdinand von Lindemann proved that π is transcendental, thus putting this problem to rest at last.

Exercises

8.1.1 Find the roots of the polynomial $x^3 - 7x^2 + 12x - 4$.

8.1.2 Find a monic quartic polynomial with rational coefficients such that $2 - 3\sqrt{2}$ and $1 - \sqrt{3}$ are roots of the polynomial.

8.1.3 Prove that $\sqrt[3]{4}$ is irrational.

8.1.4 Let $h(x) = x^4 - 3x^2 + 2$.

(a) Factor $h(x)$ completely over the rational numbers.

(b) Factor $h(x)$ completely over the reals.

8.1.5 Let $p(x)$ be a polynomial with rational coefficients. Is it true that if $p(4 + \sqrt[3]{7}) = 0$, then $p(4 - \sqrt[3]{7}) = 0$?

8.1.6★ Find a polynomial with integer coefficients that has $\sqrt{\sqrt{2} + \sqrt{3}}$ as a root. **Hints:** 153

8.2 Nonreal Roots

Problems

Problem 8.10: Find the roots of the following two polynomials:

(a) $f(x) = x^3 - 2x^2 + 3x - 18$

(b) $g(x) = 2x^4 - 8x^3 - 3x^2 + 22x - 40$

Problem 8.11: Let $p(x) = a_n x^n + a_{n-1} x^{n-1} + \cdots + a_1 x + a_0$ and $q(x) = a_m x^m$ be polynomials with real coefficients, and recall that if $z = c + di$, where c and d are real, then $\bar{z} = c - di$.

(a) Show that $\overline{q(z)} = q(\bar{z})$ for all complex numbers z.

(b) Show that $\overline{p(z)} = p(\bar{z})$ for all complex numbers z.

(c) Show that if $p(z) = 0$, then $p(\bar{z}) = 0$.

Problem 8.12:

(a) Find the monic quadratic with real coefficients that has $1 + i$ as a root.

(b) Find the monic quadratic with real coefficients that has $1 - i$ as a root.

(c) Find the monic polynomial $f(x)$ of minimal degree such that $f(1 + i) = f(-1 + i) = 0$ and all the coefficients of $f(x)$ are real.

Problem 8.13: Let $f(x) = a_n x^n + a_{n-1} x^{n-1} + \cdots + a_0$, where all the a_i are real. In this problem, we show that if r is a nonreal double root of the polynomial $f(x)$, then $\bar{r}$ is also a double root of $f(x)$.

(a) Let $r = c + di$. Find a quadratic $g(x)$ with real coefficients (in terms of c and d) such that $g(x)$ evenly divides $f(x)$.

(b) Show that when $f(x)$ is divided by $g(x)$, the quotient has real coefficients.

(c) Show that $\bar{r}$ is a double root of $f(x)$.

Problem 8.14: Prove that a nonconstant polynomial with real coefficients can be factored into a product of linear and/or quadratic polynomials with real coefficients.

Problem 8.15: Consider the equation $c_4 z^4 + i c_3 z^3 + c_2 z^2 + i c_1 z + c_0 = 0$, where c_0, c_1, c_2, c_3, and c_4 are real constants.

(a) Find a function h such that substituting $z = h(x)$ into the polynomial above produces a fourth degree polynomial in x with real coefficients.

(b) If $z = a + bi$ is a solution to the original equation, where a and b are real, then what value of x solves your equation from part (a)?

(c) What other value of x must be a solution to your equation from (a)?

(d) Prove that $z = -a + bi$ is a solution to the original equation.

(Source: AHSME)

Problem 8.10: Find the roots of the following two polynomials:

(a) $f(x) = x^3 - 2x^2 + 3x - 18$

(b) $g(x) = 2x^4 - 8x^3 - 3x^2 + 22x - 40$

Solution for Problem 8.10:

(a) We first note that because the signs of the coefficients of $f(x)$ alternate, there are no negative roots of $f(x)$. Next, we use the Rational Root Theorem to hunt for roots by testing divisors of 18 to see if they are roots of $f(x)$. We find that $f(3) = 0$ and

$$f(x) = (x - 3)(x^2 + x + 6).$$

We can't factor $x^2 + x + 6$ easily, so we turn to the quadratic formula to find that the roots of $x^2 + x + 6$ are

$$x = \frac{-1 \pm i\sqrt{23}}{2}.$$

So, the three roots of $f(x)$ are 3 and $\dfrac{-1 \pm i\sqrt{23}}{2}$.

(b) We start by searching for rational roots. We find that $x + 2$ divides $g(x)$ evenly and gives us

$$g(x) = (x + 2)(2x^3 - 12x^2 + 21x - 20).$$

We continue by searching for roots of $2x^3 - 12x^2 + 21x - 20$. First, we note that there are no negative roots because the signs of the coefficients alternate (and none of them is 0). So, we test for positive roots and eventually find that $x - 4$ is a factor. Dividing $2x^3 - 12x^2 + 21x - 20$ by $x - 4$ gives us

$$g(x) = (x + 2)(x - 4)(2x^2 - 4x + 5).$$

As in part (a), we tackle the quadratic with the quadratic formula and find that its roots are

$$x = \frac{4 \pm 2i\sqrt{6}}{4} = 1 \pm \frac{i\sqrt{6}}{2}.$$

So, the four roots of $g(x)$ are -2, 4, $1 - \frac{i\sqrt{6}}{2}$, and $1 + \frac{i\sqrt{6}}{2}$.

$\square$

Hmmm.... For both polynomials in Problem 8.10, the nonreal roots come in conjugate pairs (in other words, if $a + bi$ is a root, then so is $a - bi$). We've already seen that nonreal roots of a quadratic with real coefficients must form a conjugate pair. Let's see if we can prove a similar statement for higher degree polynomials.

As a reminder, we use $\bar{z}$ to denote the conjugate of the complex number z.

Problem 8.11: Let $p(x) = a_n x^n + a_{n-1} x^{n-1} + \cdots + a_1 x + a_0$ and $q(x) = a_m x^m$ be polynomials with real coefficients.

(a) Show that $\overline{q(z)} = q(\bar{z})$ for all complex numbers z.

(b) Show that $\overline{p(z)} = p(\bar{z})$ for all complex numbers z.

(c) Show that if $p(z) = 0$, then $p(\bar{z}) = 0$.

Solution for Problem 8.11:

(a) In Chapter 3, we showed that for all complex numbers w and z, we have $\overline{wz} = \overline{w} \cdot \overline{z}$. Using this relationship, along with the fact that a_m is real, we see that

$$\overline{a_m z^m} = \overline{a_m} \cdot \overline{z^m} = a_m \overline{z^m}.$$

Now, we note that

$$\overline{z^m} = \overline{z \cdot z^{m-1}} = \overline{z} \cdot \overline{z \cdot z^{m-2}} = \cdots = \underbrace{\overline{z} \cdot \overline{z} \cdots \overline{z}}_{m \ \overline{z}'s} = \overline{z}^m,$$

which gives us $\overline{q(z)} = \overline{a_m z^m} = a_m \overline{z^m} = a_m \overline{z}^m = q(\overline{z})$.

(b) In Chapter 3, we showed that for all complex numbers w and z, we have $\overline{w + z} = \overline{w} + \overline{z}$. Combining this with part (a), we have

$$\begin{aligned}
\overline{p(z)} &= \overline{a_n z^n + a_{n-1} z^{n-1} + \cdots + a_1 z + a_0} \\
&= \overline{a_n z^n} + \overline{a_{n-1} z^{n-1}} + \cdots + \overline{a_1 z} + \overline{a_0} \\
&= a_n \overline{z}^n + a_{n-1} \overline{z}^{n-1} + \cdots + a_1 \overline{z} + a_0 \\
&= p(\overline{z}),
\end{aligned}$$

as desired.

(c) From part (b), we have $p(\overline{z}) = \overline{p(z)}$. So, if $p(z) = 0$, then we have $p(\overline{z}) = \overline{p(z)} = \overline{0} = 0$. $\quad\blacksquare$

Important: Let $p(x)$ be a polynomial with real coefficients. If z is a root of $p(x)$, then $\overline{z}$ is also a root of $p(x)$. In other words, the nonreal roots of $p(x)$ come in conjugate pairs.

- We can sometimes use this fact to build a polynomial given some, but not all, of its roots.

Problem 8.12: Find the monic polynomial $f(x)$ of minimal degree such that $f(1 + i) = f(-1 + i) = 0$ and all the coefficients of $f(x)$ are real.

Solution for Problem 8.12: Nonreal roots of a polynomial with real coefficients come in conjugate pairs, so we know that $1 \pm i$ and $-1 \pm i$ are all roots of $f(x)$. So, the degree of f is at least four.

Now, we could find the product of four linear factors whose roots are the four known roots of $f(x)$. However, finding quadratics with the given roots is easier. The sum of $1 + i$ and $1 - i$ is 2 and the product $(1 + i)(1 - i)$ equals 2. So, Vieta's Formulas tell us that a quadratic with roots $1 \pm i$ is

$$x^2 - 2x + 2.$$

Similarly, we have $(-1 + i) + (-1 - i) = -2$ and $(-1 + i)(-1 - i) = 2$, so a quadratic with roots $-1 \pm i$ is

$$x^2 + 2x + 2.$$

The product of these quadratics contains all the known roots of $f(x)$. Multiplying these quadratics gives us

$$(x^2 - 2x + 2)(x^2 + 2x + 2) = x^4 + 4,$$

which is of degree four, so $f(x) = x^4 + 4$ is the desired polynomial. $\square$

For all the polynomials with real coefficients that we've factored so far, we've been able to write the polynomial as the product of linear and quadratic factors with real coefficients. In our next two problems, we prove that we can do so with all polynomials that have real coefficients.

Problem 8.13: Let $f(x) = a_n x^n + a_{n-1} x^{n-1} + \cdots + a_0$, where all the a_i are real. Show that if r is a nonreal double root of the polynomial $f(x)$, then $\overline{r}$ is also a double root of $f(x)$.

Solution for Problem 8.13: Let $r = c + di$ be a root of $f(x)$, where $d \neq 0$ because r is nonreal. Because the coefficients of $f(x)$ are real, $\overline{r} = c - di$ must also be a root of $f(x)$, and we have $r \neq \overline{r}$ because $d \neq 0$. Therefore, both $x - r$ and $x - \overline{r}$ are factors of $f(x)$, which means there is some polynomial $q(x)$ such that

$$f(x) = (x - r)(x - \overline{r})q(x).$$

Writing $(x - r)(x - \overline{r})$ in terms of c and d, and we have

$$(x - r)(x - \overline{r}) = x^2 - (r + \overline{r})x + r \cdot \overline{r} = x^2 - 2cx + c^2 - d^2,$$

so we have

$$f(x) = (x^2 - 2cx + c^2 - d^2)q(x).$$

In other words, $q(x)$ is the quotient when $f(x)$ is divided by a particular quadratic with real coefficients. This means all of the coefficients of $q(x)$ are real. (Consider our process in Section 6.2 for polynomial division to see why.)

We know that r is a double root of $f(x)$, so $x - r$ must appear twice as a factor in the factorization of $f(x)$. We only have one factor of $x - r$ so far in $f(x) = (x - r)(x - \overline{r})q(x)$, so $q(x)$ must contribute another factor of $x - r$. Therefore, r is a root of $q(x)$. Since all the coefficients of $q(x)$ are real but r is nonreal, we know that $\overline{r}$ is also a root of $q(x)$. Therefore, $x - \overline{r}$ is a factor of $q(x)$, which means that $\overline{r}$ is a double root of $f(x)$. $\square$

Problem 8.14: Prove that a nonconstant polynomial with real coefficients can be factored into a product of linear and/or quadratic polynomials with real coefficients.

Solution for Problem 8.14: Let $p(x)$ be a polynomial of degree n, for some positive integer n, such that all the coefficients of $p(x)$ are real. By the Fundamental Theorem of Algebra, p has exactly n complex roots. Let these roots be $r_1, r_2, \ldots, r_n$, so we have

$$p(x) = c(x - r_1)(x - r_2) \cdots (x - r_n), \tag{8.1}$$

for some nonzero real constant c.

When r_i is real, the expression $x - r_i$ is a linear factor with real coefficients. Otherwise, r_i is nonreal, so $\overline{r_i}$ is also a root of $p(x)$, because the coefficients of $p(x)$ are real. This means that $(x - \overline{r_i})$ is also one of the linear factors of p in Equation (8.1). Letting $r_i = a + bi$, we have $\overline{r_i} = a - bi$, and

$$(x - r_i)(x - \overline{r_i}) = x^2 - (r_i + \overline{r_i})x + r_i\overline{r_i}$$
$$= x^2 - 2ax + a^2 + b^2.$$

Since $2a$ and $a^2 + b^2$ are real, this quadratic has only real coefficients. So, it appears we can pair the linear factors of $p(x)$ that don't have real coefficients to create quadratic factors that do have real coefficients.

We can't yet simply deduce that all the linear factors with nonreal roots can be paired off this way! For example, suppose $f(x)$ has $(x + i)^2$ and $(x - i)$ among its factors. Then, from our work above, we can pair off either $x + i$ with the $x - i$ to produce a quadratic with real coefficients. But we can't pair off both $x + i$ factors with the same $x - i$ factor; once we pair off $x + i$ and $x - i$, we'll have one $x + i$ "left over." So, it's not enough to simply note that for every factor $x - r_i$ with r_i nonreal, there is a factor $x - \overline{r_i}$. We have to be careful about multiplicity.

> **WARNING!!** Watch out for special cases.
> ☢

This is where our work in Problem 8.13 comes into play. We showed there that if a polynomial with real coefficients has a double root that is nonreal, then the conjugate of the root is also a double root. Similarly, we can show that if a nonreal root of a polynomial with real coefficients has multiplicity k (that is, it appears k times among the roots), then the conjugate of the nonreal root also has multiplicity k. So, we can pair off linear factors with nonreal roots and not have any extra factors with nonreal roots "left over."

We're still not quite finished, because at this point, we've only explained that $p(x)$ can be written as the product of some nonzero constant, linear expressions with real coefficients, and quadratics with real coefficients. We still have to deal with the constant. Fortunately, that's easy. We can either include it in one of our linear expressions by writing $c(x - r_i) = cx - cr_i$ for any one of the real roots, or we can include it similarly in one of the quadratics.

So, every polynomial with real coefficients can be written as the product of linear and quadratic expressions with real coefficients. □

> **Problem 8.15:** Suppose $z = a + bi$ is a solution of the polynomial equation
>
> $$c_4z^4 + ic_3z^3 + c_2z^2 + ic_1z + c_0 = 0,$$
>
> where $c_0, c_1, c_2, c_3, c_4, a,$ and b are real constants. Prove that $-a + bi$ is also a solution. *(Source: AHSME)*

Solution for Problem 8.15: Let $g(z) = c_4z^4 + ic_3z^3 + c_2z^2 + ic_1z + c_0$. The coefficients of $g(z)$ are not all real, so we can't deduce that the nonreal roots come in conjugate pairs. We don't know much about roots of polynomials with nonreal coefficients, so we'd like to relate $g(z)$ to a polynomial in which all the coefficients are real. If we had such a polynomial, we'd at least then know that its nonreal roots come in conjugate pairs.

We notice that the imaginary coefficients are of the odd powers of z, so if we let $z = xi$, the resulting coefficients of x are real:

$$g(xi) = c_4x^4 + c_3x^3 - c_2x^2 - c_1x + c_0.$$

Aha! Now, we've related $g(z)$ to a polynomial with real coefficients. We let

$$f(x) = g(xi) = c_4x^4 + c_3x^3 - c_2x^2 - c_1x + c_0$$

and investigate the roots of $f(x)$. We are given that $g(a + bi) = 0$. Because $f(x) = g(xi)$, we therefore know that $f(x)$ is 0 when $xi = a + bi$. Solving for x gives us $x = (a + bi)/i = b - ai$, and we have $f(b - ai) = g((b - ai)i) = g(a + bi) = 0$. Since $f(b - ai) = 0$ and $f(x)$ has real coefficients, we know that $\overline{b - ai} = b + ai$ is also a root of $f(x)$. Now, because $g(xi) = f(x) = 0$ when $x = b + ai$, we see that $g(z) = 0$ when $z = (b + ai)i = -a + bi$, as desired. $\square$

Exercises

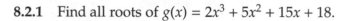

8.2.1 Find all roots of $g(x) = 2x^3 + 5x^2 + 15x + 18$.

8.2.2 Find all roots of $f(t) = 2t^4 - 23t^2 + 27t - 36$.

8.2.3 Let $g(x)$ be a polynomial with real coefficients such that $g(2 + i) = g(3 + i) = 0$, and $\deg g \le 4$. Other than $2 + i$ and $3 + i$, find all possible roots of g.

8.2.4 Prove that a fourth degree polynomial with a real root and real coefficients must have another (not necessarily distinct) real root.

8.2.5 Use the roots of $x^4 + 4$ we found in Problem 8.12 to find the four fourth roots of -1. In other words, find the four numbers whose fourth power equals -1.

8.2.6★ A polynomial of degree four with leading coefficient 1 and integer coefficients has two real zeros, both of which are integers. Which of the following can also be a zero of the polynomial?

$$\frac{1 + i\sqrt{11}}{2}, \frac{1 + i}{2}, \frac{1}{2} + i, 1 + \frac{i}{2}, \frac{1 + i\sqrt{13}}{2}.$$

(Source: AMC 12) **Hints:** 101

8.2.7★ Find all values of x such that $x^4 + 5x^2 + 4x + 5 = 0$. **Hints:** 112

8.3 Vieta's Formulas

We began our discussion of Vieta's Formulas in Section 4.2 by examining the relationship between the coefficients of a quadratic and the sum and product of the roots of the quadratic. In this section, we extend Vieta's Formulas to all polynomials.

Extra! François Vieta was a French mathematician in the late 1500's who once noted, "There ⫸⫸⫸⫸ is no problem that cannot be solved." He was likely inspired in this belief by his many mathematical successes, which were in part due to his being one of the first Europeans to use letters to represent unknown quantities in his published mathematics.

Problems ▶

Problem 8.16: Let $f(x) = x^3 - 4x^2 + 15x - 7$, and let p, q, and r be the roots of $f(x)$.

(a) Why must $f(x) = (x - p)(x - q)(x - r)$?

(b) Expand $(x - p)(x - q)(x - r)$.

(c) Compare the result in (b) to $f(x) = x^3 - 4x^2 + 15x - 7$ to find $p + q + r$.

(d) Find $pq + qr + rp$.

(e) Find pqr.

Problem 8.17: Let $f(x) = c(x - p)(x - q)(x - r)$, where c, p, q, and r are constants.

(a) What are the roots of $f(x)$?

(b) Expand the given expression for $f(x)$.

(c) Suppose we write $f(x)$ in the form $f(x) = a_3x^3 + a_2x^2 + a_1x + a_0$. Use your expansion in part (b) to write each of a_3, a_2, a_1, and a_0 in terms of c, p, q, and r.

(d) Why must $p + q + r = -a_2/a_3$?

(e) What is the product of the roots of $f(x)$ in terms of the coefficients of $f(x)$?

(f) Express $pq + qr + rp$ in terms of the coefficients of $f(x)$.

Problem 8.18: Let p, q, r, and s be the roots of $f(x) = 3x^4 + 2x^3 - 7x^2 + 9x - 10$. Find $pqr + pqs + prs + qrs$.

Problem 8.19: Let $f(x) = a_nx^n + a_{n-1}x^{n-1} + a_{n-2}x^{n-2} + \cdots + a_1x + a_0$, and let the roots of the polynomial $f(x)$ be $r_1, r_2, r_3, \ldots, r_n$.

(a) Why can we also write $f(x)$ as $f(x) = a_n(x - r_1)(x - r_2)(x - r_3) \cdots (x - r_n)$?

(b) When we expand the product in (a), what terms have x^{n-1} in them? What is the coefficient of x^{n-1} in the full expansion?

(c) When we expand the product in (a), what is the coefficient of x^{n-2}?

(d) When we expand the product in (a), what is the coefficient of x^{n-3}?

(e) When we expand the product in (a), what is the constant term?

(f) How are the coefficients of a polynomial related to its roots?

Problem 8.20: Let r, s, and t be the solutions to the equation $3x^3 - 4x^2 + 5x + 7 = 0$.

(a) Find $r + s + t$.

(b) Find $r^2 + s^2 + t^2$.

(c) Find $\frac{1}{r} + \frac{1}{s} + \frac{1}{t}$.

Extra! Vieta was well-positioned in the French government in the late 1500s. He was a favorite
▶▶▶▶ of King Henry IV, and the King relied on Vieta for a variety of mathematical matters. Vieta was so successful in decoding the messages of Henry IV's enemy, King Philip II of Spain, that Philip complained to the Pope that black magic was being used.

Problem 8.21: One of the roots of $x^3 + ax + b = 0$ is $1 + 2i$, where a and b are real numbers.

(a) Find the other nonreal root of the polynomial above.

(b) Find the sum of the roots of the polynomial above.

(c) Find a and b.

Problem 8.22: Evaluate $(2 - r)(2 - s)(2 - t)$ if r, s, and t are the roots of $f(x) = 2x^3 - 4x^2 + 3x - 9$.

Problem 8.23: The polynomial $f(x) = x^4 + 6x^3 + ax^2 - 54x + c$ has four real roots r_1, r_2, r_3, and r_4 such that $r_1 + r_2 = 0$ and $r_4 - r_3 = 4$. Find a and c.

Problem 8.24: The roots r_1, r_2, and r_3 of $x^3 - 2x^2 - 11x + a = 0$ satisfy $r_1 + 2r_2 + 3r_3 = 0$. Find all possible values of a.

Problem 8.16: Let p, q, and r be the roots of $f(x) = x^3 - 4x^2 + 15x - 7$. Determine the values of $p + q + r$, $pq + qr + rp$, and pqr.

Solution for Problem 8.16: By the Factor Theorem, since p, q, and r are the roots of $f(x)$, we have

$$f(x) = x^3 - 4x^2 + 15x - 7 = k(x - p)(x - q)(x - r)$$

for some constant k. The coefficient of the x^3 term of the expansion on the right must be 1, so $k = 1$. Expanding the right side then gives

$$x^3 - 4x^2 + 15x - 7 = x^3 - (p + q + r)x^2 + (pq + qr + rp)x - pqr.$$

Since these polynomials are the same, each coefficient on the right must match the corresponding coefficient on the left, so we have

$$p + q + r = 4,$$
$$pq + qr + rp = 15,$$
$$pqr = 7.$$

$\square$

For a set of variables $x_1, x_2, \ldots, x_n$, let s_k denote the sum of all products of the x_i taken k at a time. So for example, for two variables x_1 and x_2, we have $s_1 = x_1 + x_2$ and $s_2 = x_1 x_2$. For three variables x_1, x_2, and x_3, we have

$$s_1 = x_1 + x_2 + x_3,$$
$$s_2 = x_1 x_2 + x_2 x_3 + x_3 x_1,$$
$$s_3 = x_1 x_2 x_3.$$

The expressions $s_1, s_2, \ldots, s_n$ are known as the **elementary symmetric polynomials** in the variables $x_1, x_2, \ldots, x_n$. We will call them **symmetric sums** for short. Problem 8.16 suggests a general relationship between the coefficients and the symmetric sums of the roots of a cubic polynomial.

Problem 8.17: Let $f(x) = a_3x^3 + a_2x^2 + a_1x + a_0$, and let p, q, and r be the roots of $f(x)$. Find the symmetric sums of p, q, and r in terms of only a_0, a_1, a_2, and a_3.

Solution for Problem 8.17: Because p, q, and r are the roots of $f(x)$, we have

$$a_3x^3 + a_2x^2 + a_1x + a_0 = a_3(x-p)(x-q)(x-r)$$

for all x. The a_3 on the right side ensures that the coefficient of x^3 on the right side matches the coefficient of x^3 on the left.

Expanding the right side, we have

$$a_3x^3 + a_2x^2 + a_1x + a_0 = a_3x^3 - (a_3p + a_3q + a_3r)x^2 + (a_3pq + a_3qr + a_3rp)x - a_3pqr.$$

Because these two polynomials are identical, their coefficients must be the same. Therefore, we have

$$a_3 = a_3,$$
$$a_2 = -a_3(p + q + r),$$
$$a_1 = a_3(pq + qr + rp),$$
$$a_0 = -a_3pqr.$$

Isolating a symmetric sum of p, q, and r in each of the last three equations gives us

$$p + q + r = -\frac{a_2}{a_3},$$
$$pq + qr + rp = \frac{a_1}{a_3},$$
$$pqr = -\frac{a_0}{a_3}.$$

□

These look a lot like the relationships we found between the roots and coefficients of a quadratic. Let's see what happens when we consider a quartic.

Problem 8.18: Let p, q, r, and s be the roots of $f(x) = 3x^4 + 2x^3 - 7x^2 + 9x - 10$. Find $pqr + pqs + prs + qrs$.

Solution for Problem 8.18: We use our work on cubics as a guide. Because p, q, r, and s are the roots of $f(x)$, and the leading coefficient of $f(x)$ is 3, we can write

$$f(x) = 3(x-p)(x-q)(x-r)(x-s).$$

Each term in the expansion of the right side consists of the product of 3 and one term from each of the four linear factors. We seek expressions like pqr in this expansion. Such terms will occur when we take x from one linear factor, and take the constant from each of the other three factors. There are 4 factors from which to choose the x, and these generate all four linear terms in the expansion of $f(x)$. Therefore, the coefficient of x in the expansion is

$$3(-pqr - pqs - prs - qrs).$$

The coefficient of x in $f(x)$ is 9, so we have $3(-pqr - pqs - prs - qrs) = 9$. Dividing by -3 gives us $pqr + pqs + prs + qrs = -3$. $\square$

We've found some relationships between coefficients of a polynomial and symmetric sums of the roots of the polynomials for quadratic, cubic, and quartic polynomials. This suggests we can find similar relationships between the roots and coefficients of any polynomial. Let's try.

Problem 8.19: Let $f(x) = a_n x^n + a_{n-1} x^{n-1} + a_{n-2} x^{n-2} + \cdots + a_1 x + a_0$, and let the roots of $f(x)$ be $r_1, r_2, r_3, \ldots, r_n$. Find the symmetric sums of the roots of $f(x)$ in terms of the coefficients of $f(x)$.

Solution for Problem 8.19: Since the leading coefficient of $f(x)$ is a_n, and the roots are $r_1, r_2, \ldots, r_n$, we can write

$$a_n x^n + a_{n-1} x^{n-1} + a_{n-2} x^{n-2} + \cdots + a_1 x + a_0 = a_n(x - r_1)(x - r_2)(x - r_3) \cdots (x - r_n).$$

When tackling quadratics and cubics, we expanded the product on the right side and equated the resulting coefficients in the expansion to the corresponding coefficients on the left side. However, our product with n binomials is pretty intimidating to expand. To form each term in the expansion, we choose one of the terms in each binomial, and then take the product of all the chosen terms and the leading constant, a_n. The full expansion consists of the sum of all the possible products we can thus form.

Let's see what terms appear in the expansion of

$$a_n(x - r_1)(x - r_2)(x - r_3) \cdots (x - r_n).$$

When we choose x from each of the binomials, we form the term $a_n x^n$. What about the rest of the terms in the expansion? Suppose we choose $-r_1$ from the first binomial and x from the remaining $n - 1$ binomials. Then, we form the term $-a_n r_1 x^{n-1}$. Similarly, we can form terms by choosing each $-r_i$ in turn, and choosing x from the remaining $n - 1$ binomials. All these terms together give us

$$-a_n r_1 x^{n-1} - a_n r_2 x^{n-1} - a_n r_3 x^{n-1} - \cdots - a_n r_n x^{n-1} = -a_n(r_1 + r_2 + \cdots + r_n)x^{n-1} = -a_n s_1 x^{n-1},$$

where s_1 is the symmetric sum of the roots taken one at a time.

Perhaps you see where this is going. Let's see what happens when we choose $-r_i$ from two of the binomials. If we choose $-r_1$ and $-r_2$ from the first two binomials and x from the other $n - 2$ binomials, we form the term $a_n r_1 r_2 x^{n-2}$. Continuing in this way, we choose every possible combination of two r_i along with x from the remaining $n - 2$ binomials to give

$$a_n r_1 r_2 x^{n-2} + a_n r_1 r_3 x^{n-2} + a_n r_1 r_4 x^{n-2} + \cdots + a_n r_{n-1} r_n x^{n-2} = a_n(r_1 r_2 + r_1 r_3 + \cdots + r_{n-1} r_n)x^{n-2} = a_n s_2 x^{n-2},$$

where s_2 is the symmetric sum of the roots taken two at a time.

Similarly, if we choose all possible combinations of k of the $-r_i$ and choose x from the other $n - k$ binomials, we form the term $(-1)^k a_n s_k x^{n-k}$, where s_k is the sum of the roots taken k at a time. We have $(-1)^k$ because in each term we form by choosing k of the $-r_i$, we multiply k terms of the form $-r_i$. So, if k is even, this product equals the product of the k chosen roots, but if k is odd, the product is the opposite of the product of the chosen roots.

So, when we expand $f(x) = a_n(x - r_1)(x - r_2)(x - r_3) \cdots (x - r_n)$, we have

$$f(x) = a_n x^n - a_n s_1 x^{n-1} + a_n s_2 x^{n-2} + \cdots + (-1)^k a_n s_k x^{n-k} + \cdots + (-1)^{n-1} a_n s_{n-1} x + (-1)^n a_n s_n.$$

Equating the coefficient of each term to the corresponding coefficient in

$$f(x) = a_n x^n + a_{n-1} x^{n-1} + a_{n-2} x^{n-2} + \cdots + a_1 x + a_0$$

gives us our desired relationship between the roots and coefficients of a polynomial:

> **Important:** Let $f(x) = a_n x^n + a_{n-1} x^{n-1} + a_{n-2} x^{n-2} + \cdots + a_1 x + a_0$, and let the roots of
> ⚠️ $f(x)$ be $r_1, r_2, \ldots, r_n$. Then, we have
>
> $$s_1 = \quad r_1 + r_2 + r_3 + \cdots + r_n \quad = -\frac{a_{n-1}}{a_n},$$
>
> $$s_2 = r_1 r_2 + r_1 r_3 + r_1 r_4 + \cdots + r_{n-1} r_{n-2} = \frac{a_{n-2}}{a_n},$$
>
> $$s_3 = r_1 r_2 r_3 + r_1 r_2 r_4 + \cdots + r_{n-2} r_{n-1} r_n = -\frac{a_{n-3}}{a_n},$$
>
> $$\vdots$$
>
> $$s_n = \quad r_1 r_2 r_3 \cdots r_n \quad = (-1)^n \frac{a_0}{a_n},$$
>
> where each s_k is the symmetric sum of the roots of $f(x)$ taken k at a time.
> We call these equations **Vieta's Formulas**.

□

Let's put Vieta's Formulas to work.

> **Problem 8.20:** Let $r, s,$ and t be the solutions to the equation $3x^3 - 4x^2 + 5x + 7 = 0$.
> (a) Find $r + s + t$.
> (b) Find $r^2 + s^2 + t^2$.
> (c) Find $\frac{1}{r} + \frac{1}{s} + \frac{1}{t}$.

Solution for Problem 8.20:

(a) By Vieta, we have $r + s + t = -(-4)/3 = 4/3$.

(b) We already have $r + s + t$, and we know that squaring $r + s + t$ will give us r^2, s^2, and t^2, so we expand $(r + s + t)^2 = (4/3)^2$ to find

$$r^2 + s^2 + t^2 + 2rs + 2st + 2tr = \frac{16}{9}.$$

We can isolate the sum of the squares by subtracting $2(rs + st + tr)$ from both sides:

$$r^2 + s^2 + t^2 = \frac{16}{9} - 2(rs + st + tr).$$

We recognize $rs + st + tr$ as the symmetric sum of the roots taken two at a time. By Vieta, we have $rs + st + tr = 5/3$, so $r^2 + s^2 + t^2 = \frac{16}{9} - 2 \cdot \frac{5}{3} = -\frac{14}{9}$.

(c) Writing the fractions with a common denominator, we get a ratio of symmetric sums we can evaluate with Vieta's Formulas:

$$\frac{1}{r} + \frac{1}{s} + \frac{1}{t} = \frac{st + tr + rs}{rst} = \frac{\frac{5}{3}}{-\frac{7}{3}} = -\frac{5}{7}.$$

□

A **symmetric expression** in multiple variables is an expression that is unchanged if any two of the variables are swapped. For example, $p^2r^2 + r^2q^2 + q^2p^2$ is symmetric because swapping any two of the variables results in the same expression. The last two parts of Problem 8.20 are examples of writing symmetric expressions in terms of elementary symmetric polynomials (that is, symmetric sums).

Important:	We can think of elementary symmetric polynomials as building blocks because any symmetric polynomial can be built from them.

You'll have more practice building symmetric polynomials from symmetric sums as Exercises. Now, we use Vieta's Formulas as a problem solving tool to tackle a few more challenging polynomial problems.

Problem 8.21: One of the roots of $x^3 + ax + b = 0$ is $1 + 2i$, where a and b are real numbers. Find a and b.

Solution for Problem 8.21: Since the coefficients of the cubic are real, one of the other roots must be the conjugate of $1+2i$, which is $1-2i$. Letting the third root be r, Vieta's Formulas give us $r+(1+2i)+(1-2i) = 0$, because the coefficient of x^2 in x^3+ax+b is 0. Therefore, the third root must be $r = -[(1+2i)+(1-2i)] = -2$.

Now, by Vieta, we know each of the following:

$$a = (-2)(1 + 2i) + (-2)(1 - 2i) + (1 + 2i)(1 - 2i) = 1,$$
$$b = -(-2)(1 + 2i)(1 - 2i) = 10.$$

So, the polynomial is $x^3 + x + 10$. (We could also have found the polynomial by finding the quadratic with $1 + 2i$ and $1 - 2i$ as roots, and then multiplying this quadratic by $x + 2$.) □

Problem 8.22: Evaluate $(2 - r)(2 - s)(2 - t)$ if r, s, and t are the roots of $f(x) = 2x^3 - 4x^2 + 3x - 9$.

Solution for Problem 8.22: *Solution 1: Vieta.* When we expand the desired expression, symmetric sums pop out:

$$(2 - r)(2 - s)(2 - t) = (4 - 2(r + s) + rs)(2 - t)$$
$$= 4(2 - t) - 2(r + s)(2 - t) + rs(2 - t)$$
$$= 8 - 4t - 4r + 2rt - 4s + 2st + 2rs - rst$$
$$= 8 - 4(r + s + t) + 2(rt + st + rs) - rst.$$

From Vieta's Formulas, we have $r + s + t = -(-4)/2 = 2$, $rt + st + rs = 3/2$, and $rst = -(-9)/2 = 9/2$. Therefore, we have

$$(2 - r)(2 - s)(2 - t) = 8 - 4(r + s + t) + 2(rt + st + rs) - rst = 8 - 8 + 3 - \frac{9}{2} = -\frac{3}{2}.$$

Solution 2: Recognize the desired expression. The desired expression looks a lot like the factored form of the polynomial. Because the coefficient of the leading term of $f(x)$ is 2, we have

$$f(x) = 2x^3 - 4x^2 + 3x - 9 = 2(x-r)(x-s)(x-t),$$

where r, s, and t are the roots of $f(x)$. Letting $x = 2$ gives us the desired expression on the far right, and we have

$$2(2-r)(2-s)(2-t) = f(2) = 2 \cdot 2^3 - 4 \cdot 2^2 + 3 \cdot 2 - 9 = -3.$$

Therefore, we have $(2-r)(2-s)(2-t) = -3/2$. $\square$

In each of the previous two problems, we have suggested solutions or significant steps that don't require Vieta's Formulas, in addition to providing approaches that use Vieta's Formulas.

> **WARNING!!** While Vieta's Formulas are often very useful, don't forget about the many other tools we have for working with polynomials.

> **Problem 8.23:** The polynomial $f(x) = x^4 + 6x^3 + ax^2 - 54x + c$ has four real roots r_1, r_2, r_3, and r_4 such that $r_1 + r_2 = 0$ and $r_4 - r_3 = 4$. Find a and c.

Solution for Problem 8.23: We have a problem involving the roots and coefficients of a polynomial, so we try using Vieta. We have information about the sum of two roots, so we start by focusing on the sum of all four roots. Vieta's Formulas tell us

$$r_1 + r_2 + r_3 + r_4 = -6.$$

Since $r_1 + r_2 = 0$, we have $r_3 + r_4 = -6$. Adding this to $r_4 - r_3 = 4$ gives us $2r_4 = -2$, so $r_4 = -1$ and $r_3 = -5$. However, we don't know r_1 and r_2, so how do we find a and c?

> **Concept:** When stuck on a problem, look for information you haven't used yet.

We haven't used the fact that the coefficient of x is -54, so we focus on that. Vieta's Formulas give us

$$r_1r_2r_3 + r_1r_2r_4 + r_1r_3r_4 + r_2r_3r_4 = 54.$$

Since $r_2 = -r_1$, $r_4 = -1$, and $r_3 = -5$, we have

$$5r_1^2 + r_1^2 + 5r_1 - 5r_1 = 54,$$

so $r_1^2 = 9$, which means r_1 and r_2 are 3 and -3. (It doesn't matter which is which.) Letting r_1 be 3 and r_2 be -3, we can now use Vieta's Formulas to compute

$$a = r_1r_2 + r_1r_3 + r_1r_4 + r_2r_3 + r_2r_4 + r_3r_4 = -4$$

and $c = r_1r_2r_3r_4 = -45$. Note that we could have used a little clever factorization, together with the fact that $r_1 + r_2 = 0$, to compute a more quickly:

$$a = r_1r_2 + r_1r_3 + r_1r_4 + r_2r_3 + r_2r_4 + r_3r_4$$
$$= r_1r_2 + r_1(r_3 + r_4) + r_2(r_3 + r_4) + r_3r_4$$
$$= r_1r_2 + (r_1 + r_2)(r_3 + r_4) + r_3r_4 = -9 + (0)(-6) + 5 = -4.$$

$\square$

> **Concept:** Many problems can be solved by using Vieta's Formulas in conjunction with given information about the roots or coefficients of polynomials.

Problem 8.24: The roots r_1, r_2, and r_3 of $x^3 - 2x^2 - 11x + a = 0$ satisfy $r_1 + 2r_2 + 3r_3 = 0$. Find all possible values of a.

Solution for Problem 8.24: Vieta's Formulas give us several equations involving the roots of the polynomial. Combining these with $r_1 + 2r_2 + 3r_3 = 0$ gives us the system

$$r_1 + 2r_2 + 3r_3 = 0,$$
$$r_1 + r_2 + r_3 = 2,$$
$$r_1 r_2 + r_1 r_3 + r_2 r_3 = -11,$$
$$r_1 r_2 r_3 = -a.$$

Yikes! Four equations, four variables, and only two of the equations are linear. Let's try to solve the first three equations for r_1, r_2, and r_3, and then use the result to find a.

Subtracting the second equation from the first eliminates r_1 and gives $r_2 + 2r_3 = -2$. So, we have $r_2 = -2 - 2r_3$. We can also write r_1 in terms of r_3 by subtracting twice the second equation from the first. This gives us $-r_1 + r_3 = -4$, so $r_1 = 4 + r_3$. Substituting these expressions for r_1 and r_2 into the third equation gives us an equation with only r_3,

$$(4 + r_3)(-2 - 2r_3) + (4 + r_3)(r_3) + (-2 - 2r_3)(r_3) = -11.$$

Simplifying this equation gives us $3r_3^2 + 8r_3 - 3 = 0$. Factoring the quadratic gives $(r_3 + 3)(3r_3 - 1)$, so $r_3 = 1/3$ or $r_3 = -3$. When $r_3 = 1/3$, we have $r_1 = 13/3$, $r_2 = -8/3$ and $a = -r_1 r_2 r_3 = 104/27$. When $r_3 = -3$, we have $r_1 = 1$, $r_2 = 4$, and $a = -r_1 r_2 r_3 = 12$. $\square$

The key step in solving the unusual system of equations we faced in Problem 8.24 was solving for r_1 and r_2 in terms of r_3. This allowed us to create an equation with one variable, which we could then solve for that variable.

> **Concept:** One useful approach when solving a system of equations is to solve for all but one of the variables in terms of the remaining variable. Then, use substitution to create an equation involving only this final variable.

Exercises

8.3.1 Find the product of the roots of $3x^3 - 7x^2 + 2x - 6 = 0$.

8.3.2 Let $p(x) = x^3 - 5x^2 + 12x - 19$ have roots a, b, and c. Find the value of $\dfrac{1}{ab} + \dfrac{1}{bc} + \dfrac{1}{ca}$.

8.3.3 If two factors of $x^3 - t_1 x + t_2$ are $x + 2$ and $x - 1$, find the roots of $x^2 - t_1 x + t_2$.

8.3.4 Let a and b be real numbers, such that one of the roots of $x^3 + ax^2 - 4x + b = 0$ is $1 + i$. Find the other two roots.

8.3.5 Find $(2 + r)(2 + s)(2 + t)(2 + u)$ if r, s, t, and u are the roots of $f(x) = 3x^4 - x^3 + 2x^2 + 7x + 2$.

8.3.6 Find the two values of k for which $2x^3 - 9x^2 + 12x - k$ has a double root. *(Source: ARML)*

8.3.7★ For certain real values of a, b, c, and d, the equation $x^4 + ax^3 + bx^2 + cx + d = 0$ has four nonreal roots. The product of two of these roots is $13 + i$ and the sum of the other two roots is $3 + 4i$. Find b. *(Source: AIME)*

8.4 Using Roots to Make Equations

Vieta's Formulas are not the only way to relate roots of a polynomial to the coefficients of the polynomial. An even more straightforward way is to note that if r is a root of f, then $f(r) = 0$.

Problem 8.25: Let r be a root of $x^2 - x + 7$.
  (a) Show that $r^3 = r^2 - 7r$.
  (b) Find the value of $r^3 + 6r + \pi$.

Problem 8.26: In this problem, we find the sum of the 20th powers of the roots of $z^{20} - 19z + 2$.
  (a) Suppose r_1 is a root of the polynomial. Write an equation in terms of r_1.
  (b) Solve the equation in part (a) for r_1^{20}.
  (c) Use part (b) as inspiration to find an expression for the sum of the 20th powers of the roots.
  (d) Use your expression in (c) and the original polynomial to evaluate the sum.

Problem 8.27: Let a, b, c be the roots of $x^3 + 3x^2 - 24x + 1 = 0$. All three roots are real. Prove that $\sqrt[3]{a} + \sqrt[3]{b} + \sqrt[3]{c} = 0$.
  (a) Find a polynomial $f(a)$ such that $\sqrt[3]{a} = f(a)$. **Hints:** 41, 69
  (b) Use part (a) to show that $\sqrt[3]{a} + \sqrt[3]{b} + \sqrt[3]{c} = 0$.

Problem 8.25: Let r be a root of $x^2 - x + 7$. Find the value of $r^3 + 6r + \pi$.

Solution for Problem 8.25: We could find the possible values of r, and then evaluate the given expression for each of them. But that would take a lot of time. Instead, we note that r being a root of $x^2 - x + 7$ means that $r^2 - r + 7 = 0$, and then relate this equation to the expression $r^3 + 6r + \pi$.

  Solution 1: From $r^2 - r + 7 = 0$, we have $r^2 = r - 7$, so we can substitute $r - 7$ for r^2 until we reduce $r^3 + 6r + \pi$ to an equivalent linear or constant expression:

$$r^3 + 6r + \pi = r\left(r^2\right) + 6r + \pi = r(r - 7) + 6r + \pi = r^2 - r + \pi = (r - 7) - r + \pi = \pi - 7.$$

So, we have $r^3 + 6r + \pi = \pi - 7$. Notice that all our steps hold for any r that satisfies $r^2 - r + 7 = 0$. In

253

other words, we have $r^3 + 6r + \pi = \pi - 7$ no matter which root of $x^2 - x + 7$ we take to be r.

Solution 2: Looking for a way to relate $r^2 - r + 7 = 0$ to the cubic expression we're trying to evaluate, we multiply our equation by r to produce an r^3, which gives us $r^3 - r^2 + 7r = 0$. Therefore, we have $r^3 = r^2 - 7r$, so $r^3 + 6r + \pi = r^2 - r + \pi$. We know that $r^2 - r = -7$, so we have $r^3 + 6r + \pi = r^2 - r + \pi = \pi - 7$. $\square$

> **Important:** Vieta's Formulas are not the only way to use a polynomial to create equations in terms of the polynomial's roots. We can substitute the roots into the polynomial itself! Specifically, if $x = k$ is a root of $f(x)$, then $f(k) = 0$.

Problem 8.26: Find the sum of the 20th powers of the roots of $z^{20} - 19z + 2$.

Solution for Problem 8.26: Let the roots be $r_1, r_2, \ldots, r_{20}$. We seek

$$r_1^{20} + r_2^{20} + r_3^{20} + \cdots + r_{20}^{20}.$$

We can use Vieta to evaluate the symmetric sums of the roots, but expressing the sum of the 20th powers of the roots in terms of these symmetric sums is daunting. So, we look for another approach.

Because r_1 is a root of $z^{20} - 19z + 2$, we have

$$r_1^{20} - 19r_1 + 2 = 0.$$

We're interested in adding the 20th powers of the roots so we solve this equation for r_1^{20} to find that $r_1^{20} = 19r_1 - 2$. Aha! We can do the same for every root. Adding the resulting equations to get the sum of the 20th powers gives us

$$r_1^{20} + r_2^{20} + \cdots + r_{20}^{20} = 19(r_1 + r_2 + \cdots + r_{20}) - 2(20).$$

By Vieta, the sum of the roots of $z^{20} - 19z + 2$ is 0, so the sum of the 20th powers of the roots is $-2(20) = -40$. $\square$

Problem 8.27: Let a, b, c be the roots of $x^3 + 3x^2 - 24x + 1 = 0$, and all three roots are real. Prove that $\sqrt[3]{a} + \sqrt[3]{b} + \sqrt[3]{c} = 0$.

Solution for Problem 8.27: What's wrong with this solution:

> **Bogus Solution:** We start by working backwards, hoping to produce an equation we know how to prove is true. We isolate $\sqrt[3]{a}$ to get $\sqrt[3]{a} = -\sqrt[3]{b} - \sqrt[3]{c}$. Cubing both sides and rearranging gives
>
> $$a + b + c = -3\sqrt[3]{b^2c} - 3\sqrt[3]{bc^2} = 3\sqrt[3]{bc}(-\sqrt[3]{b} - \sqrt[3]{c}).$$
>
> Aha! We recognize that $-\sqrt[3]{b} - \sqrt[3]{c} = \sqrt[3]{a}$ from above, and we have $a + b + c = 3\sqrt[3]{abc}$. Turning to our polynomial, we use Vieta to see that $a + b + c = -3$ and $abc = -1$, so we do indeed have $a + b + c = 3\sqrt[3]{abc}$. Then, reversing our steps above, we find that $\sqrt[3]{a} + \sqrt[3]{b} + \sqrt[3]{c} = 0$, as desired.

The error in this argument is a subtle one, and had we written our proof forwards, we would have found it.

> **WARNING!!** When you solve a problem by working backwards, always write your solution forwards. This will help you catch flaws in your reasoning.

Here, when we write our solution forwards, we see that we cannot go from $a + b + c = 3\sqrt[3]{abc}$ to

$$a + b + c = 3\sqrt[3]{bc}(-\sqrt[3]{b} - \sqrt[3]{c}),$$

because we haven't shown that $\sqrt[3]{a} = -\sqrt[3]{b} - \sqrt[3]{c}$. So, our steps are not reversible! We'll have to find another way.

The main complication in this problem is the cube roots. It would be nice if we could get rid of them.

> **Concept:** Focus on removing the main complication in difficult problems.

We get rid of cube roots by cubing, so we look for cubes in the problem. We first might rewrite the polynomial equation as $x^3 = -3x^2 + 24x - 1$, but this doesn't seem to help. Then, we notice that the polynomial itself is almost a cube. Specifically, the polynomial equals $(x + 1)^3 - 27x$, and 27 itself is a cube! So, now our equation is $(x + 1)^3 - 27x = 0$, and we have a path to our solution. Adding $27x$ to both sides, and then dividing by 27, gives

$$x = \frac{(x + 1)^3}{27}.$$

We know that a, b, and c are solutions to this equation, so we have $a = \frac{(a+1)^3}{27}$. Taking the cube root of both sides gives

$$\sqrt[3]{a} = \sqrt[3]{\frac{(a + 1)^3}{27}} = \frac{a + 1}{3},$$

and likewise for b and c. No more cube roots! We now have

$$\sqrt[3]{a} + \sqrt[3]{b} + \sqrt[3]{c} = \frac{a + 1}{3} + \frac{b + 1}{3} + \frac{c + 1}{3} = \frac{a + b + c}{3} + 1.$$

Using Vieta, we have $a + b + c = -3$, so we have the desired $\sqrt[3]{a} + \sqrt[3]{b} + \sqrt[3]{c} = 0$. $\square$

Exercises

8.4.1 Let t be a root of $f(x) = x^3 - x + 2$. Evaluate $t^6 - t^2 + 4t$.

8.4.2 Let r be a root of $x^2 + 7x + 17 = 0$. Compute each of the following:

(a) $(r + 3)(r + 4)$.

(b) $(r - 2)(r + 9)$.

(c)$\star$ $(r - 1)(r + 2)(r + 8)(r + 5)$. **Hints:** 84

8.4.3★ Find the sum of the 20th powers of the roots of $z^{20} - 19z^2 + 2 = 0$. **Hints:** 52

8.4.4★ Let a, b, c, and d be four distinct real numbers such that a and b are the roots of the equation $x^2 - 3cx - 8d = 0$, and c and d are the roots of the equation $x^2 - 3ax - 8b = 0$. Find $a + b + c + d$. *(Source: Costa Rica)*

8.5 Summary

Important:	Let $p(x)$ be a polynomial with real coefficients. If z is a root of $p(x)$, then $\bar{z}$ is also a root of $p(x)$. In other words, the nonreal roots of $p(x)$ come in conjugate pairs.

Important:	If $b + c\sqrt{d}$, where b, c, and d are rational and $\sqrt{d}$ is irrational, is the root of a polynomial with rational coefficients, then $b - c\sqrt{d}$ is also a root of the polynomial.

Important:	We can sometimes prove a statement by showing that if we assume the statement is false, then an impossible conclusion results. We call this strategy **proof by contradiction** or **indirect proof**.

Important:	Let $f(x) = a_n x^n + a_{n-1} x^{n-1} + a_{n-2} x^{n-2} + \cdots + a_1 x + a_0$, and let the roots of $f(x)$ be $r_1, r_2, \ldots, r_n$. Then, we have $$s_1 = r_1 + r_2 + r_3 + \cdots + r_n = -\frac{a_{n-1}}{a_n},$$ $$s_2 = r_1 r_2 + r_1 r_3 + r_1 r_4 + \cdots + r_{n-1} r_{n-2} = \frac{a_{n-2}}{a_n},$$ $$s_3 = r_1 r_2 r_3 + r_1 r_2 r_4 + \cdots + r_{n-2} r_{n-1} r_n = -\frac{a_{n-3}}{a_n},$$ $$\vdots$$ $$s_n = r_1 r_2 r_3 \cdots r_n = (-1)^n \frac{a_0}{a_n},$$ where each s_k is the symmetric sum of the roots of $f(x)$ taken k at a time. We call these equations **Vieta's Formulas**.

Important:	Vieta's Formulas are not the only way to use a polynomial to create equations in terms of the polynomial's roots. We can substitute the roots into the polynomial itself! Specifically, if $x = k$ is a root of $f(x)$, then $f(k) = 0$.

Things To Watch Out For!

- Watch out for special cases.

- When you make a substitution to help with algebra, don't forget what your variables stand for.

- When you solve a problem by working backwards, always write your solution forwards. This will help you catch flaws in your reasoning.

Problem Solving Strategies

Concepts:

- When facing a new type of problem, comparing it to similar problems you have already solved can guide you to a solution.

- There's nothing wrong with guessing! Guessing has a long and glorious tradition in math and science. Guessing, and then confirming that your guess is correct, is a perfectly valid way to solve problems.

- When in doubt, factor. Factoring often simplifies expressions.

- When stuck on a problem, look for information you haven't used yet.

- Many problems can be solved by using Vieta's Formulas in conjunction with given information about the roots or coefficients of polynomials.

- One useful approach when solving a system of equations is to solve for all but one of the variables in terms of the remaining variable. Then, use substitution to create an equation involving only this final variable.

- Focus on removing the main complication in difficult problems.

REVIEW PROBLEMS

8.28 Find the roots of each of the following polynomials:

(a) $f(r) = 2r^3 + 7r^2 - 4r - 21$ (b) $g(s) = s^3 + 8s + 24$

8.29 One of the roots of $x^4 + 9x^3 + 48x^2 + 78x - 136 = 0$ is $-3 + 5i$. Find the other three roots.

8.30 Let $f(x)$ be a polynomial with degree 6 and rational coefficients. The polynomial has five distinct roots: $1, 2 - i, 2 + i, 1 - i$, and $1 + i$.

(a) Explain why the polynomial must have a double root.

(b) Which root is the double root?

8.31 Let $r_1 = a + bi$ and $r_2 = a - bi$, where a and b are nonzero real numbers. Find the minimum degree of a polynomial with real coefficients and roots $r_1 i$ and $r_2 i$.

8.32 One root of $x^3 - 11x^2 + 29x - 7 = 0$ is $2 + \sqrt{3}$. Find the other two roots.

8.33 Find the monic polynomial with minimal degree and integer coefficients such that $3 - 2i\sqrt{5}$ and $\sqrt{3} - 1$ are roots of the polynomial.

8.34 Prove that if c is a positive integer that is not a perfect square, then $\sqrt{c}$ is irrational.

8.35 Prove that $\sqrt{2 - \sqrt{2}}$ is not rational.

8.36 Factor $8 - 2y + 4y^3 - y^4$ completely

(a) over the integers.

(b) over the reals.

(c)⋆ over the complex numbers.

8.37 Let $p(x)$ be a cubic polynomial with real coefficients.

(a) Use the graph of $p(x)$ to give an intuitive explanation for why $p(x)$ must have at least one real root.

(b) Find an algebraic proof that $p(x)$ has at least one real root using the tools we developed in this chapter. **Hints:** 170

8.38 Let $f(x)$ be a polynomial with rational coefficients.

(a) If $\sqrt{2}$ is a root of $f(x)$, must $-\sqrt{2}$ also be a root of $f(x)$?

(b) If $\sqrt[3]{2}$ is a root of $f(x)$, must $-\sqrt[3]{2}$ also be a root of $f(x)$?

8.39 Let $g(x) = x^4 - 5x^3 + 2x^2 + 7x - 11$, and let the roots of $g(x)$ be p, q, r, and s. Find the following:

(a) $pqr + pqs + prs + qrs$

(c) $p^2 + q^2 + r^2 + s^2$

(b) $\dfrac{1}{p} + \dfrac{1}{q} + \dfrac{1}{r} + \dfrac{1}{s}$

(d) $p^2qrs + pq^2rs + pqr^2s + pqrs^2$

8.40 What is the sum of the reciprocals of the roots of the equation $\dfrac{2003}{2004}x + 1 + \dfrac{1}{x} = 0$? *(Source: AMC 10)*

8.41 If $a \pm bi$ are complex roots of the equation $x^3 + qx + r = 0$, where a, b, q, and r are real numbers and $b \neq 0$, then find q in terms of a and b. *(Source: AHSME)*

8.42 A cubic polynomial $p(x)$ has leading coefficient 1, all real coefficients, and $p(3-2i) = 0$. If $p(0) = -52$, find $p(x)$.

8.43 The equation $x^3 - 4x^2 + 5x - 1.9 = 0$ has real roots r, s, and t. Find the length of an interior diagonal of a box with sides r, s, and t. *(Source: Mandelbrot)*

8.44 Let x be a real number such that $x^3 + 4x = 8$. Determine the value of $x^7 + 64x^2$. *(Source: HMMT)*

Challenge Problems ▶

8.45 Let r be a root of $x^2 - 2x + 3$. Find the value of $r^4 - 4r^3 + 2r^2 + 4r + 3$.

8.46 Let $f(x) = x^n + a_1 x^{n-1} + a_2 x^{n-2} + 4x + 7$, where a_1 and a_2 are constant. Prove that the sum of the squares of the roots of $f(x)$ is the same for all integers $n \geq 4$.

8.47 Find the polynomial P of smallest degree with rational coefficients and leading coefficient 1 such that $P\left(\sqrt[3]{49} + \sqrt[3]{7}\right) = 4$. *(Source: NYSML)* **Hints:** 62

8.48 Find all roots of the polynomial $iy^3 - 8y^2 - 22iy + 21$. **Hints:** 43

8.49 Let x_1 and x_2 be the roots of the equation $x^2 - (a + d)x + (ad - bc) = 0$. Show that x_1^3 and x_2^3 are the roots of the equation

$$y^2 - (a^3 + d^3 + 3abc + 3bcd)y + (ad - bc)^3 = 0.$$

(Source: Hungary)

8.50 Let r_1, r_2, r_3 be the 3 zeroes of the cubic polynomial $x^3 - x - 1 = 0$. Then, the expression

$$r_1(r_2 - r_3)^2 + r_2(r_3 - r_1)^2 + r_3(r_1 - r_2)^2$$

equals a rational number. Find this number. *(Source: ARML)*

8.51 Let $f(x)$ be a polynomial with rational coefficients. If $\sqrt[4]{2}$ is a root of $f(x)$, must $-\sqrt[4]{2}$ also be a root of $f(x)$?

8.52 Determine $(r + s)(s + t)(t + r)$ if r, s, and t are the three real roots of the polynomial $x^3 + 9x^2 - 9x - 8$. *(Source: Mandelbrot)* **Hints:** 147

8.53 Show that if the roots of $x^3 + ax + b = 0$ are rational numbers m, n, and p, then the roots of $mx^2 + nx + p$ are also rational. **Hints:** 172

8.54 Let p, q, and r be the distinct roots of $x^3 - x^2 + x - 2 = 0$. Find $p^3 + q^3 + r^3$. *(Source: AHSME)* **Hints:** 28

8.55 Let n be a positive integer. Prove that

$$\frac{n}{1} + \frac{n}{2} + \cdots + \frac{n}{n} - \frac{n^2}{1 \cdot 2} - \frac{n^2}{1 \cdot 3} - \cdots - \frac{n^2}{(n-1)n} + \cdots + (-1)^{n-1}\frac{n^n}{n!} = 1,$$

where each term consists of $(-1)^{k+1}n^k$ divided by a product of k of positive integers from 1 to n, and all possible such terms are included in the sum. *(Source: Mandelbrot)* **Hints:** 167, 148

8.56 Consider all lines that meet the graph of $y = 2x^4 + 7x^3 + 3x - 5$ in four distinct points, (x_1, y_1), (x_2, y_2), (x_3, y_3), and (x_4, y_4). Show that
$$\frac{x_1 + x_2 + x_3 + x_4}{4}$$
is independent of the line and find its value. *(Source: Putnam)*

8.57 Let $f(x) = a_n x^n + a_{n-1} x^{n-1} + \cdots + a_1 x + a_0$, and let t_k be the sum of the kth powers of the roots of $f(x)$. Prove that if $k \geq n$, then $a_n t_k + a_{n-1} t_{k-1} + a_{n-2} t_{k-2} + \cdots + a_0 t_{k-n} = 0$. **Hints:** 149

8.58 If α, β, γ are the roots of $x^3 - x - 1 = 0$, compute $\dfrac{1+\alpha}{1-\alpha} + \dfrac{1+\beta}{1-\beta} + \dfrac{1+\gamma}{1-\gamma}$. *(Source: Canada)* **Hints:** 179

8.59★ Find the remainder when $x^{81} + x^{48} + 2x^{27} + x^6 + 3$ is divided by $x^3 + 1$.

8.60★ Let a, b, c be the roots of $x^3 - 9x^2 + 11x - 1 = 0$, and let $s = \sqrt{a} + \sqrt{b} + \sqrt{c}$. Find $s^4 - 18s^2 - 8s$. *(Source: HMMT)*

8.61★ Let α_1 and α_2 be the roots of the quadratic $x^2 - 5x - 2 = 0$, and let β_1, β_2, and β_3 be the roots of the cubic $x^3 - 3x - 1 = 0$. Compute $(\alpha_1 + \beta_1)(\alpha_1 + \beta_2)(\alpha_1 + \beta_3)(\alpha_2 + \beta_1)(\alpha_2 + \beta_2)(\alpha_2 + \beta_3)$. *(Source: WOOT)* **Hints:** 164, 105

8.62★ Let $f(x)$ be a polynomial in x of degree greater than 1. Define $g_i(x)$ by $g_1(x) = f(x)$, and $g_{k+1}(x) = f(g_k(x))$. Let r_k be the average of the roots of g_k. Determine r_{89} if $r_{19} = 89$. *(Source: Mathematical Mayhem)* **Hints:** 273

8.63★ Let α be a root of the cubic $x^3 - 21x + 35 = 0$. Prove that $\alpha^2 + 2\alpha - 14$ is a root of the cubic. **Hints:** 74, 7

Extra! **Descartes Rule of Signs** is one more tool that helps us search for polynomial roots.
⟫⟫⟫⟫ To use it, we order the nonzero coefficients of $f(x)$ from the coefficient of the highest degree term to the coefficient of the lowest degree term. We similarly order the nonzero coefficients of $f(-x)$. Let the number of sign changes in the list for $f(x)$ be i and the number of sign changes in the list for $f(-x)$ be j. Descartes Rule of Signs then tells us the following:

- The number of positive roots of $f(x)$ is $i - 2k$ for some nonnegative integer k.

- The number of negative roots of $f(x)$ is $j - 2k$ for some nonnegative integer k.

So, for example, let $f(x) = 3x^5 - 2x^2 - x + 1$. There are two sign changes when reading the coefficients from the highest degree term to the lowest, so there are either 2 positive roots or 0 positive roots. (That is, we cannot have exactly 1, 3, 4, or 5 positive roots.) Since $f(-x) = -3x^5 - 2x^2 + x + 1$, there is only one sign change when we read the coefficients of $f(-x)$ from the highest degree term to the lowest. Therefore, there is exactly 1 negative root. Combining our possibilities for the numbers of positive and negative roots with the fact that 0 is not a root of $f(x)$, we see that we can have at most 3 real roots. Since $\deg f = 5$, the polynomial has 5 roots, so there must be at least 2 nonreal roots.

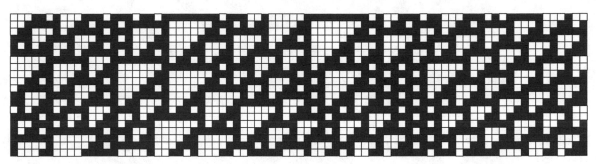

Ideas are the factors that lift civilization. They create revolutions. There is more dynamite in an idea than in many bombs. – Bishop Vincent

CHAPTER 9

Factoring Multivariable Polynomials

An expression or function with more than one variable is a **multivariable polynomial** if it is a sum of terms such that each term is a product of a constant and a nonnegative integer power of each variable. For example, the expressions

$$x^3 y^2 - 3y + 2t, \qquad \frac{r^2 t^5}{2} + r + rst - 7, \qquad \text{and} \qquad a^2 b + b^2 c + c^2 a$$

are multivariable polynomials, while

$$\frac{x}{y} + \frac{y}{x}, \qquad \sqrt{x^2 + y^3 - 3}, \qquad \text{and} \qquad r\sqrt{s} - 3$$

are not polynomials.

We extend the idea of degree to multivariable polynomials by defining the **degree** of a term to be the sum of the exponents of the variables in the term. For example, the degree of $2x^3 y^7$ is $3 + 7 = 10$. The degree of a multivariable polynomial is defined as the maximum degree of the terms in the polynomial. For example, the degree of $a^5 b + a^4 b^2 + a^3 b^3$ is 6 and the degree of $x^9 + 2xy^3 + xy$ is 9.

In this chapter, we use some of our strategies for working with single-variable polynomials to develop strategies for dealing with multivariable polynomials. We also turn around and use some of these new strategies to tackle problems with one variable.

9.1 Grouping

In this section, we factor expressions by splitting them into groups of terms, factoring each group, and then discovering that the factored groups have factors in common.

Problems

Problem 9.1:
(a) Find a common factor of $x^3 - x^2$ and $4x - 4$.
(b) Factor $x^3 - x^2 + 4x - 4$ completely into factors with real coefficients.

Problem 9.2: In this problem, we find all ordered pairs of positive integers (x, y) such that we have $6xy + 4x + 9y + 6 = 253$.
(a) Factor $6xy + 4x$.
(b) Factor $9y + 6$.
(c) Factor $6xy + 4x + 9y + 6$.
(d) Use your factorization from part (c) together with possible factorizations of 253 to solve the problem.

Problem 9.3: Write each of the following expressions as the product of two binomials, plus (or minus) a constant. For example, we have $xy + x + y = (x + 1)(y + 1) - 1$.
(a) $ab + 3a - 5b$ (c) $6x^2y^2 - 3x^2 - 2y^2$
(b) $2mn + 14m - 3n$

Problem 9.4: Find all positive integers m and n, where n is odd, that satisfy $\frac{1}{m} + \frac{4}{n} = \frac{1}{12}$. *(Source: Great Britain)*

Problem 9.5: Factor $2x^5 + x^4y - 21x^3y^2 + 16x^2 + 8xy - 168y^2$ as the product of linear and quadratic polynomials with real coefficients.

Problem 9.6: Factor $ab(c^2 + d^2) + cd(a^2 + b^2)$.

Problem 9.7: In this problem, we find all values of x that satisfy

$$\frac{6}{\sqrt{x-8}-9} + \frac{1}{\sqrt{x-8}-4} + \frac{7}{\sqrt{x-8}+4} + \frac{12}{\sqrt{x-8}+9} = 0.$$

(a) Let $y = \sqrt{x-8}$. Solve the resulting equation for y.
(b) Find all values of x that satisfy the original equation. *(Source: ARML)*

Problem 9.1: Factor $x^3 - x^2 + 4x - 4$ completely into factors with real coefficients.

Solution for Problem 9.1: We note that $x^3 - x^2 = x^2(x - 1)$ and $4x - 4 = 4(x - 1)$ share $x - 1$ as a common factor. So, we have
$$x^3 - x^2 + 4x - 4 = x^2(x - 1) + 4(x - 1) = (x^2 + 4)(x - 1).$$

The roots of x^2+4 are imaginary, so we cannot factor it as the product of linear terms with real coefficients.
□

In Problem 9.1, we applied the technique of **grouping** in order to factor $x^3 - x^2 + 4x - 4$. We looked for groups of terms that shared a common factor, factored those groups, and then factored the common factor, $x - 1$, out of the two groups.

Problem 9.2: Find all ordered pairs of positive integers (x, y) such that $6xy + 4x + 9y + 6 = 253$.

Solution for Problem 9.2: We could attempt trial and error, but that would take quite a while, and it would be hard to be sure that we found all the solutions. We've solved many equations with factorization, so we try factorization here. We start by factoring $2x$ out of our first two terms, which gives

$$2x(3y + 2) + 9y + 6 = 253.$$

We notice that factoring 3 out of $9y + 6$ also produces a factor of $3y + 2$:

$$2x(3y + 2) + 3(3y + 2) = 253.$$

So, we have

$$(2x + 3)(3y + 2) = 253.$$

We seek integer solutions (x, y), so $2x + 3$ and $3y + 2$ must be integers. Since x and y must be positive, neither $2x+3$ nor $3y+2$ can equal 1. So, the two factors can't be 1 and 253. We find the prime factorization of 253 to find other pairs of integers that multiply to 253. Since $253 = 11 \cdot 23$, the values of x and y must satisfy either the system of equations $2x + 3 = 11$, $3y + 2 = 23$, or the system $2x + 3 = 23$, $3y + 2 = 11$. Solving these gives us the solutions $(x, y) = (4, 7)$ and $(x, y) = (10, 3)$. □

We call equations for which we seek integer solutions **Diophantine equations**. Problem 9.2 highlights a powerful problem solving strategy for tackling Diophantine equations.

Concept: Factorization is a powerful tool for solving Diophantine equations. If we can write a Diophantine equation as a product of expressions that equal an integer, such as
$$(2x + 3)(3y + 2) = 253,$$
we can then use the divisors of the integer to find solutions to the equation.

Our factorization of $6xy + 4x + 9y + 6$ is an example of this factorization:

Important:

$$ap + aq + bp + bq = (a + b)(p + q).$$

This really isn't something new; it's just the reverse of expanding the product of binomials $(a+b)(p+q)$.

This factorization is so useful that we sometimes "force" it into problems. For example, if Problem 9.2 had asked us to find all positive integer pairs such that $6xy + 4x + 9y = 247$, we would add 6 to both sides to give us $6xy + 4x + 9y + 6 = 253$, because we could then factor the left side to give $(2x + 3)(3y + 2) = 253$. We call this process **Simon's Favorite Factoring Trick**.

> **Sidenote:** Simon's Favorite Factoring Trick is named after Art of Problem Solving
> Community member Simon Rubinstein-Salzedo, who described this clever
> manipulation as his favorite factoring trick when discussing the problem
> of finding all pairs of positive integers (x, y) such that
>
> $$\frac{1}{x} - \frac{1}{y} = \frac{1}{97}.$$
>
> See if you can solve the problem!

Problem 9.3:   Write each of the following expressions as the product of two binomials, plus (or minus) a constant. For example, we have $xy + x + y = (x + 1)(y + 1) - 1$.

(a) $ab + 3a - 5b$

(c) $6x^2y^2 - 3x^2 - 2y^2$

(b) $2mn + 14m - 3n$

Solution for Problem 9.3:

(a)   We guess that our product of binomials will be of the form $(a + m)(b + n)$ for some constants m and n, because this product will include an ab term in its expansion. Since we need a $3a$ term and a $-5b$ term in our expansion, we let $m = -5$ and $n = 3$, which gives us

$$(a - 5)(b + 3) = ab + 3a - 5b - 15.$$

Adding 15 to both sides, we have $ab + 3a - 5b = (a - 5)(b + 3) + 15$.

(b)   We know that a product of binomials of the form $(m + c)(n + d)$, where c and d are constants, won't do, since this will produce an mn term instead of a $2mn$ term. Instead, we try $(2m + c)(n + d)$, where we give m a coefficient of 2 rather than n because the coefficient of m in $2mn + 14m - 3n$ is even. In order to produce $14m$ and $-3n$ terms, we let $c = -3$ and $d = 7$, and we have

$$(2m - 3)(n + 7) = 2mn + 14m - 3n - 21.$$

Adding 21 to both sides, we have $2mn + 14m - 3n = (2m - 3)(n + 7) + 21$.

We could also have found this result by using grouping. We start by factoring $2m$ out of the first two terms to give

$$2m(n + 7) - 3n.$$

We need a factor of $n + 7$ in the second term of this expression in order to factor further. Adding $-3(7)$ gives us such a factor, but we also need to subtract $-3(7)$ to keep the value of expression unchanged. We have

$$2m(n + 7) - 3n = 2m(n + 7) - 3n - 3(7) - (-3(7)) = 2m(n + 7) - 3(n + 7) + 21 = (2m - 3)(n + 7) + 21.$$

(c)   We factor $3x^2$ out of the first two terms to give $3x^2(2y^2 - 1) - 2y^2$. Adding $+1 - 1$ will allow us to factor further:

$$3x^2(2y^2 - 1) - 2y^2 = 3x^2(2y^2 - 1) - 2y^2 + 1 - 1$$
$$= 3x^2(2y^2 - 1) - 1(2y^2 - 1) - 1 = (3x^2 - 1)(2y^2 - 1) - 1.$$

□

Problem 9.4: Find all positive integers m and n, where n is odd, that satisfy $\frac{1}{m} + \frac{4}{n} = \frac{1}{12}$. *(Source: Great Britain)*

Solution for Problem 9.4: We start by getting rid of the fractions. Multiplying both sides by $12mn$ leaves us $12n + 48m = mn$. Rearranging then gives us $mn - 12n - 48m = 0$. We apply Simon's Favorite Factoring Trick by adding $(12)(48)$ to both sides, which gives us $mn - 12n - 48m + (12)(48) = (12)(48)$. We can now factor the left side, to give $(m - 12)(n - 48) = 576$.

Before we go through listing all the divisors of 576 (there are a lot of them!), we note that n must be odd. Therefore, the quantity $n - 48$ must be an *odd* divisor of 576. Moreover, since m and n are positive, we must have $\frac{4}{n} < \frac{1}{12}$, which means that $n > 48$. This greatly limits our search. The prime factorization of 576 is $2^6 \cdot 3^2$, so all the odd divisors of 576 are divisors of $3^2 = 9$. This means $n - 48$ must divide 9. We have three possibilities. First, $n - 48 = 9$ and $m - 12 = 576/9 = 64$, which gives us $(m, n) = (76, 57)$. Second, $n - 48 = 3$ and $m - 12 = 576/3 = 192$, which gives us $(m, n) = (204, 51)$. Third, $n - 48 = 1$ and $m - 12 = 576$, which gives us $(m, n) = (588, 49)$. $\square$

Problem 9.5: Let $f(x, y) = 2x^5 + x^4 y - 21x^3 y^2 + 16x^2 + 8xy - 168y^2$. Factor $f(x, y)$ completely into linear and quadratic factors with real coefficients.

Solution for Problem 9.5: It's not obvious how we might find an expression that evenly divides $f(x, y)$. We try to group terms with common factors to get us started. Looking for clues, we note that $\gcd(16, 168) = 8$, so we can factor 8 out of the last three terms of $f(x, y)$:

$$16x^2 + 8xy - 168y^2 = 8(2x^2 + xy - 21y^2).$$

The factor $2x^2 + xy - 21y^2$ looks a lot like the first few terms of $f(x, y)$. Indeed, we note that

$$2x^5 + x^4 y - 21x^3 y^2 = x^3(2x^2 + xy - 21y^2).$$

So, we see that $16x^2 + 8xy - 168y^2$ and $2x^5 + x^4 y - 21x^3 y^2$ have a common factor:

$$2x^5 + x^4 y - 21x^3 y^2 + 16x^2 + 8xy - 168y^2 = x^3(2x^2 + xy - 21y^2) + 8(2x^2 + xy - 21y^2)$$
$$= (x^3 + 8)(2x^2 + xy - 21y^2).$$

Next, we check if we can factor either $x^3 + 8$ or $2x^2 + xy - 21y^2$. We factor $2x^2 + xy - 21y^2$ as a quadratic to find $2x^2 + xy - 21y^2 = (2x + 7y)(x - 3y)$. Using our tactics from Chapter 7, we find that $x + 2$ is a factor of $x^3 + 8$, and we have $x^3 + 8 = (x + 2)(x^2 - 2x + 4)$. The discriminant of $x^2 - 2x + 4$ is negative, so we can't factor $x^2 - 2x + 4$ over the real numbers, and our factorization of $f(x, y)$ is:

$$f(x, y) = 2x^5 + x^4 y - 21x^3 y^2 + 16x^2 + 8xy - 168y^2 = (x + 2)(x^2 - 2x + 4)(2x + 7y)(x - 3y).$$

$\square$

Problem 9.6: Factor $ab(c^2 + d^2) + cd(a^2 + b^2)$.

Solution for Problem 9.6: We begin by expanding the given expression to get

$$ab(c^2 + d^2) + cd(a^2 + b^2) = abc^2 + abd^2 + a^2cd + b^2cd.$$

Looking for similarities between terms to guide us, we note a couple of partial factorizations not present in the original expression we expanded:

$$abc^2 + a^2cd = ac(bc + ad),$$
$$abd^2 + b^2cd = bd(ad + bc).$$

Both right sides include the factor $ad + bc$. Putting these equations together, we have factors of the original expression:

$$ab(c^2 + d^2) + cd(a^2 + b^2) = abc^2 + abd^2 + a^2cd + b^2cd$$
$$= abc^2 + a^2cd + abd^2 + b^2cd = ac(bc + ad) + bd(ad + bc) = (ac + bd)(ad + bc).$$

$\square$

Problem 9.7: Find all values of x that satisfy

$$\frac{6}{\sqrt{x-8}-9} + \frac{1}{\sqrt{x-8}-4} + \frac{7}{\sqrt{x-8}+4} + \frac{12}{\sqrt{x-8}+9} = 0.$$

(Source: ARML)

Solution for Problem 9.7: Let $y = \sqrt{x-8}$. Then the equation becomes

$$\frac{6}{y-9} + \frac{1}{y-4} + \frac{7}{y+4} + \frac{12}{y+9} = 0,$$

which is considerably less intimidating. We multiply both sides by all four denominators to find

$$6(y-4)(y+4)(y+9) + 1(y-9)(y+4)(y+9) + 7(y-9)(y-4)(y+9) + 12(y-9)(y-4)(y+4) = 0.$$

Multiplying out all those products doesn't look easy. Instead of diving into expanding all those products, we notice that many of the products have factors in common. Maybe we can use clever factorization to simplify the equation.

> **Concept:** Look for ways to use factorization to avoid having to expand complicated products.

We group the two terms with $(y-4)(y+4)$ as factors, since expanding this product gives us a binomial, $y^2 - 16$, instead of a 3-term quadratic. We find

$$6(y-4)(y+4)(y+9) + 12(y-9)(y-4)(y+4) = (6(y+9) + 12(y-9))(y-4)(y+4)$$
$$= 18(y-3)(y^2 - 16).$$

Grouping the two terms with $(y-9)(y+9)$ gives

$$1(y-9)(y+4)(y+9) + 7(y-9)(y-4)(y+9) = (y+4 + 7(y-4))(y-9)(y+9)$$
$$= 8(y-3)(y^2 - 81).$$

So, the left side of the equation we had after multiplying by the denominators can be simplified as:

$$18(y-3)(y^2-16) + 8(y-3)(y^2-81) = (y-3)(18(y^2-16) + 8(y^2-81))$$
$$= (y-3)(2)(9(y^2-16) + 4(y^2-81))$$
$$= 2(y-3)(13y^2-468)$$
$$= 2(y-3)(13)(y^2-36)$$
$$= 26(y-3)(y-6)(y+6).$$

So, our equation now is $26(y-3)(y-6)(y+6) = 0$. Our careful factorization spared us a lot of work.

Now, we have $y = 3$, $y = 6$, and $y = -6$ as solutions. However, we must find x. Since $y = \sqrt{x-8}$, the solution $y = -6$ is extraneous, because a square root cannot be negative. The solution $y = 3$ gives us $3 = \sqrt{x-8}$, so $x = 17$, and $y = 6$ gives us $6 = \sqrt{x-8}$, so $x = 44$. Therefore, 17 and 44 are the only values of x that satisfy the original equation. $\square$

In general, grouping can be very helpful on problems that allow us to rearrange an equation as a product of factors equal to 0, and on Diophantine equations. However, it isn't always successful. For example, suppose we must find all the solutions to the equation $x^3 - 4x^2 + 8x - 32 = -27$. We might be think we have an opportunity to use grouping on the left side by noting that

$$x^3 - 4x^2 + 8x - 32 = x^2(x-4) + 8(x-4) = (x^2+8)(x-4),$$

so we have $(x^2+8)(x-4) = -27$. But now what? This isn't a Diophantine equation, and we don't have a product of factors equal to 0. We might proceed with the factors of 27, but this will only succeed if all the possible solutions are integers. Instead, we're better off rearranging the equation as $x^3 - 4x^2 + 8x - 5 = 0$ and using our usual techniques for finding the roots of polynomials.

> **WARNING!!**
> ☢
> There's an old saying that "When all you have is a hammer, every problem looks like a nail." Keep your problem solving toolbox well-stocked, so you won't find yourself swinging a hammer when you really need a wrench. In other words, don't get so attached to one strategy that you can't let it go and try something else when it isn't working.

Exercises

9.1.1 Factor each of the following polynomials:

    (a) $xy + 3x - 4y - 12$ (b) $2ab - 14a + 5b - 35$

9.1.2 Find all ordered pairs of integers (x, y) such that $\frac{1}{x} + \frac{1}{y} = \frac{1}{7}$.

9.1.3 Factor $x^3 + 2xy^2 + 5x^2 + 10y^2$ completely over the reals.

9.1.4 Find all ordered pairs of real numbers (x, y) such that $4x^2y^2 + 36x^2 = y^2 + 9$.

9.1.5 If x and y are positive integers and $x + y + xy = 54$, then find $x + y$.

9.1.6 Find all real roots of $8 - 2y + 4y^3 - y^4$.

9.1.7   Factor $x^2y + xy^2 + x^2 + 2xy + y^2 + x + y$.

9.1.8★   Find all solutions to the system of equations

$$xy + 2x + 3y = 4,$$
$$yz - y + 2z = 5,$$
$$xz - x + 3z = 33.$$

9.2   Sums and Differences of Powers

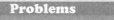

Problem 9.8: Use the Factor Theorem to find a factor of each of the following. Rewrite each as the product of a linear polynomial and another polynomial.

(a) $x^3 - 1$

(b) $x^4 - 1$

(c) $x^5 - 1$

(d) $x^n - 1$, where n is an integer with $n > 1$

Problem 9.9: Factor $x^5 - 2^5$.

Problem 9.10: Let $f(x) = x^n - y^n$ for some constant y and some integer n, where $n > 1$.

(a)   What is $f(y)$?

(b)   Factor $x^n - y^n$ into a product of two polynomial factors such that one factor is linear in x and y and the coefficient of x in this factor is 1.

Problem 9.11:

(a)   Can you factor $x^2 + y^2$ as the product of two linear expressions with real coefficients?

(b)   Factor $x^3 + y^3$ as the product of two polynomials with nonzero degree.

(c)   Let n be a positive integer. Factor $x^{2n+1} + y^{2n+1}$ as the product of two polynomials with nonzero degrees.

Problem 9.12: In this problem, we show that $3^{50} + 8^{50}$ is divisible by 73.

(a)   Use the factorization of $x^5 + y^5$ to explain why $3^5 + 8^5$ is divisible by 11.

(b)   Explain why $3^{50} + 8^{50}$ is divisible by $3^{10} + 8^{10}$.

(c)   Show that $3^{50} + 8^{50}$ is divisible by 73.

Extra! French mathematician **Marie-Sophie Germain** found a factorization during her study
➡➡➡➡ of number theory that still bears her name. **Sophie Germain's Identity** is a factorization
of $x^4 + 4y^4$ over the integers. See if you can find the factorization!

Germain carried on a notable mathematical correspondence with the great mathematician Carl Gauss. She used a pseudonym when writing Gauss to hide her gender.
Gauss would learn her identity when she asked a French commander to ensure Gauss's
safety during a French occupation of Gauss's hometown in Germany.

Problem 9.13: In this problem, we compute the number of ordered triples of integers (x, y, z) with $1729 < x, y, z < 1999$ such that

$$x^2 + xy + y^2 = y^3 - x^3 \quad \text{and} \quad yz + 1 = y^2 + z.$$

(Source: ARML)

(a) Factor the right side of the first equation. How can you simplify the resulting equation?

(b) Rearrange the second equation until you find an expression in the equation that you can factor. (This expression need not involve every term in the equation.)

(c) There are several ways to accomplish part (b). Keep experimenting until you find a way to rewrite the second equation with a product of two factors on one side and 0 on the other.

(d) Combine parts (a) and (c) to find restrictions on x, y, and z that you can use to solve the problem.

Problem 9.14:
(a) Find a positive integer n such that $x^4 + x^2 + 1$ is a factor of $x^n - 1$.

(b) Factor $x^n - 1$ as a difference of squares for your choice of n in part (a).

(c) Factor the two factors produced in part (b).

(d) Factor $x^4 + x^2 + 1$ completely over the real numbers.

Problem 9.15: In this problem, we factor

$$f(x) = x^{10} + 2x^9 + 3x^8 + 4x^7 + 5x^6 + 6x^5 + 5x^4 + 4x^3 + 3x^2 + 2x + 1.$$

(a) Expand each of the following: $(x+1)^2$, $(x^2+x+1)^2$, $(x^3+x^2+x+1)^2$. Notice anything interesting?

(b) Express $f(x)$ as the product of linear and quadratic expressions with real coefficients.

Problem 9.16: Let $N = 4^8 + 6^8 + 9^8$. In this problem, we find two 4-digit numbers whose product is N.
(a) What expression must be added to $x^2 + xy + y^2$ to give a square of a binomial?

(b) Why is $N + 6^8$ a perfect square? (Note that $6 = 2 \cdot 3$.)

(c) Write $N + 6^8 - 6^8$ as the product of two 4-digit numbers.

Problem 9.17: In this problem, we evaluate

$$\frac{(10^4 + 324)(22^4 + 324)(34^4 + 324)(46^4 + 324)(58^4 + 324)}{(4^4 + 324)(16^4 + 324)(28^4 + 324)(40^4 + 324)(52^4 + 324)}$$

without a calculator. (Source: AIME)
(a) Let $f(x) = x^4 + 324$. Find a function $g(x)$ such that $f(x) + g(x)$ is the square of a binomial.

(b) Use part (a) to factor $x^4 + 324$ as the product of polynomials with integer coefficients.

(c) Solve the problem.

Problem 9.8: Express each of the following as the product of a linear polynomial and another polynomial.

(a) $x^3 - 1$

(b) $x^4 - 1$

(c) $x^5 - 1$

(d) $x^n - 1$, where n is an integer with $n > 1$

Solution for Problem 9.8: We see that $x = 1$ is a root of each of the given polynomials. So, by the Factor Theorem, $x - 1$ is a factor of each. We divide $x - 1$ out of each to get another factor.

(a) $x^3 - 1 = (x - 1)(x^2 + x + 1)$.

(b) $x^4 - 1 = (x - 1)(x^3 + x^2 + x + 1)$.

(c) $x^5 - 1 = (x - 1)(x^4 + x^3 + x^2 + x + 1)$.

(d) Our first three parts suggest that $x^n - 1 = (x - 1)(x^{n-1} + x^{n-2} + x^{n-3} + \cdots + x + 1)$. We confirm this by expanding the product on the right side:

$$(x - 1)(x^{n-1} + x^{n-2} + \cdots + x + 1) = x(x^{n-1} + x^{n-2} + \cdots + x + 1) - 1(x^{n-1} + x^{n-2} + \cdots + x + 1)$$
$$= x^n + x^{n-1} + \cdots + x^2 + x - x^{n-1} - x^{n-2} - \cdots - x - 1$$
$$= x^n - 1.$$

$\square$

Maybe the pattern we see in Problem 9.8 is just a quirk that happens when we subtract 1 from x^n. What if we try another constant? We might try factoring $x^3 - 2$ over the rational numbers, but quickly see that we get nowhere, since $x^3 - 2$ has no rational roots. But what if our constant is raised to the same power as x is?

Problem 9.9: Factor $x^5 - 2^5$.

Solution for Problem 9.9: Let $f(x) = x^5 - 2^5$. Clearly, we have $f(2) = 0$, so $x - 2$ is a factor. We divide $x - 2$ into $x^5 - 2^5$ and find that

$$x^5 - 2^5 = (x - 2)(x^4 + 2x^3 + 4x^2 + 8x + 16) = (x - 2)(x^4 + 2^1 x^3 + 2^2 x^2 + 2^3 x + 2^4).$$

$\square$

Now, we're ready to generalize our results from Problems 9.8 and 9.9.

Problem 9.10: Let n be a positive integer. Factor $x^n - y^n$ into a product of two polynomial factors such that one factor is linear in x and y and the coefficient of x in this factor is 1.

Solution for Problem 9.10: We use the previous two problems as a guide. Suppose $f(x) = x^n - y^n$, where y is a constant. We have $f(y) = y^n - y^n = 0$, so by the Factor Theorem, $x - y$ is a factor of $x^n - y^n$. Our work in Problem 9.9 strongly suggests that we then have

$$x^n - y^n = (x - y)\left(x^{n-1} + x^{n-2}y + x^{n-3}y^2 + \cdots + xy^{n-2} + y^{n-1}\right).$$

As before, expanding the right side confirms this factorization:

$$(x - y)(x^{n-1} + x^{n-2}y + \cdots + xy^{n-2} + y^{n-1})$$

$$= x\left(x^{n-1} + x^{n-2}y + \cdots + xy^{n-2} + y^{n-1}\right) - y\left(x^{n-1} + x^{n-2}y + \cdots + xy^{n-2} + y^{n-1}\right)$$

$$= x^n + x^{n-1}y + \cdots + x^2y^{n-2} + xy^{n-1} - x^{n-1}y - x^{n-2}y^2 - \cdots - xy^{n-1} - y^n$$

$$= x^n - y^n.$$

$\square$

> **Important:**
> $$x^n - y^n = (x - y)\left(x^{n-1} + x^{n-2}y + x^{n-3}y^2 + \cdots + xy^{n-2} + y^{n-1}\right)$$

The most common instances of this factorization are the **difference of squares** ($n = 2$) and **difference of cubes** ($n = 3$) factorizations:

$$x^2 - y^2 = (x - y)(x + y),$$
$$x^3 - y^3 = (x - y)(x^2 + xy + y^2).$$

What about sums of nth powers? If we try factoring $x^2 + 1$, we immediately get stuck. The roots of $x^2 + 1$ are nonreal, so we can't factor $x^2 + 1$ as a product of linear factors with real coefficients. However, if we look at $x^3 + 1$, we see that -1 is a root, so $x + 1$ is a factor of $x^3 + 1$. Similarly, -1 is a root of $x^5 + 1$, $x^7 + 1$, $x^9 + 1$, and so on. This gives us an idea:

Problem 9.11: Let n be a positive integer. Factor $x^{2n+1} + y^{2n+1}$ as the product of a linear factor and another polynomial.

Solution for Problem 9.11: Before jumping into factoring $x^{2n+1} + y^{2n+1}$, we try a few simple examples. First, we let $y = 1$, and try small values of n. Since $x = -1$ is a root of $x^{2n+1} + 1$ for every nonnegative integer n, we quickly find

$$x^3 + 1 = (x + 1)(x^2 - x + 1),$$
$$x^5 + 1 = (x + 1)(x^4 - x^3 + x^2 - x + 1),$$
$$x^7 + 1 = (x + 1)(x^6 - x^5 + x^4 - x^3 + x^2 - x + 1).$$

Aha! We see a pattern, so we guess that

$$x^{2n+1} + 1 = (x + 1)(x^{2n} - x^{2n-1} + x^{2n-2} - \cdots + x^2 - x + 1).$$

We can multiply the right side out to confirm it is true. Let's see what happens if y is something more interesting than 1. If $x = -y$, then no matter what y is, we have $x^{2n+1} + y^{2n+1} = 0$. Using this, we can quickly find a linear factor for each of our examples. Here's what we find:

$$x^3 + 3^3 = (x + 3)(x^2 - 3^1x + 3^2),$$
$$x^5 + 2^5 = (x + 2)(x^4 - 2^1x^3 + 2^2x^2 - 2^3x + 2^4),$$
$$x^7 + 2^7 = (x + 2)(x^6 - 2^1x^5 + 2^2x^4 - 2^3x^3 + 2^4x^2 - 2^5x + 2^6).$$

Our examples strongly suggest that

$$x^{2n+1} + y^{2n+1} = (x + y)(x^{2n} - x^{2n-1}y + x^{2n-2}y^2 - \cdots + x^2y^{2n-2} - xy^{2n-1} + y^{2n}).$$

As before, we expand the right side to confirm this factorization:

$$(x + y)(x^{2n} - x^{2n-1}y + x^{2n-2}y^2 - \cdots + x^2y^{2n-2} - xy^{2n-1} + y^{2n})$$
$$= x(x^{2n} - x^{2n-1}y + x^{2n-2}y^2 - \cdots + x^2y^{2n-2} - xy^{2n-1} + y^{2n})$$
$$+ y(x^{2n} - x^{2n-1}y + x^{2n-2}y^2 - \cdots + x^2y^{2n-2} - xy^{2n-1} + y^{2n})$$
$$= x^{2n+1} - x^{2n}y + x^{2n-1}y^2 - \cdots + x^3y^{2n-2} - x^2y^{2n-1} + xy^{2n}$$
$$+ x^{2n}y - x^{2n-1}y^2 + \cdots - x^3y^{2n-2} + x^2y^{2n-1} - xy^{2n} + y^{2n+1}.$$

All terms but the first and last cancel, and we are left with

> **Important:**
>
> $$x^{2n+1} + y^{2n+1} = (x + y)(x^{2n} - x^{2n-1}y + x^{2n-2}y^2 - \cdots + x^2y^{2n-2} - xy^{2n-1} + y^{2n})$$

□

The most common instance of this factorization is the **sum of cubes** factorization:

$$x^3 + y^3 = (x + y)(x^2 - xy + y^2).$$

You might wonder why we focus on sums of odd powers. To see why we cannot factor $x^2 + y^2$ over the reals, notice that if $x^2 + y^2 = 0$, then $x^2 = -y^2$. Taking the square root of both sides gives $x = \pm iy$. In other words, if we view $x^2 + y^2$ as a polynomial in x, its roots are iy and $-iy$, which means we can factor it as $x^2 + y^2 = (x + iy)(x - iy)$. That is, we can factor it over complex numbers, but not over real numbers.

However, as we'll see in our next problem, we can sometimes still use our sum of powers factorization when the powers are even.

Problem 9.12: Show that $3^{50} + 8^{50}$ is divisible by 73.

Solution for Problem 9.12: We certainly aren't going to compute $3^{50} + 8^{50}$. Instead, we'll look for a way to factor it, hoping we can eventually write it as the product of 73 and an integer. We don't have a factorization for the sum of 50th powers, but any 50th power is also a 5th power and a 25th power. So, we factor $3^{50} + 8^{50}$ as a sum of 5th powers and as a sum of 25th powers:

$$3^{50} + 8^{50} = (3^{10})^5 + (8^{10})^5 = (3^{10} + 8^{10})((3^{10})^4 - (3^{10})^3(8^{10}) + (3^{10})^2(8^{10})^2 - (3^{10})(8^{10})^3 + (8^{10})^4),$$
$$3^{50} + 8^{50} = (3^2)^{25} + (8^2)^{25} = (3^2 + 8^2)((3^2)^{24} - (3^2)^{23}(8^2) + \cdots - (3^2)(8^2)^{23} + (8^2)^{24}).$$

The first factorization requires more work to be useful (try factoring $3^{10} + 8^{10}$), but the second is immediately a winner, because $3^2 + 8^2 = 73$. Our second factorization tells us that $3^{50} + 8^{50}$ equals 73 times an integer, which means $3^{50} + 8^{50}$ is divisible by 73. □

> **Concept:** Factorization can be useful in number theory problems involving divisibility.

Notice that we were able to apply our factorization for $x^{2n+1} + y^{2n+1}$ to a sum of even powers in Problem 9.12 because the even power, 50, is divisible by odd numbers besides 1 (namely, 5 and 25). So, we can view the 50th powers as 5th or 25th powers and use the factorization of $x^{2n+1} + y^{2n+1}$.

Problem 9.13: Compute the number of ordered triples of integers (x, y, z) with $1729 < x, y, z < 1999$ such that
$$x^2 + xy + y^2 = y^3 - x^3 \qquad \text{and} \qquad yz + 1 = y^2 + z.$$
(Source: ARML)

Solution for Problem 9.13: We hope to simplify the system of equations, so we look for helpful factorizations or manipulations. We note that the right side of the first equation is a difference of cubes. Factoring the right side of $x^2 + xy + y^2 = y^3 - x^3$ gives us
$$x^2 + xy + y^2 = (y - x)(y^2 + xy + x^2).$$

Since x and y are positive, the factor $x^2 + xy + y^2$ is nonzero. So, we can divide both sides of our equation by $x^2 + xy + y^2$ to find $1 = y - x$. Therefore, $y = x + 1$.

Unfortunately, the relationship we discovered between x and y doesn't help us with the second equation. Neither side factors and nothing jumps out of the expressions on either side that seems helpful, so we move all the terms to one side of the equation to give ourselves another perspective:
$$y^2 - yz + z - 1 = 0.$$

Concept: Experiment with equations that you're unsure how to handle. Rearrange them and look for familiar expressions or forms. Don't just stare at them!

We could factor y out of the first two terms, but this doesn't help much. However, we also notice that we have a difference of squares, $y^2 - 1$. Factoring this gives us
$$(y + 1)(y - 1) - yz + z = 0.$$

Factoring z out of the last two terms gives $(y+1)(y-1) - z(y-1) = 0$. Aha! We can factor out $y - 1$ to get
$$(y + 1 - z)(y - 1) = 0.$$

Since $y > 1729$, we know that $y - 1$ is nonzero. Dividing by $y - 1$ gives us $y + 1 - z = 0$, so $y = z - 1$.

From $y = x + 1$, we have $x = y - 1$, and $y = z - 1$ gives $z = y + 1$. So, for each possible value of y, the triple $(x, y, z) = (y - 1, y, y + 1)$ is the only possible solution to the system with that value of y. Since $1729 < x, y, z < 1999$, we must have $1730 < y < 1998$ in order to keep $x = y - 1$ and $z = y + 1$ in the permissible range of values. Since there are $1998 - 1730 - 1 = 267$ integers between 1730 and 1998, there are 267 triples of integers (x, y, z) that are solutions to the given equations. $\square$

Problem 9.14: Factor $x^4 + x^2 + 1$ over the real numbers.

Solution for Problem 9.14: The expression we must factor has the form $y^2 + y + 1$ for $y = x^2$. We recognize $y^2 + y + 1$ as a factor of $y^3 - 1$. Since $y^3 - 1 = (y - 1)(y^2 + y + 1)$, we have $y^2 + y + 1 = (y^3 - 1)/(y - 1)$.

CHAPTER 9. FACTORING MULTIVARIABLE POLYNOMIALS

Letting $y = x^2$, we have

$$x^4 + x^2 + 1 = \frac{x^6 - 1}{x^2 - 1}.$$

Now we factor the numerator, first using the difference of squares factorization, then both the sum and difference of cubes factorizations:

$$x^4 + x^2 + 1 = \frac{x^6 - 1}{x^2 - 1} = \frac{(x^3 + 1)(x^3 - 1)}{x^2 - 1} = \frac{(x + 1)(x - 1)(x^2 + x + 1)(x^2 - x + 1)}{x^2 - 1}.$$

Since $x^2 - 1 = (x + 1)(x - 1)$, we have

$$x^4 + x^2 + 1 = (x^2 + x + 1)(x^2 - x + 1).$$

Neither of the quadratics on the right side has real roots, so we cannot factor any further over the reals. $\square$

In Chapter 10, we'll take another look at expressions like $1 + x^2 + x^4$ when we discuss geometric series.

Problem 9.15: Factor

$$f(x) = x^{10} + 2x^9 + 3x^8 + 4x^7 + 5x^6 + 6x^5 + 5x^4 + 4x^3 + 3x^2 + 2x + 1$$

as the product of linear and quadratic expressions with real coefficients.

Solution for Problem 9.15:   Before tackling this tenth degree polynomial, we try to think of similar polynomials that we know how to deal with. We notice that the coefficients of the polynomial go up by 1 until they reach 6, then back down by 1 until they reach 1. The expansion $(x + 1)^2 = x^2 + 2x + 1$ is similar in this regard: the coefficients go up to 2, then back down. So, does this give us any insight into, say, $x^4 + 2x^3 + 3x^2 + 2x + 1$?

Because the leading term of $x^4 + 2x^3 + 3x^2 + 2x + 1$ is x^4, the constant term is 1, and all the coefficients are positive, we guess that

$$x^4 + 2x^3 + 3x^2 + 2x + 1 = (x^2 + ax + 1)(x^2 + bx + 1),$$

for some constants a and b. We need the coefficients of x^3 and of x in the expansion of the right side both to be 2, so we try $a = b = 1$. Sure enough, we find that

$$(x^2 + x + 1)^2 = x^4 + 2x^3 + 3x^2 + 2x + 1.$$

Hmmm.... Comparing this to our expansion of $(x + 1)^2$, we start to see a pattern. We guess that $f(x)$ equals $(x^5 + x^4 + x^3 + x^2 + x + 1)^2$. We expand

$$(x^5 + x^4 + x^3 + x^2 + x + 1)(x^5 + x^4 + x^3 + x^2 + x + 1)$$

to check if it works. To expand this product, we must add every possible product of a term in the first factor and a term in the second. The expansion clearly has x^{10} as its first term. There are two x^9 terms, $(x^5)(x^4)$ and $(x^4)(x^5)$. There are three x^8 terms, $(x^5)(x^3)$, $(x^4)(x^4)$, and $(x^3)(x^5)$. Continuing in this manner, we find that

$$x^{10} + 2x^9 + 3x^8 + 4x^7 + 5x^6 + 6x^5 + 5x^4 + 4x^3 + 3x^2 + 2x + 1 = (x^5 + x^4 + x^3 + x^2 + x + 1)^2.$$

274

We're not finished. We have to try to factor $x^5 + x^4 + x^3 + x^2 + x + 1$. We turn to our strategy from Problem 9.14. We recognize this polynomial as a factor of $x^6 - 1$, and we have

$$x^5 + x^4 + x^3 + x^2 + x + 1 = \frac{x^6 - 1}{x - 1} = \frac{(x^3 - 1)(x^3 + 1)}{x - 1} = \frac{(x - 1)(x^2 + x + 1)(x + 1)(x^2 - x + 1)}{x - 1}$$
$$= (x + 1)(x^2 + x + 1)(x^2 - x + 1).$$

The quadratics in this final expression have nonreal roots, so we cannot factor any further. So, we finally have

$$f(x) = (x^5 + x^4 + x^3 + x^2 + 1)^2 = (x + 1)^2(x^2 + x + 1)^2(x^2 - x + 1)^2.$$

□

Notice that we did a lot of guessing on the way to finding a solution. These weren't blind, lucky guesses; each was based on a few observations or patterns. Moreover, we confirmed each guess by working out our hunches to see if they were correct. This is not only a perfectly valid approach to problems, it's also an important problem solving strategy.

> **Concept:** Don't be afraid to guess. A great many problems can be solved by looking for patterns or clues, guessing an answer, then showing that your guess is indeed the correct answer.

Problem 9.16: Find two 4-digit numbers whose product is $4^8 + 6^8 + 9^8$.

Solution for Problem 9.16: None of our factorizations is immediately useful. We notice that the bases in the expression are 2^2, $2 \cdot 3$, and 3^2, so we write the expression in terms of 2s and 3s:

$$4^8 + 6^8 + 9^8 = (2^2)^8 + (2 \cdot 3)^8 + (3^2)^8 = 2^{16} + (2 \cdot 3)^8 + 3^{16}.$$

This is very close to the expansion of $(2^8 + 3^8)^2$. If the middle term were $2(2 \cdot 3)^8$, then the expression would be the expansion of $(2^8 + 3^8)^2$. This gives us an idea. Let's "complete the square":

$$2^{16} + (2 \cdot 3)^8 + 3^{16} = 2^{16} + (2 \cdot 3)^8 + 3^{16} + (2 \cdot 3)^8 - (2 \cdot 3)^8$$
$$= 2^{16} + 2(2 \cdot 3)^8 + 3^{16} - (2 \cdot 3)^8$$
$$= (2^8 + 3^8)^2 - (2 \cdot 3)^8.$$

We have a difference of squares! Factoring it gives

$$4^8 + 6^8 + 9^8 = (2^8 + 3^8)^2 - (2 \cdot 3)^8 = (2^8 + 3^8 + (2 \cdot 3)^4)(2^8 + 3^8 - (2 \cdot 3)^4) = 8113 \cdot 5521.$$

□

Problem 9.16 gives us both another method of completing the square, and another use for doing so.

> **Important:** We can sometimes factor the sum of even powers of expressions by creating a difference of squares. We do so by adding and subtracting an appropriate perfect square so that the result is the square of a binomial minus the perfect square.

Here's another problem that illustrates this strategy:

Problem 9.17: Evaluate

$$\frac{(10^4 + 324)(22^4 + 324)(34^4 + 324)(46^4 + 324)(58^4 + 324)}{(4^4 + 324)(16^4 + 324)(28^4 + 324)(40^4 + 324)(52^4 + 324)}$$

without a calculator. (Source: AIME)

Solution for Problem 9.17:   We can't cancel anything, but we do notice that each factor in the numerator and denominator has the form $n^4 + 324$. So, if we find a way to factor $n^4 + 324$, we may be able to use this factorization to do some canceling and simplify the fraction. Since $324 = 18^2$, we have $n^4 + 18^2$. This is a sum of squares, which gets us thinking about using our new "building a difference of squares" strategy.

We seek a term we can add to $n^4 + 18^2$ to make the result the square of a binomial. This binomial must be $n^2 + 18$, since squaring this will give us both an n^4 term and an 18^2 term. We find

$$(n^2 + 18)^2 = n^4 + 36n^2 + 18^2,$$

so $n^4 + 18^2 = (n^2 + 18)^2 - 36n^2$. Success! We can factor the right side as a difference of squares, to find

$$n^4 + 18^2 = (n^2 + 18)^2 - 36n^2 = (n^2 - 6n + 18)(n^2 + 6n + 18).$$

So, now our fraction is

$$\frac{(10^2 - 6 \cdot 10 + 18)(10^2 + 6 \cdot 10 + 18)(22^2 - 6 \cdot 22 + 18)(22^2 + 6 \cdot 22 + 18) \cdots (58^2 - 6 \cdot 58 + 18)(58^2 + 6 \cdot 58 + 18)}{(4^2 - 6 \cdot 4 + 18)(4^2 + 6 \cdot 4 + 18)(16^2 - 6 \cdot 16 + 18)(16^2 + 6 \cdot 16 + 18) \cdots (52^2 - 6 \cdot 52 + 18)(52^2 + 6 \cdot 52 + 18)}.$$

We might think we're stuck again, but we notice that

$$10^2 - 6 \cdot 10 = (10 - 6)(10) = 4(10) = 4(4 + 6) = 4^2 + 6 \cdot 4,$$

so the factors $(10^2 - 6 \cdot 10 + 18)$ and $(4^2 + 6 \cdot 4 + 18)$ are equal, and therefore cancel. Maybe it's a coincidence. Let's see if there's a factor in the denominator equal to $(10^2 + 6 \cdot 10 + 18)$. There is—the very next factor, $(16^2 - 6 \cdot 16 + 18)$, because

$$10^2 + 6 \cdot 10 = (10 + 6)(10) = (16)(10) = 16(16 - 6) = 16^2 - 6 \cdot 16.$$

Doing a little more factorization reveals why these factors in the numerator match factors in the denominator. We write each factor of the form $n^2 - 6n + 18$ as $n(n - 6) + 18$ and each factor of the form $n^2 + 6n + 18$ as $n(n + 6) + 18$:

$$\frac{(10 \cdot 4 + 18)(10 \cdot 16 + 18)(22 \cdot 16 + 18)(22 \cdot 28 + 18) \cdots (58 \cdot 52 + 18)(58 \cdot 64 + 18)}{(4 \cdot (-2) + 18)(4 \cdot 10 + 18)(16 \cdot 10 + 18)(16 \cdot 22 + 18) \cdots (52 \cdot 46 + 18)(52 \cdot 58 + 18)}.$$

Now we see that everything cancels except the last term in the numerator and the first term in the denominator, and we have

$$\frac{58 \cdot 64 + 18}{4 \cdot (-2) + 18} = \frac{3730}{10} = 373.$$

$\square$

> **Concept:** Experimentation and searching for patterns are critical factoring skills.

Exercises

9.2.1 Find all 6 roots of $f(x) = x^6 - 1$.

9.2.2 Compute all ordered pairs (x, y) such that $xy + 9 = y^2$ and $xy + 7 = x^2$. *(Source: ARML)*

9.2.3 Show that $2^{27} + 3^{27}$ is divisible by 35.

9.2.4 Let $x^3 = 8x + y$ and $y^3 = x + 8y$. If $|x| \neq |y|$, then find $x^2 + y^2$.

9.2.5 Find $(a/c)^3$ if $\dfrac{1}{a+c} = \dfrac{1}{a} + \dfrac{1}{c}$. *(Source: Mandelbrot)*

9.2.6 Given that 7,999,999,999 has at most two prime factors, find its largest prime factor. *(Source: HMMT)* **Hints:** 347

9.2.7 Compute the real value of x such that $(x^2 + x + 1)(x^6 + x^3 + 1) = \dfrac{10}{x-1}$. *(Source: ARML)*

9.2.8★ Show that $n^4 - 20n^2 + 4$ is composite for all integers n, where $n > 4$. *(Source: Crux Mathematicorum)* **Hints:** 283

9.3★ The Factor Theorem for Multivariable Polynomials

Problems

> **Problem 9.18:** In this problem, we factor the polynomial $f(a, b) = b^3 - ab^2 - ab - b + a^2 + a$.
>
> (a) Start by viewing $f(a, b)$ as a polynomial in a, with b being just a constant. In other words, let $g(a) = a^2 - ab^2 - ab + a + b^3 - b$. Find a function $h(b)$ such that $g(h(b)) = 0$ for all values of b.
>
> (b) Find the polynomial $q(a)$ such that $g(a) = (a - h(b))q(a)$. In other words, use your result from part (a) to factor $g(a)$.

> **Problem 9.18:** Factor the polynomial $f(a, b) = b^3 - ab^2 - ab - b + a^2 + a$.

Solution for Problem 9.18: We could try grouping, but it's not at all clear which terms to group. For example, we could factor b from the first 2 terms and a from the last two, and we have

$$f(a, b) = b(b^2 - ab - a - 1) + a(a + 1).$$

That's not too helpful.

We have 3 positive terms and 3 negative terms in our original $f(a, b)$, and each coefficient is 1 or -1. If we just let $a = b$, all the terms will cancel:

$$f(a, a) = a^3 - a(a^2) - a(a) - a + a^2 + a = a^3 - a^3 - a^2 - a + a^2 + a = 0.$$

Hmmm.... We have $f(a,b) = 0$ whenever $a = b$. Does this mean $a - b$ is a factor, just as $x - r$ is a factor of a polynomial $p(x)$ if $p(r) = 0$?

To see why $a - b$ is a factor of $f(a,b)$, let's take a different view of $f(a,b)$. We know a lot about factoring one-variable polynomials, so we view $f(a,b)$ as a polynomial in a, and treat b as a constant. We let

$$g(a) = b^3 - ab^2 - ab - b + a^2 + a = a^2 + (-b^2 - b + 1)a + (b^3 - b),$$

where b is a constant. Notice that we organize the terms of $g(a)$ by the degree of a in each term, in order to focus on the fact that it is a polynomial in a.

As shown above, we have $g(b) = 0$, so b is a root of $g(a)$. Therefore, $a - b$ is a factor of the one-variable polynomial $g(a)$, and we can write

$$g(a) = (a - b)q(a)$$

for some polynomial $q(a)$. Note that we only know that $q(a)$ is a polynomial in a at this point. We can find $q(a)$ with long division, or with a little logic. We must have

$$(a - b)q(a) = g(a) = a^2 + (-b^2 - b + 1)a + (b^3 - b),$$

Since the far right side is quadratic in a, the polynomial $q(a)$ must be linear in a. The leading term of $g(a)$ is a^2, so the leading term of $q(a)$ is simply a, which means that $q(a) = a + k$ for some constant k (where the constant k may be in terms of b, which we are treating as a constant). We therefore have

$$(a - b)(a + k) = a^2 - ab^2 - ab + a + b^3 - b.$$

Expanding the left gives $a^2 - ab + ak - bk = a^2 - ab^2 - ab + a + b^3 - b$. Removing common terms leaves us with $ak - bk = -ab^2 + a + b^3 - b$. Since k must be a constant, we equate the coefficients of a on both sides to find $k = -b^2 + 1$. (Remember, we are treating b as a constant.) So, we have

$$g(a) = (a - b)(a - b^2 + 1).$$

But this holds for any constant b, so we have our factorization:

$$f(a,b) = (a - b)(a - b^2 + 1).$$

As a check, we can multiply it out:

$$(a - b)(a - b^2 + 1) = a(a - b^2 + 1) - b(a - b^2 + 1) = a^2 - ab^2 + a - ab + b^3 - b,$$

as desired. $\square$

Taking a closer look at our solution to Problem 9.18, our key step was finding a polynomial $h(b)$ such that $f(h(b), b) = 0$ for all b. Specifically, we found that if $h(b) = b$, then $f(h(b), b) = f(b, b) = 0$ for all b. This observation guided us to guess that $a - h(b)$ (that is, $a - b$) is a factor of $f(a,b)$. Viewing $f(a,b)$ as a one-variable polynomial in a and applying the Factor Theorem made it particularly clear that we could write $f(a,b)$ as the product of $a - b$ and another factor. We then found that this other factor is also a polynomial.

In much the same way, we can show that:

> **Important:** If $f(a,b)$ is a polynomial and there is a polynomial $h(b)$ such that $f(h(b),b) = 0$ for all b, then we can write $f(a,b) = (a - h(b))g(a,b)$, where $g(a,b)$ is a polynomial.

We can also extend this to a polynomial in any number of variables. For the rest of this section, we'll focus on applying the Factor Theorem for multivariable polynomials to various problems. You'll also get a lot of practice in dealing with complicated expressions and expressions involving several variables.

Problems

Problem 9.19: Let $f(a,b,c) = (a + b + c)^3 - a^3 - b^3 - c^3$.

(a) Show that $f(-b,b,c) = 0$. What expression must therefore be a factor of $f(a,b,c)$?

(b) Show that $a + c$ and $b + c$ are factors of $f(a,b,c)$.

(c) Why are the factors found in parts (a) and (b) the only nonconstant factors of $f(a,b,c)$?

(d) Factor $f(a,b,c)$. Check your factorization by expanding $f(a,b,c)$ and the factorization you found.

Problem 9.20: Factor $a^2(b - c) + b^2(c - a) + c^2(a - b)$.

Problem 9.21: Factor $(a + b + c)^3 - (b + c - a)^3 - (c + a - b)^3 - (a + b - c)^3$.

Problem 9.19: Factor $f(a,b,c) = (a + b + c)^3 - a^3 - b^3 - c^3$.

Solution for Problem 9.19: We notice that if $a = -b$, then we have

$$f(a,b,c) = f(-b,b,c) = (-b + b + c)^3 - (-b)^3 - b^3 - c^3 = c^3 - c^3 = 0.$$

Since $f(-b,b,c) = 0$, the Factor Theorem tells us that $a + b$ is a factor of $f(a,b,c)$. Similarly, if $a = -c$, we have

$$f(-c,b,c) = (-c + b + c)^3 - (-c)^3 - b^3 - c^3 = b^3 - b^3 = 0,$$

so $a + c$ is a factor of $f(a,b,c)$. Finally, if $b = -c$, we have $f(a,-c,c) = 0$, so $b + c$ is a factor of $f(a,b,c)$.

Since $\deg f = 3$, and $a + b$, $a + c$, and $b + c$ are factors of $f(a,b,c)$, we have

$$f(a,b,c) = k(a + b)(a + c)(b + c)$$

for some constant k. There can be no more factors, since the product of $(a + b)(a + c)(b + c)$ and any nonconstant polynomial has a degree greater than 3.

Since this factorization must equal $f(a,b,c)$ for all a, b, and c, we can choose specific values of a, b, and c to find k. So, we let $a = b = c = 1$, and we have

$$(1 + 1 + 1)^3 - 1^3 - 1^3 - 1^3 = k(1 + 1)(1 + 1)(1 + 1).$$

Solving this equation gives us $k = 3$, so we have

$$(a + b + c)^3 - a^3 - b^3 - c^3 = 3(a + b)(a + c)(b + c).$$

We can spot-check our factorization by trying a few triples (a, b, c), and making sure the two sides are equal. For example, if $(a, b, c) = (1, 1, 0)$, then both sides equal 6.

> **Sidenote:** We can also recognize that both $(a+b+c)^3 - a^3 - b^3 - c^3$ and $3(a+b)(b+c)(c+a)$ are symmetric. By **symmetric**, we mean that if we exchange any 2 variables in the expression, then the expression is unchanged. For example, exchanging any 2 of the variables in the expression abc leaves the expression unchanged, so it is symmetric. On the other hand, the a^2bc is not symmetric, because exchanging a and b gives ab^2c, which is not the same as a^2bc. A symmetric expression cannot be equivalent to a non-symmetric expression, so noting that our two expressions are symmetric gives us a quick check that we haven't made a pretty obvious mistake (but it doesn't show the two are the same).

The best check is to multiply out both sides. The $(a + b + c)^3$ on the left side presents a bit of a challenge. We start by writing it as

$$(a + b + c)(a + b + c)(a + b + c).$$

We could repeatedly use the distributive property, but we can also carefully reason our way to the expansion. Each term in the expansion results from taking the product of one term from each of the factors in the product. Because all the terms in each factor are linear, and we multiply three of them to form each term in the product, each term in the expansion has degree 3. So, we have three types of terms in the expansion:

- *Terms of the form a^3.* If we take a from each expression, we get a^3. Similarly, we get b^3 by taking b from each expression and c^3 by taking c from each expression. Each of these can only be done in one way, so $a^3 + b^3 + c^3$ is part of the expansion.

- *Terms of the form a^2b.* If we take a from two expressions and b from the other, we form a^2b. We can do so in 3 ways, since there are 3 ways to choose the expression from which we take the b. So, there is a $3a^2b$ term in the expansion. Similarly, we see that the expansion must include

$$3a^2b + 3ab^2 + 3a^2c + 3ac^2 + 3b^2c + 3bc^2.$$

- *Terms of the form abc.* If we take a from one expression, b from another, and c from the remaining expression, we form the product abc. There are $3 \cdot 2 \cdot 1 = 6$ ways to form abc in the expansion, so we must have $6abc$ in our expansion.

So, we have

$$(a + b + c)^3 = a^3 + b^3 + c^3 + 3a^2b + 3ab^2 + 3a^2c + 3ac^2 + 3b^2c + 3bc^2 + 6abc.$$

Notice that both sides are symmetric, and they both equal 27 when $a = b = c = 1$. (This is just a check; it doesn't prove our expansion is correct.)

> **Important:** Don't memorize the expansion of $(a + b + c)^3$. Just understand how we came up with it, and you'll be able to re-derive it if you need it. You'll also be able to use this expansion method on other complicated expansions.

Using our expansion of $(a + b + c)^3$ gives us the expansion of the left side of

$$(a + b + c)^3 - a^3 - b^3 - c^3 = 3(a + b)(a + c)(b + c)$$

as

$$3a^2b + 3ab^2 + 3a^2c + 3ac^2 + 3b^2c + 3bc^2 + 6abc.$$

We could use the same approach to expand $3(a + b)(a + c)(b + c)$, or we could just use the distributive property:

$$\begin{aligned}
3(a + b)(a + c)(b + c) &= 3(a^2 + ac + ab + bc)(b + c) \\
&= 3(a^2b + a^2c + abc + ac^2 + ab^2 + abc + b^2c + bc^2) \\
&= 3a^2b + 3ab^2 + 3a^2c + 3ac^2 + 3b^2c + 3bc^2 + 6abc.
\end{aligned}$$

So, our factorization is correct. $\square$

Problem 9.20: Factor $a^2(b - c) + b^2(c - a) + c^2(a - b)$.

Solution for Problem 9.20: Let the polynomial be $f(a, b, c)$. We have $f(b, b, c) = b^2(b-c)+b^2(c-b)+c^2(b-b) = 0$. Since $f(a, b, c) = 0$ when $a = b$, the expression $a - b$ is a factor of $f(a, b, c)$. Similarly, we have $f(a, b, c) = 0$ for $c = a$ and $b = c$, so $c - a$ and $b - c$ are factors of $f(a, b, c)$. Because $\deg f = 3$ and we have found 3 linear factors of f, we have

$$f(a, b, c) = a^2(b - c) + b^2(c - a) + c^2(a - b) = k(a - b)(b - c)(c - a)$$

for some constant k. This must hold for all a, b, and c, so we let $a = 3$, $b = 2$, and $c = 1$, and we have

$$3^2(2 - 1) + 2^2(1 - 3) + 1^2(3 - 2) = k(3 - 2)(2 - 1)(1 - 3),$$

which gives us $k = -1$. So, we have

$$a^2(b - c) + b^2(c - a) + c^2(a - b) = -(a - b)(b - c)(c - a).$$

Note that we can write the expression $-(a - b)(b - c)(c - a)$ in many different ways. For example, because $-(a - b) = b - a$, we can write our final factorization as $(b - a)(b - c)(c - a)$. $\square$

Problem 9.21: Factor $(a + b + c)^3 - (b + c - a)^3 - (c + a - b)^3 - (a + b - c)^3$.

Solution for Problem 9.21: Let the polynomial be $f(a, b, c)$. We try $f(b, b, c)$ to see if $a - b$ is a factor, but this gives us

$$f(b, b, c) = (2b + c)^3 - c^3 - c^3 - (2b - c)^3.$$

However, this expression doesn't equal 0 for all b and c. (We can see this by multiplying out the cubes, or simply by noting that if $b = c = 1$, the expression equals 24, not 0.) Similarly, we test $f(-b, b, c)$ to see if $a + b$ is a factor, but this fails, too.

We could start trying more complicated expressions, but first, maybe we should try something simpler. Instead of letting a equal some expression involving b and/or c, we check what happens when $a = 0$. Then, we have

$$f(0, b, c) = (b + c)^3 - (b + c)^3 - (c - b)^3 - (b - c)^3,$$

which does indeed equal 0 for all b and c because $(b + c)^3 - (b + c)^3 = 0$ and $-(c - b)^3 - (b - c)^3 = (b - c)^3 - (b - c)^3 = 0$. Since $f(a, b, c) = 0$ when $a = 0$, the expression $a - 0$ is a factor of $f(a, b, c)$. That is, a itself is a factor of $f(a, b, c)$. Similarly, b and c are factors of $f(a, b, c)$. Since the degree of $f(a, b, c)$ is clearly no greater than 3, and we have 3 factors of $f(a, b, c)$, we have

$$f(a, b, c) = (a + b + c)^3 - (b + c - a)^3 - (c + a - b)^3 - (a + b - c)^3 = kabc$$

for some constant k. We find k by letting $a = b = c = 1$, which gives us $k = 24$, so

$$(a + b + c)^3 - (b + c - a)^3 - (c + a - b)^3 - (a + b - c)^3 = 24abc.$$

□

Exercises

9.3.1 Show that $(a - b)^3 + (b - c)^3 + (c - a)^3 = 3(a - b)(b - c)(c - a)$.

9.3.2 Prove that

$$(a + b + c + d)^4 + (a + b - c - d)^4 + (a - b + c - d)^4 + (a - b - c + d)^4$$
$$- (a + b + c - d)^4 - (a + b - c + d)^4 - (a - b + c + d)^4 - (-a + b + c + d)^4 = 192abcd.$$

9.3.3 Factor $abc - ab - bc - ca + a + b + c - 1$.

9.3.4 Show that $\dfrac{b - c}{a} + \dfrac{c - a}{b} + \dfrac{a - b}{c} = \dfrac{(c - b)(a - c)(b - a)}{abc}$ if $abc \neq 0$.

9.3.5 Suppose that $x^{-1} + y^{-1} + z^{-1} = (x + y + z)^{-1}$ for nonzero numbers x, y, and z such that $x + y + z \neq 0$. Prove that $x^n + y^n + z^n = (x + y + z)^n$ for all odd integers n. **Hints:** 99

9.3.6★ Factor $(x + y)^7 - (x^7 + y^7)$ over the reals.

9.4 Summary

We can sometimes factor expressions by splitting them into groups of terms, factoring the groups, and then discovering that the factored groups have factors in common. One common example is

$$ap + aq + bp + bq = a(p + q) + b(p + q) = (a + b)(p + q).$$

This factorization is so useful that we sometimes "force" it into problems. For example, if we must find all positive integer pairs such that $6xy + 4x + 9y = 247$, we would add 6 to both sides to give us $6xy + 4x + 9y + 6 = 253$, because we could then factor the left side to give $(2x + 3)(3y + 2) = 253$. We call this process **Simon's Favorite Factoring Trick**.

Two useful factorizations of sums and differences of powers are

$$x^n - y^n = (x - y)\left(x^{n-1} + x^{n-2}y + x^{n-3}y^2 + \cdots + xy^{n-2} + y^{n-1}\right),$$
$$x^{2n+1} + y^{2n+1} = (x + y)(x^{2n} - x^{2n-1}y + x^{2n-2}y^2 - \cdots + x^2y^{2n-2} - xy^{2n-1} + y^{2n}).$$

We can sometimes factor the sum of even powers of expressions by creating a difference of squares. We do so by adding and subtracting an appropriate perfect square so that the result is the square of a binomial minus the perfect square.

The Factor Theorem can be extended to multivariable polynomials. For example:

> **Important:** If $f(a,b)$ is a polynomial and there is a polynomial $h(b)$ in b such that $f(h(b),b) = 0$ for all b, then we can write $f(a,b) = (a - h(b))g(a,b)$, where $g(a,b)$ is a polynomial.

We can also apply this to polynomials with more than 2 variables.

Problem Solving Strategies

Concepts:
- A **Diophantine equation** is an equation for which we must find integer solutions. Factorization is a powerful tool for solving Diophantine equations. If we can write a Diophantine equation as a product of expressions set equal to a constant integer, we can then use the divisors of the integer to find solutions to the equation.

- Look for ways to use factorization to avoid having to expand complicated products.

- Factorization can be useful in number theory problems about divisibility.

- Experiment with equations that you're unsure how to handle. Rearrange them and look for familiar expressions or forms. Don't just stare at them!

- Don't be afraid to guess. A great many problems can be solved by looking for patterns or clues, guessing an answer, then showing that your guess is indeed the correct answer.

REVIEW PROBLEMS

9.22 Let $x = 2001^{1002} - 2001^{-1002}$ and $y = 2001^{1002} + 2001^{-1002}$. Find $x^2 - y^2$. *(Source: HMMT)*

9.23 Factor both $16y^4 - 1$ and $243x^5 + 32z^{10}$ over the integers.

9.24 Find all the roots of $x^7 - x^4 - x^3 + 1$.

9.25 Find all 8 roots of $x^8 - 1$.

9.26 If $a \neq b$, $a^3 - b^3 = 19x^3$, and $a - b = x$, then find a in terms of x. *(Source: AHSME)*

9.27 Find all ordered pairs such that $x - y = 2$ and $x^3 - y^3 = 98$.

9.28 Call a pair of numbers whose product is equal to their sum *Simonish*. For instance, 1.25 and 5 are a *Simonish* pair: $5 + 1.25 = 5 \cdot 1.25 = 6.25$. The number 2 is a *Simonish* pair with itself: $2 + 2 = 2 \cdot 2 = 4$. Find all *Simonish* pairs.

9.29 Graph all solutions to the equation $6xy + 1 = 3x + 2y$.

9.30 Factor each of the following as much as possible:

(a) $p^2 - 2p + 1 - q^2 - 2qr - r^2$

(b) $4x^2y^2 - (x^2 - z^2 + y^2)^2$

(c) $(a^2 - b^2)(b^2 - c^2) + (a - b)(b - c)$

(d)★ $x^2 + 5xy + 6y^2 + x + 3y$

9.31 Find all ordered pairs of integers (c, d) such that $6cd + 4c = 79 + 3d$.

9.32 Factor $9a^5 - 4a^3 - 81a^2 + 36$ completely

(a) over the integers.

(b) over the rationals.

(c) over the reals.

(d) over the complex numbers.

9.33 Suppose a, b, and c are distinct numbers such that $a^2 - bc = b^2 - ca$. Show that $a = -(b + c)$.

9.34 Factor $abc + abe + acd + ade + bcf + bef + cdf + def$ completely.

9.35 Factor $x^{12} - y^{12}$ completely:

(a) over the rationals.

(b)★ over the reals.

9.36 Show that $\dfrac{b+c}{(a-b)(a-c)} + \dfrac{a+c}{(b-c)(b-a)} + \dfrac{a+b}{(c-a)(c-b)} = 0$ for all distinct a, b, and c.

9.37 Factor $x^4 + 4y^4$ over real coefficients.

9.38 Factor $(x-a)^3(b-c)^3 + (x-b)^3(c-a)^3 + (x-c)^3(a-b)^3$ into a product of linear terms.

9.39 Let a, b, and c be three real numbers such that $a + b$, $b + c$, and $c + a$ are nonzero. Show that

$$\frac{2a}{a+b} + \frac{2b}{b+c} + \frac{2c}{c+a} + \frac{(b-c)(c-a)(a-b)}{(b+c)(c+a)(a+b)} = 3.$$

Challenge Problems

9.40 The distinct real numbers x and y satisfy $x^2 = 33y + 907$ and $y^2 = 33x + 907$. Find xy. *(Source: WOOT)*

9.41 If $x^3 = 1$ and $x \neq 1$, find the value of $(1 - x + x^2)(1 + x - x^2)$. **Hints:** 188

9.42 Find all ordered pairs (x, y) such that $x - xy^3 = 7$ and $xy^2 - xy = 3$.

9.43 If $x \neq y$ and $\dfrac{x}{y} + x = \dfrac{y}{x} + y$, what is the value of $\dfrac{1}{x} + \dfrac{1}{y}$? **Hints:** 246

9.44 Let $g(x) = x^5 + x^4 + x^3 + x^2 + x + 1$. What is the remainder when the polynomial $g(x^{12})$ is divided by the polynomial $g(x)$? *(Source: AHSME)* **Hints:** 109, 284

9.45 Let $Q(x)$ be the quotient when $37x^{73} - 73x^{37} + 36$ is divided by $x - 1$. Compute the sum of the coefficients of $Q(x)$. *(Source: ARML)* **Hints:** 73

9.46 Given $x + \dfrac{1}{x} = 1$, calculate $x^7 + \dfrac{1}{x^7}$.

9.47 Factor each of the following as completely as possible over the reals.

(a) $a^3(c - b) + b^3(a - c) + c^3(b - a)$.

(b)$\star$ $a^4(c - b) + b^4(a - c) + c^4(b - a)$.

9.48 Let x and y be real numbers that differ in absolute value and satisfy the equations

$$x^3 = 13x + 3y,$$
$$y^3 = 3x + 13y.$$

Find $(x^2 - y^2)^2$. *(Source: AIME)* **Hints:** 266

9.49 Let n be a positive integer. Show that $(x - 1)^2$ is a factor of $x^n - n(x - 1) - 1$. **Hints:** 287

9.50 Let z be such that $z^7 = 1$ and $z \neq 1$. Evaluate $z^{100} + \dfrac{1}{z^{100}} + z^{300} + \dfrac{1}{z^{300}} + z^{500} + \dfrac{1}{z^{500}}$. **Hints:** 70

9.51 Factor $x^5 + x^4y + x^3y^2 + x^2y^3 + xy^4 + y^5$ completely over the reals.

9.52 Find two positive integers greater than 1 whose product is $6^6 + 8^4 + 27^4$. **Hints:** 117

9.53 The expression $\sqrt{\sqrt[3]{4} - 1} + \sqrt{\sqrt[3]{16} - \sqrt[3]{4}}$ is equal to $\sqrt[3]{k}$ for some integer k. Find k.

9.54$\star$ Factor $p^3 + q^3 + r^3 - 3pqr$. **Hints:** 85

9.55$\star$ Write $\left(1 + \dfrac{1}{3}\right)\left(1 + \dfrac{1}{3^2}\right)\left(1 + \dfrac{1}{3^{2^2}}\right)\cdots\left(1 + \dfrac{1}{3^{2^{100}}}\right)$ in the form $a(1 - b^c)$, where a, b, and c are constants. **Hints:** 94

9.56$\star$ Factor $x^2(y + z) + y^2(z + x) + z^2(x + y) - x^3 - y^3 - z^3 - 2xyz$. **Hints:** 127

9.57$\star$ Simplify the expression $\dfrac{a^3(b + c)}{(a - b)(a - c)} + \dfrac{b^3(c + a)}{(b - c)(b - a)} + \dfrac{c^3(a + b)}{(c - a)(c - b)}$.

9.58$\star$ Determine all the solutions to $x^4 - 4x = 1$. **Hints:** 354

9.59$\star$ Find the two real roots of $2000x^6 + 100x^5 + 10x^3 + x - 2 = 0$. *(Source: AIME)* **Hints:** 276

We are continually faced with a series of great opportunities brilliantly disguised as insoluble problems.

– John W. Gardner

CHAPTER **10**

Sequences and Series

A **sequence** is an ordered *list* of numbers, while a **series** is an ordered *sum* of numbers. For instance, $10, 11, 12, \ldots, 99$ is a sequence, while $10 + 11 + 12 + \cdots + 99$ is a series. The order of the numbers in a sequence or series matters. For example, the sequence 1, 2, 3, 4 is not the same as the sequence 3, 1, 2, 4. Sequences and series can also have infinitely many terms. For example, both the sequence $1, 2, 3, \ldots$ and the series $1 + \frac{1}{2} + \frac{1}{4} + \frac{1}{8} + \frac{1}{16} + \cdots$ have infinitely many terms.

10.1 Arithmetic Sequences

Definition: An **arithmetic sequence** is a sequence in which the difference between any two consecutive terms is the same. This fixed difference is known as the **common difference** of the sequence.

For example,
$$28, 33, 38, 43, 48, 53, 58, 63, 68$$
is an arithmetic sequence, because each term is 5 greater than the previous term.

Arithmetic sequences are sometimes referred to as **arithmetic progressions**, and terms that form an arithmetic sequence are said to be "in arithmetic progression." For example, 7, 9, 11, and 13 are in arithmetic progression.

Problems

Problem 10.1: The first term of an arithmetic sequence is 5 and the sixth term is 40.
 (a) Find the common difference of the sequence.
 (b) Find a formula for the nth term of the sequence.
 (c) Find the 717th term of the sequence.

Problem 10.2: Let a be the first term of an arithmetic sequence with common difference d. Find a formula for the nth term of the sequence in terms of a, d, and n.

Problem 10.3: Prove that if a, b, and c are in arithmetic progression, then for any constant k, the numbers ak, bk, and ck are also in arithmetic progression.

Problem 10.4: An arithmetic sequence has first and last terms of 42 and -138 respectively. Its common difference is -10.

   (a) Find a formula for the kth term of the sequence in terms of k.

   (b) How many terms are there in the sequence?

Problem 10.5: The first four terms of an arithmetic sequence, in order, are $x + y, x - y, xy$, and x/y.

   (a) Using the first two terms, express the common difference of the sequence in terms of y.

   (b) Using the common difference, find new expressions for the third and fourth terms in the sequence.

   (c) Using your expressions from part (b), build and solve a system of equations to find x and y.

   (d) Find the fifth term in the sequence. *(Source: AMC 10)*

Problem 10.6: The roots of $64x^3 - 144x^2 + 92x - 15 = 0$ are in arithmetic progression. Find them. *(Source: AHSME)*

Problem 10.7: Let a, b, and c be in arithmetic progression.

   (a) Prove that $2b = a + c$.

   (b) Prove that $a^2 - bc$, $b^2 - ac$, and $c^2 - ab$ are in arithmetic progression.

Problem 10.1: The first term of an arithmetic sequence is 5 and the sixth term is 40. Find the 717th term of the sequence.

Solution for Problem 10.1: Finding the common difference will allow us to calculate any term of the sequence. In going from the first term to the sixth, we add the common difference 5 times. So, if the common difference is d, then $5 + 5d = 40$. Therefore, we have $d = \frac{40-5}{5} = 7$.

For any positive integer n, the nth term in the arithmetic sequence is produced by adding the common difference to the first term of the sequence $n - 1$ times. Therefore, the nth term of the sequence is

$$5 + (n - 1)7 = 7n - 2.$$

We can test this formula with the terms we already know. Letting $n = 1$ and $n = 6$ in our formula gives $7(1) - 2 = 5$ and $7(6) - 2 = 40$. These are indeed the first and sixth terms.

We've done all the heavy lifting. Now we simply substitute $n = 717$ into our formula and find that $7(717) - 2 = 5017$ is the 717th term of the sequence. $\square$

Problem 10.2: Let a be the first term of an arithmetic sequence with common difference d. Find a formula for the nth term of the arithmetic sequence in terms of a, d, and n.

Solution for Problem 10.2: The nth term of an arithmetic progression is the result of adding the common difference to the first term exactly $n - 1$ times. The resulting n^{th} term then is $a + (n - 1)d$. $\square$

> **Important:** The nth term of an arithmetic sequence that has first term a and common difference d is
> $$a + (n - 1)d.$$

Problem 10.3: Prove that if a, b, and c are in arithmetic progression, then for any constant k, the numbers ak, bk, and ck are also in arithmetic progression.

Solution for Problem 10.3: Because a, b, and c are in arithmetic progression, the difference between b and a equals the difference between c and b. Therefore, we have

$$b - a = c - b.$$

To show that ak, bk, and ck are in arithmetic progression, we must show that the difference between bk and ak equals the difference between ck and bk. In other words, we must show that $bk - ak = ck - bk$. Factoring k out of both sides of this equation gives $k(b - a) = k(c - b)$. This is just k times our equation $b - a = c - b$ from above. We can now write our proof as follows:

> **Proof:** Because a, b, and c are in arithmetic progression, we have
> $$b - a = c - b.$$
> Multiplying this equation by k gives $k(b - a) = k(c - b)$. Expanding both sides gives us $bk - ak = ck - bk$. Therefore, ak, bk, and ck are in arithmetic progression.

$\square$

In Problem 10.3, we multiplied an arithmetic sequence by a constant to generate another arithmetic sequence. As an Exercise, you'll show that adding a constant to each term in an arithmetic sequence produces another arithmetic sequence.

Problem 10.4: An arithmetic sequence has first and last terms of 42 and -138 respectively. How many terms are there in the sequence if it has a common difference of -10?

Solution for Problem 10.4: Let n be the number of terms in the sequence. To get from the first term to the last, we add the common difference $n - 1$ times. Since the common difference is -10, we must therefore have $42 + (n - 1)(-10) = -138$. Solving this equation, we find that $n = 19$ is the number of terms in the sequence. $\square$

> **Concept:** Each term in an arithmetic sequence can be expressed in terms of the first term of the sequence (usually denoted a), the common difference of the sequence (usually denoted d), and the position of the term in the sequence (often denoted n). When a problem does not provide some of these values, we can introduce variables for them to create equations.

For example, in Problem 10.4, we are given $a = 42$, $d = -10$, and the nth term, so we can quickly write an equation for n.

> **Problem 10.5:** The first four terms of an arithmetic sequence, in order, are $x + y$, $x - y$, xy, and x/y, in that order. What is the fifth term? *(Source: AMC 10)*

Solution for Problem 10.5:   The first two terms tell us that the common difference of the sequence is $(x - y) - (x + y) = -2y$. Starting from our second term, $x - y$, and adding this common difference gives us third and fourth terms of $x - 3y$ and $x - 5y$ respectively. These must match the given expressions for the third and fourth terms of the sequence, so we have the system of equations

$$x - 3y = xy,$$
$$x - 5y = \frac{x}{y}.$$

We solve the first equation for x in terms of y to find

$$x = \frac{3y}{1 - y}.$$

We could go ahead and substitute for x in the second equation. However, we can first manipulate the second equation a bit to get rid of the fraction, making substitution easier. Multiplying both sides by y gives $xy - 5y^2 = x$, and rearranging this gives $5y^2 + x - xy = 0$, from which we have

$$5y^2 + (1 - y)x = 0.$$

Substituting for x, we get $5y^2 + 3y = 0$, which gives us the possible solutions $y = 0$ and $y = -3/5$. However, because y is in the denominator of one of the terms of the sequence, we must have $y \neq 0$. Therefore, the only possible value of y is $-3/5$, which gives us

$$x = \frac{3y}{1 - y} = \frac{-\frac{9}{5}}{\frac{8}{5}} = -\frac{9}{8}.$$

Now, using the common difference $-2y$ again, the fifth term is

$$(x - 5y) + (-2y) = x - 7y = -\frac{9}{8} - 7\left(-\frac{3}{5}\right) = -\frac{9}{8} + \frac{21}{5} = \frac{123}{40}.$$

$\square$

> **Problem 10.6:** The roots of $64x^3 - 144x^2 + 92x - 15 = 0$ are in arithmetic progression. Find them. *(Source: AHSME)*

Solution for Problem 10.6: We could assign variables to the values of the roots in many ways. We might write them as a, $a + d$, and $a + 2d$, and then use Vieta's Formulas to note that

$$a + (a + d) + (a + 2d) = -\frac{-144}{64} = \frac{9}{4}.$$

However, we can also write the roots as $r - d$, r, and $r + d$. In other words, we write them in terms of the *middle term* of the sequence instead of the first term. The advantage of this approach becomes clear when we add the roots, which gives

$$(r - d) + r + (r + d) = 3r.$$

By Vieta, this sum must equal 9/4, so we have $3r = 9/4$, which gives us $r = 3/4$.

Vieta's Formulas helped us find r, so we try it again to find d. By Vieta, the product of the roots is

$$(r - d)r(r + d) = \frac{15}{64}.$$

Since $r = 3/4$, we have

$$\frac{3}{4}\left(\frac{9}{16} - d^2\right) = \frac{15}{64}.$$

Solving this equation gives $d = \pm 1/2$. The sign of d does not matter since the roots are r and $r \pm d$. From our values of r and d, the roots of the cubic are 1/4, 3/4, and 5/4. $\square$

Note that in Problem 10.6, we set the middle term in the arithmetic sequence equal to a variable.

> **Concept:** Setting the middle term of an arithmetic sequence with an odd number of terms equal to a variable often introduces a symmetry that allows us to solve problems more easily.

If the sequence has an even number of terms, we can often accomplish the same feat by assigning a variable to the average of the two middle terms.

Problem 10.7: Prove that if the numbers a, b, and c are in arithmetic progression, then so are $a^2 - bc$, $b^2 - ac$, and $c^2 - ab$.

Solution for Problem 10.7: The fact that a, b, and c are in arithmetic progression gives us $b - a = c - b$. Solving this equation for b in terms of a and c gives

$$b = \frac{a + c}{2}.$$

In other words, b is the average of a and c.

To show that the sequence $a^2 - bc$, $b^2 - ac$, $c^2 - ab$ is an arithmetic sequence, we must show that

$$(c^2 - ab) - (b^2 - ac) = (b^2 - ac) - (a^2 - bc).$$

We rearrange this to put the terms with b on one side, since we know b in terms of a and c:

$$a^2 + 2ac + c^2 = 2b^2 + ab + bc.$$

Aha! We have the square of a binomial on the left. Factoring b out of the right side leaves us the equation

$$(a + c)^2 = b(2b + a + c)$$

to prove. Substituting $b = (a + c)/2$ makes the right side

$$\left(\frac{a + c}{2}\right)\left(2\left(\frac{a + c}{2}\right) + a + c\right) = \frac{1}{2}(a + c)(a + c + a + c) = (a + c)^2,$$

as desired.

We started from an equation we wish to show true,

$$(c^2 - ab) - (b^2 - ac) = (b^2 - ac) - (a^2 - bc),$$

and manipulated it to produce an equation that we proved is true. Make sure you see why this isn't yet a complete proof! We must make sure that all our steps are reversible. That is, we must show that we can start from true equations and work forwards to the desired equation. Using our work above as a guide, we write our proof below:

Proof: Because a, b, and c are in arithmetic progression, we have $c - b = b - a$, so $2b = a + c$. Multiplying both sides of $2b = a + c$ by $a + c$ gives $2b(a + c) = (a + c)^2$, so

$$(a + c)^2 = b(2a + 2c) = b(a + c + a + c) = b(2b + a + c) = 2b^2 + ab + bc.$$

Expanding the left side of $(a + c)^2 = 2b^2 + ab + bc$ gives

$$a^2 + 2ac + c^2 = 2b^2 + ab + bc,$$

and rearranging this equation gives us

$$(c^2 - ab) - (b^2 - ac) = (b^2 - ac) - (a^2 - bc),$$

so $a^2 - bc$, $b^2 - ac$, and $c^2 - ab$ are in arithmetic progression.

□

WARNING!! Working backwards is a great way to find a solution. However, if you work backwards to find a solution, you must make sure that all your steps are reversible. The best way to do so is to write your proof forwards, starting from what is known and ending with what you wanted to prove.

At the beginning of our solution, we showed that if a, b, and c are in arithmetic progression, then b is the average of a and c. As you'll show in the Exercises, the converse is also true, so that we have:

Important: The quantities a, b, and c are in arithmetic progression if and only if b is the average of a and c.

10.1.1 The fourth term of a particular infinite arithmetic sequence is 203 and the thirteenth term is 167. What is the smallest value of n such that the nth term of the sequence is negative?

10.1.2 The fifth term of an arithmetic sequence is 10 and the tenth term is 4.

(a) Find the twentieth term in the sequence.

(b) Find a formula for the nth term of the sequence.

10.1.3 Show that if b is the average of a and c, then a, b, and c are in arithmetic progression.

10.1.4 Prove that when a constant k is added to each term in an arithmetic progression, the resulting sequence is also arithmetic.

10.1.5 In an attempt to copy down a sequence of six positive integers in arithmetic progression, a student wrote down the five numbers 113, 137, 149, 155, 173, accidentally omitting one. He later discovered that he also miscopied one of them. Can you help him recover the original sequence? *(Source: USAMTS)*

10.1.6★ Find all values of k such that $x^4 - (3k + 4)x^2 + k^2 = 0$ has 4 real roots in arithmetic progression. **Hints:** 265

10.2 Arithmetic Series

Now that we've solved a few problems involving arithmetic sequences, we move on to problems that involve summing the terms of arithmetic sequences.

> **Definition:** The sum of the terms of an arithmetic sequence is called an **arithmetic series**.

Problems

> **Problem 10.8:** Consider the arithmetic sequence with first term a, last term b, and common difference d, and also the sequence produced by reversing the order of the terms:
>
> $$a, a + d, a + 2d, \ldots, b - 2d, b - d, b,$$
> $$b, b - d, b - 2d, \ldots, a + 2d, a + d, a.$$
>
> (a) First, we find a formula for the sum of the terms in the original sequence.
>
> (i) Add the first term of the first sequence to the first term of the second sequence. Do the same with the second term of each sequence, and the third term of each sequence. Notice anything interesting?
>
> (ii) Suppose there are n terms in each sequence. Find the sum of all $2n$ terms in both sequences, then use your result to find a formula for the sum of the terms in the original sequence.
>
> (b) Show that the average of all the terms in an arithmetic sequence is equal to the average of the first and last terms.

Problem 10.9: Compute the sum of the arithmetic series $5 + 11 + 17 + \cdots + 107$.

Problem 10.10: Find a formula for the sum of the terms in an n-term arithmetic series with first term a and common difference d.

Problem 10.11: The pages of a book are numbered 1 through n. When the page numbers of the book are added, one of the page numbers was mistakenly added twice, resulting in the incorrect sum of 1986. What was the number of the page that was added twice? *(Source: AIME)*

Problem 10.12: Find all integers k such that $46 + 44 + 42 + \cdots + k = 510$.

Problem 10.13: Let $a_1, a_2, a_3, \ldots$ be an arithmetic progression with common difference 1 and let

$$S_1 = a_1 + a_2 + a_3 + \cdots + a_{98},$$
$$S_2 = a_2 + a_4 + a_6 + \cdots + a_{98}.$$

(a) Find an equation relating S_1 to S_2.

(b) Find the value of S_2 given that $S_1 = 137$. *(Source: AIME)*

Sidenote: German mathematician **Carl Friedrich Gauss** (1777-1855) is widely considered one of the greatest mathematicians ever. Due to his significant contributions to so many fields, he is sometimes called the "prince of mathematicians."

Gauss was a child prodigy, amazing adults with his mathematical prowess at a young age in the way Mozart as a child wowed audiences with his mastery of music. One account of Gauss has him discovering an error in his father's payroll calculations at the age of three. Another story recounts how one of Gauss's elementary school teachers attempted to occupy students in class by having them compute the sum of the integers from 1 to 100. Gauss immediately wrote down the sum, 5050, having recognized a convenient pairing of the terms:

$$1 + 2 + 3 + 4 + \cdots + 97 + 98 + 99 + 100$$
$$= (1 + 100) + (2 + 99) + (3 + 98) + (4 + 97) + \cdots + (50 + 51)$$
$$= 101 + 101 + 101 + 101 + \cdots + 101$$
$$= 50(101) = 5050.$$

Expanding on Gauss's clever work, we now derive a formula for an arithmetic series given its first term, last term, and total number of terms.

Extra! *I mean the word proof not in the sense of the lawyers, who set two half proofs equal to a whole one, but in the sense of a mathematician, where half proof equals nothing, and it is demanded for proof that every doubt becomes impossible.*

– Carl Gauss

Problem 10.8: Consider an arithmetic series with n terms whose first term is a and whose last term is b:

$$a + (a + d) + (a + 2d) + \cdots + (b - 2d) + (b - d) + b.$$

(a) Find a formula for the series (that is, a formula for the sum of all the terms) in terms of a, b, and n.

(b) Compute the average of the terms in the series.

Solution for Problem 10.8:

(a) Inspired by the creativity of Gauss, we write the series both forward and backward. If S is the sum of the series, then

$$S = a + (a + d) + (a + 2d) + \cdots + (b - 2d) + (b - d) + b,$$
$$S = b + (b - d) + (b - 2d) + \cdots + (a + 2d) + (a + d) + a.$$

Instead of trying to sum across in each row, we add the equations by adding the terms in each column to get

$$2S = (a + b) + (a + b) + (a + b) + \cdots + (a + b) + (a + b) + (a + b).$$

There are n terms in each of the series we just added, so the sum consists of n instances of $a + b$. Therefore, we have $2S = n(a + b)$. Dividing this equation by 2, we have $S = n(a + b)/2$.

(b) Since the sum of all n terms is $\dfrac{n(a + b)}{2}$, their average is $\dfrac{n(a + b)}{2} \div n = \dfrac{a + b}{2}$.

$\square$

In Problem 10.8 we proved the following highly useful fact:

> **Important:** The average of the terms in an arithmetic sequence is equal to the average of the first and last terms of the sequence.

Therefore, we see that:

> **Important:** The sum of the terms in an arithmetic sequence equals the product of the number of terms and the average of the first and last terms.

Problem 10.9: Compute the sum of the arithmetic series $5 + 11 + 17 + 23 + 29 + \cdots + 107$.

Solution for Problem 10.9: We have the first and last terms, so all we need is the number of terms to sum the series. We first note that the common difference of the series is $11 - 5 = 6$. So, we take $(107 - 5)/6 = 17$ steps of 6 going from 5 to 107. Therefore, there are 18 terms in the series—the first term plus one for each step.

> **WARNING!!** The term that is k terms after the 1st term of a sequence is the $(k + 1)$th term of the sequence, not the kth term.

The average of the first and the last terms is $(5 + 107)/2 = 112/2 = 56$. This is also the average of all the terms, so the sum of the arithmetic series is $18 \cdot 56 = 1008$. $\square$

> **Concept:** While we often use the sum of a group of numbers to compute their average, we can also use the average of a group of numbers to compute their sum.

> **Problem 10.10:** Find a formula for the sum of the terms in an n-term arithmetic series with first term a and common difference d.

Solution for Problem 10.10: The first term of the arithmetic series is a and the last term is $a + (n-1)d$. The average of these terms, which is also the average of all the terms in the series, is

$$\frac{a + [a + (n-1)d]}{2} = \frac{2a + (n-1)d}{2}.$$

Therefore, we find

> **Important:** The sum of an n-term arithmetic series with first term a and common difference d is
> $$\frac{n(2a + (n-1)d)}{2}.$$

Don't bother memorizing this formula. It's just another way of expressing the fact that the sum of an arithmetic series is the product of the number of terms in the series and the average of the first and last terms. $\square$

The simple formula $\dfrac{n(2a + (n-1)d)}{2}$ is much easier to work with than the equivalent series

$$a + (a + d) + (a + 2d) + \cdots + (a + (n-1)d).$$

We call this simple formula a **closed form** for the series. A closed form is basically an expression that doesn't involve a summation with "$\cdots$" or some other repeated operation.

> **Problem 10.11:** The pages of a book are numbered 1 through n. When the page numbers of the book are added, one of the page numbers was mistakenly added twice, resulting in the incorrect sum of 1986. What was the number of the page that was added twice? *(Source: AIME)*

Solution for Problem 10.11: The sum of the page numbers in the book is $1 + 2 + \cdots + n$. This is an arithmetic series with first term 1, last term n, and n terms, so its sum is

> **Important:**
> $$1 + 2 + 3 + \cdots + n = \frac{n(n+1)}{2}$$

But one page was added twice to get our sum of 1986. Suppose page k was added twice, so we have $\frac{n(n+1)}{2} + k = 1986$. Solving for k gives us $k = 1986 - \frac{n(n+1)}{2}$. We don't know n, but we do know that

$1 \le k \le n$, because k is a page number of the book. So, now we can simply guess values of n, evaluate $1986 - \frac{n(n+1)}{2}$, and see if this expression gives us a positive integer from 1 to n.

> **Concept:** Don't overlook trial and error as a problem-solving strategy. Some problems are most easily solved by guessing solutions and checking if they work.

Letting $n = 60$ gives $1986 - \frac{n(n+1)}{2} = 156$, so we need a larger value of n. Letting $n = 61$ gives $k = 95$, $n = 62$ gives $k = 33$, and $n = 63$ gives $k = -30$. Larger values of n give negative values of k, and values of n smaller than 60 give values of k larger than 60. So, the only value of n that gives a value of k from 1 to n is $n = 62$, which gives $k = 33$, the number of the page that was added twice. $\square$

Notice that in our solution, we used trial and error to find a value of n that solves the problem, and we also explained why no other values of n work. The latter step is very important, as our next problem exhibits.

> **Problem 10.12:** Find all integers k such that $46 + 44 + 42 + \cdots + k = 510$.

Solution for Problem 10.12: A complete understanding of arithmetic series helps us avoid mistakes like the following:

> **Bogus Solution:** We start with $46 + 44$, then keep adding successive terms until we have a sum of 510. We eventually find that $46 + 44 + 42 + \cdots + 16 + 14 = 510$, so we know that $k = 14$.

While $k = 14$ is a solution to the problem, it's not the only one!

> **WARNING!!** If you must find all possible solutions to a given problem, simply finding one solution isn't enough. You must not only find the solutions to the problem, but you must also show that these are the only solutions. Using "trial and error" or "guess and check" isn't enough to do the latter—you might overlook a solution.

We'll use an algebraic approach to the problem to make sure we find all possible solutions. We take $(k - 46)/(-2)$ steps of size -2 in going from 46 down to k, so the number of terms in the series $46 + 44 + 42 + \cdots + k$ is

$$\frac{k - 46}{-2} + 1 = \frac{48 - k}{2}.$$

The average of these terms is $\dfrac{46 + k}{2}$, so their sum is

$$\frac{48 - k}{2} \cdot \frac{46 + k}{2} = \frac{2208 + 2k - k^2}{4}.$$

This sum must be 510, so we have

$$\frac{2208 + 2k - k^2}{4} = 510.$$

Rearranging this equation gives $k^2 - 2k - 168 = 0$, and solving this quadratic gives us $k = 14$ and $k = -12$ as solutions. Notice how these two solutions are related. The series for $k = -12$ equals the series for $k = 14$ plus the sum $12 + 10 + \cdots + (-10) + (-12)$. This latter sum obviously equals 0, since each nonzero term is canceled by its opposite. $\square$

> **Problem 10.13:** Find the value of $a_2 + a_4 + a_6 + \cdots + a_{98}$ if $a_1, a_2, a_3, \ldots$ is an arithmetic progression with common difference 1, and $a_1 + a_2 + a_3 + \cdots + a_{98} = 137$. *(Source: AIME)*

Solution for Problem 10.13: Solution 1: Write everything in terms of a_1. Because we know the common difference of the arithmetic sequence, we can write the left side of the given equation entirely in terms of a_1. Specifically, we have $a_2 = a_1 + 1$, $a_3 = a_1 + 2$, $a_4 = a_1 + 3$, and so on, so our equation is

$$a_1 + (a_1 + 1) + (a_1 + 2) + \cdots + (a_1 + 97) - 137.$$

> **Concept:** We can often tackle problems involving arithmetic series by writing all the given information in terms of the first term, the common difference, and the number of terms in the series.

There are 98 terms, and the sum of the first 97 positive integers is $97(98)/2 = 4753$, so we find that $98a_1 + 4753 = 137$. This gives us $a_1 = -2308/49$. Therefore, the series $a_2 + a_4 + a_6 + \cdots + a_{98}$ has first term $a_2 = a_1 + 1 = -2259/49$, common difference 2, and 49 terms, so its sum is

$$\frac{49(2a_2 + (48)(2))}{2} = 93.$$

There's a lot of computation in this solution, and a lot of ways to make careless arithmetic mistakes. As soon as we saw that a_1 is a pretty ugly fraction, we might instead take a few minutes to look for a simpler solution.

> **Concept:** Before wading through excessive computation or algebraic manipulation in a problem, look for ways to solve the problem more creatively.

Solution 2: Clever manipulation. We have the sum of one arithmetic series and we want to find the sum of a related series. Instead of finding a_1, we look for a way to use the relationship between these two series to manipulate the series we know into the one we want. To form the series we want, we must eliminate all the odd-numbered terms from $a_1 + a_2 + a_3 + \cdots + a_{98}$. Each odd-numbered term in the series is 1 less than the next even-numbered term, so we have

$$
\begin{aligned}
137 &= a_1 + a_2 + a_3 + \cdots a_{98} \\
&= (a_2 - 1) + a_2 + (a_4 - 1) + a_4 + \cdots + (a_{98} - 1) + a_{98} \\
&= (2a_2 - 1) + (2a_4 - 1) + \cdots + (2a_{98} - 1) \\
&= 2(a_2 + a_4 + \cdots + a_{98}) - 49.
\end{aligned}
$$

We now have a nice equation, $2(a_2 + a_4 + a_6 + \cdots + a_{98}) - 49 = 137$. Adding 49 to both sides and dividing by 2, we see that $a_2 + a_4 + a_6 + \cdots + a_{98} = 93$. $\square$

Notice that we didn't solve for any variable in our second solution.

> **Concept:** Algebraic relationships between sequences and series can help us compute
> some series without evaluating any of the terms in the series.

Exercises

10.2.1 Find the sum of the arithmetic series $639 + 631 + 623 + \cdots + (-97)$.

10.2.2 Mrs. White has 13 grandchildren. If the youngest is 5 years old and they are spaced 2 years apart, what is the sum of the ages of all of Mrs. White's grandchildren?

10.2.3 Let n be a positive integer.

(a) Find the sum of the n smallest positive odd integers.

(b) Find the sum of the n smallest positive odd integers greater than 100.

10.2.4 Let a and b be positive integers with $a \le b$. Find the sum of the positive integers from a to b, including a and b, in terms of a and b.

10.2.5 Let $a_1, a_2, \ldots,$ and $b_1, b_2, \ldots$ be arithmetic progressions such that $a_1 = 25$, $b_1 = 75$, and $a_{100} + b_{100} = 100$. Find the sum of the first 100 terms of the progression $a_1 + b_1, a_2 + b_2, \ldots$. *(Source: AHSME)*

10.2.6 For every n, the sum of the first n terms of a certain arithmetic progression is $2n + 3n^2$. Find a formula for the nth term in the progression. *(Source: AHSME)*

10.2.7 If the sum of the first $3n$ positive integers is 150 more than the sum of the first n positive integers, then find the sum of the first $4n$ positive integers. *(Source: AHSME)*

10.2.8★ In a certain arithmetic progression, the ratio of the sum of the first r terms to the sum of the first s terms is r^2/s^2 for any r and s. Find the ratio of the 8th term to the 23rd term. *(Source: ARML)*

10.3 Geometric Sequences

> **Definition:** A **geometric sequence** is a sequence for which there exists a constant r such that each term (except for the first term) is r times the previous term. This constant r is called the **common ratio** of the sequence.

For example,
$$1, 3, 9, 27, 81, 243$$
is a geometric sequence, because each term is 3 times the previous term. Geometric sequences are also known as **geometric progressions**.

Problems

> **Problem 10.14:** The first term of a geometric sequence is a, and the common ratio of the sequence is r. Find the nth term of the sequence in terms of $a, r,$ and n.

Problem 10.15: The first term of a geometric sequence of positive numbers is 12, and the fourth term is 24.

   (a) Find the common ratio of the geometric sequence.

   (b) Find a formula for the nth term of the geometric sequence.

   (c) Find the 10th term of the geometric sequence.

Problem 10.16: The **geometric mean** of n positive numbers is the nth root of the product of the numbers. Let a be the first term of a geometric sequence with n terms, and let r be the common ratio of the sequence. Show that if a and r are positive, the geometric mean of the terms in the sequence is equal to the geometric mean of the first term and the last term.

Problem 10.17: Suppose x, y, z is a geometric sequence with common ratio r and $x \neq y$. If $x, 2y, 3z$ is an arithmetic sequence, find the value of r. *(Source: AHSME)*

Problem 10.18: Prove that if a, b, c, and d are four consecutive terms in a geometric sequence, then

$$(b - c)^2 + (c - a)^2 + (d - b)^2 = (a - d)^2.$$

(Source: CEMC)

Problem 10.14: The first term of a geometric sequence is a, and the common ratio of the sequence is r. Find the nth term of the sequence in terms of a, r, and n.

Solution for Problem 10.14: The first term of the geometric sequence is a. The second is ar. The third is ar^2. In general, we get the nth term of the sequence by multiplying the first term of the sequence by r a total of $n - 1$ times, so the nth term is ar^{n-1}. $\square$

> **Important:** If a geometric sequence has first term a and common ratio r, then the nth term of the sequence is equal to ar^{n-1}.

Problem 10.15: The first term of a geometric sequence of positive numbers is 12, and the fourth term is 24. Find the 10th term of the geometric sequence.

Solution for Problem 10.15: *Solution 1: Find the common ratio.* Let r be the common ratio. We multiply the first term by r three times to get the fourth term, so $12r^3 = 24$. Dividing by 12 and taking the cube root of both sides, we find that $r = \sqrt[3]{2}$ is the common ratio of the sequence. The tenth term of the sequence is therefore $12 \cdot (\sqrt[3]{2})^{10-1} = 12 \cdot (2^{1/3})^9 = 12 \cdot 2^3 = 96$.

Solution 2: Think about it! Rather than finding the common ratio, we notice that in addition to multiplying by the common ratio three times to get from the first term to the fourth, we also multiply by the common ratio three times to get from the fourth term to the seventh, and to get from the seventh term to the tenth. Since multiplying 12 by the common ratio three times gives us 24, the result of multiplying a number by the common ratio three times is doubling the number. So, the seventh term of the sequence is $24(2) = 48$ and the tenth term is $48(2) = 96$. $\square$

In Section 10.2, we saw that the average of all the terms in an arithmetic sequence equals the average of the first and last term. Is there a similar relationship for the terms in a geometric sequence?

Problem 10.16: The **geometric mean** of n positive numbers is the nth root of the product of the numbers. Prove that for any finite geometric sequence of positive numbers, the geometric mean of the terms in the sequence is equal to the geometric mean of the first term and the last term.

Solution for Problem 10.16: Let n be the number of terms of a geometric sequence with first term a and common ratio r. The last term of this sequence is ar^{n-1}, so the geometric mean of the first and last terms is

$$\sqrt{a \cdot ar^{n-1}} = \sqrt{a^2 r^{n-1}} = ar^{(n-1)/2}.$$

The geometric mean of all n terms is

$$\sqrt[n]{a(ar)(ar^2)\cdots(ar^{n-1})} = \sqrt[n]{a^n \cdot r^{[1+2+\cdots+(n-1)]}}$$
$$= \sqrt[n]{a^n \cdot r^{(n-1)n/2}}$$
$$= ar^{(n-1)/2},$$

which matches our expression for the geometric mean of the first and last terms of the sequence. $\square$

Problem 10.17: Suppose x, y, z is a geometric sequence with common ratio r and $x \neq y$. If $x, 2y, 3z$ is an arithmetic sequence, find the value of r. *(Source: AHSME)*

Solution for Problem 10.17: We can start by reducing the number of variables we have to deal with. From our geometric sequence, we know that $y = rx$ and $z = r^2 x$. Now that we can write y and z in terms of r and x, we can write our arithmetic sequence in terms of just x and r. So, we know that $x, 2rx$, and $3r^2 x$ are in arithmetic progression, which gives us

$$2rx - x = 3r^2 x - 2rx.$$

Rearranging this equation gives us $3r^2 x - 4rx + x = 0$. Factoring this equation gives

$$x(r-1)(3r-1) = 0.$$

We cannot have $x = 0$ or $r = 1$, since both of these give us $x = y$, which is forbidden in the problem statement. This leaves us with $r = 1/3$ as the only possible value of r. $\square$

Problem 10.18: Prove that if a, b, c, and d are four consecutive terms in a geometric sequence, then

$$(b-c)^2 + (c-a)^2 + (d-b)^2 = (a-d)^2.$$

(Source: CEMC)

Solution for Problem 10.18: Let r be the common ratio of the geometric sequence, so that $b = ar$, $c = ar^2$, and $d = ar^3$. Then, on the left side of the relationship we must prove, we have

$$(b - c)^2 + (c - a)^2 + (d - b)^2 = (ar - ar^2)^2 + (ar^2 - a)^2 + (ar^3 - ar)^2$$
$$= a^2 r^2 (1 - r)^2 + a^2 (r^2 - 1)^2 + a^2 r^2 (r^2 - 1)^2$$
$$= a^2 ((r^4 - 2r^3 + r^2) + (r^4 - 2r^2 + 1) + (r^6 - 2r^4 + r^2))$$
$$= a^2 (r^6 - 2r^3 + 1)$$
$$= a^2 (r^3 - 1)^2.$$

Over on the right side of the desired equation, we have $(a - d)^2 = (a - ar^3)^2 = a^2 (1 - r^3)^2 = a^2 (r^3 - 1)^2$, so we see that $(b - c)^2 + (c - a)^2 + (d - b)^2 = (a - d)^2$. $\square$

> **Concept:** Many problems involving geometric sequences can be tackled by writing the sequence in terms of the sequence's first term and its common ratio.

Exercises

10.3.1 The second term in a geometric sequence of positive numbers is 6 and the sixth term is $\frac{3}{8}$.

(a) Find the common ratio of the sequence.

(b) Find a formula for the mth term of the sequence.

(c) Find the value of a if the 1977th term of the sequence is $\frac{3}{2^a}$.

10.3.2 The Steering Committee of the EmailTag Club sends a message to all the members of the club as follows: On Day 1, each member of the Steering Committee emails three other people in the club. On each subsequent day, each person who received an email the day before emails three other people in the club. If no one ever gets two copies of the email on the same day, and there are 972 more emails on Day 5 than on Day 4, then how many people are on the Steering Committee?

10.3.3 A sequence of three real numbers forms an arithmetic progression with a first term of 9. If 2 is added to the second term and 20 is added to the third term, the three resulting numbers form a geometric progression. What is the smallest possible value for the third term in the geometric progression? (*Source: AMC 12*)

10.3.4

(a) Show that nonzero numbers x, y, and z are in geometric progression if and only if $y^2 = xz$.

(b) Is it true that nonzero numbers x, y and z are in geometric progression if and only if $y = \sqrt{xz}$?

10.3.5 The roots of $2x^3 - 19x^2 + kx - 54 = 0$ are in geometric progression for some constant k. Find k.

10.3.6★ The sum of three consecutive terms in a geometric sequence is 39, and the sum of their squares is 741. Find the three terms. **Hints:** 51

> **Extra!** *Great things are not done by impulse, but by a series of small things brought together.*
> ⤍⤍⤍⤍   – Vincent van Gogh

10.4 Geometric Series

> **Definition:** The sum of the numbers in a geometric sequence is called a **geometric series**.

For example, the series $4 + 8 + 16 + 32 + 64$ is a geometric series. In this section, we will also work with **infinite geometric series**, which are geometric series with infinitely many terms. For example, the infinite geometric series that starts with 1 and has common ratio $\frac{1}{2}$ is

$$1 + \frac{1}{2} + \frac{1}{4} + \frac{1}{8} + \cdots,$$

where the "$\cdots$" indicates that the series continues infinitely.

 **Problems**

Problem 10.19: In this problem, we evaluate the sum $S = a + ar + ar^2 + \cdots + ar^{n-1}$, where $r \neq 1$, in terms of a, r, and n.

(a) By what expression can we multiply our expression for S to produce another series that has many terms in common with S?

(b) Compute S in terms of a, r, and n.

(c) Explain the relationship between your result from (b) and a factoring method we've discussed in a previous chapter.

Problem 10.20: Note that $3^8 = 6561$. Compute the geometric series $1 + 3 + 9 + 27 + \cdots + 2187$.

Problem 10.21: Write the following expression as a single polynomial:

$$\frac{(t^2 + t + 1)(t^{12} + t^9 + t^6 + t^3 + 1)}{t^{10} + t^5 + 1}.$$

(Source: Mandelbrot)

Problem 10.22: In this problem we learn how to evaluate the infinite geometric series

$$\frac{1}{2} + \frac{1}{4} + \frac{1}{8} + \frac{1}{16} + \cdots.$$

We start by letting $S = \frac{1}{2} + \frac{1}{4} + \frac{1}{8} + \frac{1}{16} + \cdots + \frac{1}{2^n}$.

(a) Explain why S cannot be greater than 1.

(b) By what number can we multiply S to create another series that has nearly all of its terms in common with S?

(c) Use part (b) to explain why $S = 1 - \frac{1}{2^n}$.

(d) Evaluate S for $n = 5$, $n = 10$, and $n = 20$. (Yes you can use a calculator!)

(e) Use parts (c) and (d) to explain why $\frac{1}{2} + \frac{1}{4} + \frac{1}{8} + \frac{1}{16} + \cdots = 1$.

Problem 10.23: Find a formula for the sum of an infinite geometric series with first term a and common ratio r, where $|r| < 1$. Why must we have the $|r| < 1$ restriction?

Problem 10.24: Evaluate the sum $4 - \dfrac{8}{3} + \dfrac{16}{9} - \dfrac{32}{27} + \cdots$.

Problem 10.25: Find all infinite geometric series that have 4 as the second term and a sum of 18.

Problem 10.26: The sum of the geometric series $a + ar + ar^2 + ar^3 + \cdots$ is 15, and the sum of the geometric series $a + ar^2 + ar^4 + ar^6 + \cdots$ is 9. Find the sum of the geometric series $a + ar^3 + ar^6 + ar^9 + \cdots$.

Problem 10.27: Let P be the product of n numbers in geometric progression, S their sum, and S' the sum of their reciprocals. In this problem, we find P in terms of S, S', and n. *(Source: AHSME)*

(a) Let the first term of the progression be a and the common ratio be r. Find P, S, and S' in terms of a, r, and n.

(b) Combine your expressions from part (a) to write P in terms of S and S'.

Problem 10.19: Find the sum of the geometric series $a + ar + ar^2 + \cdots + ar^{n-1}$, where $r \neq 1$, in terms of a, r, and n.

Solution for Problem 10.19: We set the series equal to S, hoping to perform a manipulation similar to the one we used to find a formula for arithmetic series:

$$S = a + ar + ar^2 + \cdots + ar^{n-1}.$$

We can't simply write this series backwards and add it to the original series to get our answer. However, if we multiply this series by r, we create another series that has many terms in common with S:

$$rS = \quad ar + ar^2 + \cdots + ar^{n-1} + ar^n,$$
$$S = a + ar + ar^2 + \cdots + ar^{n-1}.$$

Subtracting the second equation from the first cancels most of the terms and leaves

$$rS - S = ar^n - a,$$

so $S(r - 1) = a(r^n - 1)$. Dividing by $r - 1$ gives us our formula, $S = a(r^n - 1)/(r - 1)$. To avoid dividing by 0, we must note that we cannot have $r = 1$. If we have an n-term geometric series in which $r = 1$, the series simply consists of n copies of the first term. So, if we call the first term a, the sum is just an.

> **Important:** A geometric series with n terms, first term a, and common ratio r (with $r \neq 1$) has sum
> $$\frac{a(r^n - 1)}{r - 1}.$$

We might also have found this formula by factoring a out of the whole series:

$$ar^{n-1} + ar^{n-2} + \cdots + ar^2 + ar + a = a(r^{n-1} + r^{n-2} + \cdots + r^2 + r + 1).$$

We recognize $r^{n-1} + r^{n-2} + \cdots + r^2 + r + 1$ as one of the factors of $r^n - 1$:

$$r^n - 1 = (r-1)(r^{n-1} + r^{n-2} + \cdots + r^2 + r + 1).$$

Dividing both sides by $r - 1$ gives $r^{n-1} + r^{n-2} + \cdots + r^2 + r + 1 = (r^n - 1)/(r-1)$, so we have

$$ar^{n-1} + ar^{n-2} + \cdots + ar^2 + ar + a = a(r^{n-1} + r^{n-2} + \cdots + r^2 + r + 1) = \frac{a(r^n - 1)}{r - 1}.$$

$\square$

Problem 10.20: Note that $3^8 = 6561$. Compute the geometric series $1 + 3 + 9 + 27 + \cdots + 2187$.

Solution for Problem 10.20: We simply apply the formula we just found: $1 + 3^1 + 3^2 + \cdots + 3^7 = \dfrac{3^8 - 1}{3 - 1} = \dfrac{6561 - 1}{2} = 3280.$ $\square$

Problem 10.21: Write the following expression as a single polynomial:

$$\frac{(t^2 + t + 1)(t^{12} + t^9 + t^6 + t^3 + 1)}{t^{10} + t^5 + 1}.$$

(Source: Mandelbrot)

Solution for Problem 10.21: We could multiply out the numerator and wade through long division of the product by the denominator. But each factor is a geometric series, so maybe we can use our understanding of geometric series to avoid long division. Writing $t^2 + t + 1$ as $1 + t + t^2$, we recognize it as a geometric series with first term 1 and common ratio t, so we have

$$1 + t + t^2 = \frac{1(t^3 - 1)}{t - 1} = \frac{t^3 - 1}{t - 1}.$$

You might also have recognized this relationship by noting that $t^2 + t + 1$ is a factor of $t^3 - 1$.

Similarly, the series $1 + t^3 + t^6 + t^9 + t^{12}$ is geometric with first term 1 and common ratio t^3, and the series $1 + t^5 + t^{10}$ is geometric with first term 1 and common ratio t^5. So, we have

$$1 + t^3 + t^6 + t^9 + t^{12} = \frac{1((t^3)^5 - 1)}{t^3 - 1} = \frac{t^{15} - 1}{t^3 - 1},$$
$$1 + t^5 + t^{10} = \frac{1((t^5)^3 - 1)}{t^5 - 1} = \frac{t^{15} - 1}{t^5 - 1}.$$

We see a lot of common factors among our new expressions. So, we expect we can do a lot of cancellation:

$$\frac{(t^2 + t + 1)(t^{12} + t^9 + t^6 + t^3 + 1)}{t^{10} + t^5 + 1} = \frac{\frac{t^3 - 1}{t - 1} \cdot \frac{t^{15} - 1}{t^3 - 1}}{\frac{t^{15} - 1}{t^5 - 1}} = \frac{(t^3 - 1)(t^{15} - 1)(t^5 - 1)}{(t^{15} - 1)(t - 1)(t^3 - 1)}$$

$$= \frac{t^5 - 1}{t - 1} = \frac{(t - 1)(t^4 + t^3 + t^2 + t + 1)}{t - 1} = t^4 + t^3 + t^2 + t + 1.$$

Note that if $t = 1$, we have division by 0 in many steps of this solution. So, our solution only shows that the original expression equals $t^4 + t^3 + t^2 + t + 1$ if $t \neq 1$. We take care of the case $t = 1$ by noting that when $t = 1$, the original expression equals $(3)(5)/3 = 5$, and $t^4 + t^3 + t^2 + t + 1 = 5$. Therefore, the original expression does indeed equal $t^4 + t^3 + t^2 + t + 1$ when $t = 1$, as well. $\square$

Concept: When $x \neq 0$ and $x \neq y$, the factorization

$$x^n - y^n = (x - y)(x^{n-1} + x^{n-2}y + x^{n-3}y^2 + \cdots + xy^{n-2} + y^{n-1})$$

is the same as the relationship found by summing the geometric series with first term x^{n-1} and common ratio y/x:

$$x^{n-1} + x^{n-2}y + x^{n-3}y^2 + \cdots + xy^{n-2} + y^{n-1} = \frac{x^{n-1}\left(\left(\frac{y}{x}\right)^n - 1\right)}{\frac{y}{x} - 1}$$

$$= \frac{\frac{y^n}{x} - x^{n-1}}{\frac{y}{x} - 1}$$

$$= \frac{y^n - x^n}{y - x} = \frac{x^n - y^n}{x - y}.$$

We most frequently use this relationship when confronted with a polynomial that is a geometric series, such as $t^2 + t + 1$, $t^{10} + t^5 + 1$, or $t^{12} + t^9 + t^6 + t^3 + 1$.

Some geometric series never end. These are called **infinite geometric series**. Despite the fact that these series are infinite, we can still find the sum of some of them.

Problem 10.22: Evaluate the infinite geometric series $\dfrac{1}{2} + \dfrac{1}{4} + \dfrac{1}{8} + \cdots$.

Solution for Problem 10.22: We can view our series as starting from 0 on the number line, and taking steps to the right of length $\frac{1}{2}$, then $\frac{1}{4}$, then $\frac{1}{8}$, and so on. Our first step, $\frac{1}{2}$, takes us half the distance from 0 to 1. We are therefore $\frac{1}{2}$ away from 1. The next step, $\frac{1}{4}$, takes us half the distance from $\frac{1}{2}$ to 1. We're then at $\frac{3}{4}$, which is still $\frac{1}{4}$ from 1. The next step, $\frac{1}{8}$, covers half the distance from $\frac{3}{4}$ to 1. Continuing in this way, each step halves our distance from 1.

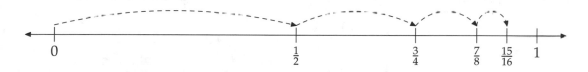

Since each step just halves our distance from 1, it appears that we never quite get to 1 and we certainly can't get beyond it.

It appears that the series

$$\frac{1}{2} + \frac{1}{4} + \frac{1}{8} + \cdots$$

never quite gets to 1. But what number does it equal?

To answer that question, we start with something that we know how to deal with: a finite geometric series.

> **Concept:** When stuck on a problem, compare it to a similar problem you know how to solve.

Suppose we instead try to find an expression for the sum

$$S(n) = \frac{1}{2} + \frac{1}{4} + \frac{1}{8} + \frac{1}{16} + \cdots + \frac{1}{2^{n-1}} + \frac{1}{2^n}.$$

This is a finite geometric series with first term $\frac{1}{2}$, common ratio $\frac{1}{2}$, and n terms, so we can use our earlier formula to find a closed form for it:

$$S(n) = \frac{\frac{1}{2}\left(\left(\frac{1}{2}\right)^n - 1\right)}{\frac{1}{2} - 1} = \frac{\frac{1}{2}\left(\left(\frac{1}{2}\right)^n - 1\right)}{-\frac{1}{2}} = 1 - \frac{1}{2^n}.$$

Now, we can easily evaluate $S(n)$ for different values of n. A few such values are shown in the table below. (The last two are approximations.)

n	$S(n)$
1	0.5
5	0.96875
10	0.9990234
20	0.9999990

We see that the larger n gets, the closer $S(n)$ gets to 1. Let's see why. Since

$$S(n) = 1 - \frac{1}{2^n},$$

the value of $S(n)$ cannot ever be larger than 1. However, the value of $\frac{1}{2^n}$ gets closer to 0 as n gets larger, so $S(n)$ will get closer to 1 as n gets larger.

If "n is infinite," meaning our series never stops, we have our original infinite series,

$$\frac{1}{2} + \frac{1}{4} + \frac{1}{8} + \cdots.$$

Moreover, when "n is infinite," the value of $\frac{1}{2^n}$ is 0, so $S(n) = 1 - \frac{1}{2^n}$ equals 1 when n is infinite. Therefore, we have

$$\frac{1}{2} + \frac{1}{4} + \frac{1}{8} + \cdots = 1.$$

This explanation omits several details about exactly what we mean by "n is infinite," but it should give you an intuitive understanding why our infinite series has a finite sum.

> **Sidenote:** The details we omitted are some of the beginning steps of calculus. You'll learn much more about these details later when you study the mathematical concept of a **limit**.

Once we know our infinite series has a finite sum, we can find that sum even more quickly using the same manipulation we used in Problem 10.19.

Let our infinite geometric series equal T. We multiply T by 1/2, producing a series that has all of its terms in common with T:

$$T = \frac{1}{2} + \frac{1}{4} + \frac{1}{8} + \cdots,$$
$$\frac{T}{2} = \frac{1}{4} + \frac{1}{8} + \cdots.$$

Subtracting the second equation from the first gives $\frac{T}{2} = \frac{1}{2}$. All the other terms cancel! Therefore, $T = 1$.
$\square$

A key step in our investigation of this infinite series was looking at sums of a finite number of terms of the series. These sums of just part of a series are referred to as **partial sums**. For example, $\frac{1}{2} + \frac{1}{4}$, $\frac{1}{2} + \frac{1}{4} + \frac{1}{8}$, and $\frac{1}{2} + \frac{1}{4} + \frac{1}{8} + \frac{1}{16}$ are all partial sums of the infinite geometric series $\frac{1}{2} + \frac{1}{4} + \frac{1}{8} + \cdots$.

> **Concept:** We can often learn more about an infinite series by investigating partial sums of the series.

Let's find a formula for computing the sum of an infinite geometric series.

> **Problem 10.23:** Find a formula for the sum of an infinite geometric series with first term a and common ratio r, where $|r| < 1$. Why must we have the $|r| < 1$ restriction?

Solution for Problem 10.23: The series we wish to evaluate is

$$a + ar + ar^2 + ar^3 + \cdots.$$

We use our previous problem as a guide, and let $T(n)$ be the finite series

$$T(n) = a + ar + ar^2 + ar^3 + \cdots + ar^{n-1}.$$

This finite geometric series has first term a, common ratio r, and n terms, so we can use our formula for the sum of a finite geometric series to give

$$T(n) = \frac{ar^n - a}{r - 1} = \frac{ar^n}{r - 1} - \frac{a}{r - 1}.$$

As we did before, we evaluate the infinite series by thinking about what happens to $T(n)$ when n is infinite. If $|r| < 1$, then r^n becomes 0 when n is infinite. This is true no matter how close to 1 that r is.

(Grab your calculator and calculate $0.999^{1000000}$.) So, when $|r| < 1$ and n is infinite, the term $ar^n/(r-1)$ equals 0, and we have

$$a + ar + ar^2 + ar^3 + \cdots = \frac{-a}{r-1} = \frac{a}{1-r}.$$

If $r = 1$, then our infinite series is just $a + a + a + \cdots$, which we clearly cannot evaluate. If $r > 1$, then r^n just gets bigger and bigger as n gets larger, so $T(n)$ either keeps getting bigger (if $a > 0$) or it keeps getting smaller (if $a < 0$). In either case, we say that the series **diverges**, and it does not have a finite sum.

> **Sidenote:** If $r = -1$, we have a different problem. Then, our infinite series is
> ♪
> $$a - a + a - a + a - a + a - a + \cdots.$$
>
> After 2 terms, the sum is 0, but after 3 terms, the sum is a. After 4 terms, the sum is 0 again, and after 5 terms, the sum is a. So, if the series has an even number of terms, then the sum is 0. If the sum has an odd number of terms, then the sum is a. But our infinite series doesn't have an even or an odd number of terms.
>
> It turns out that there is no way to define what the sum of this series is, and we call such a series an **indeterminate series**.
> If $r < -1$, then we have similar problems. As we add more and more terms of the series, the sum goes back and forth from being much less than 0 to being much greater than 0. Therefore, we can't even say whether or not the sum is negative or positive, so we certainly can't evaluate it.

Once we know that an infinite series does have a sum, we could also use the shortcut we used in our last solution to Problem 10.19. We let our infinite series equal S:

$$S = a + ar + ar^2 + ar^3 + \cdots.$$

We multiply this equation by r to produce a series that has all of its terms in common with S:

$$
\begin{aligned}
S &= a + ar + ar^2 + \cdots, \\
rS &= \phantom{a +{}} ar + ar^2 + \cdots.
\end{aligned}
$$

We then subtract, forcing nearly all the terms on the right to cancel and leaving

$$S - rS = a.$$

Factoring out S and dividing by $1 - r$ gives us:

> **Important:** The sum of an infinite geometric series with first term a and common ratio r, where $|r| < 1$, is
> $$\frac{a}{1-r}.$$

> **Concept:** While the formula for an infinite geometric series is very handy, pay more attention to the method we used to derive it. We'll use this method again, and you can always use it to rederive the formula if you forget it.

As we saw earlier, we must include the restriction $|r| < 1$ because otherwise the geometric series cannot be evaluated. An infinite series that we can evaluate is called **convergent**, and we say that such a series **converges**. On the other hand, an infinite series that doesn't converge, such as $1+2+4+8+16+\cdots$, is called **divergent**, or is said to **diverge**. □

Let's try out our new formula.

> **Problem 10.24:** Evaluate the sum $4 - \dfrac{8}{3} + \dfrac{16}{9} - \dfrac{32}{27} + \cdots$.

Solution for Problem 10.24: We have an infinite geometric series with first term 4. The common ratio between the terms is the ratio of the second term to the first, or

$$\frac{-\frac{8}{3}}{4} = -\frac{2}{3}.$$

We have an infinite geometric series with first term 4 and a common ratio with absolute value less than 1, so we can use our formula to find that

$$4 - \frac{8}{3} + \frac{16}{9} - \frac{32}{27} + \cdots = \frac{4}{1 - \left(-\frac{2}{3}\right)} = \frac{4}{\frac{5}{3}} = \frac{12}{5}.$$

□

Just as we drew a picture on page 305 for an infinite geometric series with a positive ratio, we illustrate below the addition of the first few terms of the geometric series with negative ratio from Problem 10.24.

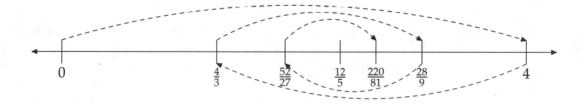

The first long dashed arrow on top represents the initial 4. Each subsequent dashed arrow represents adding another term in the series. For example, the term after the 4, which is $-\frac{8}{3}$, is the longest arrow on the bottom. As we add more and more terms, we'll continue to get closer and closer to $\frac{12}{5}$. Each positive term we add makes our sum a little greater than $\frac{12}{5}$, and each negative term we add makes our sum a little less than $\frac{12}{5}$. In both cases, we get closer to $\frac{12}{5}$ with each term we add.

> **Problem 10.25:** Find all infinite geometric series that have 4 as the second term and a sum of 18.

Solution for Problem 10.25: Let a be the first term of the series, and let r be the common ratio. Since the second term is 4 and the sum is 18, we have the system of equations

$$ar = 4,$$
$$\frac{a}{1-r} = 18.$$

Multiplying the second equation by $1 - r$ gives us $a = 18 - 18r$. From the first equation, we have $a = 4/r$. We equate these two expressions for a to give $18 - 18r = \frac{4}{r}$. Rearranging this equation gives $9r^2 - 9r + 2 = 0$, from which we find the solutions $r = 1/3$ and $r = 2/3$. If $r = 1/3$, we have $a = 12$, so the series is

$$12 + 4 + \frac{4}{3} + \frac{4}{9} + \cdots.$$

If $r = 2/3$, then $a = 4/r = 6$, and our series is

$$6 + 4 + \frac{8}{3} + \frac{16}{9} + \cdots.$$

We can use our formula for the sum of an infinite geometric series to quickly confirm that each of these sums equals 18, as required. $\square$

Problem 10.26: If $a + ar + ar^2 + ar^3 + \cdots = 15$ and $a + ar^2 + ar^4 + ar^6 + \cdots = 9$, then what is the sum of the geometric series $a + ar^3 + ar^6 + ar^9 + \cdots$?

Solution for Problem 10.26: Both of the given series are infinite geometric series, one with common ratio r and the other with common ratio r^2. Applying our formula for infinite geometric series to both series gives us

$$\frac{a}{1-r} = 15,$$
$$\frac{a}{1-r^2} = 9.$$

There are a variety of ways to solve this system of equations. Perhaps the most straightforward is to notice that we can factor the denominator of the left side in the second equation to give us

$$\frac{a}{(1+r)(1-r)} = 9.$$

We already know that $a/(1-r) = 15$, so we have

$$9 = \frac{a}{(1+r)(1-r)} = \frac{a}{1-r} \cdot \frac{1}{1+r} = 15 \cdot \frac{1}{1+r}.$$

Solving $9 = 15/(1+r)$ gives $r = 2/3$. From $a/(1-r) = 15$, we find that $a = 15(1-r) = 5$, so we have

$$a + ar^3 + ar^6 + ar^9 + \cdots = \frac{a}{1-r^3} = \frac{5}{1-(2/3)^3} = \frac{135}{19}.$$

$\square$

> **Problem 10.27:** If P is the product of n numbers in geometric progression, S their sum, and S' the sum of their reciprocals, then find P in terms of S, S', and n. *(Source: AHSME)*

Solution for Problem 10.27: At first glance, it seems unlikely that there's a simple relationship between P, S, S', and n. To search for one, we take our usual approach for working with geometric sequences: we let the first term be a and the common ratio be r. Then, we write the quantities in the problem in terms of a, r, and n. First, we take care of the product of the n terms:

$$P = a \cdot ar \cdot ar^2 \cdots ar^{n-2} \cdot ar^{n-1}.$$

Since there are n terms, our product equals a^n times all the r terms:

$$P = a \cdot ar \cdot ar^2 \cdots ar^{n-2} \cdot ar^{n-1} = a^n r^{1+2+3+\cdots+(n-1)} = a^n r^{n(n-1)/2}.$$

We also know how to write S in closed form:

$$S = a + ar + ar^2 + \cdots + ar^{n-1} = \frac{a(r^n - 1)}{r - 1}.$$

Writing S' in closed form is a little trickier. We start with factoring:

$$S' = \frac{1}{a} + \frac{1}{ar} + \frac{1}{ar^2} + \cdots + \frac{1}{ar^{n-1}} = \frac{1}{a}\left(1 + \frac{1}{r} + \frac{1}{r^2} + \cdots + \frac{1}{r^{n-1}}\right).$$

The series on the right is a geometric series with first term 1 and common ratio $\frac{1}{r}$, so we have

$$S' = \frac{1}{a}\left(1 + \frac{1}{r} + \frac{1}{r^2} + \cdots + \frac{1}{r^{n-1}}\right) = \frac{1}{a}\left(\frac{1\left(\left(\frac{1}{r}\right)^n - 1\right)}{\frac{1}{r} - 1}\right).$$

That's not too pretty, but after a little algebra, we can get the expression in a much nicer form:

$$S' = \frac{1}{a}\left(\frac{1\left(\left(\frac{1}{r}\right)^n - 1\right)}{\frac{1}{r} - 1}\right) = \frac{1}{a}\left(\frac{\frac{1}{r^n} - 1}{\frac{1}{r} - 1} \cdot \frac{r^n}{r^n}\right) = \frac{1}{a}\left(\frac{r^n - 1}{r^n - r^{n-1}}\right) = \frac{r^n - 1}{ar^{n-1}(r - 1)}.$$

Next, we write our expressions for P, S, and S' all together, hoping we see some way to relate them:

$$P = a^n r^{n(n-1)/2},$$
$$S = \frac{a(r^n - 1)}{r - 1},$$
$$S' = \frac{r^n - 1}{ar^{n-1}(r - 1)}.$$

Our expression for P doesn't have $r^n - 1$ or $r - 1$, so we have to combine S and S' in a way that cancels these. Fortunately, we see that dividing S by S' (or vice versa) eliminates both $r^n - 1$ and $r - 1$:

$$\frac{S}{S'} = \frac{\frac{a(r^n-1)}{r-1}}{\frac{r^n-1}{ar^{n-1}(r-1)}} = \frac{a(r^n - 1)}{r - 1} \cdot \frac{ar^{n-1}(r - 1)}{r^n - 1} = a^2 r^{n-1} \cdot \frac{r^n - 1}{r - 1} \cdot \frac{r - 1}{r^n - 1} = a^2 r^{n-1}.$$

Aha! We have $S/S' = a^2 r^{n-1}$ and $P = a^n r^{n(n-1)/2}$. To get from $a^2 r^{n-1}$ to $a^n r^{n(n-1)/2}$, we only have to raise the former to the $n/2$ power. So, we have

$$\left(\frac{S}{S'}\right)^{n/2} = (a^2 r^{n-1})^{n/2} = a^n r^{n(n-1)/2} = P.$$

We now have the desired (and surprising) relationship: $P = (S/S')^{n/2}$.

Seeing such a simple relationship, we check to see if we could have derived it by manipulating the series for S and S'.

> **Concept:** If you derive a simple algebraic relationship after complex algebraic manipulation, use the simple relationship as a guide to look for a simpler proof.

Now that we know what to look for, we start with S/S':

$$\begin{aligned}
\frac{S}{S'} &= \frac{a + ar + ar^2 + \cdots + ar^{n-1}}{\frac{1}{a} + \frac{1}{ar} + \frac{1}{ar^2} + \cdots + \frac{1}{ar^{n-1}}} \\
&= \frac{a(1 + r + r^2 + \cdots + r^{n-1})}{\frac{1}{a}\left(1 + \frac{1}{r} + \frac{1}{r^2} + \cdots + \frac{1}{r^{n-1}}\right)} \\
&= a^2 \cdot \frac{1 + r + r^2 + \cdots + r^{n-1}}{\frac{r^{n-1} + r^{n-2} + \cdots + r^2 + r + 1}{r^{n-1}}} \\
&= a^2 r^{n-1}.
\end{aligned}$$

Aha! Comparing this to $P = a^n r^{n(n-1)/2}$, we have $P = (a^2 r^{n-1})^{n/2} = (S/S')^{n/2}$, as before. (Notice also that $S/S' = a \cdot ar^{n-1}$, which is the product of the first and last terms of the sequence. This also is a clue that there is a simple relationship between S/S' and P.) $\square$

Exercises

10.4.1 Find the sum of each of the following series:

(a) $2 + 8 + 32 + \cdots + 2048$

(b) $54 - 18 + 6 - 2 + \cdots + \dfrac{2}{243}$

(c)$\star$ $1 + 2 + 3 + 6 + 9 + 18 + \cdots + 729 + 1458$

10.4.2 Sum each of the following infinite geometric series:

(a) $\dfrac{1}{6^1} + \dfrac{1}{6^2} + \dfrac{1}{6^3} + \cdots$

(b) $192 + 144 + 108 + \cdots$

(c) $2 - \sqrt{2} + 1 - \dfrac{\sqrt{2}}{2} + \dfrac{1}{2} - \cdots$

10.4.3 Find the sum $\dfrac{1}{7} + \dfrac{2}{7^2} + \dfrac{1}{7^3} + \dfrac{2}{7^4} + \cdots$. *(Source: AHSME)*

10.4.4 Find the value of x that satisfies the equation $1 + x + x^2 + \cdots = 4$. *(Source: Mandelbrot)*

10.4.5 Find all real values of a such that $\dfrac{(a^8 + a^4 + 1)(a^3 + a^2 + a + 1)}{a^9 + a^6 + a^3 + 1} = 21$.

10.4.6 The first term of a finite geometric series of real numbers is positive, and the sum of the series is negative. Show that the series has an even number of terms.

10.4.7 Note the following relationship between sums of powers of 5 and other powers of 5:

$$
\begin{array}{rlcrcl}
5^1 + 5^0 &= 6 & \qquad & 6 \cdot 100 + 25 &= 625 &= 5^4 \\
5^2 + 5^1 + 5^0 &= 31 & \qquad & 31 \cdot 100 + 25 &= 3125 &= 5^5 \\
5^3 + 5^2 + 5^1 + 5^0 &= 156 & \qquad & 156 \cdot 100 + 25 &= 15625 &= 5^6
\end{array}
$$

Explain the pattern and why it works.

10.5 Sequence, Summation, and Product Notation

We have special notations that allow us to write sequences, series, and products more compactly. We introduce these notations with a few examples. We start with sequences:

$$\{3n + 5\}_{n=1}^{5} = 8, 11, 14, 17, 20$$
$$\{2^k + 1\}_{k=0}^{\infty} = 2, 3, 5, 9, 17, \ldots$$

In the first row, we write the sequence $8, 11, 14, 17, 20$ by enclosing the formula $3n + 5$ in curly braces. The variable in our formula is a **dummy variable**, also called the **index**, that takes on the range of values indicated by the indices outside the braces to the right. The lower index represents the lower bound (starting point) for the dummy variable. The upper index represents the upper bound (endpoint) for the dummy variable. Note that these indices are *not* necessarily the same as the least and greatest numbers in the sequence.

The second sequence is an infinite sequence, so we denote the upper bound of the dummy variable with the symbol for infinity: ∞.

Sometimes a sequence will be defined without indices. We should only write a sequence with braces without indices when it's clear what the dummy variable is and at what value of the dummy variable we should start the sequence. For example, if we say that $F_0 = 0$, $F_1 = 1$, and $F_n = F_{n-1} + F_{n-2}$ for all $n \geq 2$, we can then comfortably refer to the sequence $\{F_n\}$, where it is implied that this refers to the infinite sequence $F_0, F_1, F_2, \ldots$.

To express some series, we use the uppercase Greek letter Σ, which is called **sigma**. For example, we have

$$\sum_{n=1}^{10} 2n = 2 + 4 + 6 + 8 + 10 + 12 + 14 + 16 + 18 + 20$$

$$\sum_{k=0}^{\infty} 2^{-k} = 1 + \frac{1}{2} + \frac{1}{4} + \frac{1}{8} + \cdots$$

We call this **summation notation**. Each Σ symbol tells us to sum the sequence whose general formula immediately follows the sigma. Below each sigma is the starting point for the dummy variable in the sum, and above each sigma is the dummy variable's upper bound. Note that the first sum is a finite arithmetic series and the second is an infinite geometric series.

Finally, we use the uppercase Greek letter Π for products in much the same way we use Σ for sums:

$$\prod_{k=2}^{7}(k+1) = 3\cdot 4\cdot 5\cdot 6\cdot 7\cdot 8$$

$$\prod_{n=1}^{\infty}\frac{n}{n+1} = \frac{1}{2}\cdot\frac{2}{3}\cdot\frac{3}{4}\cdots$$

You might notice the resemblance of Π to π. This isn't a coincidence; Π is the uppercase version of the Greek letter pi, while π is the lowercase version. Just as with summation notation, the index below the pi represents the starting point of the dummy variable in the product while the number on top represents the endpoint.

Series are extremely important in mathematics and most sciences (including, if not especially, computer science). You'll encounter Σ frequently, so it's important to understand what it means and how to manipulate expressions involving it.

Problems

Problem 10.28: Rewrite each of the following sequences, series, and products by expanding them term by term. You do not need to calculate the sums and products, but feel free to do so if you like.

(a) $\{7k+12\}_{k=1}^{8}$

(b) $\{2^n - n\}_{n=2}^{6}$

(c) $\displaystyle\sum_{n=1}^{6} n^3$

(d) $\displaystyle\sum_{k=1}^{8}(2k-1)$

(e) $\displaystyle\sum_{n=0}^{\infty} x^n$

(f) $\displaystyle\sum_{i=3}^{\infty}\left(-\frac{2}{3}\right)^i$

(g) $\displaystyle\prod_{i=1}^{10} i$

(h) $\displaystyle\prod_{n=1}^{9}\frac{n}{n+1}$

Problem 10.29: Write each of the following using sequence, summation, or product notation. Assume that each sequence or series in parts (a) through (f) is arithmetic or geometric in nature.

(a) $4, -2, 1, -\frac{1}{2}, \frac{1}{4}, -\frac{1}{8}$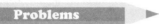

(b) $\ldots, -10, -7, -4, -1, 2, 5, \ldots$

(c) $1 + 3 + 5 + \cdots + 101$

(d) $83 + 79 + 75 + \cdots + (-105) + (-109)$

(e) $2 - 6 + 18 - 54 + \cdots + 118098$

(f) $3 + 0.9 + 0.27 + 0.081 + \cdots$

(g) $\frac{1}{3}\cdot\frac{2}{4}\cdot\frac{3}{5}\cdots\frac{8}{10}$

(h) $\frac{2^2-1}{2^2+1}\cdot\frac{3^2-1}{3^2+1}\cdot\frac{4^2-1}{4^2+1}\cdots\frac{18^2-1}{18^2+1}$

Problem 10.30: In each of the following parts, state whether or not the two sums are the same, and explain your answer.

(a) $\displaystyle\sum_{i=2}^{21} 3i$ and $\displaystyle\sum_{k=-3}^{16}(3k+15)$

(b) $\displaystyle\sum_{k=2}^{10}\frac{1}{k}$ and $\displaystyle\sum_{k=4}^{11}\frac{1}{k-2}$

(c) $\displaystyle\sum_{j=-2}^{15}(-3)^j$ and $\displaystyle\sum_{j=0}^{17}(-3)^{j+2}$

(d) $\displaystyle\sum_{i=0}^{\infty}(0.5)^i$ and $\displaystyle\sum_{i=1}^{\infty}(0.5)^{i-1}$

Problem 10.31:

(a) Express the sum $\displaystyle\sum_{i=4}^{12} \frac{2i}{i-3}$ as a summation in which the dummy variable ranges from 7 to 15.

(b) Express the sum $\displaystyle\sum_{j=0}^{\infty} 0.1^j$ as a summation in which the dummy variable starts at -5.

Problem 10.32: Evaluate the summation $\displaystyle\sum_{j=1}^{100}(5j+7)$.

Problem 10.33: Suppose $a_1, a_2, \ldots, a_{10}$ and $b_1, b_2, \ldots, b_{10}$ are numbers such that $\displaystyle\sum_{k=1}^{10} a_k = 6$ and $\displaystyle\sum_{k=1}^{10} b_k = 13$.

Find $\displaystyle\sum_{k=1}^{10}(3a_k - 2b_k + 1)$.

Problem 10.34:

(a) Find a closed form expression for $\displaystyle\sum_{n=1}^{k} 2n$. In other words, find a formula in terms of k that equals

$\displaystyle\sum_{n-1}^{k} 2n$ for all integers k.

(b) Find a closed form expression for $\displaystyle\sum_{n-1}^{k} 3$.

(c) Find a closed form expression for $\displaystyle\sum_{n=1}^{k}(2n+3)$.

Problem 10.35: Consider the sum $\displaystyle\sum_{k=0}^{\infty} \frac{1+4^k}{5^k}$.

(a) Write the first five terms of the sum without adding the terms in each numerator.

(b) Separate the sum into two sums that you can evaluate, and then compute these sums.

(c) Compute $\displaystyle\sum_{k=0}^{\infty} \frac{1+4^k}{5^k}$.

Problem 10.36: Find the polynomial $f(x)$ such that $\displaystyle\sum_{k=1}^{n} f(k) = n^2 - 3n$ for all positive integers n.

We begin by getting a little practice interpreting these new notations.

Problem 10.28: Rewrite each of the following sequences, series, and products by expanding them term by term. (You do not need to calculate the sums and products.)

(a) $\{7k + 12\}_{k=1}^{8}$

(b) $\{2^n - n\}_{n=2}^{6}$

(c) $\displaystyle\sum_{n=1}^{6} n^3$

(d) $\displaystyle\sum_{k=1}^{8}(2k - 1)$

(e) $\displaystyle\sum_{n=0}^{\infty} x^n$

(f) $\displaystyle\sum_{i=3}^{\infty}\left(-\frac{2}{3}\right)^i$

(g) $\displaystyle\prod_{i=1}^{10} i$

(h) $\displaystyle\prod_{n=1}^{9}\frac{n}{n+1}$

Solution for Problem 10.28:

(a) To produce the terms in the sequence $\{7k + 12\}_{k=1}^{8}$, we evaluate $7k + 12$ for each value of k from $k = 1$ to $k = 8$. This gives us $19, 26, 33, 40, 47, 54, 61, 68$. Notice that this is an arithmetic sequence.

(b) To produce the terms in the sequence $\{2^n - n\}_{n=2}^{6}$, we evaluate $2^n - n$ for each value of n from $n = 2$ to $n = 6$. This gives us $2, 5, 12, 27, 58$.

(c) $\displaystyle\sum_{n=1}^{6} n^3 = 1^3 + 2^3 + 3^3 + 4^3 + 5^3 + 6^3.$

(d) $\displaystyle\sum_{k=1}^{8}(2k - 1) = 1 + 3 + 5 + 7 + 9 + 11 + 13 + 15.$ This is an arithmetic series.

(e) $\displaystyle\sum_{n=0}^{\infty} x^n = 1 + x + x^2 + x^3 + \cdots.$ We might be tempted to simplify this by noting that it is an infinite geometric series with first term 1 and common ratio x. However, we don't know whether or not $|x| < 1$, so we don't know if the sum converges to $1/(1 - x)$ or not. So, we must leave it as the sum shown.

(f) We have

$$\sum_{i=3}^{\infty}\left(-\frac{2}{3}\right)^i = \left(-\frac{2}{3}\right)^3 + \left(-\frac{2}{3}\right)^4 + \left(-\frac{2}{3}\right)^5 + \cdots.$$

This is an infinite geometric series with first term $(-2/3)^3 = -8/27$ and common ratio $-2/3$, so its sum is

$$\frac{-\frac{8}{27}}{1 - \left(-\frac{2}{3}\right)} = \frac{-\frac{8}{27}}{\frac{5}{3}} = -\frac{8}{45}.$$

(g) $\displaystyle\prod_{i=1}^{10} i = 1 \cdot 2 \cdot 3 \cdot 4 \cdot 5 \cdot 6 \cdot 7 \cdot 8 \cdot 9 \cdot 10 = 10!.$

(h) $\displaystyle\prod_{n=1}^{9}\frac{n}{n+1} = \frac{1}{2} \cdot \frac{2}{3} \cdot \frac{3}{4} \cdot \frac{4}{5} \cdot \frac{5}{6} \cdot \frac{6}{7} \cdot \frac{7}{8} \cdot \frac{8}{9} \cdot \frac{9}{10} = \frac{1}{10}.$

$\square$

Now, let's get a little practice writing sequences, series, and products using our new notation.

Problem 10.29: Write each of the following using sequence, summation, or product notation. Assume that each sequence or series in parts (a) through (f) is arithmetic or geometric in nature.

(a) $4, -2, 1, -\frac{1}{2}, \frac{1}{4}, -\frac{1}{8}$

(b) $\ldots, -10, -7, -4, -1, 2, 5, \ldots$

(c) $1 + 3 + 5 + \cdots + 101$

(d) $83 + 79 + 75 + \cdots + (-105) + (-109)$

(e) $2 - 6 + 18 - 54 + \cdots + 118098$

(f) $3 + 0.9 + 0.27 + 0.081 + \cdots$

(g) $\frac{1}{3} \cdot \frac{2}{4} \cdot \frac{3}{5} \cdots \frac{8}{10}$

(h) $\frac{2^2-1}{2^2+1} \cdot \frac{3^2-1}{3^2+1} \cdot \frac{4^2-1}{4^2+1} \cdots \frac{18^2-1}{18^2+1}$

Solution for Problem 10.29: If your answers don't match the answers we come up with below, don't panic! Your answers may still be correct, as we'll discuss shortly.

(a) The first term is 4 and the common ratio is $-1/2$, so the general formula for each term is $4(-1/2)^{n-1}$. There are six terms, so we write the sequence as $\left\{4\left(-\frac{1}{2}\right)^{n-1}\right\}_{n=1}^{6}$.

(b) There is no beginning or end to this arithmetic sequence. We do know that the common difference is 3. From this we could create any number of general formulas. We choose $3k + 2$ here, but we could just as easily choose $3k - 1$, $3k + 11$, or any other that produces all the numbers that are two more than a multiple of 3. Using $3k + 2$ as our formula, we can write the sequence as $\{3k + 2\}_{k=-\infty}^{\infty}$.

(c) What's wrong with this solution:

> **Bogus Solution:** This series consists of the first 51 positive odd numbers. We can write odd numbers as $2n + 1$, and we want the first 51 positive odd numbers, so we can write the series as $\sum_{n=1}^{51}(2n + 1)$.

If we write out the sum represented by $\sum_{n=1}^{51}(2n + 1)$, the first term is $2(1) + 1 = 3$, and the last term is $2(51) + 1 = 103$. So, the resulting series is $3 + 5 + 7 + \cdots + 103$, not $1 + 3 + 5 + \cdots + 101$. Our starting and ending values of the dummy variable are 1 too large! We can fix this by subtracting 1 from each, so the sum can be written as $\sum_{n=0}^{50}(2n + 1)$.

> **WARNING!!** Pay close attention to the starting and ending values of the dummy variable when writing a series in summation notation. The easiest way to make sure you have the correct values is to simply substitute them into the formula in your summation. The starting value should produce the first term of the sum, and the ending value should produce the last term.

We also could have noted that the common difference of this arithmetic series is 2, and a general formula for each term is $1 + 2(n - 1) = 2n - 1$. Since $2n - 1 = 101$ when $n = 51$, our series consists of the sum of $2n - 1$ from $n = 1$ to $n = 51$, which we write as $\sum_{n=1}^{51}(2n-1)$. While this summation does not look the same as $\displaystyle\sum_{n=0}^{50}(2n+1)$, it is indeed the same series. Both represent the sum $1+3+5+\cdots+101$.

(d) The common difference of this arithmetic series is -4, and a general formula for each term is $83-4(n-1)=87-4n$. Looking at the last term, we see that $87-4n=-109$ when $n=49$, so our

series consists of the sum of $87-4n$ from $n=1$ to $n=49$. We can write this sum as $\displaystyle\sum_{n=1}^{49}(87-4n)$.

(e) The common ratio of this geometric series is -3, and a general formula for each term is $2\cdot(-3)^{n-1}$. Since $2\cdot(-3)^{n-1}=118098=2\cdot3^{10}$ when $n=11$, the series is the sum of $2\cdot(-3)^{n-1}$ as n goes from 1

to 11. We can write this sum as $\displaystyle\sum_{n=1}^{11}(2\cdot(-3)^{n-1})$.

(f) The common ratio of this infinite geometric series is 0.3, and a general formula for each term is

$3(0.3)^{n-1}$. We write the sum as $\displaystyle\sum_{n=1}^{\infty}3(0.3)^{n-1}$.

(g) As we did with the other parts, we look for a formula that generates each term. The denominator is two more than the numerator in each fraction, so each term can be written as $n/(n+2)$. As we let n go from $n=1$ to $n=8$ in $n/(n+2)$, we produce all the terms in the product, so the product

can be written as $\displaystyle\prod_{n=1}^{8}\frac{n}{n+2}$.

(h) One formula that generates the terms in our product is pretty clearly $(n^2-1)/(n^2+1)$. However, we must be careful to note that our product does not start with $n=1$. It starts instead with $n=2$

and goes through $n=18$, so we write it as $\displaystyle\prod_{n=2}^{18}\frac{n^2-1}{n^2+1}$.

$\square$

You might have different answers from those above, but your answers might still be correct. Here are a couple of reasons why you might have different answers:

- You used a different variable as your dummy variable. For example, the sum $\displaystyle\sum_{n=1}^{49}(87-4n)$ is the

same as the sum $\displaystyle\sum_{j=1}^{49}(87-4j)$.

- You used a different formula that produces the same sequence, sum, or product. For example, instead of using $2n-1$ to express each odd number in the sum $1+3+5+7+\cdots+101$, you might use $2n+1$. However, if you do so, you must be careful about the bounds of the dummy variable. If we use $2n+1$ to represent each term in the sum $1+3+5+\cdots+101$, then n goes from $n=0$ to

$n=50$, so our sum is $\displaystyle\sum_{n=0}^{50}(2n+1)$. Notice that this is the same sum as $\displaystyle\sum_{n=1}^{51}(2n-1)$, and as $\displaystyle\sum_{n=-1}^{49}(2n+3)$, and as $\displaystyle\sum_{n=3}^{53}(2n-5)$, and so on.

Let's take a closer look at this latter point.

Problem 10.30: In each of the following parts, state whether or not the two sums are the same, and explain your answer.

(a) $\displaystyle\sum_{i=2}^{21}3i$ and $\displaystyle\sum_{k=-3}^{16}(3k+15)$

(c) $\displaystyle\sum_{j=-2}^{15}(-3)^j$ and $\displaystyle\sum_{j=0}^{17}(-3)^{j+2}$

(b) $\displaystyle\sum_{k=2}^{10}\frac{1}{k}$ and $\displaystyle\sum_{k=4}^{11}\frac{1}{k-2}$

(d) $\displaystyle\sum_{i=0}^{\infty}(0.5)^i$ and $\displaystyle\sum_{i=1}^{\infty}(0.5)^{i-1}$

Solution for Problem 10.30:

(a) One way to see that two summations represent the same series is to write both summations as series without summation notation. We have

$$\sum_{i=2}^{21}3i = 6+9+12+15+\cdots+63,$$

$$\sum_{k=-3}^{16}(3k+15) = 6+9+12+15+\cdots+63.$$

Yep, these are the same series. Now, let's think about how we might have seen this without writing out the series. Let

$$A = \sum_{i=2}^{21}3i \qquad \text{and} \qquad B = \sum_{k=-3}^{16}(3k+15).$$

We first note that both series A and series B have 20 terms. We could also see that the two series have the same number of terms by noting that each limit of A is 5 more than the corresponding limit of B.

Since the limits of A are 5 more than the limits of B, the dummy variable in each term of A is 5 more than the dummy variable in the corresponding term in B. This means that for each value of i in A and the corresponding value of k in B, we have $i = k + 5$. In A, we add $3i$ for each value of i. Letting $i = k + 5$, adding $3i$ for each value of i from 2 to 21 is the same as adding $3(k+5)$, which equals $3k + 15$, for each value of k from -3 to 16. (We get the range -3 to 16 for k by noting that $k = i - 5$, and i ranges from 2 to 21.) Therefore, series A and series B are the same.

(b) The first series has $10 - 2 + 1 = 9$ terms and the second has $11 - 4 + 1 = 8$ terms, so these two series are not the same. (Their sums may or may not be equal, but the series are not the same.)

(c) Both series have 18 terms. To make the two series easier to discuss, we assign a variable to each, and change the dummy variable in the second series to k. (Changing the dummy variable doesn't

---

change the sum.) So, we write

$$C = \sum_{j=-2}^{15} (-3)^j \qquad \text{and} \qquad D = \sum_{k=0}^{17} (-3)^{k+2}.$$

What's wrong here:

> **Bogus Solution:** Each k is 2 more than the corresponding j, so we let $j = k + 2$ in the term $(-3)^j$ of C to find the corresponding term $(-3)^{k+2}$ of D. Therefore, the two series are the same.

Be careful! Because each k is 2 more than the corresponding j, we let $j = k - 2$, not $j = k + 2$. Letting $j = k - 2$ in the expression we sum in C gives us $(-3)^{k-2}$, which is not the expression we sum in D. Specifically, we have

$$\sum_{j=-2}^{15} (-3)^j = \sum_{k=0}^{17} (-3)^{k-2}.$$

We also could have caught our mistake by writing out a few terms of each summation:

$$\sum_{j=-2}^{15} (-3)^j = (-3)^{-2} + (-3)^{-1} + (-3)^0 + \cdots + (-3)^{15},$$

$$\sum_{k=0}^{17} (-3)^{k+2} = (-3)^2 + (-3)^3 + (-3)^4 + \cdots + (-3)^{19}.$$

These are clearly not the same.

> **Concept:** When in doubt with summation notation, write out the series without summation notation.

(d) At first, we might think that the two series are different because the second series appears to have one fewer term than the first. But, if we write both series without summation notation, we have:

$$\sum_{i=0}^{\infty} (0.5)^i = (0.5)^0 + (0.5)^1 + (0.5)^2 + (0.5)^3 + \cdots,$$

$$\sum_{i=1}^{\infty} (0.5)^{i-1} = (0.5)^0 + (0.5)^1 + (0.5)^2 + (0.5)^3 + \cdots.$$

These are the same series. Because the series are infinite, we can't count the number of terms. In other words, even though the dummy variable starts at a higher value in the second summation, we can't say the second series has "more" terms than the first.

□

> **Extra!** Here's a surprising summation:
>
> $$\sum_{k=1}^{\infty} \frac{1}{k^2} = \frac{\pi^2}{6}.$$

Problem 10.31:

(a) Express the sum $\sum_{i=4}^{12} \dfrac{2i}{i-3}$ as a summation in which the dummy variable ranges from 7 to 15.

(b) Express the sum $\sum_{j=0}^{\infty} 0.1^j$ as a summation in which the dummy variable starts at -5.

Solution for Problem 10.31:

(a) Let the dummy variable of our new summation be k. We wish to have k range from 7 to 15, so we already have the limits on our summation. We need only to find the expression that we sum in the new series. Each i in the old series is 3 less than the corresponding k in the new series, so to find our new expression, we let $i = k - 3$ in $\frac{2i}{i-3}$. This gives us $\frac{2(k-3)}{k-3-3} = \frac{2k-6}{k-6}$. So our new summation is $\sum_{k=7}^{15} = \dfrac{2k-6}{k-6}$. As a check, we can write out a few terms in each summation:

$$\sum_{i=4}^{12} \frac{2i}{i-3} = \frac{8}{1} + \frac{10}{2} + \frac{12}{3} + \cdots + \frac{24}{9},$$

$$\sum_{k=7}^{15} \frac{2k-6}{k-6} = \frac{8}{1} + \frac{10}{2} + \frac{12}{3} + \cdots + \frac{24}{9}.$$

Indeed, these two series are the same.

(b) Let the dummy variable of our new summation again be k. Our new summation starts with the dummy variable 5 less than the dummy variable in our old summation. So, for each k and corresponding j, we have $k = j - 5$, or $j = k + 5$. This means that the expression we sum in the new summation is 0.1^{k+5}. But what are the limits of the new summation? We know the new summation starts at $k = -5$, but where does it end?

The new summation must still be an infinite series, so the upper limit is still ∞:

$$\sum_{j=0}^{\infty} 0.1^j = \sum_{k=-5}^{\infty} 0.1^{k+5}.$$

When we write a few terms of each series, we see that each equals $(0.1)^0 + (0.1)^2 + (0.1)^3 + \cdots$.

$\square$

When we write $\sum_{i=4}^{12} \dfrac{2i}{i-3}$ as the equivalent summation $\sum_{k=7}^{15} \dfrac{2k-6}{k-6}$, we say we **shift the index** of the original summation. Shifting the index of a summation does not produce a new series; it merely produces a different way to represent the series. As we saw in our second example, when we shift the index of an infinite series, only the starting limit changes.

Now that we're comfortable with summation notation, let's try evaluating a sum expressed in summation notation.

Problem 10.32: Evaluate the summation $\displaystyle\sum_{j=1}^{100}(5j+7)$.

Solution for Problem 10.32: Solution 1: We write out the first few terms of the series. We see that increasing j by 1 at each step increases $5j+7$ by 5. In other words, our sum is an arithmetic series with common difference 5. We get the first term by letting $j=1$ in $5j+7$ and we get the last term by letting $j=100$ in $5j+7$. This gives us the arithmetic series

$$\sum_{j=1}^{100}(5j+7) = 12 + 17 + 22 + \cdots + 507.$$

We know that there are 100 terms, so our sum is $\displaystyle\sum_{j=1}^{100}(5j+7) = 100 \cdot \frac{12+507}{2} = 100 \cdot \frac{519}{2} = 25950.$

Solution 2: Writing each term in the form of the general term, we see that we can group and factor in a way that makes each group easy to evaluate:

$$\sum_{j=1}^{100}(5j+7) = (5\cdot 1 + 7) + (5\cdot 2 + 7) + (5\cdot 3 + 7) + \cdots + (5\cdot 99 + 7) + (5\cdot 100 + 7)$$

$$= (5\cdot 1 + 5\cdot 2 + 5\cdot 3 + \cdots + 5\cdot 99 + 5\cdot 100) + \underbrace{(7 + 7 + 7 + \cdots + 7 + 7)}_{100\ 7\text{'s}}$$

$$= 5(1 + 2 + 3 + \cdots + 99 + 100) + \underbrace{(7 + 7 + 7 + \cdots + 7 + 7)}_{100\ 7\text{'s}}$$

$$= 5\sum_{j=1}^{100} j + \sum_{j=1}^{100} 7.$$

We could have evaluated the expression

$$5(1 + 2 + 3 + \cdots + 99 + 100) + \underbrace{(7 + 7 + 7 + \cdots + 7 + 7)}_{100\ 7\text{'s}}$$

as $5(100 \cdot 101/2) + 7 \cdot 100 = 25950$, but instead we rewrote it in terms of two other summations above to make a point. We've shown how to break the original sum into simpler sums, and in the future we can jump straight to this sum of simpler sums in order to calculate similar sums more easily. Specifically, we break the initial summation into two sums, then factor the 5 out of the first sum to find:

$$\sum_{j=1}^{100}(5j+7) = \sum_{j=1}^{100}(5j) + \sum_{j=1}^{100} 7 = 5\sum_{j=1}^{100} j + \sum_{j=1}^{100} 7 = 5 \cdot \frac{100(100+1)}{2} + 100 \cdot 7 = 5 \cdot 5050 + 700 = 25950.$$

While the second solution may not look simpler at first, these sorts of summation manipulations will become more natural as you gain experience working with summation notation, and they become more important as summation problems get harder. □

Our second solution is an example of the following two manipulations of summation notation:

Important: For any real numbers $a_1, a_2, \ldots, a_n, b_1, b_2, \ldots, b_n$, and constant c,

$$\sum_{k=1}^{n}(a_k + b_k) = \sum_{k=1}^{n}a_k + \sum_{k=1}^{n}b_k,$$

$$\sum_{k=1}^{n}ca_k = c\sum_{k=1}^{n}a_k.$$

Here's a little more practice using these facts.

Problem 10.33: Suppose $\displaystyle\sum_{k=1}^{10}a_k = 6$ and $\displaystyle\sum_{k=1}^{10}b_k = 13$. Find $\displaystyle\sum_{k=1}^{10}(3a_k - 2b_k + 1)$.

Solution for Problem 10.33: We separate the sum into parts we can evaluate, using the given summation values where possible:

$$\sum_{k=1}^{10}(3a_k - 2b_k + 1) = \sum_{k=1}^{10}3a_k + \sum_{k=1}^{10}(-2b_k) + \sum_{k=1}^{10}1$$

$$= 3\sum_{k=1}^{10}a_k - 2\sum_{k=1}^{10}b_k + 10$$

$$= 3\cdot 6 - 2\cdot 13 + 10 = 18 - 26 + 10 = 2.$$

$\square$

Problem 10.34: Find a closed form expression for $\displaystyle\sum_{n=1}^{k}(2n + 3)$.

Solution for Problem 10.34: We separate the summation into two sums, then sum the results to get a closed form:

$$\sum_{n=1}^{k}(2n + 3) = \sum_{n=1}^{k}(2n) + \sum_{n=1}^{k}3 = 2\sum_{n=1}^{k}n + 3k = 2\cdot\frac{k(k+1)}{2} + 3k = k^2 + 4k.$$

Notice that our final answer gives us a nice, simple closed form in terms of k for the summation. Closed forms are usually much easier to deal with than summations. For example, which would you rather have to compute for some value of k,

$$\sum_{n=1}^{k}(2n + 3) \qquad \text{or} \qquad k^2 + 4k?$$

$\square$

Problem 10.35: Compute $\displaystyle\sum_{k=0}^{\infty}\frac{1 + 4^k}{5^k}$.

Solution for Problem 10.35: Let's start by taking a look at the first few terms of the sum:

$$\sum_{k=0}^{\infty} \frac{1+4^k}{5^k} = \frac{1+1}{1} + \frac{1+4}{5} + \frac{1+16}{25} + \frac{1+64}{125} + \frac{1+256}{625} + \cdots$$

$$= \frac{1}{1} + \frac{1}{1} + \frac{1}{5} + \frac{4}{5} + \frac{1}{25} + \frac{16}{25} + \frac{1}{125} + \frac{64}{125} + \cdots.$$

Starting with the first term and taking every other term, we see that we have the infinite geometric series $1 + \frac{1}{5} + \frac{1}{25} + \frac{1}{125} + \cdots$. The remaining fractions also form an infinite geometric series, with first term 1 and common ratio 4/5. We could have seen this directly by separating the initial summation into a pair of summations:

$$\sum_{k=0}^{\infty} \frac{1+4^k}{5^k} = \sum_{k=0}^{\infty} \left(\frac{1}{5}\right)^k + \sum_{k=0}^{\infty} \left(\frac{4}{5}\right)^k = \frac{1}{1-\frac{1}{5}} + \frac{1}{1-\frac{4}{5}} = \frac{5}{4} + 5 = \frac{25}{4}.$$

$\square$

Problem 10.36: Find the polynomial $f(x)$ such that $\sum_{k=1}^{n} f(k) = n^2 - 3n$ for all positive integers n.

Solution for Problem 10.36: Letting $n = x$ in our summation gives us

$$f(1) + f(2) + f(3) + \cdots + f(x-1) + f(x) = x^2 - 3x.$$

If only we could get rid of the first $x - 1$ terms on the left.

We can! We use our summation again, this time letting $n = x - 1$. This gives us

$$f(1) + f(2) + f(3) + \cdots + f(x-1) = (x-1)^2 - 3(x-1).$$

Subtracting this from our $n = x$ equation eliminates every term on the left except $f(x)$, and we have

$$f(x) = x^2 - 3x - (x-1)^2 + 3(x-1) = 2x - 4.$$

$\square$

Concept: If we have a summation with a variable among its bounds, we can often learn more about the summation by choosing different values for the variable.

Exercises

10.5.1 Rewrite each of the following arithmetic and geometric sequences using sequence notation:

(a) $-91, -81, -71, \ldots, 1999$

(b) $\frac{1}{8}, \frac{1}{2}, 2, 8, 32, \ldots, 8192$

(c) $15, 11, 7, 3, -1, \ldots$

(d) $54, 36, 24, 16, \frac{32}{3}, \ldots$

10.5.2 Write the following using summation notation:

(a) $3 + 3 + 3 + 3 + 3 + 3 + 3$

(b) $\dfrac{1}{2} + 1 + \dfrac{3}{2} + 2 + \dfrac{5}{2} + 3$

(c) $n + (n-1) + (n-2) + \cdots + 1$

(d) $\dfrac{1}{2} + \dfrac{2}{3} + \dfrac{3}{4} + \cdots + \dfrac{n}{n+1}$

(e) $\dfrac{x}{1!} + \dfrac{x^2}{2!} + \dfrac{x^3}{3!} + \cdots$

(f) $1 - x + x^2 - x^3 + \cdots$

10.5.3 Express each of the following as a summation in which the dummy variable starts at 5.

(a) $\displaystyle\sum_{i=-3}^{71} \dfrac{2^i}{i+5}$

(b) $\displaystyle\sum_{i=7}^{\infty} \dfrac{1}{i^2}$

10.5.4 Compute or express in closed form each of the following sums:

(a) $\displaystyle\sum_{i=1}^{15} (3i - 4)$

(b) $\displaystyle\sum_{i-1}^{n} (2 - i)$

(c) $\displaystyle\sum_{n=1}^{\infty} 4 \cdot \dfrac{1}{2^n}$

(d) $\displaystyle\sum_{k=0}^{n} x^k$

(e) $\displaystyle\sum_{n=1}^{6} (3^n - 3n)$

(f)★ $\displaystyle\sum_{k=1}^{100} (-1)^k k^2$

10.5.5 If $\displaystyle\sum_{k=1}^{99} x_k = 7$ and $\displaystyle\sum_{k=1}^{99} y_k = 17$, find the value of $\displaystyle\sum_{n=1}^{99} (107 x_n - 74 y_n)$.

10.5.6 Evaluate the sum $\displaystyle\sum_{k=0}^{\infty} \dfrac{3^k + 6^k}{9^k}$.

10.5.7★ Express, as concisely as possible, the value of the product

$$(0^3 - 350)(1^3 - 349)(2^3 - 348)(3^3 - 347) \cdots (349^3 - 1)(350^3 - 0).$$

(Source: HMMT) **Hints:** 18

10.6 Nested Sums and Products

We know how to tackle a summation of a simple expression, like $\displaystyle\sum_{i=1}^{10} (2i)$. We simply let i equal each of 1 through 10 in the expression $2i$, then sum the results. But what if we must find the summation not of a simple expression, but of another summation?

Extra! *We should be taught not to wait for inspiration to start a thing. Action always generates inspiration. Inspiration seldom generates action.*

– Frank Tibolt

Problems

Problem 10.37: Evaluate each of the following:

(a) $\displaystyle\sum_{i=1}^{4}\left(\sum_{j=1}^{6} i\right)$

(b) $\displaystyle\sum_{i=1}^{4}\left(\sum_{j=1}^{6} j\right)$

(c) $\displaystyle\sum_{i=1}^{4}\left(\sum_{j=1}^{6} (i+j)\right)$

Problem 10.38:

(a) Evaluate $\displaystyle\left(\sum_{i=1}^{5} i\right)\left(\sum_{j=1}^{5} j\right)$.

(b) Evaluate $\displaystyle\sum_{i=1}^{5}\left(\sum_{j=1}^{5} ij\right)$.

(c) Explain the relationship between your answers.

(d) Evaluate $\displaystyle\sum_{i=1}^{\infty}\left(\sum_{j=1}^{\infty} \frac{1}{2^{i+j}}\right)$.

Problem 10.39:

(a) Find a closed form for $\displaystyle\sum_{j=1}^{n}(1)$.

(c) Find a closed form for $\displaystyle\sum_{j=i}^{\infty}\left(\frac{1}{3}\right)^{j}$.

(b) Evaluate $\displaystyle\sum_{i=1}^{10}\left(\sum_{j=1}^{i}(1)\right)$.

(d) Evaluate $\displaystyle\sum_{i=0}^{\infty}\left(\sum_{j=i}^{\infty} \frac{1}{3^{j}}\right)$.

Problem 10.40: Bill thinks of a number that equals $\frac{1}{2}$ raised to a positive integer power. Betty thinks of a number that equals $\frac{1}{3}$ raised to a positive integer power. Bob then multiplies Bill's number and Betty's number. Find the sum of all the possible products Bob can form.

We call the summation of a summation, such as

$$\sum_{i=1}^{10}\left(\sum_{j=1}^{5}(2ij)\right),$$

a **nested sum**. We will often leave out the parentheses around the inner sum and simply write such a nested summation as

$$\sum_{i=1}^{10}\sum_{j=1}^{5}(2ij).$$

Problem 10.37: Evaluate $\displaystyle\sum_{i=1}^{4}\sum_{j=1}^{6}(i+j)$.

Solution for Problem 10.37: We begin by breaking the expression being summed (and summed again) into pieces:

$$\sum_{i=1}^{4}\sum_{j=1}^{6}(i+j) = \sum_{i=1}^{4}\left(\sum_{j=1}^{6}i + \sum_{j=1}^{6}j\right) = \sum_{i=1}^{4}\sum_{j=1}^{6}i + \sum_{i=1}^{4}\sum_{j=1}^{6}j.$$

Now we evaluate the inner sum for each of our pieces:

$$\sum_{i=1}^{4}\sum_{j=1}^{6}i = \sum_{i=1}^{4}\left(\sum_{j=1}^{6}i\right) = \sum_{i=1}^{4}6i,$$

$$\sum_{i=1}^{4}\sum_{j=1}^{6}j = \sum_{i=1}^{4}\left(\sum_{j=1}^{6}j\right) = \sum_{i=1}^{4}21.$$

Next, we evaluate each of these summations:

$$\sum_{i=1}^{4}\sum_{j=1}^{6}i = \sum_{i=1}^{4}6i = 6\sum_{i=1}^{4}i = 6\left(\frac{4\cdot 5}{2}\right) = 60,$$

$$\sum_{i=1}^{4}\sum_{j=1}^{6}j = \sum_{i=1}^{4}21 = 4(21) = 84.$$

Finally, we add the pieces to evaluate the whole sum:

$$\sum_{i=1}^{4}\sum_{j=1}^{6}(i+j) = \sum_{i=1}^{4}\sum_{j=1}^{6}i + \sum_{i=1}^{4}\sum_{j=1}^{6}j = 60 + 84 = 144.$$

$\square$

Problem 10.38:

(a) Evaluate $\displaystyle\sum_{i=1}^{5}\sum_{j=1}^{5}ij$.

(b) Evaluate $\displaystyle\sum_{i=1}^{\infty}\left(\sum_{j=1}^{\infty}\frac{1}{2^{i+j}}\right)$.

Solution for Problem 10.38:

(a) We focus first on the inner sum, $\displaystyle\sum_{j=1}^{5}ij$. In evaluating this sum, we treat i as constant, and let j range from 1 to 5. Therefore, we can factor i out of the sum $\displaystyle\sum_{j=1}^{5}ij$:

$$\sum_{j=1}^{5}ij = i\sum_{j=1}^{5}j.$$

This is true for any value of i, so we can write our nested sum as

$$\sum_{i=1}^{5}\sum_{j=1}^{5} ij = \sum_{i=1}^{5}\left(i\sum_{j=1}^{5} j\right).$$

Now, the inner sum doesn't depend on the variable of the outer sum, so we evaluate the inner sum:

$$\sum_{i=1}^{5}\left(i\sum_{j=1}^{5} j\right) = \sum_{i=1}^{5} i(1+2+3+4+5).$$

We can now factor the constant $(1+2+3+4+5)$ out of the remaining sum, and then evaluate what remains:

$$\sum_{i=1}^{5} i(1+2+3+4+5) = (1+2+3+4+5)\sum_{i=1}^{5} i = (1+2+3+4+5)(1+2+3+4+5),$$

which is $15^2 = 225$.

(b) With our first part as inspiration, we try to separate the nested sum into two summations. First, we must separate the variables in the inner expression:

$$\sum_{i=1}^{\infty}\left(\sum_{j=1}^{\infty} \frac{1}{2^{i+j}}\right) = \sum_{i=1}^{\infty}\left(\sum_{j=1}^{\infty} \frac{1}{2^i}\cdot\frac{1}{2^j}\right).$$

Now, we can factor $1/2^i$ out of the inner sum:

$$\sum_{i=1}^{\infty}\left(\sum_{j=1}^{\infty} \frac{1}{2^i}\cdot\frac{1}{2^j}\right) = \sum_{i=1}^{\infty}\left(\frac{1}{2^i}\sum_{j=1}^{\infty} \frac{1}{2^j}\right) = \sum_{i=1}^{\infty}\frac{1}{2^i}\left(\sum_{j=1}^{\infty} \frac{1}{2^j}\right).$$

Since the sum of the infinite geometric series $\sum_{j=1}^{\infty}\frac{1}{2^j}$ is $\frac{1/2}{1-(1/2)} = 1$, we have

$$\sum_{i=1}^{\infty}\left(\sum_{j=1}^{\infty}\frac{1}{2^{i+j}}\right) = \sum_{i=1}^{\infty}\frac{1}{2^i}\left(\sum_{j=1}^{\infty}\frac{1}{2^j}\right) = \sum_{i=1}^{\infty}\frac{1}{2^i} = 1.$$

□

We found in Problem 10.38 that each nested sum was the product of two simpler sums. For example, in the first part, we found

$$\sum_{i=1}^{5}\sum_{j=1}^{5} ij = \sum_{i=1}^{5}\left(i\sum_{j=1}^{5} j\right).$$

So, our outer summation now consists of summing the product of i and another summation *that does not depend on i*. So, we can factor this other summation out of the outer summation, and write the expression as the product of two summations, instead of as a nested sum:

$$\sum_{i=1}^{5}\sum_{j=1}^{5} ij = \sum_{i=1}^{5}\left(i\sum_{j=1}^{5} j\right) = \left(\sum_{i=1}^{5} i\right)\left(\sum_{j=1}^{5} j\right).$$

Similarly, in the second part, we saw that

$$\sum_{i=1}^{\infty}\left(\sum_{j=1}^{\infty}\frac{1}{2^{i+j}}\right) = \sum_{i=1}^{\infty}\left(\frac{1}{2^i}\left(\sum_{j=1}^{\infty}\frac{1}{2^j}\right)\right) = \left(\sum_{i=1}^{\infty}\frac{1}{2^i}\right)\left(\sum_{j=1}^{\infty}\frac{1}{2^j}\right).$$

> **Concept:** When evaluating the inner sum of a nested sum, the outer variable can be treated as a constant. It (or any function of it) can sometimes be factored out of the inner sum to separate a nested sum into the product of two simple summations.

Now that we know how to deal with the nested summation of an expression involving the dummy variables of both sums, let's take a look at what happens if the dummy variable of the outer summation is part of the *limits* of the inner summation.

Problem 10.39:

(a)   Evaluate $\displaystyle\sum_{i=1}^{10}\left(\sum_{j=1}^{i}(1)\right)$.

(b)   Evaluate $\displaystyle\sum_{i=0}^{\infty}\left(\sum_{j=i}^{\infty}\frac{1}{3^j}\right)$.

Solution for Problem 10.39:

(a)   As usual, we work from the inside out, focusing first on $\displaystyle\sum_{j=1}^{i}(1)$. We can forget for the moment that i is itself the dummy variable of a summation and just treat it as a variable, so we can write this summation in closed form in terms of i. The sum is the sum of i copies of 1, so we have

$$\sum_{j=1}^{i}(1) = i.$$

Therefore, we have $\displaystyle\sum_{i=1}^{10}\left(\sum_{j=1}^{i}(1)\right) = \sum_{i=1}^{10}i = 1 + 2 + 3 + \cdots + 10 = \frac{10(11)}{2} = 55.$

(b)   Once again, we work from the inside out, seeking a closed form for $\displaystyle\sum_{j=i}^{\infty}\frac{1}{3^j}$. This is an infinite geometric series with first term $1/3^i$ and ratio $1/3$. Therefore, the series equals

$$\sum_{j=i}^{\infty}\frac{1}{3^j} = \frac{\frac{1}{3^i}}{1-\frac{1}{3}} = \frac{\frac{1}{3^i}}{\frac{2}{3}} = \frac{3}{2}\cdot\frac{1}{3^i}.$$

This means that

$$\sum_{i=0}^{\infty}\left(\sum_{j=i}^{\infty}\frac{1}{3^j}\right) = \sum_{i=0}^{\infty}\frac{3}{2}\cdot\frac{1}{3^i} = \frac{3}{2}\sum_{i=0}^{\infty}\frac{1}{3^i}.$$

This is just another geometric series! The first term is 1 and the common ratio is 1/3, so we have

$$\sum_{i=0}^{\infty}\left(\sum_{j=i}^{\infty}\frac{1}{3^j}\right) = \frac{3}{2}\sum_{i=0}^{\infty}\frac{1}{3^i} = \frac{3}{2}\cdot\frac{1}{1-\frac{1}{3}} = \frac{9}{4}.$$

□

> **Concept:** Evaluating nested summations sometimes requires writing the inner summation in closed form in terms of the dummy variable of the outer summation.

> **Problem 10.40:** Bill thinks of a number that equals $\frac{1}{2}$ raised to a positive integer power. Betty thinks of a number that equals $\frac{1}{3}$ raised to a positive integer power. Bob then multiplies Bill's number and Betty's number. Find the sum of all the possible products Bob can form.

Solution for Problem 10.40: Bill's number is in the form $\left(\frac{1}{2}\right)^i$ for some positive integer i. Betty's is in the form $\left(\frac{1}{3}\right)^j$ for some positive integer j. Therefore, the desired sum is the sum of $\left(\frac{1}{2}\right)^i\left(\frac{1}{3}\right)^j$ for all pairs of positive integers i and j. It's not immediately clear how to find this sum, but we can write it in summation notation. We write "Add $\left(\frac{1}{2}\right)^i\left(\frac{1}{3}\right)^j$ for all integers $i, j \geq 1$" as

$$\sum_{i=1}^{\infty}\sum_{j=1}^{\infty}\left(\frac{1}{2}\right)^i\left(\frac{1}{3}\right)^j.$$

Make sure you see why this is the desired sum; it produces a term $\left(\frac{1}{2}\right)^i\left(\frac{1}{3}\right)^j$ for every pair of positive integers i and j. Moreover, it tells us how to evaluate the sum. We can factor out $\left(\frac{1}{2}\right)^i$ from the inner summation, and we have

$$\sum_{i=1}^{\infty}\sum_{j=1}^{\infty}\left(\frac{1}{2}\right)^i\left(\frac{1}{3}\right)^j = \sum_{i=1}^{\infty}\left(\left(\frac{1}{2}\right)^i\sum_{j=1}^{\infty}\left(\frac{1}{3}\right)^j\right) = \left(\sum_{i=1}^{\infty}\left(\frac{1}{2}\right)^i\right)\left(\sum_{j=1}^{\infty}\left(\frac{1}{3}\right)^j\right).$$

In other words, our desired sum is the product of two infinite geometric series! We evaluate each infinite geometric series to find that the sum is

$$\left(\sum_{i=1}^{\infty}\left(\frac{1}{2}\right)^i\right)\left(\sum_{j=1}^{\infty}\left(\frac{1}{3}\right)^j\right) = \left(\frac{\frac{1}{2}}{1-\frac{1}{2}}\right)\left(\frac{\frac{1}{3}}{1-\frac{1}{3}}\right) = \frac{1}{2}.$$

Now that we know what to look for, we can see why this is true without using summation notation. Consider the product of these two infinite geometric series:

$$\left(\frac{1}{2} + \frac{1}{2^2} + \frac{1}{2^3} + \cdots\right)\left(\frac{1}{3} + \frac{1}{3^2} + \frac{1}{3^3} + \cdots\right).$$

When we expand this product, each term is the product of a power of $\frac{1}{2}$ from the first sum and a product of $\frac{1}{3}$ from the second, and every such product is formed. □

Our key step in this solution was writing the sum in summation notation. Then, it was clear how to evaluate it.

> **Concept:** Sometimes expressing a sum in summation notation will tell us how to evaluate the sum.

Exercises

10.6.1 Evaluate each of the following as completely as possible:

(a) $\displaystyle\sum_{i=1}^{5}\sum_{j=1}^{5}(i+j)$

(b) $\displaystyle\sum_{i=1}^{4}\sum_{j=1}^{7}3^{i+j}$

(c) $\displaystyle\sum_{i=1}^{4}\sum_{j=1}^{5}\sum_{k=1}^{6}ijk$

10.6.2 Evaluate each of the following:

(a) $\displaystyle\sum_{i=1}^{5}\sum_{j=1}^{i}ij$

(b) $\displaystyle\sum_{i=0}^{5}\sum_{j=0}^{i}2^{j}$

10.6.3 Evaluate $\displaystyle\sum_{i=1}^{5}\sum_{j=i}^{12}(2j-7)$.

10.6.4 Show that $\displaystyle\sum_{i=1}^{n}\sum_{j=1}^{i}x_j = \sum_{i=1}^{n}(n+1-i)x_i$ for any sequence $x_1, x_2, \ldots, x_n$.

10.7 Summary

A **sequence** is a *list* of numbers, while a **series** is a *sum* of numbers.

In an **arithmetic sequence**, the difference between any two consecutive terms is the same. This fixed difference is known as the **common difference** of the sequence. When we add the terms of an arithmetic sequence, we form an **arithmetic series**.

Here are some useful facts about arithmetic sequences and series:

- The nth term of an arithmetic sequence that has first term a and common difference d is $a+(n-1)d$.

- The sequence a, b, c is an arithmetic sequence if and only if b is the average of a and c.

- The average of the terms in an arithmetic sequence is equal to the average of the first and last terms of the sequence.

- The sum of the terms in an arithmetic sequence equals the product of the number of terms and the average of the first and last terms.

- The sum of an n-term arithmetic series with first term a and common difference d is $\dfrac{n(2a + (n-1)d)}{2}$.

- The sum of the first n positive integers is $\dfrac{n(n+1)}{2}$.

A **geometric sequence** is a sequence for which there exists a constant r such that each term (except for the first term) is r times the previous term. This constant r is called the **common ratio** of the sequence. When we add the terms in a geometric sequence, we form a **geometric series**.

Here are some important facts about geometric sequences and series:

- If a geometric sequence has first term a and common ratio r, then the nth term of the sequence is equal to ar^{n-1}.

- A geometric series with n terms, first term a, and common ratio r (with $r \neq 1$) has sum $\dfrac{a(r^n - 1)}{r - 1}$.

- The sum of an infinite geometric series with first term a and common ratio r, where $|r| < 1$, is $\dfrac{a}{1-r}$.

We have special notations for sequences, series, and products. We use curly braces for sequences, the capital Greek letter sigma, Σ, for sums, and the capital Greek letter pi, Π, for products.

Problem Solving Strategies

Concepts:
- Setting the middle term of an arithmetic sequence with an odd number of terms equal to a variable often introduces a symmetry that allows us to solve problems more easily.

- We can often tackle problems involving arithmetic series by writing all the given information in terms of the first term, the common difference, and the number of terms in the series.

- Before wading through excessive computation or algebraic manipulation in a problem, look for ways to solve the problem more creatively.

- Don't overlook trial and error as a problem-solving strategy. Some equations are most easily solved by guessing solutions and checking if they work.

- Many problems involving geometric sequences can be tackled by writing the terms of the sequence in terms of the first term of the sequence and the common ratio of the sequence.

Continued on the next page. . .

Concepts: ... *continued from the previous page*

- When stuck on a problem, compare it to a similar problem you know how to solve.

- We can often learn more about an infinite series by investigating the sums of first n terms of the series for small values of n.

- If you derive a simple algebraic relationship after complex algebraic manipulation, use the simple relationship as a guide to look for a simpler proof.

- When in doubt with summation notation, write out the series without summation notation.

- If we have a summation with a variable among its bounds, we can often learn more about the summation by choosing different values for the variable.

REVIEW PROBLEMS

10.41 The first term of an arithmetic sequence is 4 and the tenth term is 85. Find the sum of the first fifty terms in the sequence.

10.42 Evaluate each of the following series:

(a) $12 + 15 + 18 + \cdots + 1002$

(c) $\displaystyle\sum_{k=0}^{\infty} \frac{2}{3^k}$

(b) $\displaystyle\sum_{n=1}^{40}(-3n + 4)$

10.43 Evaluate the following geometric series:

(a) $1 + \dfrac{1}{4} + \dfrac{1}{16} + \cdots$

(c) $-\sqrt{6} + \sqrt{2} - \dfrac{\sqrt{6}}{3} + \cdots$

(b) $-3 + 6 - 12 + 24 - \cdots - 768$

10.44 Write the following using summation or product notation:

(a) $\sqrt{1} + \sqrt{2} + \sqrt{3} + \cdots + \sqrt{n}$

(c) $5 \cdot 6 \cdot 7 \cdot 8 \cdot 9 \cdot 10 \cdot 11 \cdot 12$

(b) $1 + 4 + 7 + \cdots + 100$

(d) $(x - x_1)(x - x_2) \cdots (x - x_n)$

10.45 Find all ordered pairs (x, y) such that $3, x, y$ is a geometric sequence and $x, y, 9$ is an arithmetic sequence.

10.46 Find the value of $\frac{3}{2} + \frac{5}{4} + \frac{9}{8} + \frac{17}{16} + \frac{33}{32} + \frac{65}{64} - 7$. *(Source: AHSME)*

10.47 Find all real numbers a such that the roots of $x^3 - 6x^2 + 21x + a = 0$ are not all real and form an arithmetic progression.

10.48 The sum of all the terms in an infinite geometric progression is 6. The sum of the first two terms is 9/2. Find the possible first terms of the progression. *(Source: AHSME)*

10.49 In an arithmetic sequence $t_1, t_2, t_3, \ldots, t_{47}$, the sum of the odd numbered terms is 1272. What is the sum of all 47 terms in the sequence? *(Source: COMC)*

10.50 Prove that for any linear function $f(x)$, the sequence $f(0), f(1), f(2), f(3), \ldots$ is an arithmetic sequence.

10.51 In an arithmetic sequence, the pth term is q and the qth term is p, where $p \neq q$. Find the $(p+q)^{\text{th}}$ term.

10.52

(a) Find the arithmetic sequence with first term 1 and common difference not equal to 0 such that the second, tenth, and thirty-fourth terms are the first three terms of a geometric sequence.

(b) The fourth term in the geometric sequence in part (a) appears as the nth term in the arithmetic sequence in part (a). Find the value of n.

(Source: CEMC)

10.53 Find the sum of the series $1 + \frac{1}{2} + \frac{1}{10} + \frac{1}{20} + \frac{1}{100} + \cdots$, where we alternately multiply by 1/2 and 1/5 to get successive terms.

10.54 Find a positive integer solution to the equation $\dfrac{1+3+5+\cdots+(2n-1)}{2+4+6+\cdots+2n} = \dfrac{115}{116}$. *(Source: AHSME)*

10.55 Consider the ten numbers $ar, ar^2, ar^3, \ldots, ar^{10}$. If their sum is 18 and the sum of their reciprocals is 6, determine their product. *(Source: COMC)*

10.56 Show that if $|r| < 1$, then the product of $a + ar + ar^2 + \cdots$ and $a - ar + ar^2 - \cdots$ is equal to $a^2 + a^2r^2 + a^2r^4 + \cdots$.

10.57 The sum of an infinite geometric series with common ratio r such that $|r| < 1$ is 15, and the sum of the squares of the terms of this series is 45. Find the first term of this series. *(Source: AHSME)*

10.58 If $\displaystyle\sum_{k=1}^{n} g(k) = \dfrac{n}{3n-2}$, find a formula for $g(k)$. *(Source: ARML)*

10.59 In each of the following parts, state whether or not the two summations are equal, and explain why or why not the sums are equal.

(a) $\displaystyle\sum_{i=-4}^{19}(2-i^2)$ and $\displaystyle\sum_{k=-2}^{21}(-2+4k-k^2)$ (c) $\displaystyle\sum_{j=1}^{20}\dfrac{j}{22-j}$ and $\displaystyle\sum_{k=1}^{20}\dfrac{21-k}{k+1}$

(b) $\displaystyle\sum_{k=0}^{\infty}\dfrac{1}{2^k+4}$ and $\displaystyle\sum_{k=2}^{\infty}\dfrac{1}{2^k+1}$

10.60 Show that if c is a positive constant, then $\frac{1}{c-x} = \frac{1}{c} + \frac{x}{c^2} + \frac{x^2}{c^3} + \cdots$ for all x such that $|x/c| < 1$.

10.61 Evaluate each of the following:

(a) $\displaystyle\sum_{i=1}^{4}\sum_{j=1}^{4}(i-j)^2$

(b) $\displaystyle\sum_{i=1}^{5}\sum_{j=-3}^{5}j\cdot 2^i$

10.62 Evaluate each of the following:

(a) $\displaystyle\sum_{j=0}^{5}\sum_{k=0}^{j}3$

(b) $\displaystyle\sum_{i=1}^{8}\sum_{j=1}^{2i}\frac{2j}{i}$

10.63 Find a closed form for $\displaystyle\sum_{j=0}^{k}\sum_{i=0}^{k}2j.$

10.64 A **harmonic sequence** (or **harmonic progression**) is a sequence in which the reciprocals of the terms form an arithmetic sequence. A **harmonic series** is a sum of the numbers in a harmonic sequence.

(a) Let S_n denote the sum of the first n terms of a harmonic sequence whose first two terms are 3 and 4. Compute S_4.

(b) The first term of a harmonic sequence is 1/3, and the fifth term is 1. Find the second, third, and fourth terms.

(c) Find all real numbers x such that $x+4, x+1, x$ is a harmonic progression.

(d) Let a, b, and c be positive real numbers. Prove that if a, b, c is a harmonic sequence, then so is $a/(b+c), b/(a+c), c/(a+b)$.

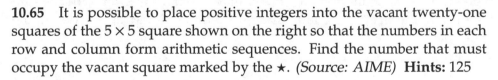

Challenge Problems

10.65 It is possible to place positive integers into the vacant twenty-one squares of the 5×5 square shown on the right so that the numbers in each row and column form arithmetic sequences. Find the number that must occupy the vacant square marked by the $\star$. *(Source: AIME)* **Hints:** 125

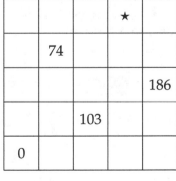

10.66 Write $x_1(x_1+2x_2)(x_1+2x_2+3x_3)\cdots(x_1+2x_2+3x_3+\cdots+nx_n)$ using product and summation notation.

10.67 A set of consecutive positive integers beginning with 1 is written on a blackboard. One number is erased. The average (arithmetic mean) of the remaining numbers is $35\frac{7}{17}$. What number was erased? *(Source: AHSME)* **Hints:** 181

10.68 If a and b are the roots of $11x^2-4x-2=0$, then compute the product

$$(1+a+a^2+a^3+\cdots)(1+b+b^2+b^3+\cdots).$$

(Source: ARML) **Hints:** 235

10.69   Evaluate $\displaystyle\sum_{i=1}^{5}\sum_{j=1}^{i}\sum_{k=1}^{j}(1)$.

10.70   Let a, b, and c be positive real numbers. Prove that if a^2, b^2, c^2 is an arithmetic sequence, then so is $1/(b + c)$, $1/(a + c)$, $1/(a + b)$.

10.71   Let $p(x_1, x_2, x_3, \ldots, x_m) = \displaystyle\prod_{n=1}^{m}\sum_{k=1}^{n} x_k$. Find the sum of the coefficients of the expanded form of the m-variable polynomial p.

10.72   The roots of the equation $x^5 - 80x^4 + Px^3 + Qx^2 + Rx + S = 0$ are in geometric progression. The sum of their reciprocals is 5. Compute $|S|$. **Hints:** 243

10.73   Consider the triangular array of numbers formed when the numbers 0, 1, 2, 3, $\ldots$ are placed along the sides and interior numbers are obtained by adding the two adjacent numbers in the previous row. Rows 1 through 6 are shown below.

$$
\begin{array}{ccccccccccc}
 & & & & & 0 & & & & & \\
 & & & & 1 & & 1 & & & & \\
 & & & 2 & & 2 & & 2 & & & \\
 & & 3 & & 4 & & 4 & & 3 & & \\
 & 4 & & 7 & & 8 & & 7 & & 4 & \\
5 & & 11 & & 15 & & 15 & & 11 & & 5 \\
\end{array}
$$

Let $f(n)$ denote the sum of the numbers in row n. What is the remainder when $f(100)$ is divided by 100? *(Source: AHSME)* **Hints:** 130, 218

10.74   If the integer k is added to each of the numbers 36, 300, and 596, one obtains the squares of three consecutive terms of an arithmetic sequence. Find k. *(Source: AIME)* **Hints:** 237

10.75 S_1 is the sum of an infinite geometric series with first term a and common ratio r. S_2 is the sum of another geometric series with first term a^2 and common ratio r^2. Similarly, for all $n \geq 2$, let S_n be the sum of the infinite geometric series with first term a^n and common ratio r^n. If $|r| < 1$ and $|a| > 1$, determine the sum of the infinite series

$$\frac{1}{S_1} + \frac{1}{S_2} + \frac{1}{S_3} + \cdots$$

in terms of a and r. *(Source: CIMC)* **Hints:** 110

10.76   Define $p = \displaystyle\sum_{k=1}^{\infty}\frac{1}{k^2}$ and $q = \displaystyle\sum_{k=1}^{\infty}\frac{1}{k^3}$. Write $\displaystyle\sum_{j=1}^{\infty}\sum_{k=1}^{\infty}\frac{1}{(j+k)^3}$ in terms of p and q. *(Source: Mandelbrot)* **Hints:** 262, 13

10.77   Set A consists of m consecutive integers whose sum is $2m$, and set B consists of $2m$ consecutive integers whose sum is m. The absolute value of the difference between the greatest element of A and the greatest element of B is 99. Find m. *(Source: AIME)*

10.78   An infinite geometric series has sum 2005. A new series, obtained by squaring each term of the original series, has sum 10 times the sum of the original series. Find the common ratio of the original series. *(Source: AIME)*

10.79★ Find all pairs of integers n and k, with $2 < k < n$, such that the binomial coefficients

$$\binom{n}{k-1}, \quad \binom{n}{k}, \quad \binom{n}{k+1}$$

form an increasing arithmetic sequence. *(Source: USAMTS)* **Hints:** 23, 150

10.80★ If a, b, and c form an arithmetic progression, and

$$a = x^2 + xy + y^2,$$
$$b = x^2 + xz + z^2,$$
$$c = y^2 + yz + z^2,$$

where $x + y + z \neq 0$, prove that x, y, and z also form an arithmetic progression. **Hints:** 194

10.81★ A sequence of positive integers with $a_1 = 1$ and $a_9 + a_{10} = 646$ is formed so that the first three terms are in geometric progression, the second, third, and fourth terms are in arithmetic progression, and, in general, for all $j \geq 1$, the terms $a_{2j-1}, a_{2j}, a_{2j+1}$ are in geometric progression, and the terms $a_{2j}, a_{2j+1},$ and a_{2j+2} are in arithmetic progression. Find the greatest term in this sequence that is less than 1000. *(Source: AIME)* **Hints:** 159, 333

10.82★ In an increasing sequence of four positive integers, the first three terms form an arithmetic progression, the last three terms form a geometric progression, and the first and fourth terms differ by 30. Find the sum of the four terms. *(Source: AIME)* **Hints:** 295

10.83 Let $|x| < 1$. Express $(1+x)(1+x^2)(1+x^4)(1+x^8)\cdots$ as $f(x)/g(x)$, where $f(x)$ and $g(x)$ are polynomials with finite degree. **Hints:** 26

10.84★ For $0 < x < 1$, let $f(x) = (1+x)(1+x^4)(1+x^{16})(1+x^{64})(1+x^{256})\cdots$. Compute: $f^{-1}\left(\dfrac{8}{5f\left(\frac{3}{8}\right)}\right)$.

(Source: ARML) **Hints:** 228

10.85★ For $0 < x < 1$, express $\displaystyle\sum_{n=0}^{\infty} \frac{x^{2^n}}{1 - x^{2^{n+1}}}$ as $f(x)/g(x)$, where $f(x)$ and $g(x)$ are polynomials with finite degree. *(Source: Putnam)* **Hints:** 53

Extra! Mathematicians have closely studied a great many different sequences, such as

$$1, 1, 2, 5, 14, 42, 132, \ldots.$$

If you come across a sequence of numbers that you'd like to learn more about, you can visit **The On-Line Encyclopedia of Integer Sequences**, which is a collaborative project to collect information about sequences. A link to The On-Line Encyclopedia of Integer Sequences can be found on the Links page cited on page vi. In case you're wondering, the sequence 1, 1, 2, 5, 14, 42, 132, ... consists of the **Catalan numbers**, which you can learn more about in Art of Problem Solving's *Intermediate Counting & Probability*.

Science is nothing but the finding of analogy, identity, in the most remote parts. – Ralph Waldo Emerson

CHAPTER **11**

Identities, Manipulations, and Induction

You've seen many identities already. The difference of squares factorization

$$x^2 - y^2 = (x + y)(x - y)$$

is an identity because it's true for any ordered pair (x, y). On the other hand, the equation

$$x^2 - y^2 = 0$$

is not an identity, because it is not true for all values of x and y.

> **Definition:** An **identity** is an equation that is true for all values of all variables for which the equation is defined.

We've already seen how useful many algebraic identities can be in our discussion of factorization. In this chapter, we develop more methods for finding algebraic identities. Most of these identities themselves are not terribly important (we'll highlight the ones that are), but the procedures we use to develop them are.

11.1 Brute Force

We can prove the difference of squares identity, $x^2 - y^2 = (x - y)(x + y)$, by simply multiplying out the right side:

$$(x - y)(x + y) = x(x + y) - y(x + y) = x^2 + xy - xy - y^2 = x^2 - y^2.$$

Many identities can be proved similarly, though often the necessary algebra is more complicated. We call this approach **brute force**.

Problems

Problem 11.1: Show that $(a + b + c)(a^2 + b^2 + c^2 - ab - bc - ca) = a^3 + b^3 + c^3 - 3abc$.

Problem 11.2:

(a) Prove that $(a_1^2 + a_2^2)(b_1^2 + b_2^2) - (a_1 b_1 + a_2 b_2)^2 = (a_1 b_2 - a_2 b_1)^2$.

(b) Prove that $(a_1^2 + a_2^2)(b_1^2 + b_2^2) \geq (a_1 b_1 + a_2 b_2)^2$ for all real numbers a_1, a_2, b_1, and b_2.

Problem 11.3:

(a) Find an identity that relates the expressions $x^2 + y^2 + z^2$, $\frac{1}{x^2} + \frac{1}{y^2} + \frac{1}{z^2}$, and $\left(\frac{x}{y} + \frac{y}{x}\right)^2 + \left(\frac{y}{z} + \frac{z}{y}\right)^2 + \left(\frac{z}{x} + \frac{x}{z}\right)^2$.

(b) Given that $x^2 + y^2 + z^2 = 6$ and $\left(\frac{x}{y} + \frac{y}{x}\right)^2 + \left(\frac{y}{z} + \frac{z}{y}\right)^2 + \left(\frac{z}{x} + \frac{x}{z}\right)^2 = 16.5$, what is $\frac{1}{x^2} + \frac{1}{y^2} + \frac{1}{z^2}$? *(Source: Duke Math Meet)*

Problem 11.4: In this problem, we evaluate $\sqrt{50 \cdot 51 \cdot 52 \cdot 53 + 1}$ without a calculator. *(Source: MATH-COUNTS)*

(a) Experiment. Evaluate $\sqrt{1 \cdot 2 \cdot 3 \cdot 4 + 1}$, $\sqrt{2 \cdot 3 \cdot 4 \cdot 5 + 1}$, $\sqrt{3 \cdot 4 \cdot 5 \cdot 6 + 1}$, and $\sqrt{4 \cdot 5 \cdot 6 \cdot 7 + 1}$. Can you guess the answer to the problem?

(b) Let $x = 50$ and write the expression we wish to evaluate in terms of x. Let the expression inside the radical be $f(x)$. What is $\deg f$?

(c) If there is a polynomial $g(x)$ such that $(g(x))^2 = f(x)$, then what is $\deg g$?

(d) Find $\sqrt{50 \cdot 51 \cdot 52 \cdot 53 + 1}$ without a calculator.

Problem 11.1: Show that $(a + b + c)(a^2 + b^2 + c^2 - ab - bc - ca) = a^3 + b^3 + c^3 - 3abc$.

Solution for Problem 11.1: We have

$$(a + b + c)(a^2 + b^2 + c^2 - ab - bc - ca) = a(a^2 + b^2 + c^2 - ab - bc - ca)$$
$$+ b(a^2 + b^2 + c^2 - ab - bc - ca)$$
$$+ c(a^2 + b^2 + c^2 - ab - bc - ca)$$
$$= a^3 + ab^2 + ac^2 - a^2 b - abc - a^2 c$$
$$+ a^2 b + b^3 + bc^2 - ab^2 - b^2 c - abc$$
$$+ a^2 c + b^2 c + c^3 - abc - bc^2 - ac^2$$
$$= a^3 + b^3 + c^3 - 3abc.$$

Make sure you see how all the other terms cancel out. $\square$

Our work in Problem 11.1 doesn't require any particular insight; it's simply careful algebraic manipulation.

> **Concept:** When performing complicated algebra, stay organized and write neatly.

Problem 11.2:

(a) Prove that $(a_1^2 + a_2^2)(b_1^2 + b_2^2) - (a_1b_1 + a_2b_2)^2 = (a_1b_2 - a_2b_1)^2$.

(b) Prove that $(a_1^2 + a_2^2)(b_1^2 + b_2^2) \ge (a_1b_1 + a_2b_2)^2$ for all real numbers $a_1, a_2, b_1,$ and b_2.

Solution for Problem 11.2:

(a) We have

$$(a_1^2 + a_2^2)(b_1^2 + b_2^2) - (a_1b_1 + a_2b_2)^2 = a_1^2b_1^2 + a_1^2b_2^2 + a_2^2b_1^2 + a_2^2b_2^2 - a_1^2b_1^2 - 2a_1a_2b_1b_2 - a_2^2b_2^2$$
$$= a_1^2b_2^2 - 2a_1a_2b_1b_2 + a_2^2b_1^2$$
$$= (a_1b_2 - a_2b_1)^2.$$

This is yet another example of how useful it is to be able to recognize the square of a binomial.

(b) Our identity from part (a) gives us

$$(a_1^2 + a_2^2)(b_1^2 + b_2^2) - (a_1b_1 + a_2b_2)^2 = (a_1b_2 - a_2b_1)^2.$$

Because $a_1b_2 - a_2b_1$ is a real number, we have $(a_1b_2 - a_2b_1)^2 \ge 0$. Combining this with the equation above gives us

$$(a_1^2 + a_2^2)(b_1^2 + b_2^2) - (a_1b_1 + a_2b_2)^2 \ge 0.$$

Adding $(a_1b_1 + a_2b_2)^2$ to both sides of this inequality gives us the desired

$$(a_1^2 + a_2^2)(b_1^2 + b_2^2) \ge (a_1b_1 + a_2b_2)^2.$$

□

The identity we proved in part (a) of Problem 11.2 is a special case of **Lagrange's Identity**. The inequality we proved in part (b) is a similar special case of the **Cauchy-Schwarz Inequality**, which we discuss in much greater detail in Chapter 12.

We continue with a couple problems in which we must discover an identity to evaluate expressions.

Problem 11.3: Given that $x^2 + y^2 + z^2 = 6$ and $\left(\dfrac{x}{y} + \dfrac{y}{x}\right)^2 + \left(\dfrac{y}{z} + \dfrac{z}{y}\right)^2 + \left(\dfrac{z}{x} + \dfrac{x}{z}\right)^2 = 16.5$, what is $\dfrac{1}{x^2} + \dfrac{1}{y^2} + \dfrac{1}{z^2}$?
(Source: Duke Math Meet)

Solution for Problem 11.3: We start by expanding the left side of the second equation, since doing so will produce squares and reciprocals of squares. This gives us

$$\frac{x^2}{y^2} + 2 + \frac{y^2}{x^2} + \frac{y^2}{z^2} + 2 + \frac{z^2}{y^2} + \frac{z^2}{x^2} + 2 + \frac{x^2}{z^2} = 16.5,$$

so

$$\frac{x^2}{y^2} + \frac{y^2}{x^2} + \frac{y^2}{z^2} + \frac{z^2}{y^2} + \frac{z^2}{x^2} + \frac{x^2}{z^2} = 10.5.$$

Our hope is to relate this equation to the first given equation and to the expression we must evaluate. The left side looks like some of what we expect from expanding the product of $x^2 + y^2 + z^2$ (the left side of the first given equation) and $\frac{1}{x^2} + \frac{1}{y^2} + \frac{1}{z^2}$ (the expression we're trying to evaluate). So, we take a look at that product:

$$(x^2 + y^2 + z^2)\left(\frac{1}{x^2} + \frac{1}{y^2} + \frac{1}{z^2}\right) = 3 + \frac{x^2}{y^2} + \frac{y^2}{x^2} + \frac{y^2}{z^2} + \frac{z^2}{y^2} + \frac{z^2}{x^2} + \frac{x^2}{z^2}.$$

Aha—we recognize many of the expressions in this identity! We know that $x^2 + y^2 + z^2 = 6$, and the expression on the right equals $3 + 10.5 = 13.5$. Substituting these into our new identity, we have

$$6\left(\frac{1}{x^2} + \frac{1}{y^2} + \frac{1}{z^2}\right) = 13.5,$$

so $\dfrac{1}{x^2} + \dfrac{1}{y^2} + \dfrac{1}{z^2} = 2.25.$ $\square$

Recall from our discussion of Vieta's Formulas that expressions like $x^2 + y^2 + z^2$ and $\frac{1}{x^2} + \frac{1}{y^2} + \frac{1}{z^2}$ are called **symmetric**, because exchanging any two of the variables in the expression does not change the expression. Our solution to Problem 11.3 is an example that:

> **Concept:** We can often relate symmetric expressions with identities.

As we saw in Problem 11.3, we'll sometimes have to develop identities as we need them. Often these identities won't be immediately obvious, so we'll have to experiment a little to find them, as we did when we multiplied $x^2 + y^2 + z^2$ and $\frac{1}{x^2} + \frac{1}{y^2} + \frac{1}{z^2}$.

> **Concept:** If it's not immediately obvious how to solve a problem, don't just sit and stare at it. Experiment!

Problem 11.4: Evaluate $\sqrt{50 \cdot 51 \cdot 52 \cdot 53 + 1}$ without a calculator. *(Source: MATHCOUNTS)*

Solution for Problem 11.4: Multiplying that out doesn't look like much fun. Then, finding the square root of the result looks like even less fun. Let's find a more clever approach. We start by making the problem simpler by evaluating similar expressions for smaller numbers.

> **Concept:** When asked to evaluate an expression involving large numbers, you can get a feel for the problem by evaluating similar expressions with small numbers.

We find:

$$\sqrt{0 \cdot 1 \cdot 2 \cdot 3 + 1} = 1,$$
$$\sqrt{1 \cdot 2 \cdot 3 \cdot 4 + 1} = 5,$$
$$\sqrt{2 \cdot 3 \cdot 4 \cdot 5 + 1} = 11,$$
$$\sqrt{3 \cdot 4 \cdot 5 \cdot 6 + 1} = 19.$$

There's a subtle pattern here. The second result is 4 more than the first, the third is 6 more than the second, and the fourth is 8 more than the third. However, it's not clear how to prove this pattern continues. To see if it does, we replace 50 with x in $\sqrt{50 \cdot 51 \cdot 52 \cdot 53 + 1}$, and see if we can find the square root of the resulting expression:

$$\sqrt{x(x+1)(x+2)(x+3) + 1} = \sqrt{(x^2 + x)(x^2 + 5x + 6) + 1} = \sqrt{x^4 + 6x^3 + 11x^2 + 6x + 1}.$$

Our examples make us suspect that this quartic is the perfect square of some polynomial. If it is, then that polynomial must be a quadratic. Because the first and last terms of the quartic are x^4 and 1, and all the coefficients of the quartic are positive, we guess that the square root is of the form $x^2 + ax + 1$ for some positive value of a. Rather than square $x^2 + ax + 1$, we note that the x^3 term of the expansion of $(x^2 + ax + 1)(x^2 + ax + 1)$ is $2ax^3$, so we guess that $a = 3$ to make this term equal to $6x^3$. Testing, we find that

$$(x^2 + 3x + 1)^2 = x^4 + 6x^3 + 11x^2 + 6x + 1,$$

so we have

$$\sqrt{x(x+1)(x+2)(x+3) + 1} = \sqrt{x^4 + 6x^3 + 11x^2 + 6x + 1} = x^2 + 3x + 1.$$

Letting $x = 50$, we find $\sqrt{50 \cdot 51 \cdot 52 \cdot 53 + 1} = 50^2 + 3 \cdot 50 + 1 = 2651.$ □

You might be wondering if it's possible to perform expansions like

$$(x^2 + y^2 + z^2)\left(\frac{1}{x^2} + \frac{1}{y^2} + \frac{1}{z^2}\right) = 3 + \frac{x^2}{y^2} + \frac{y^2}{x^2} + \frac{y^2}{z^2} + \frac{z^2}{y^2} + \frac{z^2}{x^2} + \frac{x^2}{z^2}$$

and

$$(x^2 + 3x + 1)(x^2 + 3x + 1) = x^4 + 6x^3 + 11x^2 + 6x + 1$$

in your head without filling up a lot of paper with intermediate steps. It is possible, but you have to be careful.

For example, consider

$$(x^2 + y^2 + z^2)\left(\frac{1}{x^2} + \frac{1}{y^2} + \frac{1}{z^2}\right) = 3 + \frac{x^2}{y^2} + \frac{y^2}{x^2} + \frac{y^2}{z^2} + \frac{z^2}{y^2} + \frac{z^2}{x^2} + \frac{x^2}{z^2}.$$

When expanding the left side, we are multiplying two expressions with 3 terms each. So, the expansion must have $3 \times 3 = 9$ terms. Because each term in the expansion is the product of one term of $x^2 + y^2 + z^2$ and one term of $\frac{1}{x^2} + \frac{1}{y^2} + \frac{1}{z^2}$, we see that each of the possible ratios of one of x^2, y^2, or z^2 to one of x^2, y^2, or z^2 is formed in the expansion. Three of these ratios equal 1, which gives us the 3 on the right side. The remaining possible ratios give us the other terms. We can quickly test our expansion by checking if the identity holds for $x = y = z = 1$. We find $(1 + 1 + 1)(1 + 1 + 1) = 3 + 1 + 1 + 1 + 1 + 1 + 1$, which is true.

> **Concept:** Because an identity must be true for all permissible values of the variables in the identity, we can test an identity we develop by substituting convenient values for the variables.

To expand $(x^2 + 3x + 1)(x^2 + 3x + 1)$, we focus on each power of x in the expansion in turn.

- x^4. There's only one way to form x^4: $(x^2)(x^2) = x^4$.

- x^3. There are two ways to form x^3: $(x^2)(3x) + (3x)(x^2) = 6x^3$.

- x^2. There are three ways to form x^2: $(x^2)(1) + (3x)(3x) + (1)(x^2) = 11x^2$.

- x. There are two ways to form x: $(3x)(1) + (1)(3x) = 6x$.

- Constant. There's only one way to form a constant: $(1)(1) = 1$.

Notice that across all five of these cases, we form a total of 9 different products of terms in the two expressions we are multiplying. Combining all five cases gives us

$$(x^2 + 3x + 1)(x^2 + 3x + 1) = x^4 + 6x^3 + 11x^2 + 6x + 1.$$

We can quickly test this by letting $x = 1$, which gives us $(5)(5)$ on the left and $1 + 6 + 11 + 6 + 1$ on the right. These both equal 25, so the expansion passes our quick test.

Exercises

11.1.1 Show that $a^2(b - c) + b^2(c - a) + c^2(a - b) = bc(b - c) + ca(c - a) + ab(a - b)$ for all a, b, and c.

11.1.2 Show that $2(x - y)^2 - 3x^2 + 3xy = (y - x)(x + 2y)$ for all values of x and y.

11.1.3 Consider the following equations:

$$1^2 + 2^2 + 2^2 = 3^2,$$
$$2^2 + 3^2 + 6^2 = 7^2,$$
$$3^2 + 4^2 + 12^2 = 13^2,$$
$$4^2 + 5^2 + 20^2 = 21^2.$$

State and prove a one-variable identity that is suggested by these examples.

11.1.4 If $xyz = x + y + z = \dfrac{1}{x} + \dfrac{1}{y} + \dfrac{1}{z} = 8$, then find $\left(x - \dfrac{1}{y}\right)\left(y - \dfrac{1}{z}\right)\left(z - \dfrac{1}{x}\right)$.

11.2 Ratios

In this section, we explore some very handy manipulations involving ratios.

Problems

Problem 11.5: Let $\dfrac{a}{b} = \dfrac{c}{d}$.

(a) Let $\dfrac{a}{b} = k$. Use this to show that $\dfrac{a}{b} = \dfrac{c}{d} = \dfrac{a+c}{b+d} = \dfrac{a-c}{b-d}$ for those values of b and d for which the denominators are nonzero.

(b) Show that $\dfrac{a+b}{a-b} = \dfrac{c+d}{c-d}$ if $a \ne b$ and $c \ne d$.

Problem 11.6: Find the numerical value of k for which $\dfrac{7}{x+y} = \dfrac{k}{x+z} = \dfrac{11}{z-y}$ for all x, y, and z such that the denominators of the fractions are nonzero. *(Source: ARML)*

Problem 11.7: In this problem, we show that if $\dfrac{x+y}{a^2+b^2} = \dfrac{y+z}{b^2+c^2} = \dfrac{x+z}{a^2+c^2}$ and $abc \ne 0$, then we have $\dfrac{x}{a^2} = \dfrac{y}{b^2} = \dfrac{z}{c^2}$.

(a) Show that $\dfrac{x+y}{a^2+b^2} = \dfrac{x+y+z}{a^2+b^2+c^2}$.

(b) Show that $\dfrac{x}{a^2} = \dfrac{y}{b^2} = \dfrac{z}{c^2}$.

Problem 11.8: If $\dfrac{a}{b} = \dfrac{c}{d}$, and we have $\dfrac{a}{b} > 1$ and $b > d > 0$, show that $a^2 + d^2 > b^2 + c^2$.

Problem 11.5: Let $\dfrac{a}{b} = \dfrac{c}{d}$.

(a) Show that $\dfrac{a}{b} = \dfrac{c}{d} = \dfrac{a+c}{b+d} = \dfrac{a-c}{b-d}$ for those values of b and d for which the denominators are nonzero.

(b) Show that $\dfrac{a+b}{a-b} = \dfrac{c+d}{c-d}$ if $a \ne b$ and $c \ne d$.

Solution for Problem 11.5:

(a) *Solution 1: Set the given equal ratios equal to a variable.* We'd like to get rid of the fractions. One way to do so is to let

$$k = \frac{a}{b} = \frac{c}{d}.$$

This gives us $a = bk$ and $c = dk$, so we have

$$\frac{a+c}{b+d} = \frac{bk+dk}{b+d} = \frac{k(b+d)}{b+d} = k \qquad \text{and} \qquad \frac{a-c}{b-d} = \frac{bk-dk}{b-d} = \frac{k(b-d)}{b-d} = k.$$

Therefore, we have $\dfrac{a}{b} = \dfrac{c}{d} = \dfrac{a+c}{b+d} = \dfrac{a-c}{b-d}$.

> **Concept:** When given equal ratios, setting the ratios equal to a variable helps eliminate the fractions.

Solution 2: Brute force. We also could have used brute force to show that $(a+c)/(b+d) = a/b$. It's not so clear how to start with $a/b = c/d$ and prove $(a+c)/(b+d) = a/b$, so we instead work backwards. We start with $(a+c)/(b+d) = a/b$ and try to work from here to an equation we know to be true.

We start by multiplying both sides of this equation by $b(b+d)$ to get rid of the fractions. This gives us $b(a+c) = a(b+d)$, so $ab + bc = ab + ad$, which means $bc = ad$. Dividing both sides of this by bd gives us $c/d = a/b$, which we know to be true. This gives us a path to prove $(a+c)/(b+d) = a/b$ starting from $a/b = c/d$. We simply reverse our "working backwards" steps:

Proof: We multiply $a/b = c/d$ by bd to give $ad = bc$. Adding ab to both sides gives the equation ✓ $ad + ab = bc + ab$. Factoring gives $a(b+d) = b(a+c)$. Because b and $b+d$ are nonzero, we can divide both sides by $b(b+d)$ to give the desired

$$\frac{a}{b} = \frac{a+c}{b+d}.$$

We can tackle $a/b = (a-c)/(b-d)$ in much the same way.

Our key step in finding this brute force solution was working backwards.

Concept: Working backwards from a relationship you wish to prove is a powerful problem-solving method.

WARNING!! If you do work backwards to find a solution, make sure you write your solution "forwards" as we've done in the Proof box above. This will make your work easier to read, and help you catch any mistakes you might have made.

Also, some steps are not reversible. For example, it's true that if $x = y$, then $x^2 = y^2$. But we can't reverse this and say that "If $x^2 = y^2$, then $x = y$." The step going from $x = y$ to $x^2 = y^2$ is not reversible; if we know that $x^2 = y^2$, we cannot deduce that $x = y$ because it's possible that $x = -y$ instead. (Look back to Problem 8.27 on page 254 to see an example in which writing our solution forwards saves us from a subtle error that might occur when we work backwards.)

(b) *Solution 1: Set the ratios equal to k.* Letting $a/b = c/d = k$ as before, we have $a = bk$ and $c = dk$, so

$$\frac{a+b}{a-b} = \frac{bk+b}{bk-b} = \frac{k+1}{k-1} \quad \text{and} \quad \frac{c+d}{c-d} = \frac{dk+d}{dk-d} = \frac{k+1}{k-1},$$

so $(a+b)/(a-b) = (c+d)/(c-d)$.

Solution 2: Clever manipulation. We wish to prove an equation involving $a+b$ and $c+d$. Adding 1 to $a/b = c/d$ gives us

$$\frac{a+b}{b} = \frac{c+d}{d}.$$

Similarly, we need $a-b$ and $c-d$, so we subtract 1 from $a/b = c/d$, which gives

$$\frac{a-b}{b} = \frac{c-d}{d}.$$

Dividing $\frac{a+b}{b} = \frac{c+d}{d}$ by $\frac{a-b}{b} = \frac{c-d}{d}$ gives us the desired $\frac{a+b}{a-b} = \frac{c+d}{c-d}$. $\square$

This problem gives us examples of the following useful ratio manipulations:

> **Important:** Suppose $a/b = c/d$. Then, for any constant k for which $b + kd \neq 0$, we have
> $$\frac{a}{b} = \frac{c}{d} = \frac{a + kc}{b + kd}.$$
> For any constant k for which $a - kb$ and $c - kd$ are nonzero, we have
> $$\frac{a + kb}{a - kb} = \frac{c + kd}{c - kd}.$$

You'll have a chance to prove these as Exercises. We can also extend the fact that $\frac{a}{b} = \frac{c}{d}$ implies $\frac{a+c}{b+d}$ to any number of equal ratios:

> **Important:** If $\frac{x_1}{y_1} = \frac{x_2}{y_2} = \cdots = \frac{x_n}{y_n}$ and $y_1 + y_2 + \cdots + y_n \neq 0$, then
> $$\frac{x_1}{y_1} = \frac{x_2}{y_2} = \cdots = \frac{x_n}{y_n} = \frac{x_1 + x_2 + \cdots + x_n}{y_1 + y_2 + \cdots + y_n}.$$

You'll have a chance to prove this as an Exercise.

Using these relationships allows us to combine equal ratios to produce other ratio equations. Let's take a look at how.

> **Problem 11.6:** Find the numerical value of k for which $\dfrac{7}{x + y} = \dfrac{k}{x + z} = \dfrac{11}{z - y}$ for all x, y, and z such that the denominators of the fractions are nonzero. *(Source: ARML)*

Solution for Problem 11.6: We notice that the sum of the denominators of the left and right fractions is the denominator of the middle fraction. So, we can use the result of the previous problem to determine that
$$\frac{7}{x + y} = \frac{11}{z - y} = \frac{7 + 11}{x + y + z - y} = \frac{18}{x + z}.$$

Therefore, we have $k = 18$. $\square$

> **Problem 11.7:** Show that if $\dfrac{x + y}{a^2 + b^2} = \dfrac{y + z}{b^2 + c^2} = \dfrac{x + z}{a^2 + c^2}$ and $abc \neq 0$, then $\dfrac{x}{a^2} = \dfrac{y}{b^2} = \dfrac{z}{c^2}$.

Solution for Problem 11.7: Brute-forcing this doesn't look like much fun. Let's try applying the identities we just learned. Because
$$\frac{x + y}{a^2 + b^2} = \frac{y + z}{b^2 + c^2},$$
we can apply the relationship we just proved to find
$$\frac{x + y}{a^2 + b^2} = \frac{y + z}{b^2 + c^2} = \frac{(x + y) + (y + z)}{(a^2 + b^2) + (b^2 + c^2)} = \frac{x + 2y + z}{a^2 + 2b^2 + c^2}.$$

Hmmm.... These ratios also equal $(x + z)/(a^2 + c^2)$, so we have

$$\frac{x+y}{a^2+b^2} = \frac{y+z}{b^2+c^2} = \frac{x+z}{a^2+c^2} = \frac{x+2y+z}{a^2+2b^2+c^2} = \frac{(x+z)+(x+2y+z)}{(a^2+c^2)+(a^2+2b^2+c^2)} = \frac{2x+2y+2z}{2a^2+2b^2+2c^2}.$$

So, we have

$$\frac{x+y}{a^2+b^2} = \frac{y+z}{b^2+c^2} = \frac{x+z}{a^2+c^2} = \frac{x+y+z}{a^2+b^2+c^2}.$$

In part (a) of Problem 11.5, we showed that if $\frac{p}{q} = \frac{r}{s}$, then we have $\frac{p}{q} = \frac{r}{s} = \frac{p-r}{q-s}$. Applying this to the first and last equal ratios in our equality chain above, we have

$$\frac{x+y+z}{a^2+b^2+c^2} = \frac{x+y}{a^2+b^2} = \frac{x+y+z-(x+y)}{a^2+b^2+c^2-(a^2+b^2)} = \frac{z}{c^2}.$$

Similarly, we can show that $\frac{x+y+z}{a^2+b^2+c^2}$ equals both $\frac{x}{a^2}$ and $\frac{y}{b^2}$, so $\frac{x+y+z}{a^2+b^2+c^2} = \frac{x}{a^2} = \frac{y}{b^2} = \frac{z}{c^2}$. $\square$

We can also use our ratio manipulation tactics on inequalities.

Problem 11.8: If $\frac{a}{b} = \frac{c}{d}$, and we have $\frac{a}{b} > 1$ and $b > d > 0$, show that $a^2 + d^2 > b^2 + c^2$.

Solution for Problem 11.8: We try letting $k = a/b = c/d$, as before. This gives us $a = bk$ and $c = dk$. Now, we work backwards from the inequality we wish to prove. Letting $a = bk$ and $c = dk$ in $a^2 + d^2 > b^2 + c^2$, we must now show that $b^2k^2 + d^2 > b^2 + d^2k^2$. We subtract b^2 and d^2 to give

$$b^2k^2 - b^2 > d^2k^2 - d^2.$$

Factoring then gives us $b^2(k^2 - 1) > d^2(k^2 - 1)$ to prove. We're told that $b > d > 0$, so $b^2 > d^2$. (Make sure you see why it's important that b and d are positive for this step!) We also are given that $a/b > 1$, so $k > 1$. This means that $k^2 - 1$ is positive, so multiplying $b > d$ by $k^2 - 1$ gives us the desired $b^2(k^2 - 1) > d^2(k^2 - 1)$.

Our work above is not a great way to write a solution, but it is a great way to *find* a solution. Specifically, we worked backwards from $a^2 + d^2 > b^2 + c^2$ to produce an inequality, $b^2(k^2 - 1) > d^2(k^2 - 1)$, that we know how to prove from the given information. So, we figured out how to solve the problem using a combination of working forwards and working backwards. As our last step, we write our solution entirely forwards, starting from the given information and ending at the desired inequality.

Proof: Let $k = a/b = c/d$. Because $a/b > 1$, we have $k > 1$, so $k^2 - 1 > 0$. Because $b > d > 0$, we have $b^2 > d^2$. Multiplying this inequality by $k^2 - 1$ gives $b^2(k^2 - 1) > d^2(k^2 - 1)$. Since $k = a/b = c/d$, we have

$$b^2(k^2 - 1) = b^2\left(\frac{a^2}{b^2} - 1\right) = a^2 - b^2.$$

Similarly, we have $d^2(k^2 - 1) = c^2 - d^2$, so our inequality is now $a^2 - b^2 > c^2 - d^2$. Adding $b^2 + d^2$ to both sides gives the desired $a^2 + d^2 > b^2 + c^2$.

$\square$

Exercises

11.2.1 Suppose that $\dfrac{a}{b} = \dfrac{c}{d}$.

(a) Show that $\dfrac{a}{b} = \dfrac{c}{d} = \dfrac{a + kc}{b + kd}$ for any constant k such that $b + kd \neq 0$.

(b) Show that $\dfrac{a + kb}{a - kb} = \dfrac{c + kd}{c - kd}$ for any constant k such that $a - kb$ and $c - kd$ are nonzero.

11.2.2 Show that if $\dfrac{x_1}{y_1} = \dfrac{x_2}{y_2} = \cdots = \dfrac{x_n}{y_n}$ and $y_1 + y_2 + \cdots + y_n \neq 0$, then

$$\frac{x_1}{y_1} = \frac{x_2}{y_2} = \cdots = \frac{x_n}{y_n} = \frac{x_1 + x_2 + \cdots + x_n}{y_1 + y_2 + \cdots + y_n}.$$

11.2.3 If $\dfrac{a}{x + y - z} = \dfrac{b}{x - y + z} = \dfrac{c}{-x + y + z}$ and $x + y + z \neq 0$, must all three fractions equal $\dfrac{a + b + c}{x + y + z}$?

11.2.4 Show that if $\dfrac{x + y}{x - y} = \dfrac{c^2 + d^2}{c^2 - d^2}$ and $yd \neq 0$, then $\dfrac{x}{y} = \dfrac{c^2}{d^2}$.

11.2.5 Let $\dfrac{x}{y} = \dfrac{y}{z}$.

(a) Show that $\dfrac{x^2 + y^2}{y^2 + z^2} = \dfrac{y^2}{z^2}$.

(b)⋆ Show that $\dfrac{x^2 + xy + y^2}{z^2 + yz + y^2} = \dfrac{x}{z}$.

11.2.6⋆ If a, b, c are nonzero real numbers such that

$$\frac{a + b - c}{c} = \frac{a - b + c}{b} = \frac{-a + b + c}{a},$$

and

$$x = \frac{(a + b)(b + c)(c + a)}{abc},$$

and $x < 0$, then find x. *(Source: AHSME)* **Hints:** 89, 199

11.2.7⋆ Suppose that $x_i > 0$ for $1 \leq i \leq 6$, and that $\dfrac{x_1}{x_2} < \dfrac{x_3}{x_4} < \dfrac{x_5}{x_6}$. Show that $\dfrac{x_1}{x_2} < \dfrac{x_1 + x_3 + x_5}{x_2 + x_4 + x_6} < \dfrac{x_5}{x_6}$.
Hints: 351

11.3 Induction

The principle of **mathematical induction** (or often just **induction**) is one of the key proof methods that we use to prove statements about positive integers. The classic metaphor that is usually used to describe mathematical induction is a row of dominos all standing on end:

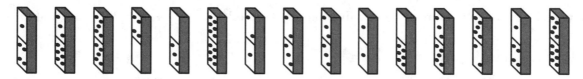

If we tip the first domino over:

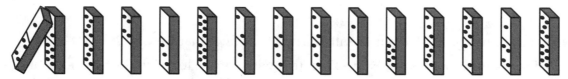

and if each domino, as it falls, knocks over the next domino in line:

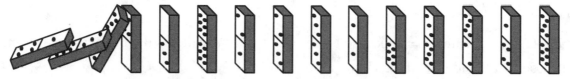

then ultimately, all of the dominos will fall over:

This is the basis of mathematical induction. Each "domino" is a mathematical statement that depends on n. The first domino is the statement where $n = 1$, the second domino is the statement where $n = 2$, and so on. If we can prove that the first statement (where $n = 1$) is true, and that each statement implies the next one, then we can knock all the dominos down, and prove that the statement is true for all n.

Obviously, this is all rather vague at the moment. We'll start by walking through a basic proof by induction to see how it works. Mathematical induction is one of the most common proof techniques, and you should master it.

Problems

Problem 11.9: In this problem, we prove the identity $1 + 2 + 2^2 + 2^3 + \cdots + 2^n = 2^{n+1} - 1$ for all positive integers n.

(a) Show that the identity holds for $n = 1$.

(b) Show that if the equation $1 + 2 + 4 + \cdots + 2^k = 2^{k+1} - 1$ is true, then we also have

$$1 + 2 + 4 + \cdots + 2^k + 2^{k+1} = 2^{k+2} - 1.$$

(c) Why can we use part (b) together with part (a) to deduce that $1 + 2^1 + 2^2 = 2^3 - 1$?

(d) Explain why we can use part (b) together with part (a) to conclude that $1 + 2 + 4 + \cdots + 2^n = 2^{n+1} - 1$, where n is any positive integer.

We'll start with an identity that you may know how to prove with other methods.

Problem 11.9: In this problem, we prove the identity $1 + 2 + 2^2 + 2^3 + \cdots + 2^n = 2^{n+1} - 1$ for all positive integers n.

  (a) Show that the identity holds for $n = 1$.

  (b) Show that if the equation $1 + 2 + 4 + \cdots + 2^k = 2^{k+1} - 1$ is true, then we also have

$$1 + 2 + 4 + \cdots + 2^k + 2^{k+1} = 2^{k+2} - 1.$$

  (c) Why can we use part (b) together with part (a) to deduce that $1 + 2^1 + 2^2 = 2^3 - 1$?

  (d) Explain why we can use $1 + 2^1 = 2^2 - 1$ together with part (b) to conclude that $1 + 2 + 4 + \cdots + 2^n = 2^{n+1} - 1$, where n is any positive integer.

Solution for Problem 11.9:

  (a) The identity holds for $n = 1$ because $1 + 2^1 = 2^{1+1} - 1$. We call this the **base case** of our induction.

  (b) Suppose that the identity holds if $n = k$, so that

$$1 + 2^1 + 2^2 + \cdots + 2^k = 2^{k+1} - 1. \tag{11.1}$$

We wish to show that
$$1 + 2 + 4 + \cdots + 2^k + 2^{k+1} = 2^{k+2} - 1. \tag{11.2}$$

By simply adding 2^{k+1} to both sides of Equation (11.1), we make the left side equal the left side of Equation (11.2), and we have

$$1 + 2 + 4 + \cdots + 2^k + 2^{k+1} = 2^{k+1} - 1 + 2^{k+1}.$$

Simplifying the right side of this equation gives $2(2^{k+1}) - 1 = 2^{k+2} - 1$, so we have

$$1 + 2 + 4 + \cdots + 2^k + 2^{k+1} = 2^{k+2} - 1,$$

as desired.

  (c) We've shown that the identity holds for $n = 1$. We've also shown that if the identity holds for $n = k$, then it holds for $n = k + 1$. So, because the identity holds for $n = 1$, it holds for $n = 2$, which means $1 + 2 + 2^2 = 2^3 - 1$.

  (d) Because the identity holds for $n = 2$, it holds for $n = 2 + 1 = 3$. Because it holds for $n = 3$, it holds for $n = 4$. Because it holds for.... You get the picture. We can continue this reasoning indefinitely, thereby showing that the identity

$$1 + 2 + 2^2 + 2^3 + \cdots + 2^n = 2^{n+1} - 1$$

holds for every positive integer n.

$\square$

Our solution above is a basic example of **proof by induction**. The proof involves two main steps:

1. *The base case.* We showed that the identity is true for $n = 1$. This is usually pretty easy.

2. *The inductive step.* We show that if the identity is true for $n = k$, then the identity holds for $n = k+1$.

Combining the base case with the inductive step tells us that because the identity is true for $n = 1$, it is true for $n = 1 + 1 = 2$. We can then use $n = 2$ together with the inductive step to deduce that the identity is true for $n = 2 + 1 = 3$. And so on.

The inductive step is usually the hardest part of proving a statement by induction. It begins with the **inductive assumption**, which is the assumption that the statement is true for some integer k. The inductive assumption is also sometimes called the **inductive hypothesis**. In Problem 11.9, our inductive assumption is our identity for $n = k$, which is

$$1 + 2^1 + 2^2 + \cdots + 2^k = 2^{k+1} - 1.$$

In the inductive step we used this assumption to prove the identity for $n = k + 1$, which is

$$1 + 2^1 + 2^2 + \cdots + 2^k + 2^{k+1} = 2^{k+2} - 1.$$

This, combined with our base case $n = 1$, completed the proof by induction.

Solve each of the following problems with induction. In each case, clearly identify the base case, the inductive assumption, and the inductive step.

Problems

Problem 11.10: Show that the sum of the first n positive integers is $\dfrac{n(n + 1)}{2}$.

Problem 11.11: Let n be a positive integer, and let $n! = n \cdot (n - 1) \cdot (n - 2) \cdots 2 \cdot 1$. Prove that

$$1 \cdot 1! + 2 \cdot 2! + 3 \cdot 3! + \cdots + n \cdot n! = (n + 1)! - 1.$$

Problem 11.12: Use induction, together with the identity established in Problem 11.10, to show that

$$1^3 + 2^3 + 3^3 + \cdots + n^3 = (1 + 2 + 3 + \cdots + n)^2$$

for all positive integers n.

Problem 11.13: Prove that

$$\frac{m!}{0!} + \frac{(m + 1)!}{1!} + \frac{(m + 2)!}{2!} + \cdots + \frac{(m + n)!}{n!} = \frac{(m + n + 1)!}{n!(m + 1)}$$

for all nonnegative integers m and n. (Note that $0! = 1$.)

Problem 11.14: Show that if n is a positive integer greater than 2, then

$$\frac{1}{n + 1} + \frac{1}{n + 2} + \frac{1}{n + 3} + \cdots + \frac{1}{2n} > \frac{3}{5}.$$

Problem 11.15: Evaluate $\displaystyle\prod_{n=1}^{\infty}\left(3-\sum_{k=0}^{2^n}2^{-k}\right)$.

Problem 11.10: Show that the sum of the first n positive integers is $\dfrac{n(n+1)}{2}$.

Solution for Problem 11.10: We wish to show that

$$1+2+3+\cdots+n=\frac{n(n+1)}{2}.$$

For our base case, we have $1=\frac{1(1+1)}{2}$, which is true.

For the inductive step, we assume the identity holds for $n=k$; that is, we assume

$$1+2+3+\cdots+k=\frac{k(k+1)}{2}. \tag{11.3}$$

We must now show that the identity holds for $n=k+1$. When $n=k+1$, the identity is

$$1+2+3+\cdots+k+(k+1)=\frac{(k+1)(k+2)}{2}. \tag{11.4}$$

We wish to use our inductive assumption, Equation (11.3), to show that Equation (11.4) is true. Adding $k+1$ to both sides of Equation (11.3) gives us

$$\begin{aligned}1+2+3+\cdots+k+(k+1)&=\frac{k(k+1)}{2}+k+1\\&=\frac{k^2+k}{2}+\frac{2k+2}{2}\\&=\frac{k^2+3k+2}{2}\\&=\frac{(k+1)(k+2)}{2},\end{aligned}$$

which shows that Equation (11.4) is true. So, we have shown that if the identity holds for $n=k$, then it also holds for $n=k+1$. Since we have also shown that the identity holds for $n=1$, our induction is complete and we have proved the identity for all positive integers. $\square$

Important: The sum of the first n positive integers is $\frac{n(n+1)}{2}$.

When writing up an induction proof, we don't have to be so long-winded. For example, we can write our solution to Problem 11.10 as:

Proof: We proceed by induction. For the base case, we have $1 = \frac{1(1+1)}{2}$, which is true. Now, we assume that the identity holds for $n = k$, which gives

$$1 + 2 + 3 + \cdots + k = \frac{k(k+1)}{2}.$$

Adding $k + 1$ to both sides of this equation gives us

$$1 + 2 + 3 + \cdots + k + k + 1 = \frac{k(k+1)}{2} + k + 1 = \frac{k^2 + k}{2} + \frac{2k + 2}{2} = \frac{(k+1)(k+2)}{2},$$

so the identity holds for $n = k + 1$. This completes the inductive step, so the identity holds for all positive integers n.

Problem 11.11: Let n be a positive integer. Prove that $1 \cdot 1! + 2 \cdot 2! + 3 \cdot 3! + \cdots + n \cdot n! = (n+1)! - 1$.

Solution for Problem 11.11: What's wrong with this solution:

Bogus Solution: We assume that

$$1 \cdot 1! + 2 \cdot 2! + 3 \cdot 3! + \cdots + k \cdot k! = (k+1)! - 1$$

for some positive integer k. Using this inductive assumption, we have

$$
\begin{aligned}
1 \cdot 1! + \cdots + k \cdot k! + (k+1) \cdot (k+1)! &= (k+1)! - 1 + (k+1) \cdot (k+1)! \\
&= (k+2) \cdot (k+1)! - 1 \\
&= (k+2)! - 1.
\end{aligned}
$$

This completes the inductive step, so our induction is complete.

The error here is that we forgot the base case! If we don't have a starting point for our induction, then showing that the identity holds for $n = k + 1$ if it holds for $n = k$ does us no good. Trying to perform an induction without establishing the base case is like trying to knock over a row of dominos without knocking over the first domino to start the dominos falling!

Fortunately, the base case is easy. For $n = 1$, we have $1 \cdot 1! = (1+1)! - 1$, which is true. Combining this with the work from our Bogus Solution gives us a complete proof by induction. $\square$

WARNING!! Don't forget the base case when using induction.

Notice that each of the identities we have proved with induction so far is an identity that holds "for all positive integers." This is usually a strong sign that induction might succeed.

> **Concept:** Mathematicians often use the letter n to stand for a nonnegative or positive integer. Because of this, you'll sometimes hear people say about a proof problem, "If it has an n, you can use induction." While this isn't always the case, it is generally true that if you must prove a mathematical statement for all positive integers, then you should consider trying induction.

Problem 11.12: Show that for any positive integer n, we have

$$1^3 + 2^3 + 3^3 + \cdots + n^3 = (1 + 2 + 3 + \cdots + n)^2.$$

Solution for Problem 11.12: That square of a huge expression is pretty intimidating. Fortunately, we have already proved that

$$1 + 2 + 3 + \cdots + n = \frac{n(n + 1)}{2},$$

so the identity we wish to prove can be written as

$$1^3 + 2^3 + 3^3 + \cdots + n^3 = \frac{n^2(n + 1)^2}{4}. \tag{11.5}$$

First, we take care of the base case: $1^3 = (1)^2(1 + 1)^2/4$, which is clearly true. Now, we assume the identity holds for $n = k$, so we have

$$1^3 + 2^3 + 3^3 + \cdots + k^3 = \frac{k^2(k + 1)^2}{4}.$$

Adding $(k + 1)^3$ to both sides of this equation gives us

$$1^3 + 2^3 + 3^3 + \cdots + k^3 + (k + 1)^3 = \frac{k^2(k + 1)^2}{4} + (k + 1)^3 = \frac{k^2(k + 1)^2}{4} + \frac{4(k + 1)(k + 1)^2}{4}$$

$$= \frac{k^2(k + 1)^2 + (4k + 4)(k + 1)^2}{4} = \frac{(k^2 + 4k + 4)(k + 1)^2}{4}$$

$$= \frac{(k + 1)^2(k + 2)^2}{4} = \frac{(k + 1)^2((k + 1) + 1)^2}{4}.$$

This is our identity in Equation (11.5) for $n = k + 1$, so we have completed our induction. $\square$

> **Important:** For all positive integers n, we have
>
> $$1^3 + 2^3 + 3^3 + \cdots + n^3 = (1 + 2 + 3 + \cdots + n)^2 = \frac{n^2(n + 1)^2}{4}.$$

So far, we have used induction on problems in which there is only one variable. We can also use induction to prove identities involving multiple variables.

Problem 11.13: Prove that

$$\frac{m!}{0!} + \frac{(m + 1)!}{1!} + \frac{(m + 2)!}{2!} + \cdots + \frac{(m + n)!}{n!} = \frac{(m + n + 1)!}{n!(m + 1)}$$

for all nonnegative integers m and n. (Note that $0! = 1$.)

Solution for Problem 11.13: In this problem, we have our choice of which variable to use for our induction. We'll focus on n, since there are fewer terms that involve n in the identity we wish to prove. First, we tackle our base case, $n = 0$. When $n = 0$, we have $\frac{m!}{0!} = m!$ on the left and $\frac{(m+0+1)!}{0!(m+1)} = m!$ on the right, so the identity holds when $n = 0$. (Notice that our base case works for any m, as well.)

> **WARNING!!** Note that we must start this induction with a base case of 0, not 1, because we must prove the identity for all nonnegative integers, not just all positive integers.

Next, we assume the identity holds for $n = k$, so we have

$$\frac{m!}{0!} + \frac{(m+1)!}{1!} + \frac{(m+2)!}{2!} + \cdots + \frac{(m+k)!}{k!} = \frac{(m+k+1)!}{k!(m+1)}$$

for all nonnegative integers m. To show that the identity holds for $n = k+1$, we add $\dfrac{(m+k+1)!}{(k+1)!}$ to both sides:

$$
\begin{aligned}
\frac{m!}{0!} + \frac{(m+1)!}{1!} + \frac{(m+2)!}{2!} + \cdots + \frac{(m+k)!}{k!} + \frac{(m+k+1)!}{(k+1)!} &= \frac{(m+k+1)!}{k!(m+1)} + \frac{(m+k+1)!}{(k+1)!} \\
&= \frac{(k+1)(m+k+1)!}{(k+1)k!(m+1)} + \frac{(m+1)(m+k+1)!}{(m+1)(k+1)!} \\
&= \frac{(k+1)(m+k+1)! + (m+1)(m+k+1)!}{(k+1)!(m+1)} \\
&= \frac{(m+k+2)(m+k+1)!}{(k+1)!(m+1)} \\
&= \frac{(m+k+2)!}{(k+1)!(m+1)}.
\end{aligned}
$$

This is our identity for $n = k+1$, so our induction is complete. Notice that every part of the induction holds for all nonnegative values of m. So, even though we must prove this statement for all nonnegative integers m and n, we can treat m as a constant and use induction on n. Because all of our steps are valid for any value of m, our induction shows that for all nonnegative values of m, the identity holds for all nonnegative values of n. $\square$

We can also use induction on inequalities that we wish to prove for positive integers.

> **Problem 11.14:** Show that if n is a positive integer greater than 2, then
> $$\frac{1}{n+1} + \frac{1}{n+2} + \frac{1}{n+3} + \cdots + \frac{1}{2n} > \frac{3}{5}.$$

Solution for Problem 11.14: What's wrong with this solution:

> **Bogus Solution:** For our base case, we have $n = 1$, which gives us $\frac{1}{2}$ on the left. This is not greater than $\frac{3}{5}$, so the inequality is not true.

Read the question carefully: we have to prove this inequality for all integers n *greater than* 2. So, our base case is $n = 3$, for which we have a left side of

$$\frac{1}{4} + \frac{1}{5} + \frac{1}{6} = \frac{15 + 12 + 10}{60} = \frac{37}{60},$$

which is indeed greater than $3/5$, because $3/5 = 36/60$.

Our next step is to assume that the inequality holds for $n = k$, so we have

$$\frac{1}{k+1} + \frac{1}{k+2} + \frac{1}{k+3} + \cdots + \frac{1}{2k} > \frac{3}{5}. \qquad (11.6)$$

We wish to use this to show that the inequality holds for $n = k + 1$; that is, we wish to show

$$\frac{1}{k+2} + \frac{1}{k+3} + \frac{1}{k+4} + \cdots + \frac{1}{2k+2} > \frac{3}{5}. \qquad (11.7)$$

To produce the left side of (11.7) from the left side of (11.6), we must add $\frac{1}{2k+1} + \frac{1}{2k+2} - \frac{1}{k+1}$ to both sides of (11.6). This gives us

$$\begin{aligned}
\frac{1}{k+2} + \frac{1}{k+3} + \frac{1}{k+4} + \cdots + \frac{1}{2k+2} &> \frac{3}{5} + \frac{1}{2k+1} + \frac{1}{2k+2} - \frac{1}{k+1} \\
&= \frac{3}{5} + \frac{1}{2k+1} + \frac{1}{2k+2} - \frac{2}{2k+2} \\
&= \frac{3}{5} + \frac{1}{2k+1} - \frac{1}{2k+2} \\
&= \frac{3}{5} + \frac{(2k+2) - (2k+1)}{(2k+1)(2k+2)} \\
&= \frac{3}{5} + \frac{1}{(2k+1)(2k+2)}.
\end{aligned}$$

Because $\dfrac{1}{(2k+2)(2k+1)}$ is positive, we have

$$\frac{1}{k+2} + \frac{1}{k+3} + \frac{1}{k+4} + \cdots + \frac{1}{2k+2} > \frac{3}{5} + \frac{1}{(2k+2)(2k+1)} > \frac{3}{5},$$

so the inequality holds for $n = k + 1$. This completes our induction. $\square$

Problem 11.15: Evaluate $\displaystyle\prod_{n=1}^{\infty} \left(3 - \sum_{k=0}^{2^n} 2^{-k} \right)$.

Solution for Problem 11.15: We begin by evaluating the inner sum, which is a geometric series with common ratio 2^{-1}:

$$\sum_{k=0}^{2^n} 2^{-k} = \frac{1 - 2^{-(2^n+1)}}{1 - 2^{-1}} = 2 - 2^{-2^n},$$

so

$$\prod_{n=1}^{\infty} \left(3 - \sum_{k=0}^{2^n} 2^{-k} \right) = \prod_{n=1}^{\infty} [3 - (2 - 2^{-2^n})] = \prod_{n=1}^{\infty} \left(1 + 2^{-2^n} \right).$$

Now what? We might be tempted to try this step:

Bogus Solution: We separate the product into two products:

$$\prod_{n=1}^{\infty}\left(1 + 2^{-2^n}\right) = \prod_{n=1}^{\infty} 1 \cdot \prod_{n=1}^{\infty} 2^{-2^n}.$$

The first product is just 1, so we focus on the second product.

But before we focus on that second product, we hopefully notice that we can't separate products of sums like this. For example, it is *not* generally true that $(1 + x)(1 + y)(1 + z) = 1 \cdot 1 \cdot 1 + xyz$. There are a bunch more terms in the expansion of $(1 + x)(1 + y)(1 + z)$.

So, back to the drawing board. To get a guess at what the infinite product is, let's try looking at the product of the first few terms.

Concept: When faced with a complicated summation or product, work out the result for small numbers of terms and look for a pattern.

The one-term product is easy: $1 + 2^{-2}$.

If we have two terms in the product, we get $(1 + 2^{-2})(1 + 2^{-4}) = 1 + 2^{-2} + 2^{-4} + 2^{-6}$.

The product of the first three terms in the product is

$$
\begin{aligned}
(1 + 2^{-2})(1 + 2^{-4})(1 + 2^{-8}) &= (1 + 2^{-2} + 2^{-4} + 2^{-6})(1 + 2^{-8}) \\
&= 1(1 + 2^{-2} + 2^{-4} + 2^{-6}) + 2^{-8}(1 + 2^{-2} + 2^{-4} + 2^{-6}) \\
&= 1 + 2^{-2} + 2^{-4} + 2^{-6} + 2^{-8} + 2^{-10} + 2^{-12} + 2^{-14}.
\end{aligned}
$$

Aha! Our experimentation suggests that our product equals the infinite geometric series with first term 1 and ratio $\frac{1}{4}$. Moreover, our experimentation suggests a way to prove it. We used the product of the first two terms to get the product of the first three terms. Similarly, we can use the product of the first k terms to get the product of the first $k + 1$ terms. Yep—it's induction time! We wish to show that

$$\prod_{n=1}^{j}(1 + 2^{-2^n}) = 1 + 2^{-2} + 2^{-4} + \cdots + 2^{-2^{j+1}+2}.$$

We've already taken care of the base case. For the inductive step, we assume that

$$\prod_{n=1}^{m}(1 + 2^{-2^n}) = 1 + 2^{-2} + 2^{-4} + \cdots + 2^{-2^{m+1}+2}.$$

Multiplying both sides of our inductive assumption by $1 + 2^{-2^{m+1}}$ gives us

$$
\begin{aligned}
\prod_{n=1}^{m+1}(1 + 2^{-2^n}) &= (1 + 2^{-2^{m+1}})(1 + 2^{-2} + 2^{-4} + \cdots + 2^{-2^{m+1}+2}) \\
&= 1 + 2^{-2} + 2^{-4} + \cdots + 2^{-2^{m+1}+2} + 2^{-2^{m+1}}(1 + 2^{-2} + 2^{-4} + \cdots + 2^{-2^{m+1}+2}) \\
&= 1 + 2^{-2} + 2^{-4} + \cdots + 2^{-2^{m+1}+2} + 2^{-2^{m+1}} + 2^{-2^{m+1}-2} + 2^{-2^{m+1}-4} + \cdots + 2^{-2^{m+2}+2}.
\end{aligned}
$$

This completes our induction, thus showing that for all positive integers j, we have

$$\prod_{n=1}^{j}(1+2^{-2^n}) = 1 + 2^{-2} + 2^{-4} + \cdots + 2^{-2^{j+1}+2}.$$

So, the infinite product $\displaystyle\prod_{n=1}^{\infty}(1+2^{-2^n})$ equals the infinite geometric series with first term 1 and common ratio $\frac{1}{4}$. The sum of this geometric series is $1/(1 - \frac{1}{4}) = \frac{4}{3}$. $\square$

> **Concept:** Many problems, particularly those involving sequences or series, can be solved with the strategy:
>
> Experiment $\rightarrow$ See Pattern $\rightarrow$ Prove Pattern with Induction

Exercises

11.3.1 Use induction to prove each of the following for all positive integers n:

(a) $1 + 4 + 7 + 10 + \cdots + (3n-2) = \dfrac{n(3n-1)}{2}$

(b) $1^2 + 4^2 + 7^2 + \cdots + (3n-2)^2 = \dfrac{n(6n^2 - 3n - 1)}{2}$

(c) $2^2 + 5^2 + 8^2 + \cdots + (3n-1)^2 = \dfrac{n(6n^2 + 3n - 1)}{2}$

11.3.2 The kth triangular number is given by the formula $\frac{k(k+1)}{2}$. Prove that the sum of the first n triangular numbers is $\frac{n(n+1)(n+2)}{6}$.

11.3.3 Use induction to prove that for any complex number z, we have $|z^n| = |z|^n$ for all positive integers n.

11.3.4 Use induction to show that $\dfrac{1}{1\cdot3} + \dfrac{1}{3\cdot5} + \dfrac{1}{5\cdot7} + \cdots + \dfrac{1}{(2n-1)(2n+1)} = \dfrac{n}{2n+1}$ for all positive integers n.

11.3.5★ Show that if n is a positive integer, then $1 + \dfrac{1}{\sqrt{2}} + \dfrac{1}{\sqrt{3}} + \cdots + \dfrac{1}{\sqrt{n}} < 2\sqrt{n}$.

11.3.6★ The sequence $x_1, x_2, x_3, \ldots$ is defined by $x_1 = 2$ and $x_{k+1} = x_k^2 - x_k + 1$ for all $k \geq 1$. Find $\displaystyle\sum_{k=1}^{\infty}\dfrac{1}{x_k}$.

Hints: 144

11.4 Binomial Theorem

If you are not familiar with combinations or the notation $\binom{n}{r}$, you may wish to review them in our *Introduction to Counting & Probability* text before studying this section.

Definition: Let $n! = n \cdot (n-1) \cdot (n-2) \cdots 2 \cdot 1$, and define $0! = 1$. For any nonnegative integers n and r with $n \geq r$, let

$$\binom{n}{r} = \frac{n!}{r!(n-r)!}.$$

The numbers $\binom{n}{r}$ are sometimes called **binomial coefficients** for their relationship to the expansions of powers of binomials. That relationship is:

Important: The **Binomial Theorem** states that

$$(x+y)^n = \sum_{i=0}^{n} \binom{n}{n-i} x^{n-i} y^i.$$

Problems

Problem 11.16:

(a) Use our definition of $\binom{n}{r}$ and some brute force to show that $\binom{n}{r} + \binom{n}{r+1} = \binom{n+1}{r+1}$.

(b) Show that if

$$(x+y)^k = \binom{k}{k} x^k + \binom{k}{k-1} x^{k-1} y + \binom{k}{k-2} x^{k-2} y^2 + \cdots + \binom{k}{2} x^2 y^{k-2} + \binom{k}{1} xy^{k-1} + \binom{k}{0} y^k,$$

then

$$(x+y)^{k+1} = \binom{k+1}{k+1} x^{k+1} + \binom{k+1}{k} x^k y + \binom{k+1}{k-1} x^{k-1} y^2 + \cdots + \binom{k+1}{2} x^2 y^{k-1} + \binom{k+1}{1} xy^k + \binom{k+1}{0} y^{k+1}.$$

(c) Prove the Binomial Theorem.

Problem 11.17: Use the Binomial Theorem to expand each of the following:

(a) $(x+y)^3$

(b) $(x+y)^4$

(c) $(2z-1)^5$

(d) $(x^2 - x)^4$

Problem 11.18: Find the real solution to the equation $x^3 + 3x^2 + 3x = 3$. (*Source: Mandelbrot*)

Problem 11.19: In this problem, we find the remainder when 65^{20} is divided by 512.

(a) Consider the expansion of $(64+1)^{20}$, and notice that $512 = 2^9$. Which terms in the expansion are not divisible by 512?

(b) Find the desired remainder.

Problem 11.20: In the expansion of $(ax + b)^{2000}$, where a and b are relatively prime positive integers (meaning their greatest common divisor is 1), the coefficients of x^2 and x^3 are equal. Find a and b. (*Source: AIME*)

Problem 11.21: Find the sum of all the roots, real and nonreal, of the equation

$$x^{2001} + \left(\frac{1}{2} - x\right)^{2001} = 0,$$

given that there are no multiple roots. *(Source: AIME)*

Problem 11.22:

(a) Show that $0 < \sqrt{3} - \sqrt{2} < 1$.

(b) Compute $(\sqrt{3} + \sqrt{2})^6 + (\sqrt{3} - \sqrt{2})^6$.

(c) What is the smallest integer larger than $(\sqrt{3} + \sqrt{2})^6$? *(Source: AHSME)*

Problem 11.16:

(a) Show that $\binom{n}{r} + \binom{n}{r+1} = \binom{n+1}{r+1}$.

(b) Show that

$$(x + y)^n = \binom{n}{n}x^n + \binom{n}{n-1}x^{n-1}y + \binom{n}{n-2}x^{n-2}y^2 + \cdots + \binom{n}{2}x^2y^{n-2} + \binom{n}{1}xy^{n-1} + \binom{n}{0}y^n$$

for all positive integers n.

Solution for Problem 11.16:

(a) A little brute force goes a long way:

$$\binom{n}{r} + \binom{n}{r+1} = \frac{n!}{r!(n-r)!} + \frac{n!}{(r+1)!(n-r-1)!}$$

$$= \frac{(r+1)n!}{(r+1)r!(n-r)!} + \frac{(n-r)n!}{(r+1)!(n-r)(n-r-1)!}$$

$$= \frac{(r+1)n! + (n-r)n!}{(r+1)!(n-r)!}$$

$$= \frac{(n+1)n!}{(r+1)!(n-r)!} = \frac{(n+1)!}{(r+1)!(n-r)!}$$

$$= \binom{n+1}{r+1}.$$

This useful identity is known as **Pascal's Identity**.

(b) We must prove a statement for all positive integers n, so we try using induction. For our base case, we must show $(x + y)^1 = \binom{1}{1}x^1 + \binom{1}{0}y^1$. Both sides of this equation equal $x + y$, so we have established the base case. Next, we let $n = k$, and we assume that

$$(x + y)^k = \binom{k}{k}x^k + \binom{k}{k-1}x^{k-1}y + \binom{k}{k-2}x^{k-2}y^2 + \cdots + \binom{k}{2}x^2y^{k-2} + \binom{k}{1}xy^{k-1} + \binom{k}{0}y^k.$$

Multiplying both sides by $x + y$ gives us

$$(x + y)^{k+1} = (x + y)\left(\binom{k}{k}x^k + \binom{k}{k-1}x^{k-1}y + \cdots + \binom{k}{1}xy^{k-1} + \binom{k}{0}y^k\right)$$

$$= x\left(\binom{k}{k}x^k + \binom{k}{k-1}x^{k-1}y + \cdots + \binom{k}{0}y^k\right) + y\left(\binom{k}{k}x^k + \binom{k}{k-1}x^{k-1}y + \cdots + \binom{k}{0}y^k\right)$$

$$= \binom{k}{k}x^{k+1} + \binom{k}{k-1}x^k y + \cdots + \binom{k}{0}xy^k + \binom{k}{k}x^k y + \binom{k}{k-1}x^{k-1}y^2 + \cdots + \binom{k}{0}y^{k+1}$$

$$= \binom{k}{k}x^{k+1} + \left(\binom{k}{k-1} + \binom{k}{k}\right)x^k y + \left(\binom{k}{k-2} + \binom{k}{k-1}\right)x^{k-1}y^2 + \cdots + \left(\binom{k}{0} + \binom{k}{1}\right)xy^k + \binom{k}{0}y^{k+1}.$$

Since $\binom{k}{k} = \binom{k+1}{k+1}$ and $\binom{k}{0} = \binom{k+1}{0}$ because all four of these expressions equal 1, we have

$$(x+y)^{k+1} = \binom{k+1}{k+1}x^{k+1} + \left(\binom{k}{k-1} + \binom{k}{k}\right)x^k y + \left(\binom{k}{k-2} + \binom{k}{k-1}\right)x^{k-1}y^2 + \cdots + \left(\binom{k}{0} + \binom{k}{1}\right)x^1 y^k + \binom{k+1}{0}y^{k+1}.$$

For all the other terms, we apply Pascal's Identity from part (a). For example, we have the equations $\binom{k}{k-1} + \binom{k}{k} = \binom{k+1}{k}$, and $\binom{k}{k-2} + \binom{k}{k-1} = \binom{k+1}{k-1}$, and so on down to $\binom{k}{0} + \binom{k}{1} = \binom{k+1}{1}$, which gives us the desired

$$(x + y)^{k+1} = \binom{k+1}{k+1}x^{k+1} + \binom{k+1}{k}x^k y + \binom{k+1}{k-1}x^{k-1}y^2 + \cdots + \binom{k+1}{2}x^2 y^{k-1} + \binom{k+1}{1}xy^k + \binom{k+1}{0}y^{k+1}.$$

This completes our induction. $\square$

As shown at the beginning of this section, we can write the Binomial Theorem more succinctly with summation notation:

$$(x + y)^n = \binom{n}{n}x^n + \binom{n}{n-1}x^{n-1}y + \binom{n}{n-2}x^{n-2}y^2 + \cdots + \binom{n}{2}x^2 y^{n-2} + \binom{n}{1}xy^{n-1} + \binom{n}{0}y^n = \sum_{i=0}^{n} \binom{n}{n-i}x^{n-i}y^i.$$

There are other ways we can express the Binomial Theorem. For example, because

$$\binom{n}{n-i} = \frac{n!}{(n-i)!(n-(n-i))!} = \frac{n!}{(n-i)!i!} = \binom{n}{i},$$

we can write it as $(x + y)^n = \sum_{i=0}^{n} \binom{n}{i}x^{n-i}y^i$.

Let's put the Binomial Theorem to work.

Problem 11.17: Expand each of the following:

(a) $(x + y)^3$

(b) $(x + y)^4$

(c) $(2z - 1)^5$

(d) $(x^2 - x)^4$

Solution for Problem 11.17:

(a) $(x + y)^3 = \binom{3}{3}x^3 + \binom{3}{2}x^2y + \binom{3}{1}xy^2 + \binom{3}{0}y^3 = x^3 + 3x^2y + 3xy^2 + y^3$.

(b) $(x + y)^4 = \binom{4}{4}x^4 + \binom{4}{3}x^3y + \binom{4}{2}x^2y^2 + \binom{4}{1}xy^3 + \binom{4}{0}y^4 = x^4 + 4x^3y + 6x^2y^2 + 4xy^3 + y^4$.

(c) We have

$$(2z - 1)^5 = \binom{5}{5}(2z)^5 + \binom{5}{4}(2z)^4(-1) + \binom{5}{3}(2z)^3(-1)^2 + \binom{5}{2}(2z)^2(-1)^3 + \binom{5}{1}(2z)(-1)^4 + \binom{5}{0}(-1)^5$$

$$= 32z^5 - 80z^4 + 80z^3 - 40z^2 + 10z - 1.$$

(d) $(x^2 - x)^4 = \binom{4}{4}(x^2)^4 + \binom{4}{3}(x^2)^3(-x) + \binom{4}{2}(x^2)^2(-x)^2 + \binom{4}{1}(x^2)(-x)^3 + \binom{4}{0}(-x)^4 = x^8 - 4x^7 + 6x^6 - 4x^5 + x^4$.

Notice that we're careful to keep our signs correct in parts (c) and (d). $\square$

For the rest of this section, we apply the Binomial Theorem to more challenging algebra problems.

Problem 11.18: Find the real solution to the equation $x^3 + 3x^2 + 3x = 3$. *(Source: Mandelbrot)*

Solution for Problem 11.18: The left side looks a lot like the expansion of $(x + 1)^3$, which is $x^3 + 3x^2 + 3x + 1$. So, we add 1 to both sides of the equation, which gives us $x^3 + 3x^2 + 3x + 1 = 4$, so we have

$$(x + 1)^3 = 4.$$

Taking the cube root of both sides of the equation, we get $x + 1 = \sqrt[3]{4}$, so $x = \sqrt[3]{4} - 1$ is the real root. (There are two nonreal numbers whose cube is 4; see if you can find them, and thereby find the other two solutions to the equation $x^3 + 3x^2 + 3x = 3$.) $\square$

The key to our solution to Problem 11.18 was recognizing the resemblance between $x^3 + 3x^2 + 3x$ and the expansion of $(x + 1)^3$.

Concept: Keep an eye out for expressions that are expansions of powers of binomials, or are very similar to expansions of powers of binomials.

Problem 11.19: Find the remainder when 65^{20} is divided by 512.

Solution for Problem 11.19: We surely don't want to evaluate 65^{20}. We notice that 512 is an integer power of 2, namely $2^9 = 512$. Next, we see that 65 is very close to 64, which equals 2^6. So, we consider the binomial expansion of $(64 + 1)^{20}$:

$$(64 + 1)^{20} = \binom{20}{20}64^{20} + \binom{20}{19}64^{19}(1) + \cdots + \binom{20}{2}(64^2)(1^{18}) + \binom{20}{1}(64^1)(1^{19}) + \binom{20}{0}1^{20}.$$

Because $64^2 = (2^6)^2 = 2^{12}$, we see that all but the last two terms of the expansion are divisible by 2^9. Therefore, the remainder when 65^{20} is divided by 512 equals the remainder when the sum of these last two terms is divided by 512. The sum of these last two terms is

$$(20)(64) + 1 = 1280 + 1 = 1281.$$

Since 1281 divided by 512 is 2 with a remainder of 257, our answer is 257. $\square$

> **Concept:** Whenever faced with a problem that involves large powers of expressions or numbers, consider looking for a way to apply the Binomial Theorem.

Problem 11.20: In the expansion of $(ax + b)^{2000}$, where a and b are relatively prime positive integers, the coefficients of x^2 and x^3 are equal. Find a and b. *(Source: AIME)*

Solution for Problem 11.20: Using the Binomial Theorem, the first few terms of $(ax + b)^{2000}$ are

$$b^{2000} + \binom{2000}{1999} b^{1999} \cdot ax + \binom{2000}{1998} b^{1998} \cdot a^2 x^2 + \binom{2000}{1997} b^{1997} \cdot a^3 x^3 + \cdots .$$

Since $\binom{2000}{1998} = \binom{2000}{2} = \frac{2000 \cdot 1999}{2 \cdot 1}$ and $\binom{2000}{1997} = \binom{2000}{3} = \frac{2000 \cdot 1999 \cdot 1998}{3 \cdot 2 \cdot 1}$, setting the coefficients of x^2 and x^3 equal gives us

$$\frac{2000 \cdot 1999}{2} b^{1998} a^2 = \frac{2000 \cdot 1999 \cdot 1998}{6} b^{1997} a^3,$$

which simplifies to $3b^{1998} a^2 = 1998 b^{1997} a^3$. Dividing both sides by $3a^2 b^{1997}$ gives $b = 666a$, so a divides b evenly. Since the greatest common divisor of a and b is 1 and a divides b, we must have $a = 1$, so $b = 666$. $\square$

Problem 11.21: Find the sum of all the roots, real and nonreal, of the equation

$$x^{2001} + \left(\frac{1}{2} - x\right)^{2001} = 0,$$

given that there are no multiple roots. *(Source: AIME)*

Solution for Problem 11.21: *Solution 1:* We can use Vieta's Formulas to find the sum of the roots of a polynomial, so we expand $\left(\frac{1}{2} - x\right)^{2001}$ with the Binomial Theorem:

$$\left(\frac{1}{2} - x\right)^{2001} = \left(\frac{1}{2}\right)^{2001} + \binom{2001}{2000}\left(\frac{1}{2}\right)^{2000}(-x)^1 + \cdots + \binom{2001}{2}\left(\frac{1}{2}\right)^2 (-x)^{1999} + \binom{2001}{1}\left(\frac{1}{2}\right)^1 (-x)^{2000} + (-x)^{2001}.$$

Since $\binom{2001}{2000} = \binom{2001}{1} = 2001$ and $\binom{2001}{2} = \frac{2001 \cdot 2000}{2 \cdot 1}$, we have

$$\left(\frac{1}{2} - x\right)^{2001} = \frac{1}{2^{2001}} - \frac{2001}{2^{2000}} x + \cdots - \frac{2001 \cdot 2000}{2 \cdot 2^2} x^{1999} + \frac{2001}{2} x^{2000} - x^{2001}.$$

Adding x^{2001} gives us

$$x^{2001} + \left(\frac{1}{2} - x\right)^{2001} = \frac{1}{2^{2001}} - \frac{2001}{2^{2000}} x + \cdots - \frac{2001 \cdot 2000}{2 \cdot 2^2} x^{1999} + \frac{2001}{2} x^{2000}.$$

By Vieta's Formulas, the sum of the roots is $\dfrac{\frac{2001 \cdot 2000}{2 \cdot 2^2}}{\frac{2001}{2}} = 500$.

Solution 2: We focus on the expressions that are raised to the same power, x and $(\frac{1}{2} - x)$. Noting that the average of x and $\frac{1}{2} - x$ is $\frac{1}{4}$ allows us to make a clever substitution that simplifies the problem. Letting $y = x - \frac{1}{4}$, we have the equation

$$\left(\frac{1}{4} + y\right)^{2001} + \left(\frac{1}{4} - y\right)^{2001} = 0.$$

The expansions of $\left(\frac{1}{4} + y\right)^{2001}$ and $\left(\frac{1}{4} - y\right)^{2001}$ are the same except that the signs of the terms with odd exponents are opposites. When the expansions are added, these terms with odd exponents cancel, leaving a degree 2000 polynomial that has only terms with even powers of y. Now when we apply Vieta's Formulas, the coefficient of y^{1999} is 0, which means that the sum of the 2000 solutions for y is 0. Since each value of y is 1/4 less than a solution for x in the original equation, the sum of the values of x that satisfy the original equation is $2000(\frac{1}{4}) = 500$. $\square$

Our second solution is a clever use of symmetry that makes applications of the Binomial Theorem and Vieta's Formulas much easier. In this solution, we make use of the fact that the terms with odd powers of b in the expansions of $(a + b)^n$ and $(a - b)^n$ cancel each other out. This can be a very useful problem solving strategy.

Problem 11.22: What is the smallest integer larger than $(\sqrt{3} + \sqrt{2})^6$? *(Source: AHSME)*

Solution for Problem 11.22: When we expand $\left(\sqrt{3} + \sqrt{2}\right)^6$, we have terms with square roots still in them. We could work hard to estimate them, but, inspired by our second solution to the previous problem, we can also use a related binomial expansion to help us. Note that $0 < \sqrt{3} - \sqrt{2} < 1$, so $\left(\sqrt{3} - \sqrt{2}\right)^6 < 1$. Now, we note that

$$\left(\sqrt{3} + \sqrt{2}\right)^6 = \binom{6}{6}\left(\sqrt{3}\right)^6 + \binom{6}{5}\left(\sqrt{3}\right)^5\left(\sqrt{2}\right)^1 + \binom{6}{4}\left(\sqrt{3}\right)^4\left(\sqrt{2}\right)^2 + \binom{6}{3}\left(\sqrt{3}\right)^3\left(\sqrt{2}\right)^3$$

$$+ \binom{6}{2}\left(\sqrt{3}\right)^2\left(\sqrt{2}\right)^4 + \binom{6}{1}\left(\sqrt{3}\right)^1\left(\sqrt{2}\right)^5 + \binom{6}{0}\left(\sqrt{2}\right)^6,$$

and

$$\left(\sqrt{3} - \sqrt{2}\right)^6 = \binom{6}{6}\left(\sqrt{3}\right)^6 - \binom{6}{5}\left(\sqrt{3}\right)^5\left(\sqrt{2}\right)^1 + \binom{6}{4}\left(\sqrt{3}\right)^4\left(\sqrt{2}\right)^2 - \binom{6}{3}\left(\sqrt{3}\right)^3\left(\sqrt{2}\right)^3$$

$$+ \binom{6}{2}\left(\sqrt{3}\right)^2\left(\sqrt{2}\right)^4 - \binom{6}{1}\left(\sqrt{3}\right)^1\left(\sqrt{2}\right)^5 + \binom{6}{0}\left(\sqrt{2}\right)^6.$$

Adding these expansions rids us of those pesky square roots, leaving an integer to work with:

$$\left(\sqrt{3} + \sqrt{2}\right)^6 + \left(\sqrt{3} - \sqrt{2}\right)^6 = 2\left(3^3\right) + 30\left(3^2\right)\left(2^1\right) + 30\left(3^1\right)\left(2^2\right) + 2\left(2^3\right)$$

$$= 54 + 540 + 360 + 16 = 970.$$

So, we have $\left(\sqrt{3} + \sqrt{2}\right)^6 = 970 - \left(\sqrt{3} - \sqrt{2}\right)^6$. Since $0 < \left(\sqrt{3} - \sqrt{2}\right)^6 < 1$, we see that $969 < \left(\sqrt{3} + \sqrt{2}\right)^6 < 970$, which means the smallest integer greater than $\left(\sqrt{3} + \sqrt{2}\right)^6$ is 970. $\square$

Exercises ▶

11.4.1 Expand each of the following:

(a) $(1 - x)^5$

(b) $(2x^2 + 3y)^4$

11.4.2 Find the constant term in the expansion of $\left(x + \frac{2}{x^2}\right)^9$.

11.4.3

(a) Without using a calculator, show that $31^{19} + 33^{99}$ is divisible by 64.

(b)★ Without using a calculator, show that $31^{19} + 33^{99}$ is not divisible by 128.

11.4.4

(a) Calculate $\binom{10}{0} + \binom{10}{1} + \binom{10}{2} + \cdots + \binom{10}{10}$.

(b)★ Find $\displaystyle\sum_{k=0}^{49} \binom{99}{2k}$. **Hints:** 326

11.4.5 Let n be a positive integer and $x \geq 0$. Show that $(1 + x)^n \geq 1 + nx$.

11.4.6★ Let P be the product of the nonreal roots of $x^4 - 4x^3 + 6x^2 - 4x = 2005$. Find $\lfloor P \rfloor$, where $\lfloor x \rfloor$ stands for the greatest integer that is less than or equal to x. *(Source: AIME)*

11.5 Summary

An **identity** is an equation that is true for all values of all variables for which the equation is defined. In this chapter, we discussed a few useful identities and covered several methods for proving specific identities.

First, we discussed **brute force**, which is essentially expanding both sides of a proposed identity to see if the two sides match.

Next, we worked with ratios. We proved a particularly useful pair of ratio manipulations:

> **Important:** Suppose $a/b = c/d$. Then, for any constant k for which $b + kd \neq 0$, we have
> $$\frac{a}{b} = \frac{c}{d} = \frac{a + kc}{b + kd}.$$
> For any constant k for which $a - kb$ and $c - kd$ are nonzero, we have
> $$\frac{a + kb}{a - kb} = \frac{c + kd}{c - kd}.$$

We then moved on to an extremely powerful tool, **mathematical induction**. While we primarily used induction to prove identities in this chapter, you will see it in many more contexts throughout your study of mathematics. To use induction to prove an identity holds for all positive integers n, we must perform two steps:

1. *The base case.* Show that the identity is true for $n = 1$. This is usually pretty easy.

2. *The inductive step.* Show that if the identity is true for $n = k$, then the identity holds for $n = k + 1$.

We used induction to prove several identities, including:

> **Important:** The sum of the first n positive integers is $n(n + 1)/2$.

> **Important:** For all positive integers n, we have
> $$1^3 + 2^3 + 3^3 + \cdots + n^3 = (1 + 2 + 3 + \cdots + n)^2 = \frac{n^2(n + 1)^2}{4}.$$

Finally, we used induction to prove the **Binomial Theorem**:

> **Important:** The **Binomial Theorem** states that
> $$(x + y)^n = \sum_{i=0}^{n} \binom{n}{n - i} x^{n-i} y^i.$$

Problem Solving Strategies

> **Concepts:**
> - When performing complicated algebra, stay organized and write neatly.
> - We can often relate symmetric expressions with identities.
> - If it's not immediately obvious how to solve a problem, don't just sit and stare at it. Experiment!
> - When asked to evaluate an expression involving large numbers, you can get a feel for the problem by evaluating similar expressions with small numbers.
> - Because an identity must be true for all permissible values of the variables in the identity, we can test an identity we develop by substituting convenient values for the variables.
>
> *Continued on the next page. . .*

Concepts: ... *continued from the previous page*

- When given equal ratios, setting the ratios equal to a variable helps eliminate the fractions.

- Working backwards from a relationship you wish to prove is a powerful problem-solving method.

- If you must prove a mathematical statement for all positive integers, then you should probably consider trying induction.

- When faced with a complicated summation or product, work out the result for small numbers of terms and look for a pattern.

- Many problems, particularly those involving sequences or series, can be solved with the strategy:

  Experiment → See Pattern → Prove Pattern with Induction

- Keep an eye out for expressions that are expansions of powers of binomials, or are very similar to expansions of powers of binomials.

- Whenever faced with a problem that involves large powers of expressions or numbers, consider looking for a way to apply the Binomial Theorem.

REVIEW PROBLEMS

11.23 In each of the following parts, state whether or not the equation is an identity. If it is an identity, prove that it is an identity. If it is not an identity, find values of the variables for which the equation is not true.

(a) $a(b - c) + b(c - a) + c(a - b) = 0$

(b) $(x - y)^4 + (x + y)^4 = 2(x^2 + y^2)^2 + 8x^2y^2$

(c) $\dfrac{a}{b} = \dfrac{a - \sqrt{a}}{b - \sqrt{b}}$, where $b \neq 0$ and $b \neq 1$

(d) $x^2(x - y) + y^2(y - x) = x^3 + y^3$

11.24 Find $x + y + z$ if $\dfrac{1}{xy} + \dfrac{1}{yz} + \dfrac{1}{zx} = 12$ and $xyz = \dfrac{1}{18}$.

11.25 Find $x + y + z$ if $\dfrac{x}{3 - x} = \dfrac{y}{5 - y} = \dfrac{z}{16 - z} = 2$.

11.26 Suppose $\dfrac{a}{b} = \dfrac{c}{d} = \dfrac{e}{f}$, where $a, b, c, d, e, f > 0$. Show that $\dfrac{a}{b} = \dfrac{\sqrt[n]{a^n + c^n + e^n}}{\sqrt[n]{b^n + d^n + f^n}}$.

11.27 Let $\dfrac{p + q - r}{p - q + r} = \dfrac{u + v}{u - v}$. Prove that $pv = u(q - r)$.

11.28 Suppose $\dfrac{a}{b} = \dfrac{x}{y}$, where $a, b, x, y > 0$. Show that $\dfrac{a-x}{\sqrt{a^2+x^2}} = \dfrac{b-y}{\sqrt{b^2+y^2}}$.

11.29 Use induction to prove each of the following for all positive integers n:

(a) $\quad 1 + 3 + 5 + \cdots + (2n-1) = n^2$ $\qquad$ (b) $\quad 1 + 4 + 9 + \cdots + n^2 = \dfrac{n(n+1)(2n+1)}{6}$

11.30 Show that $\dfrac{0}{1!} + \dfrac{1}{2!} + \dfrac{2}{3!} + \dfrac{3}{4!} + \cdots + \dfrac{n-1}{n!} = 1 - \dfrac{1}{n!}$.

11.31 Let $a_1, a_2, a_3, \ldots, a_n$ and $b_1, b_2, b_3, \ldots, b_n$ be two sequences of numbers. Prove the identity

$$a_1b_1 + a_2b_2 + a_3b_3 + \cdots + a_nb_n = (a_1 - a_2)b_1 + (a_2 - a_3)(b_1 + b_2) + (a_3 - a_4)(b_1 + b_2 + b_3) + \cdots$$
$$+ (a_{n-1} - a_n)(b_1 + b_2 + b_3 + \cdots + b_{n-1}) + a_n(b_1 + b_2 + b_3 + \cdots + b_n),$$

where $n > 1$. This identity is sometimes referred to as the **Abel Summation**.

11.32 Find the sum of the digits of the number $2003^4 - 1997^4$. *(Source: Mandelbrot)*

11.33 State and prove a one-variable identity that is suggested by the examples below:

$$1 + 2 = 3,$$
$$4 + 5 + 6 = 7 + 8,$$
$$9 + 10 + 11 + 12 = 13 + 14 + 15,$$
$$16 + 17 + 18 + 19 + 20 = 21 + 22 + 23 + 24.$$

11.34 Show that for all x, y, and z, we have

$$x^2(2x^2 - y^2 - z^2) + y^2(2y^2 - x^2 - z^2) + z^2(2z^2 - x^2 - y^2) = (x-y)^2(x+y)^2 + (y-z)^2(y+z)^2 + (z-x)^2(z+x)^2.$$

11.35 Find the coefficient of x^{14} in the expansion of $(2x^3 - x)^8$.

11.36 Show that $\dfrac{2^3 - 1}{2^3 + 1} \cdot \dfrac{3^3 - 1}{3^3 + 1} \cdot \dfrac{4^3 - 1}{4^3 + 1} \cdot \cdots \cdot \dfrac{n^3 - 1}{n^3 + 1} = \dfrac{2(n^2 + n + 1)}{3n(n+1)}$ for any positive integer $n \geq 2$.

11.37

(a) Verify that for all $x \neq 0$, we have $x^{n+1} + \dfrac{1}{x^{n+1}} = \left(x + \dfrac{1}{x}\right)\left(x^n + \dfrac{1}{x^n}\right) - \left(x^{n-1} + \dfrac{1}{x^{n-1}}\right)$.

(b)★ Let $x + \dfrac{1}{x} = a$, for some integer a. Prove that $x^n + \dfrac{1}{x^n}$ is an integer for all $n \geq 0$. **Hints:** 241

11.38 Without using a calculator, prove that $2005 \cdot 2007^3 - 2006 \cdot 2004^3$ is the cube of an integer .

11.39 In Problem 11.4.5, we showed that $(1 + x)^n \geq 1 + nx$ if $x \geq 0$ and n is a positive integer. In this problem, we show that this inequality still holds for all positive integers n even if $-1 \leq x < 0$.

(a) Explain why we cannot use the exact same Binomial Theorem approach we used in Problem 11.4.5.

(b) What's another tool we can use to prove a statement "for all positive integers n"? Prove that $(1 + x)^n \geq 1 + nx$ for all positive integers n even if $-1 \leq x < 0$.

11.40 Find x if $(x^2 - 3x + 4)/(x^2 - 5x + 5) = (x^2 - 3x + 9)/(x^2 - 5x + 10)$.

11.41 Let a, b, c, d, and e be real numbers such that $\frac{a}{b} = \frac{b}{c} = \frac{c}{d} = \frac{d}{e} = \frac{e}{a}$. Find the value of $\frac{ab^2}{c^2d}$.

Challenge Problems

11.42 Find all real values x that satisfy $x^4 - 4x^3 + 5x^2 - 4x + 1 = 0$.

11.43 Prove that $n^n > (n+1)^{n-1}$ for all positive integers n. **Hints:** 175, 195

11.44 Let $\frac{x}{a+b} = \frac{y}{a-b}$, where $x \neq y$ and $b \neq 0$. Show that $\frac{x+y}{x-y} = \frac{a}{b}$ and $\frac{x^3+y^3}{a^2+3b^2} = \frac{x^3-xy^2}{2b(a+b)}$.
Hints: 305

11.45 Find the units digit of the decimal expansion of $\left(15 + \sqrt{220}\right)^{19} + \left(15 + \sqrt{220}\right)^{82}$. *(Source: AHSME)*
Hints: 64

11.46 State and prove a one-variable identity that is suggested by the examples below:

$$2^2 + 3^2 + 4^2 + 14^2 = 15^2,$$
$$4^2 + 5^2 + 6^2 + 38^2 = 39^2,$$
$$6^2 + 7^2 + 8^2 + 74^2 = 75^2,$$
$$8^2 + 9^2 + 10^2 + 122^2 = 123^2$$

Hints: 214

11.47 Because $\binom{n}{i} = \frac{n!}{i!(n-i)!}$, we can write the Binomial Theorem as

$$(x+y)^n = \sum_{\substack{0 \leq i,j \leq n \\ i+j=n}} \frac{n!}{i!j!} x^i y^j,$$

where the $0 \leq i, j \leq n$ and $i + j = n$ under the summation means, "Sum the expression for all possible i and j such that i and j sum to n, and both i and j are nonnegative integers no greater than n." Express the expansion of $(x + y + z)^9$ using a similar summation notation.

11.48 Prove that $n^4 \leq 4^n$ for all positive integers n greater than 3.

11.49 Let $(1 + x + x^2)^n = a_0 + a_1x + a_2x^2 + \cdots + a_{2n}x^{2n}$ for all x. Find $a_0 + a_2 + a_4 + \cdots + a_{2n}$ in terms of n. *(Source: AHSME)* **Hints:** 129

11.50 Show that if $a_1x + b_1y + c_1z = a_2x + b_2y + c_2z = 0$, then

$$\frac{x}{b_1c_2 - b_2c_1} = \frac{y}{c_1a_2 - c_2a_1} = \frac{z}{a_1b_2 - a_2b_1}$$

for all values of the variables for which the three fractions are defined. **Hints:** 120

11.51 Prove that, for each positive integer n, we have

$$1 - \frac{1}{2} + \frac{1}{3} - \cdots + \frac{1}{2n-1} - \frac{1}{2n} = \frac{1}{n+1} + \frac{1}{n+2} + \cdots + \frac{1}{2n}.$$

Hints: 118

11.52 Find all ordered pairs of real numbers (x, y) such that $x^3 - y^3 = 19$ and $x^2 y - xy^2 = 6$.

11.53 The polynomial $1 - x + x^2 - x^3 + \cdots + x^{16} - x^{17}$ may be written in the form

$$a_0 + a_1 y + a_2 y^2 + a_3 y^3 + \cdots + a_{16} y^{16} + a_{17} y^{17},$$

where $y = x + 1$ and the a_i's are constants. Find the value of a_2. *(Source: AIME)* **Hints:** 192

11.54 Factor $(x + y)^5 - (x^5 + y^5)$ completely into a product of polynomials with real coefficients.

11.55 Find the smallest positive integer n for which the expansion of $(xy - 3x + 7y - 21)^n$, after like terms have been collected, has at least 1996 terms. *(Source: AIME)* **Hints:** 182

11.56 Prove that $\dfrac{bc(b+c)}{(a-b)(a-c)} + \dfrac{ca(c+a)}{(b-c)(b-a)} + \dfrac{ab(a+b)}{(c-a)(c-b)} = a + b + c$.

11.57 The Binomial Theorem is valid for exponents that are not integers. That is, for all real numbers x, y, and r with $|x| > |y|$, we have

$$(x + y)^r = x^r + rx^{r-1}y + \frac{r(r-1)}{2!}x^{r-2}y^2 + \frac{r(r-1)(r-2)}{3!}x^{r-3}y^3 + \cdots.$$

What are the first three digits to the right of the decimal point in the decimal representation of the number $(10^{2002} + 1)^{10/7}$? *(Source: AIME)* **Hints:** 310

11.58★ Let n be a positive integer. Prove that $\dfrac{1}{n+1} + \dfrac{1}{n+2} + \cdots + \dfrac{1}{2n} \geq \dfrac{2n}{3n+1}$.

11.59★ Prove that if

$$A = \frac{b^2 + c^2 - a^2}{2bc}, \quad B = \frac{a^2 + c^2 - b^2}{2ac}, \quad C = \frac{a^2 + b^2 - c^2}{2ab},$$

and $A + B + C = 1$, then $ABC = -1$. **Hints:** 275

11.60★ Write $(x^2 - yz)^3 + (y^2 - zx)^3 + (z^2 - xy)^3 - 3(x^2 - yz)(y^2 - zx)(z^2 - xy)$ as the square of a polynomial. **Hints:** 100

11.61★ Prove that for all positive integers n, we have $\displaystyle\sum_{k=1}^{n} \frac{1}{k^2} < 2$. **Hints:** 10

11.62★

(a) Find a polynomial $f(a, b, c, d)$ in four variables such that

$$f(x_1, x_2, x_3, x_4)f(y_1, y_2, y_3, y_4) = (x_1 y_1 + x_2 y_2 + x_3 y_3 + x_4 y_4)^2 + (x_1 y_2 - x_2 y_1 + x_3 y_4 - x_4 y_3)^2$$
$$+ (x_1 y_3 - x_3 y_1 + x_4 y_2 - x_2 y_4)^2 + (x_1 y_4 - x_4 y_1 + x_2 y_3 - x_3 y_2)^2$$

for all values of $x_1, x_2, x_3, x_4, y_1, y_2, y_3, y_4$. **Hints:** 102

(b) Every prime number can be written as the sum of four squares. Combine this fact with part (a) to prove that every positive integer can be written as the sum of four perfect squares.

Nature is unfair? So much the better. Inequality is the only bearable thing; the monotony of equality can only lead us to boredom. – Francis Picabia

CHAPTER **12**

_____

Inequalities

So far in this book, much of our work has involved solving equations or inequalities by finding those specific values of variables that make these equations or inequalities true. We have also proved some equations are always true; we called these identities. In this chapter, we turn to proving that certain inequalities are always true.

12.1 Manipulating Inequalities

We'll sometimes impose certain conditions on the variables in an inequality we wish to prove, and we'll occasionally use a shorthand for writing some of these conditions. For example, rather than writing, "$a > 0$, $b > 0$, and $c > 0$," we can write "$a, b, c > 0$." Similarly, we can write $0 < w, x, y, z < 1$ to indicate that w, x, y, and z are all between 0 and 1.

> **Problems**

Problem 12.1: Suppose that $\dfrac{a}{b} > \dfrac{c}{d}$ and we wish to prove that $\dfrac{a+b}{b} > \dfrac{c+d}{d}$.

(a) Suppose we start with, "We multiply $\dfrac{a}{b} > \dfrac{c}{d}$ by bd to get $ad > bc$." Why is this invalid?

(b) Prove that $\dfrac{a+b}{b} > \dfrac{c+d}{d}$.

Problem 12.2: In this problem, we show that if $a \geq b$ and $c \geq d$, then $ac + bd \geq ad + bc$.

(a) Move all the terms of $ac + bd \geq ad + bc$ to one side and factor the result.

(b) Explain why the inequality you produced in part (a) must be true if $a \geq b$ and $c \geq d$.

(c) In parts (a) and (b), we worked backwards to figure out why $ac + bd \geq ad + bc$. Write a complete solution forwards, starting with $a \geq b$ and $c \geq d$ and ending with $ac + bd \geq ad + bc$.

Problem 12.3:

(a) Find some positive numbers a, b, c, and d such that $\frac{a}{b} < \frac{c}{d} < 1$.

(b) Arrange the quantities $\frac{b}{a}$, $\frac{d}{c}$, $\frac{bd}{ac}$, and 1 from greatest to least according to your values from part (a).

(c) Show that the ordering found in part (b) is the same for all positive numbers a, b, c, and d.

Problem 12.4: In this problem, we will determine which is greater, $\dfrac{445721}{9902923}$ or $\dfrac{445725}{9902927}$.

(a) Prove that if $b > a > 0$ and $x > 0$, then $\dfrac{a}{b} < \dfrac{a+x}{b+x}$.

(b) Without using a calculator or computer, determine which is greater, $\dfrac{445721}{9902923}$ or $\dfrac{445725}{9902927}$.

Problem 12.5: Let $0 < x, y \le c$. Prove that $\dfrac{x+y}{1 + \frac{xy}{c^2}} \le c$.

Problem 12.6: Show that if $0 < a < c$ and $0 < b < d$, then $\dfrac{ab}{a+b} < \dfrac{cd}{c+d}$.

Problem 12.1: Suppose that $\dfrac{a}{b} > \dfrac{c}{d}$. Show that $\dfrac{a+b}{b} > \dfrac{c+d}{d}$.

Solution for Problem 12.1: What's wrong with this solution:

> **Bogus Solution:** Multiplying both sides of $\frac{a}{b} > \frac{c}{d}$ by bd gives us $ad > bc$. Adding bd to both sides gives us $ad + bd > bc + bd$. Factoring both sides gives $d(a + b) > b(c + d)$. Dividing both sides by b gives $\frac{d}{b}(a + b) > c + d$, and dividing both sides by d gives the desired $\frac{a+b}{b} > \frac{c+d}{d}$.

Our first step, multiplying $\frac{a}{b} > \frac{c}{d}$ by bd to get $ad > bc$, is only valid if bd is positive. But we aren't guaranteed that this is the case; if bd is negative, then multiplying $\frac{a}{b} > \frac{c}{d}$ by bd produces $ad < bc$. Each of our divisions have similar difficulties, as well.

> **WARNING!!** Pay special attention to whether or not the variables can be negative in inequalities you wish to prove. Some common inequality manipulations are only valid if the variables are positive.

We could fix our Bogus Solution with some casework, but instead, let's look for manipulations that don't depend on the signs of the variables.

Comparing the left side of the inequality we are given, a/b, to the left side of the inequality we want to prove, $(a + b)/b$, we see a very simple relationship:

$$\frac{a}{b} + 1 = \frac{a+b}{b}.$$

So, we add 1 to both sides of $\frac{a}{b} > \frac{c}{d}$, which gives us $\frac{a}{b} + 1 > \frac{c}{d} + 1$. Writing each side of the inequality with a common denominator gives the desired $\frac{a+b}{b} > \frac{c+d}{d}$. □

> **Concept:** Look for ways to manipulate one side of a given inequality to match the corresponding side of the desired inequality. Sometimes simple manipulations will be enough to prove the desired inequality.

Problem 12.2: Show that if $a \geq b$ and $c \geq d$, then $ac + bd \geq ad + bc$.

Solution for Problem 12.2: What's wrong with starting as follows:

> **Bogus Solution:** We need an ac and a bd, so we start by multiplying $a \geq b$ by $c \geq d$ to get $ac \geq bd$.

This is bad for two reasons. First, it doesn't get us any closer to what we want to prove, since $ac \geq bd$ has bd on the lesser side, but the desired inequality has bd on the greater side. Even subtracting bd from both sides doesn't help, since that gives $ac - bd \geq 0$, which is quite different from the $ac + bd$ we want on the larger side.

> **Concept:** Keep your eye on the ball! Don't just blindly manipulate inequalities—try to make sure you're moving in the right direction.

Even worse than the fact that we're moving in the wrong direction is the fact that the step "multiplying $a \geq b$ by $c \geq d$ gives $ac \geq bd$" is just plain invalid. What if $a = 2$, $b = -1$, $c = -3$, and $d = -4$? Then, we have $ac < bd$, not $ac \geq bd$.

> **WARNING!!** If a, b, c, and d can be any real numbers, then we cannot multiply $a \geq b$ by $c \geq d$ to get $ac \geq bd$. If a, b, c, and d are positive, then this manipulation is valid.

So, back to the drawing board. We might note that we need ac on the larger side and bc on the smaller, so we could try multiplying $a \geq b$ by c to give $ac \geq bc$. But that's invalid too! Since c might be negative, we can't say that $ac \geq bc$. Now, we seem stuck. There are two ways we might proceed from here:

Solution 1: Focus on why we are stuck. We are stuck because every time we try multiplying inequalities, we run into the problem of possibly multiplying by a negative number. So, we try to find a way to multiply inequalities such that we are guaranteed to multiply by a nonnegative number.

> **Concept:** When stuck on a problem, focus on why you are stuck, and try to find ways to remove that obstacle.

We can turn our given inequalities into ones involving nonnegative expressions by subtracting b from $a \geq b$ and subtracting d from $c \geq d$. This gives us $a - b \geq 0$ and $c - d \geq 0$. Aha! Nonnegative numbers. The product of nonnegative expressions $a - b$ and $c - d$ must be nonnegative, and it will produce all the products in our desired inequality. We therefore have $(a - b)(c - d) \geq 0$. Expanding the left side gives $ac - ad - bc + bd \geq 0$, and rearranging gives the desired $ac + bd \geq ad + bc$.

Solution 2: Work backwards. We're stuck going forwards, so we try manipulating the desired inequality to produce an inequality we know how to prove.

> **Concept:** Working backwards is a very powerful method for proving inequalities.

We start by moving everything to one side, which gives us $ac + bd - ad - bc \geq 0$. We see pairs of terms we can factor, so we try factoring, which gives us $c(a - b) + d(b - a) \geq 0$. Aha! We can factor some more. We have

$$c(a - b) + d(b - a) = c(a - b) - d(a - b) = (c - d)(a - b),$$

so our desired inequality is now $(c - d)(a - b) \geq 0$. Now, we see the path to a solution. The given inequalities $a \geq b$ and $c \geq d$ give us $a - b \geq 0$ and $c - d \geq 0$, and multiplying these gives us $(c - d)(a - b) \geq 0$.

This second solution is a combination of working backwards and forwards. As usual when we find a solution wholly or partially by working backwards, we finish by writing the whole solution forwards:

> **Proof:** Since $a \geq b$ and $c \geq d$, we have $a - b \geq 0$ and $c - d \geq 0$. Therefore, we have $(a - b)(c - d) \geq 0$.
> ✓ Expanding the product on the left gives $ac - ad - bc + bd$, and rearranging gives the desired $ac + bd \geq ad + bc$.

□

> **Problem 12.3:** Let $a, b, c,$ and d be positive numbers such that $\frac{a}{b} < \frac{c}{d} < 1$. Arrange the four quantities $\frac{b}{a}, \frac{d}{c}, \frac{bd}{ac},$ and 1 from greatest to least, and show that the ordering is the same for all positive numbers $a, b, c,$ and d.

Solution for Problem 12.3: Before diving into manipulating the given inequalities, we might simply choose some positive numbers that fit the description of the problem. Since $\frac{a}{b}$ and $\frac{c}{d}$ must be less than 1, we must choose $a < b$ and $c < d$, and we must also choose the numbers so that $\frac{a}{b} < \frac{c}{d}$. Letting $a = 1$, $b = 2$, $c = 3$, and $d = 4$ works just fine.

> **Concept:** When possible, testing numbers that fit an inequality problem can help you get started with a solution.

For our choice of values for the variables, we have $\frac{b}{a} = 2, \frac{d}{c} = \frac{4}{3},$ and $\frac{bd}{ac} = \frac{8}{3}$, so the order from greatest to least is $\frac{bd}{ac}, \frac{b}{a}, \frac{d}{c}, 1$.

We now have a target. We'd like to prove that

$$\frac{bd}{ac} > \frac{b}{a} > \frac{d}{c} > 1.$$

We see that the last three terms in this inequality chain are reciprocals of the terms in the given inequality, so we wonder what happens when we take the reciprocal of both sides of an inequality.

If x and y are positive and $x \geq y$, then dividing both sides by xy gives $\frac{1}{y} \geq \frac{1}{x}$. So, when we take the reciprocal of both sides of an inequality (and both sides are positive), we reverse the direction of the inequality sign.

Applying this fact to our given inequality, $\frac{a}{b} < \frac{c}{d} < 1$, gives us

$$\frac{b}{a} > \frac{d}{c} > 1.$$

So, all we have left to prove is that $\frac{bd}{ac} > \frac{b}{a}$. We note that the left side of this inequality is $\frac{d}{c}$ times the right side. But we already know that $\frac{d}{c} > 1$. Multiplying this by $\frac{b}{a}$ gives us $\frac{bd}{ac} > \frac{b}{a}$. Combining this with our earlier inequality chain gives us

$$\frac{bd}{ac} > \frac{b}{a} > \frac{d}{c} > 1,$$

as desired. □

Not all problems that involve inequality proofs say "Prove this inequality."

Problem 12.4: Without using a calculator or computer, determine which is greater, $\dfrac{445721}{9902923}$ or $\dfrac{445725}{9902927}$.

Solution for Problem 12.4: We're certainly not interested in performing the long division. So, we look for clues in the fractions that will help us find a slicker solution. We notice that the numerator and denominator in the second fraction are each 4 larger than in the first fraction. So, we can recast our problem as comparing

$$\frac{a}{b} \quad \text{to} \quad \frac{a+4}{b+4},$$

where $a = 445721$ and $b = 9902923$. Working with a and b will probably be a lot easier than working with those huge numbers.

Concept: If a problem is given in terms of gigantic numbers, it's often useful to replace those numbers with variables.

In fact, we could even go further and replace 4 with x, so that we are comparing

$$\frac{a}{b} \quad \text{to} \quad \frac{a+x}{b+x}.$$

We can experiment to get a sense for which is larger. For example, if we let $a = 1$, $b = 2$, and $x = 3$, we have $a/b = 1/2$ and $(a+x)/(b+x) = 4/5$. So, we think we have

$$\frac{a}{b} < \frac{a+x}{b+x}.$$

We're dealing with positive numbers in the original problem, so we let $a, b, x > 0$. It's not clear how to get from $a, b, x > 0$ to $a/b < (a + x)/(b + x)$, so we work backwards. Multiplying both sides of the desired inequality by $b(b + x)$ to get rid of the fractions gives

$$ab + ax < ab + bx.$$

Rearranging this inequality gives us $(a - b)x < 0$. This inequality is only true if $b > a$. However, for the two numbers we wish to compare, we do indeed have $b > a$.

So, now we're ready to prove an inequality that we can use to solve our problem. If $b > a > 0$ and $x > 0$, then we have $(a - b)x < 0$, so $ax < bx$. Adding ab to both sides and factoring gives $a(b + x) < b(a + x)$. Dividing both sides by $b(b + x)$ gives

$$\frac{a}{b} < \frac{a + x}{b + x}.$$

Letting $a = 445721$, $b = 9902923$, and $x = 4$ gives us $\dfrac{445721}{9902923} < \dfrac{445725}{9902927}$. $\square$

Problem 12.5: Let $0 < x, y \le c$. Prove that $\dfrac{x + y}{1 + \frac{xy}{c^2}} \le c$.

Solution for Problem 12.5: We don't have any clear starting point for working forwards from $0 \le x, y \le c$ to the desired inequality, so we'll have to try working backwards.

We'd like to start by getting rid of all the ugly fractions in the desired inequality. Multiplying both sides by the denominator of the left side would be easiest, though we must first check that this expression is positive. Because x, y, and c are positive, the expression $1 + \frac{xy}{c^2}$ is positive, too. Therefore, multiplying both sides of the desired inequality by $1 + \frac{xy}{c^2}$ does not change the direction of the inequality. This gives us

$$x + y \le c + \frac{xy}{c}.$$

Similarly, since $c > 0$, we can multiply both sides of this inequality by c to get

$$(x + y)c \le c^2 + xy.$$

That's much nicer than the original inequality. We move all terms to the left side, which gives

$$-c^2 + (x + y)c - xy \le 0.$$

We can view the left side as a quadratic in c. Multiplying the inequality by -1 (and reversing the direction of the inequality sign) gives

$$c^2 - (x + y)c + xy \ge 0,$$

which makes the quadratic, and its factorization, much clearer. We now factor the left side, and we have

$$(c - x)(c - y) \ge 0$$

as the inequality we wish to prove. Since $c - x \ge 0$ and $c - y \ge 0$, this inequality is true.

Now, we can walk backwards through our steps to write our proof:

Proof: Since $x, y \leq c$, we have $c - x \geq 0$ and $c - y \geq 0$. Multiplying these two gives $(c - x)(c - y) \geq 0$.
☑ Expanding the left side and rearranging gives $c^2 + xy \geq c(x + y)$. Since $c > 0$, dividing both sides by c gives

$$c + \frac{xy}{c} \geq x + y.$$

Factoring c out of the left side gives

$$c\left(1 + \frac{xy}{c^2}\right) \geq x + y.$$

Since x, y, and c are positive, the expression $1 + \frac{xy}{c^2}$ is positive. So, we can divide both sides by this expression to give the desired

$$c \geq \frac{x + y}{1 + \frac{xy}{c^2}}.$$

☐

Problem 12.6: Show that if $0 < a < c$ and $0 < b < d$, then $\dfrac{ab}{a + b} < \dfrac{cd}{c + d}$.

Solution for Problem 12.6: We're definitely going to try working backwards.

Solution 1: Because $a + b$ and $c + d$ are positive, we can multiply both sides of the desired inequality by $(a + b)(c + d)$ to produce the equivalent inequality

$$abc + abd < acd + bcd.$$

We can produce some of these terms from our given inequalities. Multiplying $b < d$ by ac gives $abc < acd$ and multiplying $a < c$ by bd gives $abd < bcd$. Adding $abc < acd$ and $abd < bcd$ gives

$$abc + abd < acd + bcd.$$

We know how to finish from here; we retrace our initial "working backwards" steps. Factoring gives $ab(c + d) < cd(a + b)$ and dividing both sides by $(a + b)(c + d)$ gives $\frac{ab}{a+b} < \frac{cd}{c+d}$.

Solution 2: Multiplying by denominators is not the only way to deal with fractions. Sometimes, reciprocals of fractions are much nicer than the original fractions. (This is particularly true when the denominator of a fraction is a sum and the numerator is a product.)

Concept: Dealing with reciprocals of complicated fractions is often nicer than dealing with the original fractions.

Taking the reciprocal of both sides of the desired inequality (and therefore reversing the direction of the inequality) gives us

$$\frac{a + b}{ab} > \frac{c + d}{cd}.$$

Splitting both sides into the sum of two fractions, we have

$$\frac{1}{b} + \frac{1}{a} > \frac{1}{d} + \frac{1}{c}.$$

Because we have $0 < a < c$, we have $\frac{1}{a} > \frac{1}{c}$. Similarly, $0 < b < d$ gives $\frac{1}{b} > \frac{1}{d}$. Adding these reciprocal inequalities puts us on track to walk backwards through our steps above:

Proof: Because we have $0 < a < c$ and $0 < b < d$, we have $\frac{1}{a} > \frac{1}{c}$ and $\frac{1}{b} > \frac{1}{d}$. Adding these gives

$$\frac{1}{b} + \frac{1}{a} > \frac{1}{d} + \frac{1}{c}.$$

Writing each side with a common denominator, we have

$$\frac{a+b}{ab} > \frac{c+d}{cd}.$$

Because $a, b, c, d > 0$, we can take the reciprocal of both sides of this inequality to give

$$\frac{ab}{a+b} < \frac{cd}{c+d}.$$

WARNING!! Many of the manipulations we performed in our solution to Problem 12.6 are valid only because the variables are all positive.

Exercises

12.1.1 Let $x > y > 0$ and $z \neq 0$. In each of the parts below, determine whether or not the inequality *always* holds. If it does, prove that it does. If it doesn't, provide an example in which it fails.

(a) $x + z > z + y$

(b) $(x + y)z > 0$

(c) $x - z > y - z$

(d) $xz > yz$

12.1.2 Let $-1 < x < 0$. Arrange x, x^2, and x^3 in order from least to greatest.

12.1.3 Show that if $0 < x \leq y \leq c$, then $y - x \leq c\left(1 - \frac{x}{y}\right)$.

12.1.4 Suppose that a, b, c, and d are positive real numbers such that $\frac{a}{b} \leq \frac{c}{d}$. Show that $\frac{a}{a+b} \leq \frac{c}{c+d}$.

12.1.5 Show that if a, b, x, y are constants such that $a > b > 0$, and $x > y > 0$, then $\frac{a+y}{b+y} > \frac{a+x}{b+x}$.

12.1.6★ Let $0 \leq a \leq b \leq c$ and $a + b + c = 1$. Show that $a^2 + 3b^2 + 5c^2 \geq 1$. **Hints:** 244

12.2 The Trivial Inequality

Already throughout this book, we have used the fact that the square of any real number is nonnegative. This fact is so simple, you might even consider it trivial.

> **Important:** The **Trivial Inequality** states that if x is a real number, then $x^2 \geq 0$. Equality holds if and only if $x = 0$.

The second sentence above seems rather simple, but, in fact, it's very important. We call the condition under which equality holds in a nonstrict inequality the **equality condition** of the inequality. Don't overlook the equality conditions of inequalities—we'll see that they can be very helpful in solving problems.

We've already seen in our discussion of finding maxima or minima of quadratic functions that this Trivial Inequality is not so trivial. But it's useful for much more than just quadratics; it's the foundation of many of the inequalities we will study throughout the rest of this chapter and in Chapter 18.

Problems

Problem 12.7: Show that $\dfrac{x^2 + 2}{\sqrt{x^2 + 1}} \geq 2$ for all real numbers x.

Problem 12.8: In this problem, we prove that if x and y are positive, then $\dfrac{1}{x} + \dfrac{1}{y} \geq \dfrac{4}{x + y}$.

(a) Manipulate the desired inequality to produce an inequality that is equivalent to the Trivial Inequality.

(b) Reverse your steps in part (a) to prove the desired inequality.

Problem 12.9:
(a) Prove that $a^2 + b^2 \geq 2ab$ for all real numbers a and b. When does equality occur?
(b) Prove that $a^2 + b^2 + c^2 \geq ab + ac + bc$ for all real numbers a, b, and c. When does equality occur?
(c) Prove that $a^4 + b^4 + c^4 \geq abc(a + b + c)$ for all real numbers a, b, and c.

Problem 12.10: Let a and b be positive real numbers. Show that $\dfrac{b}{a^2} + \dfrac{a}{b^2} \geq \dfrac{1}{a} + \dfrac{1}{b}$.

Problem 12.11: In this problem, we find all ordered pairs of real numbers (x, y) such that we have $(4x^2 + 4x + 3)(y^2 - 6y + 13) = 8$.

(a) Complete the square in both quadratics.

(b) Solve the problem.

Problem 12.7: Show that $\dfrac{x^2 + 2}{\sqrt{x^2 + 1}} \geq 2$ for all real numbers x.

Solution for Problem 12.7: As usual, we start by working backwards. We multiply by $\sqrt{x^2 + 1}$ to get rid of the fractions, and we have $x^2 + 2 \geq 2\sqrt{x^2 + 1}$. Because $x^2 \geq 0$ for all real x, we know that both sides of

$x^2 + 2 \geq 2\sqrt{x^2+1}$ are positive. So, we can square both sides to give

$$x^4 + 4x^2 + 4 \geq 4(x^2 + 1).$$

Rearranging this inequality gives us simply $x^4 \geq 0$, which by the Trivial Inequality is always true.

Now we have our path to a solution:

Proof: Because x^2 is a real number, the Trivial Inequality gives us $(x^2)^2 \geq 0$, so $x^4 \geq 0$. Adding $4x^2 + 4$ to both sides gives

$$x^4 + 4x^2 + 4 \geq 4(x^2 + 1),$$

so we have $(x^2+2)^2 \geq 4(x^2+1)$. Because $x^2 \geq 0$, we know that both x^2+2 and x^2+1 are positive. So, we can take the square root of both sides of $(x^2+2)^2 \geq 4(x^2+1)$ to give $x^2+2 \geq 2\sqrt{x^2+1}$. Dividing both sides by $\sqrt{x^2+1}$ gives the desired

$$\frac{x^2+2}{\sqrt{x^2+1}} \geq 2$$

for all real numbers x.

While the Trivial Inequality is stated in terms of one variable, it's a powerful tool for multivariable inequalities, too.

Problem 12.8: Prove that if x and y are positive, then $\dfrac{1}{x} + \dfrac{1}{y} \geq \dfrac{4}{x+y}$.

Solution for Problem 12.8: Again, we work backwards. Because x and y are positive, multiplying both sides of the desired inequality by $xy(x + y)$ to get rid of the fractions gives us

$$y(x + y) + x(x + y) \geq 4xy.$$

Rearranging this gives us $x^2 - 2xy + y^2 \geq 0$. Factoring the left side gives us $(x - y)^2 \geq 0$, which is true by the Trivial Inequality. So, now we can write our solution forwards:

Proof: The Trivial Inequality gives us $(x - y)^2 \geq 0$. Expanding the left side gives $x^2 - 2xy + y^2 \geq 0$. Adding $4xy$ to both sides gives $x^2 + xy + y^2 + xy \geq 4xy$. Factoring on the left side gives $x(x + y) + y(x + y) \geq 4xy$. Since x and y are positive, we can divide both sides by $xy(x + y)$ to give the desired

$$\frac{1}{x} + \frac{1}{y} \geq \frac{4}{x+y}.$$

For many of the problems in the rest of this chapter, we will not be writing "forwards" solutions after discovering solutions by working backwards. After each such problem, you should try closing the book and re-creating the forwards solution on your own.

Problem 12.9:

(a) Prove that $a^2 + b^2 \geq 2ab$ for all real numbers a and b. When does equality occur?

(b) Prove that $a^2 + b^2 + c^2 \geq ab + ac + bc$ for all real numbers a, b, and c. When does equality occur?

(c) Prove that $a^4 + b^4 + c^4 \geq abc(a + b + c)$ for all real numbers a, b, and c.

Solution for Problem 12.9:

(a) Rearranging the desired inequality gives $a^2 - 2ab + b^2 \geq 0$. We recognize that $a^2 - 2ab + b^2 = (a-b)^2$, so the given inequality is just an instance of the Trivial Inequality. Equality occurs if and only if $(a - b)^2 = 0$, which is equivalent to $a = b$.

(b) Rearranging the desired inequality gives

$$a^2 + b^2 + c^2 - ab - ac - bc \geq 0.$$

The expression on the left side is not a square, nor can we easily factor it. However, individual parts of the expression look familiar. The terms a^2, b^2, and ab should remind us of the inequality we just derived, $a^2 + b^2 \geq 2ab$.

> **Concept:** Don't treat every problem as if it is brand-new. Try to relate problems or parts of problems to other problems that you already know how to do.

Notice that $a^2 + b^2 + c^2 - ab - ac - bc$ is symmetric, so it may help to write inequalities of the form $a^2 + b^2 \geq 2ab$ for all three pairs of variables:

$$a^2 + b^2 \geq 2ab,$$
$$a^2 + c^2 \geq 2ac,$$
$$b^2 + c^2 \geq 2bc.$$

Adding these inequalities gives us twice our desired sum of squares on the greater side:

$$2a^2 + 2b^2 + 2c^2 \geq 2ab + 2ac + 2bc.$$

Aha! Dividing by 2, we get the desired inequality, $a^2 + b^2 + c^2 \geq ab + ac + bc$.

For equality to occur, we must have equality in each of the three inequalities we added, and this occurs if and only if $a = b = c$.

> **Concept:** We can sometimes prove a complicated inequality by proving simpler inequalities that can be combined to give us the desired complicated one.

(c) We just proved an inequality that involved a sum of squares. The sum $a^4 + b^4 + c^4$ is a sum of squares, too. Letting $x = a^2$, $y = b^2$, and $z = c^2$ in

$$x^2 + y^2 + z^2 \geq xy + yz + zx \tag{12.1}$$

gives us

$$a^4 + b^4 + c^4 \geq a^2b^2 + b^2c^2 + c^2a^2. \tag{12.2}$$

We're not quite there, but if we can show that the right side of (12.2) is at least $abc(a + b + c)$, we'll have our desired inequality. The right side of (12.2) is also a sum of three squares. We let $x = ab$, $y = bc$, and $z = ca$ in (12.1), and we have

$$a^2b^2 + b^2c^2 + c^2a^2 \geq ab^2c + abc^2 + a^2bc.$$

Success! Factoring the right side of this inequality gives us

$$a^2b^2 + b^2c^2 + c^2a^2 \geq abc(a + b + c). \tag{12.3}$$

Combining inequalities (12.2) and (12.3) gives us $a^4 + b^4 + c^4 \geq a^2b^2 + b^2c^2 + c^2a^2 \geq abc(a + b + c)$.

□

Our solutions to parts (b) and (c) show how we can build on inequalities we have already proved in order to prove more complicated inequalities. First, in part (b), we recognized that the greater side of the desired inequality was similar to that of the inequality in part (a). Moreover, the lesser side of the desired inequality was also similar to the lesser side in the inequality of part (a). So, we added three applications of part (a) to prove the desired inequality in part (b).

> **Concept:** Once we have proved an inequality, it becomes a tool to prove other inequalities, just like we can use proven identities to find new identities.

In going from part (b) to part (c), we first recognized that the desired greater side in (c) had the same form as the greater side in (b)—they were both sums of three squares. So, we applied part (b) to $a^4 + b^4 + c^4$. Notice that we used a little wishful thinking here; we hadn't even thought about the lesser side yet.

> **Concept:** In more complicated inequality problems, we often just focus on one side of the desired inequality, hoping to be able to manipulate it into or compare it to something that we can show is greater than the lesser side.

In other words, if we want to show that $P \geq R$, we sometimes first find an expression Q such that we can prove $P \geq Q$. Then, we try to show that $Q \geq R$, so we can deduce that $P \geq Q \geq R$. Above, we did this by showing that $a^4 + b^4 + c^4 \geq a^2b^2 + b^2c^2 + c^2a^2$ and $a^2b^2 + b^2c^2 + c^2a^2 \geq abc(a + b + c)$, so $a^4 + b^4 + c^4 \geq abc(a + b + c)$.

> **Sidenote:** An old legend has it that a man once willed his herd of camels to his three children, giving the oldest 1/2 of the herd to the oldest, 1/3 to the middle child, and 1/9 to the youngest child. Unfortunately, the herd only had 17 camels, so the children did not know how to divide the camels among them. They asked their wise uncle what to do. He responded, "I will give you a magic camel to help you divide the herd. But after you divide the herd, you must return the camel."
>
> When they added the magic camel to the herd, the herd then had 18 camels. The oldest then took 1/2 the herd, 9 camels. The middle child took 1/3 of the herd, 6 camels. And the youngest took 1/9 of the herd, 2 camels. Together, then, they took $9 + 6 + 2 = 17$ camels, and returned the 18th, the magic camel, to their wise uncle.

Our intermediary expression $a^2b^2 + b^2c^2 + c^2a^2$ is a magic camel—it doesn't exist in the problem when we start, and it's not there in the final result. But it's a crucial step from the beginning to the end.

> **Concept:** Be on the lookout for magic camels.

Problem 12.10: Let a and b be positive real numbers. Show that $\dfrac{b}{a^2} + \dfrac{a}{b^2} \geq \dfrac{1}{a} + \dfrac{1}{b}$.

Solution for Problem 12.10: First, we get rid of the fractions by multiplying by a^2b^2, which gives

$$a^3 + b^3 \geq a^2b + ab^2.$$

Now, we move all the terms to one side and try to factor. This gives us $a^3 - a^2b - ab^2 + b^3 \geq 0$. Factoring a^2 out of the first two terms on the left and b^2 out of the last two terms gives

$$a^3 - a^2b - ab^2 + b^3 = a^2(a - b) + b^2(-a + b) = a^2(a - b) - b^2(a - b) = (a^2 - b^2)(a - b) = (a - b)^2(a + b).$$

So, our inequality now is $(a - b)^2(a + b) \geq 0$. We now have a path to a solution.

Since $a > 0$ and $b > 0$, we have $a + b > 0$. The Trivial Inequality gives us $(a - b)^2 \geq 0$. Multiplying these two gives $(a - b)^2(a + b) \geq 0$, and working backwards through our steps above gives us the desired inequality. $\square$

> **Concept:** Many inequality problems are essentially manipulation and factorization problems.

Problem 12.11: Find all ordered pairs of real numbers (x, y) such that $(4x^2 + 4x + 3)(y^2 - 6y + 13) = 8$.

Solution for Problem 12.11: We certainly aren't interested in expanding the product of those quadratics, so we look for other ways to work with the quadratics.

> **Concept:** When stuck on a problem involving quadratics, try completing the square.

We complete the square in the two quadratics, because we know plenty about perfect squares. The resulting equation is
$$((2x + 1)^2 + 2)((y - 3)^2 + 4) = 8.$$
The Trivial Inequality tells us that $(2x + 1)^2 \geq 0$ and $(y - 3)^2 \geq 0$, so we have

$$(2x + 1)^2 + 2 \geq 2,$$
$$(y - 3)^2 + 4 \geq 4.$$

The product of these inequalities looks a lot like our equation:

$$((2x+1)^2 + 2)((y-3)^2 + 4) \geq 8. \tag{12.4}$$

This inequality holds for all real x and y. In order for equality to hold, we must have equality in both of the inequalities we multiplied together to produce line (12.4). In other words, equality only holds if $(2x+1)^2 = (y-3)^2 = 0$. Therefore, we have $((2x+1)^2 + 2)((y-3)^2 + 4) = 8$ if and only if $(x,y) = (-\frac{1}{2}, 3)$. $\square$

> **Concept:** Some equations can be solved by viewing them as the equality case of a nonstrict inequality. We can then use the equality condition of the inequality to find the solutions to the equation.

Exercises

12.2.1 Show that $x^2 + x + 1 \geq 3x$ for all real x.

12.2.2 Show that $(a^2 + 1)(b^2 + 1) \geq (ab + 1)^2$ for all real a and b.

12.2.3 We wish to show that if $a > b$, then $a^3 + a^2 b \geq b^3 + ab^2$.

(a) What is wrong with this solution: "Because $a > b$, we have $a^2 > b^2$. Multiplying both sides by $a+b$ gives $a^3 + a^2 b > b^3 + ab^2$, so $a^3 + a^2 b \geq b^3 + ab^2$, as desired."

(b) Prove that if $a > b$, then $a^3 + a^2 b \geq b^3 + ab^2$. When does equality hold?

12.2.4 Show that for all nonzero x, y, and z, we have $x^2 + y^2 + z^2 + \frac{1}{x^2} + \frac{1}{y^2} + \frac{1}{z^2} \geq 2\left(\frac{x}{y} + \frac{y}{z} + \frac{z}{x}\right)$. **Hints:** 186

12.2.5★ Prove that if a and b are nonnegative real numbers, then $(a^7 + b^7)(a^2 + b^2) \geq (a^5 + b^5)(a^4 + b^4)$. **Hints:** 212

12.2.6★ Prove that $\frac{x^2}{4} + y^2 + z^2 \geq xy - xz + 2yz$ for all real numbers x, y, and z. **Hints:** 189

12.3 AM-GM Inequality with Two Variables

While the Trivial Inequality is the most fundamental of the inequalities we will study, there are many others that are powerful problem solving tools. As we'll see, the Trivial Inequality is a key tool in proving many of these other inequalities.

The first new inequality we will explore is the **Arithmetic Mean-Geometric Mean Inequality**, which is usually more succinctly called the **AM-GM Inequality**.

> **Definition:** The **arithmetic mean (AM)** of the nonnegative numbers $a_1, a_2, \ldots, a_n$ is
>
> $$\frac{a_1 + a_2 + \cdots + a_n}{n},$$
>
> and the **geometric mean (GM)** of these numbers is
>
> $$\sqrt[n]{a_1 a_2 \cdots a_n}.$$

You probably already know the arithmetic mean of a group of numbers as the **average** of the numbers.

We'll start by comparing the arithmetic mean and geometric mean of pairs of nonnegative numbers.

Problems

Problem 12.12: Show that $\dfrac{a+b}{2} \geq \sqrt{ab}$ for all $a, b \geq 0$. For what values of a and b are the two sides equal?

Problem 12.13: Let x and y be positive real numbers.
(a) Find the geometric mean of $\frac{x}{y}$ and $\frac{y}{x}$.
(b) Show that $\frac{x}{y} + \frac{y}{x} \geq 2$. When does equality occur?

Problem 12.14: Show that if $x, y > 0$, then $\dfrac{1}{x^2} + \dfrac{1}{y^2} \geq \dfrac{2}{xy}$.

Problem 12.15: Show that if x, y, and z are nonnegative, then $xy + yz + zx \geq x\sqrt{yz} + y\sqrt{zx} + z\sqrt{xy}$.

Problem 12.16: Show that $(x+y)(y+z)(z+x) \geq 8xyz$ for all nonnegative numbers x, y, and z.

Problem 12.12: Show that
$$\frac{a+b}{2} \geq \sqrt{ab}$$
for all $a, b \geq 0$. For what values of a and b are the two sides equal?

Solution for Problem 12.12: Working backwards, we start by multiplying both sides by 2 to give $a + b \geq 2\sqrt{ab}$. Because $a, b \geq 0$, we can square both sides to give $(a+b)^2 \geq 4ab$. Expanding the left side and rearranging gives $a^2 - 2ab + b^2 \geq 0$. Factoring then gives $(a-b)^2 \geq 0$, which is true by the Trivial Inequality. We can now walk backwards through our steps to write our proof:

Proof: Let $a, b \geq 0$. The Trivial Inequality gives us $(a-b)^2 \geq 0$, and expanding the left side gives $a^2 - 2ab + b^2 \geq 0$. Adding $4ab$ to both sides gives $a^2 + 2ab + b^2 \geq 4ab$, so we have $(a+b)^2 \geq 4ab$. Because both sides are positive, we can take the square root of both sides to give $a + b \geq 2\sqrt{ab}$. Dividing both sides by 2 gives us $(a+b)/2 \geq \sqrt{ab}$.

We only have equality in our first step, $(a-b)^2 \geq 0$, if $a = b$. Otherwise, the left side is strictly larger than the right. If we have $a = b$, we have $(a+b)/2 = \sqrt{ab}$, so equality holds in $(a+b)/2 \geq \sqrt{ab}$ if and only if $a = b$. $\square$

We have now proved the AM-GM Inequality for two variables:

Important: The arithmetic mean of two nonnegative numbers is greater than or equal to the geometric mean of the numbers. The two means are equal if and only if the two numbers are equal. In other words, if a and b are nonnegative, then

$$\frac{a+b}{2} \ge \sqrt{ab},$$

and equality holds if and only if $a = b$.

Don't overlook the equality condition of the AM-GM Inequality. The sentence, "The two means are equal if and only if the two numbers are equal," not only tells us that the two means are the same if the two numbers are equal, but it also tells us that the two numbers must be equal if the two means are the same.

Let's try AM-GM on some problems.

Problem 12.13: Let x and y be positive real numbers. Show that $\frac{x}{y} + \frac{y}{x} \ge 2$. When does equality occur?

Solution for Problem 12.13: Because x and y are positive, so are x/y and y/x. So, we can apply in the AM-GM Inequality to find

$$\frac{x/y + y/x}{2} \ge \sqrt{\frac{x}{y} \cdot \frac{y}{x}} = 1.$$

Multiplying the far left side and the far right side by 2, we get

$$\frac{x}{y} + \frac{y}{x} \ge 2.$$

Equality occurs if and only if $\frac{x}{y} = \frac{y}{x}$. Cross-multiplying gives $x^2 = y^2$. So, because x and y are positive, equality occurs if and only if $x = y$. $\square$

One of our biggest clues to try AM-GM in Problem 12.13 is the fact that $\frac{x}{y}$ and $\frac{y}{x}$ have a constant product. This means that if we apply AM-GM to these quantities, we will have a constant on the lesser side of the inequality, just as our desired inequality has.

Similarly, if we have nonnegative expressions with a constant sum, we can apply AM-GM to produce an inequality in which the larger side is constant.

Concept: When trying to prove an inequality involving nonnegative expressions that have a constant sum or product, consider using AM-GM.

Problem 12.14: Show that if $x, y > 0$, then $\frac{1}{x^2} + \frac{1}{y^2} \ge \frac{2}{xy}$.

Solution for Problem 12.14: We have a sum on the greater side and a product on the lesser side. This looks like a job for AM-GM. Applying AM-GM to $\frac{1}{x^2}$ and $\frac{1}{y^2}$ gives

$$\frac{\frac{1}{x^2} + \frac{1}{y^2}}{2} \geq \sqrt{\left(\frac{1}{x^2}\right)\left(\frac{1}{y^2}\right)}.$$

Multiplying both sides by 2 and simplifying the right side gives the desired $\frac{1}{x^2} + \frac{1}{y^2} \geq \frac{2}{xy}$. $\square$

Problem 12.15: Show that if x, y, and z are nonnegative, then $xy + yz + zx \geq x\sqrt{yz} + y\sqrt{zx} + z\sqrt{xy}$.

Solution for Problem 12.15: We're not sure where to start, but we see sums of expressions, so we try AM-GM. What's wrong with the following start:

Bogus Solution: We try applying AM-GM to the sum $x\sqrt{yz} + y\sqrt{zx}$. This gives us

$$\frac{x\sqrt{yz} + y\sqrt{zx}}{2} \geq \sqrt{(x\sqrt{yz})(y\sqrt{zx})}.$$

Not only is the expression on the right a mess, but the result is an inequality in which $x\sqrt{yz}$ is on the greater side, not the lesser side.

WARNING!! If you want to show that $A \geq B$, then inequalities like $C \geq A$ or $B \geq D$ are not going to help you. You need to work with inequalities in which A is on the greater side or B is on the lesser side.

The $\sqrt{yz}$ term is a big clue how to use AM-GM, particularly because the square root is on the lesser side of the inequality. We produce the square roots by applying AM-GM to each pair of x, y, and z:

$$\frac{x+y}{2} \geq \sqrt{xy},$$
$$\frac{y+z}{2} \geq \sqrt{yz},$$
$$\frac{z+x}{2} \geq \sqrt{zx}.$$

Multiplying these by z, x, and y, respectively, gives us all the terms we want on the lesser side:

$$z \cdot \frac{x+y}{2} \geq z\sqrt{xy},$$
$$x \cdot \frac{y+z}{2} \geq x\sqrt{yz},$$
$$y \cdot \frac{z+x}{2} \geq y\sqrt{zx}.$$

Aha! This gives us the terms we want on the greater side, too!

> **Concept:** Focus on one side of an inequality and the other will often fall into place.

Adding all three of these inequalities gives the desired $xy + yz + zx \geq x\sqrt{yz} + y\sqrt{zx} + z\sqrt{xy}$. $\square$

Addition is not the only way we can combine inequalities.

> **Problem 12.16:** Show that $(x + y)(y + z)(z + x) \geq 8xyz$ for all nonnegative numbers x, y, and z.

Solution for Problem 12.16: We have a product of sums on the greater side of the inequality. Sums on the greater side—that sounds like AM-GM. Let's try it! We use AM-GM to produce each sum on the desired left side, which gives us

$$\frac{x + y}{2} \geq \sqrt{xy},$$
$$\frac{y + z}{2} \geq \sqrt{yz},$$
$$\frac{z + x}{2} \geq \sqrt{zx}.$$

We want the product of these sums, so we multiply all three inequalities to get

$$\frac{(x + y)(y + z)(z + x)}{8} \geq \sqrt{x^2 y^2 z^2}.$$

We have $\sqrt{x^2 y^2 z^2} = xyz$ because $x, y, z \geq 0$, so multiplying both sides by 8 gives the desired inequality $(x + y)(y + z)(z + x) \geq 8xyz$. $\square$

Exercises

12.3.1 Show that the sum of any positive number and its reciprocal is no less than 2.

12.3.2 Show that $a^6 + b^4 \geq 2a^3 b^2$ for all positive real numbers a and b.

12.3.3 Let $a, b, c,$ and d be nonnegative real numbers. Show that $(ab + cd)(ac + bd) \geq 4abcd$.

12.3.4 Show that if $x, y > 0$, then $\sqrt{x + \frac{1}{y}} + \sqrt{y + \frac{1}{x}} \geq 2\sqrt{2}$.

12.3.5 Let x and y be positive real numbers such that $x + y = 1$. Show that $\left(1 + \frac{1}{x}\right)\left(1 + \frac{1}{y}\right) \geq 9$. *(Source: CMO)*

12.3.6 Use AM-GM to show that $\dfrac{x^2 + 2}{\sqrt{x^2 + 1}} \geq 2$ for all real x.

12.3.7★ Show that for all nonnegative $a, b, c,$ and d, we have $\dfrac{a + b + c + d}{4} \geq \sqrt[4]{abcd}$. **Hints:** 254

> **Extra!** *The beginning of knowledge is the discovery of something we do not understand.*
> – Frank Herbert

12.4 AM-GM with More Variables

We've seen that AM-GM with two variables is basically just a routine application of the Trivial Inequality. If that were all there were to AM-GM, it probably wouldn't even have its own name. But having seen that AM-GM holds for two variables, we should wonder if it holds for more variables.

The AM-GM Inequality does, indeed, hold for more variables.

> **Important:** The **AM-GM Inequality** tells us that the arithmetic mean of any n non-negative numbers is greater than or equal to the geometric mean of the numbers. That is, if $a_1, a_2, \ldots, a_n$ are nonnegative, then
>
> $$\frac{a_1 + a_2 + \cdots + a_n}{n} \geq \sqrt[n]{a_1 a_2 \cdots a_n}.$$
>
> Equality holds if and only if all a_i are equal.

The proof is very intricate (and beautiful!). We'll give you a chance to prove the general AM-GM Inequality as a series of Challenge Problems. Through this series of problems, you'll walk in the footsteps of the great mathematician **Augustin Louis Cauchy**.

> **Sidenote:** The French mathematician and engineer Augustin Louis Cauchy was one of the most prolific mathematicians of the 1800s. He made numerous contributions relating to inequalities, one of which bears his name. He also made important contributions to number theory, geometry, calculus, and even physics. It's hard to go very far in math and science without reading about his many accomplishments. However, the name Cauchy would probably be more well-known were it not for Cauchy's abrasiveness and the many contentious rifts he had with other mathematicians of his day.

Fortunately, using AM-GM for any number of variables is a lot easier than proving it works! Try it and see.

Problems

> **Problem 12.17:** Let x, y, and z be positive real numbers. Show that
>
> $$(x + y + z)\left(\frac{1}{x} + \frac{1}{y} + \frac{1}{z}\right) \geq 9.$$

> **Problem 12.18:** In this problem, we show that that $\left(\frac{n+1}{2}\right)^n > n!$ for all positive integers $n \geq 2$.
>
> (a) Take the nth root of both sides. Why does the result make us think of the AM-GM Inequality?
>
> (b) Solve the problem.

Problem 12.19: Prove that for all nonnegative real numbers x, y, z, we have

$$x^2(1 + y^2) + y^2(1 + z^2) + z^2(1 + x^2) \geq 6xyz.$$

When does equality occur?

Problem 12.17: Let x, y, and z be positive real numbers. Show that $(x + y + z)\left(\dfrac{1}{x} + \dfrac{1}{y} + \dfrac{1}{z}\right) \geq 9$.

Solution for Problem 12.17: We have two sums on the greater side, so we apply AM-GM to both. We find

$$\frac{x + y + z}{3} \geq \sqrt[3]{xyz},$$

$$\frac{\frac{1}{x} + \frac{1}{y} + \frac{1}{z}}{3} \geq \sqrt[3]{\frac{1}{x} \cdot \frac{1}{y} \cdot \frac{1}{z}} = \frac{1}{\sqrt[3]{xyz}}.$$

Multiplying these inequalities, we have

$$\frac{(x + y + z)\left(\frac{1}{x} + \frac{1}{y} + \frac{1}{z}\right)}{9} \geq \frac{\sqrt[3]{xyz}}{\sqrt[3]{xyz}} = 1.$$

Multiplying both sides by 9 gives the desired result. $\square$

Problem 12.18: Prove that $\left(\dfrac{n + 1}{2}\right)^n > n!$ for all positive integers $n \geq 2$.

Solution for Problem 12.18: First, we take the nth root of both sides to get

$$\frac{n + 1}{2} > \sqrt[n]{n!}.$$

We now have an nth root of a product of n numbers on the lesser side, which is a strong indication to use the AM-GM Inequality. So, we apply AM-GM to the n numbers $1, 2, \ldots, n$, which gives

$$\frac{1 + 2 + \cdots + n}{n} > \sqrt[n]{1 \cdot 2 \cdots n} = \sqrt[n]{n!}.$$

(Note that this is a strict inequality, since the numbers $1, 2, \ldots, n$ are not all equal, and therefore, they do not satisfy the equality condition of AM-GM.) Writing the left side in closed form gives us

$$\frac{1 + 2 + \cdots + n}{n} = \frac{\frac{n(n+1)}{2}}{n} = \frac{n + 1}{2},$$

so we have $\frac{n+1}{2} > \sqrt[n]{n!}$. Raising both sides to the nth power gives the desired result. $\square$

Problem 12.19: Prove that for all nonnegative real numbers x, y, z, we have

$$x^2(1 + y^2) + y^2(1 + z^2) + z^2(1 + x^2) \geq 6xyz.$$

When does equality occur?

Solution for Problem 12.19: We have a sum on the greater side and a product on the lesser, which suggests using the AM-GM Inequality. However, when we apply AM-GM to the three added terms on the left, their geometric mean is a very messy expression. Going the other direction, we could view xyz as the geometric mean of x^3, y^3, and z^3, but that doesn't seem to help much, either.

> **Concept:** It's often necessary to manipulate a desired inequality into a form for which tools like AM-GM apply. Factor! Expand! Regroup terms!

Here, the 6 is a big clue. Expanding all the terms on the left gives us 6 terms:

$$x^2 + x^2y^2 + y^2 + y^2z^2 + z^2 + z^2x^2 \geq 6xyz.$$

The larger side is now the sum of six terms, and the smaller is a product, so the inequality is a natural candidate for AM-GM. All six terms are nonnegative, so by the AM-GM Inequality, we have

$$\frac{x^2 + x^2y^2 + y^2 + y^2z^2 + z^2 + z^2x^2}{6} \geq \sqrt[6]{x^2 \cdot x^2y^2 \cdot y^2 \cdot y^2z^2 \cdot z^2 \cdot z^2x^2}$$

$$= \sqrt[6]{x^6y^6z^6}$$

$$= xyz.$$

Multiplying both sides by 6 gives the desired $x^2 + x^2y^2 + y^2 + y^2z^2 + z^2 + z^2x^2 \geq 6xyz$.

We must also determine when equality occurs. Equality occurs if and only if $x^2 = x^2y^2 = y^2 = y^2z^2 = z^2 = z^2x^2$. Since x, y, and z are nonnegative and $x^2 = y^2 = z^2$, we get $x = y = z$. Then $x^2 = x^4$, so $x = 0$ or $x = 1$. Therefore, equality occurs only for $(x, y, z) = (0, 0, 0)$ and $(x, y, z) = (1, 1, 1)$. $\square$

Exercises

12.4.1 Let $a, b, c \geq 0$. Show that $a^2b + ab^2 + a^2c + ac^2 + b^2c + bc^2 \geq 6abc$.

12.4.2 Show that if a, b, and c are positive, then $\dfrac{a}{b} + \dfrac{b}{c} + \dfrac{c}{a} \geq 3$.

12.4.3 Find the minimum value of $xy + \dfrac{2}{x} + \dfrac{4}{y}$ for positive real numbers x and y.

12.4.4 Show that, for all positive real numbers p, q, r, s, we have

$$(p^2 + p + 1)(q^2 + q + 1)(r^2 + r + 1)(s^2 + s + 1) \geq 81pqrs.$$

12.4.5★ Show that $\dfrac{a_2 + a_3}{a_1} + \dfrac{a_3 + a_4}{a_2} + \cdots + \dfrac{a_n + a_1}{a_{n-1}} + \dfrac{a_1 + a_2}{a_n} \geq 2n$ for all positive reals $a_1, a_2, \ldots, a_n$.
Hints: 210

12.4.6★ If $a > b > 0$, find the minimum value of $a + \dfrac{1}{(a - b)b}$. **Hints:** 345

12.5 The Cauchy-Schwarz Inequality

The AM-GM Inequality isn't the only inequality that Cauchy proved. He also developed a powerful inequality that still bears his name.

Problems

Problem 12.20: Let a, b, c, and d be real numbers. Prove that $(a^2 + b^2)(c^2 + d^2) \geq (ac + bd)^2$. When does equality occur?

Problem 12.21: Let $a_1, a_2, \ldots, a_n, b_1, b_2, \ldots, b_n$ be real numbers. In this problem, we find two proofs that
$$(a_1^2 + a_2^2 + \cdots + a_n^2)(b_1^2 + b_2^2 + \cdots + b_n^2) \geq (a_1b_1 + a_2b_2 + \cdots + a_nb_n)^2.$$
We call this the **Cauchy-Schwarz Inequality**.

 (a) *Proof 1:* Consider the expansion of both sides. What terms have no variables besides a_1, a_2, b_1, and b_2?

 (b) Apply the inequality from the previous problem to the terms in part (a).

 (c) Use your observations in parts (a) and (b) to prove the Cauchy-Schwarz Inequality.

 (d) *Proof 2:* Consider the quadratic $f(x) = (a_1x - b_1)^2 + (a_2x - b_2)^2 + \cdots + (a_nx - b_n)^2$. What is the discriminant of this quadratic?

 (e) Why must the discriminant of $f(x)$ be nonpositive?

 (f) Find the equality condition of the Cauchy-Schwarz Inequality.

Problem 12.22: Show that $a^2 + b^2 + c^2 \geq \dfrac{(a + b + c)^2}{3}$.

Problem 12.23: Let x, y, and z be positive real numbers. Use the Cauchy-Schwarz Inequality to prove that
$$(x + y + z)\left(\frac{1}{x} + \frac{1}{y} + \frac{1}{z}\right) \geq 9.$$

Problem 12.24: Let a and b be positive real numbers with $a + b = 1$. Prove that $\dfrac{a^2}{a + 1} + \dfrac{b^2}{b + 1} \geq \dfrac{1}{3}$. (*Source: Hungary*)

Problem 12.25: If a, b, c, x, y, and z are nonzero real numbers and $a^2 + b^2 + c^2 = 25$, $x^2 + y^2 + z^2 = 36$, and $ax + by + cz = 30$, compute
$$\frac{a + b + c}{x + y + z}.$$
(*Source: ARML*)

Just as we did with AM-GM, we start with the 2-variable version of the Cauchy-Schwarz Inequality.

Problem 12.20: Let a, b, c, and d be real numbers. Prove that $(a^2 + b^2)(c^2 + d^2) \geq (ac + bd)^2$. When does equality occur?

Solution for Problem 12.20: *Solution 1: Bash it out.* We start by working backwards. We expand both sides, which gives

$$a^2c^2 + a^2d^2 + b^2c^2 + b^2d^2 \geq a^2c^2 + 2abcd + b^2d^2.$$

Simplifying this gives $a^2d^2 + b^2c^2 \geq 2abcd$. What's wrong with this finish:

> **Bogus Solution:** Because a^2d^2 and b^2c^2 are positive, we can apply AM-GM to find
>
> $$\frac{a^2d^2 + b^2c^2}{2} \geq \sqrt{(a^2d^2)(b^2c^2)} = abcd.$$
>
> Multiplying both sides by 2 gives $a^2d^2 + b^2c^2 \geq 2abcd$. Adding $a^2c^2 + b^2d^2$ to both sides, and then factoring the result, gives the desired
> $$(a^2 + b^2)(c^2 + d^2) \geq (ac + bd)^2.$$
>
> We have equality when we have equality in our application of AM-GM, which is when $a^2d^2 = b^2c^2$.

We've overlooked one detail, and that is the fact that a, b, c, or d might be negative, so $abcd$ might be negative above. Therefore, we must write $\sqrt{(a^2d^2)(b^2c^2)} = |abcd|$. We can still patch our proof with a little careful casework, or we can remember that 2-variable AM-GM has its roots in the Trivial Inequality. Subtracting $2abcd$ from both sides of $a^2d^2 + b^2c^2 \geq 2abcd$ leads us to the following solution:

> **Proof:** The Trivial Inequality gives us $(ad - bc)^2 \geq 0$. Expanding the left side and rearranging gives
> ☑ $a^2d^2 + b^2c^2 \geq 2abcd$. Adding $a^2c^2 + b^2d^2$ to both sides, and then factoring the result, gives the desired
> $$(a^2 + b^2)(c^2 + d^2) \geq (ac + bd)^2.$$
>
> We have equality when we have equality in our initial application of the Trivial Inequality, which was $(ad - bc)^2 \geq 0$. Therefore, our equality condition is $ad = bc$.

Solution 2: Recall a similar identity. Back on page 340, we proved the identity

$$(a^2 + b^2)(c^2 + d^2) = (ac + bd)^2 + (ad - bc)^2.$$

By the Trivial Inequality, we have $(ad - bc)^2 \geq 0$, so we have

$$(a^2 + b^2)(c^2 + d^2) - (ac + bd)^2 = (ad - bc)^2 \geq 0,$$

which means that

$$(a^2 + b^2)(c^2 + d^2) \geq (ac + bd)^2.$$

Equality holds when $(ad - bc)^2 = 0$, which occurs if and only if $ad = bc$. $\square$

Don't worry if you didn't remember the identity we used in our second solution. Our first solution shows that, when all else fails, some algebra and the Trivial Inequality will often do the trick.

We were able to extend AM-GM to any number of variables. Can we do the same with the inequality we proved in Problem 12.20?

Problem 12.21: Let $a_1, a_2, \ldots, a_n, b_1, b_2, \ldots, b_n$ be real numbers. Prove that

$$(a_1^2 + a_2^2 + \cdots + a_n^2)(b_1^2 + b_2^2 + \cdots + b_n^2) \geq (a_1 b_1 + a_2 b_2 + \cdots + a_n b_n)^2.$$

Solution for Problem 12.21: Solution 1: Extend our previous solution. If we expand the products on both sides, we have terms of the form $a_i^2 b_i^2$ on both sides. On the left, we also have all the possible terms $a_i^2 b_j^2$ for which $i \neq j$ and $1 \leq i, j \leq n$, which we'll write as $\displaystyle\sum_{i \neq j, 1 \leq i, j \leq n} a_i^2 b_j^2$, and on the right, we also have all possible terms $2 a_i b_i a_j b_j$ for which $1 \leq i < j \leq n$, which we'll write as $\displaystyle\sum_{1 \leq i < j \leq n} 2 a_i b_i a_j b_j$. We can therefore write the expansions as

$$\sum_{i=1}^{n} a_i^2 b_i^2 + \sum_{i \neq j, 1 \leq i, j \leq n} a_i^2 b_j^2 \geq \sum_{i=1}^{n} a_i^2 b_i^2 + \sum_{1 \leq i < j \leq n} 2 a_i b_i a_j b_j.$$

The first sum on the left matches the first sum on the right, so we don't have to worry about those. Next, rather than work with the remaining complicated summations, we look at one $2 a_i b_i a_j b_j$ on the right and the corresponding $a_i^2 b_j^2$ and $a_j^2 b_i^2$ on the left. This is exactly the same problem we faced in Problem 12.20! The Trivial Inequality gives us

$$(a_i b_j - a_j b_i)^2 \geq 0.$$

Expanding the left side and rearranging gives $a_i^2 b_j^2 + a_j^2 b_i^2 \geq 2 a_i b_i a_j b_j$. If we do this for all pairs i and j such that $1 \leq i < j \leq n$ and $i \neq j$, and then add the results, we have

$$\sum_{i \neq j, 1 \leq i, j \leq n} a_i^2 b_j^2 \geq \sum_{1 \leq i < j \leq n} 2 a_i b_i a_j b_j.$$

Adding $\displaystyle\sum_{i=1}^{n} a_i^2 b_i^2$ to both sides gives

$$\sum_{i=1}^{n} a_i^2 b_i^2 + \sum_{i \neq j, 1 \leq i, j \leq n} a_i^2 b_j^2 \geq \sum_{i=1}^{n} a_i^2 b_i^2 + \sum_{1 \leq i < j \leq n} 2 a_i b_i a_j b_j,$$

which is our expansion above of the desired inequality. Factoring both sides gives

$$(a_1^2 + a_2^2 + \cdots + a_n^2)(b_1^2 + b_2^2 + \cdots + b_n^2) \geq (a_1 b_1 + a_2 b_2 + \cdots + a_n b_n)^2.$$

We have equality if and only if we have equality in all of our applications of the Trivial Inequality, which were the inequalities $(a_i b_j - a_j b_i)^2 \geq 0$, where $i \neq j$. In other words, for any i and j such that $i \neq j$, we must have $a_i b_j = a_j b_i$. If b_i and b_j are nonzero, we can rearrange this as $a_i/b_i = a_j/b_j$. Since this relationship must hold for any i and j, the ratio a_k/b_k must be constant for all k. There is however a subtlety to this equality condition. Some of the variables can equal 0. For example, equality occurs if $(a_1, a_2, a_3) = (0, 2, 3)$ and $(b_1, b_2, b_3) = (0, 4, 6)$. Also, equality occurs if either all the a_i or all the b_i are zero. So, to be completely accurate, equality occurs if and only if either all a_i are 0, or there is a constant t such that $b_i = ta_i$ for all i. (The case that all the b_i are 0 is addressed by letting $t = 0$.)

Solution 2: Use the discriminant. Discriminant? What discriminant? To see how we might think to use the discriminant, we let

$$P = a_1^2 + a_2^2 + \cdots + a_n^2,$$
$$Q = a_1 b_1 + a_2 b_2 + \cdots + a_n b_n,$$
$$R = b_1^2 + b_2^2 + \cdots + b_n^2.$$

This allows us to write our desired inequality as simply $PR \geq Q^2$, or $Q^2 - PR \leq 0$. A square minus a product less than or equal to zero; this might make us think of the discriminant of a quadratic that has no more than one real root. (Yes, this is a leap, but great insights come from such creative jumps.)

But what quadratic might have such a discriminant? The coefficients of its quadratic term and constant term must be $a_1^2 + a_2^2 + \cdots + a_n^2$ and $b_1^2 + b_2^2 + \cdots + b_n^2$, so we can try to produce a suitable quadratic $f(x)$ by writing

$$f(x) = (a_1 x - b_1)^2 + (a_2 x - b_2)^2 + \cdots + (a_n x - b_n)^2.$$

This gives us the appropriate quadratic and constant terms. But what about the linear term? When we expand each of the squares, we have

$$f(x) = (a_1^2 + a_2^2 + \cdots + a_n^2)x^2 - 2(a_1 b_1 + a_2 b_2 + \cdots + a_n b_n)x + (b_1^2 + b_2^2 + \cdots + b_n^2).$$

Now, we're in business. We have $f(x) = Px^2 - 2Qx + R$, and we wish to prove that $PR \geq Q^2$.

Before we dive into using our quadratic, we have to address the possibility that $P = 0$. If $P = a_1^2 + a_2^2 + \cdots + a_n^2 = 0$, then each a_i must be equal to 0. Then we have $Q = 0$, so the statement $PR \geq Q^2$ is obviously true. Otherwise, we have $P > 0$, so $f(x) = Px^2 - 2Qx + R$ is a quadratic.

We thought of building this quadratic by thinking about discriminants. To create an inequality from the discriminant, we must learn more about the roots of $f(x)$. Because $f(x)$ is a sum of squares, we have $f(x) \geq 0$ for any real number x. Since $f(x)$ is never negative, it cannot have two distinct real roots. Therefore, its discriminant must be nonpositive. This gives us $(-2Q)^2 - 4PR \leq 0$, from which we have the desired $Q^2 \leq PR$, or $PR \geq Q^2$.

Substituting back for $P, Q,$ and R, we have the Cauchy-Schwarz Inequality:

$$(a_1^2 + a_2^2 + \cdots + a_n^2)(b_1^2 + b_2^2 + \cdots + b_n^2) \geq (a_1 b_1 + a_2 b_2 + \cdots + a_n b_n)^2.$$

We have equality if and only if the discriminant of our quadratic is 0. There are a couple ways this could happen. First, we might have all the a_i or all the b_i equal to 0. (If all the a_i equal 0, then $f(x)$ isn't even a quadratic, but we clearly still have equality in the Cauchy-Schwarz Inequality.) Second, if some of the a_i are nonzero, the discriminant of $f(x)$ is still 0 if $f(x)$ has a double root. Let's take another look at $f(x)$:

$$f(x) = (a_1 x - b_1)^2 + (a_2 x - b_2)^2 + \cdots + (a_n x - b_n)^2.$$

Because $f(x)$ is a sum of squares, the Trivial Inequality tells us that $f(x)$ can never be negative. It can equal 0, but only if all of the squares equal 0 for the same value of x. In other words, we must have $(a_i x - b_i)^2 = 0$ for the same value of x for all values of i. This means that there is some x such that $b_i = a_i x$ for all i. In this case, this value of x is the only value of x such that $f(x) = 0$. Since $f(x)$ is a quadratic, this means that this value of x is a double root of the quadratic, so the discriminant of the quadratic equals 0, as desired.

Therefore, the equality condition for the Cauchy-Schwarz Inequality is that either all the a_i equal 0, or there exists a constant t such that $b_i = ta_i$ for all i. $\square$

Important: The **Cauchy-Schwarz Inequality** states that for any two sequences of n real numbers $a_1, a_2, \ldots, a_n$ and $b_1, b_2, \ldots, b_n$, we have

$$(a_1^2 + a_2^2 + \cdots + a_n^2)(b_1^2 + b_2^2 + \cdots + b_n^2) \geq (a_1 b_1 + a_2 b_2 + \cdots + a_n b_n)^2.$$

Equality occurs if and only if either $a_i = 0$ for all i, or there exists a constant t such that $b_i = ta_i$ for all i.

Sidenote: The Cauchy-Schwarz Inequality is also sometimes called the Cauchy-Schwarz-Buniakowsky Inequality.

Rather than writing Cauchy-Schwarz Inequality (or Cauchy-Schwarz-Buniakowsky Inequality) over and over, we will often simply refer to the inequality as "Cauchy." From example, we might write, "By Cauchy, we have $(a^2 + 1)(b^2 + 1) \geq (a \cdot b + 1 \cdot 1)^2$, so $(a^2 + 1)(b^2 + 1) \geq (ab + 1)^2$."

Sidenote: If you ever have trouble remembering which side of the Cauchy-Schwarz Inequality is the larger side and you don't have time to re-derive it, just compute a simple case. For example, let $a_1 = 1$, $a_2 = 2$, $b_1 = 3$, and $b_2 = 4$. Then, we have $(a_1^2 + a_2^2)(b_1^2 + b_2^2) = (5)(25) = 125$, and $(a_1 b_1 + a_2 b_2)^2 = (3 + 8)^2 = 121$, so $(a_1^2 + a_2^2)(b_1^2 + b_2^2)$ is on the larger side.

If you like to remember things with words, you can think of Cauchy as, "The product of the sum of the squares is greater than or equal to the square of the sum of the products."

Or perhaps you like symbols: $\prod \sum \square \geq \square \sum \prod$. (Remember, $\prod$ is the symbol we use for products and $\sum$ is the symbol for sums.)

It should not be a surprise that the Cauchy-Schwarz Inequality is particularly useful for inequalities that involve sums of squares.

Problem 12.22: Show that $a^2 + b^2 + c^2 \geq \dfrac{(a + b + c)^2}{3}$.

Solution for Problem 12.22: We have a sum of squares on the greater side, and the square of a sum on the lesser side. This may be a job for Cauchy. We isolate the square of the sum by multiplying both sides by 3:

$$3(a^2 + b^2 + c^2) \geq (a + b + c)^2.$$

To use Cauchy here, one of our sequences clearly must be a, b, c, since we have $a^2 + b^2 + c^2$ on the greater side. What is the other sequence? If we don't see it right away, we can let the other sequence be x, y, z, so that Cauchy gives us

$$(x^2 + y^2 + z^2)(a^2 + b^2 + c^2) \geq (xa + yb + zc)^2.$$

Comparing this to our desired inequality, we see that if we let $x = y = z = 1$, then we have the desired

$$3(a^2 + b^2 + c^2) \geq (a + b + c)^2.$$

□

Concept: If an inequality involves a sum of squares or a product of two sums with the same number of terms, try using the Cauchy-Schwarz Inequality to prove it.

We also saw another useful inequality manipulation in Problem 12.22 when we let one of the sequences be 1, 1, 1 in the Cauchy-Schwarz Inequality:

Concept: Some inequalities result from setting variables or even entire sequences of variables equal to carefully chosen constants, and then applying known inequalities to those variables or sequences.

Just as sums of squares are a tip-off to consider Cauchy, so, too, are products of sums:

Problem 12.23: Let x, y, and z be positive real numbers. Show that $(x + y + z)\left(\dfrac{1}{x} + \dfrac{1}{y} + \dfrac{1}{z}\right) \geq 9$.

Solution for Problem 12.23: We tackled this once already in Problem 12.17, but Cauchy now offers us an even faster solution. The product of sums on the greater side is our clue to try Cauchy. We first write the Cauchy-Schwarz Inequality for three variables:

$$(a_1^2 + a_2^2 + a_3^2)\left(b_1^2 + b_2^2 + b_3^2\right) \geq (a_1 b_1 + a_2 b_2 + a_3 b_3)^2.$$

In the case of the desired inequality, we let $a_1 = \sqrt{x}$, $a_2 = \sqrt{y}$, $a_3 = \sqrt{z}$, and $b_1 = \frac{1}{\sqrt{x}}$, $b_2 = \frac{1}{\sqrt{y}}$, $b_3 = \frac{1}{\sqrt{z}}$ to get

$$(x + y + z)\left(\frac{1}{x} + \frac{1}{y} + \frac{1}{z}\right) \geq \left(\sqrt{x} \cdot \frac{1}{\sqrt{x}} + \sqrt{y} \cdot \frac{1}{\sqrt{y}} + \sqrt{z} \cdot \frac{1}{\sqrt{z}}\right)^2$$
$$= (1 + 1 + 1)^2 = 9,$$

which is the desired inequality. □

Problem 12.23 is an example of how Cauchy can help deal with variables in denominators. While the product of sums was our first clue to use Cauchy, a more experienced problem-solver might see the denominators themselves as a clue to try Cauchy.

Problem 12.24: Let a and b be positive real numbers with $a + b = 1$. Prove that

$$\frac{a^2}{a+1} + \frac{b^2}{b+1} \geq \frac{1}{3}.$$

(Source: Hungary)

Solution for Problem 12.24: The inconvenient denominators make the expression on the left side difficult to deal with. We just saw an example of the Cauchy-Schwarz Inequality eliminating denominators, so we try that here. However, in this problem, we aren't given a second sequence that eliminates the denominators for us. So, we introduce it. Specifically, we pair the denominator $a + 1$ with another appearance of $a + 1$, and likewise for $b + 1$:

$$[(a+1) + (b+1)]\left(\frac{a^2}{a+1} + \frac{b^2}{b+1}\right) \geq \left(\sqrt{a+1} \cdot \sqrt{\frac{a^2}{a+1}} + \sqrt{b+1} \cdot \sqrt{\frac{b^2}{b+1}}\right)^2$$
$$= (a+b)^2,$$

so we have $\dfrac{a^2}{a+1} + \dfrac{b^2}{b+1} \geq \dfrac{(a+b)^2}{a+b+2} = \dfrac{1}{3}.$ $\square$

> **Concept:** Cauchy can help us prove inequalities that have a sum of fractions on the greater side. To try using Cauchy on such an inequality, let the sum of the fractions be one sum of squares, and the sum of the denominators be the other sum of squares:
>
> $$\left(\frac{x_1^2}{y_1^2} + \frac{x_2^2}{y_2^2} + \cdots + \frac{x_n^2}{y_n^2}\right)(y_1^2 + y_2^2 + \cdots + y_n^2) \geq (x_1 + x_2 + \cdots + x_n)^2.$$
>
> This strategy should particularly come to mind when the numerators of the fractions are perfect squares.

Problem 12.25: If a, b, c, x, y, and z are nonzero real numbers and $a^2 + b^2 + c^2 = 25$, $x^2 + y^2 + z^2 = 36$, and $ax + by + cz = 30$, compute

$$\frac{a+b+c}{x+y+z}.$$

(Source: ARML)

Solution for Problem 12.25: The given equations include two sums of squares, and a sum of products of the variables that are squared in the other two equations. These sums remind us of the Cauchy-Schwarz Inequality. Applying Cauchy, we get

$$(a^2 + b^2 + c^2)(x^2 + y^2 + z^2) \geq (ax + by + cz)^2.$$

But from the given equations, we have

$$(a^2 + b^2 + c^2)(x^2 + y^2 + z^2) = 900 = (ax + by + cz)^2.$$

Therefore, the information in the problem fits the case in which equality holds for the Cauchy-Schwarz Inequality. This tells us that the equality condition for the Cauchy-Schwarz Inequality must be met. Specifically, it means that for some real number t, we must have $a = tx$, $b = ty$, and $c = tz$. Substituting into $ax + by + cz = 30$, the left side becomes

$$ax + by + cz = tx^2 + ty^2 + tz^2 = t(x^2 + y^2 + z^2),$$

so $t(x^2 + y^2 + z^2) = 30$. Since $x^2 + y^2 + z^2 = 36$, we find that $t = 5/6$. Finally, we note that

$$\frac{a+b+c}{x+y+z} = \frac{tx + ty + tz}{x+y+z} = t = \frac{5}{6}.$$

$\square$

Concept: Some equations are really inequality problems in disguise. If an equation tells us that equality holds in some inequality, then we can use the equality condition of the inequality to solve the equation.

Sidenote: The Cauchy-Schwarz Inequality is a special case of the **Hölder Inequality**, which states that if p and q are positive numbers such that $\frac{1}{p} + \frac{1}{q} = 1$ then

$$x_1 y_1 + x_2 y_2 + \cdots + x_n y_n \le \left(x_1^p + x_2^p + \cdots + x_n^p\right)^{\frac{1}{p}} \left(y_1^q + y_2^q + \cdots + y_n^q\right)^{\frac{1}{q}}$$

for all nonnegative numbers $x_1, x_2, \ldots, x_n$ and $y_1, y_2, \ldots, y_n$. What happens when $p = q = 2$?

Exercises

12.5.1 Show that $(a_1 b_1 + a_2 b_2 + \cdots + a_n b_n)\left(\dfrac{a_1}{b_1} + \dfrac{a_2}{b_2} + \cdots + \dfrac{a_n}{b_n}\right) \ge (a_1 + a_2 + \cdots + a_n)^2$, where $a_i, b_i > 0$ for all i.

12.5.2 Use the Cauchy-Schwarz Inequality to prove that $a^2 + b^2 + c^2 \ge ab + ac + bc$ for all positive a, b, and c.

12.5.3 Jane has just drawn 6 rectangles, and each rectangle has a different length and width. The sum of the squares of the lengths is 40 and the sum of the squares of the widths is 20. What is largest possible value of the sum of the areas of the rectangles?

12.5.4 Suppose that a, b, c, and d are positive. Prove that $2\sqrt{a+b+c+d} \ge \sqrt{a} + \sqrt{b} + \sqrt{c} + \sqrt{d}$. **Hints:** 8

12.5.5 Let a, b, c, and d be positive real numbers. Show that $\dfrac{1}{a} + \dfrac{1}{b} + \dfrac{4}{c} + \dfrac{16}{d} \ge \dfrac{64}{a+b+c+d}$. *(Source: Leningrad)* **Hints:** 126

12.5.6 Let $x, y > 0$. Find the maximum value of $\dfrac{(3x + 4y)^2}{x^2 + y^2}$.

12.6   Maxima and Minima

In this section, we use our inequality tools to find the minimum or maximum value of various expressions.

Important: Maximization and minimization problems are two-part problems. We must show that the maximum or minimum value is attainable, and that no better value (no higher for maximization, no lower for minimization) can be attained.

We sometimes refer to maximization or minimization problems as **optimization** problems.

Problems

Problem 12.26: Suppose that $a + b \le c + d + 3$ and $c - b \le 3$. Find the maximum possible value of $a - d$.

Problem 12.27: Let $x^2 + 3xy + y^2 = 60$, where x and y are real. In this problem, we determine the maximum possible value of xy. *(Source: CEMC)*

(a) Isolate xy in the equation, and then maximize the expression that xy equals.

(b) Confirm that your maximum value in part (a) can be attained. If it cannot, return to part (a), and find a different way to isolate xy.

Problem 12.28: Let $\mathcal{B}$ be a rectangle with perimeter 36. In this problem, we find the largest possible area of $\mathcal{B}$.

(a) Use one of the inequalities in this chapter to relate the perimeter and the area of the rectangle.

(b) Use your inequality from part (a) to maximize the area of the rectangle. What must be true of the rectangle if it has this maximum area?

Problem 12.29: BigBox Company is making big rectangular cardboard boxes. They don't want to use any more cardboard than they have to. A customer requests boxes that have a volume of 1000 cubic inches.

(a) Suppose the customer wants each box to have a top, a bottom, and all four sides. Show that the BigBox Company should make boxes that are cubes in order to minimize the amount of cardboard they have to use.

(b) Suppose the customer wants each box to have a bottom and all four sides, but no top. What dimensions should BigBox Company use to minimize the amount of cardboard they use on each box?

Problem 12.30: For $x > 0$, let $f(x) = \dfrac{x^2 + 16}{3x}$. Find the minimum possible value of $f(x)$.

Extra!   *Which facts are likely to reappear? The simple facts. How to recognize them? Choose those that*
▸▸▸▸   *seem simple.*

– Robert M. Pirsig

Problem 12.31: Let $x_1, x_2, \ldots, x_n$ be a sequence of integers such that

(i) $-1 \leq x_i \leq 2$, for $i = 1, 2, 3, \ldots, n$,

(ii) $x_1 + x_2 + \cdots + x_n = 19$, and

(iii) $x_1^2 + x_2^2 + \cdots + x_n^2 = 99$.

Find the minimal and maximal possible values of $x_1^3 + x_2^3 + \cdots + x_n^3$. *(Source: AHSME)*

Problem 12.26: Suppose that $a + b \leq c + d + 3$ and $c - b \leq 3$. Find the maximum possible value of $a - d$.

Solution for Problem 12.26: What's wrong with the following:

> **Bogus Solution:** If equality holds in our inequalities, we have $c - b = 3$ and $a + b = c + d + 3$. Solving for c in the first equation gives us $c = b + 3$, and substituting this in the second equation gives us $a + b = b + 3 + d + 3$. Simplifying this equation gives us $a - d = 6$, so the maximum value of $a - d$ is 6.
>
>

As we'll see, the numerical answer is correct, but we haven't proved that $a - d$ cannot possibly equal anything higher.

> **WARNING!!** We can't be sure we have the maximum or minimum value of an expression until we show that no better value can be achieved.

We want to maximize $a - d$, so we want to find an inequality of the form

$$a - d \leq \text{ some number.}$$

If we also show that $a - d$ can equal this number, then we have found the maximum of $a - d$.

Since we want $a - d$ on the lesser side of an inequality, we subtract b and d from both sides of $a + b \leq c + d + 3$ to get $a - d \leq c - b + 3$. We also know that $c - b \leq 3$, so we have

$$a - d \leq c - b + 3 \leq 3 + 3 = 6.$$

Therefore, we have $a - d \leq 6$. We now want to show that it is possible to have $a - d = 6$. Our Bogus Solution above shows us how. We choose values such that equality holds in all the inequalities. For example, if $(a, b, c, d) = (6, 2, 5, 0)$, then both inequalities are satisfied and $a - d = 6$. $\square$

Problem 12.27: If $x^2 + 3xy + y^2 = 60$, where x and y are real, then determine the maximum possible value of xy. *(Source: CEMC)*

Solution for Problem 12.27: Here's another common error in maximization and minimization problems:

> **Bogus Solution:** We recognize the left side as $xy+(x+y)^2$, so we have $xy+(x+y)^2 = 60$.
> Isolating xy gives $xy = 60 - (x+y)^2$. By the Trivial Inequality, we have $(x+y)^2 \geq 0$, so $60 - (x+y)^2 \leq 60$. Therefore, we have $xy \leq 60$, which means the maximum possible value of xy is 60.

What we haven't done in this Bogus Solution is shown that the value 60 can be achieved. We only have $xy = 60$ if $60 - (x+y)^2 = 60$. That is, we achieve our maximum when $(x+y)^2 = 0$. We have $(x+y)^2 = 0$ if and only if $x = -y$. But if $x = -y$, then we have $-x^2 = 60$, so $x^2 = -60$. There are no real solutions to this equation! So, we see that we cannot have $xy = 60$.

> **WARNING!!** To show that an expression has a specific maximum value, there are two steps:
>
> 1. Show that the expression cannot be greater than this value.
>
> 2. Show that the expression can be equal to this value.
>
> Both steps are necessary! Similarly, if we wish to show that an expression has a minimum value, we show both that the expression cannot be less than this value, and that the expression can equal this value.

Back to the drawing board. We might next try to isolate xy directly, and we find that

$$xy = 20 - \frac{x^2}{3} - \frac{y^2}{3}.$$

Here, we have $x^2 \geq 0$ and $y^2 \geq 0$, so $xy \leq 20$. But once again, showing that this value can be attained stymies us. We only have $xy = 20$ in our equation if $x^2/3 = y^2/3 = 0$. But in this case, we have $x = y = 0$, so $xy = 0$, not 20.

Each of our false leads has failed us when we looked at the equality condition of the Trivial Inequality. In the first failed solution, the equality condition had $x = -y$, which failed. The second failed solution offered $x = 0$ and $y = 0$ as the equality condition for the Trivial Inequality. That's a clear loser. What might a successful equality condition be? Due to symmetry, a reasonable guess is $x = y$. So, we think about what Trivial Inequality application has $x = y$ as the equality condition. This leads us to $(x - y)^2$.

The Trivial Inequality gives us $(x - y)^2 \geq 0$. Expanding the left side gives $x^2 - 2xy + y^2 \geq 0$. Adding $5xy$ to both sides of this inequality to match the left side of the given equation yields $x^2 + 3xy + y^2 \geq 5xy$. Because $x^2 + 3xy + y^2 = 60$, we have $60 \geq 5xy$, so $12 \geq xy$. We've been fooled before, so we check our equality condition again. Here, the equality condition is $x = y$. Substituting this in our given equation gives us $5y^2 = 60$, so $y^2 = 12$. If $y^2 = 12$ and $x = y$, we have $xy = y^2 = 12$, so 12 is the maximum. $\square$

Problem 12.28: Let $\mathcal{B}$ be a rectangle with perimeter 36. Find the largest possible area of $\mathcal{B}$.

Solution for Problem 12.28: Here's a common mistake in geometric optimization problems:

> **Bogus Solution:** The maximum area occurs when the rectangle is a square. Since the
> perimeter is 36, each side has length 9, so the area is $9^2 = 81$.

The answer is correct, but we've skipped the most important step: we haven't proved the statement "The maximum area occurs when the rectangle is a square."

To do so, we let the side lengths of the rectangle be x and y. From the information about the perimeter, we have $2x + 2y = 36$, so $x + y = 18$. We wish to maximize the area, which is xy. We have an inequality that relates $x + y$ and xy. The AM-GM Inequality gives us

$$\frac{x + y}{2} \geq \sqrt{xy}.$$

Since $x + y = 18$, we have $\sqrt{xy} \leq 9$. Squaring both sides gives $xy \leq 81$. This shows that the area of the rectangle cannot be greater than 81. To show that it can possibly equal 81, we note that the equality condition of AM-GM gives us $x = y$, which describes the case in which the rectangle is a square with perimeter 36 and area 81. □

With Problem 12.28 as a guide, we're ready to tackle a tougher geometric problem.

Problem 12.29: BigBox Company is making big rectangular cardboard boxes. They don't want to use any more cardboard than they have to. A customer requests boxes that have a volume of 1000 cubic inches.

(a) Suppose the customer wants each box to have a top, a bottom, and all four sides. Show that the BigBox Company should make boxes that are cubes in order to minimize the amount of cardboard they have to use.

(b) Suppose the customer wants each box to have a bottom and all four sides, but no top. What dimensions should BigBox Company use to minimize the amount of cardboard they use on each box?

Solution for Problem 12.29: We let the dimensions of the box be x, y, and z. The volume of the box is $xyz - 1000$.

(a) We wish to minimize the surface area of the box, which equals $2(xy + yz + zx)$. This means we want $xy + yz + zx$ on the greater side of an inequality. A sum on the greater side of an inequality—this is a job for AM-GM. Applying AM-GM to the nonnegative quantities xy, yz, and zx gives

$$\frac{xy + yz + zx}{3} \geq \sqrt[3]{(xy)(yz)(zx)} = \sqrt[3]{x^2y^2z^2}.$$

We have $xyz = 1000$, so $\sqrt[3]{x^2y^2z^2} = 100$. Multiplying our inequality above by 6 therefore gives $2(xy + yz + zx) \geq 6\sqrt[3]{x^2y^2z^2} = 600$. The equality condition of AM-GM tells us that this surface area occurs when $xy = yz = zx$. From $xy = yz$, we have $x = z$, and from $yz = zx$, we have $y = x$. So, our minimum surface area occurs when $x = y = z$, which means that BigBox Company should make cubes. If we have $x = y = z = 10$, then we do indeed have a box with volume 1000 in^3 and surface area 600 in^2.

(b) The results of Problem 12.28 and part (a) of this problem are intuitively obvious. However, if BigBox Company must make boxes *without tops*, then the answer is not so intuitively obvious. They could go ahead and still make cubes of volume 1000 in^3. Without the top, such a cubical box would have surface area 500 in^2. But can they do better?

Let x be the "height" of the box, so that y and z are the dimensions of the top and bottom. So, the surface area of the box is $2xy + 2xz + yz$. Again, we have a sum that we must minimize, so we try AM-GM:

$$\frac{2xy + 2xz + yz}{3} \geq \sqrt[3]{(2xy)(2xz)(yz)} = \sqrt[3]{4x^2y^2z^2} = 100\sqrt[3]{4}.$$

Multiplying by 3 gives us $2xy + 2xz + yz \geq 300\sqrt[3]{4}$. Since $300\sqrt[3]{4} \approx 476.2$, this does indeed suggest we can make a box with less than 500 square inches of cardboard. To find out how to build the boxes, we use the equality condition of AM-GM, which gives us $2xy = 2xz = yz$. From $2xy = 2xz$, we have $y = z$. From $2xz = yz$, we have $y = 2x$. Therefore, we must have $2x = y = z$. So, the dimensions of the bottom must each be twice the height of the box.

To find out what these dimensions are exactly, we let $y = 2x$ and $z = 2x$ in $xyz = 1000$ to find $4x^3 = 1000$, so $x^3 = 250$. This gives us $x = 5\sqrt[3]{2}$, so $y = z = 10\sqrt[3]{2}$. A topless box whose bottom is a square with side length $10\sqrt[3]{2}$ inches and whose sides have height $5\sqrt[3]{2}$ inches does indeed have volume 1000 in^3 and surface area $300\sqrt[3]{4}$, so our minimum surface area is $300\sqrt[3]{4}$ in^2.

$\square$

So far, all of our optimization problems have involved expressions with more than one variable. We can also sometimes optimize one-variable functions using the inequalities we have studied in this chapter. We've already seen examples of this when we studied quadratic optimization in Section 5.3.

Problem 12.30: For $x > 0$, let $f(x) = \dfrac{x^2 + 16}{3x}$. Find the minimum possible value of $f(x)$.

Solution for Problem 12.30: As written, it's not clear how we can minimize $f(x)$. So, we look for other ways in which we can write $f(x)$. We can rewrite the function as a sum of two expressions, one with the variable in the numerator, and one with the variable in the denominator:

$$f(x) = \frac{x}{3} + \frac{16}{3x}.$$

We do so because this produces two expressions whose product is a constant. This suggests we can use AM-GM to compare $f(x)$ to a constant. (Note that we can only use AM-GM because $x > 0$, so both $\frac{x}{3}$ and $\frac{16}{3x}$ are positive.) AM-GM gives us

$$\frac{\frac{x}{3} + \frac{16}{3x}}{2} \geq \sqrt{\frac{x}{3} \cdot \frac{16}{3x}} = \sqrt{\frac{16}{9}} = \frac{4}{3}.$$

Multiplying by 2 gives us $f(x) \geq 8/3$. We have equality when $x/3 = 16/(3x)$. Solving this equation (and remembering that we must have $x > 0$) gives us $x = 4$. We confirm that

$$f(4) = \frac{4^2 + 16}{3 \cdot 4} = \frac{8}{3},$$

so the minimum value $8/3$ can be achieved. $\square$

We finish our study of maximization and minimization with a problem that doesn't quite fit the mold of any of the inequalities we have studied in this chapter.

Problem 12.31: Let $x_1, x_2, \ldots, x_n$ be a sequence of integers such that

(i) $-1 \le x_i \le 2$, for $i = 1, 2, 3, \ldots, n$,

(ii) $x_1 + x_2 + \cdots + x_n = 19$, and

(iii) $x_1^2 + x_2^2 + \cdots + x_n^2 = 99$.

Find the minimal and maximal possible values of $x_1^3 + x_2^3 + \cdots + x_n^3$. *(Source: AHSME)*

Solution for Problem 12.31: The sum of squares suggests Cauchy, but before we go too far with Cauchy, we note the odd restrictions on the x_i. First, all the x_i must be integers. Second, each must be $-1, 0, 1$, or 2. This doesn't fit the restrictions of any of our inequalities from this chapter. Furthermore, it looks pretty difficult to express the sum of the cubes of the x_i in terms of the sum of the x_i and the sum of the squares of the x_i. So, we need to try something different than just applying our toolbox of inequalities.

We focus on the most unusual restriction, which is that each x_i is either $-1, 0, 1$, or 2.

Concept: When stuck on an unusual problem, try focusing on the what makes the problem unusual.

We can use this idea to recast the problem by counting the number of x_i that equal each of $-1, 0, 1$, or 2. We let $a, b, c,$ and d denote the number of appearances among the x_i of $-1, 0, 1,$ and 2, respectively. Now, we have 4 variables instead of n variables. Writing our two given equations in terms of $a, b, c,$ and d gives

$$-a + c + 2d = 19,$$
$$a + c + 4d = 99.$$

We want the minimum and maximum values of $(-1)^3 a + 0^3 b + 1^3 c + 2^3 d = -a + c + 8d$. From the first of our two linear equations above, we have $a = c + 2d - 19$, so the expression we want to maximize and minimize is $-a + c + 8d = -c - 2d + 19 + c + 8d = 19 + 6d$. Now, our task is simply to find the minimum and maximum possible values of d.

The only restrictions we have on d are the two linear equations in our system above. We must choose d such that both a and c are nonnegative integers. So, we solve the linear equations for a and c in terms of d. Subtracting the first equation from the second gives $2a + 2d = 80$, so $a = 40 - d$. Adding the equations gives $2c + 6d = 118$, so $c = 59 - 3d$. For a to be nonnegative, we therefore must have $d \le 40$. For c to be nonnegative, we must have $d \le \frac{59}{3} = 19\frac{2}{3}$. Since d is an integer, this means $d \le 19$.

Now, we're ready to maximize and minimize $19 + 6d$. We know that d must be a nonnegative integer, so the smallest we can allow d to be is 0, which gives us $19 + 6d = 19$. To show that this case is possible, we note that it gives us $a = 40 - d = 40$ and $c = 59 - 3d = 59$, so the x_i must have 40 -1s and 59 1s. If we let these be all the x_i, then the sum of the x_i is indeed 19 and the sum of their squares is $40(-1)^2 + 59(1)^2 = 99$.

To maximize $19 + 6d$, we must have d as large as possible. Above, we found that $d \le 19$ and $d \le 40$. Clearly, the former is more restrictive. So, we let $d = 19$, which gives us $a = 40 - d = 21$ and $c = 59 - 3d = 2$. If the x_i consist of 21 -1s, 2 1s, and 19 2s, then the sum of the x_i is $21(-1) + 2(1) + 19(2) = 19$

and the sum of their squares is $21(-1)^2 + 2(1)^2 + 19(2)^2 = 99$, as required. So, the maximum value of $19 + 6d = 19 + 6(19) = 133$ can be achieved.

To summarize, the minimum is 19 and the maximum is 133. $\square$

Because the Trivial Inequality, AM-GM, and Cauchy can be used to solve so many problems, some people fall into the trap of forgetting the rest of their problem solving toolbox when faced with an inequality problem.

> **WARNING!!** Not all optimization and inequality problems are straightforward applications of the inequalities we have studied in this chapter. Don't forget all your basic problem solving skills when faced with an optimization or inequality problem. The Trivial Inequality, AM-GM, and Cauchy are nice tools, but they won't solve every problem. Be ready to use all your problem solving strategies: experiment, recast the problem in different forms, focus on what makes the problem unusual, compare the problem to others you know how to do, and so on.

Exercises

12.6.1 A box contains chips, each of which is red, white, or blue. The number of blue chips is at least half the number of white chips, and at most one third the number of red chips. The number of chips that are white or blue is at least 55. What is the minimum possible number of red chips? *(Source: AHSME)*

12.6.2 If $a - 7 = b + 3 = c - 9 = d + 22$, which of $a, b, c,$ and d is largest?

12.6.3 What is the largest possible volume of a rectangular box whose diagonal length is 12? Prove that your answer is correct.

12.6.4 Find the minimum value of $\dfrac{x^2 - 2x + 2}{2x - 2}$ for $x > 1$. *(Source: AHSME)*

12.6.5★ Find the minimum value of $\sqrt{x^2 + y^2}$ if $5x + 12y = 60$. *(Source: AHSME)*

12.7 Summary

In this chapter, we covered many strategies for proving that certain inequalities are always true, or that these inequalities are true under specific conditions.

Many inequalities can be proved by using basic inequality manipulations.

> **WARNING!!** Pay special attention to whether or not the variables can be negative in inequalities you wish to prove. Some common inequality manipulations are only valid if the variables are always positive.

Three powerful tools for proving inequalities are:

- The **Trivial Inequality**. If x is a real number, then $x^2 \geq 0$. Equality holds if and only if $x = 0$.

- The **AM-GM Inequality**. The arithmetic mean of any n nonnegative numbers is greater than or equal to the geometric mean of the numbers. That is, if $a_1, a_2, \ldots, a_n$ are nonnegative, then

$$\frac{a_1 + a_2 + \cdots + a_n}{n} \geq \sqrt[n]{a_1 a_2 \cdots a_n}.$$

Equality holds if and only if all a_i are equal.

- The **Cauchy-Schwarz Inequality**. For any two sequences of n real numbers $a_1, a_2, \ldots, a_n$ and $b_1, b_2, \ldots, b_n$, we have

$$(a_1^2 + a_2^2 + \cdots + a_n^2)(b_1^2 + b_2^2 + \cdots + b_n^2) \geq (a_1 b_1 + a_2 b_2 + \cdots + a_n b_n)^2.$$

Equality occurs if and only if either $a_i = 0$ for all i, or there exists a constant t such that $b_i = t a_i$ for all i.

We then used these strategies to solve maximization and minimization problems:

> **Important:** Maximization and minimization problems are two-part problems. We must show that the maximum or minimum value is attainable, and that no better value (no higher for maximization, no lower for minimization) can be attained.

Problem Solving Strategies

We encountered so many useful strategies in this chapter, that we split the concepts summary into inequality-specific strategies and general strategies.

> **Concepts:** Useful general strategies discussed in this chapter include:
>
>
>
> - When stuck on a problem, focus on why you are stuck, and try to find ways to remove that obstacle.
>
> - If a problem is given in terms of gigantic numbers, it's often useful to replace those numbers with variables.
>
> - Dealing with reciprocals of complicated fractions is often nicer than dealing with the original fractions.
>
> - Don't treat every problem as if it is brand-new. Try to relate problems or parts of problems to other problems you already know how to do.
>
> - When stuck on a problem involving quadratics, try completing the square.

Concepts: Useful inequality strategies discussed in this chapter include:

- Keep your eye on the ball! Don't just blindly manipulate inequalities—try to make sure you're moving in the right direction.

- When possible, testing numbers that fit an inequality problem can help you get started with a solution.

- Working backwards is a very powerful method for proving inequalities.

- We can sometimes prove a complicated inequality by proving simpler inequalities that can be combined to give us the desired complicated one.

- Some equations can be solved by viewing them as the equality case of a nonstrict inequality. We can then use the equality condition of the inequality to find the solutions to the equation.

- When trying to prove an inequality involving nonnegative expressions that have a constant sum or product, consider using AM-GM.

- Focus on one side of an inequality and the other will often fall into place.

- It's often necessary to manipulate a desired inequality into a form for which tools like AM-GM apply. Factor! Expand! Regroup terms!

- If an inequality involves a sum of squares or a product of two sums with the same number of terms, try using the Cauchy-Schwarz Inequality to prove it.

- Some inequalities result from setting variables or even entire sequences of variables equal to carefully chosen constants, and then applying known inequalities to those variables or sequences.

- Cauchy can help us prove inequalities that have a sum of fractions on the greater side. To try using Cauchy on such an inequality, let the sum of the fractions be one sum of squares, and the sum of the denominators be the other sum of squares.

 REVIEW PROBLEMS

12.32 Let $x > y > 0$ and $z \neq 0$. Determine whether each of the following inequalities *always* holds. If it does, then prove that it always holds. If it doesn't, then give an example for which it doesn't hold.

(a) $xz^2 > yz^2$

(b) $\dfrac{x}{z^2} > \dfrac{y}{z^2}$

12.33 Show that if $0 < b < a < 1$, then we must have $\dfrac{1 - ab}{a - b} > 1$.

12.34 Show that if $x < a < 0$, then $x^2 > ax > a^2$. *(Source: AHSME)*

12.35 Show that if $x - y > x$ and $x + y < y$, then both x and y must be negative. *(Source: AHSME)*

12.36 Show that if $a + b \geq 0$ then $a^3 + b^3 \geq a^2b + ab^2$.

12.37 If $\frac{4}{2001} < \frac{a}{a+b} < \frac{5}{2001}$, compute the number of possible integer values of $\frac{b}{a}$. *(Source: ARML)*

12.38 Let $s = x + y$ and $p = xy$ for real numbers x and y. Show that $s^2 \geq 4p$.

12.39 Let a, b, and c be nonnegative real numbers. Prove that $a^2b + b^2c + c^2a \geq 3abc$.

12.40 Let n be a positive integer, and let $a, b \geq 0$. Show that $\dfrac{a^{n+1} + nb^{n+1}}{n + 1} \geq ab^n$.

12.41 Let a, b, c and d be integers with $a < 2b$, $b < 3c$, and $c < 4d$. If $d < 100$, find the largest possible value for a. *(Source: AHSME)*

12.42 Let a, b, c, and d be positive numbers such that $\frac{a}{b} < \frac{c}{d}$. Show that $\frac{a}{b} < \frac{a+c}{b+d} < \frac{c}{d}$.

12.43 A student attempted to compute the average, A, of x, y and z by computing the average of x and y, and then computing the average of that result with z. The student's final result was B. If $x < y < z$, is either A or B necessarily larger? *(Source: AHSME)*

12.44 Determine the minimum value of the sum $\dfrac{a}{2b} + \dfrac{b}{4c} + \dfrac{c}{8a}$, where a, b, and c are positive real numbers. *(Source: Mandelbrot)*

12.45 A three-dimensional box has a surface area of 96 cm^2. What is the maximum volume of the box?

12.46 For $x \geq 0$, find the minimum value of $\dfrac{4x^2 + 8x + 13}{6(1 + x)}$. *(Source: AHSME)*

12.47 In counting n colored balls, some red and some black, it was found that 49 of the first 50 counted were red. Thereafter, the balls were counted in groups of 8, and 7 balls in each of these groups were red. Find the maximum possible value of n if 90% or more of the balls counted were red. *(Source: AHSME)*

12.48 A dealer bought n radios for d dollars total (where d is a positive integer). He sold two radios to a community bazaar at half their cost. The rest he sold at a profit of \$8 on each radio sold. If the overall profit was \$72, find the least possible value of n. *(Source: AHSME)*

12.49 Let $a + b + c + d = 1$. Prove that $a^2 + b^2 + c^2 + d^2 \geq \frac{1}{4}$.

12.50 A gambler played the following game with a friend. The gambler bet half the money in his pocket on the toss of a coin; he won on heads and lost on tails. The coin was tossed and the money handed over. The game was repeated, each time for half the money held by the gambler. At the end, the number of times the gambler lost was equal to the number of times he won. Did he gain, lose, or break even?

12.51 What's the smallest possible value of $k + m + n$ if k, m, and n are positive integers such that $\frac{1}{k} + \frac{1}{m} + \frac{1}{n} < 1$?

12.52 Show that for all real numbers a, b, c, d we have $(a^2 - b^2)(c^2 - d^2) \leq (ac - bd)^2$.

12.53 Show that for all positive numbers a, b, c, we have $\dfrac{a^2 + b^2 + c^2}{a^3 + b^3 + c^3} \geq \dfrac{a^3 + b^3 + c^3}{a^4 + b^4 + c^4}$.

12.54 Let P, A, G, and Q be points on a line, in that order, with $PG = a$ and $GQ = b$. Let A be be the midpoint of $\overline{PQ}$. Construct a semicircle with diameter $\overline{PQ}$, and let M be a point on the semicircle such that $\overline{MG}$ is perpendicular to $\overline{PQ}$.

(a) Compute AM and GM in terms of a and b.

(b) Find a geometric proof of the 2-variable AM-GM Inequality.

12.55 Let x, y, and z be real numbers such that $x^2 + y^2 + z^2 = 1$. Find the maximum value of $3x + 4y + 12z$.

Challenge Problems

12.56 Prove that the inequality $3a^4 - 4a^3b + b^4 \geq 0$ holds for all real numbers a and b. **Hints:** 338

12.57 Let x, y, and z be positive real numbers such that $xy^2z^3 = 108$. What is the minimum value of $x + y + z$? **Hints:** 355, 296

12.58 Let $a_1, a_2, \ldots, a_n > 0$, and $a_1 a_2 \cdots a_n = 1$. Show that $(1 + a_1)(1 + a_2) \cdots (1 + a_n) \geq 2^n$.

12.59 Find all ordered triples (x, y, z) such that $3x^2 + y^2 + z^2 = 2x(y + z)$.

12.60 Find all ordered triples of real numbers (a, b, c) that satisfy the system of equations

$$a^2 + 10b = -34,$$
$$b^2 + 4c = 17,$$
$$c^2 + 8a = -28.$$

12.61 Show that $a^2 + ab + b^2 \geq 0$ for all real numbers a and b. When does equality occur? **Hints:** 14

12.62 Find the ordered pair of real numbers (x, y) that satisfies the equation below, and demonstrate that it is unique:

$$\frac{36}{\sqrt{x}} + \frac{9}{\sqrt{y}} = 42 - 9\sqrt{x} - \sqrt{y}.$$

(Source: USAMTS) **Hints:** 215

12.63 Show that for all positive x, y, and z, we have $\dfrac{x^2 + y^2}{x + y} + \dfrac{y^2 + z^2}{y + z} + \dfrac{z^2 + x^2}{z + x} \geq x + y + z$.

12.64 The sum of the following seven numbers is exactly 19: $a_1 = 2.56$, $a_2 = 2.61$, $a_3 = 2.65$, $a_4 = 2.71$, $a_5 = 2.79$, $a_6 = 2.82$, $a_7 = 2.86$. It is desired to replace each a_i by an integer approximation A_i, $1 \leq i \leq 7$, so that the sum of the A_i's is also 19, and so that M, the maximum of the "errors" $|A_i - a_i|$, is as small as possible. For this minimum M, what is $100M$? *(Source: AIME)*

12.65   Show that for all positive real numbers $x \neq 1$ and nonnegative integers n, we have

$$\frac{x^{2n+1} - 1}{x^{n+1} - x^n} \geq 2n + 1.$$

Hints: 338

12.66   Let $0 \leq x \leq 4$. Find the maximum value of $x^3(4 - x)$. **Hints:** 257

12.67   Show that if $a + b + c = 3$, then $ab + bc + ca \leq 3$. **Hints:** 282

12.68   Arrange the following products in increasing order from left to right:

$$1000! \quad (400!)(400!)(200!) \quad (500!)(500!) \quad (600!)(300!)(100!) \quad (700!)(300!).$$

(Source: ARML)

12.69   Show that for all positive integers n with $n \geq 2$, we have

$$\frac{1}{n} + \frac{1}{n+1} + \cdots + \frac{1}{2n - 1} > n(2^{1/n} - 1).$$

Hints: 145, 4, 279

12.70   Let $0 \leq x \leq 1$. Prove that $\dfrac{1 - x^{n+1}}{n + 1} \leq \dfrac{1 - x^n}{n}$. **Hints:** 338, 289

12.71   Let S be a list of positive integers—not necessarily distinct—in which the number 68 appears. The average (arithmetic mean) of the numbers in S is 56. However, if 68 is removed, the average of the remaining numbers drops to 55. What is the largest number that can appear in S? *(Source: AIME)* **Hints:** 152

12.72   Let a, b, c, be positive real numbers. Prove that $\left(1 + \frac{a}{b}\right)\left(1 + \frac{b}{c}\right)\left(1 + \frac{c}{a}\right) > 2\left(1 + \frac{a+b+c}{\sqrt[3]{abc}}\right)$. **Hints:** 11

12.73★   At a wedding reception, n guests have assembled into m groups to converse. (The groups are not necessarily equal sized.) The host is preparing m square cakes, each with an ornate ribbon adorning its perimeter, to serve the m groups. No guest is allowed to have more than 25 cm^2 of cake. Prove that no more than $20\sqrt{mn}$ cm of ribbon is needed to embellish the m cakes. *(Source: Mandelbrot)* **Hints:** 233, 32

12.74★   Find all values of k such that $36x^2 + 25y^2 \geq kxy$ for all real values of x and y. **Hints:** 27

12.75★   Prove that for all real numbers a, b, x, and y, we have $\sqrt{a^2 + b^2} + \sqrt{x^2 + y^2} \geq \sqrt{(a - x)^2 + (b - y)^2}$. **Hints:** 341

12.76★   Show that for positive numbers a, b, c, we have $abc(a + b + c) \leq ac^3 + cb^3 + ba^3$. **Hints:** 281

12.77★   Given that a, b, c, d, e are real numbers such that

$$a + b + c + d + e = 8,$$
$$a^2 + b^2 + c^2 + d^2 + e^2 = 16,$$

determine the maximum value of e. *(Source: USAMO)* **Hints:** 81, 107, 216

12.78★   Prove that if $x_i > 0$ for all i then

$$(x_1^{19} + x_2^{19} + \cdots + x_n^{19})(x_1^{93} + x_2^{93} + \cdots + x_n^{93}) \geq (x_1^{20} + x_2^{20} + \cdots + x_n^{20})(x_1^{92} + x_2^{92} + \cdots + x_n^{92}).$$

(Source: Mandelbrot) **Hints:** 230, 309

12.79★   Suppose that a, b, and c are positive numbers and $ab + bc + ca = 1/3$. Show that

$$\frac{a}{a^2 - bc + 1} + \frac{b}{b^2 - ca + 1} + \frac{c}{c^2 - ab + 1} \geq \frac{1}{a + b + c}.$$

Hints: 311, 176

12.80★   Let a, b, and c be positive real numbers such that $abc = 1$. Prove that

$$\frac{1}{a^3(b + c)} + \frac{1}{b^3(a + c)} + \frac{1}{c^3(a + b)} \geq \frac{3}{2}.$$

(Source: IMO)   **Hints:** 226, 221

In the following problems, we prove the general AM-GM Inequality. You can use AM-GM for two variables in the following problems, but you cannot use any other AM-GM inequalities until you prove them.

12.81   Prove that $\frac{w+x+y+z}{4} \geq \sqrt[4]{wxyz}$ for all nonnegative numbers w, x, y, and z. Under what conditions are the two sides equal?

12.82   Prove that the arithmetic mean of any 2^n nonnegative numbers, where $n > 1$, is greater than or equal to the geometric mean of those numbers. Under what conditions are these two means equal? **Hints:** 340

12.83★   Show that if a, b, and c are nonnegative, then $\frac{a + b + c}{3} \geq \sqrt[3]{abc}$. When does equality hold? **Hints:** 16, 5

12.84★   Show that the arithmetic mean of any n nonnegative numbers, where $n > 1$, is greater than or equal to the geometric mean of those numbers. Under what conditions does equality hold? **Hints:** 19, 72

Extra!   Here are a few more special inequalities:

- **Chebyshev's Inequality.** If $a_1 \geq a_2 \geq \cdots \geq a_n$ and $b_1 \geq b_2 \geq \cdots \geq b_n$, then

$$n \sum_{i=1}^{n} a_i b_i \geq \left(\sum_{i=1}^{n} a_i \right) \left(\sum_{i=1}^{n} b_i \right).$$

- **Schur's Inequality.** If x, y, and z are nonnegative real numbers and t is positive, then

$$x^t(x - y)(x - z) + y^t(y - z)(y - x) + z^t(z - x)(z - y) \geq 0.$$

- **Minkowski's Inequality.** If $p > 1$ and the sequences $\{a_i\}$ and $\{b_i\}$ consist of positive numbers, then

$$\left(\sum_{i=1}^{n} (a_i + b_i)^p \right)^{1/p} \leq \left(\sum_{i=1}^{n} (a_i)^p \right)^{1/p} + \left(\sum_{i=1}^{n} (b_i)^p \right)^{1/p}.$$

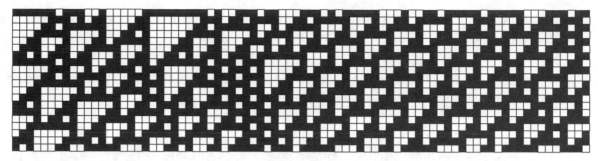

Exponential growth looks like nothing is happening, and then suddenly you get this explosion at the end.

— Ray Kurzweil

Exponents and Logarithms

Most of the expressions and functions we have worked with so far either are polynomials or are closely related to polynomials. In the next four chapters, we will explore several other common types of expressions and functions. These present some complexities that are quite different from those we faced when working with polynomials. While it may at first feel like we are presenting a bunch of unrelated tricks, where each trick is associated with a particular type of expression, we are really presenting many different examples of a very important general strategy:

> **Concept:** When working with an unusual expression in a problem, try to find ways to convert the problem into one involving expressions you know how to handle.

Think of most of the tactics you learn in the next four chapters as putting this strategy into practice. Then, you won't have to memorize a bunch of different tricks—you'll have a strategy that you can use to re-create whatever tool you need whenever you need it. Even better, you'll develop the ability to create new tools when you encounter new unusual expressions in problems.

13.1 Exponential Function Basics

In the expression 2^3, the 3 is called the **exponent** and the 2 is called the **base**. We call a function $f(x) = a^x$, where a is a positive constant, an **exponential function** with base a.

Suppose $g(x) = 2^x$. At right are values of $g(x)$ for several values of x. Notice that as x gets larger and larger, so does $g(x)$.

In general, if $f(x) = a^x$, where $a > 1$, then $f(x)$ grows as x grows. More formally, we say that a function f is a **monotonically increasing** function if $f(x) > f(y)$ whenever $x > y$. Often we will just say the function is "increasing" rather than "monotonically increasing." All exponential functions $f(x) = a^x$ such that $a > 1$ are monotonically increasing. Showing

x	$g(x)$
-2	0.25
0	1
2	4
5	32
8	256

that exponential functions with base greater than 1 are increasing functions requires more advanced tools than we study in this book.

> **Sidenote:** Rigorously defining the output of exponential functions requires more advanced tools, too. For example, consider $g(x) = 2^x$. It's very clear that $g(2) = 4$ and $g(5) = 32$. We can also easily define the result when we input rational numbers to $g(x)$, such as $g(3/2) = 2^{3/2} = (2^3)^{1/2} = 8^{1/2} = 2\sqrt{2}$. But what about $g(\pi)$? We have $g(\pi) = 2^\pi$, but how do we raise 2 to the power of π? We could estimate it with $2^{3.14}$, which we can evaluate because 3.14 is rational. But this doesn't tell us how to define 2^π. We have to rely on more advanced tools to define $g(x) = 2^x$ for all real numbers x. You'll learn these tools when you study calculus.

 Problems

Problem 13.1: Let $f(x) = 2^x$ and $g(x) = \left(\frac{1}{2}\right)^x$. Graph f and g on the same Cartesian plane. Notice anything interesting? If so, explain the relationship you see between the graphs.

Problem 13.2: Suppose that a is a constant greater than 1 and that $a^x = a^y$. Must it be true that $x = y$? Why or why not?

Problem 13.3:
(a) Find x if $5^{x^2-3} = 25^x$.
(b) Find t if $3^{27^t} = 27^{3^t}$.

Problem 13.1: Let $f(x) = 2^x$ and $g(x) = \left(\frac{1}{2}\right)^x$. Graph f and g on the same Cartesian plane. Notice anything interesting? If so, explain the relationship you see between the graphs.

Solution for Problem 13.1: By choosing various values of x, we produce the graphs of $y = 2^x$ and $y = \left(\frac{1}{2}\right)^x$ at right. The graph of $y = 2^x$ is bold. We see that the graph of $y = 2^x$ is the graph of $y = \left(\frac{1}{2}\right)^x$ reflected over the y-axis. In other words, for any point (a, b) on the graph of $y = \left(\frac{1}{2}\right)^x$, there is a point $(-a, b)$ on the graph of $y = 2^x$, and vice-versa. To see why this is the case, we note that $g(x) = \left(\frac{1}{2}\right)^x = (2^{-1})^x = 2^{-x} = f(-x)$. So, for each point $(a, g(a))$ on the graph of g, there is a point $(-a, f(-a))$ on the graph of f (and vice versa) with the same y-coordinate, but opposite x-coordinate. $\square$

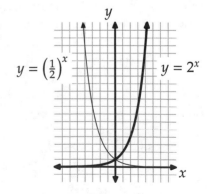

The graph of $y = 2^x$ above exhibits what we call **exponential growth**. When x is negative, the value of 2^x increases very slowly as x increases. But when x is positive, we reach a point where the value of 2^x skyrockets as x increases. Every exponential function a^x with $a > 1$ exhibits this behavior. Even if a is very close to 1, the value of a^x will appear to "explode" at some point as x increases.

On the other hand, if $0 < a < 1$ in a^x, we have the behavior depicted by the graph of $y = \left(\frac{1}{2}\right)^x$. In this case, as x increases, the value of a^x decreases. As x gets very large, the value of a^x gets very close to 0, but never equals 0. This behavior is called **exponential decay**, which we'll discuss more in Section 13.6.

Problem 13.2: Suppose that a is a constant greater than 1 and that $a^x = a^y$. Must it be true that $x = y$? Why or why not?

Solution for Problem 13.2: Let $f(x) = a^x$. Because f is increasing, we know that $f(x) > f(y)$ for any x and y such that $x > y$. So, to test if we must have $x = y$ when $f(x) = f(y)$, we consider what happens in each of the three possible cases, $x > y$, $x < y$, and $x = y$.

If $x > y$, then we must have $f(x) > f(y)$ because f is increasing. Therefore, if $x > y$, then $a^x > a^y$.

If $x < y$, then we must have $f(x) < f(y)$ because f is increasing. Therefore, if $x < y$, then $a^x < a^y$.

If $x = y$, then $f(x) = f(y)$, so $a^x = a^y$.

We must have either $x > y$, $x < y$, or $x = y$, and in the first two cases, we cannot also have $a^x = a^y$. So, if $a^x = a^y$, the only possible relationship between x and y is $x = y$. (Notice that this is not true if $a = 1$. However, if $a = 1$, then the function $f(x) = a^x$ is just $f(x) = 1$.) $\square$

We can use similar reasoning to our solution in Problem 13.2 to show that if $0 < a < 1$ and $a^x = a^y$, then $x = y$. (You'll have a chance to do this on your own as an Exercise.)

Important: If a is a positive constant such that $a \neq 1$ and $a^x = a^y$, then $x = y$.

Let's see how this simple fact can be used to solve equations in which the variables are in exponents.

Problem 13.3:

(a) Find all x such that $5^{x^2-3} = 25^x$.

(b) Find all t such that $3^{27^t} = 27^{3^t}$.

Solution for Problem 13.3:

(a) We don't know much about dealing with variables in exponents, so we'd like to convert this equation into one in which the variables are not in exponents. We know that $a^x = a^y$ implies $x = y$. So, if we can rewrite the equation such that the base is the same on both sides, then we can produce a simpler equation. Since $25 = 5^2$, we have $25^x = (5^2)^x = 5^{2x}$, so our equation now is

$$5^{x^2-3} = 5^{2x}.$$

Since the base is the same on both sides, the exponents must be equal. So, we have $x^2 - 3 = 2x$. Rearranging gives $x^2 - 2x - 3 = 0$ and factoring gives $(x-3)(x+1) = 0$, so the two values of x that satisfy the equation are $x = 3$ and $x = -1$.

> **Concept:** If you have an equation with constants raised to powers, it's often helpful to write all those constants with the same base if possible.

(b) Our first step is to note that $27 = 3^3$, so we can write each side as a power of 3. Specifically, on the left we have $3^{27^t} = 3^{(3^3)^t} = 3^{3^{3t}}$ and on the right, we have $27^{3^t} = (3^3)^{3^t} = 3^{3 \cdot 3^t} = 3^{3^{t+1}}$, so our equation is

$$3^{3^{3t}} = 3^{3^{t+1}}.$$

Because the base of both sides is 3, the exponents of both sides must be equal, which gives us $3^{3t} = 3^{t+1}$. Again, the base of both sides is 3, so the exponents must be equal. This gives us $3t = t + 1$, from which we find $t = 1/2$.

$\square$

Exercises

13.1.1 Find all t such that $16^t = 8^{2-t}$.

13.1.2 The equation $2^{x^2} = 32^{3x+8}$ has two real solutions. Find their product.

13.1.3 A function f is **monotonically decreasing** if $f(x) < f(y)$ whenever $x > y$. Let $g(x) = a^x$, where $0 < a < 1$.

(a) Show that g is monotonically decreasing.
(b) Show that if $g(x) = g(y)$, then $x = y$.

13.1.4 Find the maximum value of 3^{-x^2+2x+3}.

13.1.5 Find all values of y such that $4^y - 12 \cdot 2^y = -32$. **Hints:** 93

13.2 Introduction to Logarithms

In addition to helping us solve equations, the fact that exponential functions with base greater than 1 are monotonically increasing tells us something very important about them.

Problems

Problem 13.4: Suppose that $f(x) = a^x$, where a is constant such that $a > 1$. Show that $f(x)$ has an inverse. (If you don't remember what an inverse function is, read Section 2.4, then try this problem again.)

Problem 13.5: Let $g(x) = 2^x$. Find $g^{-1}(32)$.

Problem 13.4: Suppose that $f(x) = a^x$, where a is constant such that $a > 1$. Show that $f(x)$ has an inverse.

Solution for Problem 13.4: A function has an inverse if it does not ever give the same output for two different inputs. Because f is increasing, we know that if $x_1 > x_2$, then $f(x_1) > f(x_2)$. In other words, if two inputs to f are not equal, then their outputs from f cannot be equal. So, f must have an inverse. $\square$

Problem 13.5: Let $g(x) = 2^x$. Find $g^{-1}(32)$.

Solution for Problem 13.5: The previous problem tells us that g has an inverse. The value of $g^{-1}(32)$ is the input to $g(x)$ such that $g(x) = 32$. Therefore, we seek the value of x such that $2^x = 32$. In other words, $g^{-1}(32)$ is the power to which we must raise 2 in order to get 32. Since $32 = 2^5$, we find that $g^{-1}(32) = 5$. $\square$

The inverses of exponential functions are so common that they have their own name: **logarithms**. To get a feel for logarithms, we'll look at a specific exponential function and its inverse. We write

$$x = \log_2 32$$

to indicate that x is the power to which we must raise 2 in order to get 32. In other words, the equations

$$x = \log_2 32 \quad \text{and} \quad 2^x = 32$$

are equivalent. Since $2^5 = 32$, we know that

$$\log_2 32 = 5.$$

Using this notation, we can now write the inverse of $f(x) = 2^x$ as $f^{-1}(x) = \log_2 x$.

Just as 2 is the base of the expression 2^5, we say that 2 is the **base** of the logarithm $\log_2 32$. The base of a logarithm must be a positive number, and it cannot equal 1. In the expression $\log_2 32$, the 32 is sometimes referred to as the **argument** of the logarithm. When speaking, we say $\log_2 32 = 5$ as "The logarithm base 2 of 32 is 5," and this means "The exponent to which we must raise 2 to get 32 is 5."

As we will see, much of understanding logarithms requires being able to convert between **logarithmic form**, $\log_a b = c$, and **exponential form**, $a^c = b$.

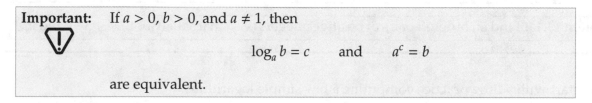

Important: If $a > 0$, $b > 0$, and $a \neq 1$, then

$$\log_a b = c \quad \text{and} \quad a^c = b$$

are equivalent.

In this book, we only define logarithms for positive bases, and we exclude 1. (To see why we exclude 1, try to figure out what $\log_1 1$ is. We have $1^y = 1$ for all real y, so $\log_1 x$ would not be a function.) We also will restrict each logarithm to having only real numbers in its domain and range.

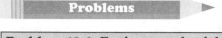

Problems

Problem 13.6: Evaluate each of the following:

(a) $\log_2 4$

(b) $\log_2 8$

(c) $\log_2 16$

(d) $\log_a a^p$, where $a > 0$ and $a \neq 1$

Problem 13.7:

(a) Find the domain and range of $f(x) = 2^x$.

(b) Find the domain and range of $g(x) = \log_2 x$.

(c) Graph f and g on the same Cartesian plane.

Problem 13.8: If $\log_5 a = \log_5 b$, then must we have $a = b$?

Problem 13.9: Solve the following inequalities:

(a) $\log_4(2x - 3) > 2$

(b) $\log_{1/5} x > 3$

Problem 13.10:

(a) Find the domain of $h(x) = \log_{1/2} x$.

(b) Find the domain of $g(x) = \log_{1/2} (\log_3 x)$.

(c) Find the domain of $f(x) = \log_{1/2} (\log_5 (\log_{1/3} x))$.

Problem 13.11: Let $x = \log_4 128$.

(a) Evaluate 4^x.

(b) Evaluate 2^x.

(c) Find x.

Problem 13.12: Evaluate each of the following:

(a) $\log_{16} 8$ (b) $\log_{3/2} 4/9$ (c) $\log_{3\sqrt{3}} 243$

Problem 13.13: Suppose that a and b are constants such that $f(x) = ax^b$, $f(3) = 14$, and $f(27) = 686$. Find a and b.

Problem 13.14: Find all ordered pairs of positive integers (x, y) that satisfy both $y \leq 5 - x$ and $y \geq \log_2 x$.

We start with a little practice computing a few simple logarithms.

Problem 13.6: Evaluate each of the following:

(a) $\log_2 4$ (c) $\log_2 16$

(b) $\log_2 8$ (d) $\log_a a^p$, where $a > 0$ and $a \neq 1$

Solution for Problem 13.6:

(a) Because $2^2 = 4$, we have $\log_2 4 = \log_2 2^2 = 2$.

(b) Because $2^3 = 8$, we have $\log_2 8 = \log_2 2^3 = 3$.

(c) Because $2^4 = 16$, we have $\log_2 16 = \log_2 2^4 = 4$.

(d) We raise a to the power p to get a^p, so $\log_a a^p = p$.

$\square$

> **Important:** If $a > 0$ and $a \neq 1$, then
> $$\log_a a^p = p.$$

One way to learn more about a new function is to find the domain, range, and graph of the function.

> **Problem 13.7:**
> (a) Find the domain and range of $f(x) = 2^x$.
>
> (b) Find the domain and range of $g(x) = \log_2 x$.
>
> (c) Graph f and g on the same Cartesian plane.

Solution for Problem 13.7:

(a) We can raise 2 to any real power, so the domain of $f(x) = 2^x$ is all real numbers. All powers of 2 are positive, so there are no nonpositive numbers in the range of 2^x. We can make 2^x as small a positive number as we like by choosing a suitably small negative number for x. For example, when $x = -20$, we have $2^x = 2^{-20} = 1/2^{20} \approx 0.000000954$. That's pretty small. Imagine what 2^x is if $x = -200$. So, the range of f is all positive real numbers.

(b) Because $g(x) = \log_2 x$ is the inverse of $f(x)$, the domain of g is the range of f, and vice versa. Therefore, the domain of g is all positive real numbers and the range of g is all real numbers. We can also see this by noting that $\log_2 x$ equals the power to which we must raise 2 in order to get x. So, the range is all real numbers because we can raise 2 to any real power. Similarly, the domain of $\log_2 x$ consists of all numbers that equal 2 raised to some power. As we just discussed, all positive real numbers can be expressed as 2 raised to some power. So, the domain of g is all positive real numbers.

(c) By choosing a few values for x, we can plot $y = 2^x$ and $y = \log_2 x$ as shown at right. Notice that the two graphs are reflections of each other over the line $y = x$. This is as expected, because $f(x) = 2^x$ and $g(x) = \log_2 x$ are inverses of each other.

$\square$

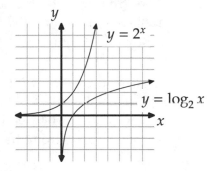

Problem 13.8: If $\log_5 a = \log_5 b$, then must we have $a = b$?

Solution for Problem 13.8: Let both logarithms equal x. The equation $x = \log_5 a$ means that $5^x = a$. Similarly, $x = \log_5 b$ means that $5^x = b$. Therefore, we have $a = 5^x = b$. $\square$

Problem 13.9: Solve the following inequalities:

(a) $\log_4(2x - 3) > 2$

(b) $\log_{1/5} x > 3$

Solution for Problem 13.9:

(a) If $\log_4(2x - 3)$ is greater than 2, then the expression $2x - 3$ equals 4 raised to a power greater than 2. Because the function 4^t is increasing, when we raise 4 to any power greater than 2, the result is greater than 4^2. Therefore, we have $2x - 3 > 4^2$. Solving this inequality gives $x > 19/2$.

   We also could have tackled this problem in much the same way we solved Problem 13.8. We start by assigning a variable to the logarithm, letting $y = \log_4(2x - 3)$. Writing this in exponential form gives $4^y = 2x - 3$. Because $y > 2$, we have $4^y > 16$. Therefore, we have $2x - 3 > 16$, so $x > 19/2$, as before.

(b) In part (a), we rewrote $\log_4(2x - 3) > 2$ as $2x - 3 > 4^2$. So, we might be inspired to do this:

> **Bogus Solution:** Because $\log_{1/5} x > 3$, we have $x > (1/5)^3$, so $x > 1/125$.

At first glance, this looks fine. But then we test it. If the solution really is $x > 1/125$, then $x = 1$ should satisfy the inequality. However, we have $\log_{1/5} 1 = 0$, which clearly is not greater than 3. Where did we go wrong?

   Our mistake was rewriting the logarithm inequality in exponential form without thinking about what the inequality means.

> **WARNING!!** Don't just perform algebraic manipulations blindly. Think about each step and make sure you know why it is valid.

If $\log_{1/5} x > 3$, then we must raise $1/5$ to a power greater than 3 to get x. Because $1/5$ is less than 1, when we raise $1/5$ to a power greater than 3, we get a number that is *smaller than* $(1/5)^3$, not greater than $(1/5)^3$. Therefore, we must have $x < (1/5)^3$, or $x < 1/125$.

   We're not quite finished, and testing our answer will reveal why. When we consider trying numbers less than $1/125$ in the original inequality, $\log_{1/5} x > 3$, we see that the logarithm is not defined if x is nonpositive, so we must have $x > 0$. Therefore, the full solution to the inequality is $0 < x < 1/125$.

   As with the previous part, we can also solve this problem by assigning a variable to the logarithm. Letting $y = \log_{1/5} x$, we have $(1/5)^y = x$, and the given inequality is $y > 3$. Since $1/5$ is between 0 and 1, we have $(1/5)^y < (1/5)^3$, so $x < (1/5)^3 = 1/125$. Combining this with the restriction $x > 0$ from the domain of the logarithm, we have $0 < x < 1/125$, as before.

□

Problem 13.10: Find the domain of $f(x) = \log_{1/2}(\log_5(\log_{1/3} x))$.

Solution for Problem 13.10: We can either work from the inside out, by starting with the base $1/3$ logarithm, or we can work from the outside in, by starting with the base $1/2$ logarithm. We'll work from the outside in; see if you can find a solution working from the inside out.

We let $h(x) = \log_{1/2} x$, which has all positive numbers as its domain. Next, we look at $g(x) = \log_{1/2}(\log_5 x)$. We know that the domain of $\log_{1/2} x$ is all positive numbers, so the output of $\log_5 x$ must be positive in order for $\log_{1/2}(\log_5 x)$ to be defined. The output of $\log_5 x$ is positive if and only if x is a positive power of 5. We have $5^0 = 1$, and 5^t is an increasing function, so all real numbers greater than 1 are positive powers of 5. Therefore, the domain of $g(x) = \log_{1/2}(\log_5 x)$ is $(1, +\infty)$.

Finally, we consider $f(x) = \log_{1/2}(\log_5(\log_{1/3} x))$, so that we have $f(x) = g(\log_{1/3} x)$. We know that the domain of g is $(1, +\infty)$, so, in order to be able to evalute $g(\log_{1/3} x)$, we must restrict x to those values for which $\log_{1/3} x > 1$. The statement $\log_{1/3} x > 1$ means that the power to which we raise 1/3 to get x is greater than 1. Raising 1/3 to a power greater than 1 gives a number less than $(1/3)^1$, so $x < \frac{1}{3}$. Therefore, the domain of f is $(0, 1/3)$.

We can check our answer by noting that if $0 < x < \frac{1}{3}$, then $\log_{1/3} x > 1$, so $\log_5(\log_{1/3} x) > 0$. Because $\log_5(\log_{1/3} x)$ is positive, we can take its logarithm with any positive base. $\square$

Now that we have a feel for how logarithmic functions behave, let's compute a few more specific logarithms.

Problem 13.11: Compute $\log_4 128$.

Solution for Problem 13.11: Let $x = \log_4 128$. We don't know much yet about dealing with logarithms in equations, so we convert this equation to exponential form:

$$4^x = 128.$$

We know how to tackle this equation; we write 4 and 128 with a common base, 2, and we have $(2^2)^x = 2^7$, so $2^{2x} = 2^7$. The bases on both sides are the same, so the exponents must be the same. Therefore, we have $2x = 7$, so $x = 7/2$. $\square$

Concept: We can often evaluate logarithms by setting the logarithm equal to x, then writing the resulting equation in exponential form.

After we do so, we use the same tactic we used to solve exponential equations; we write constants in the problem as powers of a common base. We could also have done this from the beginning with our logarithm problem, so we have

$$\log_4 128 = \log_{2^2} 2^7.$$

Now, it's easy to see that 2^2 must be raised to the 7/2 power to get 2^7, so we have $\log_4 128 = \log_{2^2} 2^7 = 7/2$.

Concept: Finding a common exponential base between two numbers a and b often helps compute $\log_a b$ more easily.

Here's a little more practice.

Problem 13.12: Evaluate each of the following:

(a) $\log_{16} 8$ (b) $\log_{3/2} 4/9$ (c) $\log_{3\sqrt{3}} 243$

Solution for Problem 13.12:

(a) We let $x = \log_{16} 8$, so we have $16^x = 8$. Writing the constants with a common base gives $(2^4)^x = 2^3$, so $2^{4x} = 2^3$. Therefore, we have $4x = 3$, so $x = 3/4$.

   We also could have noted that $\log_{16} 8 = \log_{2^4} 2^3$, and we must raise 2^4 to the 3/4 power to get 2^3, so $\log_{16} 8 = 3/4$.

(b) We let $x = \log_{3/2} 4/9$, so $(3/2)^x = 4/9 = (2/3)^2 = (3/2)^{-2}$. Therefore, $x = -2$.

(c) We let $x = \log_{3\sqrt{3}} 243$, so $(3\sqrt{3})^x = 243$. We have $3\sqrt{3} = 3 \cdot 3^{1/2} = 3^{3/2}$ and $243 = 3^5$, so our equation is $3^{3x/2} = 3^5$. Therefore, we have $3x/2 = 5$, so $x = 10/3$.

□

> **Problem 13.13:** Suppose that a and b are constants such that $f(x) = ax^b$, $f(3) = 14$, and $f(27) = 686$. Find a and b.

Solution for Problem 13.13: Because $f(3) = 14$ and $f(27) = 686$, we have the system of equations

$$a \cdot 3^b = 14,$$
$$a \cdot 27^b = 686.$$

Dividing the second equation by the first eliminates a and leaves $27^b/3^b = 49$. The bases on the left are both powers of 3, so we can simplify the left side by noting $27^b/3^b = (3^3)^b/3^b = 3^{3b}/3^b = 3^{2b}$, which makes our equation $3^{2b} = 49$. Taking the square root of both sides gives $3^b = \pm 7$. We can't have $3^b = -7$, so our equation is now just $3^b = 7$. Now what? All this tells us is that b is the power to which we raise 3 in order to get 7. We can't find this power without a calculator or computer, but we can write it as a logarithm: $b = \log_3 7$.

   Since $3^b = 7$, we can use the equation $a \cdot 3^b = 14$ from above to find $a = 14/3^b = 2$. □

> **Problem 13.14:** Find all ordered pairs of positive integers (x, y) that satisfy both $y \le 5 - x$ and $y \ge \log_2 x$.

Solution for Problem 13.14: Solution 1: Casework. There aren't many pairs of positive integers that satisfy the inequality $y \le 5 - x$, because if $x \ge 5$, there are no positive integers y such that $y \le 5 - x$. So, we only have to try the cases $x = 1, 2, 3,$ and 4.

   When $x = 1$, we have $y \le 4$ and $y \ge \log_2 1 = 0$, so there are 4 solutions for this case with y as a positive integer, $(x, y) = (1, 1), (1, 2), (1, 3),$ and $(1, 4)$.

   When $x = 2$, we have $y \le 3$ and $y \ge 1$ as our inequalities, so there are 3 solutions for this case, $(x, y) = (2, 1), (2, 2),$ and $(2, 3)$.

   When $x = 3$, we have $y \le 2$ and $y \ge \log_2 3$ as our inequalities. Because $2^1 = 2$ and $2^2 = 4$, we know that we must raise 2 to a power that is between 1 and 2 to get 3. So, we have $2 > \log_2 3$, which means $(x, y) = (3, 2)$ satisfies both inequalities. That gives us 1 solution for this case.

   When $x = 4$, we have $y \le 1$ and $y \ge 2$ as our inequalities. There are no solutions for this case.

   This gives us a total of $4 + 3 + 1 = 8$ solutions.

> **Concept:** Not all problems have nice neat solutions. Sometimes a little casework is required to find a solution.

Solution 2: Graphing. We graph the two inequalities as shown at right. The graph of $y \le 5 - x$ consists of the diagonal line shown and the region below this line. The graph of $y \ge \log_2 x$ consists of the graph of $y = \log_2 x$ and the shaded region above this graph. The points that satisfy both these inequalities are the dark shaded region and the solid portions of the graphs of $y = 5 - x$ and $y = \log_2 x$.

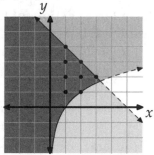

We see that there are 8 lattice points (points with integer coordinates) with positive coordinates that are either in the darkest shaded region, or on the solid lines in the diagram. These points are in bold in the diagram. So, there are 8 ordered pairs of positive integers that satisfy both inequalities. □

Our first solution is more rigorous than our second solution. While graphs can be excellent guides to finding an answer, they are not proofs. For example, our graph above of $y = \log_2 x$ appears to pass through $(2, 1)$. But we can't just trust what the graph appears to be in order to be completely confident in our answer. We should confirm that it passes through $(2, 1)$ by testing that $(2, 1)$ satisfies the equation $y = \log_2 x$. Similarly, our graph of $y = \log_2 x$ appears to pass between the points $(3, 1)$ and $(3, 2)$. While this looks obvious, our casework solution is more thorough in explaining why this is the case.

> **Concept:** Graphing can be a useful tool when working with unusual equations or inequalities.

> **WARNING!!** Graphs are excellent for providing intuition about a problem, but they shouldn't be used as a key part of a proof or a rigorous solution to a problem.

Exercises

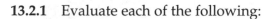

13.2.1 Evaluate each of the following:

(a) $\log_4 256$ (b) $\log_3 \frac{1}{9}$ (c) $\log_{125} 5$

13.2.2 Evaluate each of the following:

(a) $\log_{125} 25$ (b) $\log_{2\sqrt{2}} 16$ (c) $\log_{1/3} 9$

13.2.3 Solve the equation $3 \cdot 2^x - 9 = 14$.

13.2.4 Solve the equation $\log_2(x^2 - 2x - 7) = 3$.

13.2.5 Find the domain and range of each of the following:

(a) $f(x) = \log_{1/5} x$ (b) $g(x) = \log_6(\log_{1/5} x)$

13.2.6★ Let $f(x) = \log_3 (x - 1)$ and $g(x) = \sqrt{\frac{x}{x-1}}$. Find the domain of $g(f(x))$. **Hints:** 68

13.3 Logarithmic Identities

At right are exponential relationships sometimes referred to as **exponent laws**. In the next two sections, we use these laws to develop an analogous set of identities for logarithms, then put those identities to use.

(i) $x^m \cdot x^n = x^{m+n}$ (v) $\left(\frac{x}{y}\right)^n = \frac{x^n}{y^n}$

(ii) $\frac{x^m}{x^n} = x^{m-n}$ (vi) $x^{-n} = \frac{1}{x^n}$

(iii) $(x^m)^n = x^{mn}$ (vii) $\left(\frac{x}{y}\right)^{-n} = \left(\frac{y}{x}\right)^n$

(iv) $(xy)^m = x^m y^m$

Problems

Problem 13.15: In this problem, we find all positive values of x such that $(x)\left(x^{1/x}\right)^3 = \frac{x^x}{x}$. *(Source: NYSML)*

(a) Use the exponent laws to simplify both sides, writing the equation in the form $x^{f(x)} = x^{g(x)}$.

(b) Use your equation from part (a) to find all values of x that satisfy the original equation. Don't forget to check special cases!

Problem 13.16:

(a) Evaluate $\log_2 8$, $\log_2 8^2$, $\log_2 8^3$, and $\log_2 8^4$.

(b) How do you think $\log_a b$ and $\log_a b^c$ are related?

(c) Let $\log_a b = x$ and $\log_a b^c = y$. Write these equations in exponential form.

(d) Use the exponential equations in part (c) to show that $y = cx$. Does this prove your intuition from part (b)?

Problem 13.17: Prove that if a and b are positive and n is nonzero, then $\log_{a^n} b^n = \log_a b$.

Problem 13.18:

(a) Evaluate $\log_3 9 + \log_3 3$ and $\log_3 27$.

(b) Evaluate $\log_4 2 + \log_4 64$ and $\log_4 128$.

(c) Take a guess as to how we might simplify $\log_a b + \log_a c$.

(d) Let $\log_a b = x$ and $\log_a c = y$. Write these equations in exponential form.

(e) Prove that $\log_a b + \log_a c = \log_a bc$.

(f) Evaluate $\log_6 2 + \log_6 27 + \log_6 4$.

Problem 13.19: Prove that $\log_a b - \log_a c = \log_a \frac{b}{c}$.

Problem 13.20:

(a) Evaluate $(\log_2 8)(\log_8 64)$.

(b) Evaluate $(\log_3 9)\left(\log_9 \frac{1}{3}\right)$.

(c) Use the first two parts as a guide to guess how we might simplify $(\log_a b)(\log_b c)$.

(d) Prove your guess.

Problem 13.21: Prove that $\dfrac{\log_a b}{\log_a c} = \log_c b$.

Problem 13.22: Evaluate $\dfrac{4}{\log_{10} 5} - \dfrac{2}{\log_4 5}$.

Problem 13.23: Evaluate $3^{\log_3 4}$.

We start off by practicing with some of the exponent laws.

Problem 13.15: Find all positive values of x such that $(x)\left(x^{1/x}\right)^3 = \dfrac{x^x}{x}$. *(Source: NYSML)*

Solution for Problem 13.15: We start by simplifying both sides using the exponent laws. On the left, we have

$$x\left(x^{1/x}\right)^3 = x^1 \cdot x^{\frac{3}{x}} = x^{\frac{3}{x}+1}.$$

On the right, we have $\dfrac{x^x}{x} = x^{x-1}$, so our equation now is $x^{\frac{3}{x}+1} = x^{x-1}$. What's wrong with the following solution:

> **Bogus Solution:** Since the base on both sides is x, the exponents must be equal, so we must have $\frac{3}{x} + 1 = x - 1$. Multiplying both sides by x, and then rearranging, gives $x^2 - 2x - 3 = 0$, which has solutions $x = 3$ and $x = -1$. Since x must be positive, we discard the solution $x = -1$, so $x = 3$ is the only solution to the original equation.

The Bogus Solution does correctly identify those values of x for which the exponents are equal, and $x = 3$ is a solution to the original equation. However, the solution is incomplete. There is a special case we must consider. Specifically, if x is 1, then the equation $x^a = x^b$ does not require a and b to be equal. If $x = 1$ in the original equation, then both sides equal 1, so $x = 1$ is a solution. Therefore, the two solutions to our equation are $x = 1$ and $x = 3$. $\square$

In general, the case $x = 1$ is not the only special case to worry about when dealing with an equation of the form $x^a = x^b$. If $x = 0$, then $x^a = x^b$ for all nonzero numbers a and b. If we restrict a and b to being integers, then if $x = -1$, we have $x^a = x^b$ when a and b are either both even or both odd.

> **WARNING!!** When working with equations involving exponential expressions, watch out for special cases in which -1, 0, or 1 is the base.

Problem 13.16: Let $\log_a b = x$ and $\log_a b^c = y$. How are x and y related?

Solution for Problem 13.16: We start by exploring various logarithms to look for a pattern:

$$\begin{aligned}
\log_2 8 &= \log_2 2^3 &&= 3, \\
\log_2 8^2 &= \log_2 2^6 &&= 6, \\
\log_2 8^3 &= \log_2 2^9 &&= 9, \\
\log_2 8^4 &= \log_2 2^{12} &&= 12.
\end{aligned}$$

> **Concept:** When trying to find a relationship between two expressions involving variables, experiment by choosing simple values for the variables. Sometimes, a clear pattern will emerge.

We see that in each case, $\log_2 8^c$ is c times $\log_2 8$. So, we guess that

$$\log_a b^c = c \log_a b.$$

But how do we prove it? We don't have a lot of tools for working with logarithms, but we know a lot about exponents. So, we let

$$x = \log_a b \qquad \text{and} \qquad y = \log_a b^c,$$

and then write these equations in exponential form:

$$a^x = b \qquad \text{and} \qquad a^y = b^c.$$

We believe that $y = cx$. We can relate x to y by substituting our expression for b in the first equation, $b = a^x$, into the second equation. This gives us

$$a^y = (a^x)^c = a^{cx}.$$

Since the bases are the same in the equation $a^y = a^{cx}$, we have $y = cx$, as desired. (Note that we do not have to worry about the possibilities of a being -1, 0, or 1 in the equation $a^y = a^{cx}$, because a is the base of a logarithm in the problem.) □

We have our first logarithmic identity:

> **Important:** $\qquad\qquad\qquad\qquad \log_a b^c = c \log_a b.$

This identity is a direct result of the exponent laws.

> **Concept:** Most logarithmic identities are the result of the exponent laws. Therefore, we can usually prove logarithmic identities by converting them to exponential form and using our exponent laws.

Problem 13.17: Prove that if a and b are positive and n is nonzero, then $\log_{a^n} b^n = \log_a b$.

Solution for Problem 13.17: This problem looks a lot like the last one, so we try the same strategy.

> **Concept:** When facing a problem that is very similar to a problem that you have already solved, try using the tactics you used on the solved problem to tackle the new problem.

We start with

$$x = \log_a b \quad \text{and} \quad y = \log_{a^n} b^n.$$

Converting these to exponential notation gives

$$a^x = b \quad \text{and} \quad (a^n)^y = b^n.$$

As we did in the previous problem, we substitute our expression for b from the first equation into the second. We then have $(a^n)^y = (a^x)^n$. Because $(a^n)^y = a^{yn}$ and $(a^x)^n = a^{xn}$, we have $a^{xn} = a^{yn}$. Therefore, we have $xn = yn$, which means $x = y$.

We also could have written this proof by raising both sides of $a^x = b$ to the nth power to give $(a^x)^n = b^n$, so $a^{xn} = b^n$, which tells us that $(a^n)^x = b^n$. The equation $(a^n)^x = b^n$ means that raising a^n to the power of x gives us b^n, which means $\log_{a^n} b^n = x$, so we have $\log_{a^n} b^n = \log_a b$. $\square$

> **Concept:** Your first solution to a problem often will not be the simplest solution. Reviewing your first solution will sometimes reveal new, quicker methods.

Problem 13.18:

(a) Prove that $\log_a b + \log_a c = \log_a bc$.

(b) Evaluate $\log_6 2 + \log_6 27 + \log_6 4$.

Solution for Problem 13.18:

(a) Exponential form helped us so much in the last two problems that we try it again here. We let

$$x = \log_a b \quad \text{and} \quad y = \log_a c.$$

We wish to show that $\log_a bc = x + y$. We write our equations for x and y in exponential form:

$$a^x = b \quad \text{and} \quad a^y = c.$$

Since we are interested in the product bc, we multiply these two equations:

$$a^{x+y} = bc.$$

Writing this equation in logarithmic form gives the desired

$$\log_a bc = x + y = \log_a b + \log_a c.$$

Notice that this identity is just another way of writing the exponent law $a^{x+y} = a^x a^y$.

(b) As an example of how this identity helps us, consider the expression

$$\log_6 2 + \log_6 27 + \log_6 4.$$

We can't easily evaluate any one of these logarithms. But when we use our new identity to combine them, we have

$$\log_6 2 + \log_6 27 + \log_6 4 = \log_6(2)(27) + \log_6 4 = \log_6(2)(27)(4) = \log_6 6^3 = 3.$$

$\square$

You won't be surprised that there is an identity for the difference of two logarithms.

Problem 13.19: Prove that $\log_a b - \log_a c = \log_a \dfrac{b}{c}$.

Solution for Problem 13.19: Solution 1: Use exponential form. This looks so much like our previous identity that we figure we can use the same approach. We let

$$x = \log_a b \qquad \text{and} \qquad y = \log_a c.$$

We wish to show that $\log_a \dfrac{b}{c} = x - y$. We write our equations for x and y in exponential form:

$$a^x = b \qquad \text{and} \qquad a^y = c.$$

We want b/c, so we divide:

$$\frac{a^x}{a^y} = \frac{b}{c}.$$

Since $\dfrac{a^x}{a^y} = a^{x-y}$, we have $a^{x-y} = \dfrac{b}{c}$. Writing this in logarithmic form, we have

$$\log_a \frac{b}{c} = x - y = \log_a b - \log_a c.$$

Solution 2: Use our other logarithm identities.

Concept: Once we have proved an identity, we can use it to prove new identities.

We've already proved an identity that involves adding logarithms, and we wish to prove that

$$\log_a b - \log_a c = \log_a \frac{b}{c}.$$

Therefore, if we can turn the subtraction sign in front of $\log_a c$ into an addition sign, we can use our addition identity. Fortunately, we can use our very first identity to write

$$-\log_a c = (-1)\log_a c = \log_a c^{-1} = \log_a \frac{1}{c}.$$

So, we have

$$\log_a b - \log_a c = \log_a b + \log_a \frac{1}{c} = \log_a (b)\left(\frac{1}{c}\right) = \log_a \frac{b}{c}.$$

We also could have used the addition identity to deduce that $\log_a \frac{b}{c} + \log_a c = \log_a \left(\frac{b}{c} \cdot c\right) = \log_a b$, so $\log_a b - \log_a c = \log_a \frac{b}{c}$. $\square$

We can now combine sums and differences of logarithms that have the same base:

Important:

$$\log_a b + \log_a c = \log_a bc,$$

$$\log_a b - \log_a c = \log_a \frac{b}{c}.$$

WARNING!! The logarithms in our sum and difference identities *must have the same base*. For example, we cannot use the addition identity to combine the logarithms in the sum

$$\log_2 5 + \log_3 6.$$

We have identities for adding and subtracting logs; what about multiplying them?

Problem 13.20:

(a) Evaluate $(\log_2 8)(\log_8 64)$.

(b) Evaluate $(\log_3 9)\left(\log_9 \frac{1}{3}\right)$.

(c) Use the first two parts as a guide to guess how we might simplify $(\log_a b)(\log_b c)$.

(d) Prove your guess.

Solution for Problem 13.20:

(a) Since $\log_2 8 = 3$ and $\log_8 64 = 2$, we have $(\log_2 8)(\log_8 64) = (2)(3) = 6$.

(b) Since $\log_3 9 = 2$ and $\log_9 \frac{1}{3} = -\frac{1}{2}$, we have $(\log_3 9)\left(\log_9 \frac{1}{3}\right) = (2)\left(-\frac{1}{2}\right) = -1$.

(c) Our final result in the first part equals $\log_2 64$, and the answer in the second part equals $\log_3 \frac{1}{3}$:

$$(\log_2 8)(\log_8 64) = \log_2 64,$$

$$(\log_3 9)\left(\log_9 \frac{1}{3}\right) = \log_3 \frac{1}{3}.$$

Interesting! These both follow the pattern $(\log_a b)(\log_b c) = \log_a c$. Let's see if we can prove this identity.

(d) As usual, we shift to exponential notation. We let

$$x = \log_a b \qquad \text{and} \qquad z = \log_b c.$$

Therefore,

$$a^x = b \qquad \text{and} \qquad b^z = c.$$

We wish to show that $\log_a c = (\log_a b)(\log_b c) = xz$. Writing $\log_a c = xz$ in exponential form gives us $a^{xz} = c$. Since we want a^{xz}, we raise both sides of the equation $a^x = b$ to the z power:

$$a^{xz} = b^z.$$

However, we already know that $b^z = c$, so we have $a^{xz} = b^z = c$. Because $a^{xz} = c$, we have

$$\log_a c = xz = (\log_a b)(\log_b c),$$

as desired.

$\square$

One step that helped guide our proof was working backwards, writing the equation we wanted to prove, $\log_a c = xz$, in exponential form, $a^{xz} = c$. This gave us an equation to aim for when working with our exponential equations.

> **Concept:** Working backwards can be very useful in guiding your exploration when trying to find a proof.

We can use our multiplication identity to deal with dividing logarithms with the same base.

Problem 13.21: Prove that $\dfrac{\log_a b}{\log_a c} = \log_c b$.

Solution for Problem 13.21: We know how to deal with the product of logarithms, so we multiply both sides of the identity we'd like to prove by $\log_a c$, which gives

$$\log_a b = (\log_c b)(\log_a c).$$

But this is just the identity we proved in the previous problem! Because we have already proved that

$$(\log_c b)(\log_a c) = \log_a b,$$

we can divide this equation by $\log_a c$ to get the desired $\dfrac{\log_a b}{\log_a c} = \log_c b$. $\square$

Once again, we worked backwards to find our solution. Then, we wrote our solution forwards, starting from the multiplication identity and ending with the division identity we wanted to prove.

Problem 13.22: Evaluate $\dfrac{4}{\log_{10} 5} - \dfrac{2}{\log_4 5}$.

Solution for Problem 13.22: We might start by writing the expression with a common denominator, which gives us

$$\frac{4}{\log_{10} 5} - \frac{2}{\log_4 5} = \frac{4 \log_4 5 - 2 \log_{10} 5}{(\log_{10} 5)(\log_4 5)}.$$

Unfortunately, we don't have a good way to deal with the numerator or the denominator. None of our identities gets us anywhere. One reason we can't simplify the difference in the numerator is that the bases of the logarithms are different. However, in both cases, the argument (the number we are taking the logarithm of) is 5.

We know how to simplify sums and differences of logarithms if their bases are the same, so we put our wishful thinking hat on and try to find a way to turn the arguments of our logarithms into bases.

> **Concept:** Wishful thinking is an excellent problem solving tool. When stuck on a problem, ask yourself what, if it were true, would make the problem easier.

We do have an identity that turns the argument of a logarithm into a base, $\dfrac{\log_a b}{\log_a c} = \log_c b$. This turns the argument of the denominator on the left side into the base of the right side. So, if we let $a = 10$ and $c = 5$, we have

$$\frac{\log_{10} b}{\log_{10} 5} = \log_5 b.$$

This is true for any value of b. Choosing $b = 10$ gives us $\dfrac{1}{\log_{10} 5} = \log_5 10$ because $\log_{10} 10 = 1$. Similarly, letting $a = b$ in $\dfrac{\log_a b}{\log_a c} = \log_c b$ gives us

$$\frac{1}{\log_b c} = \log_c b.$$

Applying this to our problem, we have $\dfrac{1}{\log_{10} 5} = \log_5 10$ and $\dfrac{1}{\log_4 5} = \log_5 4$, so our expression is

$$\frac{4}{\log_{10} 5} - \frac{2}{\log_4 5} = 4\log_5 10 - 2\log_5 4.$$

Now, we can use some of our other identities to evaluate the expression:

$$\frac{4}{\log_{10} 5} - \frac{2}{\log_4 5} = 4\log_5 10 - 2\log_5 4 = \log_5 10^4 - \log_5 4^2 = \log_5 \left(\frac{10^4}{4^2}\right) = \log_5 625 = 4.$$

□

Note that we discovered a new identity on the way to our solution:

> **Important:**
> $$\frac{1}{\log_b c} = \log_c b.$$

Actually, this isn't a new identity at all! As we saw, it's just a special case of an identity we already knew, $\dfrac{\log_a b}{\log_a c} = \log_c b$, which itself is just a rearrangement of another identity we already proved, $(\log_a c)(\log_c b) = \log_a b$. So, there's really no need to memorize all three identities. If you know this last identity, you can quickly produce the others when you need them.

We finish our discovery of identities with an identity that involves both exponents and logarithms.

Problem 13.23: Evaluate $3^{\log_3 4}$.

Solution for Problem 13.23: What makes this expression so complicated is the very unusual exponent. Rather than diving into our exponent and logarithm tools right away, we stop and think about what this expression means.

> **Concept:** Don't be intimidated by unusual notation. Often, it represents a very simple idea.

Here, we focus on what the exponent means. The expression $\log_3 4$ equals the power to which we must raise 3 in order to get 4. Therefore, when we raise 3 to this power, by definition, the result is 4. $\square$

Similarly, for any positive numbers a and x with $a \neq 1$, the expression $\log_a x$ is the power to which we must raise a to get x, so $a^{\log_a x} = x$.

Putting all the identities we proved in this section in one place, we have:

> **Important:**
> $$\log_a b^c = c \log_a b$$
> $$\log_{a^n} b^n = \log_a b$$
> $$\log_a b + \log_a c = \log_a bc$$
> $$\log_a b - \log_a c = \log_a \frac{b}{c}$$
> $$(\log_a b)(\log_b c) = \log_a c$$
> $$\frac{\log_a b}{\log_a c} = \log_c b$$
> $$a^{\log_a x} = x$$

Hopefully you understand these identities so well that you do not need to memorize them. After working with them for a while, and proving them on your own, they should become as clear to you as the exponent laws we listed at the start of the section. In fact, most of these are simply restatements of exponent laws in different forms.

> **WARNING!!** Be careful when you apply the laws of logarithms. Here are some common mistakes that students make when working with logarithms:
> $$\log_b(x + y) = \log_b x + \log_b y,$$
> $$\frac{\log_b x}{\log_b y} = \log_b \frac{x}{y},$$
> $$(\log_b x)(\log_b y) = \log_b xy.$$
> **None of these are true in general!**

Exercises ▶

13.3.1 Evaluate each of the following without a calculator:

(a) $\log_3\left(27^{2007}\right)$

(b) $6^{\log_6 418}$

(c) $\log_2 \frac{2}{3} + \log_2 6$

(d) $(\log_2 5)(\log_5 12) + (\log_2 7)(\log_7 \frac{8}{3})$

(e) $\dfrac{\log_2 125}{\log_2 25}$

(f)★ $\dfrac{2^{\log_4 108}}{2^{\log_4 3}}$

13.3.2 Prove that $\log_b \frac{1}{a} = -\log_b a$.

13.3.3 If $c = \log_y b$, $c \neq 0$, and $d = 2\log_{y^3}(b^3)$, then find the value of $\frac{d}{c}$.

13.3.4 Prove that if $|a| \neq 1$, $|c| \neq 1$, $ac \neq 0$, and we have both $a^x = c^q$ and $c^y = a^z$, then $xy = qz$.

13.3.5 To the nearest hundredth, $\log_{10} 2$ is .301 and $\log_{10} 3$ is 0.477. Which of the following is the best approximation of $\log_5 10$?

$$\frac{8}{7}, \quad \frac{9}{7}, \quad \frac{10}{7}, \quad \frac{11}{7}, \quad \frac{12}{7}.$$

(Source: AHSME)

13.3.6★ Prove that $(\log_a b)(\log_c d) = (\log_c b)(\log_a d)$. **Hints:** 249

13.4 Using Logarithm Identities

In this section, we use our logarithm identities to solve more challenging problems. In most of these problems, our main strategy is to use our identities to simplify expressions and equations. In fact, that's the main strategy for solving a great deal of algebra problems:

> **Concept:** Use basic tools to simplify expressions and to convert complicated equations into simpler equations you know how to solve.

Try using this strategy on the following problems.

Problems ▶

> **Problem 13.24:** In this problem, we find all values of x such that $2\log_3(x + 4) - \log_3(4x - 11) = 2$.
> (a) Use logarithm identities to write the left side as a single logarithm.
> (b) Convert your equation from part (a) to exponential form, and solve the resulting equation.
> (c) Confirm that your solutions satisfy the original equation.

Problem 13.25: Find all ordered pairs of real numbers (x, y) that satisfy the system of equations

$$\log_{10}\left(\frac{x^3}{y^4}\right) = 5,$$

$$\log_{10}(x^2 y^5) = 11.$$

Problem 13.26: Suppose that $3^x = 12$. Find 9^{x-1}.

Problem 13.27: Suppose I have a calculator that only computes base 10 logarithms. How can I use my calculator to compute $\log_2 3$?

Problem 13.28: In this problem, we find the largest real value of b such that the solutions to the following equation are integers:

$$\left(\log_{2^{10}} x^{2b}\right)^2 = \log_{2^{10}} x^4.$$

(a) Rewrite the equation so that the logarithm terms are all $\log_2 x$ instead of $\log_{2^{10}} x^{2b}$ and $\log_{2^{10}} x^4$.

(b) Solve the problem. *(Source: ARML)*

Problem 13.29: The solutions to the system of equations

$$\log_{225} x + \log_{64} y = 4,$$

$$\log_x 225 - \log_y 64 = 1,$$

are (x_1, y_1) and (x_2, y_2). Find $\log_{30}(x_1 y_1 x_2 y_2)$. *(Source: AIME)* **Hints:** 47, 169

Problem 13.24: Find all values of x such that $2\log_3(x + 4) - \log_3(4x - 11) = 2$.

Solution for Problem 13.24: We use our logarithm identities to simplify the left side. First, we have $2\log_3(x + 4) = \log_3(x + 4)^2$, so our equation is $\log_3(x + 4)^2 - \log_3(4x - 11) = 2$. We now have a difference of logarithms with a common base, which we can write as

$$\log_3(x + 4)^2 - \log_3(4x - 11) = \log_3 \frac{(x + 4)^2}{4x - 11}.$$

This makes our original equation $\log_3 \dfrac{(x + 4)^2}{4x - 11} = 2$. Writing this in exponential form, we have

$$\frac{(x + 4)^2}{4x - 11} = 3^2.$$

We know how to solve this equation. We multiply both sides by $4x - 11$ and expand $(x + 4)^2$ to give

$$x^2 + 8x + 16 = 9(4x - 11).$$

Rearranging this equation gives $x^2 - 28x + 115 = 0$. Factoring the quadratic yields $(x - 23)(x - 5) = 0$, so the solutions are $x = 23$ and $x = 5$.

We test these solutions by substituting them in the original equation. For $x = 23$, we have

$$2\log_3 27 - \log_3 81 = 2(3) - 4 = 2,$$

so this solution works. For $x = 5$, we have

$$2\log_3 9 - \log_3 9 = 2(2) - 2 = 2,$$

so this solution works, too. □

Our first example exhibited two powerful strategies for solving equations involving logarithms:

> **Concept:** Equations involving several logarithms are often solved by using logarithm identities to write the equation with a single logarithm.

> **Concept:** Many logarithmic equations can be solved by converting them into exponential equations, then solving the exponential equations.

However, we must be careful:

> **WARNING!!** When you solve a logarithmic equation, be on the lookout for extraneous solutions. The best way to avoid extraneous solutions is to check that your solutions satisfy the original equation.

For example, consider the equation $\log_2 x + \log_2(x + 2) = 3$. Combining the logarithms on the left gives $\log_2(x^2 + 2x) = 3$, and converting to exponential form gives $x^2 + 2x = 2^3 = 8$. Solving this equation gives $x = 2$ and $x = -4$ as solutions. While $x = 2$ does satisfy the original equation, the value $x = -4$ does not, because $\log_2(-4)$ and $\log_2(-2)$ are not defined. This extraneous solution is produced when we add $\log_2 x$ and $\log_2(x + 2)$ to get $\log_2(x^2 + 2x)$. The expression $\log_2(x^2 + 2x)$ is defined for $x = -4$, even though $\log_2 x$ and $\log_2(x + 2)$ are not.

Another way we can be on the lookout for extraneous solutions when working with an equation like $\log_2 x + \log_2(x + 2) = 3$ is to start by determining those values of x for which the expressions in the equation are defined. Since $\log_2 x$ is defined only for $x > 0$ and $\log_2(x + 2)$ is defined only for $x > -2$, we know that any valid solution must satisfy $x > 0$.

> **Problem 13.25:** Find all ordered pairs of real numbers (x, y) that satisfy the system of equations
>
> $$\log_{10}\left(\frac{x^3}{y^4}\right) = 5,$$
> $$\log_{10}(x^2 y^5) = 11.$$

Solution for Problem 13.25: Solution 1: Logarithm identities. The mix of variables inside the logarithms is the chief obstacle, so we start by rewriting the system with the variables separated:

$$\log_{10} x^3 - \log_{10} y^4 = 5,$$
$$\log_{10} x^2 + \log_{10} y^5 = 11.$$

> **Concept:** Just as our sum and difference identities can be used to combine logarithms, they can also be used to separate logarithms of products and quotients into easier to handle logarithms.

Unfortunately, we can't yet combine these equations to eliminate x or y. However, we can use the identity $\log_a b^c = c \log_a b$ to write both equations in terms of $\log_{10} x$ and $\log_{10} y$. This gives us

$$3 \log_{10} x - 4 \log_{10} y = 5,$$
$$2 \log_{10} x + 5 \log_{10} y = 11.$$

Adding 5 times the first equation to 4 times the second eliminates $\log_{10} y$ and gives us $23 \log_{10} x = 69$, so $\log_{10} x = 3$. Therefore, we have $x = 10^3 = 1000$. Substituting this into either of our original equations gives $\log_{10} y = 1$, so $y = 10$. Therefore, the solution to our system of equations is $(x, y) = (1000, 10)$.

Solution 2: Convert to exponential form. Exponential equations are often easier to work with than logarithmic equations, so we rewrite the system as

$$\frac{x^3}{y^4} = 10^5,$$
$$x^2 y^5 = 10^{11}.$$

Now, we can eliminate y by taking the first equation to the fifth power and multiplying it by the second equation raised to the fourth power:

$$\left(\frac{x^3}{y^4}\right)^5 (x^2 y^5)^4 = (10^5)^5 \cdot (10^{11})^4.$$

Simplifying both sides of this equation gives us $x^{23} = 10^{69}$, so $x = 10^3$. Substituting this value into the equation $x^2 y^5 = 10^{11}$ and solving for y, we get $(x, y) = (1000, 10)$. $\square$

> **Problem 13.26:** Suppose that $3^x = 12$. Find 9^{x-1}.

Solution for Problem 13.26: *Solution 1: Exponential manipulation.* We are working with two different powers of 3, so we start by writing everything as a power of 3. We already have $3^x = 12$. We write 9^{x-1} as $9^{x-1} = (3^2)^{x-1} = 3^{2(x-1)} = 3^{2x-2}$. If we can write this in terms of 3^x, we can substitute $3^x = 12$ to evaluate it. Fortunately, it's not hard to do so:

$$3^{2x-2} = \frac{3^{2x}}{3^2} = \frac{(3^x)^2}{9} = \frac{(12)^2}{9} = 16.$$

Solution 2: x is a logarithm. From $3^x = 12$, we have $x = \log_3 12$. Substituting this into 9^{x-1} gives us

$$9^{\log_3 12 - 1}.$$

We only have one tool in our bag of tricks for dealing with logarithms as exponents:

$$a^{\log_a b} = b.$$

In order to use this identity, we must rewrite $9^{\log_3 12 - 1}$ so that the complicated exponential expression is in the form $a^{\log_a b}$. We must therefore have only a logarithm in the exponent, and the base of the exponential must be the same as the base of the logarithm. We could isolate the logarithm in the exponent first:

$$9^{\log_3 12 - 1} = \frac{9^{\log_3 12}}{9^1} = \frac{(3^2)^{\log_3 12}}{9} = \frac{3^{2\log_3 12}}{9} = \frac{3^{\log_3 144}}{9} = \frac{144}{9} = 16.$$

We could also note that $\log_3 12 = \log_{3^2} 12^2 = \log_9 144$, so

$$9^{\log_3 12 - 1} = \frac{9^{\log_3 12}}{9^1} = \frac{9^{\log_{3^2} 12^2}}{9} = \frac{9^{\log_9 144}}{9} = \frac{144}{9} = 16.$$

Or, we could use identities to write the exponent as a single logarithm:

$$9^{\log_3 12 - 1} = 9^{\log_3 12 - \log_3 3} = 9^{\log_3 4} = (3^2)^{\log_3 4} = 3^{2\log_3 4} = 3^{\log_3 16} = 16.$$

We also could have evaluated $9^{\log_3 4}$ by noting that $\log_3 4 = \log_{3^2} 4^2 = \log_9 16$, so $9^{\log_3 4} = 9^{\log_9 16} = 16.$ □

Problem 13.27: Suppose I have a calculator that only computes base 10 logarithms. How can I use my calculator to compute $\log_2 3$?

Solution for Problem 13.27: Because my calculator only understands base 10 logarithms, I can only compute numbers of the form $\log_{10} x$. However, I need to compute something that doesn't have any 10's at all! So, we look through all our logarithm tools in search of an identity that has a base on one side of the equation but doesn't have that base on the other side. Our division identity stands out:

$$\log_c b = \frac{\log_a b}{\log_a c}.$$

We want $\log_2 3$, and our identity gives us the way to get it:

$$\log_2 3 = \frac{\log_{10} 3}{\log_{10} 2}.$$

So, to compute $\log_2 3$, I compute $\log_{10} 3$ and $\log_{10} 2$ with my calculator, then divide the former by the latter. □

Problem 13.28: Compute the largest real value of b such that the solutions to the following equation are integers:

$$\left(\log_{2^{10}} x^{2b}\right)^2 = \log_{2^{10}} x^4.$$

(Source: ARML)

Solution for Problem 13.28: The two logarithmic expressions are pretty complicated, so we start by looking for ways to simplify them. First, we can remove the exponents from the arguments, x^{2b} and x^4, and we have

$$\left(2b \log_{2^{10}} x\right)^2 = 4 \log_{2^{10}} x.$$

Now, the two logarithm expressions are the same. We can simplify them both even further by noting that

$$\log_{2^{10}} x = \log_2 x^{1/10} = \frac{1}{10} \log_2 x,$$

so our equation becomes

$$\left(\frac{b}{5} \log_2 x \right)^2 = \frac{2}{5} \log_2 x.$$

We multiply both sides by 25 to get rid of the fractions, which gives us

$$b^2 (\log_2 x)^2 = 10 \log_2 x.$$

Since the logarithm expressions are the same, we bring all terms to the left and factor the resulting expression to give

$$(b^2 \log_2 x - 10) \log_2 x = 0.$$

> **WARNING!!**
> ☢
> Notice that we did not divide both sides by $\log_2 x$ to cancel out a factor of $\log_2 x$ from both sides of $b^2 (\log_2 x)^2 = 10 \log_2 x$. This is because $\log_2 x$ might be 0, in which case we cannot divide by it. Instead, we rearrange and factor, so that we don't lose track of the fact that the original equation is true when $\log_2 x = 0$.

The equation $(b^2 \log_2 x - 10) \log_2 x = 0$ tells us that either $\log_2 x = 0$, which gives $x = 1$ no matter what b is, or $b^2 \log_2 x = 10$. Isolating the logarithm gives $\log_2 x = 10/b^2$, and converting this to exponential form gives

$$x = 2^{10/b^2}.$$

In order for $2^{10/b^2}$ to be an integer, the expression $10/b^2$ must be a positive integer. The largest value of b for which $10/b^2$ is an integer is $\sqrt{10}$. For any value of b with $b > \sqrt{10}$, we have $10/b^2 < 1$, so we have $1 < 2^{10/b^2} < 2$. Therefore, the largest value of b for which all solutions to the original equation are integers is $b = \sqrt{10}$. □

The key to our solution is using logarithm identities to simplify the logarithm expressions in the equation. In fact, this has been the key to most of the problems in this section:

> **Concept:**
> ⊙═🔑
> When working with a complicated equation involving logarithms, try using logarithm identities to simplify the expressions involved. Common ways to do so are:
>
> - Simplify the bases or arguments of logarithms using the identities $\log_a b^c = c \log_a b$ and $\log_{a^n} b^n = \log_a b$.
>
> - Combine logarithms of the same base with the sum, difference, or quotient identities.
>
> - Break logarithms of products or quotients into simpler pieces with the sum or difference identities.

Our final two points seem to contradict each other. Here's where a little problem-solving savvy is required. We combine logarithms with our identities when the result is a simpler expression. For example, if we have $\log_2 \frac{32}{7} + \log_2 \frac{7}{4}$, then using our addition identity cancels the 7s and gets rid of the fractions, leaving

$$\log_2 \frac{32}{7} + \log_2 \frac{7}{4} = \log_2\left(\frac{32}{7} \cdot \frac{7}{4}\right) = \log_2 8 = 3.$$

On the other hand, when confronted with $\log_{10} \frac{x^3}{y^4}$ in Problem 13.25, breaking the logarithm up gave us $3\log_{10} x - 4\log_{10} y$, which has much simpler logarithms. Notice that we can also use our "breaking logarithms up for simplification" strategy to evaluate $\log_2 \frac{32}{7} + \log_2 \frac{7}{4}$:

$$\log_2 \frac{32}{7} + \log_2 \frac{7}{4} = \log_2 32 - \log_2 7 + \log_2 7 - \log_2 4 = 5 - 2 = 3.$$

Problem 13.29: The solutions to the system of equations

$$\log_{225} x + \log_{64} y = 4,$$
$$\log_x 225 - \log_y 64 = 1,$$

are (x_1, y_1) and (x_2, y_2). Find $\log_{30}(x_1 y_1 x_2 y_2)$. *(Source: AIME)*

Solution for Problem 13.29: First, we note that the logarithms in the second equation are the reciprocals of those in the first equation, so our system of equations is

$$\log_{225} x + \log_{64} y = 4,$$
$$\frac{1}{\log_{225} x} - \frac{1}{\log_{64} y} = 1.$$

Now, at least the logarithms are the same. Rather than continuing to write $\log_{225} x$ and $\log_{64} y$ over and over, we let $a = \log_{225} x$ and $b = \log_{64} y$ to make the algebraic manipulation simpler. Our system is now

$$a + b = 4,$$
$$\frac{1}{a} - \frac{1}{b} = 1.$$

This is a lot less difficult to deal with.

Concept: Substituting variables for complex expressions in equations can make the equations easier to solve.

We solve the first equation for a to find $a = 4 - b$. Substituting this in the second equation gives

$$\frac{1}{4-b} - \frac{1}{b} = 1.$$

Multiplying both sides by $(4-b)b$ gives $b - (4-b) = (4-b)b$, and rearranging gives $b^2 - 2b - 4 = 0$. If we find b, then we can find y, since the equation $b = \log_{64} y$ tells us that $y = 64^b$. Unfortunately, the roots of $b^2 - 2b - 4 = 0$ are irrational. However, looking at the expression we seek,

$$\log_{30} x_1 y_1 x_2 y_2,$$

we see that we don't actually have to find y_1 and y_2. We only need their product.

> **Concept:** Keep your eye on the ball. Sometimes we don't need to find the values of all the variables in a problem to solve the problem.

Fortunately, we can find this product without ever finding the values of b that satisfy $b^2 - 2b - 4 = 0$. Suppose b_1 and b_2 are the solutions to this equation, so we have $y_1 = 64^{b_1}$ and $y_2 = 64^{b_2}$. Then, we have

$$y_1 y_2 = 64^{b_1} \cdot 64^{b_2} = 64^{b_1 + b_2}.$$

Since b_1 and b_2 are the solutions to the equation $b^2 - 2b - 4 = 0$, Vieta tells us that $b_1 + b_2 = -(-2)/1 = 2$. So, we have $y_1 y_2 = 64^2$.

Similarly, from $a = \log_{225} x$, we have $x = 225^a$. Since $a = 4 - b$, we have $x = 225^{4-b}$. So, from our two values of b, namely b_1 and b_2, we get two values of x, which we call x_1 and x_2, and we have

$$x_1 x_2 = 225^{4-b_1} \cdot 225^{4-b_2} = 225^{8-(b_1+b_2)} = 225^6.$$

So, we have $\log_{30} x_1 y_1 x_2 y_2 = \log_{30}(64^2)(225^6) = \log_{30}(2^{12} \cdot 3^{12} \cdot 5^{12}) = \log_{30} 30^{12} = 12$. $\square$

Exercises

13.4.1 If r and s are the roots of $3x^2 - 16x + 12 = 0$, then find $\log_2 r + \log_2 s$.

13.4.2 Given that $\log_{10} 17 = r$ and $\log_{10} 2 = s$, find the value of $\dfrac{\log_{10} \frac{1600}{17}}{\log_{10} 136}$ in terms of r and s.

13.4.3 Find all t such that $2\log_3(1 - 5t) = \log_3(2t + 5) + 2$.

13.4.4 Let $x_1, x_2, \ldots, x_n$ be a geometric sequence of positive real numbers. Prove that for any positive number b, with $b \neq 1$, the sequence $\log_b x_1, \log_b x_2, \ldots, \log_b x_n$ is an arithmetic sequence.

13.4.5 Suppose the ordered pair (x, y) satisfies $\dfrac{\log_{10}(xy)}{\log_{10}\left(\frac{x}{y}\right)} = \dfrac{1}{2}$. If y is increased by 50%, by what fraction must x be multiplied to keep this equation true? *(Source: ARML)*

13.4.6 A line $x = k$ intersects the graph of $y = \log_5 x$ and the graph of $y = \log_5 (x + 4)$. The distance between the points of intersection is 0.5. Find k. *(Source: AHSME)*

13.4.7 For all integers n greater than 1, define

$$a_n = \frac{1}{\log_n 2002}.$$

Let $b = a_2 + a_3 + a_4 + a_5$ and $c = a_{10} + a_{11} + a_{12} + a_{13} + a_{14}$. What is $b - c$? *(Source: AMC 12)* **Hints: 219**

13.4.8 Compute the integer k, where $k > 2$, for which

$$\log_{10}(k - 2)! + \log_{10}(k - 1)! + 2 = 2\log_{10} k!.$$

(Source: ARML) **Hints: 86**

13.5 Switching Between Logs and Exponents

We've already switched from logarithmic notation to exponential notation as a step in several problems. In this section, we highlight this strategy with some more challenging problems. We also introduce the reverse strategy: using logarithms to simplify complicated exponential expressions.

Problems

Problem 13.30: If $\log_{10} 5 \approx 0.699$, how many digits are in 5^{2006}?

Problem 13.31: In this problem, we prove that $a^{\log_b c} = c^{\log_b a}$.

  (a) Simplify $\log_a a^{\log_b c}$. Simplify $\log_a c^{\log_b a}$.

  (b) Use part (a) to prove that $a^{\log_b c} = c^{\log_b a}$.

Problem 13.32: Suppose that p and q are positive numbers for which $\log_9 p = \log_{12} q = \log_{16}(p + q)$. What is the value of q/p? *(Source: AHSME)*

Problem 13.30: If $\log_{10} 5 \approx 0.699$, how many digits are in 5^{2006}?

Solution for Problem 13.30: There's no way we're going to compute 5^{2006}, so we need to find some way to count the digits of a number without knowing the number exactly. There are some large numbers whose digits are easy to count: powers of 10. For example, 10^{54} is a 1 followed by 54 zeroes, so it has 55 digits. We can use this observation to count the digits in other numbers, as well. For example, 3.5×10^{63} starts with "35," and then has 62 zeroes (one power of 10 turns 3.5 into 35: $3.5(10^{63}) = 35(10^{62})$).

So, we'd like to write 5^{2006} in terms of powers of 10. This gives us the idea of writing the equation $\log_{10} 5 \approx 0.699$ in exponential form, since this will give us 5 as a power of 10:

$$10^{0.699} \approx 5.$$

Therefore, we have

$$5^{2006} \approx \left(10^{0.699}\right)^{2006} = 10^{(0.699)(2006)} = 10^{1402.19}.$$

This tells us that 5^{2006} is between 10^{1402} and 10^{1403}. Since 10^{1402} has 1403 digits and 10^{1403} is the smallest positive base 10 number with 1404 digits, the number 5^{2006} must have 1403 digits.

We could also have solved this problem by taking the base 10 logarithm of 5^{2006}:

$$\log_{10} 5^{2006} = 2006 \log_{10} 5 \approx (2006)(0.699) = 1402.19.$$

Putting this in exponential form gives us $5^{2006} \approx 10^{1402.19}$, and the rest of our solution proceeds as before. $\square$

Problem 13.30 is an example of how we can take the base 10 logarithm of a number to count the digits in the decimal expansion of the number.

Problem 13.31: Prove that $a^{\log_b c} = c^{\log_b a}$.

Solution for Problem 13.31:   If the base of the logarithm in $a^{\log_b c}$ were a instead of b, we'd know how to simplify it. But it isn't, so we'll have to find some other way to deal with such a complicated exponent. We do at least have an identity that gives us a way to turn exponents into factors of a product: $\log_x y^z = z \log_x y$. So, we can get $\log_b c$ out of the exponent of $a^{\log_b c}$ by taking the logarithm of $a^{\log_b c}$. But what base should we use?

We try using a, since we have

$$\log_a a^{\log_b c} = (\log_b c)(\log_a a) = \log_b c,$$

which is pretty simple. Let's see what happens when we take the base a logarithm of $c^{\log_b a}$:

$$\log_a c^{\log_b a} = (\log_b a)(\log_a c).$$

Back in Problem 13.20, we showed that $(\log_b a)(\log_a c) = \log_b c$, so we have

$$\log_a a^{\log_b c} = \log_b c = \log_a c^{\log_b a}.$$

We have $\log_a x = \log_a y$ if and only if $x = y$, so the equation above gives us the desired $a^{\log_b c} = c^{\log_b a}$.

Our proof does overlook the fact that a could be 1, in which case we can't take the logarithm base a. If $a = 1$, we have $a^{\log_b c} = 1$ and $c^{\log_b a} = c^{\log_b 1} = c^0 = 1$, so we still have $a^{\log_b c} = c^{\log_b a}$. We also could have gotten around this difficulty by seeing that we could use any base when we took the logarithms of $a^{\log_b c}$ and $c^{\log_b a}$. Using the multiplication identity in Exercise 13.3.6, we have

$$\log_d a^{\log_b c} = (\log_b c)(\log_d a) = (\log_d c)(\log_b a) = \log_d c^{\log_b a}.$$

In our last step, we used the logarithm identity $\log_x y^n = n \log_x y$ *in reverse*, writing the product $(\log_b a)(\log_d c)$ as $\log_d c^{\log_b a}$. Because the bases are the same in $\log_d a^{\log_b c} = \log_d c^{\log_b a}$, so are the arguments: $a^{\log_b c} = c^{\log_b a}$. $\square$

Problem 13.31 gives us a strategy for dealing with complicated exponents:

> **Concept:**   If you have an expression containing a complicated exponent, take the logarithm of it to turn the exponent into a factor.

> **WARNING!!**   This "take the logarithm to remove complicated exponents" strategy does not work on sums or differences of expressions. For example,
>
> $$\log_a(b^x + c^y) \quad \textbf{does not equal} \quad x \log_a b + y \log_a c.$$

Problem 13.32: Suppose that p and q are positive numbers for which $\log_9 p = \log_{12} q = \log_{16}(p + q)$. What is the value of q/p? *(Source: AHSME)*

Solution for Problem 13.32: All the bases are different, as are the arguments p, q, and $p + q$. Moreover, the bases aren't all powers of the same integer. So, there's no obvious way to use our identities.

In other problems, when we've been stuck working with logarithms, we've tried using exponential notation instead. Let's try this strategy here. We start with $\log_9 p = \log_{12} q$. Writing this in exponential form gives

$$9^{\log_{12} q} = p \qquad \text{or} \qquad 12^{\log_9 p} = q.$$

Yuck. It might be possible to find a solution from here, but those are pretty forbidding equations.

We've introduced new variables to simplify complicated equations before. Let's try that here:

$$x = \log_9 p = \log_{12} q = \log_{16}(p + q).$$

This gives us much simpler exponential equations:

$$9^x = p, \qquad 12^x = q, \qquad \text{and} \qquad 16^x = p + q.$$

We might then note that this makes our original equation $9^x + 12^x = 16^x$, but this doesn't seem to help us with q/p. Instead, we divide the second equation by the first to find a simple expression for q/p:

$$\frac{q}{p} = \frac{12^x}{9^x} = \left(\frac{4}{3}\right)^x.$$

But how can we find $(4/3)^x$?

We haven't used the equation $16^x = p + q$, and we expect we'll have to. Comparing this to our other equations, we note that $16^x/12^x = (4/3)^x$, which is the quantity we seek. So, we divide $16^x = p + q$ by $12^x = q$ to give

$$\left(\frac{4}{3}\right)^x = \frac{p + q}{q} = \frac{p}{q} + 1.$$

So, we have

$$\frac{q}{p} = \frac{p}{q} + 1.$$

Letting $t = \frac{q}{p}$, this means we have $t = \frac{1}{t} + 1$, so we have $t^2 - t - 1 = 0$. Applying the quadratic formula gives

$$t = \frac{1 \pm \sqrt{5}}{2}.$$

We must have $t > 0$, since $\frac{q}{p} = \left(\frac{4}{3}\right)^x > 0$. Since $\frac{1 - \sqrt{5}}{2}$ is negative, the only possible value of $\frac{q}{p}$ is $t = \frac{1 + \sqrt{5}}{2}$.

We could also have divided our equation $9^x + 12^x = 16^x$ by 9^x to get $1 + \left(\frac{4}{3}\right)^x = \left(\frac{4}{3}\right)^{2x}$, because $\left(\frac{16}{9}\right)^x = \left(\frac{4}{3}\right)^{2x}$. Letting $t = \left(\frac{4}{3}\right)^x$ leads to the same quadratic as before. □

Recall our story about the magic camel from page 382. In Problem 13.32, the variable x is a magic camel—we don't ever find x, but we use it to recast the problem in a more convenient form that allows us to solve the problem.

Exercises

13.5.1 Suppose that $\log_{10} xy^3 = 1$ and $\log_{10} x^2 y = 1$. What is $\log_{10} xy$? *(Source: AMC 12)*

13.5.2 If $60^a = 3$ and $60^b = 5$, then find $12^{[(1-a-b)/2(1-b)]}$. *(Source: AHSME)*

13.5.3 If $\log_2 p = \log_3 r = \log_{36} 17$, then determine pr. *(Source: Mandelbrot)* **Hints:** 67

13.5.4 Suppose that $a > b > 0$. Solve the equation $(a^4 - 2a^2b^2 + b^4)^{y-1} = (a - b)^{2y}(a + b)^{-2}$ for y in terms of a and b.

13.5.5 Find all x such that $x^{\log_{10} x} = \frac{x^3}{100}$. *(Source: AHSME)*

13.6 Natural Logarithms and Exponential Decay

Problems

Problem 13.33: Let $f(n) = \left(1 + \dfrac{1}{n}\right)^n$ and $g(n) = 1 + \dfrac{1}{1!} + \dfrac{1}{2!} + \dfrac{1}{3!} + \cdots + \dfrac{1}{n!}$.

(a) Evaluate $f(1)$, $f(2)$, $f(5)$, and $f(10)$. (Yes, you can use a calculator on this problem.)

(b) Evaluate $g(1)$, $g(2)$, $g(5)$, and $g(10)$.

(c) Use a calculator or computer to evaluate $f(n)$ and $g(n)$ for large values of n. Notice anything interesting?

We start this section by introducing a special constant that, much like π, has a great many uses in mathematics and science.

Problem 13.33: Let $f(n) = \left(1 + \dfrac{1}{n}\right)^n$ and $g(n) = 1 + \dfrac{1}{1!} + \dfrac{1}{2!} + \dfrac{1}{3!} + \cdots + \dfrac{1}{n!}$. Use a calculator or computer to evaluate $f(n)$ and $g(n)$ for large values of n. Notice anything interesting?

Solution for Problem 13.33: The table below shows $f(n)$ and $g(n)$ to the nearest 10^{-8} for several values of n.

n	$f(n)$	$g(n)$
1	2.00000000	2.00000000
2	2.25000000	2.50000000
5	2.48832000	2.71666668
10	2.59374246	2.71828180
100	2.70481383	2.71828183
1000	2.71692393	2.71828183
100000	2.71826824	2.71828183
10000000	2.71828169	2.71828183

It looks like both $f(n)$ and $g(n)$ approach constant values as n gets very large. Moreover, it looks like they might approach the *same* constant value. □

In fact, $f(n)$ and $g(n)$ in Problem 13.33 do indeed approach the same constant value as n gets very large. We call this constant e, a label given to this number by the great mathematician **Leonhard Euler**. (Euler is pronounced "oiler.") Not until we have the tools of calculus can we really appreciate the many fascinating properties of e, but e does appear in a great many scientific contexts. The base e logarithm is also frequently used in mathematics and science. We call the base e logarithm the **natural logarithm**. Rather than writing $\log_e$, we usually write ln, so $\ln x = \log_e x$.

> **WARNING!!** Many mathematicians will also write log without a base to indicate the
> ☢ natural logarithm. However, it is also common to write log without a
> base to indicate the base 10 logarithm. In fact, it's so common that such
> a logarithm is sometimes called the **common logarithm**. Because of
> the ambiguity surrounding the use of log without an indicated base,
> we will avoid doing so in this book. When you see it used in other
> sources, you'll have to use context to decide which of $\log_e$ and $\log_{10}$
> is intended.

The letter e and exponential functions are frequently used to describe natural processes. One example is **radioactive decay**. A radioactive substance decays into non-radioactive substances over time. It does so following a process called **exponential decay**. Under exponential decay, the same proportion of a substance decays in any fixed time t. In other words, the fraction of a 100 gram sample of a radioactive substance that decays in 3 seconds is the same as the fraction of a 30 kilogram sample of the same substance that decays in 3 seconds.

The **half-life** of a radioactive substance is the amount of time it takes for exactly half the radioactive substance to decay. Because the same proportion of a substance decays in any fixed time t, the half life of a substance is the same no matter how much of the radioactive substance is present. So, if the half-life of a substance is 40 days and initially there are 400 grams of the substance, then after 40 days there will be 200 grams of radioactive substance remaining, and after another 40 days, there will be 100 grams, and so on.

> **Problems**

Problem 13.34: Initially there are 8 grams of a radioactive material in a container. The half-life of the material is 2 days.

(a) After how many days will there be 0.5 grams of radioactive material left in the container?

(b) Suppose there are x grams of the radioactive material remaining after 5 days. Use the amounts of radioactive material remaining after 4 days and after 6 days to determine x. (Warning: The answer is *not* 1.5.)

(c) Let $f(x)$ be the number of grams of the substance that has not decayed after x days. Graph $y = f(x)$.

(d) Let t_1, t_2, and t_3 be three amounts of time after the material is placed in the container such that $t_2 - t_1 = t_3 - t_2$. (For example, we could have $t_1 = 2$ days, $t_2 = 3.5$ days, and $t_3 = 5$ days.) Let $f(x)$ be the function you found in part (c). Show that $f(t_2)/f(t_1) = f(t_3)/f(t_2)$.

Problem 13.35: Suppose that a radioactive element has a half-life of $t_{1/2}$, and that we run an experiment in which we initially place N_o atoms of this element in a container. After a time t, suppose there are N_n atoms of the radioactive element remaining in the container. In this problem, we will show that

$$t \approx -t_{1/2} \cdot \frac{\ln(N_n/N_o)}{0.693}.$$

(This formula frequently appears in science textbooks.)

(a) Explain why $\left(\frac{1}{2}\right)^{t/t_{1/2}} = \frac{N_n}{N_o}$.

(b) Take the natural logarithm of both sides of part (a) to show that $(t/t_{1/2})\ln(1/2) = \ln(N_n/N_o)$.

(c) Show that $t = -t_{1/2} \cdot \dfrac{\ln(N_n/N_o)}{\ln 2}$.

Problem 13.36: Explain why the formula $\left(\frac{1}{2}\right)^{t/t_{1/2}} = \frac{N_n}{N_o}$ can be written as $N_n = N_o e^{-\lambda t}$, where $\lambda = \dfrac{\ln 2}{t_{1/2}}$.

Problem 13.37: Dr. Curie took two measurements of a radioactive sample exactly 100 seconds apart. There were 403 grams of the sample in the first measurement and 83 grams remained undecayed in the second measurement.

(a) Find the half-life of the sample.

(b) If she measures the sample a third time, 38 seconds after the second measurement, how many grams of undecayed material will there be in the sample?

Problem 13.34: Initially there are 8 grams of a radioactive material in a container. The half-life of the material is 2 days.

(a) After how many days will there be 0.5 grams of radioactive material left in the container?

(b) How much of the radioactive substance will remain after 5 days?

(c) Let $f(x)$ be the number of grams of the substance that has not decayed after x days. Graph $y = f(x)$.

(d) Let t_1, t_2, and t_3 be three amounts of time after the material is placed in the container such that $t_2 - t_1 = t_3 - t_2$. Let $f(x)$ be the function you found in part (c). Show that $f(t_2)/f(t_1) = f(t_3)/f(t_2)$.

Solution for Problem 13.34:

(a) Because the material has a half-life of 2 days, exactly half the material decays in 2 days. So, after two days, there are $8/2 = 4$ grams of the radioactive material remaining. Over the next two days, half the remaining radioactive material decays, leaving $4/2 = 2$ grams of the radioactive material. Two days later (6 days from the beginning), there is $2/2 = 1$ gram of radioactive material. Finally, two days later, which is 8 days from the beginning, there are $1/2 = 0.5$ grams of radioactive material.

(b) Above, we saw that after 4 days, there are 2 grams of radioactive material, and after 6 days, there is 1 gram of radioactive material. Let there be x grams of radioactive material after 5 days. The same fraction of the remaining radioactive substance decays each day. So, the fraction of the radioactive substance that decays from day 4 to day 5, which is $(2 - x)/2$, must equal the fraction of the radioactive substance that decays from day 5 to day 6, which is $(x - 1)/x$. Therefore, we have

$$\frac{2 - x}{2} = \frac{x - 1}{x}.$$

Cross-multiplying gives $2x - x^2 = 2x - 2$, from which we find $x = \sqrt{2}$.

(c) Every 2 days, the amount of radioactive substance remaining is divided by 2. So, after x days, the amount of radioactive substance remaining has been divided by 2 exactly $x/2$ times, and we have

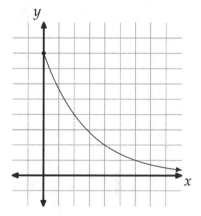

$$f(x) = 8\left(\frac{1}{2}\right)^{\frac{x}{2}}.$$

This is an exponential function! Its graph is shown at right.

(d) We can check that the function $f(x)$ satisfies our definition of exponential decay. Specifically, suppose that times t_1, t_2, and t_3 are equally spaced with $t_1 < t_2 < t_3$. Then, we have $t_2 - t_1 = t_3 - t_2$, and

$$\frac{f(t_2)}{f(t_1)} = \frac{8\left(\frac{1}{2}\right)^{\frac{t_2}{2}}}{8\left(\frac{1}{2}\right)^{\frac{t_1}{2}}} = \left(\frac{1}{2}\right)^{\frac{t_2 - t_1}{2}},$$

$$\frac{f(t_3)}{f(t_2)} = \frac{8\left(\frac{1}{2}\right)^{\frac{t_3}{2}}}{8\left(\frac{1}{2}\right)^{\frac{t_2}{2}}} = \left(\frac{1}{2}\right)^{\frac{t_3 - t_2}{2}}.$$

Because $t_2 - t_1 = t_3 - t_2$, we have $f(t_2)/f(t_1) = f(t_3)/f(t_2)$. We can subtract both sides of this equation from 1 to give

$$1 - \frac{f(t_2)}{f(t_1)} = 1 - \frac{f(t_3)}{f(t_2)},$$

which means that the fraction of the sample that decays from time t_1 to time t_2 is the same as the fraction of the sample that decays from time t_2 to time t_3.

□

We still haven't seen e. But if you open a science text, you'll usually find e in any discussion of exponential decay.

Problem 13.35: Suppose that a radioactive element has a half-life of $t_{1/2}$, and that we run an experiment in which we initially place N_o atoms of this element in a container. After a time t, suppose there are N_n atoms of the element remaining in the container. Show that

$$t \approx -t_{1/2} \cdot \frac{\ln(N_n/N_o)}{0.693}.$$

Solution for Problem 13.35: We start with the definition of half-life. In any time period of length $t_{1/2}$, the number of atoms of the radioactive element is halved. In a time period of length t, there are $t/t_{1/2}$ time periods of length $t_{1/2}$. (For example, if $t_{1/2} = 3$ days and $t = 12$ days, then $12/3 = 4$ half-lives have transpired.) Therefore, the number of atoms of the radioactive element is halved $t/t_{1/2}$ times in a time period of t. So, we must have

$$\frac{N_n}{N_0} = \left(\frac{1}{2}\right)^{t/t_{1/2}}.$$

Taking the natural logarithm of both sides gets the $t/t_{1/2}$ out of the exponent, and introduces the $\ln(N_n/N_0)$ term that's in the formula we want to prove:

$$\ln\left(\frac{N_n}{N_0}\right) = \ln\left(\frac{1}{2}\right)^{t/t_{1/2}} = \frac{t}{t_{1/2}}\ln\frac{1}{2}.$$

Solving for t gives us

$$t = t_{1/2}\frac{\ln(N_n/N_0)}{\ln\frac{1}{2}}.$$

This looks close, but our desired formula has both a negative sign and that 0.693. Where did those come from? We pull out our calculator and find that $\ln\frac{1}{2} \approx -0.693$. Aha! Now, we have our desired

$$t \approx -t_{1/2}\cdot\frac{\ln(N_n/N_0)}{0.693}.$$

$\square$

You might be wondering why science books report a formula that relies on taking the natural logarithm of both sides of

$$\frac{N_n}{N_0} = \left(\frac{1}{2}\right)^{t/t_{1/2}}.$$

We might instead write the equation as $\frac{N_n}{N_0} = 2^{-t/t_{1/2}}$, then take the base 2 logarithm of both sides to find

$$\log_2\left(\frac{N_n}{N_0}\right) = -\frac{t}{t_{1/2}}.$$

This gives us a formula for t without any constant to remember like the 0.693 in the previous formula we found.

> **WARNING!!** Formulas, particularly those involving logarithms, can be written in many different forms.

The reason we use e and $\ln$ in so many formulas in math and science is because of many nice properties of the function $f(x) = e^x$ that we need calculus to explain.

You certainly don't want to memorize all the different forms in which a formula can be written, but it's very important to be able to manipulate a formula from one form to another.

Problem 13.36: Explain why the formula $\left(\dfrac{1}{2}\right)^{t/t_{1/2}} = \dfrac{N_n}{N_0}$ can be written as $N_n = N_0 e^{-\lambda t}$, where $\lambda = \dfrac{\ln 2}{t_{1/2}}$.

Solution for Problem 13.36: We can isolate N_n by multiplying both sides by N_0 to give

$$N_n = N_0 \left(\frac{1}{2}\right)^{t/t_{1/2}}.$$

Now, our task is to write the $\left(\dfrac{1}{2}\right)^{t/t_{1/2}}$ term with e as the base. We want e as a base, so we write $1/2$ as a power of e. Specifically, we have $1/2 = e^{\ln 1/2}$, so we have

$$N_n = N_0 \left(\frac{1}{2}\right)^{t/t_{1/2}} = N_0 \left(e^{\ln(1/2)}\right)^{t/t_{1/2}}.$$

We need a negative sign and $\ln 2$, so we note that $\ln(1/2) = \ln 1 - \ln 2 = 0 - \ln 2 = -\ln 2$, and we have

$$N_n = N_0 \left(e^{-\ln 2}\right)^{t/t_{1/2}} = N_0 e^{-t(\ln 2)/t_{1/2}}.$$

Letting $\lambda = (\ln 2)/t_{1/2}$, we have $N_n = N_0 e^{-\lambda t}$. $\square$

Notice that taking the natural logarithm of this final formula gives us

$$\ln N_n = \ln(N_0 e^{-\lambda t}) = \ln N_0 + \ln e^{-\lambda t} = \ln N_0 - \lambda t,$$

so

$$t = \frac{\ln N_n - \ln N_0}{-\lambda} = -\frac{t_{1/2}\ln(N_n/N_0)}{\ln 2},$$

which is the formula we found back in Problem 13.35. Hopefully, this convinces you of the futility of memorizing all the different forms of these types of formulas. All of them can be derived from:

Important: If the half-life of a substance is $t_{1/2}$, and there is initially N_0 of the substance, then the amount remaining after a time t is given by the function

$$f(t) = N_0 \left(\frac{1}{2}\right)^{t/t_{1/2}}.$$

Problem 13.37: Dr. Curie took two measurements of a radioactive sample exactly 100 seconds apart. There were 403 grams of the radioactive sample in the first measurement and 83 grams remained undecayed in the second measurement. If she measures the sample a third time, 38 seconds after the second measurement, how many grams will remain undecayed in the sample?

Solution for Problem 13.37: If we can find the half-life of the sample, we can determine what fraction of the sample remains radioactive 138 seconds after the first measurement (which is 38 seconds after the second measurement). We know that in 100 seconds, the initial sample of 403 grams of radioactive material decays to 83 grams. So, if $t_{1/2}$ is the half-life of the sample, then

$$\frac{83}{403} = \left(\frac{1}{2}\right)^{100/t_{1/2}}.$$

To get the variable out of the exponent, we take the natural logarithm of both sides (we could use any base for the logarithm). This gives us

$$\ln\frac{83}{403} = \ln\left(\frac{1}{2}\right)^{100/t_{1/2}} = \frac{100}{t_{1/2}}\ln\frac{1}{2}.$$

Solving for $t_{1/2}$, we find

$$t_{1/2} = \frac{100\ln\frac{1}{2}}{\ln\frac{83}{403}} \approx 43.9 \text{ seconds.}$$

Now that we have the half-life, we can find the amount of the sample left in the third measurement. This measurement occurs when $t = 100 + 38 = 138$, so the amount remaining is

$$403\left(\frac{1}{2}\right)^{138/43.9} \approx 45.6 \text{ grams.}$$

We can quickly sanity-check this answer. The half-life is approximately 43.9 seconds, so in the 38 seconds after Dr. Curie measured 83 radioactive grams remaining, we should expect a little less than half of this amount to decay. Indeed, the 45.6 grams we expect to remain is a bit more than half of 83 grams.

Notice that we didn't have to rely on memorizing any formulas to solve the problem. We simply used the definition of half-life. □

Exercises

13.6.1   There are 20 grams of a radioactive substance at the start of an experiment. In the first 10 minutes of the experiment, 15 grams of the substance decays.

(a)   What is the half-life of the substance?

(b)   How much of the substance remains radioactive 24 minutes after the experiment starts?

13.6.2   Solve the equation $\dfrac{A}{B + Ce^{-kt}} = D$ for t in terms of A, B, C, D, and k.

13.6.3   400 kilograms of a radioactive element is buried. The half-life of the element is 120.3 years. At the end of the year 2314, the buried material is unearthed, and 96.2 kilograms of the radioactive element remains. In what year is the material buried?

13.6.4   Find a real-valued function $f(t)$ that satisfies $f(t + 2) = \frac{1}{9}f(t)$ for all real values of t.

13.6.5   Three grams of a radioactive substance is placed in a vial. Find the half-life of the substance if the number of grams of the substance remaining after t seconds is $3e^{-5t}$.

13.6.6   Solve $y = \ln(2 - e^y) - \ln 3$ for y.

13.7 Summary

We call a function $f(x) = a^x$, where a is a positive constant, an **exponential function** with base a. We say that a function f is a **monotonically increasing** function if $f(x) > f(y)$ whenever $x > y$. If $a > 1$, then the exponential function $f(x) = a^x$ is monotonically increasing.

> **Important:** If a is a positive constant such that $a \neq 1$ and $a^x = a^y$, then $x = y$.

We write $x = \log_a b$, where a and b are positive and $a \neq 1$, to indicate that x is the power to which we must raise a to get b. In other words, the equations

$$x = \log_a b \qquad \text{and} \qquad a^x = b$$

are equivalent. We call the expression $\log_a b$ a **logarithm**, where a is the **base** of the logarithm and b is the **argument** of the logarithm. The logarithmic function $f(x) = \log_a x$ and the exponential function $g(x) = a^x$ are inverses of each other.

Logarithms satisfy the following identities:

> **Important:**
> $$\log_a a^p = p$$
> $$\log_a b^c = c \log_a b$$
> $$\log_{a^n} b^n = \log_a b$$
> $$\log_a b + \log_a c = \log_a bc$$
> $$\log_a b - \log_a c = \log_a \frac{b}{c}$$
> $$(\log_a b)(\log_b c) = \log_a c$$
> $$\frac{\log_a b}{\log_a c} = \log_c b$$
> $$a^{\log_a x} = x$$

We use the letter e to denote a special constant that equals the value that the expression $\left(1 + \frac{1}{n}\right)^n$ approaches when n is very large. This constant appears in many mathematical and scientific contexts, but we require calculus to fully understand many of the interesting properties of e. Since e is so common in exponential and logarithmic expressions, the special notation ln is often used to denote the base e logarithm.

One place that e often appears in science is in the study of radioactive decay. The **half-life** of a radioactive substance is the amount of time it takes for exactly half the radioactive substance to decay.

> **Important:** If the half-life of a substance is $t_{1/2}$, and there is initially N_0 of the substance, then the amount remaining after a time t is given by the function
> $$f(t) = N_0 \left(\frac{1}{2}\right)^{t/t_{1/2}}.$$

Things To Watch Out For!

- Don't just perform algebraic manipulations blindly. Think about each step and make sure you know why it is valid.

- Graphs are excellent for providing intuition about a problem, but they shouldn't be used as a key part of a proof or a rigorous solution to a problem.

- When working with equations involving exponential expressions, watch out for special cases in which $-1, 0$, or 1 is the base.

- The logarithms in our sum and difference identities *must have the same base*. For example, we cannot use the addition identity to combine the logarithms in the sum $\log_2 5 + \log_3 6$.

- Be careful when you apply the laws of logarithms. Here are some common mistakes that students make when working with logarithms:

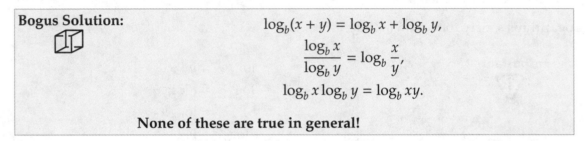

Bogus Solution:

$$\log_b(x + y) = \log_b x + \log_b y,$$
$$\frac{\log_b x}{\log_b y} = \log_b \frac{x}{y},$$
$$\log_b x \log_b y = \log_b xy.$$

None of these are true in general!

- When you solve a logarithmic equation, be on the lookout for extraneous solutions. The best way to avoid extraneous solutions is to check that your solutions satisfy the original equation.

- Formulas, particularly those involving logarithms, can be written in many different forms.

Problem Solving Strategies

Concepts:

- When working with an unusual expression in a problem, try to find ways to convert the problem into one involving expressions you know how to handle.

- If you have an equation with constants raised to powers, it's often helpful to write all those constants with the same base if possible.

- We can often evaluate logarithms by setting the logarithm equal to x, then writing the resulting equation in exponential form.

- Finding a common exponential base between two numbers a and b often helps compute $\log_a b$ more easily.

Continued on the next page. . .

> **Concepts:** ... *continued from the previous page*
>
>
> - Wishful thinking is an excellent problem solving tool. When stuck on a problem, ask yourself what, if it were true, would make the problem easier.
>
> - Equations involving several logarithms are often solved by using logarithm identities to write the equation with a single logarithm.
>
> - Many logarithmic equations can be solved by converting them into exponential equations, then solving the exponential equations.
>
> - Just as our sum and difference identities can be used to combine logarithms, they can also be used to separate logarithms of products and quotients into easier to handle logarithms.
>
> - If you have an expression containing a complicated exponent, take the logarithm of it to turn the exponent into a factor.

REVIEW PROBLEMS

13.38 Find all values of n such that $6^2 \cdot (6^n)^n = 6^n \cdot 6^n \cdot 6^n$.

13.39 Find the real value of t such that $2^{t+1} + 2^t = 2^3 + 2^5 + 2^{7-t}$.

13.40 Find the value of a given that $4 \cdot 9^a - 21 = 25 \cdot 3^a$.

13.41 Evaluate each of the following:

(a) $\log_3 243$ (b) $\log_4 \frac{1}{32}$ (c) $\log_{10} \sqrt[3]{0.01}$

13.42 Evaluate the following:

(a) $\log_4 12 + 2\log_4 3 - 3\log_4 6$ (b) $\log_2 7 \cdot \log_7 2$

13.43 Find x such that $\log_2 (\log_2 (\log_2 x)) = 2$. *(Source: AMC 12)*

13.44 Find the value of b if $4^a = 8^b$ and $3^b = 2 \cdot 3^a$.

13.45 Solve the following inequalities:

(a) $\log_5(3 - x) \geq 7$ (b) $\log_{1/2}(2x) > 3$

13.46 Let n be a positive integer. If the number of integers in the domain of $f(x) = \log((1 - x)(x - n))$ equals $2n - 6$, compute n. *(Source: ARML)*

13.47 Compute 3^A, where $A = \dfrac{(\log_3 1 - \log_3 4)(\log_3 9 - \log_3 2)}{(\log_3 1 - \log_3 9)(\log_3 8 - \log_3 4)}$. *(Source: Mandelbrot)*

13.48 Prove that $\log_{a^n} b = \dfrac{1}{n} \log_a b$.

13.49 Find all ordered pairs (a, b) such that $\log_b(9a) = 3\log_a b = -3$.

13.50 Compute the numerical value of $5^{\log_{10} 2} \cdot 2^{\log_{10} 3} \cdot 2^{\log_{10} 6} \cdot 5^{\log_{10} 9}$. *(Source: NYSML)*

13.51 If $\log_{10} 5 = x$, find the value of $\log_2 5$ in terms of x.

13.52 At what point do the graphs of $y = \log_2(x + 3)$ and $y = 3 - \log_2(x - 1)$ intersect?

13.53 What is the value of the expression

$$\frac{1}{\log_2 100!} + \frac{1}{\log_3 100!} + \frac{1}{\log_4 100!} + \cdots + \frac{1}{\log_{100} 100!},$$

where $100! = 100 \cdot 99 \cdot 98 \cdot 97 \cdots \cdot 2 \cdot 1$? *(Source: AHSME)*

13.54 The value of $\log_3 2$ is 0.631, rounded to the nearest thousandth. Compute the smallest positive integer n for which $3^n > 2^{102}$. *(Source: Mandelbrot)*

13.55 Solve $2^{41} = 2 + \displaystyle\sum_{k=0}^{39} \log_{10} x^{(2^k)}$ for all real value(s) of x. *(Source: NYSML)*

13.56 Solve the equation $a + b \ln c = d$ for c in terms of a, b, and d.

13.57 A certain radioactive isotope has a half-life of 4.3 seconds. If 4.2 ounces of the isotope remain 12 seconds after an experiment starts, how much was present when the experiment began?

13.58 My clock reads 1:01 PM when I take a magic pill. Every time the minutes number on my clock changes from 01 to 02, my height increases 50%. Every time the minutes number changes from 31 to 32, my height decreases 40%. I was 6 feet tall when I took the pill. What time is it when I am first less than 1 foot tall?

13.59 Jamie forgot to take a measurement of how much of a radioactive sample she had at the start of an experiment. She also doesn't know what the half-life of the sample is. She did note that there were 12.2 grams exactly 63 seconds after she started the experiment, and 4.1 grams remained radioactive exactly 107 seconds after the experiment started. How many grams in the sample were radioactive at the start of the experiment?

Challenge Problems

13.60 Solve the equation $(x + 2)^{2y} - x^2 = 3$ for y in terms of x. For what values of x is there no value of y that satisfies the equation?

13.61 A triangle with sides $a \le b \le c$ is *log-right* if $\log_{10}(a^2) + \log_{10}(b^2) = \log_{10}(c^2)$. Compute the largest possible value of a in a right triangle that is also log-right. *(Source: ARML)* **Hints:** 322

13.62 The graph, G, of $y = \log_{10} x$ is rotated 90° counterclockwise about the origin to obtain a new graph, G'. Find an equation whose graph is G'. *(Source: AMC 12)* **Hints:** 121, 318

13.63 If $x, y > 0$, $\log_y x + \log_x y = 10/3$, and $xy = 144$, then find $(x + y)/2$. *(Source: AHSME)* **Hints:** 222

13.64 For integers x and y with $1 < x, y \le 100$, compute the number of ordered pairs (x, y) such that $\log_x y + \log_y x^2 = 3$. *(Source: ARML)*

13.65 Find the product of the positive roots of $\sqrt{1995}\, x^{\log_{1995} x} = x^2$. *(Source: AIME)* **Hints:** 335

13.66 If $\frac{\log_b a}{\log_c a} = \frac{19}{99}$, then $\frac{b}{c} = c^k$. Compute k. *(Source: ARML)* **Hints:** 302

13.67 Let $a \ge b > 1$. What is the largest possible value of $\log_a \frac{a}{b} + \log_b \frac{b}{a}$? *(Source: AMC 12)* **Hints:** 183, 344

13.68 The points $A = (x_1, y_1)$ and $B = (x_2, y_2)$ are two points on the graph of $y = \log_2 x$. Through the midpoint of the line segment $\overline{AB}$, a horizontal line is drawn to intersect the graph at $C = (x_3, y_3)$. Prove that $x_3^2 = x_1 x_2$. *(Source: CEMC)*

13.69 Find x in terms of a and y given that $y = \frac{a^x + a^{-x}}{2}$. **Hints:** 229

13.70 Let $x = 2^{\log_b 3}$ and $y = 3^{\log_b 2}$. Find $x - y$. **Hints:** 335

13.71 If $\log_2 a + \log_2 b \ge 6$, then find the smallest possible value of $a + b$. *(Source: AHSME)* **Hints:** 344, 40

13.72 The equation $2^{333x-2} + 2^{111x+2} = 2^{222x+1} + 1$ has three real roots. Find their sum. *(Source: AIME)* **Hints:** 245

13.73 How many digits does it take to write the base 10 integer 4^{18} in base 3, given that $\log_3 2 \approx .631$.

13.74 Let x, y, and z all exceed 1 and let w be a positive number such that $\log_x w = 24$, $\log_y w = 40$, and $\log_{xyz} w = 12$. Find $\log_z w$. *(Source: AIME)*

13.75 Let S be the set of ordered triples (x, y, z) of real numbers for which

$$\log_{10}(x + y) = z \qquad \text{and} \qquad \log_{10}(x^2 + y^2) = z + 1.$$

There are real numbers a and b such that for all ordered triples (x, y, z) in S, we have $x^3 + y^3 = a \cdot 10^{3z} + b \cdot 10^{2z}$. What is the value of $a + b$? *(Source: AMC 12)* **Hints:** 174

13.76 Find $(\log_2 x)^2$ if $\log_2(\log_8 x) = \log_8(\log_2 x)$. *(Source: AIME)* **Hints:** 33

13.77★ Let x and y be positive numbers such that $x^{\log_y x} = 2$ and $y^{\log_x y} = 16$. Find x. *(Source: Bay Area Math Meet)*

13.78★ Find both ordered triplets (x, y, z) that satisfy the system of equations

$$\log_{10}(2000xy) - (\log_{10} x)(\log_{10} y) = 4,$$
$$\log_{10}(2yz) - (\log_{10} y)(\log_{10} z) = 1,$$
$$\log_{10}(zx) - (\log_{10} z)(\log_{10} x) = 0.$$

(Source: AIME) **Hints:** 353

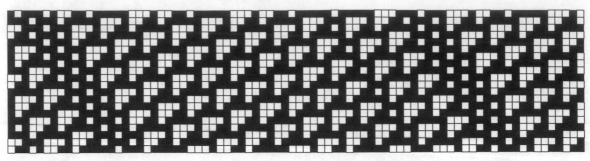

I am a radical in thought and a conservative in method. – Rutherford B. Hayes

CHAPTER **14**

Radicals

At the start of Chapter 13, we noted that the key to working with unusual expressions or inconvenient notation in a problem is to convert the problem into one involving expressions that we understand better.

In this chapter, we'll explore some common methods for dealing with expressions that have radicals. The goal is almost always to remove the radicals from the problem entirely, so that we're left with less complicated notation.

14.1 Raising Radicals to Powers

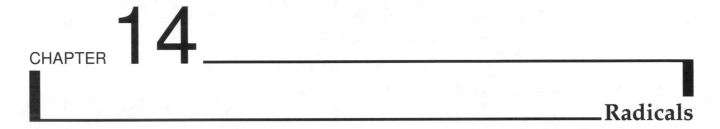

Problems

Problem 14.1: Consider the equation $x = \sqrt{2x + 24}$.

(a) Square the given equation to produce a new equation. Solve your new equation.

(b) Do all your solutions from (a) satisfy $x = \sqrt{2x + 24}$?

Problem 14.2: Find all x such that $\sqrt{6 - x} - \sqrt{10 + 3x} = 2$.

Problem 14.3: In this problem, we solve the equation $\sqrt{x^2 - 5x + 4} - \sqrt{x^2 - 10x + 9} = x - 1$.

(a) Factor the quadratics.

(b) Solve the original equation.

(c) Check your solutions. You should have 2 solutions. If you don't, try the problem again.

Problem 14.4: Find all real numbers x such that $3\sqrt{x} \geq x + 2$.

Problem 14.5:

(a) Find all r such that $\sqrt{r+2} \geq -r$.

(b) Test your solution by substituting a variety of values of r into the inequality. If you find values of r that satisfy the inequality but are not among the solutions you found in part (a), do part (a) over again.

Problem 14.6: Prove that $\dfrac{\sqrt{3n+1}}{\sqrt{3n+4}} \geq \dfrac{2n+1}{2n+2}$ for all nonnegative integers n.

Problem 14.7: Find all x such that $\sqrt[3]{2x^2 + 3x - 26} = x - 2$.

Problem 14.8: In this problem, we solve the equation $\sqrt[3]{13x + 37} - \sqrt[3]{13x - 37} = \sqrt[3]{2}$.

(a) Let $a = \sqrt[3]{13x + 37}$ and $b = \sqrt[3]{13x - 37}$. Find two equations relating a and b.

(b) Find ab.

(c) Find x.

Problem 14.1: Find all solutions to the equation $x = \sqrt{2x + 24}$.

Solution for Problem 14.1: The radical on the right side of our equation is the major complication in this problem. We get rid of it by squaring the equation.

> **Concept:** One way to get rid of a radical in an equation is to raise both sides to a power that gets rid of the radical.

Squaring both sides of the equation gives us $x^2 = 2x + 24$. Rearranging and factoring gives $(x-6)(x+4) = 0$, which has solutions $x = 6$ and $x = -4$.

Because we squared the equation as a step in our solution, we must check that our solutions are not extraneous. When $x = 6$, our equation is $6 = \sqrt{2(6) + 24}$, which is true. When $x = -4$, we have $\sqrt{2x + 24} = 4$, so we do not have $x = \sqrt{2x + 24}$. Our original equation is not satisfied by $x = -4$, but our squared equation is; therefore, the solution $x = -4$ is extraneous.

> **WARNING!!** Whenever we square an equation as a step in solving it, we have to check that our solutions are valid. If a solution to the squared equation does not satisfy our original equation, then the solution is called **extraneous**. An extraneous solution is not a valid solution to the original equation.

In conclusion, the only solution to $x = \sqrt{2x + 24}$ is $x = 6$. $\square$

Let's try our radical-removal strategy on some more complicated problems.

Problem 14.2: Find all x such that $\sqrt{6-x} - \sqrt{10+3x} = 2$.

Solution for Problem 14.2: *Solution 1: Square first.* Squaring both sides of the equation to get rid of the radicals gives us $6 - x - 2\sqrt{(6-x)(10+3x)} + 10 + 3x = 4$, so

$$2x + 12 = 2\sqrt{(6-x)(10+3x)}.$$

Dividing both sides by 2 and then squaring again, we have

$$x^2 + 12x + 36 = -3x^2 + 8x + 60,$$

so $4x^2 + 4x - 24 = 0$. Dividing by 4 gives us $x^2 + x - 6 = 0$, and factoring gives $(x+3)(x-2) = 0$. Our solutions appear to be $x = -3$ and $x = 2$. Because we squared the equation (twice) on the way to our solution, we have to check if our solutions are extraneous.

For $x = -3$, we have $\sqrt{6-(-3)} - \sqrt{10-9} = 3 - 1 = 2$, so this solution is valid. For $x = 2$, we have $\sqrt{6-2} - \sqrt{10+6} = 2 - 4 = -2$ on the left side, which does not equal the right side, 2. Therefore, $x = 2$ is an extraneous solution, and $x = -3$ is the only solution to the original equation.

Solution 2: We first move one of the radicals to the right side of the equation to give the equation $\sqrt{6-x} = \sqrt{10+3x} + 2$, and then we square both sides to find:

$$6 - x = 10 + 3x + 4\sqrt{10+3x} + 4.$$

Simplifying this equation gives $-8 - 4x = 4\sqrt{10+3x}$. Dividing by 4 gives us $-2 - x = \sqrt{10+3x}$. Squaring both sides of this equation gives $x^2 + 4x + 4 = 10 + 3x$, so $x^2 + x - 6 = 0$. From here, the rest of the solution is the same as before. $\square$

Either solution to Problem 14.2 is fine, but the second solution has the advantage of avoiding quadratic expressions until the final equation.

> **Concept:** Rearranging an equation with radicals before raising it to a power may simplify the algebra you have to perform to solve the equation.

Problem 14.3: Solve the equation $\sqrt{x^2 - 5x + 4} - \sqrt{x^2 - 10x + 9} = x - 1$.

Solution for Problem 14.3: We first factor the quadratics inside the radicals to see if there are any useful relationships to exploit:

$$\sqrt{(x-1)(x-4)} - \sqrt{(x-1)(x-9)} = x - 1.$$

We immediately see that if $x = 1$, then both sides equal 0, so $x = 1$ is one solution. What's wrong with the following:

> **Extra!** *Inspiration is the impact of a fact on a well-prepared mind.*
> — Louis Pasteur

Bogus Solution: If $x \neq 1$, we can divide both sides by $\sqrt{x-1}$, to give

$$\sqrt{x-4} - \sqrt{x-9} = \sqrt{x-1}.$$

We could start squaring and manipulating, and so on, but instead we can do something clever. We note that $x-4$ is smaller than $x-1$ for all x, so $\sqrt{x-4}$ is less than $\sqrt{x-1}$. Moreover, $\sqrt{x-4} - \sqrt{x-9}$ is less than $\sqrt{x-4}$, so $\sqrt{x-4} - \sqrt{x-9}$ has to be less than $\sqrt{x-1}$. Therefore, the equation $\sqrt{x-4} - \sqrt{x-9} = \sqrt{x-1}$ has no solutions. So, the only solution to the original equation is $x = 1$.

Pretty convincing. But maybe you noticed that the constants in the radicals, 4 and 9, are perfect squares. If we let $x = 0$ in the original equation, we have $\sqrt{4} - \sqrt{9} = -1$, which is true! So, $x = 0$ is also a solution to the original equation. However, $x = 0$ is not a solution to the equation $\sqrt{x-4} - \sqrt{x-9} = \sqrt{x-1}$ that we produced by dividing the original equation by $\sqrt{x-1}$. This suggests that it's the "divide the original equation by $\sqrt{x-1}$" step that's the problem. Looking more closely at this step, we realize that we can only divide the equation by $\sqrt{x-1}$ if $x > 1$. If $x < 1$, then $\sqrt{x-1}$ is undefined, and if $x = 1$, then dividing by $\sqrt{x-1}$ is dividing by 0.

Blindly manipulating the equation misled us into discarding the valid solution $x = 0$. But did we miss any other solutions? We'll have to find a valid solution to the problem to figure that out. Rather than dividing the equation

$$\sqrt{(x-1)(x-4)} - \sqrt{(x-1)(x-9)} = x-1$$

by $\sqrt{x-1}$, let's try squaring it to get rid of the radicals. This gives us

$$(x-1)(x-4) - 2\sqrt{(x-1)^2(x-4)(x-9)} + (x-1)(x-9) - (x-1)^2.$$

Once again, we don't blindly multiply everything out. But we still have to be careful. For example, we can't do this:

Bogus Solution: Since $x-1$ divides every term, we divide both sides by $x-1$ to give

$$x - 4 - 2\sqrt{(x-4)(x-9)} + x - 9 = x - 1.$$

This Bogus Solution assumes that

$$\frac{2\sqrt{(x-1)^2(x-4)(x-9)}}{x-1} = 2\sqrt{(x-4)(x-9)},$$

but this isn't true! If $x < 1$, then $\sqrt{(x-1)^2}/(x-1) = -1$, not 1.

We could resort to casework, and handle the cases $x < 1$, $x = 1$, and $x > 1$ separately, or we could avoid the difficulty altogether by rearranging the equation and squaring again. With a little organization, the latter approach is easier than casework. We start with

$$(x-1)(x-4) - 2\sqrt{(x-1)^2(x-4)(x-9)} + (x-1)(x-9) = (x-1)^2.$$

We isolate the radical to give

$$2\sqrt{(x-1)^2(x-4)(x-9)} = (x-1)(x-4) + (x-1)(x-9) - (x-1)^2.$$

We then factor $(x-1)$ out of each term on the right to give us

$$2\sqrt{(x-1)^2(x-4)(x-9)} = (x-1)((x-4) + (x-9) - (x-1)) = (x-1)(x-12).$$

Now, squaring both sides of the equation $2\sqrt{(x-1)^2(x-4)(x-9)} = (x-1)(x-12)$ gives us the equation $4(x-1)^2(x-4)(x-9) = (x-1)^2(x-12)^2$. At this point, we might make yet another mistake:

> **Bogus Solution:** 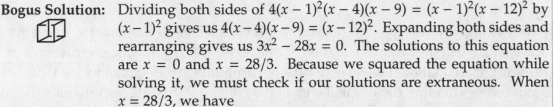 Dividing both sides of $4(x-1)^2(x-4)(x-9) = (x-1)^2(x-12)^2$ by $(x-1)^2$ gives us $4(x-4)(x-9) = (x-12)^2$. Expanding both sides and rearranging gives us $3x^2 - 28x = 0$. The solutions to this equation are $x = 0$ and $x = 28/3$. Because we squared the equation while solving it, we must check if our solutions are extraneous. When $x = 28/3$, we have
>
> $$\sqrt{(x-1)(x-4)} - \sqrt{(x-1)(x-9)} = \sqrt{\left(\frac{25}{3}\right)\left(\frac{16}{3}\right)} - \sqrt{\left(\frac{25}{3}\right)\left(\frac{1}{3}\right)} = 5,$$
>
> but $x - 1 = 25/3$. So, $x = 28/3$ is extraneous. As we've already seen, $x = 0$ is a valid solution. Therefore, the only solution to the original equation is $x = 0$.

This Bogus Solution overlooks the solution $x = 1$. The step where this Bogus Solution "loses" the solution $x = 1$ occurs when we divide both sides of $4(x-1)^2(x-4)(x-9) = (x-1)^2(x-12)^2$ by $(x-1)^2$. If $x = 1$, we cannot divide by $(x-1)^2$, since this would be division by 0. Indeed, if $x = 1$, then both sides of the equation equal 0, so $x = 1$ is a solution to the equation. If $x \neq 1$, then we can divide both sides by $(x-1)^2$. From there, we can follow our Bogus Solution above to find the solution $x = 0$. So, the two solutions to the equation are $x = 0$ and $x = 1$.

> **WARNING!!** If you divide a common factor from two sides of an equation, you cannot forget about the solutions that might arise if the factor equals 0.

We also could have avoided forgetting the solution $x = 1$ by continuing to rearrange and factor. Subtracting $(x-1)^2(x-12)^2$ from both sides of $4(x-1)^2(x-4)(x-9) = (x-1)^2(x-12)^2$ gives a left side of

$$4(x-1)^2(x-4)(x-9) - (x-1)^2(x-12)^2 = (x-1)^2(4(x-4)(x-9) - (x-12)^2) = (x-1)^2(3x^2 - 28x).$$

This expression is 0 when $x = 1$, $x = 0$, or $x = 28/3$. As before, $x = 28/3$ is extraneous, but the other two are valid. Again, the only solutions to the equation are $x = 0$ and $x = 1$.

This solution shows yet again the benefit of not immediately bashing through equations. Here, we factor relentlessly rather than multiply everything out right away. As a result, the necessary algebraic manipulations are much simpler. Factoring is your friend—imagine trying to solve the problem by squaring the original equation without even factoring the quadratics in the radicals!

> **Concept:** Don't be hasty to jump into lengthy computations or algebraic manipula-
> tions. First check and see if there is a more clever approach.

□

Take some extra time to think about all the Bogus Solutions we pursued while finding the solution to Problem 14.3. These Bogus Solutions all stem from blindly pursuing algebraic manipulations without thinking about what these manipulations mean. Each *seems* valid, but each exhibits an error that stems from overlooking subtle, but critical, concepts.

We now turn from equations to inequalities. Can we square to remove square roots from inequalities?

Problem 14.4: Find all real numbers x such that $3\sqrt{x} \geq x + 2$.

Solution for Problem 14.4: First, for $\sqrt{x}$ to be defined, x must be nonnegative. So, we must have $x \geq 0$, which means that $x + 2 \geq 0$. Since we also have $\sqrt{x} \geq 0$, both sides of the inequality are nonnegative, which means we can square both sides to get $9x \geq x^2 + 4x + 4$. Rearranging gives $x^2 - 5x + 4 \leq 0$, and factoring gives $(x - 1)(x - 4) \leq 0$. This inequality holds for all $x \in [1, 4]$. □

Just as with equations, we must be careful when we square inequalities. Above, we made sure that both sides are nonnegative before squaring. Here's an example of what can go wrong if you don't:

Problem 14.5: Find all r such that $\sqrt{r + 2} \geq -r$.

Solution for Problem 14.5: What's wrong here:

> **Bogus Solution:** Squaring both sides gives $r + 2 \geq r^2$. Rearranging gives $r^2 - r - 2 \leq 0$,
> and factoring gives us $(r - 2)(r + 1) \leq 0$. This gives us the solutions
> $r \in [-1, 2]$.

If we test $r = -1$, $r = 0$, and $r = 2$, we see that all of these values satisfy the original inequality, so we might think our answer is correct. However, then we might try $r = 7$, and see that this value of r satisfies the inequality, too!

Our mistake is that we aren't careful about our signs. Our solution only tells us when $(\sqrt{r + 2})^2$ is greater than or equal to $(-r)^2$, not when $\sqrt{r + 2}$ is greater than or equal to $(-r)$. Considering signs first, we see that for *all positive values of r*, the left side of $\sqrt{r + 2} \geq -r$ is positive and the right side is negative. So, for all positive values of r, the left side is larger than the right.

To finish, we also analyze what happens when $r \leq 0$. Since $\sqrt{r + 2}$ is only defined for $r \geq 2$, we only need to consider $r \in [-2, 0]$. When $r \in [-2, 0]$, both sides of the inequality are nonnegative. Therefore, we can follow the steps in the Bogus Solution to see that the inequality is satisfied when $r \in [-1, 0]$ and not satisfied when $r \in [-2, -1)$. So, our full solution to the inequality is $r \in [-1, +\infty)$. □

This is yet another example of the importance of thinking about your algebraic manipulations before performing them.

> **WARNING!!** ☢ Don't just blindly manipulate equations and inequalities. Doing so can introduce subtle errors and significant oversights.

> **Sidenote:** ♪ See if you can find the flaw in this proof that $1 = 2$. Suppose x and y are positive numbers, and that $x = y$. Multiplying by y gives $xy = y^2$. Subtracting x^2 from both sides gives $xy - x^2 = y^2 - x^2$. Factoring then gives $x(y - x) = (y - x)(x + y)$. Dividing by $y - x$ gives $x = x + y$. Since $x = y$, we now have $x = 2x$, and dividing by x gives us $1 = 2$.

Problem 14.6: Prove that $\dfrac{\sqrt{3n + 1}}{\sqrt{3n + 4}} \geq \dfrac{2n + 1}{2n + 2}$ for all nonnegative integers n.

Solution for Problem 14.6: We'll deal with the radicals before dealing with the fractions. Since both sides of the inequality are positive, we can simply square both sides to get an equivalent inequality:

$$\frac{3n + 1}{3n + 4} \geq \frac{4n^2 + 4n + 1}{4n^2 + 8n + 4}.$$

Because both denominators are positive, we can multiply both sides by both denominators without changing the direction of the inequality. This gives us

$$12n^3 + 28n^2 + 20n + 4 \geq 12n^3 + 28n^2 + 19n + 4.$$

Simplifying this inequality gives us $n \geq 0$, which is true for all nonnegative integers n, so the original inequality holds.

We found our solution by working backwards. It's important to note that all of our steps are reversible, so we can start from $n \geq 0$ and derive the desired inequality. We can check that all steps are indeed reversible by writing a "forwards" solution:

> **Proof:** ☑ Since n is nonnegative, we have $n \geq 0$. Adding $12n^3 + 28n^2 + 19n + 4$ to both sides gives
>
> $$12n^3 + 28n^2 + 20n + 4 \geq 12n^3 + 28n^2 + 19n + 4.$$
>
> Factoring gives $(3n + 1)(4n^2 + 8n + 4) \geq (3n + 4)(4n^2 + 4n + 1)$, and dividing both sides by $(3n + 4)(4n^2 + 8n + 4)$ gives
>
> $$\frac{3n + 1}{3n + 4} \geq \frac{4n^2 + 4n + 1}{4n^2 + 8n + 4}.$$
>
> Because both sides are positive, we can take the square root of both sides to give the desired
>
> $$\frac{\sqrt{3n + 1}}{\sqrt{3n + 4}} \geq \frac{2n + 1}{2n + 2}.$$

□

Our forwards proof shows that all our steps working backwards are reversible. However, it's very hard to tell from reading the forwards proof how we found the solution—why did we think to add $12n^3 + 28n^2 + 19n + 4$ to both sides in our first step?

In general, writing forwards is best when trying to convey that your solution is correct. It is very clear in our forwards proof that every step is valid. However, it's not always clear from such solutions how you would think of such a solution. Sometimes, you'll read a solution that will appear to be almost magic. It'll be clear that the solution is correct, but it won't be clear how the writer ever thought of it. In these cases, working through the solution backwards can sometimes reveal how the writer found the solution.

Now that you've mastered square roots, let's try cube roots.

Problem 14.7: Find all x such that $\sqrt[3]{2x^2 + 3x - 26} = x - 2$.

Solution for Problem 14.7: We squared to get rid of square roots, so here, we'll cube to get rid of cube roots. Cubing gives us $2x^2 + 3x - 26 = (x - 2)^3$. Expanding the right side gives $2x^2 + 3x - 26 = x^3 - 6x^2 + 12x - 8$, and moving all terms to the same side gives us $x^3 - 8x^2 + 9x + 18 = 0$. Factoring this polynomial gives us $(x + 1)(x - 3)(x - 6) = 0$, so we appear to have the solutions $x = -1$, $x = 3$, and $x = 6$. Checking each solution, we find that all three work, so all three are valid solutions.

It's a good idea to check each solution, just to make sure we didn't make an algebra error. However, we don't have to worry about cubing the equation producing extraneous roots the way that squaring an equation might lead to extraneous roots.

Extraneous roots can occur when we square an equation because it is possible for two different real numbers to have the same square; for example, $(-1)^2 = 1^2$. So, it's possible to have $a \neq b$, but $a^2 = b^2$. While two different real numbers can have the same square, no two different real numbers have the same *cube*. So, if $a^3 = b^3$, then we must have $a = b$. In other words, the function $f(x) = x^3$ is an invertible function; its inverse is $f^{-1}(x) = \sqrt[3]{x}$. This means that if we cube an equation, we cannot introduce any extraneous roots, since any value that satisfies the cubed equation must satisfy the original equation. $\square$

> **Important:** If we raise an equation to an even power as a step in solving it, we must check for extraneous solutions. If we raise an equation to an odd power while solving it, we don't have to check for extraneous solutions. (But it wouldn't hurt to check your solution for algebra or arithmetic mistakes.)

Problem 14.8: Solve $\sqrt[3]{13x + 37} - \sqrt[3]{13x - 37} = \sqrt[3]{2}$.

Solution for Problem 14.8: We might try isolating one of the roots and cubing first, but cubing both sides of

$$\sqrt[3]{13x + 37} = \sqrt[3]{2} + \sqrt[3]{13x - 37}$$

is pretty daunting. While it will give us $13x + 37$ on the left, the right side will have 4 terms, and two of them will still have complicated cube root expressions that we can't easily combine. So, cubing this equation won't help us much.

Cubing the original equation looks pretty scary, too. So, we look for ways to simplify our task. We start by substituting variables for the ugly cube roots. We let $a = \sqrt[3]{13x + 37}$ and $b = \sqrt[3]{13x - 37}$, so our equation becomes $a - b = \sqrt[3]{2}$. That's a pretty simple equation. If we can find another simple equation for a and b, perhaps we can use the resulting system to solve the problem, because it will be easy to find x if we know a or b.

We can get rid of the cube roots in $a = \sqrt[3]{13x + 37}$ and $b = \sqrt[3]{13x - 37}$ by cubing both equations, which gives

$$a^3 = 13x + 37,$$
$$b^3 = 13x - 37.$$

Aha! Subtracting the second equation from the first eliminates x and gives us another equation for a and b, namely $a^3 - b^3 = 74$. We now have the system of equations

$$a - b = \sqrt[3]{2},$$
$$a^3 - b^3 = 74.$$

We have two attractive first steps for trying to solve this system. Let's explore both a little bit, and then decide which is worth pursuing further.

> **Concept:** When you have multiple options for how to proceed with a problem, but aren't sure which will work, try taking a few steps down each path before committing yourself to one.

- *Option 1: Factor $a^3 - b^3$.* Since we know $a - b$, and $a - b$ is a factor of $a^3 - b^3$, we factor the left side of $a^3 - b^3 = 74$ to find $(a - b)(a^2 + ab + b^2) = 74$. We therefore have

$$a^2 + ab + b^2 = \frac{74}{a - b} = \frac{74}{\sqrt[3]{2}} = \frac{74}{\sqrt[3]{2}} \cdot \frac{\sqrt[3]{4}}{\sqrt[3]{4}} = 37\sqrt[3]{4}.$$

Unfortunately, it's not so clear how to proceed from here. Rather than continuing with this option, we take a look at our other possible approach. If our second option fails, we can always come back to Option 1.

- *Option 2: Cube $a - b$.* Cubing $a - b$ will give us both an a^3 term and a $-b^3$ term, and we know that $a^3 - b^3 = 74$. So, we cube both sides of $a - b = \sqrt[3]{2}$, and we have

$$a^3 - 3a^2b + 3ab^2 - b^3 = 2.$$

Since $a^3 - b^3 = 74$, we have $-3a^2b + 3ab^2 = -72$. Dividing by 3 and factoring the left side gives $-ab(a - b) = -24$. We have $a - b = \sqrt[3]{2}$, and substituting this in our equation gives $-ab(\sqrt[3]{2}) = -24$. Therefore, we have $ab = 12\sqrt[3]{4}$.

The equation that results from Option 2 is simpler than the one from Option 1, so we proceed with that one. Substituting our expressions for a and b into $ab = 12\sqrt[3]{4}$ gives

$$\sqrt[3]{13x + 37} \cdot \sqrt[3]{13x - 37} = 12\sqrt[3]{4}, \qquad \text{so} \qquad \sqrt[3]{169x^2 - 1369} = 12\sqrt[3]{4}.$$

Cubing both sides of this last equation gives us $169x^2 - 1369 = 6912$, from which we find $x^2 = 49$. Therefore, we have $x = \pm 7$.

Checking our solutions to see if they work, we note that if $x = 7$, then we have

$$\sqrt[3]{13 \cdot 7 + 37} - \sqrt[3]{13 \cdot 7 - 37} = \sqrt[3]{128} - \sqrt[3]{54} = 4\sqrt[3]{2} - 3\sqrt[3]{2} = \sqrt[3]{2},$$

and if $x = -7$, then we have

$$\sqrt[3]{13 \cdot (-7) + 37} - \sqrt[3]{13 \cdot (-7) - 37} = \sqrt[3]{-54} - \sqrt[3]{-128} = \sqrt[3]{128} - \sqrt[3]{54} = \sqrt[3]{2}.$$

So, both solutions satisfy the given equation. $\square$

Notice that we didn't lock in to one solution tactic; we explored both factoring the left side of $a^3 - b^3 = 72$ and cubing both sides of $a - b = \sqrt[3]{2}$. Only after taking a couple steps down each path did we see which offered a more promising route to a solution. (Both paths will get us to a solution eventually, though. See if you can figure out how to finish from where we stopped on the first path.)

Exercises

14.1.1

(a) Is it true that $\sqrt{(x-1)^2} = x - 1$ for all real values of x? Why or why not?

(b) Is it true that $\sqrt[3]{(x-1)^3} = x - 1$ for all real values of x? Why or why not?

(c) Is it true that $\sqrt{(x-1)^4} = (x-1)^2$ for all real values of x? Why or why not?

14.1.2 Find all positive numbers x that satisfy the inequality $\sqrt{x} < 2x$. *(Source: AHSME)*

14.1.3 Find all solutions to the equation $\sqrt{x-1} - \sqrt{x+1} + 1 = 0$.

14.1.4 Find all solutions to the equation $\sqrt{x+8} - \dfrac{6}{\sqrt{x+8}} = 5$.

14.1.5 Find all values of x such that $9\sqrt{2 - 3x + x^2} = \sqrt{x^6 - 3x^5 + 2x^4}$.

14.1.6 Find all r such that $\sqrt[3]{5r^2 + 24r + 8} - 2 < r$.

14.1.7★ Solve: $\sqrt[3]{3x - 5} + \sqrt[3]{2x - 4} = \sqrt[3]{5x - 9}$. *(Source: ARML)* **Hints:** 252, 157

14.2 Evaluating Expressions With Radicals

We've seen that raising equations with radicals to powers is an effective way to simplify them. In this section, we use this strategy to evaluate expressions with radicals.

Extra! *Mathematics takes us still further from what is human, into the region of absolute necessity, to*
➡➡➡➡ *which not only the actual world, but every possible world, must conform.*

– Bertrand Russell

Problems

Problem 14.9:

(a) Simplify $\sqrt{3 + 2\sqrt{2}}$.

(b) Suppose $\sqrt{74 - 12\sqrt{30}} = \sqrt{a} - \sqrt{b}$, where a and b are integers. Find a and b, and simplify $\sqrt{74 - 12\sqrt{30}}$.

(c) Simplify $\sqrt[3]{45 - 29\sqrt{2}}$.

Problem 14.10: Between which two consecutive integers is $\sqrt{23} + \sqrt{39}$? (No calculators!) *(Source: Mandelbrot)*

Problem 14.11: In this problem, we simplify $\sqrt[3]{\sqrt{\dfrac{980}{27}} + 6} - \sqrt[3]{\sqrt{\dfrac{980}{27}} - 6}$.

(a) Let $a = \sqrt[3]{\sqrt{\frac{980}{27}} + 6}$ and $b = \sqrt[3]{\sqrt{\frac{980}{27}} - 6}$ and let the expression we seek to simplify be x. Write a relationship among a, b, and x.

(b) Find two equations without radical signs relating a and b.

(c) Combine your equations from (b) with your equation from (a) in a way that produces an equation you can solve for x. Find x.

Problem 14.9:

(a) Simplify $\sqrt{3 + 2\sqrt{2}}$.

(b) Suppose $\sqrt{74 - 12\sqrt{30}} = \sqrt{a} - \sqrt{b}$, where a and b are integers. Find a and b, then simplify $\sqrt{74 - 12\sqrt{30}}$.

(c) Simplify $\sqrt[3]{45 - 29\sqrt{2}}$.

Solution for Problem 14.9:

(a) We know that squaring gets rid of radicals, but we don't have an equation to square. If we start with $x = \sqrt{3 + 2\sqrt{2}}$ and square, we get $x^2 = 3 + 2\sqrt{2}$. This doesn't directly tell us x, but it does suggest what form x has. The square of a number of the form $a + b\sqrt{2}$ is $a^2 + 2b^2 + 2ab\sqrt{2}$, so we guess that $x = a + b\sqrt{2}$ for some integers a and b. This gives us

$$a^2 + 2b^2 + 2ab\sqrt{2} = 3 + 2\sqrt{2}.$$

If there are integers a and b that satisfy this equation, then $a^2 + 2b^2 = 3$ and $2ab = 2$. We could do more algebra to find a and b, or notice that $a = b = 1$ and $a = b = -1$ satisfy the equations. Our answer must be positive, so we have $\sqrt{3 + 2\sqrt{2}} = 1 + \sqrt{2}$.

(b) Our basic strategy in the previous part was to guess that the answer had the form $a + b\sqrt{2}$. Unfortunately, there are several different forms of numbers that we can square to get a number of the form $p - q\sqrt{30}$. For example, squaring $a - b\sqrt{30}$ or $a\sqrt{2} - b\sqrt{15}$ will give a number of the form $p - q\sqrt{30}$. And there are several other cases we'd have to investigate. So, rather than guessing one of these forms specifically, we try the more generic expression $\sqrt{a} - \sqrt{b}$, where a and b are integers.

If $\sqrt{74 - 12\sqrt{30}} = \sqrt{a} - \sqrt{b}$, then squaring both sides gives

$$74 - 12\sqrt{30} = a + b - 2\sqrt{ab}.$$

So, we must have $a + b = 74$ and $2\sqrt{ab} = 12\sqrt{30}$. Dividing the second equation by 2, then squaring, gives $ab = 1080$. Solving $a + b = 74$ for a gives $a = 74 - b$. Substituting this into $ab = 1080$ and rearranging gives $b^2 - 74b + 1080 = 0$. Notice that factoring this equation is exactly the same guessing game as solving the system of equations $a + b = 74$ and $ab = 1080$. We find that a and b are 54 and 20. Since $\sqrt{74 - 12\sqrt{30}}$ must be positive, we must choose a and b so that $\sqrt{a} - \sqrt{b}$ is positive. Therefore, we have

$$\sqrt{74 - 12\sqrt{30}} = \sqrt{54} - \sqrt{20} = 3\sqrt{6} - 2\sqrt{5}.$$

(c) As with the first two parts, we take a guess at the form of the answer, and let

$$\sqrt[3]{45 - 29\sqrt{2}} = a - b\sqrt{2},$$

where a and b are integers. We cube both sides to find $45 - 29\sqrt{2} = a^3 - 3a^2b\sqrt{2} + 6ab^2 - 2b^3\sqrt{2}$. Equating the integer parts of both sides, and equating the parts with $\sqrt{2}$, gives us the system of equations

$$a^3 + 6ab^2 = 45,$$
$$-3a^2b - 2b^3 = -29.$$

Factoring these two equations gives us

$$a(a^2 + 6b^2) = 45,$$
$$b(3a^2 + 2b^2) = 29.$$

If a and b are integers, then the second equation tells us that b times an integer is 29. Therefore, b must be 1 or 29. If $b = 29$, then our second equation gives us $3a^2 + 2(29^2) = 1$, which clearly doesn't have an integer solution. If $b = 1$, we have $3a^2 + 2 = 29$, which gives $a = \pm 3$. Trying both values in the first equation reveals that $(a, b) = (3, 1)$ is a solution to the system, so we have

$$\sqrt[3]{45 - 29\sqrt{2}} = 3 - \sqrt{2}.$$

□

Problem 14.10: Between which two consecutive integers is $\sqrt{23} + \sqrt{39}$? *(Source: Mandelbrot)*

Solution for Problem 14.10: Again, we have square roots. Again, we'd like to square to get rid of the square roots. And again, we don't have an equation to square. This is just like our last problem, so we try a similar strategy.

> **Concept:** When faced with the same complications as in problems you know how to solve, try the same strategies that succeeded on the problems you've already solved.

We introduce an equation to work with by letting $x = \sqrt{23} + \sqrt{39}$. Squaring x will result in a quantity with only one radical, so we give that a try to see what we can learn:

$$x^2 = 62 + 2\sqrt{897}.$$

Since $\sqrt{897}$ is just a little less than $\sqrt{900} = 30$, we see that x^2 is a little less than $62 + 2 \cdot 30 = 122$. Because $x^2 < 122$, we know that $x < 12$. But is x greater than or less than 11? We might have $x^2 = 120.9$, in which case $x < 11$. Or we might have $x^2 = 121.1$, which makes $x > 11$. So, we must determine if $62 + 2\sqrt{897}$ is greater than or less than 121. This is the same as comparing $2\sqrt{897}$ to $121 - 62 = 59$, which is the same as comparing $\sqrt{897}$ to $59/2 = 29.5$. We have $(29.5)^2 = 870.25 < 897$, so $\sqrt{897} > 29.5$. Therefore, we have

$$x^2 = 62 + 2\sqrt{897} > 62 + 2(29.5) = 121,$$

so $x > 11$, and $\sqrt{23} + \sqrt{39}$ is between 11 and 12. $\square$

Problem 14.11: Simplify $\sqrt[3]{\sqrt{\dfrac{980}{27}} + 6} - \sqrt[3]{\sqrt{\dfrac{980}{27}} - 6}$.

Solution for Problem 14.11: We start off with the same tactic that we used on the previous problem. We let

$$x = \sqrt[3]{\sqrt{\frac{980}{27}} + 6} - \sqrt[3]{\sqrt{\frac{980}{27}} - 6}.$$

We might cube both sides now, but those are pretty scary looking radicals. Back in Problem 14.8, we made a similarly scary looking problem simpler using substitution. We try the same here. We let $a = \sqrt[3]{\sqrt{\frac{980}{27}} + 6}$ and $b = \sqrt[3]{\sqrt{\frac{980}{27}} - 6}$, and we seek the value of $a - b$.

We try getting rid of the outer radicals by cubing our expressions for a and b, which gives

$$a^3 = \sqrt{\frac{980}{27}} + 6 \qquad \text{and} \qquad b^3 = \sqrt{\frac{980}{27}} - 6.$$

Subtracting the second equation from the first gets rid of the square roots and leaves $a^3 - b^3 = 12$.

This is a much simpler equation, but we still need to find $a - b$. Unfortunately, factoring $a - b$ out of the left side of $a^3 - b^3 = 12$ won't help much, since this just gives $(a - b)(a^2 + ab + b^2) = 12$. Instead, we go back to our original strategy of setting the expression we wish to evaluate equal to x. However, now we can express x in terms of a and b; we have $x = a - b$. We hope to perform some algebra that will allow us to find x. Cubing $a - b$ will give us an expression including $a^3 - b^3$, which we know is 12. So we try that:

$$x^3 = (a - b)^3 = a^3 - 3a^2b + 3ab^2 - b^3 = a^3 - b^3 - 3ab(a - b) = 12 - 3ab(a - b).$$

This is promising, because we can replace $a - b$ with x. Turning to ab, we have

$$ab = \sqrt[3]{\sqrt{\frac{980}{27}} + 6} \cdot \sqrt[3]{\sqrt{\frac{980}{27}} - 6} = \sqrt[3]{\left(\sqrt{\frac{980}{27}} + 6\right)\left(\sqrt{\frac{980}{27}} - 6\right)} = \sqrt[3]{\frac{980}{27} - 36} = \sqrt[3]{\frac{8}{27}} = \frac{2}{3}.$$

Now our equation is $x^3 = 12 - 2x$. Rearranging this equation gives $x^3 + 2x - 12 = 0$. Searching for rational roots of $x^3 + 2x - 12$, we find that $x = 2$ is a root, and factoring gives $x^2 + 2x - 12 = (x - 2)(x^2 + 2x + 6)$, so we have

$$(x - 2)(x^2 + 2x + 6) = 0.$$

But $x^2 + 2x + 6$ has no real roots, so $x = 2$ is the only real solution to $x^3 + 2x - 12 = 0$. Therefore, we have

$$\sqrt[3]{\sqrt{\frac{980}{27}} + 6} - \sqrt[3]{\sqrt{\frac{980}{27}} - 6} = x = 2.$$

$\square$

In our solution to Problem 14.11, our general strategy is the same as in all the other problems we've faced in this chapter: get rid of the radicals. In addition to cubing equations, we added a new tactic that allows us to remove radicals from a problem: substitution. We'll see more examples of using substitution to simplify expressions in Section 20.2.

Exercises

14.2.1 Simplify $\sqrt{6 + \sqrt{11}} + \sqrt{6 - \sqrt{11}}$.

14.2.2 Simplify each of the following:

(a) $\sqrt{49 + 28\sqrt{3}}$

(b) $\sqrt[4]{49 + 20\sqrt{6}}$

14.2.3 Find the ordered pair of positive integers (a, b), with $a < b$, for which

$$\sqrt{1 + \sqrt{21 + 12\sqrt{3}}} = \sqrt{a} + \sqrt{b}.$$

(Source: ARML)

14.2.4 Which is greater, $\sqrt{2} + \sqrt{3}$ or $\sqrt{10}$? (No calculators!)

14.2.5 Find the value of $(52 + 6\sqrt{43})^{3/2} - (52 - 6\sqrt{43})^{3/2}$. *(Source: AIME)*

14.3 Radical Conjugates

In our solution to Problem 14.11, we encountered the product

$$\left(\sqrt{\frac{980}{27}} + 6\right)\left(\sqrt{\frac{980}{27}} - 6\right) = \frac{980}{27} - 36 = \frac{8}{27}.$$

This might inspire us to notice that raising radicals to powers is not the only way to get rid of them. In this section, we explore ways to use the difference of squares and difference of cubes factorizations to get rid of radicals in problems.

Some problems in this section involve **rationalizing the denominators** of fractions. When we rationalize the denominator of a fraction, we write the fraction as an equivalent fraction with a rational denominator. For example, to rationalize the denominator of $\frac{1}{\sqrt{2}}$, we multiply the numerator and denominator by $\sqrt{2}$:

$$\frac{1}{\sqrt{2}} = \frac{1}{\sqrt{2}} \cdot \frac{\sqrt{2}}{\sqrt{2}} = \frac{\sqrt{2}}{2}.$$

Problems

Problem 14.12: Rationalize the denominator of $\dfrac{1}{\sqrt{7} + \sqrt{2}}$.

Problem 14.13: Find all real values of x such that $\dfrac{x + \sqrt{x^2 - 1}}{x - \sqrt{x^2 - 1}} + \dfrac{x - \sqrt{x^2 - 1}}{x + \sqrt{x^2 - 1}} = 98$.

Problem 14.14: What is the smallest positive integer n such that $\sqrt{n} - \sqrt{n-1} < 0.01$? *(Source: AHSME)*

Problem 14.15: Rationalize the denominator of $\dfrac{1}{\sqrt[3]{3} - 1}$.

Problem 14.12: Rationalize the denominator of $\dfrac{1}{\sqrt{7} + \sqrt{2}}$.

Solution for Problem 14.12: We need to get rid of the square roots in the denominator. The difference of squares factorization, $(x + y)(x - y) = x^2 - y^2$, gives us a way to square away both square roots in the denominator. Since we have $\sqrt{7} + \sqrt{2}$ in the denominator, we multiply the numerator and denominator by $\sqrt{7} - \sqrt{2}$:

$$\frac{1}{\sqrt{7} + \sqrt{2}} = \frac{1}{\sqrt{7} + \sqrt{2}} \cdot \frac{\sqrt{7} - \sqrt{2}}{\sqrt{7} - \sqrt{2}} = \frac{\sqrt{7} - \sqrt{2}}{\left(\sqrt{7}\right)^2 - \left(\sqrt{2}\right)^2} = \frac{\sqrt{7} - \sqrt{2}}{5}.$$

$\square$

Problem 14.13: Find all real values of x such that $\dfrac{x + \sqrt{x^2 - 1}}{x - \sqrt{x^2 - 1}} + \dfrac{x - \sqrt{x^2 - 1}}{x + \sqrt{x^2 - 1}} = 98$.

Solution for Problem 14.13: We can eliminate the square root from the denominator of the first fraction by multiplying the numerator and denominator by $x + \sqrt{x^2 - 1}$:

$$\frac{x + \sqrt{x^2 - 1}}{x - \sqrt{x^2 - 1}} = \frac{x + \sqrt{x^2 - 1}}{x - \sqrt{x^2 - 1}} \cdot \frac{x + \sqrt{x^2 - 1}}{x + \sqrt{x^2 - 1}} = \frac{(x + \sqrt{x^2 - 1})^2}{x^2 - (x^2 - 1)} = (x + \sqrt{x^2 - 1})^2.$$

Similarly, we can rationalize the denominator of the second fraction on the left of our original equation to give

$$(x + \sqrt{x^2 - 1})^2 + (x - \sqrt{x^2 - 1})^2 = 98.$$

Expanding the squares gives $x^2 + 2x\sqrt{x^2 - 1} + x^2 - 1 + x^2 - 2x\sqrt{x^2 - 1} + x^2 - 1 = 98$, so $4x^2 = 100$. This gives us $x = \pm 5$. $\square$

Notice that in each of the first two problems, a key step in our solution is multiplying an expression of the form $\sqrt{a} + \sqrt{b}$ by an expression of the form $\sqrt{a} - \sqrt{b}$. The resulting product has no radicals:

$$\left(\sqrt{a} + \sqrt{b}\right)\left(\sqrt{a} - \sqrt{b}\right) = a - b.$$

The expressions $\sqrt{a} + \sqrt{b}$ and $\sqrt{a} - \sqrt{b}$ are sometimes referred to as **radical conjugates**.

> **Concept:** One way to eliminate radicals in a problem involving an expression of the form $\sqrt{a} + \sqrt{b}$ or $\sqrt{a} - \sqrt{b}$ is to multiply by the radical conjugate, $\sqrt{a} - \sqrt{b}$ or $\sqrt{a} + \sqrt{b}$, respectively.

Of course, this tactic also works when the radicals have rational coefficients besides 1. For example, we can rationalize the denominator of $3/(3\sqrt{5} - 2\sqrt{2})$ by multiplying the numerator and denominator by $3\sqrt{5} + 2\sqrt{2}$:

$$\frac{3}{3\sqrt{5} - 2\sqrt{2}} = \frac{3}{3\sqrt{5} - 2\sqrt{2}} \cdot \frac{3\sqrt{5} + 2\sqrt{2}}{3\sqrt{5} + 2\sqrt{2}} = \frac{9\sqrt{5} + 6\sqrt{2}}{\left(3\sqrt{5}\right)^2 - \left(2\sqrt{2}\right)^2} = \frac{9\sqrt{5} + 6\sqrt{2}}{37}.$$

Multiplying by radical conjugates isn't just for problems with square roots in a denominator:

Problem 14.14: What is the smallest positive integer n such that $\sqrt{n} - \sqrt{n-1} < 0.01$? *(Source: AHSME)*

Solution for Problem 14.14: We don't yet have a very good method for approximating $\sqrt{n} - \sqrt{n-1}$, even if n is a perfect square. However, estimating $\sqrt{n} + \sqrt{n-1}$ isn't too hard, particularly if n is a perfect square. So, we multiply the inequality by $\sqrt{n} + \sqrt{n-1}$, because we know this will give us 1 on the lesser side. We are left with

$$1 < 0.01(\sqrt{n} + \sqrt{n-1}).$$

Multiplying both sides by 100 gives us

$$100 < \sqrt{n} + \sqrt{n-1}.$$

If $n \leq 50^2$, then $\sqrt{n} \leq 50$ and $\sqrt{n-1} < 50$, so $\sqrt{n} + \sqrt{n-1} < 100$ when $n \leq 50^2$. However, when $n = 50^2 + 1$, we have $\sqrt{n} > 50$ and $\sqrt{n-1} = 50$, so $\sqrt{n} + \sqrt{n-1} > 100$ when $n = 50^2 + 1$. Therefore, the smallest value of n that satisfies the inequality is $n = 50^2 + 1 = 2501$. $\square$

Let's see if our "multiply by the conjugate" tactic works on cube roots, too:

Problem 14.15: Rationalize the denominator of $\dfrac{1}{\sqrt[3]{3} - 1}$.

Solution for Problem 14.15: We might start by multiplying our numerator and denominator, $-1 + \sqrt[3]{3}$, by $-1 - \sqrt[3]{3}$, hoping that our "multiply by the conjugate" strategy will still work with cube roots. We find:

$$\frac{1}{-1 + \sqrt[3]{3}} = \frac{1}{-1 + \sqrt[3]{3}} \cdot \frac{-1 - \sqrt[3]{3}}{-1 - \sqrt[3]{3}} = \frac{-1 - \sqrt[3]{3}}{1 - \sqrt[3]{9}}.$$

Hmmm.... That didn't quite work. We could keep at it, multiplying next by $1 + \sqrt[3]{9}$, but that will make quite a mess in the numerator, and still won't make our denominator rational.

Instead, we take a step back and think about why multiplying by the radical conjugate works with square roots. As we saw earlier, it works because the product $\left(\sqrt{a} + \sqrt{b}\right)\left(\sqrt{a} - \sqrt{b}\right)$ is the difference of squares factorization of $a - b$.

But here, we have a cube root. The difference of squares helped us out with square roots; maybe the difference of cubes factorization will help us with cube roots. So, we try to use the difference of cubes factorization $x^3 - y^3 = (x - y)(x^2 + xy + y^2)$ just as we used the difference of squares factorization in Problem 14.12. Letting $x = \sqrt[3]{3}$ and $y = 1$ in this equation, we see that $\sqrt[3]{3} - 1$ is a factor of $3 - 1$:

$$3 - 1 = \left(\sqrt[3]{3} - 1\right)\left(\sqrt[3]{9} + \sqrt[3]{3} + 1\right).$$

We can use this to rationalize the denominator of our fraction:

$$\frac{1}{\sqrt[3]{3} - 1} = \frac{1}{\sqrt[3]{3} - 1} \cdot \frac{\sqrt[3]{9} + \sqrt[3]{3} + 1}{\sqrt[3]{9} + \sqrt[3]{3} + 1} = \frac{\sqrt[3]{9} + \sqrt[3]{3} + 1}{3 - 1} = \frac{\sqrt[3]{9} + \sqrt[3]{3} + 1}{2}.$$

$\square$

Extra! Here's a clever proof that it is possible for an irrational number raised to an irrational ⫸⫸⫸⫸ power to result in a rational number. Consider the number $x = \left(\sqrt{2}\right)^{\sqrt{2}}$. If x is rational, then we have our sought-after example. If x is irrational, then consider the number $x^{\sqrt{2}}$. This number equals $\left(\sqrt{2}\right)^{\sqrt{2} \cdot \sqrt{2}} = \left(\sqrt{2}\right)^2 = 2$. So, either x is rational, or $x^{\sqrt{2}}$ is rational. In either case, we have found an irrational number raised to an irrational power that results in a rational number.

Exercises

14.3.1 Rationalize the denominator in each of the following:

(a) $\dfrac{4}{\sqrt{11}+3}$

(c) $\dfrac{2\sqrt{6}}{\sqrt{2}+\sqrt{3}+\sqrt{5}}$ *(Source: AHSME)*

(b) $\dfrac{8}{\sqrt[3]{5}+\sqrt[3]{3}}$

(d) $\dfrac{\sqrt{4+2\sqrt{3}}+\sqrt{4-2\sqrt{3}}}{\sqrt{4+2\sqrt{3}}-\sqrt{4-2\sqrt{3}}}$

14.3.2 Evaluate the product $\left(\sqrt{5}+\sqrt{6}+\sqrt{7}\right)\left(\sqrt{5}+\sqrt{6}-\sqrt{7}\right)\left(\sqrt{5}-\sqrt{6}+\sqrt{7}\right)\left(-\sqrt{5}+\sqrt{6}+\sqrt{7}\right)$. *(Source: AIME)*

14.3.3 Write the number $\dfrac{1}{\sqrt{2}-\sqrt[3]{2}}$ as the sum of terms of the form 2^q, where q is rational. (For example, $2^1 + 2^{-1/3} + 2^{8/5}$ is a sum of this form.) *(Source: USAMTS)*

14.3.4 Show that $\dfrac{1}{10} < \sqrt{101} - \sqrt{99}$ without using a calculator.

14.4 Summary

The key idea in this chapter is to deal with problems involving radicals by getting rid of the radicals.

> **Concept:** One way to get rid of a radical in an equation is to raise both sides to a power that gets rid of the radical.

> **Concept:** One way to eliminate the radicals in a problem involving an expression of the form $\sqrt{a}+\sqrt{b}$ or $\sqrt{a}-\sqrt{b}$ is to multiply by the radical conjugate, $\sqrt{a}-\sqrt{b}$ or $\sqrt{a}+\sqrt{b}$, respectively.

We often use this idea when **rationalizing the denominator** of a fraction, which means writing the fraction as an equivalent fraction with a rational (usually integer) denominator.

Things To Watch Out For!

- Whenever we square an equation as a step in solving it, we have to check that our solutions are valid. If the solution to the squared equation does not satisfy our original equation, then the solution is called **extraneous**. An extraneous solution is not a valid solution to the original equation.

- If you divide a common factor from two sides of an equation, you cannot forget about the solutions that might arise if the factor equals 0.

Problem Solving Strategies

Concepts:	• Rearranging an equation with radicals before raising it to a power may simplify the algebra you have to perform to solve the equation.
	• Don't just manipulate equations blindly. If you think about what an equation means, you might find a fast route to solve the equation or see that it has no solutions.
	• Don't be hasty to jump into lengthy computations or algebraic manipulations. First check and see if there is a more clever approach.
	• When you have multiple options for how to proceed with a problem, but aren't sure which will work, try taking a few steps down each path before committing yourself to one.
	• When faced with the same complications as in problems you know how to solve, try the same strategies that succeeded on the problems you've already solved.

REVIEW PROBLEMS

14.16 Rationalize the denominator in each of the following:

(a) $\dfrac{1}{\sqrt{3} - \sqrt{2}}$

(b) $\dfrac{4}{3 + \sqrt{7}}$

(c) $\dfrac{1}{\sqrt[3]{7} - 1}$

(d) $\dfrac{2}{\sqrt[3]{25} + \sqrt[3]{5} + 1}$

14.17 Simplify each of the following:

(a) $\sqrt{6 + 2\sqrt{5}}$

(b) $\sqrt[4]{89 + 28\sqrt{10}}$

14.18 Simplify $\sqrt{4 + \sqrt{7}} - \sqrt{4 - \sqrt{7}}$.

14.19 Solve the equation $\sqrt{5x - 1} + \sqrt{x - 1} = 2$. *(Source: AHSME)*

14.20 Find all solutions to the equation $4\sqrt{x^3 - 2x^2 + x} = \sqrt{x^3 - x^2}$.

14.21 Find all pairs of rational numbers (x, y) such that $\sqrt{2\sqrt{3} - 3} = \sqrt{x\sqrt{3}} - \sqrt{y\sqrt{3}}$.

14.22 Simplify $\dfrac{\sqrt{\sqrt{5}+2}+\sqrt{\sqrt{5}-2}}{\sqrt{\sqrt{5}+1}}-\sqrt{3-2\sqrt{2}}$. *(Source: AHSME)*

14.23

(a) Find all values of x such that $\sqrt{x^2+7x+10}=x+2+\sqrt{x+2}$.

(b)★ Find all values of x such that $\sqrt{x^2+7x+10}>x+2+\sqrt{x+2}$.

14.24 Simplify the sum $\sqrt[3]{18+5\sqrt{13}}+\sqrt[3]{18-5\sqrt{13}}$.

14.25 For $x>1$, the equation $\sqrt{x+\sqrt{2x-1}}-\sqrt{x-\sqrt{2x-1}}=\sqrt{k}$ is an identity. Find k. *(Source: IMO)*

14.26 The number

$$\sqrt{104\sqrt{6}+468\sqrt{10}+144\sqrt{15}+2006}$$

can be written as $a\sqrt{2}+b\sqrt{3}+c\sqrt{5}$, where a, b, and c are positive integers. Find abc. *(Source: AIME)*

14.27 If $x+\sqrt{x^2-1}+\dfrac{1}{x-\sqrt{x^2-1}}=20$, then find $x^2+\sqrt{x^4-1}+\dfrac{1}{x^2+\sqrt{x^4-1}}$. *(Source: AHSME)*

14.28 If $\sqrt[3]{n+\sqrt{n^2+8}}+\sqrt[3]{n-\sqrt{n^2+8}}=8$, where n is an integer, then find n.

Challenge Problems

14.29 Solve the equation $\sqrt[3]{60-x}+\sqrt[3]{x-11}=\sqrt[3]{4}$. *(Source: Sweden)* **Hints:** 79, 106

14.30 Find the largest integer N such that $N<\left(\sqrt{33+\sqrt{128}}+\sqrt{2}-8\right)^{-1}$. *(Source: ARML)*

14.31 Find the ordered triple of positive integers (a,b,c) for which

$$\left(\sqrt{5}+\sqrt{2}-\sqrt{3}\right)\left(3\sqrt{a}+\sqrt{b}-2\sqrt{c}\right)=12.$$

(Source: NYSML)

14.32 Let $x=\dfrac{4}{\left(\sqrt{5}+1\right)\left(\sqrt[4]{5}+1\right)\left(\sqrt[8]{5}+1\right)\left(\sqrt[16]{5}+1\right)}$. Find $(x+1)^{48}$. *(Source: AIME)*

14.33 Show that $\sqrt{x-4\sqrt{x-4}}+2=\sqrt{x+4\sqrt{x-4}}-2$ for all $x\ge 8$.

14.34★ Solve the equation $(\sqrt{2}+1)^x+(\sqrt{2}-1)^x=6$. *(Source: Sweden)* **Hints:** 236

14.35★ Find a polynomial $f(x,y,z)$ in three variables, with integer coefficients, such that for all integers a,b,c, the sign of $f(a,b,c)$ (that is, positive, negative, or zero) is the same as the sign of $a+b\sqrt[3]{2}+c\sqrt[3]{4}$. *(Source: USAMTS)* **Hints:** 190, 95

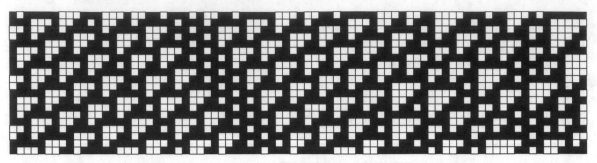

Nothing travels faster than the speed of light with the possible exception of bad news, which obeys its own special laws. – Douglas Adams

CHAPTER **15**

Special Classes of Functions

15.1 Rational Functions and Their Graphs

Any function that can be expressed as one polynomial divided by another is called a **rational function**. So, $f(x)$ is a rational function if

$$f(x) = \frac{s(x)}{t(x)},$$

where $s(x)$ and $t(x)$ are polynomials, and $t(x)$ is not the polynomial 0. For example, each of the following is a rational function:

$$f(x) = \frac{1}{x},$$

$$g(x) = \frac{x^2}{2x - 1},$$

$$h(y) = \frac{5y^3 - 7y + 13}{4y^4},$$

$$t(z) = \frac{z^{100}}{(z - 1)(z - 2)}.$$

Problems

Problem 15.1: Let $f(x) = \dfrac{1}{x}$.

  (a) Find the domain of f.   (b) Find the range of f.   (c) Graph f.

Problem 15.1: Let $f(x) = \dfrac{1}{x}$.

   (a)   Find the domain of f.   (b)   Find the range of f.   (c)   Graph f.

Solution for Problem 15.1:

   (a)   The only value of x for which $f(x)$ is not defined is $x = 0$, so the domain is all nonzero real numbers.

   (b)   We let $y = \frac{1}{x}$. Solving for x in terms of y gives $x = \frac{1}{y}$, so we cannot have $y = 0$. For any other value of y, letting $x = \frac{1}{y}$ in $f(x) = \frac{1}{x}$ gives us $f(x) = y$, so the range of f is all nonzero real numbers.

   (c)   The graph of f is the graph of the equation $y = \frac{1}{x}$. Multiplying this by x gives us $xy = 1$. We found in Section 5.6 that the graph of this equation is a hyperbola, as shown at right.

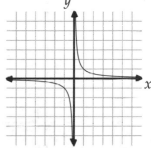

      As we noted back in Section 5.6, an **asymptote** of a graph is a line that the graph approaches but does not intersect as x or y becomes very far from 0. For example, as x gets very large, the value of y approaches, but never equals, 0. So, the graph approaches, but doesn't intersect, the x-axis (where $y = 0$) as x gets large. Similarly, as x gets much less than 0, the graph again approaches but does not intersect the x-axis. So, we call the x-axis a **horizontal asymptote** of the graph. Similarly, the y-axis is a **vertical asymptote** of the graph.

□

Problems

Problem 15.2: Let $f(x) = \dfrac{x+4}{x-3}$ and $g(x) = \dfrac{x^2 + 3x - 4}{x^2 - 4x + 3}$.

   (a)   Factor the numerator and denominator of g. For what values of x are $f(x)$ and $g(x)$ equal?

   (b)   Find the domain and range of f, and graph $y = f(x)$. Find all asymptotes of the graph.

   (c)   Find the domain and range of g, and graph $y = g(x)$. Find all asymptotes of the graph.

   (d)   Are f and g the same function? Why or why not?

Problem 15.3: Consider your graph of $f(x) = \frac{x+4}{x-3}$ from the previous problem. Notice that the graph looks like the hyperbolas of the form $(x - h)(y - k) = c$ from Section 5.6. In this problem, we learn why.

   (a)   Let $y = \frac{x+4}{x-3}$. Get rid of the fraction, and move all the variables to one side. Can you factor the side with the variables?

   (b)   Show that the graph of f is a hyperbola of the form $(x - h)(y - k) = c$.

Problem 15.4: Let $f(x) = \dfrac{p(x)}{d(x)}$, where $p(x)$ and $d(x)$ are polynomials.

   (a)   Suppose the line $x = c$, where c is a constant, is an asymptote of the graph of f. Must $d(c) = 0$?

   (b)   If c is a constant such that $d(c) = 0$, then must the line $x = c$ be a vertical asymptote of the graph of f? Why or why not?

Problem 15.5: Which of the graphs below could be the graph of $g(x) = \dfrac{(x+4)^2(x-1)}{2(x+3)(x-2)}$?

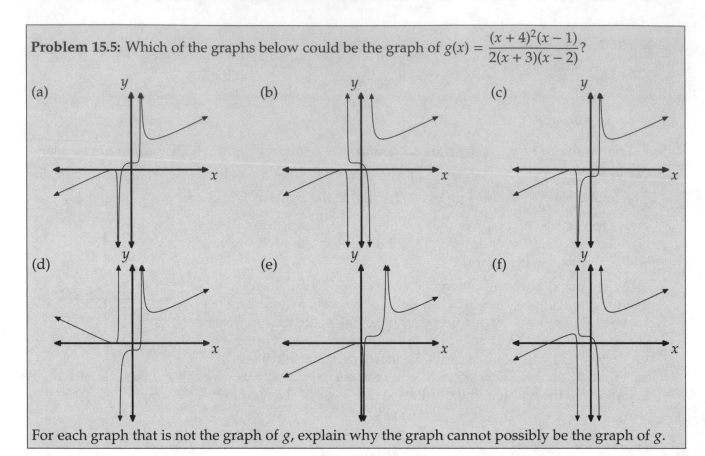

For each graph that is not the graph of g, explain why the graph cannot possibly be the graph of g.

Problem 15.6: Let $g(x) = \dfrac{2x^2 - 5x - 1}{x - 3}$.

(a) Graph $y = g(x)$.

(b) Is there a non-vertical line that appears to be an asymptote of the graph?

(c) Find the quotient and remainder when $2x^2 - 5x - 1$ is divided by $x - 3$. Use the result to explain your observations from part (b).

(d) What is the range of g?

Problem 15.7: Let $f(x) = \dfrac{2x^2 - 3}{x^4 + 3x^3 + 2x - 5}$. In this problem, we find a horizontal asymptote of the graph of f.

(a) Divide the numerator and denominator by $f(x)$ by x^4, thereby finding an expression for $f(x)$ for all $x \neq 0$ such that the highest degree term in the denominator is a constant.

(b) Use the form you found in part (a) to investigate what happens to $f(x)$ when x is very far from 0. What line is a horizontal asymptote of the graph of f?

(c) Generalize this problem. In other words, find a general statement about a certain class of rational functions that this problem represents.

Problem 15.2: Let $f(x) = \dfrac{x+4}{x-3}$ and $g(x) = \dfrac{x^2+3x-4}{x^2-4x+3}$.

(a) Factor the numerator and denominator of g. For what values of x are $f(x)$ and $g(x)$ equal?

(b) Find the domain and range of f, and graph $y = f(x)$.

(c) Find the domain and range of g, and graph $y = g(x)$.

(d) Are f and g the same function? Why or why not?

Solution for Problem 15.2:

(a) Factoring the numerator of $g(x)$ gives us $g(x) = \dfrac{(x+4)(x-1)}{(x-3)(x-1)}$. So, it appears that we have

$$g(x) = \frac{(x+4)(x-1)}{(x-3)(x-1)} = \frac{x+4}{x-3} = f(x)$$

for all x. However, while $f(1) = (1+4)/(1-3) = -5/2$, the function $g(x)$ is not defined for $x = 1$, because the denominator of $g(x)$ is 0 if $x = 1$. For all other values of x, the $x-1$ factors in the numerator and denominator of $g(x)$ do cancel. This leaves us with $f(x) = g(x)$ for all x besides $x = 1$ and $x = 3$ (both functions are undefined for $x = 3$).

(b) The domain of f is all real numbers except 3. To find the range, we let $y = f(x) = (x+4)/(x-3)$. Solving for x in terms of y gives us $x = (3y+4)/(y-1)$. For any value of y except $y = 1$, we can use this equation to find a value of x such that $f(x) = y$. Therefore, the range of f is all real numbers except 1.

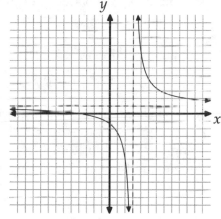

Choosing several values of x reveals the graph of $y = f(x)$ shown at right. The line $x - 3$ is a vertical asymptote, because choosing values of x closer and closer to 3 makes y much greater than 0 (when $x > 3$) or much less than 0 (when $x < 3$). Similarly, $y = 1$ is a horizontal asymptote. To see why, consider the equation $x = (3y+4)/(y-1)$, which we found by solving $y = f(x)$ for x in terms of y. As y gets very close to 1, the value of x is much greater than 0 (when $y > 1$) or much less than 0 (when $y < 1$). The asymptotes are dashed in the graph.

(c) Because the denominator of $g(x)$ is $(x-3)(x-1)$, the domain of $g(x)$ is all real numbers except 1 and 3. We've seen that $g(x) = f(x)$ for all values of x except $x = 1$ and $x = 3$. So, the range of g appears to be the same as the range of f, except that it may not include $f(1)$, because $x = 1$ is not in the domain of g. There are a couple ways to see that $f(1)$, which equals $-5/2$, is not in the range of g.

Approach 1: Solve $g(x) = -5/2$. Cross-multiplying

$$\frac{x^2+3x-4}{x^2-4x+3} = -\frac{5}{2}$$

gives $2x^2 + 6x - 8 = -5x^2 + 20x - 15$, so we have $7x^2 - 14x + 7 = 0$. This gives us $7(x-1)^2 = 0$, so we only have $g(x) = -5/2$ if $x = 1$. This is as expected, since $f(1) = -5/2$. However, because 1 is not in the domain of g, there are no values of x for which $g(x) = -5/2$. Therefore, the range of g is all real numbers except $-5/2$ and 1 (remember, except for $-5/2$, the range of g matches the range of f).

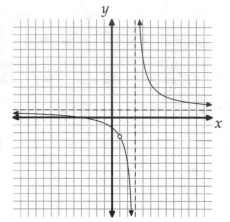

Approach 2: Graph g. Because $g(x) = f(x)$ for all x except for $x = 1$, the graph of $y = g(x)$ is the same as the graph of $y = f(x)$ everywhere except where $x = 1$. So, to graph $g(x)$, we draw the graph of $f(x)$, but place an open circle at the point $(1, -5/2)$, since this point is on $f(x)$, but not on $g(x)$. This graph is shown at right. Again, the asymptotes are shown as dashed lines. Note that we do not consider the line $x = 1$ an asymptote. The graph of g does approach this line, and does not intersect it, but the graph of g does not approach the line $x = 1$ as y gets far from 0.

We say that the graph of g has a **hole** at $(1, -5/2)$.

(d) Because $f(x)$ and $g(x)$ are not the same for $x = 1$, they are not the same function.

$\square$

Problem 15.3: Consider your graph of

$$f(x) = \frac{x + 4}{x - 3}$$

from the previous problem. Notice that it looks like the hyperbolas of the form $(x - h)(y - k) = c$ from Section 5.6. Is this a coincidence?

Solution for Problem 15.3: The graph of f is the graph of the equation $y = (x+4)/(x-3)$. Multiplying both sides by $x - 3$ gives $xy - 3y = x + 4$. Hey, this looks like an equation for Simon's Favorite Factoring Trick! Moving the variables to the left gives $xy - 3y - x = 4$. Adding 3 to both sides gives $xy - 3y - x + 3 = 7$, and factoring the left side of this gives us

$$(x - 3)(y - 1) = 7.$$

So, rearranging the equation $y = f(x)$ gives us an equation of the form $(x - h)(y - k) = c$. Therefore, the graph of f is a hyperbola with horizontal and vertical asymptotes. $\square$

Now that we've graphed a few rational functions, perhaps you have some intuition for how to find the vertical asymptotes of the graph of a rational function.

Problem 15.4: Let $f(x) = \dfrac{p(x)}{d(x)}$, where $p(x)$ and $d(x)$ are polynomials.

(a) Suppose the line $x = c$, where c is a constant, is a vertical asymptote of the graph of f. Must $d(c) = 0$?

(b) If c is a constant such that $d(c) = 0$, then must the line $x = c$ be a vertical asymptote of the graph of f? Why or why not?

Solution for Problem 15.4:

(a) If $d(c) \neq 0$, then $f(c) = p(c)/d(c)$ is defined. In other words, the graph of f passes through the point $(c, \frac{p(c)}{d(c)})$. If $x = c$ were an asymptote of the graph of f, then the graph would approach the line, but not intersect it. Instead, the graph approaches the line, then passes through a point on the line. So, $x = c$ is not an asymptote of the graph of f if $d(c) \neq 0$. This means that we must have $d(c) = 0$ if $x = c$ is a vertical asymptote of the graph of f.

(b) Consider our graph of

$$g(x) = \frac{x^2 + 3x - 4}{x^2 - 4x + 3}$$

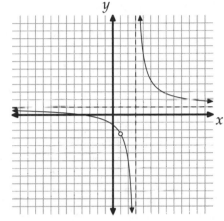

in Problem 15.2. This graph is shown at right. The denominator equals 0 when $x = 1$, but $x = 1$ is not a vertical asymptote of the graph. Why isn't it?

The numerator of $g(x)$ is also 0 when $x = 1$. As we saw in our discussion of Problem 15.2, if $x \neq 1$, then we can cancel the factors of $x - 1$ in the numerator and denominator of $g(x)$ to find $g(x) = \frac{x+4}{x-3}$. So, the graph of $g(x)$ approaches $\frac{1+4}{1-3} = -\frac{5}{2}$ as x gets close to 1. In other words, $g(x)$ is not far from 0 when x gets close to 1. This tells us that $x = 1$ is not a vertical asymptote of the graph of g.

What if the numerator is not also 0 when the denominator equals 0? We'll answer this by considering the rational function $f(x) = p(x)/d(x)$, where $d(c) = 0$ and $p(c) \neq 0$ for some constant c. Because $d(c) = 0$, one of the factors of $d(c)$ is $x - c$. Therefore, as x gets very close to 0, the value of $d(x)$ gets very close to 0. Since $p(c) \neq 0$, when we compute $f(x)$ for values of x very close to $x = c$, we are dividing a nonzero quantity by a number that is very close to 0. The result of dividing a nonzero constant by a number that is very close to 0 is a quotient that is very far from 0. This tells us that if $d(c) = 0$ and $p(c) \neq 0$, the line $x - c$ is a vertical asymptote of the graph of f.

Note that if $p(x)$ and $d(x)$ are polynomials such that $p(c) = d(c) = 0$ for some constant c, then we cannot conclude that $x = c$ is definitely not a vertical asymptote of the graph of $y = p(x)/d(x)$. For example, consider the function $f(x) = x/x^2$. Here, both the numerator and denominator are 0 for $x = 0$. However, we have $f(x) = 1/x$ for all nonzero x, so the graph of $f(x) = x/x^2$ is the same as the graph of $y = 1/x$. This latter graph clearly has a vertical asymptote at $x = 0$.

$\square$

Together, our two parts of Problem 15.4 give us some guidance for finding vertical asymptotes of the graph of a rational function:

> **Important:** Let $p(x)$ and $d(x)$ be polynomials such that $d(x)$ is not 0 for all x. If c is a constant such that $d(c) = 0$ and $p(c) \neq 0$, then the graph of $f(x) = p(x)/d(x)$ has a vertical asymptote at $x = c$.

We can use our understanding of asymptotes and polynomial roots to quickly get qualitative information about the graph of a rational function.

Problem 15.5: Which of the graphs below could be the graph of $g(x) = \dfrac{(x+4)^2(x-1)}{2(x+3)(x-2)}$?

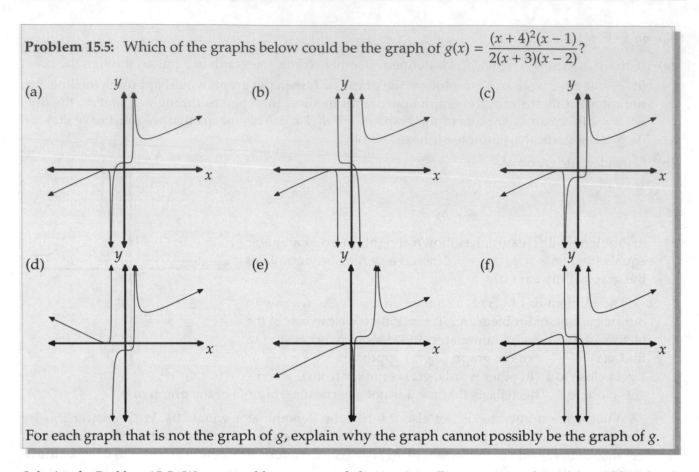

(a)

(b)

(c)

(d)

(e)

(f)

For each graph that is not the graph of g, explain why the graph cannot possibly be the graph of g.

Solution for Problem 15.5: We proceed by process of elimination. First, we note that $g(x) = 0$ if and only if $x = -4$ or $x = 1$. So, the graph of g intersects the x-axis once to the right of the y-axis and once to the left of the y-axis. That eliminates graph (a) and graph (f). It looks like it eliminates graph (e) as well, but it's not completely clear where that graph intersects the x-axis, so we should look for another way to eliminate (e).

Next, we note that the two lines $x = -3$ and $x = 2$ are vertical asymptotes of g. These two asymptotes are on opposite sides of the y-axis. This does eliminate option (e).

Comparing graphs (c) and (d), we see that the only difference is the left branch of the graph. Consider what happens when $x < -4$. This makes $(x - 1)$, $(x + 3)$, and $(x - 2)$ all negative. However, $(x + 4)^2$ is always positive. So, if $x < -4$, then our expression for $g(x)$ has a negative numerator and a positive denominator. Therefore $g(x) < 0$ for $x < -4$. This eliminates graph (d).

All we have left are graphs (b) and (c). The only difference between these two is the middle branch. There are several ways to eliminate (c). First, we could note that $g(0) = 4/3$, so $g(0) > 0$, which is not the case in graph (c). We could also note that if $-3 < x < 0$, then the numerator of $g(x)$ is negative, as is the denominator, so $g(x) > 0$ if $-3 < x < 0$. This is the case in graph (b), but not in graph (c). (Similarly, we could have examined the behaviors of the graphs near $x = 2$.)

Extra! *One factor that has remained constant through all the twists and turns of the history of physical*
➥➥➥➥ *science is the decisive importance of the mathematical imagination.*

– Freeman Dyson

We have eliminated all but graph (b). Graph (b) has roots and vertical asymptotes in the correct locations. To check if the graph is below the y-axis where g is negative and above the y-axis where g is positive, we make the table at right. We see that $g(x)$ is positive for $x > 2$ and $-3 < x < 1$, and negative for $1 < x < 2$,

	$(x+4)^2$	$x+3$	$x-1$	$x-2$	$g(x)$
$x > 2$	$+$	$+$	$+$	$+$	$+$
$1 < x < 2$	$+$	$+$	$+$	$-$	$-$
$-3 < x < 1$	$+$	$+$	$-$	$-$	$+$
$-4 < x < -3$	$+$	$-$	$-$	$-$	$-$
$x < -4$	$+$	$-$	$-$	$-$	$-$

$-4 < x < -3$, and $x < -4$. Though we don't know for sure that these exactly match the corresponding ranges in graph (b) for which the graph is above or below the x-axis, there are no points (x, y) on graph (b) at which the sign of y obviously cannot match the sign of $g(x)$. $\square$

Our graph of $g(x)$ in Problem 15.5 exhibits some interesting behavior as x becomes far from 0. Specifically, the graph doesn't appear to have a horizontal asymptote, but it does appear to "straighten out" and look more and more like a straight line. Let's take a look at what might cause this behavior.

Problem 15.6: Let $g(x) = \dfrac{2x^2 - 5x - 1}{x - 3}$.

(a) Graph $y = g(x)$.

(b) Is there a non-vertical line that appears to be an asymptote of the graph?

(c) Find the quotient and remainder when $2x^2 - 5x - 1$ is divided by $x - 3$. Use the result to explain your observations from part (b).

(d) What is the range of g?

Solution for Problem 15.6:

(a) We start with the table at left below, and use it to generate the graph of $y = g(x)$ at right below.

x	$g(x)$	x	$g(x)$
-7	-13.2	3.1	27.2
-5	-9.25	3.5	12
-3	-5.33	4	11
-1	-1.5	5	12
1	2	6	13.67
2	3	7	15.5
2.5	2	8	17.4
2.9	-13.2	9	19.33

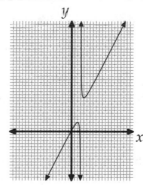

(b) The graph has two pieces, and a vertical asymptote at $x = 3$ (where the denominator of $g(x)$ is 0, as expected). The graph does not seem to have a horizontal asymptote. However, as x gets far from 0, the graph appears to become more and more like a line. Both the graph and our table suggest that the slope of the line is close to 2.

(c) We explained the hyperbola form of the graph $y = (x+4)/(x-3)$ by writing the equation in another form. We try the same here. We hope to explain the line-like behavior of the graph of g by finding another form in which to write $g(x)$. One natural way to do so is to perform polynomial division, which gives us

$$g(x) = \frac{2x^2 - 5x - 1}{x - 3} = 2x + 1 + \frac{2}{x - 3}.$$

Aha! This tells the story. When x is very large, the term $2/(x - 3)$ is very close to zero, so $g(x)$ is very close to $2x + 1$. The same is true if x is much less than 0. So, as x becomes far from 0, the graph of g gets closer and closer to the line $y = 2x + 1$. We call this line a **slant asymptote** or an **oblique asymptote** of the graph.

In fact, a horizontal asymptote can be thought of as a slant asymptote with slope 0, and we can discover horizontal asymptotes the same way we found the vertical asymptote of the graph of g. For example, we earlier found that the graph of $f(x) = (x + 4)/(x - 3)$ has a horizontal asymptote at $y = 1$. Performing the polynomial division $(x + 4)/(x - 3)$ gives us

$$f(x) = \frac{x + 4}{x - 3} = 1 + \frac{7}{x - 3}.$$

As x gets very far from 0, the term $7/(x - 3)$ gets very close to 0, so $f(x)$ gets very close to 1.

While horizontal asymptotes are essentially slant asymptotes with slope 0, when we use the term "slant asymptote" in this book, we refer specifically to asymptotes that are neither horizontal nor vertical.

(d) Our graph and table suggest that the range of g is $(-\infty, 3] \cup [11, +\infty)$. To show that this is the case, we again rewrite

$$y = \frac{2x^2 - 5x - 1}{x - 3}.$$

Multiplying both sides by $x - 3$ gives

$$xy - 3y = 2x^2 - 5x - 1,$$

and bringing all terms to one side gives us a quadratic in x:

$$2x^2 + (-5 - y)x + (3y - 1) = 0.$$

We wish to know for what values of y there is a real solution x to this equation. This is a job for the discriminant. This quadratic has real solutions for x if and only if its discriminant is nonnegative, so we must have

$$(-5 - y)^2 - 4(2)(3y - 1) \geq 0.$$

Simplifying the left side of this inequality gives us $y^2 - 14y + 33 \geq 0$. Factoring the left side gives us $(y - 11)(y - 3) \geq 0$, which has solutions $y \in (-\infty, 3] \cup [11, +\infty)$, as expected. $\square$

In Problem 15.6, we learned a lot about g and its graph by writing the equation $y = g(x)$ in different forms.

> **Concept:** Don't limit yourself to an equation or expression you're given. Different algebraic forms of the same equation or expression can offer different information, so experiment with manipulating equations and expressions into different forms.

Problem 15.7: Let $f(x) = \dfrac{2x^2 - 3}{x^4 + 3x^3 + 2x - 5}$. Find the horizontal asymptote of the graph of f.

Solution for Problem 15.7: Intuitively, it seems like $y = 0$ is a horizontal asymptote of the graph because when x is very large, the value of x^4 is much greater than that of $2x^2$. To more clearly show that this is the case, we look back to Problem 15.6 for inspiration. In that problem, we rewrote the given rational function in a more convenient form by performing polynomial division. Here, polynomial division isn't so obviously useful, since the quotient is 0 (a clue that $y = 0$ is a horizontal asymptote), which means $f(x)$ can't be rewritten with polynomial division.

We wish to analyze what happens when x is very large. We don't have good tools for comparing what happens to different powers of x as x gets large, but we do know that dividing by large values produces numbers close to 0. So, we rewrite our rational function such that in both the numerator and denominator, all the instances of x are in denominators. Specifically, for all $x \neq 0$, we can divide the numerator and denominator of $f(x)$ by the highest power of x in either, x^4, to eliminate the positive powers of x and leave

$$f(x) = \frac{2x^2 - 3}{x^4 + 3x^3 + 2x - 5} \cdot \frac{\frac{1}{x^4}}{\frac{1}{x^4}} = \frac{\frac{2}{x^2} - \frac{3}{x^4}}{1 + \frac{3}{x} + \frac{2}{x^3} - \frac{5}{x^4}}$$

for $x \neq 0$. Notice that we have a constant term in the denominator, but not the numerator. This is because the denominator of $f(x)$ has a higher degree than the numerator.

Now, the behavior of $f(x)$ when x is far from 0 is clear. Except for the constant term, 1, every term in the numerator and the denominator is very close to 0 when x is far from 0. Therefore, when x is large, the numerator is very close to 0, while the denominator is close to 1. This means $f(x)$ is very close to 0. To see that $f(x)$ is not equal to 0, we can note that the roots of $2x^2 - 3$ are $\pm\sqrt{3}/2$, so $f(x) \neq 0$ when x is very far from 0. Since $f(x)$ gets very, very close to 0, but never equals 0, as x gets far from 0, we know that $y = 0$ is a horizontal asymptote of the graph of f. $\square$

We can follow essentially the same steps to generalize Problem 15.7. Suppose that

$$g(x) = \frac{a_n x^n + a_{n-1} x^{n-1} + \cdots + a_0}{b_m x^m + b_{m-1} x^{m-1} + \cdots + b_0},$$

where a_n and b_m are nonzero. The key feature of our specific $f(x)$ above was that the degree of the denominator was greater than that of the numerator. So, we let $m > n$. Then, for $x \neq 0$, we can divide the numerator and denominator of $g(x)$ by x^m to get

$$g(x) = \frac{\frac{a_n}{x^{n-m}} + \frac{a_{n-1}}{x^{n \; m-1}} + \cdots + \frac{a_0}{x^m}}{b_m + \frac{b_{m-1}}{x} + \cdots + \frac{b_0}{x^m}}$$

for $x \neq 0$. When x is far from zero, all the terms except the b_m in the denominator are very close zero. So, $g(x)$ approaches 0 when x is far from 0.

> **Important:** If $f(x) = g(x)/h(x)$, where $g(x)$ and $h(x)$ are nonzero polynomials such that $\deg g < \deg h$, then $y = 0$ is a horizontal asymptote.

15.1.1 Find the domain and range of each of the following rational functions. Also, sketch the graph of $y = f(x)$ for each function, and identify any asymptotes.

(a) $f(x) = \dfrac{x+2}{x-1}$

(b) $f(x) = \dfrac{x^2-1}{2x+3}$

(c) $f(x) = \dfrac{x}{(x-2)(x+3)}$

(d) $f(x) = \dfrac{x^3}{x^2-1}$

15.1.2 Find a rational function that has vertical asymptotes $x = 1$ and $x = -2$. Sketch the graph of your function.

15.1.3 For what values of the constant c does the graph of $f(x) = \dfrac{x^2-x+c}{x^2-8x+15}$ have exactly one vertical asymptote?

15.1.4 Let f and g be rational functions. Suppose the graph of f has a vertical asymptote at $x = -1$ and the graph of g has a vertical asymptote at $x = 3$, and that these graphs have no other vertical asymptotes. What can we say about the vertical asymptotes of $f(x) \cdot g(x)$? Of $f(x) + g(x)$? Be careful—there are some special cases that must be considered.

15.1.5

(a) Does the graph of $f(x) = \dfrac{2x-1}{x+2}$ have a horizontal asymptote? If so, what is it?

(b) Does the graph of $f(x) = \dfrac{(x-2)(3x-1)}{x+1}$ have a horizontal asymptote? If so, what is it?

(c) Does the graph of $f(x) = \dfrac{(x-2)(3x-1)}{(x+1)(x-3)}$ have a horizontal asymptote? If so, what is it?

(d) Does the graph of $f(x) = \dfrac{x-2}{x(x+1)}$ have a horizontal asymptote? If so, what is it?

(e) Suppose $f(x) = g(x)/h(x)$, where $g(x)$ and $h(x)$ are non-constant polynomials. What must be true about $g(x)$ and $h(x)$ if the graph of f has a horizontal asymptote?

15.2 Rational Function Equations and Inequalities

In this section, we continue our work with rational functions by solving equations and inequalities involving rational functions.

Problems

Problem 15.8: Find all values of x such that $\dfrac{x}{x-1} + \dfrac{3x-2}{x-3} = \dfrac{2-5x}{4}$.

Problem 15.9: In this problem, we find all solutions to the equation

$$8x + 12 + \frac{5x^3 + 13x^2 + 6x}{x^2 + x} = 2x + \frac{x^3 - 3x + 2}{x^2 - 1}.$$

(a) Can you simplify either side of the equation before getting rid of the fractions?

(b) Solve the equation and check that your solutions satisfy the original equation.

Problem 15.10: In this problem, we solve the inequality $\frac{r-3}{r+4} + 2 \geq \frac{5-r}{r+1}$.

(a) What's wrong with simply multiplying the inequality by $(r+4)(r+1)$ and solving the resulting inequality?

(b) Move all the terms to the left side and write that side with a common denominator.

(c) Factor the numerator and denominator of the fraction you formed in part (b). Use the result to solve the inequality.

Problem 15.11: There is a unique real value A such that for all x with $1 < x < 3$ and $x \neq 2$,

$$\left| \frac{A}{x^2 - x - 2} + \frac{1}{x^2 - 6x + 8} \right| < 1999.$$

Compute A. (Source: Duke Math Meet)

Problem 15.8: Find all values of x such that $\frac{x}{x-1} + \frac{3x-2}{x-3} = \frac{2-5x}{4}$.

Solution for Problem 15.8: We first form a common denominator on the left, which gives us

$$\frac{x}{x-1} + \frac{3x-2}{x-3} = \frac{x(x-3) + (3x-2)(x-1)}{(x-1)(x-3)} = \frac{4x^2 - 8x + 2}{(x-1)(x-3)},$$

so the equation is $\frac{4x^2 - 8x + 2}{(x-1)(x-3)} = \frac{2-5x}{4}$. Multiplying both sides by $4(x-1)(x-3)$ gets rid of the fractions and gives us

$$4(4x^2 - 8x + 2) = (2 - 5x)(x - 1)(x - 3).$$

Next, we expand the product on left to give $16x^2 - 32x + 8$, and we expand the product on the right to give

$$(2-5x)(x-1)(x-3) = (2-5x)(x^2 - 4x + 3) = 2(x^2 - 4x + 3) - 5x(x^2 - 4x + 3) = -5x^3 + 22x^2 - 23x + 6.$$

So, our equation now is

$$16x^2 - 32x + 8 = -5x^3 + 22x^2 - 23x + 6.$$

Rearranging this equation gives $5x^3 - 6x^2 - 9x + 2 = 0$. We first try $x = 1$ and $x = -1$, hoping to find a root quickly. We see that $x = -1$ solves the equation, so $x + 1$ is a factor of the left side. Factoring this out gives $(x + 1)(5x^2 - 11x + 2) = 0$. Factoring this quadratic gives us $(x + 1)(5x - 1)(x - 2) = 0$, so our solutions are $x = -1$, $x = 1/5$, and $x = 2$. Testing each solution, we find that all three satisfy the original equation. $\square$

There are many other ways we could have solved Problem 15.8 that essentially amount to the same solution as above. For example, we could have multiplied by $4(x - 1)(x - 3)$ as our very first step to get rid of the fractions, rather than first finding the common denominator on the left side. We also could have moved all the terms to one side initially, and then found a common denominator on that side.

All these possible solutions have one common general strategy:

> **Concept:** When faced with a new type of equation, focus on what makes the equation difficult to solve. Try to remove this difficulty and turn the equation into one you already know how to handle.

For equations involving rational functions, the difficulty is typically the variables in the denominator. We usually take care of this the way we did in Problem 15.8, by multiplying the equation by the denominators to get rid of the fractions. However, we must be careful to check our answers at the end. Our next problem shows why.

Problem 15.9: Find all solutions to the equation $8x + 12 + \dfrac{5x^3 + 13x^2 + 6x}{x^2 + x} = 2x + \dfrac{x^3 - 3x + 2}{x^2 - 1}$.

Solution for Problem 15.9: We start by simplifying the fractions. We can factor x out of the numerator and denominator of $(5x^3 + 13x^2 + 6x)/(x^2 + x)$, then cancel these common factors to leave

$$8x + 12 + \frac{5x^2 + 13x + 6}{x + 1} = 2x + \frac{x^3 - 3x + 2}{x^2 - 1}.$$

> **Concept:** Don't make problems harder than they are! When solving equations involving rational functions, eliminate any common factors in the numerator and denominator of each rational function.

We can't simplify the fraction on the left any further, but we quickly see that $x - 1$ divides the numerator and denominator of the fraction on the right (since both the numerator and denominator equal 0 when $x = 1$). We have

$$\frac{x^3 - 3x + 2}{x^2 - 1} = \frac{(x - 1)(x^2 + x - 2)}{(x - 1)(x + 1)} = \frac{x^2 + x - 2}{x + 1},$$

and our equation is now

$$8x + 12 + \frac{5x^2 + 13x + 6}{x + 1} = 2x + \frac{x^2 + x - 2}{x + 1}.$$

Before multiplying by $x + 1$ to get rid of the fractions, we rearrange the equation to combine the fractions. We then have

$$\frac{4x^2 + 12x + 8}{x + 1} = -6x - 12.$$

Cross-multiplying gives us $4x^2 + 12x + 8 = -6x^2 - 18x - 12$, so we have $10x^2 + 30x + 20 = 0$. Solving this equation gives us $x = -1$ and $x = -2$ as solutions. However, $x = -1$ makes denominators in our original equation equal to 0, so we must discard it as extraneous.

> **WARNING!!** When you solve an equation involving rational functions, you must
> ☢ check that your solutions do not make any denominator equal to 0.

The only solution to our equation is $x = -2$. □

When we turn from equations to inequalities, we have to be more careful how we handle the complications caused by rational functions.

Problem 15.10: Find all values of r that satisfy $\dfrac{r-3}{r+4} + 2 \geq \dfrac{5-r}{r+1}$.

Solution for Problem 15.10: What's wrong with this approach:

> **Bogus Solution:** We multiply both sides by $(r + 1)(r + 4)$ to get rid of the fractions.
> ▯ On the left, this gives us
>
> $$(r-3)(r+1) + 2(r+1)(r+4) = [(r-3) + 2(r+4)](r+1) = 3r^2 + 8r + 5,$$
>
> and on the right we have
>
> $$(5-r)(r+4) = -r^2 + r + 20,$$
>
> so our inequality is now $3r^2 + 8r + 5 \geq -r^2 + r + 20$. Rearranging this inequality gives $4r^2 + 7r - 15 \geq 0$, so we have $(r+3)(4r-5) \geq 0$, which gives us $r \in (-\infty, -3] \cup [\frac{5}{4}, +\infty)$.

All the steps look OK, but if we look at our solution closely, we see one problem right away. Our final answer suggests that $r = -4$ satisfies the inequality, but it clearly doesn't, since it makes the left side undefined. A little more experimenting will also reveal that $r = -2$ satisfies the inequality, but isn't included in our answer. Uh-oh. Something went wrong, but what?

The problem comes when we multiply the inequality by $(r + 1)(r + 4)$. If this expression is negative, then we have to reverse the direction of the inequality.

> **WARNING!!** Be careful when multiplying an inequality by a variable expression.
> ☢ If the expression is sometimes negative and sometimes positive, then
> you can't just multiply the inequality by the expression and proceed
> as usual.

We can avoid multiplying the inequality by a variable expression by moving all the terms to one side and finding a common denominator. Moving all the terms of our inequality to the left gives us a left side of

$$\frac{r-3}{r+4} + 2 - \frac{5-r}{r+1} = \frac{(r-3)(r+1)}{(r+1)(r+4)} + \frac{2(r+1)(r+4)}{(r+1)(r+4)} - \frac{(5-r)(r+4)}{(r+1)(r+4)}$$

$$= \frac{(r^2 - 2r - 3) + (2r^2 + 10r + 8) - (-r^2 + r + 20)}{(r+1)(r+4)}$$

$$= \frac{4r^2 + 7r - 15}{(r+1)(r+4)}$$

$$= \frac{(r+3)(4r-5)}{(r+1)(r+4)}.$$

Our inequality is now

$$\frac{(r+3)(4r-5)}{(r+1)(r+4)} \geq 0.$$

To determine when the expression on the left is positive, we have to analyze the values of r for which each factor is positive. This is just like our previous work with quadratic inequalities in Section 4.5 and polynomial inequalities in Sections 7.2

	$r+4$	$r+3$	$r+1$	$4r-5$	$f(r)$
$r > \frac{5}{4}$	+	+	+	+	+
$-1 < r < \frac{5}{4}$	+	+	+	−	−
$-3 < r < -1$	+	+	−	−	+
$-4 < r < -3$	+	−	−	−	−
$r < -4$	−	−	−	−	+

and 7.3. We make a table as shown, letting $f(r)$ be the expression on the greater side of our inequality. We see that $f(r)$ is positive for $r \in (-\infty, -4) \cup (-3, -1) \cup (\frac{5}{4}, +\infty)$. However, this isn't the whole story!

> **WARNING!!** When solving an inequality, we have to pay special attention to whether or not the inequality is strict. If the inequality is nonstrict, we must include those values of the variable for which the two sides of the inequality are equal.

Here, the inequality is indeed nonstrict, so we must include those values of r such that $f(r) = 0$. This adds $r = -3$ and $r = \frac{5}{4}$ to the solutions we already found, and makes our final answer

$$r \in (-\infty, -4) \cup [-3, -1) \cup [\tfrac{5}{4}, +\infty).$$

□

> **Problem 15.11:** There is a unique real value A such that for all x with $1 < x < 3$ and $x \neq 2$,
>
> $$\left| \frac{A}{x^2 - x - 2} + \frac{1}{x^2 - 6x + 8} \right| < 1999.$$
>
> Compute A. *(Source: Duke Math Meet)*

Solution for Problem 15.11: We start by letting $f(x) = \frac{A}{x^2-x-2} + \frac{1}{x^2-6x+8}$ and writing $f(x)$ with a common denominator:

$$f(x) = \frac{A}{(x-2)(x+1)} + \frac{1}{(x-2)(x-4)} = \frac{A(x-4) + (x+1)}{(x-2)(x+1)(x-4)}.$$

So, our inequality now is

$$\left| \frac{A(x-4)+(x+1)}{(x-2)(x+1)(x-4)} \right| < 1999.$$

Trying to solve for A in terms of x, or for x in terms of A, looks pretty difficult. It's also not clear how doing either of these would help. Instead, we stop and think about the expression on the left side.

> **Concept:** Don't just blindly bash equations and inequalities with algebra when you're stuck on a problem. Take a little time to think about properties of the expressions in the problem; sometimes a little qualitative understanding will go a long way. Thinking about graphs is a particularly good way to look for useful properties of complicated expressions.

Considering the graph of $f(x)$ for values of x such that $1 < x < 3$, it appears that $f(x)$ has a vertical asymptote at $x = 2$ as a result of the factor $x - 2$ in the denominator. If this is the case, then $|f(x)|$ grows without bound as x gets close to 2. Specifically, if $x = 2$ is a vertical asymptote of the graph of f, then there are values of x such that $1 < x < 3$ and $|f(x)| > 1999$.

So, we must choose A such that $f(x)$ does not have a vertical asymptote at $x = 2$. The only way that $f(x)$ does not have a vertical asymptote at $x = 2$ is if the numerator of $f(x)$ is also 0 for $x = 2$. Then, $f(x)$ would have a hole in its graph at $x = 2$ rather than a vertical asymptote. Therefore, we must have

$$A(2-4)+(2+1) = 0,$$

which gives us $A = 3/2$. When $A = 3/2$ and $x \neq 2$, the left side of our inequality simplifies to

$$\left| \frac{\frac{3}{2}(x-4)+(x+1)}{(x-2)(x+1)(x-4)} \right| = \left| \frac{\frac{5}{2}x-5}{(x-2)(x+1)(x-4)} \right| = \left| \frac{\frac{5}{2}(x-2)}{(x-2)(x+1)(x-4)} \right| = \left| \frac{5}{2(x+1)(x-4)} \right|.$$

To confirm that this final expression is less than 1999 for all x between 1 and 3, we note that the expression is maximized when $2(x+1)(x-4)$ is as close as possible to 0. Completing the square, we have $2(x+1)(x-4) = 2(x-\frac{3}{2})^2 - \frac{25}{2}$. The graph of $g(x) = 2(x-\frac{3}{2})^2 - \frac{25}{2}$ is an upward-opening parabola with vertex $(\frac{3}{2}, -\frac{25}{2})$. For $1 \leq x \leq 3$, the graph reaches its highest point to the left of the vertex at $(1, g(1)) = (1, -12)$, and it reaches its highest point to the right of the vertex at $(3, g(3)) = (3, -8)$. Therefore, the closest $g(x)$ gets to 0 is -8, which means that $f(x)$ is no greater than $5/8$ for $1 \leq x \leq 3$ when $A = \frac{3}{2}$. $\square$

 Exercises

15.2.1 Find all values of t such that $\dfrac{1}{2t-3} + \dfrac{t}{3t-1} = \dfrac{7}{5}$.

15.2.2 Find all r such that $\dfrac{3r^2-12r}{r^2-3r-4} = \dfrac{2r-3}{r+7} - 1$.

15.2.3 Find all solutions to the equation $\dfrac{x-12}{x^2+3x} - \dfrac{2}{x} = \dfrac{5}{x+3}$.

15.2.4 Find all real values of x such that $\dfrac{x^2-9}{x^2-25} > 0$.

15.2.5 Find all real values of x that satisfy $\dfrac{1}{x+1} + \dfrac{6}{x+5} \geq 1$. *(Source: ARML)*

15.3   Even and Odd Functions

A function f is called an **even function** if $f(x) = f(-x)$ for all x in the domain of f, and it is called an **odd function** if $f(x) = -f(-x)$ for all x in the domain of f. As we'll see, these names are much less arbitrary than they seem!

Problems

Problem 15.12: For each of the following functions, state if the function is even, odd, or neither.

(a) $f(x) = x$

(b) $f(x) = 3^x$

(c) $f(x) = \dfrac{1}{x^4 - 2}$

(d) $f(x) = 5^{x^2}$

(e) $f(x) = \sqrt[3]{x}$

(f) $f(x) = \sqrt[3]{x^2}$

Problem 15.13:

(a)   For each of the following polynomials, determine if the polynomial is even, odd, or neither:

$$p(x) = x^3 - x,$$
$$q(x) = x^4 - 3x^3 + 2x,$$
$$r(x) = 2x^5 - 3x - 3,$$
$$s(x) = 6x^{42} + x^6 - 2x^2 + 1.$$

(b)   Suppose $f(x)$ is a polynomial such that f is an even function. Is it possible for the coefficient of x to be nonzero?

(c)   Suppose $f(x)$ is a polynomial such that f is an odd function. Is it possible for the coefficient of x^4 to be nonzero?

(d)   Explain why the name "even" is appropriate for functions f that satisfy $f(x) = f(-x)$ for all x in their domains. What about the name "odd"?

Problem 15.14: Is it possible to tell just by looking at the graph of a function whether the function is even, odd, or neither? If so, how can you tell? If not, why is it not possible?

Problem 15.15: Let f and g be even functions that are defined for all real numbers.

(a)   Must $f + g$ be even?

(b)   Must $f \cdot g$ be even?

Problem 15.16: Let f be a function with all real numbers as its domain.

(a)   Show that the function $g(x) = f(x) + f(-x)$ is even.

(b)   Show that the function $h(x) = f(x) - f(-x)$ is odd.

(c)   Show that every function with domain $\mathbb{R}$ can be written as the sum of an even function and an odd function. For example, if $f(x) = x^2 + x + 3\sqrt[3]{x} - 9$, we can write $f(x) = (x^2 - 9) + \left(x + 3\sqrt[3]{x}\right)$.

Problem 15.12: For each of the following functions, state if the function is even, odd, or neither.

(a) $f(x) = x$

(b) $f(x) = 3^x$

(c) $f(x) = \dfrac{1}{x^4 - 2}$

(d) $f(x) = 5^{x^2}$

(e) $f(x) = \sqrt[3]{x}$

(f) $f(x) = \sqrt[3]{x^2}$

Solution for Problem 15.12:

(a) We have $f(x) = x$ and $f(-x) = -x$, so $f(x) = -f(-x)$, which means $f(x) = x$ is an odd function.

(b) We have $f(x) = 3^x$ and $f(-x) = 3^{-x} = 1/3^x$. For instance, we have $f(1) = 3$ and $f(-1) = 1/3$, so it is not true that $f(x) = f(-x)$ or $f(x) = -f(-x)$ for all x.

(c) We have $f(x) = \dfrac{1}{x^4 - 2}$ and $f(-x) = \dfrac{1}{(-x)^4 - 2} = \dfrac{1}{x^4 - 2}$, so $f(x) = f(-x)$ for all x. Therefore, f is an even function.

(d) We have $f(-x) = 5^{(-x)^2} = 5^{x^2} = f(x)$ for all x, so f is an even function.

(e) We have $f(-x) = \sqrt[3]{-x} = -\sqrt[3]{x} = -f(x)$ for all x, so $f(x) = \sqrt[3]{x}$ is an odd function.

(f) We have $f(-x) = \sqrt[3]{(-x)^2} = \sqrt[3]{x^2} = f(x)$ for all x, so $f(x) = \sqrt[3]{x^2}$ is an even function.

□

Problem 15.13:

(a) For each of the following polynomials, determine if the polynomial is even, odd, or neither:

$$p(x) = x^3 - x,$$
$$q(x) = x^4 - 3x^3 + 2x,$$
$$r(x) = 2x^5 - 3x - 3,$$
$$s(x) = 6x^{42} + x^6 - 2x^2 + 1.$$

(b) Suppose $f(x)$ is a polynomial such that f is an even function. Is it possible for the coefficient of x to be nonzero?

(c) Suppose $f(x)$ is a polynomial such that f is an odd function. Is it possible for the coefficient of x^4 to be nonzero?

(d) Explain why the name "even" is appropriate for functions f that satisfy $f(x) = f(-x)$ for all x in their domains. What about the name "odd"?

Solution for Problem 15.13:

(a) We have $p(-x) = (-x)^3 - (-x) = -x^3 + x = -p(x)$, so $p(x)$ is odd.

We have $q(-x) = (-x)^4 - 3(-x)^3 + 2(-x) = x^4 + 3x^3 - 2x$, which is not the same as $q(x)$ or $-q(x)$. Therefore, $q(x)$ is neither even nor odd.

We have $r(-x) = 2(-x)^5 - 3(-x) - 3 = -2x^5 + 3x - 3$, which is not the same as $r(x)$ or $-r(x)$. Therefore, $r(x)$ is neither even nor odd.

We have $s(-x) = 6(-x)^{42} + (-x)^6 - 2(-x)^2 + 1 = 6x^{42} + x^6 - 2x^2 + 1 = s(x)$, so $s(x)$ is even.

For each of the following parts, let $f(x) = a_n x^n + a_{n-1} x^{n-1} + \cdots + a_1 x + a_0$.

(b) Two polynomials are equal for all values of x if and only if all their corresponding coefficients are the same. If f is an even function, then we must have $f(x) = f(-x)$ for all x, which means that the coefficients of $f(x)$ must be the same as the coefficients of $f(-x)$. However, the coefficient of x in $f(-x)$ is $-a_1$, while the coefficient of x in $f(x)$ is a_1. Therefore, we must have $-a_1 = a_1$, which means $a_1 = 0$. We conclude that if polynomial $f(x)$ is an even function, then the coefficient of x must be 0.

(c) If f is odd, then $f(x) = -f(-x)$, which means the coefficients of $-f(-x)$ match those of $f(x)$. The coefficient of x^4 in $f(-x)$ is a_4, so the coefficient of x^4 in $-f(-x)$ is $-a_4$. This must equal the coefficient of x^4 in $f(x)$, which is a_4. So, we have $a_4 = -a_4$, which means $a_4 = 0$. This tells us that if polynomial $f(x)$ is an odd function, then the coefficient of x^4 must be 0.

(d) The first two parts guide us. We let

$$f(x) = a_n x^n + a_{n-1} x^{n-1} + \cdots + a_3 x^3 + a_2 x^2 + a_1 x + a_0.$$

Then, we have

$$f(-x) = a_n(-1)^n x^n + a_{n-1}(-1)^{n-1} x^{n-1} + \cdots - a_3 x^3 + a_2 x^2 - a_1 x + a_0.$$

If f is even, then these two polynomials must be the same, which means that their corresponding coefficients must match. However, the coefficients of corresponding terms with odd powers of x are opposites. Since we can only have $a_i = -a_i$ if $a_i = 0$, we have $f(x) = f(-x)$ only if $a_i = 0$ for all odd i.

The coefficients of the even powers of x always match in $f(x)$ and $f(-x)$, so a polynomial $f(x)$ that only has even powers of x satisfies $f(x) = f(-x)$ for all x. Therefore, a polynomial $f(x)$ is an even function if and only if it only has even powers of x. Hence the name "even function."

We now have a pretty good idea now where the name "odd function" comes from. We have

$$-f(-x) = a_n(-1)^{n+1} x^n + a_{n-1}(-1)^n x^{n-1} + \cdots + a_3 x^3 - a_2 x^2 + a_1 x - a_0.$$

Therefore, we have $f(x) = -f(-x)$ only if $a_i = -a_i$ when i is even. This tells us that $a_i = 0$ for all even i. The coefficients of the odd powers of x always match in $f(x)$ and $-f(-x)$, so a polynomial $f(x)$ that only has odd powers of x satisfies $f(x) = -f(-x)$ for all x. Therefore, a polynomial $f(x)$ is an odd function if and only if it only has odd powers of x. Hence the name "odd function."

□

Important: A polynomial function $f(x)$ is an even function if and only if its only nonzero coefficients are coefficients of even powers of x, and $f(x)$ is an odd function if and only if its only nonzero coefficients are coefficients of odd powers of x.

So, we can tell at a glance if a polynomial is even or odd. Can we tell by a glance at a function's graph if it is even or odd?

Problem 15.14: Is it possible to tell just by looking at the graph of a function whether the function is even, odd, or neither? If so, how can you tell? If not, why is it not possible?

Solution for Problem 15.14: If a function f is even, then $f(x) = f(-x)$. Suppose (a, b) is on the graph of $y = f(x)$, so we have $b = f(a)$. Because f is even, we have $f(-a) = f(a) = b$, which means the point $(-a, b)$ must also be on the graph of $y = f(x)$. The point $(-a, b)$ is the reflection of the point (a, b) over the y-axis. Therefore, for each point on the graph of an even function, the reflection of that point over the y-axis must also be on the graph of the function. This means that the graph of an even function must have the y-axis as a line of symmetry, as shown in the graph at left below.

Conversely, suppose the graph of a function f is symmetric about the y-axis. This means that if (a, b) is on the graph, then $(-a, b)$ must be as well, which means $f(a) = f(-a)$. Therefore, the graph of any function that is symmetric about the y-axis must be the graph of an even function.

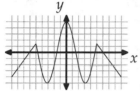

Figure 15.1: Graph of an Even Function

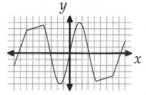

Figure 15.2: Graph of an Odd Function

If a function f is odd, then $f(x) = -f(-x)$. Suppose the point (a, b) is on the graph of $y = f(x)$, so we have $b = f(a)$, as before. Since $f(x) = -f(-x)$, we have $f(-a) = -f(-(-a)) = -f(a) = -b$, so the point $(-a, -b)$ is on the graph of f. The point $(-a, -b)$ is the image of (a, b) upon a $180°$ rotation about the origin. We say that two such points are "symmetric about the origin." So, for each point on the graph of an odd function, the rotation of that point $180°$ about the origin is also on the graph of the function. This means that the graph of an odd function must be symmetric about the origin, as shown at right above.

Conversely, suppose the graph of a function f is symmetric about the origin. This means that if (a, b) is on the graph, then $(-a, -b)$ must be as well, which means that $f(a) = -f(-a)$. So, the graph of any function that is symmetric about the origin must be the graph of an odd function. $\square$

> **Important:** A function is even if and only if its graph is symmetric about the y-axis, and a function is odd if and only if its graph is symmetric about the origin.

Problem 15.15: Let f and g be even functions that are defined for all real numbers.

(a) Must $f + g$ be even?

(b) Must $f \cdot g$ be even?

Solution for Problem 15.15:

(a) Because f and g are even, we have $f(x) = f(-x)$ and $g(x) = g(-x)$, so we have $f(-x) + g(-x) = f(x) + g(x)$. So, if $h = f + g$, we have $h(x) = h(-x)$, which means h is even.

(b) We have $f(-x) \cdot g(-x) = f(x) \cdot g(x)$ for all x, so $f \cdot g$ is even.

$\square$

In the Exercises, you'll have a chance to see what happens to $f + g$ and $f \cdot g$ if f and/or g is odd instead of even.

We'll finish our exploration with a surprising fact about functions.

Problem 15.16: Show that every function with domain $\mathbb{R}$ can be written as the sum of an even function and an odd function.

Solution for Problem 15.16: We can obviously write any polynomial as the sum of an even and an odd function, since we can break any polynomial $f(x)$ into two polynomials such that one has all the even powers of x in $f(x)$ and the other has all the odd powers. For example, if $f(x) = x^4 - 3x^3 + 7x - 13$, we have

$$f(x) = (x^4 - 13) + (-3x^3 + 7x).$$

But what about functions like $g(x) = (x^4 - x + 9)/(x^2 + 5)$ or $h(x) = |x^2 - x|$?

One way to show that any function g can be written as the sum of an even function and an odd function is to find those two functions. But g can be just about anything! How can we possibly find the even and odd functions into which we can split g? The only thing we have to work with is g, so we'll have to find an even function and an odd function *in terms of g*.

We look back to our polynomial $f(x) = x^4 - 3x^3 + 7x - 13$ for inspiration.

Concept: When trying to prove a general fact, consider specific examples. You may be able to use them as inspiration for a proof.

We know we can split $f(x)$ into $x^4 - 13$ and $-3x^3 + 7x$. We hope to find a way to write these both in terms of f. Let's focus on the even piece first. We get the even powers of x when we write $f(x)$, but we also get the odd powers of x. We'd like to cancel those odd powers out. We know how to do that! The coefficient of each odd power of x in $f(-x)$ is the opposite of the corresponding odd power of x in $f(x)$. So, $f(x) + f(-x)$ will not have any odd powers of x:

$$f(x) + f(-x) = (x^4 - 3x^3 + 7x - 13) + (x^4 + 3x^3 - 7x - 13) = 2(x^4 - 13).$$

Therefore, $f(x) + f(-x)$ equals twice the "even part" of $f(x)$. So, we can write the even part of $f(x)$ as $(f(x) + f(-x))/2$.

Now that we've found the even part, the path to the "odd part" is clear. If we want to eliminate the even powers of x in $f(x)$, we subtract $f(-x)$ instead of adding it:

$$f(x) - f(-x) = (x^4 - 3x^3 + 7x - 13) - (x^4 + 3x^3 - 7x - 13) = 2(-3x^3 + 7x).$$

So, the odd part of $f(x)$ is just $(f(x) - f(-x))/2$.

Well, that worked for our specific $f(x)$. Will it work for any function $g(x)$? Specifically, can we write $g(x)$ as the sum of an even function and an odd function by writing it as the sum of $(g(x) + g(-x))/2$ and $(g(x) - g(-x))/2$? First, we confirm that these pieces do in fact add to $g(x)$:

$$\frac{g(x) + g(-x)}{2} + \frac{g(x) - g(-x)}{2} = g(x).$$

So far, so good. Now, we must confirm that one piece is even and the other is odd. If we let $p(x) = (g(x) + g(-x))/2$, we have

$$p(-x) = \frac{g(-x) + g(x)}{2} = \frac{g(x) + g(-x)}{2} = p(x),$$

so p is even. Similarly, letting $q(x) = (g(x) - g(-x))/2$, we have

$$q(-x) = \frac{g(-x) - g(x)}{2} = -\frac{g(x) - g(-x)}{2} = -q(x),$$

so q is odd. Therefore, we can write any function g as the sum of an even function and an odd function by writing $g(x) = p(x) + q(x)$, where $p(x) = (g(x) + g(-x))/2$ and $q(x) = (g(x) - g(-x))/2$. $\square$

Exercises

15.3.1 State whether each of the following functions is odd, even, or neither:

(a) $f(x) = x^4 + x^6 + x^8 + 2008 - |x|$

(b) $g(x) = 3^{x^2 - x}$

(c) $h(x) = |3x^3 - 7x|$

(d) $f(x) = \sqrt{x^2 - 4}$

15.3.2 Suppose that f and g are both odd functions.

(a) Must $f + g$ be an odd function, an even function, or neither?

(b) Must $f \cdot g$ be odd function, an even function, or neither?

15.3.3 Suppose that $g(x)$ is a polynomial that is an odd function. Find all possible values of the product of the roots of $g(x)$.

15.3.4 Suppose that $f(x) = g(x)/h(x)$ is a rational function, where $g(x)$ and $h(x)$ are distinct polynomials. If f is an even function, must g and h be even functions?

15.4 Monotonic Functions

A function f is a **monotonically increasing** function if $f(a) > f(b)$ whenever $a > b$. Often we will just say the function is "increasing" rather than "monotonically increasing." For example, back in Section 13.1, we noted that exponential functions $f(x) = c^x$, where c is a constant, are monotonically increasing if $c > 1$. Similarly, a function is **monotonically decreasing** if $f(a) < f(b)$ whenever $a > b$, and we will often say that such a function is "decreasing" rather than "monotonically decreasing." The graph of an increasing function goes strictly upward as it goes from left to right, and the graph of a decreasing function goes strictly downward as it goes from left to right.

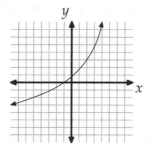

Figure 15.3: An Increasing Function

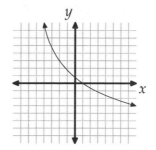

Figure 15.4: A Decreasing Function

Problems

Problem 15.17: In this problem, we show that $f(x) = x^3$ is a monotonically increasing function.

(a) Let c be a positive constant. Find $f(x + c) - f(x)$.

(b) Your answer in part (a) should be a quadratic in x. Show that this quadratic does not have real roots.

(c) Show that $f(x + c) - f(x)$ is positive for all x and all positive c.

(d) Why does part (c) tell us that $f(x)$ is a monotonically increasing function?

Problem 15.18:

(a) Is the sum of two monotonically increasing functions necessarily monotonically increasing? If so, prove it. If not, give an example of two monotonically increasing functions whose sum is not monotonically increasing.

(b) Is the product of two monotonically increasing functions necessarily monotonically increasing? If so, prove it. If not, give an example of two monotonically increasing functions whose product is not monotonically increasing.

Problem 15.19: If a function is monotonically increasing, must the function be invertible? Why or why not? What if a function is monotonically decreasing?

Problem 15.20: Show that if $f(x)$ is a monotonically increasing function, and $f(x)$ is positive for all values of x in the domain of f, then $1/f(x)$ is a monotonically decreasing function.

Problem 15.21:

(a) Show that $f(x) = \sqrt{x}$ is monotonically increasing.

(b) Find the largest possible value of $\sqrt{x + 4} - \sqrt{x}$.

Problem 15.22:

(a) Show that $p(x) = x(15 - x)$ decreases monotonically when $x \in (7.5, +\infty)$.

(b) Which is greater, π^2 or 7.5?

(c) Without using a calculator, determine if $\pi^2(15 - \pi^2)$ is greater than, less than, or equal to 50.

Problem 15.23:

(a) Use the fact that 8^x is an increasing function to show that $\log_8 x$ is monotonically increasing.

(b) Find the smallest possible value of $\log_8(x^3 - 4x^2 + x + 26) - \log_8(x + 2)$.

Problem 15.17:

(a) Suppose that $f(x + c) - f(x) > 0$ for all x and c such that $c > 0$ and both x and $x + c$ are in the domain of f. Explain why f must be monotonically increasing.

(b) Use part (a) to show that $f(x) = x^3$ increases monotonically.

Solution for Problem 15.17:

(a) Adding $f(x)$ to both sides of $f(x+c) - f(x) > 0$ gives $f(x+c) > f(x)$ for $c > 0$. Suppose that a and b are both in the domain of f, and that $b > a$. Because $f(x+c) > f(x)$ for all $c > 0$, letting $x = a$ and $c = b - a$ gives us $f(b) > f(a)$. Therefore, we have $f(b) > f(a)$ for any numbers a and b in the domain of f such that $b > a$. This means that f is monotonically increasing.

> **Important:** If $f(x+c) - f(x)$ is positive for all values of x and c such that $c > 0$ and both $x + c$ and x are in the domain of f, then f is monotonically increasing. Similarly, if $f(x+c) - f(x)$ is negative for all values of x and c such that $c > 0$ and both $x + c$ and x are in the domain of f, then f is monotonically decreasing.

(b) To use part (a), we must show that $f(x+c) - f(x) > 0$ if $c > 0$ and $f(x) = x^3$. We have

$$f(x + c) - f(x) = (x + c)^3 - x^3 = 3cx^2 + 3c^2x + c^3 = c(3x^2 + 3cx + c^2).$$

Because $c > 0$, the expression $c(3x^2 + 3cx + c^2)$ is positive whenever $3x^2 + 3cx + c^2$ is positive.

The graph of $g(x) = 3x^2 + 3cx + c^2$ is an upward-opening parabola. If it never crosses the x-axis, then $g(x)$ must be positive for all x. So, we examine its discriminant to determine if $g(x) = 0$ for any real values of x. The discriminant is

$$(3c)^2 - 4(3)(c^2) = 9c^2 - 12c^2 = -3c^2,$$

which is negative for all real values of c, so it is negative for all positive values of c. Therefore, $3x^2 + 3cx + c^2$ has no real roots. So, we know that $3x^2 + 3cx + c^2$ is positive for all values of x and all positive c, which tells us that $f(x+c) - f(x) = c(3x^2 + 3cx + c^2)$ is positive for all x and all $c > 0$. This means that $f(x) = x^3$ increases monotonically.

☐

> **Problem 15.18:**
> (a) Is the sum of two monotonically increasing functions necessarily monotonically increasing? If so, prove it. If not, give an example of two monotonically increasing functions whose sum is not monotonically increasing.
>
> (b) Is the product of two monotonically increasing functions necessarily monotonically increasing? If so, prove it. If not, give an example of two monotonically increasing functions whose product is not monotonically increasing.

Solution for Problem 15.18:

(a) Yes. Suppose f and g are monotonically increasing, and $h = f + g$. We have

$$h(x + c) - h(x) = f(x + c) + g(x + c) - (f(x) + g(x)) = f(x + c) - f(x) + g(x + c) - g(x).$$

Because f and g are increasing, when $c > 0$ we have $f(x + c) - f(x) > 0$ and $g(x + c) - g(x) > 0$. So, we have

$$h(x + c) - h(x) = f(x + c) - f(x) + g(x + c) - g(x) > 0,$$

which means that h is monotonically increasing.

(b) No. The function $f(x) = x$ is obviously monotonically increasing—as x increases, so does $f(x)$. If $f(x) = x$, $g(x) = x$, and $h(x) = f(x) \cdot g(x) = x^2$, then f and g are monotonically increasing, but h is not. For example, $h(-2) > h(0)$. (There are infinitely many other choices of f and g that we could pick such that f and g are monotonically increasing, but $f \cdot g$ is not.)

$\square$

Back in Section 13.2, we used the fact that an exponential function $f(x) = a^x$ with $a > 1$ is monotonically increasing to prove that the function is invertible. This might make us wonder if every monotonically increasing function is invertible.

Problem 15.19: If a function is monotonically increasing, must the function be invertible? Why or why not? What if a function is monotonically decreasing?

Solution for Problem 15.19: Let $f(x)$ be a monotonically increasing function. Because $f(x)$ is increasing, we know that if $a > b$, then $f(a) > f(b)$. In other words, if two inputs to f are not equal, then their outputs from f cannot be equal. So, f must have an inverse.

If $f(x)$ is monotonically decreasing, we have $f(b) < f(a)$ for all $a > b$. Again, we see that if two inputs to f are not equal, then their outputs from f are not equal, so f has an inverse. $\square$

Important: If a function is either monotonically increasing or monotonically decreasing, then it has an inverse.

Problem 15.20: Show that if $f(x)$ is a monotonically increasing function, and $f(x)$ is positive for all values of x in the domain of f, then $1/f(x)$ is a monotonically decreasing function.

Solution for Problem 15.20: Let $g(x) = \frac{1}{f(x)}$. We wish to show that if $a > b$, then $g(b) > g(a)$.

Because $f(x)$ is monotonically increasing, we have $f(a) > f(b)$ whenever $a > b$. Since $f(x)$ is positive for all x, we can divide our inequality by $f(a)$ and $f(b)$ to give $\frac{1}{f(b)} > \frac{1}{f(a)}$. Because $g(x) = \frac{1}{f(x)}$, we have $g(b) > g(a)$. Therefore, $g(x)$ is monotonically decreasing, as desired. $\square$

Let's put this principle to work on a specific function.

Problem 15.21:
(a) Show that $f(x) = \sqrt{x}$ is monotonically increasing.
(b) Find the largest possible value of $\sqrt{x+4} - \sqrt{x}$.

Solution for Problem 15.21:

(a) We wish to show that if $a > b \geq 0$, then $\sqrt{a} > \sqrt{b}$. (We must have a and b nonnegative so that both are in the domain of f.) What's wrong with this proof:

Bogus Solution: Since $a > b$, taking the square root of both sides gives us $\sqrt{a} > \sqrt{b}$. Therefore, $f(x) = \sqrt{x}$ is monotonically increasing.

This "proof" assumes that if $a > b$, then $\sqrt{a} > \sqrt{b}$. But this is the very fact that we are asked to prove! So, we have to find another way to show that f is increasing.

What's wrong with this proof:

> **Bogus Solution:** If $\sqrt{a} > \sqrt{b}$, then, because both sides are positive, we can square both sides to give $a > b$. Therefore, if $a > b$, then $\sqrt{a} > \sqrt{b}$.

This argument is backwards; we show that if $\sqrt{a} > \sqrt{b}$, then $a > b$. But we can't conclude just from this fact that "If $a > b$, then $\sqrt{a} > \sqrt{b}$."

> **WARNING!!** Let X and Y be mathematical statements. If we show that "If X, then Y," then we cannot conclude that "If Y, then X." Doing so would be like reasoning that the statement "If I live in America, then I live on the Earth," implies that "If I live on the Earth, then I live in America."

We've shown that $a > b$ if $\sqrt{a} > \sqrt{b}$, but we haven't addressed the possibility that $a > b$ if $\sqrt{a} < \sqrt{b}$ or $\sqrt{a} = \sqrt{b}$. So, we investigate these other two possibilities.

If $\sqrt{a} < \sqrt{b}$, then squaring both sides gives $a < b$. So, we can't have $a > b$ if $\sqrt{a} < \sqrt{b}$.

If $\sqrt{a} = \sqrt{b}$, then squaring both sides gives $a = b$, so we can't have $a > b$ if $\sqrt{a} = \sqrt{b}$, either.

So, of the three possibilities $\sqrt{a} > \sqrt{b}$, $\sqrt{a} < \sqrt{b}$, and $\sqrt{a} = \sqrt{b}$, only $\sqrt{a} > \sqrt{b}$ allows $a > b$. Therefore, if $a > b$, then we must also have $\sqrt{a} > \sqrt{b}$. Make sure you see why this solution is valid, but the second Bogus Solution above is not.

(b) Let $g(x) = \sqrt{x + 4} - \sqrt{x}$. We start by experimenting. We evaluate (or approximate) $g(x)$ for various values of x at right. It looks like $g(x)$ is monotonically decreasing. If we can show that it is, then we can conclude that $g(x)$ is maximized when x is minimized, which is when $x = 0$.

x	$g(x)$
0	2
1	1.24
2	1.04
5	0.76
10	0.58

If we start with $g(x + c) - g(x)$, we have

$$g(x + c) - g(x) = \left(\sqrt{x + 4 + c} - \sqrt{x + c}\right) - \left(\sqrt{x + 4} - \sqrt{x}\right),$$

which is quite a mess. Before we dive into that scary expression, we look for a faster way.

Our observation in Problem 15.20 offers us a considerably simpler approach. We note that

$$\frac{1}{g(x)} = \frac{1}{\sqrt{x + 4} - \sqrt{x}} = \frac{1}{\sqrt{x + 4} - \sqrt{x}} \cdot \frac{\sqrt{x + 4} + \sqrt{x}}{\sqrt{x + 4} + \sqrt{x}} = \frac{\sqrt{x + 4} + \sqrt{x}}{4}.$$

Because both $\sqrt{x + 4}$ and $\sqrt{x}$ are monotonically increasing, our result from Problem 15.18 tells us that their sum is monotonically increasing. Therefore, the function $1/(g(x))$ is monotonically increasing. Since $g(x) > 0$ for all x, this means that $g(x)$ is monotonically decreasing. So, the function $g(x)$ is maximized when x is minimized, which means the maximum value of $g(x)$ is $g(0) = 2$.

□

Many functions are neither monotonically increasing nor monotonically decreasing, but they are increasing or decreasing over parts of their domains.

Problem 15.22: Without using a calculator, determine if $\pi^2(15 - \pi^2)$ is greater than, less than, or equal to 50.

Solution for Problem 15.22: Multiplying out $\pi^2(15 - \pi^2)$ doesn't look like much fun. So, we look for different ways to think about this expression. We can view it as $f(\pi^2)$, where $f(x) = x(15 - x)$. This seems much simpler. Moreover, we have $f(5) = f(10) = 50$, so our problem now is to compare $f(\pi^2)$ to $f(5)$ or $f(10)$. The graph of $f(x)$ is a downward-opening parabola. We can see exactly what parabola this is by completing the square:

$$f(x) = 15x - x^2 = -\left(x - \frac{15}{2}\right)^2 + \frac{225}{4}.$$

So, the axis of symmetry of the parabola is $x = 7\frac{1}{2}$. To the right of this line, f is decreasing. Both $x = \pi^2$ and $x = 10$ are to the right of this line, so now we must compare π^2 to 10. Since $\pi^2 \approx 9.87$, we have $\pi^2 < 10$. Because $f(x)$ is strictly decreasing for all $x > 7.5$, we have $f(\pi^2) > f(10) = 50$. Therefore, $\pi^2(15 - \pi^2)$ is greater than 50. □

Note that our invoking the graph of a parabola is not a proof that $f(x)$ is decreasing for $x > 7.5$; it's merely an intuitive explanation. You'll have a chance to prove it formally as an Exercise.

Problem 15.23: Find the smallest possible value of $\log_8(x^3 - 4x^2 + x + 26) - \log_8(x + 2)$.

Solution for Problem 15.23: We first simplify the expression by using a logarithm identity:

$$\log_8(x^3 - 4x^2 + x + 26) - \log_8(x + 2) = \log_8\left(\frac{x^3 - 4x^2 + x + 26}{x + 2}\right).$$

Next, we note that $x + 2$ divides the polynomial evenly and leaves us with

$$\log_8(x^3 - 4x^2 + x + 26) - \log_8(x + 2) = \log_8\left(\frac{x^3 - 4x^2 + x + 26}{x + 2}\right) = \log_8(x^2 - 6x + 13).$$

Now what?

We might think that $\log_8(x^2 - 6x + 13)$ is minimized when $x^2 - 6x + 13$ is minimized, but this is only true if $\log_8 x$ is a monotonically increasing function. It seems intuitively obvious that it is increasing, but let's check to be sure. If $f(x) = \log_8 x$, we have

$$f(x + c) - f(x) = \log_8(x + c) - \log_8 x = \log_8 \frac{x + c}{x} = \log_8\left(1 + \frac{c}{x}\right).$$

Is the power to which we must raise 8 to get $1 + \frac{c}{x}$ positive or negative? We first note that c and x must be positive, so $1 + \frac{c}{x} > 1$. Since $8^0 = 1$ and 8^x is monotonically increasing, we must raise 8 to a positive power to get any number greater than 1. So, $\log_8\left(1 + \frac{c}{x}\right)$ is positive, which means that $f(x) = \log_8 x$ is monotonically increasing.

We now know that $\log_8(x^2 - 6x + 13)$ is minimized when $x^2 - 6x + 13$ is minimized. Since $x^2 - 6x + 13 = (x-3)^2 + 4$, the expression $x^2 - 6x + 13$ is minimized when $x = 3$. So, the minimum value of $\log_8(x^2 - 6x + 13)$ is

$$\log_8(3^2 - 6 \cdot 3 + 13) = \log_8 4 = \frac{2}{3}.$$

As a final check, we let $x = 3$ in our original expression to ensure that both logarithms in the original expression are defined for $x = 3$. They are, so we have indeed found the correct minimum. $\square$

 Exercises

15.4.1 Suppose that $f(x)$ and $g(x)$ are positive for all x. If $f(x)$ and $g(x)$ are both monotonically increasing, then must $f(x) \cdot g(x)$ be monotonically increasing?

15.4.2 Prove that $x(15 - x)$ is a decreasing function for $x > 7.5$.

15.4.3 Find all x such that $8^x(3x + 1) = 4$, and prove that you have found all values of x that satisfy this equation.

15.4.4 Show that $f(x) = x^n$ increases monotonically for any positive odd integer n.

15.4.5 For each of the following functions, find the maximum of the function, and prove that the function cannot equal any greater value.

(a) $f(x) = \sqrt{x + 3} - \sqrt{x - 3}$, for $x \geq 3$.

(b)$\star$ $g(x) = \sqrt[3]{x + 1} - \sqrt[3]{x}$, for $x \geq 0$. **Hints:** 316

15.5 Summary

Any function that can be expressed as one polynomial divided by another is called a **rational function**.

An **asymptote** of a graph is a line that the graph approaches but does not intersect as x or y becomes very far from 0. Determining the asymptotes of the graph of a rational function can be very helpful in graphing the function and understanding its behavior.

> **Important:** Let $p(x)$ and $d(x)$ be polynomials, where $d(x)$ is not 0 for all x. If c is a constant such that $d(c) = 0$ and $p(c) \neq 0$, then the graph of $f(x) = p(x)/d(x)$ has a vertical asymptote at $x = c$.

To find horizontal and slant asymptotes, we divide the denominator of a rational function into the numerator of the function. If the resulting quotient is a constant, c, then $y = c$ is a horizontal asymptote. If the quotient is a linear function with a nonzero linear coefficient, then the graph of the quotient is a slant asymptote of the graph of the rational function.

Solving equations involving rational functions usually involves getting rid of all the denominators by multiplying through by them. But we have to be careful:

> **WARNING!!** When you solve an equation involving rational functions, you must check that your solutions do not make any denominator equal to 0.

When solving inequalities involving rational functions, we usually move all the terms to one side of the inequality, form a single rational function on that side, then compare that rational function to 0. This typically involves factoring the numerator and denominator of the rational function, and analyzing the signs of the factors for different values of the variable.

> **WARNING!!** When solving an inequality, we have to pay special attention to whether or not the inequality is strict. If the inequality is nonstrict, we must include those values of the variable for which the two sides of the inequality are equal.

A function f is called an **even function** if $f(x) = f(-x)$ for all x in the domain of f, and it is called an **odd function** if $f(x) = -f(-x)$ for all x in the domain of f.

> **Important:** A function is even if and only if its graph is symmetric about the y-axis, and a function is odd if and only if its graph is symmetric about the origin.

A function f is a **monotonically increasing** function if $f(a) > f(b)$ whenever $a > b$. Similarly, a function is **monotonically decreasing** if $f(a) < f(b)$ whenever $a > b$. We often exclude the word "monotonically," and simply refer to a function as "increasing" or "decreasing." The graph of an increasing function goes strictly upward as it goes from left to right, and the graph of a decreasing function goes strictly downward as it goes from left to right.

> **Important:** If $f(x+c) - f(x)$ is positive for all values of x and c such that $c > 0$ and both $x + c$ and x are in the domain of f, then f is monotonically increasing. Similarly, if $f(x + c) - f(x)$ is negative for all values of x and c such that $c > 0$ and both $x + c$ and x are in the domain of f, then f is monotonically decreasing.

Problem Solving Strategies

> **Concepts:**
> - Don't limit yourself to an equation or expression you're given. Different algebraic forms of the same equation or expression can offer different information, so experiment with manipulating equations and expressions into different forms.
>
> - When faced with a new type of equation, focus on what makes the equation difficult to solve. Try to remove this difficulty and turn the equation into one you already know how to handle.
>
> - When you're stuck on an algebra problem, take a little time to think about properties of the expressions in the problem; sometimes a little qualitative understanding will go a long way. Thinking about graphs is a particularly good way to look for useful properties of complicated expressions.
>
> - When trying to prove a general fact, consider specific examples. You may be able to use them as inspiration for a proof.

15.24 Sketch graphs of each of the following. Identify the domain, range, and any asymptotes of each function.

(a) $f(x) = \dfrac{2x+1}{3-x}$

(b) $f(x) = \dfrac{x^2 - 4x + 3}{x - 2}$

15.25 Suppose $f(x) = p(x)/q(x)$, where $p(x) = a_n x^n + a_{n-1} x^{n-1} + \cdots + a_0$ and $q(x) = b_n x^n + b_{n-1} x^{n-1} + \cdots + b_0$ are nonzero polynomials. Suppose also that $a_n b_n \neq 0$, and that the graph of $y = f(x)$ has a horizontal asymptote. Find the equation of the horizontal asymptote in terms of the coefficients of $p(x)$ and $q(x)$.

15.26 Which of the four graphs below could be the graph of $f(x) = \dfrac{x^3 - 2x^2 - 15x}{3x^2 + 12x + 12}$? For each that cannot be the graph of f, explain why not.

(a)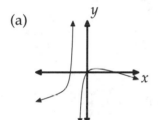

(b)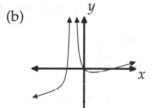

(c)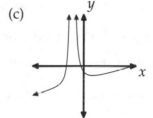

(d)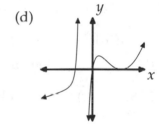

15.27 How are the graphs of $f(x) = \dfrac{x-3}{x-7}$ and $g(x) = \dfrac{x-3}{7-x}$ related?

15.28 For each of the following functions, state whether or not the graph of the function has a slant asymptote. If it does have a slant asymptote, then find it.

(a) $f(x) = \dfrac{2x-1}{x+2}$

(c) $f(x) = \dfrac{(x-2)(3x-1)}{(x+1)(x-3)}$

(e) $f(x) = \dfrac{2x^3 + 5x^2 - 4x - 3}{x^2 + 2x}$

(b) $f(x) = \dfrac{(x-2)(3x-1)}{x+1}$

(d) $f(x) = \dfrac{x-2}{x(x+1)}$

(f) $f(x) = \dfrac{2x^3 + 5x^2 - 4x - 3}{x+2}$

Suppose $f(x) = g(x)/h(x)$, where $g(x)$ and $h(x)$ are polynomials. What must be true about $g(x)$ and $h(x)$ if the graph of f has a slant asymptote?

15.29 Is it possible for the graph of a rational function to have:

(a) Two horizontal asymptotes?

(b) Two vertical asymptotes?

(c) A horizontal and a vertical asymptote?

(d) A horizontal and a slant asymptote?

(e) A slant and a vertical asymptote?

(f) Two slant asymptotes?

15.30 What is the range of the function $f(x) = \dfrac{x^2 - 2x + 3}{x^2 + 2x - 3}$?

15.31 Find all positive real values of x that satisfy $\dfrac{1}{x + \sqrt{x}} + \dfrac{1}{x - \sqrt{x}} \le 1$.

15.32 Find the solutions of the equation $\dfrac{15}{x^2 - 4} - \dfrac{2}{x - 2} = 1$. *(Source: AHSME)*

15.33 Find all x such that $\dfrac{2 - x}{x + 5} - \dfrac{3 + 2x}{1 - x} = 3x + 1$.

15.34 Find the maximum value of the function $f(x) = \dfrac{1}{4x^2 + 1}$.

15.35 Find all real numbers x such that $\dfrac{2x^2 - 16}{x^2 + 3x + 2} < 1$.

15.36 Suppose that $f(x)$ is a polynomial that is an even function. Find all possible values of the sum of the roots of $f(x)$.

15.37 Let f and g be functions such that the domain of g includes the range of f.

(a) If f is even, then must $g \circ f$ be even, odd, or neither?

(b) If f is odd, then must $g \circ f$ be even, odd, or neither?

15.38 If $f(x)$ is even and $g(x)$ is odd, identify each of the following as definitely even, definitely odd, or neither:

(a) $f(-x)$ (c) $f(x) \cdot g(x)$ (e) $f(g(2x))$

(b) $g(x^2)$ (d) $f(x) + g(x)$

15.39 Let $g(x) = |f(x)|$.

(a) If f is even, must g be an even function?

(b) If f is odd, must g be an even function, an odd function, or not necessarily either?

(c) If g is even, then must it be true that f is either even or odd?

15.40 Suppose that $f(x)$ is nonzero and monotonically increasing. Must $1/f(x)$ be monotonically decreasing? (Make sure you see why this is a different question from Problem 15.20.)

15.41 Show that $\sqrt[3]{x}$ is a monotonically increasing function.

15.42 Suppose a and b are constants greater than 1. How many solutions x are there to the equation $(\log_a x)(\log_b x) = 2007$? Prove your answer is correct.

15.43 Show that the function $f(x) = \dfrac{x}{x + 10^7}$ is increasing for all positive x.

15.44 Without using a calculator, order the following from greatest to least:

$$10 - 3\sqrt{11},\ 7 - 4\sqrt{3},\ 5\sqrt{41} - 32,\ 9 - 4\sqrt{5}.$$

15.45   Let a, b, and c be positive constants. Show that $x = 0$ is the only value of x such that

$$\sqrt{a + bx} + \sqrt{b + cx} + \sqrt{c + ax} = \sqrt{b - ax} + \sqrt{c - bx} + \sqrt{a - cx}.$$

15.46   Solve the inequality $\dfrac{x + \sqrt{x} + 4}{x - 1} < 2$. **Hints:** 306, 15

15.47   Solve the equation $\sqrt[3]{x + 2} + \sqrt[3]{x + 3} + \sqrt[3]{x + 4} = 0$. *(Source: Finland)* **Hints:** 87

15.48   Find all values of the constant A such that the equation $\dfrac{3x}{x + 2} + 2 = \dfrac{2}{Ax + 1}$ has exactly one solution.

15.49★   Find the minimum possible value of $f(x) = \log_{10}(x^3 - 5x^2 + 11x - 15) - \log_{10}(x - 3)$, or show that the function has no minimum. **Hints:** 258, 25

15.50★   Find the largest positive value attained by the function $f(x) = \sqrt{8x - x^2} - \sqrt{14x - x^2 - 48}$, where x is a real number. *(Source: AHSME)* **Hints:** 288, 45

15.51★   Find the number of real solutions to the equation $567^x + 678^x = 789^x$. **Hints:** 298

Extra!
▶▶▶▶   The **Nobel Prize** is awarded in chemistry, economics, medicine, and physics, but not mathematics. Shortly after the Nobel Prize was established in the late 1800s, Norwegian mathematician Sophus Lie began advocating that the Nobel Prize's lack of a mathematics prize be remedied by establishing an **Abel Prize**, named after the great mathematician Niels Abel. Over a century later, Lie's vision was realized, as the Norwegian government announced in August, 2001, the formation of an Abel Fund worth 200 million Norwegian krone (which is worth around 40 million dollars now) to fund an international Abel Prize.

The first Abel Prize was awarded in 2003 to Jean-Pierre Serre. He was awarded 6,000,000 Norwegian krone (worth around $850,000 at the time). In 2008, the prize was awarded to John Griggs Thompson and Jacques Tits, who shared the 6,000,000 Norwegian krone (worth around $1,200,000 now).

Another major prize given to mathematicians is the **Fields Medal**. The Fields Medal is awarded every four years to two, three, or four mathematicians not more than 40 years old. Whereas the Abel Prize has been used mainly to reward a lifetime of achievement in mathematics, the Fields Medal is designed to recognize and encourage young mathematicians.

There are even special prizes for specific problems. In 2000, the **Clay Mathematics Institute** announced it would award $1,000,000 for solutions to each of seven **Millennium Problems**. So far, only one of the seven problems has been solved, so there are still prizes to be claimed!

For more information about these prizes, and about the Millennium Problems, visit the pages linked from our links page cited on page vi.

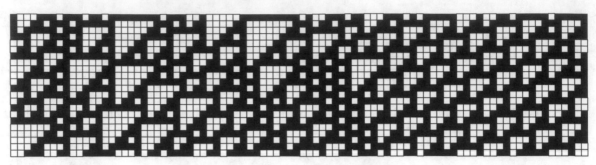

Always watch where you are going. Otherwise, you may step on a piece of the forest that was left out by mistake.
– Winnie the Pooh

CHAPTER **16**

Piecewise Defined Functions

Most of the functions we have studied so far have a single rule that is applied to every valid input. For example, the function $f(x) = 2x$ doubles any input. In this chapter, we look at functions that apply different rules to different inputs, as well as some functions that require special notation to denote the rules they represent.

While we'll see several new types of problems in this chapter, our key strategy for solving them is not new:

> **Concept:** When working with an unusual expression in a problem, try to find ways to convert the problem into one involving expressions you know how to handle.

16.1 Introduction to Piecewise Defined Functions

We've described functions as machines that take an input and produce an output according to some rule. We could also imagine a machine that applies different rules to different inputs in order to produce an output. We call such a function a **piecewise defined function**, and to describe such a function, we must both give the rules used to produce outputs, and describe when to apply each rule. Here is an example:

$$f(x) = \begin{cases} 2x & \text{if } x < 0, \\ 3x & \text{if } x > 0. \end{cases}$$

This notation describes a function that doubles negative inputs and triples positive inputs. Notice that $f(x)$ is not defined for $x = 0$, so 0 is not in the domain of f.

Problems

Problem 16.1: Let g be the function defined for all real numbers such that

$$g(x) = \begin{cases} 2x & \text{if } x < 2, \\ 2^x & \text{if } x \geq 2. \end{cases}$$

(a) Compute $g(0)$, $g(1)$, $g(2)$, $g(3)$, and $g(4)$.

(b) Graph $g(x)$.

(c) Is g an even function, an odd function, or neither?

Problem 16.2: Let

$$f(x) = \begin{cases} x + 5 & \text{if } -9 \leq x \leq -3, \\ 2x^2 - 1 & \text{if } -3 < x \leq 2, \\ -x/2 & \text{if } x > 2. \end{cases}$$

(a) What are the domain and range of f?

(b) For what values of x is $f(x) = x$?

Problem 16.3: Loosely speaking, we say that a function is **continuous** if we can draw the graph of the function between any two points without lifting our pencil from the paper. If a function is not continuous, we say it is **discontinuous**. Determine which of the following functions are continuous and which are discontinuous.

(a) $f(x) = \begin{cases} x + 5 & \text{if } -9 \leq x \leq -3, \\ 2x^2 - 1 & \text{if } -3 < x \leq 2, \\ -x/2 & \text{if } x > 2. \end{cases}$

(b) $g(x) = \begin{cases} 2x & \text{if } x < 2, \\ 2^x & \text{if } x \geq 2. \end{cases}$

(c) $h(x) = \begin{cases} 2x & \text{if } x < 2, \\ 2^x & \text{if } x > 2. \end{cases}$

Problem 16.4: Each week, Alex and Barbara get paid based on how many hours each works. Alex receives $30 per hour for each of his first 40 hours. He receives $50 per hour for each hour after his first 40 hours. (He receives the respective fraction of these amounts if he works some fraction of an hour.) Barbara receives $22 per hour for each of her first 20 hours, then $40 per hour for each of her next 10 hours, then $60 for each hour after that.

Let the function $A(x)$ represent how much Alex gets paid if he works x hours, and let $B(x)$ represent how much Barbara gets paid if she works x hours. Write $A(x)$ and $B(x)$ as piecewise defined functions in order to answer the following questions.

(a) How many hours must Alex work to earn $800?

(b) How many hours must Barbara work to earn $800?

(c) For what positive number k will they both get the same pay for working k hours each?

Problem 16.5: Determine whether each of the following functions has an inverse. If a function has an inverse, find the inverse.

(a) $t(x) = \begin{cases} 4x & \text{if } x < 5, \\ 5/x & \text{if } x \geq 5. \end{cases}$

(b) $f(x) = \begin{cases} 2x & \text{if } x < 2, \\ 2^x & \text{if } x \geq 2. \end{cases}$

Problem 16.6: The function f is defined for all integers and satisfies

$$f(n) = \begin{cases} n - 3 & \text{if } n \geq 1000, \\ f(f(n + 5)) & \text{if } n < 1000. \end{cases}$$

Find $f(84)$. *(Source: AIME)*

Problem 16.1: Let g be the function defined for all real numbers such that

$$g(x) = \begin{cases} 2x & \text{if } x < 2, \\ 2^x & \text{if } x \geq 2. \end{cases}$$

(a) Compute $g(0)$, $g(1)$, $g(2)$, $g(3)$, and $g(4)$.

(b) Graph $g(x)$.

(c) Is $g(x)$ even or odd?

Solution for Problem 16.1:

(a) Since $g(x) = 2x$ for $x < 2$, we have $g(0) = 2 \cdot 0 = 0$ and $g(1) = 2 \cdot 1 = 2$. Since $g(x) = 2^x$ for all $x \geq 2$, we have $g(2) = 2^2 = 4$, $g(3) = 2^3 = 8$, and $g(4) = 2^4 = 16$.

(b) For $x < 2$, we have $g(x) = 2x$. So, for all $x < 2$, the graph of g is the same as the graph of $y = 2x$. Similarly, for $x \geq 2$, we have $g(x) = 2^x$, so for all $x \geq 2$, the graph of g is the same as the graph of $y = 2^x$. So, we graph $y = 2x$ for $x < 2$ and $y = 2^x$ for $x \geq 2$, and we produce the graph at right.

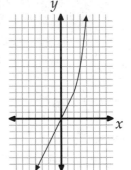

(c) Our graph at right is certainly not that of an even function, because it is not symmetric about the y-axis. We can also show that g is not even by noting that $g(-1) = -2$ and $g(1) = 2$, so $g(-1) \neq g(1)$.

To see that $g(x)$ is not odd, we note that $g(-3) = -6$ and $g(3) = 8$, so $g(-3) \neq -g(3)$.

☐

Problem 16.2: Let

$$f(x) = \begin{cases} x + 5 & \text{if } -9 \leq x \leq -3, \\ 2x^2 - 1 & \text{if } -3 < x \leq 2, \\ -x/2 & \text{if } x > 2. \end{cases}$$

(a) What are the domain and range of f?

(b) For what values of x is $f(x) = x$?

Solution for Problem 16.2:

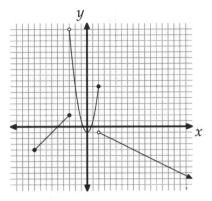

(a) First, we note that f is defined for all x greater than or equal to -9, so its domain is $[-9, +\infty)$. To determine the range, we can find the range of each piece. As an aid in doing so, we graph $y = f(x)$ at right.

In our graph, the open circles mark points that our graph approaches but never quite reaches, and the solid circles mark points that are on the graph. We can clearly see the three pieces of our function.

From the right piece of the graph, we see that f appears to continue forever downward. Since $f(x) = -x/2$ for $x > 2$, this piece has a range of $(-\infty, -1)$. The lowest point on the middle piece is $(0, -1)$, so -1 is in the range of f. The middle piece extends up to, but does not include, the boundary point on its left end. The y-coordinate of this point is $2(-3)^2 - 1 = 17$. So, the range of the middle piece is $[-1, 17)$. Make sure you see why 17 itself is not in the range. The range of the leftmost piece of f is $[-4, 2]$. We combine the ranges of all three pieces to find the range of f, so the range of f is $(-\infty, 17)$.

(b) We wish to find the solutions to the equation $f(x) = x$. First, we check for solutions where $x \leq -3$. These must satisfy the equation $x + 5 = x$ because $f(x) = x + 5$ if $x \leq -3$. There are no solutions to the equation $x + 5 = x$, so there are no values of x such that $x \leq -3$ and $f(x) = x$.

Next, we check for solutions between -3 and 2. These must satisfy $2x^2 - 1 = x$ because $f(x) = 2x^2 - 1$ when $-3 < x \leq 2$. Rearranging $2x^2 - 1 = x$ gives $2x^2 - x - 1 = 0$ and factoring gives $(2x + 1)(x - 1) = 0$. The solutions to this equation are $x = 1$ and $x = -1/2$, both of which are between -3 and 2.

Finally, we look for solutions that are greater than 2. These must satisfy $-x/2 = x$. The only solution to this equation is $x = 0$, but 0 is not greater than 2, so there are no values of x such that $x > 2$ and $f(x) = x$.

Therefore, the only values of x for which $f(x) = x$ are $x = 1$ and $x = -1/2$.

□

Loosely speaking, we say that a function is **continuous** if we can draw the graph of the function between any two points without lifting our pencil from the paper. If a function is not continuous, we say it **discontinuous**.

Problem 16.3: Determine which of the following functions are continuous and which are discontinuous.

(a) $f(x) = \begin{cases} x + 5 & \text{if } -9 \leq x \leq -3, \\ 2x^2 - 1 & \text{if } -3 < x \leq 2, \\ -x/2 & \text{if } x > 2. \end{cases}$

(c) $h(x) = \begin{cases} 2x & \text{if } x < 2, \\ 2^x & \text{if } x > 2. \end{cases}$

(b) $g(x) = \begin{cases} 2x & \text{if } x < 2, \\ 2^x & \text{if } x \geq 2. \end{cases}$

Solution for Problem 16.3:

(a) This is the function from Problem 16.2. We can see from our graph that f is not continuous because there are two "breaks" in the graph. We therefore see that f is discontinuous.

(b) This is the function from Problem 16.1. We reproduce its graph at left below. There are no "jumps" in either of the pieces of the graph of g. In other words, both $2x$ and 2^x are continuous functions. So, we know that $g(x)$ is continuous for $x < 2$, and for $x > 2$. In order to determine whether or not $g(x)$ is continuous at $x = 2$, we must check if the pieces meet when $x = 2$. Since $2x$ and 2^x both equal 4 when $x = 2$, the two pieces do indeed meet when $x = 2$. So, to graph g, we graph $y = 2x$ up to the point $(2, 4)$, and then graph $y = 2^x$ from $(2, 4)$ on. There is no jump in this graph, so the function is continuous.

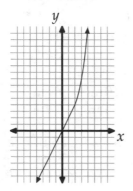

Figure 16.1: Graph of $y = g(x)$

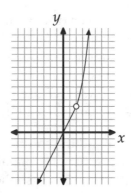

Figure 16.2: Graph of $y = h(x)$

(c) The only difference between this function and the prior one is that this function is not defined at $x = 2$. So, to graph h, we graph $y = 2x$ to the left of $x = 2$ and $y = 2^x$ to the right of $x = 2$. However, at $x = 2$ there is a **hole** in the graph because $h(x)$ is not defined for $x = 2$. We indicate this hole in the graph with an open circle, as shown in the graph at right above. Since we would have to lift our pencil at $x = 2$ to graph $y = h(x)$, the function h is discontinuous.

$\square$

You may be a little uncomfortable with our definition of continuity. When you study calculus, you'll learn a rigorous definition of continuity. However, our intuitive description of continuity based on graphing is usually enough to determine if a function is continuous or not.

Problem 16.4: Each week, Alex and Barbara get paid based on how many hours each works. Alex receives $30 per hour for each of his first 40 hours. He receives $50 per hour for each hour after his first 40 hours. (He receives the respective fraction of these amounts if he works some fraction of an hour.) Barbara receives $22 per hour for each of her first 20 hours, then $40 per hour for each of her next 10 hours, then $60 for each hour after that.

(a) How many hours must Alex work to earn $800?

(b) How many hours must Barbara work to earn $800?

(c) For what positive number k will they both get the same pay for working k hours each?

Solution for Problem 16.4: Alex's and Barbara's salaries are defined in pieces—different definitions for different amounts of hours. Piecewise defined functions help us organize piecewise defined data. We let $A(x)$ be Alex's salary in dollars if he works x hours. If Alex works no more than 40 hours, he earns $30/hr, so $A(x) = 30x$ if $0 \le x \le 40$. If he works more than 40 hours, he gets (\$30)(40) = \$1200 for the first 40 hours, plus \$50 per hour for the remaining $x - 40$ hours. So, his salary is $1200 + 50(x - 40) = 50x - 800$ dollars if $x > 40$. Putting these pieces together gives

$$A(x) = \begin{cases} 30x & \text{if } 0 \le x \le 40, \\ 50x - 800 & \text{if } x > 40. \end{cases}$$

Similarly, we let $B(x)$ be Barbara's salary in dollars if she works x hours. We consider three cases to determine $B(x)$:

- $0 \le x \le 20$. Since Barbara earns \$22/hr for her first 20 hours, we have $B(x) = 22x$ dollars if $0 \le x \le 20$.

- $20 < x \le 30$. She earns $22(20) = 440$ dollars for the first 20 hours, then she earns \$40/hr for the rest of the time. So, she earns $440 + 40(x - 20)$ dollars. Simplifying this expression, we find that $B(x) = 40x - 360$ dollars if $20 < x \le 30$.

- $x > 30$. She earns 440 dollars for the first 20 hours, then 400 dollars for the next 10 hours, then $60(x - 30)$ dollars for the time past 30 hours. This gives her a total pay of $440 + 400 + 60x - 1800 = 60x - 960$ dollars.

Putting the information from all three cases together, we have

$$B(x) = \begin{cases} 22x & \text{if } 0 \le x \le 20, \\ 40x - 360 & \text{if } 20 < x \le 30, \\ 60x - 960 & \text{if } x > 30. \end{cases}$$

We can test both $A(x)$ and $B(x)$ by trying a specific value of x. If Alex works 60 hours, he gets (\$30)(40) = \$1200 for the first 40 hours and (\$50)(20) = \$1000 for the last 20 hours, for a total of \$2200. Indeed, we find that $A(60) = 2200$. Similarly, if Barbara works 60 hours, she should make (\$22)(20) + (\$40)(10) + (\$60)(30) = \$2640, and we find $B(60) = 2640$.

(a) Alex makes \$1200 if he works 40 hours, so the amount he must work to make \$800 is less than 40 hours. So, we must have $30x = 800$, which gives us $x = 800/30 = 26\frac{2}{3}$ hours that Alex must work to make \$800.

(b) Barbara makes \$440 if she works 20 hours, so she must work more than 20 hours to make \$800. She makes $40(30) - 360 = 840$ dollars if she works 30 hours, so she must work between 20 and 30 hours to make \$800. Therefore, we must have $40x - 360 = 800$, which gives us $x = 29$ hours.

(c) We seek the positive value of k such that $A(k) = B(k)$. Since these two functions are defined in pieces, we must check for equality by pieces. We won't have equality if $0 < k \le 20$, since then we must have $22k = 30k$, which only gives $k = 0$ as a solution.

If $20 < k \le 30$, then $A(k) = 30k$ and $B(k) = 40k - 360$, which gives us $30k = 40k - 360$. Solving this gives us $k = 36$, which does not fit the $20 < k \le 30$ restriction of this case.

Next, we consider $30 < k \le 40$. Then, we have $A(k) = 30k$ and $B(k) = 60k - 960$. Setting these equal gives us $k = 32$, so they make the same amount if they both work 32 hours.

Finally, we check $k > 40$, to see if there's a second possible solution. If $k > 40$, we have $A(k) = 50k - 800$ and $B(k) = 60k - 960$. Setting these equal gives us $k = 16$, which isn't greater than 40. So, the only way they can earn the same pay for the same nonzero time is if they both work 32 hours.

□

Problem 16.5: Determine whether each of the following functions has an inverse. If a function has an inverse, find the inverse.

(a) $\quad t(x) = \begin{cases} 4x & \text{if } x < 5, \\ 5/x & \text{if } x \geq 5. \end{cases}$

(b) $\quad f(x) = \begin{cases} 2x & \text{if } x < 2, \\ 2^x & \text{if } x \geq 2. \end{cases}$

Solution for Problem 16.5:

(a) What's wrong with the following reasoning:

> **Bogus Solution:** We consider each piece separately. Let $g(x) = 4x$. Then, we have $g^{-1}(x) = x/4$ because then we have $g^{-1}(g((x)) = (4x)/4 = x$. Similarly, we let $h(x) = 5/x$, and we find that $h^{-1}(x) = 5/x$, because then we have $h^{-1}(h(x)) = 5/(5/x) = x$. Therefore, we have
>
> $$t^{-1}(x) = \begin{cases} x/4 & \text{if } x < 5, \\ 5/x & \text{if } x \geq 5. \end{cases}$$

We see that our answer is wrong by checking a few specific values. We have $t(2) = 8$, but $t^{-1}(8) = 5/8$, so we don't have $t^{-1}(t(2)) = 2$. Something went wrong.

By experimenting with a few more values, or by graphing t, we find that t does not have an inverse. For example, because $t(1/4) = t(5) = 1$, we know that t does not have an inverse. So, even though each piece of t has an inverse, t itself does not have an inverse.

(b) We've already graphed f, so we start by looking at the graph of f to see if f has an inverse. No horizontal line intersects f at more than one point, so there are no two inputs to f that give the same output. This means that f does have an inverse.

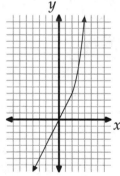

We might be tempted to think that we simply take the inverse of each piece of f to find that its inverse is

$$h(x) = \begin{cases} x/2 & \text{if } x < 2, \\ \log_2 x & \text{if } x \geq 2. \end{cases}$$

Inverting each piece failed in part (a), so we take a closer look here. Again, we try a few values. We find that $h(3) = \log_2 3$, but because $\log_2 3 < 2$, we have $f(\log_2 3) = 2 \log_2 3$. So, we don't have $f(h(3)) = 3$. Uh-oh. Back to the drawing board.

We correctly determined the inverses of both pieces; the inverse of $2x$ is $x/2$, and the inverse of 2^x is $\log_2 x$. We erred in making the domains of the two pieces in our inverse the same as the domains of the pieces of f. To see why this is a mistake, let's think about $f^{-1}(3)$. The value of $f^{-1}(3)$ is the value of x for which $f(x) = 3$. Examining both pieces of f, we see that the output of the piece with $x \geq 2$ is always at least 4, and that the output of the piece with $x < 2$ is 3 for $x = 1.5$.

Since we have $f(1.5) = 3$, we must have $f^{-1}(3) = 1.5$. Aha! In determining the domains of the pieces of $f^{-1}(x)$, we must consider the *ranges* of each piece of $f(x)$.

If $x < 2$, then we have $f(x) = 2x$, so $f(x) < 4$. In other words, the left piece of $f(x)$ takes numbers in the interval $(-\infty, 2)$ as input and outputs numbers in the interval $(-\infty, 4)$. The inverse of this piece must "undo" this process, taking numbers in the interval $(-\infty, 4)$ as input and returning numbers in the interval $(-\infty, 2)$ as output. Since the inverse of $2x$ is $x/2$, we therefore have $f^{-1}(x) = x/2$ for $x < 4$.

Similarly, if $x \geq 2$, then we have $f(x) = 2^x$, which means $f(x) \geq 4$. This piece of $f(x)$ takes the numbers in the interval $[2, +\infty)$ as input and outputs numbers in the interval $[4, +\infty)$. The inverse of 2^x is $\log_2 x$, so we have $f^{-1}(x) = \log_2 x$ for $x \geq 4$.

We therefore have

$$f^{-1}(x) = \begin{cases} x/2 & \text{if } x < 4, \\ \log_2 x & \text{if } x \geq 4. \end{cases}$$

$\square$

Problem 16.6: The function f is defined for all integers and satisfies

$$f(n) = \begin{cases} n - 3 & \text{if } n \geq 1000, \\ f(f(n+5)) & \text{if } n < 1000. \end{cases}$$

Find $f(84)$. *(Source: AIME)*

Solution for Problem 16.6: For $n \geq 1000$, the behavior of the function is easy to understand. To compute $f(84)$, we might first note that $f(84) = f(f(89))$ from our function definition. But what's $f(89)$? We can use our function definition to find that $f(89) = f(f(94))$, so $f(84) = f(f(89)) = f(f(f(94)))$. Uh-oh. It's going to take a long time to reach a number that's greater than 1000. Let's try a different strategy.

It's easy to find $f(n)$ if $n \geq 1000$. To see what happens when $n < 1000$, let's take a look at values of n that are just below 1000:

$$f(999) = f(f(1004)) = f(1001) = 998,$$
$$f(998) = f(f(1003)) = f(1000) = 997,$$
$$f(997) = f(f(1002)) = f(999) = 998,$$
$$f(996) = f(f(1001)) = f(998) = 997,$$
$$f(995) = f(f(1000)) = f(997) = 998,$$

and so on. We appear to have a pattern. It looks like $f(n) = 997$ if n is an even number less than 1000, and $f(n) = 998$ if n is an odd number less than 1000.

We've already seen that this pattern holds for all integers n such that $995 \leq n \leq 999$. Now, we want to show that it holds for all $n < 1000$. Intuitively, it seems pretty clear that this pattern will hold. It looks like we'll always have

$$f(\text{odd number}) = f(f(\text{odd number} + 5)) = f(f(\text{even number})) = f(997) = 998,$$
$$f(\text{even number}) = f(f(\text{even number} + 5)) = f(f(\text{odd number})) = f(998) = 997.$$

To prove this formally, we use induction. We assume that $f(k) = 997$ if k is even and $f(k) = 998$ if k is odd, for some specific k less than 1000. To "induct downward" to prove that $f(84) = 997$, we must use this assumption to show that $f(k-1) = 997$ if $k-1$ is even and $f(k-1) = 998$ if $k-1$ is odd.

So, what is $f(k-1)$? Since $k-1 < 1000$, we have $f(k-1) = f(f(k+4))$. But we don't know anything about $f(k+4)$. What will we do?

We have to go back and adjust our induction slightly. We cannot relate $f(k-1)$ to $f(k)$; we can only relate it to $f(k+4)$. In our exploration above, we found several values of k for which $f(k) = 997$ if k is even and $f(k) = 998$ if k is odd. So, we adjust our inductive assumption so that it includes $f(k+4)$ in addition to $f(k)$.

Specifically, we assume that $f(n) = 997$ if n is even and $f(n) = 998$ if n is odd, for all n such that $k \le n < 1000$. So, if $k < 996$, our assumption tells us that $f(n) = 997$ if n is even and $f(n) = 998$ if n is odd for both $n = k$ and $n = k+4$, instead of only telling us this for $n = k$. We've already shown that the assumption is true for all n from 995 to 999, so we can "start" our induction with $k = 995$.

We're now ready to return to $f(k-1)$, for $k \le 995$. Above, we found that $f(k-1) = f(f(k+4))$. Moreover, since $k \le 995$, we have $k+4 \le 999$, so $k < k+4 < 1000$. Because $k+4$ is between k and 1000, our inductive assumption tells us that $f(k+4) = 997$ if $k+4$ is even and $f(k+4) = 998$ if $k+4$ is odd. If $k-1$ is even, then $k+4$ is odd, so $f(k-1) = f(f(k+4)) = f(998) = 997$. If $k-1$ is odd, then $k+4$ is even, and we have $f(k-1) = f(f(k+4)) = f(997) = 998$. Combining these, we have $f(k-1) = 997$ if $k-1$ is even and $f(k-1) = 998$ if $k-1$ is odd.

This completes our inductive step, so we have shown that $f(n) = 998$ if n is an odd integer less than 1000, and $f(n) = 997$ if n is an even integer less than 1000. So, we have $f(84) = 997$.

Make sure you see why it is important that we showed that the assumption is true for all n from 995 to 999 as the base case of the induction. This is because our inductive step uses the inductive assumption for $n = k+4$ to prove that the statement also holds for $n = k-1$. In other words, we are taking steps of 5 in our induction, so we need 5 base cases. If we only showed that the inductive assumption is true for $n = 999$ as the base case, then our inductive step can only be used to show that the statement is true for $n = 994, 989, 984, 979$, etc. $\square$

Our solution to Problem 16.6 is a great example of a very common problem solving process: experiment, see a pattern, and prove the pattern with induction.

Concept: Don't just stare at a problem when you're stuck. Experiment and look for patterns.

Exercises

16.1.1 Let f be the function defined for all real numbers such that

$$f(x) = \begin{cases} -x^2, & \text{if } x < -2, \\ 2x, & \text{if } -2 \le x < 0, \\ 3x, & \text{if } 0 \le x < 2, \\ x^2 + 3, & \text{if } x \ge 2. \end{cases}$$

(a) Find $f(3/2)$. (b) Graph f. Is f continuous? (c) Find the range of $f(x)$.

16.1.2 What real value of c will make the following function continuous?

$$f(x) = \begin{cases} x^2 + 2c & \text{if } x \le 3, \\ 2cx + 1 & \text{if } x > 3. \end{cases}$$

Graph f for this choice of c.

16.1.3 Find the inverse of f if $f(x) = \begin{cases} \sqrt{2-x} & \text{if } x < 0, \\ 1 - x^2 & \text{if } x \ge 0. \end{cases}$

16.1.4 Mientka Publishing Company prices its bestseller *Where's Walter?* as follows:

$$C(n) = \begin{cases} 12n, & \text{if } 1 \le n \le 24, \\ 11n, & \text{if } 25 \le n \le 48, \\ 10n, & \text{if } 49 \le n, \end{cases}$$

where n is the number of books ordered and $C(n)$ is the cost in dollars of n books. Notice that 25 books cost less than 24 books. For how many values of n is it cheaper to buy more than n books than to buy exactly n books? *(Source: AHSME)*

16.2 Absolute Value

We can think of the absolute value of a number as the distance between that number and 0 on the number line. However, we can also view absolute value as a piecewise defined function.

> **Definition:** The **absolute value** of the real number x, denoted as $|x|$, is defined as
>
> $$|x| = \begin{cases} x & \text{if } x \ge 0, \\ -x & \text{if } x < 0. \end{cases}$$

For example, we have $|4| = 4$, $|-2| = -(-2) = 2$, and $|0| = 0$. Furthermore, because $|x|$ is positive if x is positive or negative, and $|0| = 0$, we have $|x| \ge 0$ for all x, where $|x| = 0$ if and only if $x = 0$.

▏▎ Problems ▶

Problem 16.7: Write each of the following as a piecewise defined function:
(a) $f(x) = |3 - x|$
(b) $g(x) = |x + 2| + |x + 5|$

Problem 16.8:
(a) Solve the equation $|3 - x| = 5$.
(b) Solve the equation $|x + 2| + |x + 5| = 8$.
(c) Find all solutions to $|x - 1| = |x - 2| + |x - 3|$. *(Source: HMMT)*

Problem 16.9: Find all values of x such that $|x + 3| + |x + 4| < 7$.

Problem 16.10:

(a) If $|a| + |b| = 0$, then what can be said about a and b?

(b) Find all constants c such that $|cx - 1| + |x^2 - x - 2| = 0$ has a solution in x.

Problem 16.11: How many ordered triples of integers (a, b, c) satisfy $|a + b| + c = 19$ and $ab + |c| = 97$? *(Source: AHSME)*

Many absolute value problems are solved by thinking of the absolute value expression as a piecewise defined function, so we'll start by getting a little practice writing absolute value expressions as piecewise defined functions.

Problem 16.7: Write each of the following as a piecewise defined function:

(a) $f(x) = |3 - x|$

(b) $g(x) = |x + 2| + |x + 5|$

Solution for Problem 16.7:

(a) If $3 - x \geq 0$, then $|3 - x| = 3 - x$. Rearranging $3 - x \geq 0$ gives us $x \leq 3$, so we have $|3 - x| = 3 - x$ if $x \leq 3$. Similarly, if $3 - x < 0$, then $|3 - x| = -(3 - x) = x - 3$. So, we have $|3 - x| = x - 3$ if $x > 3$. Therefore, we can write f as

$$f(x) = \begin{cases} x - 3 & \text{if } x > 3, \\ 3 - x & \text{if } x \leq 3. \end{cases}$$

(b) In order to write $|x + 2|$ and $|x + 5|$ without the absolute value notation, we consider three cases.

- *Both $x + 2$ and $x + 5$ are nonnegative.* If $x \geq -2$, then $|x + 2| = x + 2$ and $|x + 5| = x + 5$, so $|x + 2| + |x + 5| = 2x + 7$.

- *$x + 2$ is negative and $x + 5$ is nonnegative.* If $-5 \leq x < -2$, then $|x + 2| = -(x + 2) = -x - 2$ and $|x + 5| = x + 5$, so $|x + 2| + |x + 5| = 3$. (Note that we cannot have $x + 5$ negative and $x + 2$ nonnegative.)

- *Both $x + 2$ and $x + 5$ are negative.* If $x < -5$, then $|x + 2| = -(x + 2) = -x - 2$ and $|x + 5| = -x - 5$, so $|x + 2| + |x + 5| = -2x - 7$.

Putting these three cases together gives

$$g(x) = \begin{cases} 2x + 7 & \text{if } x \geq -2, \\ 3 & \text{if } -5 \leq x < -2, \\ -2x - 7 & \text{if } x < -5. \end{cases}$$

□

Let's see how we can use this technique to solve equations:

Extra! *You cannot depend on your eyes when your imagination is out of focus.*

▪▶ ▪▪▶ ▪▪▶ ▪▪▶

– Mark Twain

> **Problem 16.8:**
>
> (a) Solve the equation $|3 - x| = 5$.
>
> (b) Solve the equation $|x + 2| + |x + 5| = 8$.
>
> (c) Find all solutions to $|x - 1| = |x - 2| + |x - 3|$. *(Source: HMMT)*

Solution for Problem 16.8:

(a) For simple equations like this one, we have several straightforward approaches:

Solution 1: Use the piecewise defined function. If we let $f(x) = |3 - x|$, then our problem is to solve the equation $f(x) = 5$. In part (a) of Problem 16.7, we saw that $f(x) = |3 - x|$ can be written as

$$f(x) = \begin{cases} x - 3 & \text{if } x > 3, \\ 3 - x & \text{if } x \le 3. \end{cases}$$

So, we test each piece to see what values of x give $f(x) = 5$. If $x - 3 = 5$, we have $x = 8$, which satisfies $x > 3$ (the restriction for the first piece). If $3 - x = 5$, then $x = -2$, which satisfies $x \le 3$ (the restriction for the second piece). Checking our answers, we see that $x = -2$ and $x = 8$ both satisfy $|3 - x| = 5$.

Solution 2: Use the number line. The expression $|3 - x|$ equals the distance between 3 and x on the number line. So, if $|3 - x| = 5$, then x is 5 away from 3 on the number line, which means either $x = -2$ (5 to the left of 3) or $x = 8$ (5 to the right of 3).

Solution 3: What numbers have absolute value 5? The only two numbers with absolute value 5 are -5 and 5. So, either we must have $3 - x = -5$, which gives $x = 8$, or we must have $3 - x = 5$, which gives $x = -2$.

Solution 4: Square the absolute value signs away. Because the square of a real number is always nonnegative, we have $(|x|)^2 = |x^2| = x^2$ for all real x. So, we can square both sides of our equation to get rid of the absolute value signs. This gives $(|3 - x|)^2 = 5^2$, so $(3 - x)^2 = 25$. Expanding the square and rearranging gives us $x^2 - 6x - 16 = 0$, which has solutions $x = 8$ and $x = -2$. Because we squared the equation as a step in our solution, we must check for extraneous solutions. We find that both solutions are acceptable.

All four solutions are valid, and all four have the same key step: finding a way to get rid of the absolute value signs and turn the problem into one we know how to solve.

(b) Of our four solutions to part (a), you might find the first one the most complicated. But as absolute value problems get harder, some of our simpler methods start to fail us. For example, in solving $|x + 2| + |x + 5| = 8$, we won't get far with our third or fourth approaches to part (a). Let's give our piecewise defined function approach a shot. We saw in part (b) of Problem 16.7 that

$$|x + 2| + |x + 5| = \begin{cases} 2x + 7 & \text{if } x \ge -2, \\ 3 & \text{if } -5 \le x < -2, \\ -2x - 7 & \text{if } x < -5. \end{cases}$$

Clearly, the middle piece cannot ever equal 8. But each of the other two pieces can. For the first piece, we have $2x + 7 = 8$, so $x = 1/2$, which satisfies $x \ge -2$, as required. For the third piece, we have $-2x - 7 = 8$, so $x = -15/2$, which satisfies $x < -5$. Therefore, our two solutions are $x = -15/2$ and $x = 1/2$. (As an Exercise, you'll have a chance to use the number line to solve this problem.)

(c) We aren't too sure how to deal with an equation that has absolute values on both sides, so we move all the absolute value signs to the left, giving $|x - 1| - |x - 2| - |x - 3| = 0$. Now, our previous two parts give us a guide. We let the left side be $f(x)$ and write $f(x)$ as a piecewise defined function. We consider the cases $x < 1$, $1 \le x < 2$, $2 \le x < 3$, and $x \ge 3$, and we have

$$f(x) = |x - 1| - |x - 2| - |x - 3| = \begin{cases} x - 4 & \text{if } x < 1, \\ 3x - 6 & \text{if } 1 \le x < 2, \\ x - 2 & \text{if } 2 \le x < 3, \\ -x + 4 & \text{if } x \ge 3. \end{cases}$$

As before, we check each piece for solutions. Setting our first piece equal to 0 gives us $x = 4$, which doesn't satisfy $x < 1$, so we have no solutions for the first piece. The second piece equals 0 for $x = 2$, which doesn't satisfy the restriction $1 \le x < 2$. So, there are no solutions for the second piece. The third piece gives us $x = 2$, which satisfies $2 \le x < 3$, and the fourth piece gives us $x = 4$, which satisfies $x \ge 3$. Therefore, the only solutions to our equation are $x = 2$ and $x = 4$.

□

> **Concept:** An effective technique for solving some equations with absolute values is to break the problem into cases by viewing the absolute values as piecewise defined functions. Remember to check each solution. In each case, the solution must satisfy the restrictions of that case in order to be a valid solution.

Casework helped us with absolute value equations. Let's try an absolute value inequality.

Problem 16.9: Find all values of x such that $|x + 3| + |x + 4| < 7$.

Solution for Problem 16.9: Solution 1: Casework. When $x \ge -3$, we have $|x + 3| + |x + 4| = 2x + 7$, so our inequality is $2x + 7 < 7$ if $x \ge -3$. Solving $2x + 7 < 7$ gives $x < 0$. Combining $x < 0$ with the restriction $x \ge -3$ of this case gives us $-3 \le x < 0$.

When $-4 \le x < -3$, we have $|x+3|+|x+4| = -(x+3)+x+4 = 1$. Since $1 < 7$, we have $|x+3|+|x+4| < 7$ for all x such that $-4 \le x < -3$.

When $x < -4$, we have $|x + 3| + |x + 4| = -(x + 3) - (x + 4) = -2x - 7$, so our inequality is $-2x - 7 < 7$ if $x < -4$. Solving $-2x - 7 < 7$ gives $x > -7$. Combining $x > -7$ with the restriction $x < -4$ gives us $-7 < x < -4$.

Putting these three cases together, the values of x that satisfy the original inequality are $-3 \le x < 0$, $-4 \le x < -3$, and $-7 < x < -4$. We can combine all three of these ranges for x; our original inequality holds when $-7 < x < 0$.

Solution 2: Use the number line. The quantity $|x + 3| = |x - (-3)|$ is the distance between x and -3 on the number line. Likewise, the quantity $|x+4|$ is the distance between x and -4. Therefore, if $|x+3|+|x+4| < 7$, then the sum of the distances from x to -3 and to -4 must be no more than 7. Clearly this is the case if $-4 \le x \le -3$. For $x < -4$, we consider the following number line:

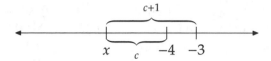

Since -3 and -4 are 1 apart, if x is c to the left of -4 as shown above, then the sum of distances from x to -3 and to -4 is $2c + 1$. We must then have $2c + 1 < 7$, so $c < 3$, which means $x > -7$. Similarly, if x is to the right of -3, it must be within 3 of -3, so $x < 0$. This gives us $-7 < x < 0$ as our solution. $\square$

Note that in the second solution to Problem 16.9, we avoided the need for any algebra by interpreting the absolute value function as a measure of distance.

> **Concept:** Viewing the absolute value of the difference between two real numbers as the distance between the numbers on the number line can help solve some absolute value problems.

Problem 16.10: Find all constants c such that the following equation has a solution in x:

$$|cx - 1| + |x^2 - x - 2| = 0.$$

Solution for Problem 16.10: We have $|cx - 1| \geq 0$ and $|x^2 - x - 2| \geq 0$ for all x, so $|cx - 1| + |x^2 - x - 2| = 0$ if and only if $|cx - 1| = |x^2 - x - 2| = 0$. So, we must have $cx - 1 = x^2 - x - 2 = 0$.

The quadratic $x^2 - x - 2$ factors as $(x - 2)(x + 1)$, so $x = 2$ or $x = -1$. If $x = 2$ is a solution of the original equation, then $|cx - 1| = 0$ gives us $|2c - 1| = 0$, so $c = 1/2$. Similarly, if $x = -1$ is a solution, then we must have $|-c - 1| = 0$, so $c = -1$. Therefore, the only possible values for c are $1/2$ and -1. $\square$

Don't forget that the absolute value of any expression must be nonnegative. This fact alone will help simplify many problems.

Problem 16.11: How many ordered triples of integers (a, b, c) satisfy $|a + b| + c = 19$ and $ab + |c| = 97$?

Solution for Problem 16.11: At first glance, it looks like we can easily eliminate c by subtracting the first equation from the second. However, the pesky absolute value signs get in the way. But we know how to take care of them! We'll consider two cases:

Case 1: $c \geq 0$. If $c \geq 0$, then $|c| = c$, so our system of equations is $|a + b| + c = 19$, $ab + c = 97$. Subtracting the first equation from the second gives us $ab - |a + b| = 78$. Once again, if it weren't for the absolute value signs, we'd know what to do with this equation: we'd use Simon's Favorite Factoring Trick. So, we get rid of the absolute value signs by considering two subcases:

- *Subcase 1a: $a + b \geq 0$, $c \geq 0$.* If $a + b \geq 0$, then $|a + b| = a + b$, so we have $ab - a - b = 78$. Adding 1 to both sides and then factoring the left side gives $(a - 1)(b - 1) = 79$. Since 79 is prime and $a + b \geq 0$, we must have $a - 1 = 79$, $b - 1 = 1$ or $a - 1 = 1$, $b - 1 = 79$. So, we have $(a, b) = (80, 2)$ or $(a, b) = (2, 80)$. In either case, we have $a + b = 82$, so we have $c = 19 - |a + b| = -63$. But we must have $c \geq 0$, so there are no solutions for this subcase.

- *Subcase 1b:* $a + b < 0$, $c \geq 0$. If $a + b < 0$, then $|a + b| = -(a + b)$, so we have $ab + a + b = 78$. Adding 1 to both sides and then factoring the left side gives $(a + 1)(b + 1) = 79$. Since $a + b < 0$, we have $(a, b) = (-80, -2)$ or $(a, b) = (-2, -80)$. In either case, we again have $c = 19 - |a + b| = -63$, and since we must have $c \geq 0$, we have no solutions for this subcase, either.

Case 2: $c < 0$. If $c < 0$, we have $|c| = -c$, so our system of equations is $|a + b| + c = 19$, $ab - c = 97$. Adding these gives $ab + |a + b| = 116$. Once again, we consider two subcases:

- *Subcase 2a:* $a + b \geq 0$, $c < 0$. If $a + b \geq 0$, then $|a + b| = a + b$, so we have $ab + a + b = 116$, so $(a + 1)(b + 1) = 117$. Since $117 = 1 \cdot 117 = 3 \cdot 39 = 9 \cdot 13$, we have three pairs of numbers whose product is 117. Each gives us two solutions (a, b). For example, from $117 = 1 \cdot 117$, we have $(a, b) = (0, 116)$ and $(a, b) = (116, 0)$. For both these solutions, we have $c = 19 - |a + b| = -97$, which satisfies the restrictions of this subcase. Similarly, $117 = 3 \cdot 39$ gives us the solutions $(a, b, c) = (2, 38, -21)$ and $(a, b, c) = (38, 2, -21)$, and $117 = 9 \cdot 13$ gives us $(a, b, c) = (8, 12, -1)$ and $(a, b, c) = (12, 8, -1)$. So, there are six solutions in this subcase.

- *Subcase 2b:* $a + b < 0$, $c < 0$. In this case, we have $ab - a - b = 116$, so $(a - 1)(b - 1) = 117$. Proceeding as in the previous subcase, we get the six solutions $(a, b, c) = (-116, 0, -97)$, $(0, -116, -97)$, $(-2, -38, -21)$, $(-38, -2, -21)$, $(-12, -8, -1)$, and $(-8, -12, -1)$.

Phew! Putting all our cases together, we have a total of 12 solutions to the original system of equations. $\square$

Exercises

16.2.1 Let x be a real number no greater than -4.

(a) Simplify $|x + 2|$.

(b) Simplify $|2 - |x + 2||$.

(c) Is $|\pi^2 - 10|$ equal to $\pi^2 - 10$ or $10 - \pi^2$?

(d) Simplify $|x + \pi^2 - 6|$.

16.2.2 Use the number line to solve the equation $|x + 2| + |x + 5| = 8$.

16.2.3 Find all x such that $1 \leq |x - 2| \leq 7$. *(Source: AHSME)*

16.2.4 Find all solutions to the equation $|x - |2x + 1|| = 3$.

16.2.5 If $|x| + x + y = 10$ and $x + |y| - y = 12$, find $x + y$. *(Source: AHSME)*

16.2.6 Find all real values of y for which $\dfrac{|y - |y||}{y} > 0$.

16.2.7★ If $f(x) = (x - 3)^2 - 1$, compute the set of real numbers such that $f(|x|) = |f(x)|$. *(Source: ARML)*
Hints: 251

16.3 Graphing Absolute Value

In the previous section, we explored several algebraic approaches for handling problems involving absolute value. In this section, we take a more geometric approach to absolute value.

Problems

Problem 16.12: Graph the function $f(x) = x + |x|$.

Problem 16.13: Let $f(x) = |x - 2| + |x - 4| + |2x - 6|$, for $2 \le x \le 8$.
 (a) Rewrite $f(x)$ as a piecewise defined function.
 (b) Graph f.
 (c) Sketch the largest and smallest possible values of $f(x)$.

Problem 16.14:
 (a) Graph $y = |x + 1| - 2$ and $y = -|x - 4| + 7$.
 (b) The graphs of $y = -|x - a| + b$ and $y = |x - c| + d$ intersect at points $(2, 5)$ and $(8, 3)$. Find $a + c$. *(Source: AHSME)*

Problem 16.15: Find the area of the region of points that satisfy the inequality $|x + y| + |x - y| \le 8$.

Problem 16.16:
 (a) Solve the equation $|x^2 + 6x - 3| = x + 3$.
 (b) Graph the equation $y = |x^2 + 6x - 3|$ and use your graph to help you solve the inequality $|x^2 + 6x - 3| \ge x + 3$.
 (c) Find the value of c if the line $y = c$ intersects the graph of $y = |x^2 + 6x - 3|$ in exactly three points.

Problem 16.12: Graph the function $f(x) = x + |x|$.

Solution for Problem 16.12: When $x \ge 0$, we have $f(x) = x + |x| = x + x = 2x$. When $x < 0$, we have $f(x) = x + |x| = x - x = 0$. So, for $x \ge 0$, the graph of $f(x)$ is the same as the graph of $y = 2x$, and for $x < 0$, the graph of f is the same as the graph of $y = 0$. Putting these together, the graph of f is shown at right. $\square$

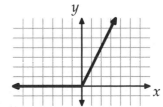

Concept: The absolute value function is a piecewise defined function, so it's often easiest to graph functions involving absolute values piece-by-piece.

Problem 16.13: Let $f(x) = |x - 2| + |x - 4| + |2x - 6|$, for $2 \le x \le 8$. Find the largest and smallest possible values of $f(x)$.

Solution for Problem 16.13: We begin by rewriting $f(x)$ as a piecewise defined function:

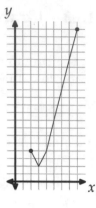

$$f(x) = \begin{cases} 8 - 2x & \text{if } 2 \le x < 3, \\ 2x - 4 & \text{if } 3 \le x < 4, \\ 4x - 12 & \text{if } 4 \le x \le 8. \end{cases}$$

The graph of f consists of three line segments, as shown at right. Each segment is non-horizontal, so we know that the highest and lowest points of the graph of each segment are the endpoints of the segment. With this in mind, we could have skipped actually graphing f, and evaluated $f(x)$ at the x values of each of these endpoints. These x values are the boundary values of our cases in the piecewise definition of f, so to calculate the minimum and maximum of f, we find

$$f(2) = 4, \quad f(3) = 2, \quad f(4) = 4, \quad f(8) = 20.$$

So, the minimum value of $f(x)$ is 2 and the maximum value is 20. $\square$

Problem 16.14: The graphs of $y = -|x - a| + b$ and $y = |x - c| + d$ intersect at points $(2, 5)$ and $(8, 3)$. Find $a + c$. *(Source: AMC)*

Solution for Problem 16.14: We start by figuring out what the graphs of equations of the forms $y = -|x-a|+b$ and $y = |x - c| + d$ look like. We do so by first graphing some example equations, by choosing specific values for $a, b, c,$ and d. On the left below, we graph $y = |x+1|-2$. For $x \ge -1$, we graph $y = x+1-2 = x-1$, and for $x < -1$, we graph $y = -(x + 1) - 2 = -x - 3$.

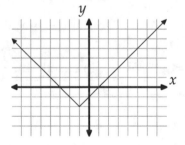

Figure 16.3: Graph of $y = |x + 1| - 2$

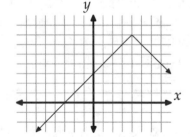

Figure 16.4: Graph of $y = -|x - 4| + 7$

On the right above, we graph $y = -|x - 4| + 7$. For $x \ge 4$, the resulting graph is the graph of $y = -(x - 4) + 7 = -x + 11$, and for $x < 4$, the graph matches the graph of $y = -(4 - x) + 7 = x + 3$.

We see that graphs of the form $y = -|x - a| + b$ have an inverted V shape. The top of the inverted V occurs when $x = a$, since this makes $-|x - a| + b$ as large as possible. For any other value of x, the expression $-|x - a|$ is negative, so $-|x - a| + b$ is less than b. When $x \ge a$, the graph of $y = -|x - a| + b$ matches that of a line with slope -1, and if $x < a$, then the graph matches a line with slope 1. Similarly, graphs of the form $y = |x + c| + d$ form an upward V shape, where the branches have slopes ± 1 and the bottommost point is (c, d).

Extra! *It has long been an axiom of mine that the little things are infinitely the most important.*

‖➡ ‖➡ ‖➡ ‖➡   – Sir Arthur Conan Doyle

Armed with these observations, we're ready to tackle the problem. We know what the graphs look like, so we start from the intersection points $(2,5)$ and $(8,3)$ and construct a V and an inverted V, using rays with slopes 1 and -1 for both branches of each V. The bottommost point of the V is $(6,1)$, so $c = 6$. The highest point of the inverted V is $(4,7)$, so $a = 4$, and $a + c = 10$.

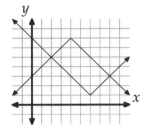

Notice that $a + c$ equals the sum of the x-coordinates of the given intersection points. Is this a coincidence? □

Problem 16.15: Find the area of the region of points that satisfy the inequality $|x + y| + |x - y| \le 8$.

Solution for Problem 16.15: We've tackled other absolute value graphs by considering cases, so we try the same here.

Case 1: $x + y \ge 0$. This makes our inequality $x + y + |x - y| \le 8$. Hmmm.... We still have an absolute value to get rid of. Let's try subcases.

- *Subcase 1a:* $x + y \ge 0$, $x - y \ge 0$. Now, we have $|x - y| = x - y$, so our inequality is $2x \le 8$, or $x \le 4$. Therefore, if both $x + y \ge 0$ and $x - y \ge 0$, then we must have $x \le 4$.

- *Subcase 1b:* $x + y \ge 0$, $x - y < 0$. This gives us $|x - y| = -(x - y) = y - x$, so our inequality is $2y \le 8$, or $y \le 4$. Therefore, if $x + y \ge 0$ and $x - y < 0$, then we have $y \le 4$.

But how do we graph these solutions? We start by splitting the Cartesian plane into regions that correspond to our subcases. At left below, we have graphed $x + y = 0$ and $x - y = 0$ with dashed lines. Region $\boxed{1}$ is where $x + y > 0$ and $x - y > 0$. So, to graph our solutions in Subcase 1a, we shade the region of points inside region $\boxed{1}$ such that $x \le 4$, as shown in the middle diagram below. Notice that we also make solid the portions of the lines $x + y = 0$ and $x - y = 0$ that border region $\boxed{1}$ for $x \le 4$, since we have $x \le 4$ for $x + y \ge 0$ and $x - y \ge 0$. This includes the appropriate points on these lines in the region we're graphing.

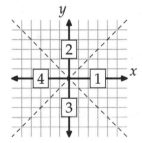

Figure 16.5: Our Regions

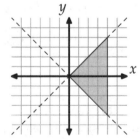

Figure 16.6: Subcase 1a

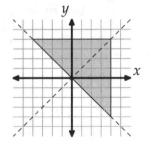

Figure 16.7: Case 1

Region $\boxed{2}$ is where $x + y > 0$ and $x - y < 0$, so to graph Subcase 1b, we shade the region of points inside region $\boxed{2}$ for which $y \le 4$ (including the appropriate portion of the line $x + y = 0$). Putting this together with our graph from Subcase 1a gives us the graph at right above, which is the graph for the entire Case 1.

We have a pretty good guess what we'll find for Case 2. Let's see:

Case 2: x + y < 0. This gives us $-x - y + |x - y| \le 8$, and we turn again to subcases.

- *Subcase 2a:* $x + y < 0, x - y \ge 0$. This gives us $-2y \le 8$, so $y \ge -4$. This subcase corresponds to region $\boxed{3}$ in our graph, so we shade $y \ge -4$ inside that region.

- *Subcase 2b:* $x + y < 0, x - y < 0$. This gives us $-2x \le 8$, so $x \ge -4$. This subcase corresponds to region $\boxed{4}$ in our graph, so we shade $x \ge -4$ inside that region.

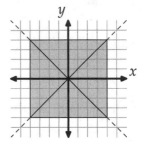

When we include our shaded regions from Case 2 with those from Case 1, we get the square at right. This square has side length 8, so its area is $8^2 = 64$.

With such a simple region as our answer, we might think that there's a simpler way to find it.

We consider the various tools we have for dealing with absolute value and see if any of them can be used to explain the result. The number line does it! We can think of $|x + y|$ as the distance between x and $-y$, and $|x - y|$ as the distance between x and y. Therefore, $|x + y| + |x - y| \le 8$ tells us that the sum of these two distances is no more than 8. If $|y| > 4$, then no matter what x is, the sum of these distances is greater than 8. To see why, consider the diagram below.

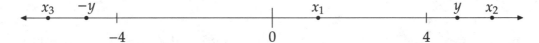

Here, we have $y > 4$, and three different possibilities for x. For x_1, which is between y and $-y$, the sum of the distances from x_1 to y and to $-y$ equals $y - (-y) = 2y$, which is greater than 8. For x_2, we have $x_2 > y$, so the distance between x_2 and $-y$ is greater than 8. Similarly, the distance between x_3 and y is greater than 8.

By similar arguments, if $|y| \le 4$ and $|x| > 4$, then the sum of the distances $x - y$ and $|x - (-y)|$ is $2|x|$, which is greater than 8. But if $|y| \le 4$ and $|x| \le 4$, then the sum of these distances is at most 8, so the square described by $-4 \le x \le 4, -4 \le y \le 4$ is the solution to $|x + y| + |x - y| \le 8$. $\square$

Don't feel bad if you didn't see the second solution right away. Often, we will work through complicated casework or computations, only to arrive at a very simple answer. When you do so, you shouldn't just immediately move on.

> **Concept:** When you get a simple answer after doing a lot of casework, look for a simple explanation for the simple answer.

Finding a simple explanation will not only provide more confidence that your answer is correct, but it may also give you greater insight into future problems.

> **Extra!** German mathematician Karl Weierstrass is credited with introducing the notation $|x|$ for absolute value in 1841. While mathematics is often regarded as a young person's pursuit, Weierstrass published a result at age 70 that still bears his name (the Weierstrass Polynomial Approximation Theorem)!

Problem 16.16:

(a) Solve the equation $|x^2 + 6x - 3| = x + 3$.

(b) Solve the inequality $|x^2 + 6x - 3| \geq x + 3$.

(c) Find the value of c if the line $y = c$ intersects the graph of $y = |x^2 + 6x - 3|$ in exactly three points.

Solution for Problem 16.16:

(a) Using casework based on where $x^2 + 6x - 3$ is positive or negative isn't too attractive (but it can be done), because the roots of this quadratic are not rational. Squaring to get rid of the absolute value signs doesn't look like much fun either. However, we can apply casework in a slightly different way. If $|x^2 + 6x - 3| = x + 3$, then either $x^2 + 6x - 3$ and $x + 3$ are equal, or they are opposites. If $x^2 + 6x - 3 = x + 3$, then $x^2 + 5x - 6 = 0$, so $x = -6$ or $x = 1$. If $x^2 + 6x - 3 = -(x + 3)$, then $x^2 + 7x = 0$, so $x = 0$ or $x = -7$.

Therefore, we appear to have four solutions: $x = -7, -6, 0$, and 1. However, if we test the solutions, something interesting happens. When $x = -7$, we have $|x^2 + 6x - 3| = |49 - 42 - 3| = 4$, but $x + 3 = -4$. In other words, we don't have $|x^2 + 6x - 3| = x + 3$ for $x = -7$. What went wrong?

Our approach to the solution overlooked an important fact: absolute values must be non-negative. When we rewrote $|x^2 + 6x - 3| = x + 3$ as the two equations $x^2 + 6x - 3 = x + 3$ and $x^2 + 6x - 3 = -(x + 3)$, we lost sight of this fact. While $x = -7$ is indeed a solution to $x^2 + 6x - 3 = -(x + 3)$, it is not a solution to $|x^2 + 6x - 3| = x + 3$. Similarly, $x = -6$ does not satisfy the original equation, but both $x = 0$ and $x = 1$ do.

> **WARNING!!** When solving an equation of the form $|f(x)| = g(x)$ by considering the cases $f(x) = g(x)$ and $f(x) = -g(x)$, you must check that the answers you find satisfy the original equation. Any solution for which $g(x)$ is negative is extraneous, since $|f(x)|$ cannot be negative.

So, the only solutions to $|x^2 + 6x - 3| = x + 3$ are $x = 0$ and $x = 1$.

(b) Unfortunately, we can't simply use the same reasoning as in part (a). It is true that if $x^2 + 6x - 3 \geq x + 3$, then $|x^2 + 6x - 3| \geq x + 3$, because $|x^2 + 6x - 3| \geq x^2 + 6x - 3$. However, if we compare $x^2 + 6x - 3$ to $-(x + 3)$ as our next step, we have a problem. It is *not true* that if $x^2 + 6x - 3 \geq -(x + 3)$, then $|x^2 + 6x - 3| \geq x + 3$. For example, if $x = \frac{1}{2}$, we have $x^2 + 6x - 3 = \frac{1}{4}$ and $-(x + 3) = -3\frac{1}{2}$, so we have $x^2 + 6x - 3 \geq -(x + 3)$. But it is not true that $|x^2 + 6x - 3| \geq x + 3$ when $x = \frac{1}{2}$. Dealing with the inequality is much trickier than dealing with the equation in (a)!

We've tried using graphs to understand complicated equations before; let's try that here. We'll graph both sides of the equation and see how the graphs are related. The right side is easy. The graph of $y = x + 3$ is just a line through $(0, 3)$ with slope 1. Graphing $y = |x^2 + 6x - 3|$ is tougher. We start by graphing $y = x^2 + 6x - 3 = (x + 3)^2 - 12$. This is an upward-opening parabola as shown at right.

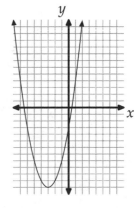

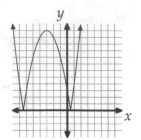

For those values of x for which $x^2 + 6x - 3$ is nonnegative, the graphs of $y = |x^2 + 6x - 3|$ and $y = x^2 + 6x - 3$ are the same. But if $x^2 - 6x + 3 < 0$, then the y values of the corresponding points on the graphs are opposites. Therefore, we get the graph of $y = |x^2 + 6x - 3|$ by copying the graph of $y = x^2 + 6x - 3$ when y is nonnegative, and by reflecting the graph of $y = x^2 + 6x - 3$ over the x-axis when y is negative. The result is the graph at right.

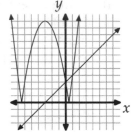

Graphing $y = x + 3$ and $y = |x^2 + 6x - 3|$ on the same Cartesian plane helps us set up our solution to the inequality. We could just read the answer off the graph, trusting that the two graphs don't intersect again somewhere off to the right of the portion of the graph shown. However, we'd like to be sure that there are no more intersection points, so we'll use our graph as a guide to an algebraic solution.

First, we note that $x + 3 < 0$ for $x < -3$. Since $|x^2 + 6x - 3|$ is nonnegative for all x, we have $|x^2 + 6x - 3| > x + 3$ for all $x < -3$. Next, we must check for solutions for which $x + 3 \geq 0$. If $|x^2 + 6x - 3| \geq x + 3 \geq 0$, then, in addition to $x \geq -3$, we must have either $x^2 + 6x - 3 \geq x + 3$ or $-(x^2 + 6x - 3) \geq x + 3$. The former gives us $(x + 6)(x - 1) \geq 0$, so $x \geq 1$ or $x \leq -6$. Since we must also have $x \geq -3$, we have $x \geq 1$ for this case. If $-(x^2 + 6x - 3) \geq x + 3$, then we have $x(x + 7) \leq 0$, so $0 \geq x \geq -7$. Since we must also have $x \geq -3$, we have $-3 \leq x \leq 0$.

Combining $x < -3$, $-3 \leq x \leq 0$, and $x \geq 1$ gives $x \in (-\infty, 0] \cup [1, +\infty)$ as our solution.

(c) *Solution 1: Graphing.* With part (b) as inspiration, we seek a horizontal line that passes through exactly 3 points on the graph of $y = |x^2 + 6x - 3|$. We use our graph from part (b). We see that any horizontal line below the x-axis misses the graph of $y = |x^2 + 6x - 3|$ entirely. Any horizontal line below the "vertex" of the middle "hump" of the graph intersects the graph in 4 points, and any horizontal line above this vertex of the middle hump intersects the graph at 2 points.

The only horizontal line that intersects the graph in 3 points is the line through the vertex. The vertex of the parabola formed by graphing $y = x^2 + 6x - 3 = (x + 3)^2 - 12$ is $(-3, -12)$. When we graph $y = |x^2 + 6x - 3|$, this point is reflected over the x-axis to give $(-3, 12)$. Therefore, the desired horizontal line is $y = 12$.

Solution 2: Algebra. If the graph of $y = c$ intersects the graph of $y = |x^2 + 6x - 3|$ at three points, then the equation $|x^2 + 6x - 3| = c$ has three real solutions. If $|x^2 + 6x - 3| = c$, then $x^2 + 6x - 3 = c$ or $x^2 + 6x - 3 = -c$. Each of these is a quadratic equation. Together, these quadratics must have 3 distinct real solutions. The only way the two quadratics can share a solution is if $c = 0$, in which case the two quadratics share both solutions, and we only have 2 distinct solutions. So, the quadratics can't share any solutions. This means that the only way they can together have 3 distinct solutions is if one of them has only 1 distinct solution.

Since the quadratic and linear terms of each quadratic equation are x^2 and $6x$, one of these equations must be equivalent to $(x + 3)^2 = 0$, which means we must have $c = \pm 12$. Clearly, c must be positive, since $|x^2 + 6x - 3| = c$ has no solutions if c is negative. So, the desired horizontal line is $y = 12$.

We can quickly check our answer by solving the equation $|x^2 + 6x - 3| = 12$. We must have either $x^2 + 6x - 3 = 12$ or $x^2 + 6x - 3 = -12$, so either $x^2 + 6x - 15 = 0$ or $x^2 + 6x + 9 = 0$. The former has solutions $x = -3 \pm 2\sqrt{6}$ and the latter only has the solution $x = -3$. Therefore, the equation $|x^2 + 6x - 3| = 12$ has exactly 3 solutions.

□

> **Concept:** Sometimes we can use the graph of a function to give us intuition about an algebraic problem.

Exercises

16.3.1 Graph the function $f(x) = |2 - 4x| - 4$, and find the x-intercepts and the y-intercepts of the graph.

16.3.2 Find all intersection points of the graphs of $f(x) = |x - 1|$ and $g(x) = |2 - x|$.

16.3.3 Find the total area of the region of points that satisfy both $|x + y| \le 8$ and $xy \ge 0$.

16.3.4 Find the area of the smallest region bounded by the graphs of $y = |x|$ and $x^2 + y^2 = 4$. *(Source: AHSME)*

16.3.5

(a) Graph $f(x) = |3 - x|$.

(b) Graph $f(x) = |x + 2| + |x + 5|$.

(c) Find another solution to part (c) of Problem 16.8 using graphing.

16.3.6 Find the minimum possible value of the function $h(x) = |x + 1| + |x - 4| + |x - 6|$.

16.3.7 Find the area of the region described by $|5x| + |6y| \le 30$.

16.4 Floor and Ceiling

> **Definitions:**
>
> - The **floor** of x is defined as the greatest integer that is less than or equal to x, and is denoted by $\lfloor x \rfloor$.
>
> - Similarly, the **ceiling** of x is defined as the least integer that is greater than or equal to x, and is denoted by $\lceil x \rceil$.
>
> - The **fractional part** of x is defined as $x - \lfloor x \rfloor$, and is denoted by $\{x\}$.

In other words, floor "rounds down" and ceiling "rounds up." The floor function is sometimes referred to as the **greatest integer function** and the ceiling function is sometimes referred to as the **least integer function**. Moreover, the notation $[x]$ is sometimes used instead of $\lfloor x \rfloor$ for the floor function.

> **Extra!** Let x and y be positive irrational numbers such that $\frac{1}{x} + \frac{1}{y} = 1$. Use a computer to compute the first 50 terms in the sequences $\lfloor x \rfloor$, $\lfloor 2x \rfloor$, $\lfloor 3x \rfloor$, ... and $\lfloor y \rfloor$, $\lfloor 2y \rfloor$, $\lfloor 3y \rfloor$, ...
> Notice anything interesting?

Here are a few examples:

$$\lfloor 3 \rfloor = 3, \qquad \lceil 3 \rceil = 3, \qquad \{3\} = 0,$$
$$\lfloor 6.25 \rfloor = 6, \qquad \lceil 6.25 \rceil = 7, \qquad \{6.25\} = 0.25,$$
$$\lfloor -6.25 \rfloor = -7, \qquad \lceil -6.25 \rceil = -6, \qquad \{-6.25\} = 0.75,$$
$$\lfloor \pi \rfloor = 3, \qquad \lceil \pi \rceil = 4, \qquad \{\pi\} = \pi - 3.$$

In a sense, $\lfloor x \rfloor$, $\lceil x \rceil$, and $\{x\}$ can all be thought of as piecewise defined functions. For example, $\lfloor x \rfloor$ equals -2 if $-2 \le x < -1$, equals -1 if $-1 \le x < 0$, equals 0 if $0 \le x < 1$, equals 1 if $1 \le x < 2$, and so on.

Most of our problems will focus on the floor function, because it is the most common of these three functions.

 Problems

Problem 16.17: Sketch the graphs of $y = \lfloor x \rfloor$, $y = \lceil x \rceil$, and $y = \{x\}$.

Problem 16.18:

(a) Find the smallest integer n such that $2^n > 1000$.

(b) Compute $\lfloor \log_2 1000 \rfloor$.

Problem 16.19: Graph each of the following:

(a) $f(x) = \lfloor x \rfloor + \lfloor -x \rfloor$ (b) $g(x) = \{x\} + \{-x\}$ (c) $f + g$

Problem 16.20:

(a) Find all x such that $\lfloor x \rfloor = 7$.

(b) Find all x such that $\lfloor 3x \rfloor = 7$.

(c) Find all x such that $\lfloor 3x + \frac{1}{2} \rfloor = 7$.

Problem 16.21: Find all y such that $3 < \lfloor 4y - 5 \rfloor \le 8$.

Problem 16.22: Find all real values of a such that $\left\lfloor \left| -a^2 + 10a - 16 \right| \right\rfloor = 1$.

Problem 16.23:

(a) Prove that $\lfloor x + n \rfloor = \lfloor x \rfloor + n$ for all real numbers x and integers n.

(b) Prove that $\{x + n\} = \{x\}$ for all real numbers x and integers n.

(c) Prove that $\lfloor x + y \rfloor \ge \lfloor x \rfloor + \lfloor y \rfloor$ for all real x, y.

Problem 16.24: Evaluate $\left\lfloor \dfrac{2007^3}{2005 \cdot 2006} - \dfrac{2005^3}{2006 \cdot 2007} \right\rfloor$ without a calculator.

Problem 16.17: Sketch the graphs of $y = \lfloor x \rfloor$, $y = \lceil x \rceil$, and $y = \{x\}$.

Solution for Problem 16.17: We start with $y = \lfloor x \rfloor$. We have $y = 0$ for all x such that $0 \le x < 1$. Similarly, $y = 1$ for $1 \le x < 2$, and $y = 2$ for $2 \le x < 3$, and so on. Looking at negative values of x, we have $y = -1$ for $-1 \le x < 0$, and $y = -2$ for $-2 \le x < -1$, and so on. Putting all this together gives us the step-like graph at left below.

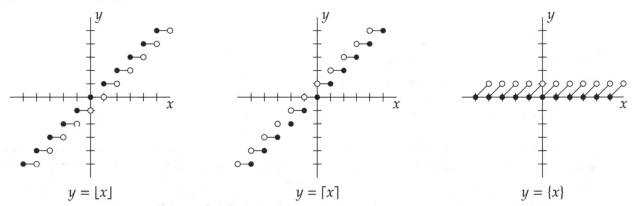

$$y = \lfloor x \rfloor \qquad\qquad y = \lceil x \rceil \qquad\qquad y = \{x\}$$

We find the graph of $y = \lceil x \rceil$ similarly, and discover the step-like graph in the middle above.

In graphing $y = \{x\}$, we note that as x goes from one integer to just less than the next larger integer, the value of $\{x\}$ increases linearly from 0 to just less than 1. This produces the repeating saw-tooth pattern at right above.

These three graphs clearly depict the piecewise nature of these three functions. $\square$

Just to make sure we have a handle on the floor function, we'll start by computing the floor of a specific number.

Problem 16.18: Compute $\lfloor \log_2 1000 \rfloor$.

Solution for Problem 16.18: If $\lfloor \log_2 1000 \rfloor = n$, then we must have $n \le \log_2 1000 < n + 1$, where n is an integer.

> **Concept:** The equation $\lfloor x \rfloor = n$, where n is an integer, is equivalent to the inequality chain $n \le x < n + 1$. Similarly, $\lceil x \rceil = n$ is equivalent to $n - 1 < x \le n$. This provides a method for getting rid of floor or ceiling notation in a problem.

Converting our inequality chain from logarithmic notation to exponential notation gives

$$2^n \le 1000 < 2^{n+1}.$$

Since $512 = 2^9 < 1000 < 2^{10} = 1024$, we have $n = \lfloor \log_2 1000 \rfloor = 9$. $\square$

Problem 16.19: Graph each of the following:

(a) $f(x) = \lfloor x \rfloor + \lfloor -x \rfloor$ (b) $g(x) = \{x\} + \{-x\}$ (c) $f + g$

Solution for Problem 16.19:

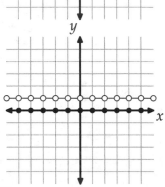

(a) When x is an integer, we have $f(x) = \lfloor x \rfloor + \lfloor -x \rfloor = x + (-x) = 0$. But what if x is not an integer? Then, we have $n < x < n+1$ for some integer n, so we have $-n > -x > -n-1$. This gives $f(x) = \lfloor x \rfloor + \lfloor -x \rfloor = n + (-n-1) = -1$. We can therefore write $f(x)$ as

$$f(x) = \begin{cases} 0 & \text{if } x \text{ is an integer,} \\ -1 & \text{if } x \text{ is not an integer.} \end{cases}$$

The graph of f is shown at right.

(b) We start with the definition of $\{x\}$, and we have

$$g(x) = \{x\} + \{-x\} = x - \lfloor x \rfloor + (-x) - \lfloor -x \rfloor = -\left(\lfloor x \rfloor + \lfloor -x \rfloor \right).$$

Aha! This is just the negative of the function in part (a)! So, we graph g by reflecting the graph of f in part (a) over the x-axis, as shown at right.

(c) Looking at the graphs of f and g in parts (a) and (b), we expect the graph of $f+g$ to simply be the x-axis. We can also see that $f(x) + g(x) = 0$ by using our definition of $\{x\}$. Since $\{x\} = x - \lfloor x \rfloor$, we have $\{x\} + \lfloor x \rfloor = x$, so

$$f(x) + g(x) = \lfloor x \rfloor + \lfloor -x \rfloor + \{x\} + \{-x\}$$
$$= \lfloor x \rfloor + \{x\} + \lfloor -x \rfloor + \{-x\} = x + (-x) = 0.$$

So, the graph of $f + g$ is the x-axis.

$\square$

Now that we have a little experience computing and graphing the floor function, let's try solving some equations and inequalities involving the floor function.

Problem 16.20: Find all x such that $\lfloor 3x + \frac{1}{2} \rfloor = 7$.

Solution for Problem 16.20: Because $\lfloor 3x + \frac{1}{2} \rfloor = 7$, we know that $3x + \frac{1}{2}$ is at least 7 and less than 8:

$$7 \le 3x + \frac{1}{2} < 8.$$

We subtract $\frac{1}{2}$ from all three parts of this inequality chain to get $\frac{13}{2} \le 3x < \frac{15}{2}$, and then divide this chain by 3 to find $\frac{13}{6} \le x < \frac{5}{2}$. $\square$

The process we used to turn the equation in Problem 16.20 into an inequality is yet another example of removing complicated notation from a problem to turn it into a problem we know how to handle.

Problem 16.21: Find all y such that $3 < \lfloor 4y - 5 \rfloor \le 8$.

Solution for Problem 16.21: Because $\lfloor 4y - 5 \rfloor$ must be an integer, we can rewrite the inequality as $4 \le \lfloor 4y - 5 \rfloor \le 8$. This tells us that $4y - 5$ is at least 4 and less than 9, so we can get rid of the floor function notation, leaving $4 \le 4y - 5 < 9$. Adding 5 to all three parts gives $9 \le 4y < 14$, and then dividing by 4 gives $\frac{9}{4} \le y < \frac{7}{2}$. $\square$

Problem 16.22: Find all real values of a such that $\lfloor |-a^2 + 10a - 16| \rfloor = 1$.

Solution for Problem 16.22: In order to solve this problem, we get rid of all the complicated notation, one step at a time. First, removing the floor function notation gives

$$1 \le |-a^2 + 10a - 16| < 2.$$

Now, removing the absolute value notation, we're looking for values of a such that either

$$1 \le -a^2 + 10a - 16 < 2 \qquad \text{or} \qquad -2 < -a^2 + 10a - 16 \le -1.$$

Rather than solve 4 different quadratic inequalities, we complete the square in $-a^2 + 10a - 16$, which gives us

$$1 \le -(a - 5)^2 + 9 < 2 \qquad \text{or} \qquad -2 < -(a - 5)^2 + 9 \le -1.$$

Isolating $(a - 5)^2$ in these inequalities gives

$$8 \ge (a - 5)^2 > 7 \qquad \text{or} \qquad 11 > (a - 5)^2 \ge 10.$$

All of the quantities in each inequality chain are positive, so we can take the square root of both inequalities to obtain

$$\sqrt{8} \ge a - 5 > \sqrt{7} \qquad \text{or} \qquad \sqrt{11} > a - 5 \ge \sqrt{10}.$$

Noting that $\sqrt{8} = 2\sqrt{2}$ and isolating a in each inequality chain gives us

$$5 + 2\sqrt{2} > a > 5 + \sqrt{7} \qquad \text{or} \qquad 5 + \sqrt{11} > a \ge 5 + \sqrt{10}.$$

$\square$

Problem 16.23:

(a) Prove that $\lfloor x + n \rfloor = \lfloor x \rfloor + n$ for all real numbers x and integers n.

(b) Prove that $\{x + n\} = \{x\}$ for all real numbers x and integers n.

(c) Prove that $\lfloor x + y \rfloor \ge \lfloor x \rfloor + \lfloor y \rfloor$ for all real x, y.

Solution for Problem 16.23:

(a) To show that $\lfloor x + n \rfloor = \lfloor x \rfloor + n$, we must show that $\lfloor x \rfloor + n \le x + n < \lfloor x \rfloor + n + 1$. Subtracting n from all three parts gives $\lfloor x \rfloor \le x < \lfloor x \rfloor + 1$, which we know is true for all x. So, we have our proof:

> **Proof:** For all x, we have $\lfloor x \rfloor \le x < \lfloor x \rfloor + 1$. Adding n to all three parts of this inequality gives
> ✓ $\lfloor x \rfloor + n \le x + n < \lfloor x \rfloor + n + 1$, so we have $\lfloor x + n \rfloor = \lfloor x \rfloor + n$.

(b) First, we write both $\{x + n\}$ and $\{x\}$ in terms of the floor function. We have $\{x + n\} = x + n - \lfloor x + n \rfloor$ and $\{x\} = x - \lfloor x \rfloor$. We can use part (a) to write $\lfloor x + n \rfloor$ as $\lfloor x \rfloor + n$, and we have

$$\{x + n\} = x + n - \lfloor x + n \rfloor = x + n - (\lfloor x \rfloor + n) = x - \lfloor x \rfloor = \{x\}.$$

(c) We need to have both $\lfloor x \rfloor$ and $\lfloor y \rfloor$ on the lesser side of an inequality. We know two inequalities with these on the lesser sides: $\lfloor x \rfloor \leq x < \lfloor x \rfloor + 1$ and $\lfloor y \rfloor \leq y < \lfloor y \rfloor + 1$. Adding these, we get

$$\lfloor x \rfloor + \lfloor y \rfloor \leq x + y < \lfloor x \rfloor + \lfloor y \rfloor + 2.$$

So, the number $x + y$ is greater than or equal to the integer $\lfloor x \rfloor + \lfloor y \rfloor$ and less than the integer $\lfloor x \rfloor + \lfloor y \rfloor + 2$. This means that $\lfloor x + y \rfloor$ is equal to $\lfloor x \rfloor + \lfloor y \rfloor$ or $\lfloor x \rfloor + \lfloor y \rfloor + 1$. In either case, we have $\lfloor x + y \rfloor \geq \lfloor x \rfloor + \lfloor y \rfloor$. $\blacksquare$

$\square$

We'll finish this section where we started: evaluating the floor of an expression. This one's a little more complicated than $\lfloor \log_2 1000 \rfloor$.

Problem 16.24: Evaluate $\left\lfloor \dfrac{2007^3}{2005 \cdot 2006} - \dfrac{2005^3}{2006 \cdot 2007} \right\rfloor$ without a calculator.

Solution for Problem 16.24: We certainly aren't going to multiply and divide everything out. We hope we can simplify the expression with some algebra, so we let $n = 2005$, write the expression inside the floor function in terms of n, and hope something good happens. We have

$$\frac{2007^3}{2005 \cdot 2006} - \frac{2005^3}{2006 \cdot 2007} = \frac{(n+2)^3}{n(n+1)} - \frac{n^3}{(n+1)(n+2)} = \frac{(n+2)^4 - n^4}{n(n+1)(n+2)}$$

$$= \frac{8n^3 + 24n^2 + 32n + 16}{n(n+1)(n+2)} = \frac{8(n+1)(n^2 + 2n + 2)}{n(n+1)(n+2)}$$

$$= \frac{8(n^2 + 2n + 2)}{n(n+2)} = \frac{8n(n+2) + 16}{n(n+2)}$$

$$= 8 + \frac{16}{n(n+2)}.$$

Something good happened! Because $0 < \dfrac{16}{n(n+2)} < 1$ for $n = 2005$, we have

$$\left\lfloor \frac{2007^3}{2005 \cdot 2006} - \frac{2005^3}{2006 \cdot 2007} \right\rfloor = 8.$$

$\square$

Concept: When you have to compute an expression involving very large numbers, assign a variable to one of the numbers in the problem. Write the expression in terms of that variable, and then simplify the expression with algebra before evaluating it.

Exercises

16.4.1 Find all values of x that satisfy $\lfloor x \rfloor + \lceil x \rceil = 9$.

16.4.2 Find all integers n such that $\lfloor n/6 \rfloor = 5$.

16.4.3 Evaluate $\left\lfloor \sqrt{\log_3 1000} \right\rfloor$ in your head. Do not use a calculator and do not use paper and pencil.

16.4.4 Graph both $f(x) = \lceil x \rceil + \lfloor x \rfloor$ and $g(x) = \lceil x \rceil - \lfloor x \rfloor$.

16.4.5 Find all x such that $-2 \le \lfloor 3x + 2 \rfloor \le 2$.

16.4.6 Find $\dfrac{\{\sqrt{3}\}^2 - 2\{\sqrt{2}\}^2}{\{\sqrt{3}\} - 2\{\sqrt{2}\}}$ without a calculator. *(Source: NYSML)*

16.4.7★ Compute $\left\lfloor \dfrac{3^{31} + 2^{31}}{3^{29} + 2^{29}} \right\rfloor$ without a calculator. *(Source: ARML)* **Hints:** 260, 22

16.5 Problem Solving with the Floor Function

In this section, we explore a couple more approaches to solving problems involving the floor function. You shouldn't view these as brand-new techniques; they have the same motivation as nearly all the problem solving approaches we've employed in the last few chapters:

> **Concept:** When faced with a problem involving a new notation or new function, try to find a way to turn the problem into one you know how to do.

Problems

Problem 16.25:
 (a) Find all values of x such that $x + \lfloor x \rfloor = 6.7$.
 (b) Find all real numbers k satisfying $\lfloor k \rfloor = 5k - 14$.

Problem 16.26: Determine the value of $10(x + y)$, given that $\lfloor x \rfloor + \lfloor y \rfloor + y = 43.8$ and $x + y - \lfloor x \rfloor = 18.4$. *(Source: AIME)*

Problem 16.27:
 (a) Find $\lfloor \sqrt{1} \rfloor, \lfloor \sqrt{2} \rfloor, \lfloor \sqrt{3} \rfloor, \lfloor \sqrt{4} \rfloor, \lfloor \sqrt{5} \rfloor, \lfloor \sqrt{6} \rfloor, \lfloor \sqrt{7} \rfloor, \lfloor \sqrt{8} \rfloor$, and $\lfloor \sqrt{9} \rfloor$.
 (b) Find $\displaystyle\sum_{k=1}^{99} \lfloor \sqrt{k} \rfloor$.

Problem 16.28: Let $f(x) = \lfloor 2x \rfloor + \lfloor 4x \rfloor + \lfloor 6x \rfloor + \lfloor 8x \rfloor$.
 (a) At what values of x does the graph of $y = \lfloor 2x \rfloor$ "step up" from one value of y to another? How about the graph of $y = \lfloor 4x \rfloor$? The graph of $y = \lfloor 6x \rfloor$? The graph of $y = \lfloor 8x \rfloor$?
 (b) How much does the graph of f increase at each of the values of x you found in part (a)?
 (c) How many of the first 1000 positive integers are in the range of $f(x)$? *(Source: AIME)*

> **Problem 16.25:**
> (a) Find all values of x such that $x + \lfloor x \rfloor = 6.7$.
> (b) Find all real numbers k satisfying $\lfloor k \rfloor = 5k - 14$.

Solution for Problem 16.25:

(a) We know that $\lfloor x \rfloor$ is an integer, so if $x + \lfloor x \rfloor$ equals 6.7, then the ".7" must be the fractional part of x. Therefore, we can write x as $\lfloor x \rfloor + 0.7$, so our equation becomes $2\lfloor x \rfloor + 0.7 = 6.7$, from which we find $\lfloor x \rfloor = 3$. This gives us $x = 3.7$.

The key step in this solution was viewing x as the sum of its floor and its fractional part: $x = \lfloor x \rfloor + \{x\}$.

> **Concept:** Writing a variable as the sum of its fractional part and its floor can help unravel problems involving the floor function.

(b) We use our previous part as inspiration, writing $k = \lfloor k \rfloor + \{k\}$. This gives us

$$\lfloor k \rfloor = 5k - 14 = 5\lfloor k \rfloor + 5\{k\} - 14.$$

Solving for $\{k\}$, we get $\{k\} = \dfrac{14 - 4\lfloor k \rfloor}{5}$. Since $0 \le \{k\} < 1$, we have $0 \le \dfrac{14 - 4\lfloor k \rfloor}{5} < 1$. Solving this inequality chain for $\lfloor k \rfloor$ gives $\frac{9}{4} < \lfloor k \rfloor \le \frac{7}{2}$. The only integer that lies in this interval is 3, so $\lfloor k \rfloor = 3$. Therefore, we have $\{k\} = (14 - 4 \cdot 3)/5 = 2/5$, so $k = \lfloor k \rfloor + \{k\} = 17/5$.

We also could have started with the "build an inequality" approach from the previous section. We first note that $\lfloor k \rfloor \le k < \lfloor k \rfloor + 1$. Solving the given equation for k gives us

$$k = \frac{\lfloor k \rfloor + 14}{5},$$

so $\lfloor k \rfloor \le \dfrac{\lfloor k \rfloor + 14}{5} < \lfloor k \rfloor + 1$. From here, we can obtain the same bounds on $\lfloor k \rfloor$ as before. $\square$

In both parts, we use the fact that:

> **Important:** The fractional part of a number is greater than or equal to 0 and less than 1.

The inequality chain $0 \le \{x\} < 1$ is really just another way of writing $\lfloor x \rfloor \le x < \lfloor x \rfloor + 1$. (If you don't see why these inequality chains are equivalent, add $\lfloor x \rfloor$ to all three parts of $0 \le \{x\} < 1$.)

Here's another problem in which viewing a number as the sum of its integer and fractional parts helps immensely.

> **Problem 16.26:** Determine the value of $10(x + y)$, given that $\lfloor x \rfloor + \lfloor y \rfloor + y = 43.8$ and $x + y - \lfloor x \rfloor = 18.4$. *(Source: AIME)*

Solution for Problem 16.26: From our first equation, we know that the fractional part of y is 0.8. This makes us think to write $x = \lfloor x \rfloor + \{x\}$ and $y = \lfloor y \rfloor + \{y\}$. Then the given equations become

$$\lfloor x \rfloor + 2\lfloor y \rfloor + \{y\} = 43.8,$$
$$\{x\} + \lfloor y \rfloor + \{y\} = 18.4.$$

The only non-integer on the left side of the first equation is $\{y\}$, so we must have $\{y\} = 0.8$ to account for the fractional part of 43.8 on the right side. Therefore, our second equation gives us $\{x\} + \lfloor y \rfloor = 18.4 - 0.8 = 17.6$. This tells us that $\{x\}$ must be the fractional part of 17.6, so $\{x\} = 0.6$. Now our system of equations is

$$\lfloor x \rfloor + 2\lfloor y \rfloor = 43,$$
$$\lfloor y \rfloor = 17.$$

Subtracting twice the second equation from the first gives $\lfloor x \rfloor = 43 - 2 \cdot 17 = 9$, so $x = \lfloor x \rfloor + \{x\} = 9.6$ and $y = \lfloor y \rfloor + \{y\} = 17.8$. Finally, we have $10(x + y) = 274$. $\square$

As we saw in Problem 16.17, the graph of $y = \lfloor x \rfloor$ looks somewhat like a series of steps. Some problems can be solved by considering where these "steps" take place.

Problem 16.27: Find $\displaystyle\sum_{k=1}^{99} \left\lfloor \sqrt{k} \right\rfloor$.

Solution for Problem 16.27: First, we get rid of the summation notation to get a look at the terms in the sum:

$$\sum_{k=1}^{99} \left\lfloor \sqrt{k} \right\rfloor = \left\lfloor \sqrt{1} \right\rfloor + \left\lfloor \sqrt{2} \right\rfloor + \left\lfloor \sqrt{3} \right\rfloor + \left\lfloor \sqrt{4} \right\rfloor + \left\lfloor \sqrt{5} \right\rfloor + \cdots + \left\lfloor \sqrt{99} \right\rfloor.$$

Let's start with the first few terms. The first term is easy: $\left\lfloor \sqrt{1} \right\rfloor = \lfloor 1 \rfloor = 1$. What is $\left\lfloor \sqrt{2} \right\rfloor$? Because $1 < 2 < 4$, we have $1 < \sqrt{2} < 2$, so $\left\lfloor \sqrt{2} \right\rfloor = 1$. Similarly, we have $\left\lfloor \sqrt{3} \right\rfloor = 1$.

The next term is then $\left\lfloor \sqrt{4} \right\rfloor = 2$. Each of $\left\lfloor \sqrt{5} \right\rfloor, \left\lfloor \sqrt{6} \right\rfloor, \left\lfloor \sqrt{7} \right\rfloor$, and $\left\lfloor \sqrt{8} \right\rfloor$ also equals 2, and then $\left\lfloor \sqrt{9} \right\rfloor = 3$.

Continuing in this manner, each term $\lfloor \sqrt{k} \rfloor$ in the sum is equal to the previous term until k becomes a perfect square, at which point the term "steps up" by 1. We see that 1 appears 3 times in the sum, and 2 appears 5 times. We might guess that 3 appears 7 times, 4 appears 9 times, and so on. Let's see if we can prove that this is the case.

We do so by finding all integers k such that $\left\lfloor \sqrt{k} \right\rfloor = m$, where m is a positive integer. If $\left\lfloor \sqrt{k} \right\rfloor = m$, then we must have $m \le \sqrt{k} < m + 1$, so $m^2 \le k < (m + 1)^2$. In other words, k must be one of the integers m^2, $m^2 + 1, m^2 + 2, \ldots, (m + 1)^2 - 1$. There are $(m + 1)^2 - 1 - m^2 + 1 = 2m + 1$ integers in this list, so there are $2m + 1$ integers k such that $\left\lfloor \sqrt{k} \right\rfloor = m$.

So, each integer m with $1 \le m \le 9$ appears $2m + 1$ times in the sum. We have to be careful to check the boundary cases. Since our sum starts at $\sqrt{1}$ and ends one term shy of $\left\lfloor \sqrt{10^2} \right\rfloor$, we do indeed have all $2 \cdot 1 + 1 = 3$ terms that equal 1 and all $2 \cdot 9 + 1 = 19$ terms that equal 9.

Now, we can evaluate our sum:

$$\sum_{k=1}^{99} \left\lfloor \sqrt{k} \right\rfloor = \sum_{m=1}^{9} (2m+1)m = 3 + 10 + 21 + 36 + 55 + 78 + 105 + 136 + 171 = 615.$$

□

In this problem, we removed the floor function by counting how many times each integer appears as a term in the sum. This approach was inspired by checking the first few terms of the sum, which revealed a pattern in the number of times that each integer appears in the sum. Then we proved that our pattern always holds.

> **Concept:** Finding patterns is an important problem solving tool. This is particularly true of problems involving sequences and series—looking at early terms in a sum can reveal clues about how to evaluate the sum.

Problem 16.28: How many of the first 1000 positive integers can be expressed in the form

$$\lfloor 2x \rfloor + \lfloor 4x \rfloor + \lfloor 6x \rfloor + \lfloor 8x \rfloor,$$

where x is a real number? *(Source: AIME)*

Solution for Problem 16.28: Let $f(x) = \lfloor 2x \rfloor + \lfloor 4x \rfloor + \lfloor 6x \rfloor + \lfloor 8x \rfloor$. We experiment with this function by examining what happens to $f(x)$ as x goes from 0 to 1. For all $x \in [0, \frac{1}{8})$, we have $f(x) = 0$, since $2x$, $4x$, $6x$, and $8x$ are positive numbers less than 1. But when we hit $x = \frac{1}{8}$, the value of $\lfloor 8x \rfloor$ steps up from 0 to 1, and we have $f(\frac{1}{8}) = 1$. For all $x \in [\frac{1}{8}, \frac{1}{6})$, the first three terms of $f(x)$ equal zero, while the last equals 1. But when we hit $x = \frac{1}{6}$, $f(x)$ steps up again, and we have $f(\frac{1}{6}) = 2$.

Continuing in this vein, we see that $f(x) = 2$ for all $x \in [\frac{1}{6}, \frac{1}{4})$. But when $x = \frac{1}{4}$, something interesting happens: $f(x)$ steps up by 2, because both $\lfloor 4x \rfloor$ and $\lfloor 8x \rfloor$ step up by 1 when we hit $x = \frac{1}{4}$. So, we find $f(\frac{1}{4}) = 0 + 1 + 1 + 2 = 4$. Now, we're starting to get a grasp on $f(x)$. We see that $f(x)$ steps up whenever at least one of its terms steps up, so we focus on those values of x at which each term steps up.

We find that

$$\lfloor 2x \rfloor \text{ steps up when } x = \frac{n}{2}, \text{ for any integer } n,$$

$$\lfloor 4x \rfloor \text{ steps up when } x = \frac{n}{4}, \text{ for any integer } n,$$

$$\lfloor 6x \rfloor \text{ steps up when } x = \frac{n}{6}, \text{ for any integer } n,$$

$$\lfloor 8x \rfloor \text{ steps up when } x = \frac{n}{8}, \text{ for any integer } n.$$

This means that steps occur at $\frac{1}{8}, \frac{1}{6}, \frac{1}{4}, \frac{1}{3}, \frac{3}{8}, \frac{1}{2}, \frac{5}{8}, \frac{2}{3}, \frac{3}{4}, \frac{5}{6}, \frac{7}{8}$, and 1. So, from $f(0)$ to $f(1)$, $f(x)$ steps up a total of 12 different times. However, $f(x)$ sometimes steps up more than 1 unit at a time, so some integers are not in its range. Because $f(0) = 0$ and $f(1) = 20$, as f takes 12 steps from 0 to 20, it hits 12 positive integers (including 20).

Now, we have a handle on how $f(x)$ behaves for $0 \leq x \leq 1$. What about other values of x? Do we need to start from scratch for $1 \leq x \leq 2$?

No! From our analysis above, we see that $f(x)$ steps up exactly 12 times as x goes from n to $n+1$ for any integer n, not just for $n = 0$. Moreover, for any integer n, we have $f(n+1) - f(n) = 20n + 20 - 20n = 20$. So, during these 12 steps, $f(x)$ hits 12 of the integers from $20n$ to $20n+20$, including $20n+20$ but excluding $20n$.

Putting all this together, we note that as x goes from 0 to 50, the value of $f(x)$ starts at 0, steps up $12 \cdot 50 = 600$ times, and ends at 1000. Each step gives us another integer than can be expressed in the form $\lfloor 2x \rfloor + \lfloor 4x \rfloor + \lfloor 6x \rfloor + \lfloor 8x \rfloor$, so we have 600 such integers. $\square$

> **Concept:** Much of the behavior of many functions that involve the floor function is determined by when and by how much the floor function part(s) step up.

Exercises

16.5.1 Find all values of x such that $x + \lfloor x \rfloor = 5.7$.

16.5.2 Compute the number of ordered pairs (x, y) with $x > 0$ and $y > 0$ that satisfy $x + \lfloor y \rfloor = 5.3$ and $y + \lfloor x \rfloor = 5.7$. *(Source: ARML)*

16.5.3 Solve the equation $7t + \lfloor 2t \rfloor = 52$.

16.5.4 For how many positive integers n is it true that $n < 1000$ and that $\lfloor \log_2 n \rfloor$ is a positive even integer? *(Source: AIME)*

16.5.5★ Let $S_n = |\sqrt{1}| + \lfloor \sqrt{2} \rfloor + \lfloor \sqrt{3} \rfloor + \cdots + \lfloor \sqrt{n} \rfloor$. Compute the largest value of $k < 1997$ such that $S_{1997} - S_k$ is a perfect square. *(Source: ARML)*

16.5.6★ Find the smallest real number x such that $\dfrac{x}{\lfloor x \rfloor} = \dfrac{2002}{2003}$. *(Source: ARML)* **Hints:** 261, 290

16.6 Summary

A **piecewise defined function** applies different rules to produce output for different inputs. To define such a function, we must describe both the rules used to produce output, and the inputs each rule is for.

In this chapter, we studied a few special classes of functions that can be thought of as piecewise defined functions.

> **Definition:** The **absolute value** of the real number x, denoted as $|x|$, is defined as
> $$|x| = \begin{cases} x & \text{if } x \geq 0, \\ -x & \text{if } x < 0. \end{cases}$$

Definitions:

- The **floor** of x is defined as the greatest integer that is less than or equal to x, and is denoted by $\lfloor x \rfloor$.

- Similarly, the **ceiling** of x is defined as the least integer that is greater than or equal to x, and is denoted by $\lceil x \rceil$.

- The **fractional part** of x is defined as $x - \lfloor x \rfloor$, and is denoted by $\{x\}$.

Problem Solving Strategies

Concepts:

- When working with an unusual expression in a problem, try to find ways to convert the problem into one involving expressions you know how to handle.

- Don't just stare at a problem when you're stuck. Experiment and look for patterns.

- An effective technique for solving some equations with absolute values is to break the problem into cases by viewing the absolute values as piecewise defined functions. Remember to check each solution. In each case, the solution must satisfy the restrictions of that case in order to be a valid solution.

- Viewing the absolute value of the difference between two real numbers as the distance between the numbers on the number line can help solve some absolute value problems.

- When you get a simple answer after doing a lot of casework, look for a simple explanation for the simple answer.

- Sometimes we can use the graph of a function to give us intuition about an algebraic problem.

- The equation $\lfloor x \rfloor = n$, where n is an integer, is equivalent to the inequality chain $n \le x < n + 1$. Similarly, $\lceil x \rceil = n$ is equivalent to $n - 1 < x \le n$. This provides a method for getting rid of floor or ceiling notation in a problem.

- When you have to compute an expression involving very large numbers, assign a variable to one of the numbers in the problem. Write the expression in terms of that variable, then simplify the expression with algebra before evaluating it.

- Writing a variable as the sum of its fractional part and its floor can help unravel problems involving the floor function.

16.29 Let $f(x) = \begin{cases} x^3 - x & \text{if } x \le 3, \\ 2\sqrt{x} & \text{if } 3 < x \le 8, \\ 2|17 - x| & \text{if } x > 8. \end{cases}$

(a) Find $f(-2)$, $f(3)$, and $f(8)$.

(b) Find all c such that $f(c) = 6$.

16.30 Find all intersection points of the graphs of f and g if

$$f(x) = \begin{cases} x^2 - 12 & \text{if } x \le -3, \\ x + 7 & \text{if } -3 < x < 2, \\ 5 & \text{if } 2 \le x \le 9, \end{cases} \quad \text{and} \quad g(x) = \begin{cases} -4x & \text{if } x \le 4, \\ x - 9 & \text{if } x > 4. \end{cases}$$

16.31 Simplify $|x - \sqrt{(x-9)^2}|$ for $x < 0$.

16.32 Find all x that satisfy the equation $x|x + 1| - 1 = x - |x + 1|$.

16.33 Given that a, b, and c are all real numbers, find all possible values of the expression

$$\frac{a}{|a|} + \frac{b}{|b|} + \frac{c}{|c|} + \frac{abc}{|abc|}.$$

(Source: AHSME)

16.34 Find the minimum possible value of $|x| + |x + 3|$, where x is a real number.

16.35 Let a and b be distinct real numbers. Find all x in terms of a and b such that $|x - a| = |x - b|$.

16.36 Let a, b, and c be real numbers such that $|a| = |b - 2|$, $|b| = |c - 2|$, and $|c| = |a - 2|$. Prove that $a + b + c = 3$.

16.37 Find all possible values of a such that there are real numbers x for which $|x - 4| + |x + 5| < a$.

16.38 Find all ordered pairs of real numbers (x, y) such that $|x| + |y| = 10$ and $xy = 24$.

16.39 Find $f^{-1}(x)$ if $f(x) = x|x| + 2$.

16.40 If x and y are nonzero real numbers such that $|x| + y = 3$ and $|x|y + x^3 = 0$, then find the integer nearest to $x - y$. *(Source: AHSME)*

16.41 Solve each of the following:

(a) $|x^2 + 8x - 3| = 19$

(b) $|z^2 - z - 18| + 3 = z$

(c) $|2 - 4r - r^2| + 3 \ge 0$

(d) $|t^2 - 4t + 5| \le 6$

16.42 A line intersects the graph of $y = |x|$ at the points A and B. The midpoint of $\overline{AB}$ is $(3,4)$. Find the length of $\overline{AB}$.

16.43 Evaluate $\lfloor \log_2 \sqrt{999} \rfloor$.

16.44 What values of r satisfy $\left\lfloor \sqrt{2r-7} \right\rfloor = 9$?

16.45 Find all real x such that $1990\lfloor x \rfloor + 1989\lfloor -x \rfloor = 1$.

16.46 Find all x such that $x^3 - \lfloor x \rfloor = 47$.

16.47 Show that $\lceil x \rceil = -\lfloor -x \rfloor$ for all real numbers x.

16.48 For each real number x, let $f(x)$ be the minimum of the numbers $4x + 1$, $x + 2$, and $-2x + 4$. Find the maximum value of $f(x)$. *(Source: AHSME)*

Challenge Problems

16.49 Suppose r is a real number for which

$$\left\lfloor r + \frac{19}{100} \right\rfloor + \left\lfloor r + \frac{20}{100} \right\rfloor + \left\lfloor r + \frac{21}{100} \right\rfloor + \cdots + \left\lfloor r + \frac{91}{100} \right\rfloor = 546.$$

Find $\lfloor 100r \rfloor$. *(Source: AIME)* **Hints:** 207

16.50 Compute $\lfloor \sqrt{n^2 - 10n + 29} \rfloor$ when $n = 19941994$. *(Source: ARML)* **Hints:** 332

16.51 Find the largest positive number δ such that $|\sqrt{x} - 2| < 0.1$ whenever $|x - 4| < \delta$.

16.52 Find the area of the region enclosed by the graph of $|x-60|+|y| = |x/4|$. *(Source: AIME)* **Hints:** 349

16.53 Find all values of y such that $\left| \frac{2-3y}{3+y} \right| \le \frac{1}{y}$. **Hints:** 248

16.54 Let $S(x) = 1$ if $x \ge 0$ and $S(x) = 0$ if $x < 0$. Express the function

$$g(x) = \begin{cases} 2 & \text{if } -3 \le x \le 3, \\ 0 & \text{otherwise,} \end{cases}$$

in terms of $S(x)$ without using piecewise notation. **Hints:** 111

16.55 Find the domain of $f(x) = \dfrac{\sqrt{64 - x^2}}{\sqrt[4]{16 - |2x + 5|}}$.

16.56 Find $100(x + y + z)$ given that

$$x + \lfloor y \rfloor + \{z\} = 1.08,$$
$$\lfloor x \rfloor + \{y\} + z = 3.31,$$
$$\{x\} + y + \lfloor z \rfloor = 2.99.$$

Hints: 291

16.57 What positive real number x has the property that x, $\lfloor x \rfloor$, and $x - \lfloor x \rfloor$ form a geometric progression?
Hints: 242, 193

16.58 Compute the smallest positive integer x greater than 9 such that

$$\lfloor x \rfloor - 19 \cdot \left\lfloor \frac{x}{19} \right\rfloor = 9 = \lfloor x \rfloor - 89 \cdot \left\lfloor \frac{x}{89} \right\rfloor.$$

(Source: ARML) **Hints:** 98

16.59 Find polynomials $f(x)$, $g(x)$, and $h(x)$, such that, for all x,

$$|f(x)| - |g(x)| + h(x) = \begin{cases} -1 & \text{if } x < -1, \\ 3x + 2 & \text{if } -1 \le x \le 0, \\ -2x + 2 & \text{if } x > 0. \end{cases}$$

(Source: Putnam) **Hints:** 88, 142

16.60 Let x be a real number selected uniformly at random between 100 and 200. If $\lfloor \sqrt{x} \rfloor = 12$, find the probability that $\lfloor \sqrt{100x} \rfloor = 120$. *(Source: AHSME)*

16.61 Prove that

$$\lfloor nx \rfloor = \lfloor x \rfloor + \left\lfloor x + \frac{1}{n} \right\rfloor + \left\lfloor x + \frac{2}{n} \right\rfloor + \cdots + \left\lfloor x + \frac{n-1}{n} \right\rfloor.$$

This relationship is called **Hermite's Identity**. **Hints:** 57, 204

16.62 Prove that $\lfloor 2x \rfloor + \lfloor 2y \rfloor \ge \lfloor x \rfloor + \lfloor y \rfloor + \lfloor x + y \rfloor$ for all real numbers x and y.

16.63★ Let S be the set of points in the Cartesian plane that satisfy $||\,|x| - 2| - 1| + ||\,|y| - 2| - 1| = 1$. If you build a model of S from wire of negligible thickness, then exactly how much wire would you need? *(Source: AIME)*

16.64★ Let $f(x) = x \cdot \lfloor x \cdot \lfloor x \cdot \lfloor x \rfloor \rfloor \rfloor$ for all positive real numbers x.

(a) Find x such that $f(x) = 2001$.

(b) Prove there are no solutions to $f(x) = 2002$.

(Source: USAMTS)

16.65★ Find the number of points of intersection of the graphs of $(x - \lfloor x \rfloor)^2 + y^2 = x - \lfloor x \rfloor$ and $y = \frac{x}{5}$. *(Source: ARML)* **Hints:** 211, 331

16.66★ Suppose that a is positive, $\{a^{-1}\} = \{a^2\}$, and $2 < a^2 < 3$. Find the value of $a^{12} - 144a^{-1}$. *(Source: AIME)* **Hints:** 158, 232

16.67★ Let $f(n)$ be the integer closest to $\sqrt[4]{n}$. Find $\displaystyle\sum_{k=1}^{1995} \frac{1}{f(k)}$. *(Source: AIME)* **Hints:** 161, 97

16.68★ Define $f(n)$ by

$$f(n) = \begin{cases} n/2, & \text{if } n \text{ is even,} \\ (n + 1023)/2, & \text{if } n \text{ is odd.} \end{cases}$$

Find the least positive integer n such that $f(f(f(f(f(n))))) = n$. *(Source: Duke Math Meet)* **Hints:** 78, 271, 131

16.69★ What is the units digit of $\left\lfloor \dfrac{10^{20000}}{10^{100} + 3} \right\rfloor$? *(Source: Putnam)* **Hints:** 156, 128

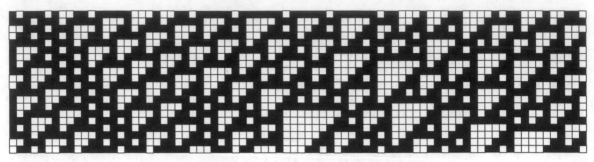

In order to understand recursion, one must first understand recursion. – Anonymous

CHAPTER 17
_____ More Sequences and Series

In this chapter, we explore more types of sequences and series, and we develop more techniques for working with them. Many of these techniques have a common thread:

> **Concept:** We can often learn a lot about a sequence or series by comparing each term
> to neighboring terms.

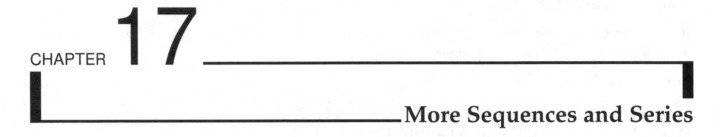

17.1 Algebra of Recursive Sequences

A **recursive sequence** is a sequence in which, after one or more initial terms, each term is defined as a function of previous terms in the sequence. Such a sequence definition is called a **recursion**. The recursion is sometimes also called the **recursive formula** or the **recurrence relation**.

The **Fibonacci sequence** is one of the most famous recursive sequences. The Fibonacci sequence begins 0, 1, and then each subsequent term is the sum of the two preceding terms. The first ten terms of the Fibonacci sequence are

$$0, 1, 1, 2, 3, 5, 8, 13, 21, 34.$$

Algebraically, we label the first two terms of this sequence F_0 and F_1, so $F_0 = 0$ and $F_1 = 1$. Each subsequent term is given by the recursion $F_n = F_{n-1} + F_{n-2}$.

In this section, we explore how to work with recursions algebraically. Art of Problem Solving's *Intermediate Counting & Probability* text includes much more information about recursions, such as how to solve counting problems with recursions.

One common way to work with recursive sequences is to find a **closed form** for the sequence. A closed form for a sequence $\{a_n\}$ is a formula that can be used to compute a_n for any n, without knowing any other terms of the sequence, and without summations or notation involving "$\cdots$". So, for example,

$a_n = 2n - 3$ is a closed form for a sequence $\{a_n\}$, but $b_n = 2b_{n-1}$ and $c_n = 1 + 2 + 3 + \cdots + (n-1)^2 + n^2$ are not closed forms.

Problems

Problem 17.1:

(a) Let $a_0 = 3$ and $a_n = a_{n-1} + 7$ for $n > 0$. What type of sequence is $\{a_n\}$? Find a closed form for a_n.

(b) Let $b_0 = 4$ and $b_n = 3b_{n-1}$ for $n > 0$. What type of sequence is $\{b_n\}$? Find a closed form for b_n.

(c) Let $c_0 = 7$ and $c_n = 2c_{n-1} + 1$ for $n > 0$. Compute several terms of $\{c_n\}$. Find a constant k such that $\{c_n + k\}$ is a sequence you recognize. Find a closed form for c_n.

Problem 17.2: Let $\{a_k\}$ be a sequence of integers such that $a_1 = 1$ and $a_{m+n} = a_m + a_n + mn$ for all positive integers m and n. In this problem, we find a_{12}. *(Source: AMC 10)*

(a) Choose a convenient value for m to reduce the number of variables in the recursion.

(b) Use the recursion that results from part (a) to find a_{12}.

Problem 17.3: For $n \geq 1$, the terms in a sequence satisfy the general relation $a_n^2 - a_{n-1}a_{n+1} = (-1)^n$.

(a) Find a formula for a_{n+1} in terms of a_n and a_{n-1}.

(b) Given $a_0 = 1$ and $a_1 = 3$, find the value of a_3. *(Source: AHSME)*

Problem 17.4: Let a be a positive constant. Consider the sequence defined recursively by $u_1 = a$ and $u_{n+1} = -1/(u_n + 1)$, for $n = 1, 2, 3, \ldots$. Find the smallest positive integer d such that $u_{n+d} = u_n$ for all positive integers n and all possible values of a.

Problem 17.5: A sequence is defined such that $a_1 = a_2 = 1$ and $a_{n+2} = \dfrac{a_{n+1} + 1}{a_n}$ for $n \geq 1$.

(a) Compute a_3, a_4, a_5, a_6, a_7, and a_8.

(b) Find the smallest positive integer d such that $a_{n+d} = a_n$ for all $n \geq 1$.

(c) Compute a_t where $t = 1998^5$. *(Source: ARML)*

Problem 17.6: A sequence of positive real numbers $a_1, a_2, a_3, \ldots$ has the property that for $i \geq 2$, each a_i is equal to the sum of all the previous terms.

(a) If $a_{19} = 99$, then what is a_{20}? *(Source: Mandelbrot)*

(b) Let $a_1 = k$. Find a closed form for a_n in terms of k for $n \geq 2$.

Problem 17.7: Define two sequences of rational numbers as follows. Let $a_0 = 2$ and $b_0 = 3$, then recursively define

$$a_{n+1} = \frac{a_n^2}{b_n} \quad \text{and} \quad b_{n+1} = \frac{b_n^2}{a_n}$$

for $n \geq 0$. Determine b_8, leaving your answer in the form $3^m/2^n$. *(Source: Mandelbrot)*

Problem 17.1:

   (a) Find a closed form for a_n given that $a_0 = 3$ and $a_n = a_{n-1} + 7$ for $n \geq 1$.

   (b) Find a closed form for b_n given that $b_0 = 4$ and $b_n = 3b_{n-1}$ for $n \geq 1$.

   (c) Find a closed form for c_n given that $c_0 = 7$ and $c_n = 2c_{n-1} + 1$ for $n \geq 1$.

Solution for Problem 17.1:

   (a) We start by computing the first few terms of the sequence $\{a_n\}$, and we have $a_0 = 3, a_1 = a_0 + 7 = 10$, $a_2 = a_1 + 7 = 17$, and $a_3 = a_2 + 7 = 24$. Our sequence starts $3, 10, 17, 24$; this looks like an arithmetic sequence. Subtracting a_{n-1} from both sides of our recursive formula gives us $a_n - a_{n-1} = 7$. This shows that the difference between consecutive terms in the sequence $\{a_n\}$ is constant, so $\{a_n\}$ is indeed an arithmetic sequence with common difference 7. To find a_n, we add the common difference n times to a_0, which gives us $a_n = 7n + a_0 = 7n + 3$. (We add the common difference n times, not $n-1$ times, because we start with a_0, not a_1.)

   (b) Computing the first few terms gives us $4, 12, 36, 108$. This looks like a geometric sequence. This shouldn't be surprising; the recursion $b_n = 3b_{n-1}$ tells us that after the first term, each term in the sequence is 3 times the previous term, so the sequence $\{b_n\}$ is geometric. To find b_n, we multiply b_0 by the common ratio n times: $b_n = 3^n \cdot b_0 = 4 \cdot 3^n$.

   (c) This one's tougher than the last two. Again, we write out a few terms in the sequence, hoping to spot something helpful. We note that the sequence is almost, but not quite, geometric:

$$7, \ 15, \ 31, \ 63, \ 127, \ 255, \ 511, \ 1023, \ 2047, \ 4095, \ \ldots$$

If we let $d_n = c_n + 1$, then $\{d_n\}$ appears to be a geometric sequence. To prove that it is, we let $c_n = d_n - 1$ in the recursion $c_n = 2c_{n-1} + 1$, and we have $d_n - 1 = 2(d_{n-1} - 1) + 1$. Rearranging this gives $d_n = 2d_{n-1}$. Therefore, after d_0, each term in $\{d_n\}$ is double the term before it, so the sequence $\{d_n\}$ is geometric. Because $d_0 = c_0 + 1 = 8 = 2^3$, we have $d_n = 2^{n+3}$. Therefore, $c_n = d_n - 1 = 2^{n+3} - 1$.

$\square$

   Problem 17.1 showed us a very common way of investigating recursions:

> **Concept:**   To learn more about a recursion, compute the first few terms and look for a pattern. If you find a pattern, use the recursion to figure out why the pattern exists.

Some recursive formulas use more than one variable to define the recursion. Here's an example:

Problem 17.2: Let $\{a_k\}$ be a sequence of integers such that $a_1 = 1$ and $a_{m+n} = a_m + a_n + mn$ for all positive integers m and n. Find a_{12}. *(Source: AMC 10)*

Solution for Problem 17.2:   We can assign any positive integer values to m and n. Because we know $a_1 = 1$, we can simplify the recursion by letting $m = 1$, which gives us

$$a_{n+1} = a_1 + a_n + n = a_n + n + 1.$$

Using this simpler recursion, we relate successive terms in the sequence:

$$a_2 = a_1 + 2,$$
$$a_3 = a_2 + 3,$$
$$a_4 = a_3 + 4,$$
$$\vdots$$
$$a_{12} = a_{11} + 12.$$

Now we can calculate terms rather easily. In particular,

$$a_{12} = 12 + a_{11} = 12 + 11 + a_{10} = \cdots = 12 + 11 + \cdots + 2 + 1 = \frac{12 \cdot 13}{2} = 78.$$

□

Problem 17.3: Given $a_0 = 1$, $a_1 = 3$, and the general relation $a_n^2 - a_{n-1}a_{n+1} = (-1)^n$ for $n \geq 1$, find a_3. (Source: AHSME)

Solution for Problem 17.3: We could jump right into substituting values for n and evaluating terms of the sequence. However, if we first solve the equation for a_{n+1}, we'll have a simple computation at each step, rather than a linear equation to solve. Isolating $a_{n-1}a_{n+1}$ gives us $a_{n-1}a_{n+1} = a_n^2 - (-1)^n$, and dividing by a_{n-1} gives

$$a_{n+1} = \frac{a_n^2 - (-1)^n}{a_{n-1}},$$

as long as $a_{n-1} \neq 0$.

Concept: Just as we can do with any other equation, we can manipulate recursive formulas into more convenient forms.

Now that we have a_{n+1} written in terms of earlier terms in the sequence, we quickly compute terms up to a_3:

$$a_2 = \frac{a_1^2 - (-1)^1}{a_0} = \frac{3^2 + 1}{1} = 10,$$

$$a_3 = \frac{a_2^2 - (-1)^2}{a_1} = \frac{10^2 - 1}{3} = 33.$$

We could continue, but we have the term we needed. □

Sidenote: Compute a few more terms of the sequence in Problem 17.3. Are all the terms of the sequence integers?

A **periodic sequence** is a sequence that repeats regularly. For example, we say the sequence

$$0, 1, 2, 0, 1, 2, 0, 1, 2, \ldots$$

has **period** 3 because it repeats every three terms. More formally, we say that the sequence $\{a_n\}$ is periodic with period k, where $k > 0$, if $a_n = a_{n+k}$ for all n for which a_n and a_{n+k} are defined.

> **Problem 17.4:** Let a be a positive number. Consider the sequence defined recursively by $u_1 = a$ and $u_{n+1} = -1/(u_n + 1)$, for $n = 1, 2, 3, \ldots$. Find the smallest number d such that $\{u_n\}$ is periodic with period d for all values of a.

Solution for Problem 17.4: We start by computing the first few terms of $\{u_n\}$ in terms of a. We are trying to determine where the sequence begins to repeat.

We note that $u_2 = -1/(a + 1)$ and then compute the next couple of terms:

$$u_3 = -\frac{1}{u_2 + 1} = -\frac{1}{-\dfrac{1}{a+1} + 1} = -\frac{a+1}{a},$$

$$u_4 = -\frac{1}{u_3 + 1} = -\frac{1}{-\dfrac{a+1}{a} + 1} = a.$$

We've determined that $u_4 = a = u_1$. Because the recursion relates each term only to the term immediately prior, once we have found that $u_4 = u_1$, we know that $u_{n+3} = u_n$ for all $n \geq 1$. For example, because $u_4 = u_1$, we know that $u_5 = -1/(u_4 + 1) = -1/(u_1 + 1) = u_2$. Similarly, because $u_2 = u_5$, we know that $u_6 = u_3$, and so on. Therefore, regardless of the value of a, we have $u_{n+3} = u_n$ for all positive integers n, so the sequence $\{u_n\}$ has a period of 3. We now also confirm that the sequence does not have a smaller period. If $a = 1$, then the first three terms are $1, -\frac{1}{2}, -2$, so the period cannot be smaller than 3. Therefore, the desired value of d is 3. $\square$

Here's an example of periodic sequences in action:

> **Problem 17.5:** If $a_1 = a_2 = 1$ and $a_{n+2} = \dfrac{a_{n+1} + 1}{a_n}$ for $n \geq 1$, compute a_t where $t = 1998^5$. *(Source: ARML)*

Solution for Problem 17.5: We begin by computing a few terms of the sequence. We sure aren't going to write out 1998^5 terms, but hopefully we'll find a pattern.

> **Concept:** When you're unsure how to begin with a sequence problem, consider writing out a few terms to see if anything interesting happens.

> **Extra!** Consider the recursion
>
> $$x_n = \frac{x_{n-1} + \dfrac{9}{x_{n-1}}}{2}$$
>
> Let $x_0 = 4$, and compute several terms in the sequence $\{x_n\}$. Do the same with different positive values of x_0. Notice anything interesting? Try replacing the 9 in the recursion with different positive constants and computing the first few terms in the sequence $\{x_n\}$ for various positive values of x_0.

We find:

$$a_1 = 1,$$
$$a_2 = 1,$$
$$a_3 = \frac{a_2 + 1}{a_1} = \frac{1 + 1}{1} = 2,$$
$$a_4 = \frac{a_3 + 1}{a_2} = \frac{2 + 1}{1} = 3,$$
$$a_5 = \frac{a_4 + 1}{a_3} = \frac{3 + 1}{2} = 2,$$
$$a_6 = \frac{a_5 + 1}{a_4} = \frac{2 + 1}{3} = 1,$$
$$a_7 = \frac{a_6 + 1}{a_5} = \frac{1 + 1}{2} = 1,$$
$$a_8 = \frac{a_7 + 1}{a_6} = \frac{1 + 1}{1} = 2,$$

and so on.

Each term depends only on the previous two terms, so once a pair of consecutive terms repeats, the whole sequence repeats. From our work above, we see that the consecutive terms a_1, a_2 are the same as the consecutive terms a_6, a_7. Since a_3 depends on a_1 and a_2 in the same way as a_8 depends on a_6 and a_7, we have $a_3 = a_8$. Similarly, $a_2 = a_7$ and $a_3 = a_8$ give $a_4 = a_9$, and so on. In other words, the sequence has period 5.

To figure out a_t when $t = 1998^5$, we must find the remainder when 1998^5 is divided by 5. To do so, we can write

$$1998^5 = (1995 + 3)^5 = 1995^5 + 5(1995^4)(3) + 10(1995^3)(3^2) + 10(1995^2)(3^3) + 5(1995)(3^4) + 3^5.$$

All terms but the last are divisible by 5. Since $3^5 = 243$, we therefore know that 1998^5 is 3 more than a multiple of 5. We could also have used modular arithmetic: $t = 1998^5 \equiv 3^5 \equiv 243 \equiv 3 \pmod 5$. Therefore, the 1998^5th term in the sequence is the same as the 3rd term, and we have $a_t = a_3 = 2$. $\square$

> **Concept:** When looking for useful relationships to help solve sequence problems (recursion problems in particular), keep an eye out for periodicity.

Problem 17.6: A sequence of positive real numbers $a_1, a_2, a_3, \ldots$ has the property that for $i \geq 2$, each a_i is equal to the sum of all the previous terms.

(a) If $a_{19} = 99$, then what is a_{20}? *(Source: Mandelbrot)*

(b) Let $a_1 = k$. Find a closed form for a_n in terms of k for $n \geq 2$.

Solution for Problem 17.6:

(a) Our sequence definition tells us that for $n \geq 2$, we have

$$a_n = a_{n-1} + a_{n-2} + \cdots + a_2 + a_1.$$

Since $a_{19} = 99$, we have

$$a_{20} = a_{19} + a_{18} + \cdots + a_2 + a_1 = 99 + a_{18} + \cdots + a_2 + a_1.$$

So, if we can find $a_{18} + \cdots + a_2 + a_1$, we can find a_{20}. From our sequence definition, this sum simply equals a_{19}! Therefore, we have

$$a_{20} = 99 + a_{18} + \cdots + a_2 + a_1 = 99 + a_{19} = 198.$$

(b) We use part (a) as a guide. If we can simplify $a_{n-2} + \cdots + a_2 + a_1$, then we can simplify our recursion for a_n. From our sequence definition, this sum is simply a_{n-1}:

$$a_{n-1} = a_{n-2} + \cdots + a_2 + a_1,$$

so we have

$$a_n = a_{n-1} + (a_{n-2} + \cdots + a_2 + a_1) = a_{n-1} + (a_{n-1}) = 2a_{n-1}.$$

Aha! Each term a_n is double the preceding term. We might conclude that the sequence is a geometric sequence with first term k and common ratio 2, and deduce that $a_n = k2^{n-1}$. But testing this formula reveals our error.

> **WARNING!!** After deriving a closed form for a sequence or sum, test it with a few simple cases.

Our formula gives $a_1 = k$, as expected, but it also gives $a_2 = 2k$. However, our recursion tells us that $a_2 = a_1$. Where did we go wrong?

Our recursive formula $a_n = a_{n-1} + a_{n-2} + \cdots + a_2 + a_1$ only holds for $n \geq 2$. Therefore, we can only have $a_{n-1} = a_{n-2} + \cdots + a_2 + a_1$ when $n - 1 \geq 2$, which means we must have $n \geq 3$. In other words, we only have $a_n = 2a_{n-1}$ for $n \geq 3$. So, *from the second term on*, our sequence is geometric with common ratio 2 and first term k, which means $a_n = k2^{n-2}$ for $n \geq 2$.

> **WARNING!!** When developing a general formula for a sequence or series, be careful about the initial terms. Sometimes a general relationship that holds for most terms will not hold for the first few terms.

□

Our key step in Problem 17.6 is rewriting the recursion for a_n,

$$a_n = a_{n-1} + a_{n-2} + \cdots + a_2 + a_1,$$

by replacing n with $n - 1$ to get

$$a_{n-1} = a_{n-2} + \cdots + a_2 + a_1.$$

We call this "**shifting the index**" of the recursion.

> **Concept:** Shifting the index of a recursion can be an effective way to learn more about the recursion.

> **Extra!** *There is a difference between not knowing and not knowing yet.*
>
> – Shelia Tobias

Problem 17.7: Define two sequences of rational numbers as follows. Let $a_0 = 2$ and $b_0 = 3$, then recursively define

$$a_{n+1} = \frac{a_n^2}{b_n} \quad \text{and} \quad b_{n+1} = \frac{b_n^2}{a_n}$$

for $n \geq 0$. Determine b_8, leaving your answer in the form $3^m/2^n$. *(Source: Mandelbrot)*

Solution for Problem 17.7: We first notice that simply multiplying the two given recursions allows us to cancel some terms and thereby obtain a simpler relationship among the a's and b's.

Concept: Seek simplicity.

This gives us

$$a_{n+1}b_{n+1} = \frac{a_n^2}{b_n} \cdot \frac{b_n^2}{a_n} = a_n b_n.$$

Since $a_0 b_0 = 6$ and $a_{n+1}b_{n+1} = a_n b_n$, we have $a_n b_n = 6$ for all n. Therefore, we have $a_n = 6/b_n$, so we can write our recursion for b_{n+1} solely in terms of b_n:

$$b_{n+1} = \frac{b_n^2}{a_n} = \frac{b_n^3}{6}.$$

From here, we can just crank out b_8:

$$b_8 = \frac{b_7^3}{6} = \frac{(b_6^3/6)^3}{6} = \frac{b_6^{3^2}}{6^3 \cdot 6} = \frac{(b_5^3/6)^{3^2}}{6^3 \cdot 6} = \frac{b_5^{3^3}}{6^{3^2} \cdot 6^3 \cdot 6^1} = \cdots = \frac{b_0^{3^8}}{6^{3^7} \cdot 6^{3^6} \cdots 6^{3^1} \cdot 6^1}.$$

The exponents of 6 in the denominator form a geometric sequence with sum $1 + 3^1 + 3^2 + \cdots + 3^7 = (3^8 - 1)/(3 - 1) = 6560/2 = 3280$, so we have

$$b_8 = \frac{b_0^{3^8}}{6^{3^7 + 3^6 + \cdots + 3^1 + 3^0}} = \frac{3^{6561}}{6^{3280}} = \frac{3^{3281}}{2^{3280}}.$$

$\square$

A key step in our solution was finding a recursive definition for b_n that did not include terms in the sequence $\{a_n\}$.

Concept: If a sequence $\{x_n\}$ is defined in terms of numbers in another sequence, try to find a way to define $\{x_n\}$ without the other sequence.

Exercises

17.1.1 Find a closed form for a_n if $a_0 = 4$ and $a_n = a_{n-1} + \frac{2}{3}$.

17.1.2 The sequence $a_1, a_2, a_3, \ldots$ satisfies $a_1 = 19$, $a_9 = 99$, and, for all $n \geq 3$, a_n is the arithmetic mean of the first $n - 1$ terms. Find a_2. *(Source: AHSME)*

17.1.3   The sequence of numbers $t_1, t_2, t_3, \ldots$ is defined by $t_1 = 2$ and

$$t_{n+1} = \frac{t_n - 1}{t_n + 1}$$

for every positive integer n. Determine the value of t_{999}. *(Source: COMC)*

17.1.4   Let $\{t_n\}$ be a sequence of positive real numbers, and let $S_n = t_1 + t_2 + \cdots + t_n$.

(a) Find t_n if $S_n = 1 - (1/2)^n$.

(b) Find t_n and S_n if $t_n + 2S_n = 3$ for all positive integers n.

17.1.5   The increasing sequence of positive integers $a_1, a_2, a_3, \ldots$ has the property that $a_{n+2} = a_n + a_{n+1}$ for all $n \geq 1$. If $a_7 = 120$, then find a_8. *(Source: AHSME)*

17.1.6★   A survey is given to a number of "sheeple" to determine how often sheeple go out to dinner. Sheeple will always lie to make themselves seem as socially acceptable as possible. Sheeple all feel like it would be best to go out to dinner slightly more than the average. Therefore, when asked how often they go out to dinner on average per month, each sheeple determines his answer by averaging all of the previous answers and then adding 0.6 (answers need not be integers). If the 7th sheeple claims to go out to dinner 5.47 times per month, how often does the first sheeple claim to go out to dinner per month? **Hints: 141**

17.1.7   The sequence $a_0, a_1, \ldots$ satisfies $a_n = ka_{n-1} - a_{n-2}$ for all $n \geq 2$. Show that $\{a_n\}$ is periodic for each of the following values of k:

(a) -1

(b) $\sqrt{2}$

(c)★ $\dfrac{1 + \sqrt{5}}{2}$ **Hints: 259**

17.2   Telescoping

In certain sums and products, the terms can be manipulated so that most of them cancel, vastly simplifying calculations. This technique is called **telescoping** after the collapsible telescopes that sea captains once used.

Problems

Problem 17.8: Evaluate $\displaystyle\sum_{n=1}^{99} \left(\frac{1}{n} - \frac{1}{n+1} \right)$.

Problem 17.9: Simplify the product $\left(1 - \frac{1}{3}\right)\left(1 - \frac{1}{4}\right)\left(1 - \frac{1}{5}\right)\cdots\left(1 - \frac{1}{n}\right)$. *(Source: AHSME)*

Problem 17.10: Let $T_n = 1 + 2 + 3 + \cdots + n$ and

$$P_n = \frac{T_2}{T_2 - 1} \cdot \frac{T_3}{T_3 - 1} \cdot \frac{T_4}{T_4 - 1} \cdots \frac{T_n}{T_n - 1}$$

for $n = 2, 3, 4, \ldots$. Find P_{1991}. *(Source: AHSME)*

Problem 17.11: In this problem we evaluate the sum

$$\frac{1}{1 \cdot 2} + \frac{1}{2 \cdot 3} + \frac{1}{3 \cdot 4} + \frac{1}{4 \cdot 5} + \cdots + \frac{1}{99 \cdot 100}.$$

(a) Find the sums of the first 2 terms, the first 3 terms, the first 4 terms, and the first 5 terms. Conjecture a formula for the sum of the first n terms.

(b) Use your formula to compute a guess for the given sum. Is your answer the same as you found in Problem 17.8? If not, do both that problem and this one again. If so, explain why.

Problem 17.8: Evaluate $\displaystyle\sum_{n=1}^{99} \left(\frac{1}{n} - \frac{1}{n+1} \right)$.

Solution for Problem 17.8: Writing out the summation reveals a pattern that greatly simplifies the sum:

$$\sum_{n=1}^{99} \left(\frac{1}{n} - \frac{1}{n+1} \right) = \left(\frac{1}{1} - \frac{1}{2} \right) + \left(\frac{1}{2} - \frac{1}{3} \right) + \left(\frac{1}{3} - \frac{1}{4} \right) + \cdots + \left(\frac{1}{98} - \frac{1}{99} \right) + \left(\frac{1}{99} - \frac{1}{100} \right).$$

Each negative term except the last cancels with the next positive term, leaving only

$$\sum_{n=1}^{99} \left(\frac{1}{n} - \frac{1}{n+1} \right) = \frac{1}{1} - \frac{1}{100} = \frac{99}{100}.$$

$\square$

The type of cancellation that we used to solve Problem 17.8 is the key step in tackling many series problems. We call any sum that can be evaluated with this technique a **telescoping series**.

As well as telescoping sums, there are telescoping products.

Problem 17.9: Simplify the product $\left(1 - \frac{1}{3}\right)\left(1 - \frac{1}{4}\right)\left(1 - \frac{1}{5}\right)\cdots\left(1 - \frac{1}{n}\right)$. *(Source: AHSME)*

Solution for Problem 17.9: When we rewrite each term in the product as a proper fraction, it becomes clear that most of the terms cancel:

$$\frac{2}{3} \cdot \frac{3}{4} \cdot \frac{4}{5} \cdots \frac{n-1}{n} = \frac{2 \cdot (3 \cdot 4 \cdot 5 \cdots (n-1))}{(3 \cdot 4 \cdot 5 \cdots (n-1)) \cdot n} = \frac{2}{n}.$$

$\square$

It sometimes takes a bit of algebraic manipulation to see how terms will cancel in a telescoping sum or product:

Problem 17.10: Let $T_n = 1 + 2 + 3 + \cdots + n$ and

$$P_n = \frac{T_2}{T_2 - 1} \cdot \frac{T_3}{T_3 - 1} \cdot \frac{T_4}{T_4 - 1} \cdots \frac{T_n}{T_n - 1}$$

for $n = 2, 3, 4, \ldots$. Find P_{1991}. *(Source: AHSME)*

Solution for Problem 17.10: First, we note that $T_k = k(k+1)/2$. Our hope is that the product telescopes, so we look for a way to factor the denominators, which are of the form $T_k - 1$. We find that

$$T_k - 1 = \frac{k^2 + k}{2} - 1 = \frac{k^2 + k - 2}{2} = \frac{(k-1)(k+2)}{2}.$$

Now we write out the product to see which factors cancel, and which remain:

$$
\begin{aligned}
P_{1991} &= \frac{T_2}{T_2 - 1} \cdot \frac{T_3}{T_3 - 1} \cdot \frac{T_4}{T_4 - 1} \cdots \frac{T_{1990}}{T_{1990} - 1} \cdot \frac{T_{1991}}{T_{1991} - 1} \\
&= \frac{2 \cdot 3/2}{1 \cdot 4/2} \cdot \frac{3 \cdot 4/2}{2 \cdot 5/2} \cdot \frac{4 \cdot 5/2}{3 \cdot 6/2} \cdots \frac{1990 \cdot 1991/2}{1989 \cdot 1992/2} \cdot \frac{1991 \cdot 1992/2}{1990 \cdot 1993/2} \\
&= \frac{2 \cdot 3}{1 \cdot 4} \cdot \frac{3 \cdot 4}{2 \cdot 5} \cdot \frac{4 \cdot 5}{3 \cdot 6} \cdots \frac{1990 \cdot 1991}{1989 \cdot 1992} \cdot \frac{1991 \cdot 1992}{1990 \cdot 1993} \\
&= \frac{2 \cdot 3^2 \cdot 4^2 \cdots 1991^2 \cdot 1992}{1 \cdot 2 \cdot 3 \cdot 4^2 \cdot 5^2 \cdots 1990^2 \cdot 1991 \cdot 1992 \cdot 1993} \\
&= \frac{3 \cdot 1991}{1993} = \frac{5973}{1993}.
\end{aligned}
$$

$\square$

Problem 17.11: Evaluate the sum $\dfrac{1}{1 \cdot 2} + \dfrac{1}{2 \cdot 3} + \dfrac{1}{3 \cdot 4} + \dfrac{1}{4 \cdot 5} + \cdots + \dfrac{1}{99 \cdot 100}$.

Solution for Problem 17.11: Adding term-by-term would take a long time, so we'd like to find a nicer approach. We begin by adding the first few terms of the series to see if partial sums of the series shed any light on the problem:

$$\frac{1}{1 \cdot 2} + \frac{1}{2 \cdot 3} = \frac{1}{2} + \frac{1}{2 \cdot 3} = \frac{2}{3},$$

$$\frac{1}{1 \cdot 2} + \frac{1}{2 \cdot 3} + \frac{1}{3 \cdot 4} = \frac{2}{3} + \frac{1}{3 \cdot 4} = \frac{3}{4},$$

$$\frac{1}{1 \cdot 2} + \frac{1}{2 \cdot 3} + \frac{1}{3 \cdot 4} + \frac{1}{4 \cdot 5} = \frac{3}{4} + \frac{1}{4 \cdot 5} = \frac{4}{5},$$

$$\frac{1}{1 \cdot 2} + \frac{1}{2 \cdot 3} + \frac{1}{3 \cdot 4} + \frac{1}{4 \cdot 5} + \frac{1}{5 \cdot 6} = \frac{4}{5} + \frac{1}{5 \cdot 6} = \frac{5}{6}.$$

> **Concept:** Simplifying a problem and looking for patterns is a great way to learn more about the problem.

It looks like we have a pattern, and if we had to guess, we'd guess that

$$\frac{1}{1 \cdot 2} + \frac{1}{2 \cdot 3} + \frac{1}{3 \cdot 4} + \frac{1}{4 \cdot 5} + \cdots + \frac{1}{99 \cdot 100} = \frac{99}{100}.$$

Hey, that's the same answer we found when adding the 99 terms in the summation of Problem 17.8. Maybe this is the same problem as that one! Let's take a closer look.

First, we write down the general statement we think is true:

$$\frac{1}{1 \cdot 2} + \frac{1}{2 \cdot 3} + \frac{1}{3 \cdot 4} + \frac{1}{4 \cdot 5} + \cdots + \frac{1}{(n) \cdot (n+1)} = \frac{n}{n+1}.$$

> **Concept:** When trying to prove that a pattern exists, write a general form that represents the pattern.

We want to compare this equation to what we found in Problem 17.8. So, we write a general result for Problem 17.8:

$$\sum_{k=1}^{n} \left(\frac{1}{k} - \frac{1}{k+1} \right) = \left(\frac{1}{1} - \frac{1}{2} \right) + \left(\frac{1}{2} - \frac{1}{3} \right) + \cdots + \left(\frac{1}{n-1} - \frac{1}{n} \right) + \left(\frac{1}{n} - \frac{1}{n+1} \right) = 1 - \frac{1}{n+1} = \frac{n}{n+1}.$$

Now, we should be very suspicious!

> **Concept:** There aren't many coincidences in mathematics. If two types of problems consistently give the same answer, look for a reason why.

Specifically, we suspect that the sum of the first n terms of the form $\frac{1}{k(k+1)}$ equals the sum of the first n terms of the form $\left(\frac{1}{k} - \frac{1}{k+1} \right)$. Now that we know what to look for, the answer is staring us right in the face. We find a common denominator in the latter expression, and we have

$$\frac{1}{k} - \frac{1}{k+1} = \frac{k+1}{k(k+1)} - \frac{k}{k(k+1)} = \frac{1}{k(k+1)}.$$

Aha! Each term in the sum

$$\frac{1}{1 \cdot 2} + \frac{1}{2 \cdot 3} + \frac{1}{3 \cdot 4} + \frac{1}{4 \cdot 5} + \cdots + \frac{1}{(n) \cdot (n+1)} = \sum_{k=1}^{n} \frac{1}{k(k+1)}$$

equals the corresponding term in the sum $\sum_{k=1}^{n} \left(\frac{1}{k} - \frac{1}{k+1} \right)$. So, we have

$$\sum_{k=1}^{n} \frac{1}{k(k+1)} = \sum_{k=1}^{n} \left(\frac{1}{k} - \frac{1}{k+1} \right).$$

We know that the latter sum telescopes to give $n/(n+1)$, so the answer to our problem is indeed $99/100$.
$\square$

A key step in our solution to Problem 17.11 was writing the rational function $\frac{1}{k(k+1)}$ in terms of rational functions with simpler denominators:

$$\frac{1}{k(k+1)} = \frac{1}{k} - \frac{1}{k+1}.$$

We call this the **partial fraction decomposition** of $\frac{1}{k(k+1)}$.

When we find the partial fraction decomposition of a rational function, we express it as the sum of rational functions with linear or quadratic denominators. As we saw in Problem 17.11, partial fraction decomposition sometimes helps us evaluate complicated series by turning them into telescoping series.

We'll find the partial fraction decomposition of one more rational function, and then put this technique to work on a few more series.

Problems

Problem 17.12: Find constants A and B such that $\dfrac{12x+11}{(x-2)(2x+3)} = \dfrac{A}{x-2} + \dfrac{B}{2x+3}$.

Problem 17.13:

(a) Evaluate $\displaystyle\sum_{k=1}^{100} \frac{1}{k(k+2)}$.

(b) Evaluate $\displaystyle\sum_{k=1}^{\infty} \frac{1}{k(k+2)}$.

Problem 17.14: Find the value of the sum $\displaystyle\sum_{n=1}^{\infty} \frac{(7n+32)\cdot 3^n}{n(n+2)\cdot 4^n}$. *(Source: Thomas Mildorf)*

Problem 17.15: Let $a_1, a_2, \ldots, a_n$ be an arithmetic sequence with nonzero common difference d. Find a closed form in terms of a_1, n, and a_n for

$$\sum_{i=1}^{n-1} \frac{1}{a_i a_{i+1}}.$$

Problem 17.12: Find constants A and B such that $\dfrac{12x+11}{(x-2)(2x+3)} = \dfrac{A}{x-2} + \dfrac{B}{2x+3}$.

Solution for Problem 17.12: Multiplying both sides by $(x-2)(2x+3)$ to get rid of the fractions gives

$$12x + 11 = A(2x+3) + B(x-2).$$

We could expand the products on the right side and equate coefficients of like terms to produce a system of equations, but we have a much faster way to find A and B. The equation must hold for all x; we can choose values of x that enable us to find A and B quickly. Specifically, letting $x = 2$ eliminates the B term and leaves $35 = 7A$, so $A = 5$. Similarly, letting $x = -3/2$ eliminates A and leaves $-7 = -7B/2$, so $B = 2$. This gives us the partial fraction decomposition

$$\frac{12x + 11}{2x^2 - x - 6} = \frac{5}{x - 2} + \frac{2}{2x + 3}.$$

We can check our work by writing the right side with a common denominator:

$$\frac{5}{x - 2} + \frac{2}{2x + 3} = \frac{5(2x + 3)}{(x - 2)(2x + 3)} + \frac{2(x - 2)}{(2x + 3)(x - 2)} = \frac{12x + 11}{2x^2 - x - 6}.$$

□

We've already seen partial fraction decomposition help us telescope one series. Let's try it on a few more:

Problem 17.13: Evaluate $\displaystyle\sum_{k=1}^{100} \frac{1}{k(k + 2)}$ and $\displaystyle\sum_{k=1}^{\infty} \frac{1}{k(k + 2)}$.

Solution for Problem 17.13: We first let $\frac{1}{k(k+2)} = \frac{A}{k} + \frac{B}{k+2}$. Multiplying by $k(k + 2)$ gives $1 = A(k + 2) + Bk$. Letting $k = 0$ gives $A = \frac{1}{2}$ and letting $k = -2$ gives $B = -\frac{1}{2}$, so we have

$$\sum_{k=1}^{100} \frac{1}{k(k + 2)} = \sum_{k=1}^{100} \left(\frac{1}{2} \cdot \frac{1}{k} - \frac{1}{2} \cdot \frac{1}{k + 2}\right) = \frac{1}{2}\left(\sum_{k=1}^{100} \frac{1}{k} - \sum_{k=1}^{100} \frac{1}{k + 2}\right).$$

Now, we see that the last 98 terms of the first sum cancel with the first 98 terms of the second sum, leaving:

$$\frac{1}{2}\left(\sum_{k=1}^{100} \frac{1}{k} - \sum_{k=1}^{100} \frac{1}{k + 2}\right) = \frac{1}{2}\left(1 + \frac{1}{2} - \frac{1}{101} - \frac{1}{102}\right) = \frac{7625}{10302}.$$

We also could have seen the cancellation by rewriting the second summation as a summation of $\frac{1}{k}$, to match the expression we sum in the first summation. We do so by letting k range from 3 to 102 instead of 1 to 100. Then, we clearly see which terms cancel and which are left over:

$$\frac{1}{2}\sum_{k=1}^{100} \frac{1}{k} - \frac{1}{2}\sum_{k=1}^{100} \frac{1}{k + 2} = \frac{1}{2}\sum_{k=1}^{100} \frac{1}{k} - \frac{1}{2}\sum_{k=3}^{102} \frac{1}{k} = \frac{1}{2}\left(1 + \frac{1}{2} - \frac{1}{101} - \frac{1}{102}\right) = \frac{7625}{10302}.$$

The shift of index made it easy to see exactly which terms would be left.

To examine what happens when the sum has infinitely many terms, we first let there be n terms, and find the sum in terms of n. From our work above, we see that

$$\sum_{k=1}^{n} \frac{1}{k(k + 2)} = \frac{1}{2}\left(\sum_{k=1}^{n} \frac{1}{k} - \sum_{k=1}^{n} \frac{1}{k + 2}\right) = \frac{1}{2}\left(1 + \frac{1}{2} - \frac{1}{n + 1} - \frac{1}{n + 2}\right).$$

As n becomes infinite, the terms $\frac{1}{n+1}$ and $\frac{1}{n+2}$ equal 0, so the infinite series has sum $\frac{1}{2}\left(1 + \frac{1}{2}\right) = \frac{3}{4}$. □

We also might have tried a shift of index to see that the infinite series telescopes infinitely:

$$\sum_{k=1}^{\infty}\frac{1}{k(k+2)}=\frac{1}{2}\sum_{k=1}^{\infty}\frac{1}{k}-\frac{1}{2}\sum_{k=1}^{\infty}\frac{1}{k+2}=\frac{1}{2}\sum_{k=1}^{\infty}\frac{1}{k}-\frac{1}{2}\sum_{k=3}^{\infty}\frac{1}{k}=\frac{1}{2}\sum_{k=1}^{2}\frac{1}{k}=\frac{1}{2}\left(1+\frac{1}{2}\right)=\frac{3}{4}.$$

We have to be careful about this type of approach, for it can be misleading. For example, consider the following:

Bogus Solution: $\displaystyle\sum_{k=1}^{\infty}1=\sum_{k=1}^{\infty}((k+1)-k)=\sum_{k=1}^{\infty}(k+1)-\sum_{k=1}^{\infty}k=\sum_{k=2}^{\infty}k-\sum_{k=1}^{\infty}k=-1.$

Clearly something is wrong there. We see the mistake when we consider the corresponding finite sum:

$$\sum_{k=1}^{n}1=\sum_{k=1}^{n}((k+1)-k)=\sum_{k=1}^{n}(k+1)-\sum_{k=1}^{n}k=\sum_{k=2}^{n+1}k-\sum_{k=1}^{n}k=(n+1)-1=n.$$

Now, we see that as n becomes infinite, so does the sum.

WARNING!! Be careful with infinity!

Problem 17.14: Find the value of the sum $\displaystyle\sum_{n=1}^{\infty}\frac{(7n+32)\cdot 3^n}{n(n+2)\cdot 4^n}$. *(Source: Thomas Mildorf)*

Solution for Problem 17.14: We're not so sure what to do about the 3^n and the 4^n, but we have a pretty good idea what to do about the rest of the expression we are summing. We let

$$\frac{7n+32}{n(n+2)}=\frac{A}{n}+\frac{B}{n+2}.$$

Multiplying both sides by $n(n+2)$ gives $7n+32=A(n+2)+Bn$. Letting $n=0$ gives $A=16$ and letting $n=-2$ gives $B=9$. Aha! We have $A=4^2$ and $B=3^2$. More 3's and 4's. We must be on the right track. Our partial fraction decomposition is

$$\frac{7n+32}{n(n+2)}=\frac{4^2}{n}-\frac{3^2}{n+2}.$$

We now have

$$\sum_{n=1}^{\infty}\frac{(7n+32)\cdot 3^n}{n(n+2)\cdot 4^n}=\sum_{n=1}^{\infty}\left(\frac{4^2}{n}-\frac{3^2}{n+2}\right)\frac{3^n}{4^n}.$$

Extra! *The infinite! No other question has ever moved so profoundly the spirit of man.*

– David Hilbert

To get a feel for the sum, we write out a few terms:

$$\sum_{n=1}^{\infty}\left(\frac{4^2}{n}-\frac{3^2}{n+2}\right)\frac{3^n}{4^n} = \left(\frac{4^2}{1}\cdot\frac{3}{4}-\frac{3^2}{3}\cdot\frac{3}{4}\right)+\left(\frac{4^2}{2}\cdot\frac{3^2}{4^2}-\frac{3^2}{4}\cdot\frac{3^2}{4^2}\right)$$

$$+\left(\frac{4^2}{3}\cdot\frac{3^3}{4^3}-\frac{3^2}{5}\cdot\frac{3^3}{4^3}\right)+\left(\frac{4^2}{4}\cdot\frac{3^4}{4^4}-\frac{3^2}{6}\cdot\frac{3^4}{4^4}\right)+\cdots$$

$$=\left(\frac{1}{1}\cdot 3\cdot 4-\frac{1}{3}\cdot\frac{3^3}{4}\right)+\left(\frac{1}{2}\cdot 3^2-\frac{1}{4}\cdot\frac{3^4}{4^2}\right)$$

$$+\left(\frac{1}{3}\cdot\frac{3^3}{4}-\frac{1}{5}\cdot\frac{3^5}{4^3}\right)+\left(\frac{1}{4}\cdot\frac{3^4}{4^2}-\frac{1}{6}\cdot\frac{3^6}{4^4}\right)+\cdots.$$

We expect to see terms cancel, and we're not disappointed. Because the denominator of the positive term in the summation is n and the denominator of the negative term is $n+2$, we expect to see each negative term cancel with a positive term two positive expressions later in the series. That is, we expect the negative term for $n=k$ to cancel with the positive term for $n=k+2$. Indeed, when $n=k$, we have

$$-\frac{3^2}{k+2}\cdot\frac{3^k}{4^k}=-\frac{3^{k+2}}{4^k(k+2)}=-\frac{4^2 3^{k+2}}{4^{k+2}(k+2)}=-\frac{4^2}{k+2}\cdot\frac{3^{k+2}}{4^{k+2}},$$

which is the opposite of the positive term when $n=k+2$, as expected. This makes it even more clear how our sum telescopes, leaving us only with the first two terms in the positive summation above, which give us

$$\frac{1}{1}\cdot 3\cdot 4+\frac{1}{2}\cdot 3^2=\frac{33}{2}.$$

We can confirm that the infinite series does indeed converge by finding the first k terms, then considering what happens when n is infinite. We first break our summation into two summations:

$$\sum_{n=1}^{k}\left(\frac{4^2}{n}-\frac{3^2}{n+2}\right)\frac{3^n}{4^n}=\sum_{n=1}^{k}\left(\frac{4^2}{n}\cdot\frac{3^n}{4^n}\right)-\sum_{n=1}^{k}\left(\frac{3^2}{n+2}\cdot\frac{3^n}{4^n}\right)=\sum_{n=1}^{k}\left(\frac{3^n}{n\cdot 4^{n-2}}\right)-\sum_{n=1}^{k}\left(\frac{3^{n+2}}{(n+2)\cdot 4^n}\right).$$

We then shift the index by 2 in the summation we are subtracting to get

$$\sum_{n=1}^{k}\left(\frac{3^n}{n\cdot 4^{n-2}}\right)-\sum_{n=1}^{k}\left(\frac{3^{n+2}}{(n+2)\cdot 4^n}\right)=\sum_{n=1}^{k}\left(\frac{3^n}{n\cdot 4^{n-2}}\right)-\sum_{n=3}^{k+2}\left(\frac{3^n}{n\cdot 4^{n-2}}\right).$$

All the terms in the first summation except the first two are subtracted in the second summation, and all terms in the second summation are eliminated except the last two terms. We're left with

$$\sum_{n=1}^{k}\left(\frac{3^n}{n\cdot 4^{n-2}}\right)-\sum_{n=3}^{k+2}\left(\frac{3^n}{n\cdot 4^{n-2}}\right)=\frac{3}{4^{-1}}+\frac{3^2}{2\cdot 4^0}-\frac{3^{k+1}}{(k+1)\cdot 4^{k-1}}-\frac{3^{k+2}}{(k+2)\cdot 4^k}.$$

The last two terms become 0 as k becomes infinite (note that $3^{k+1}/4^{k-1}=9\cdot(3/4)^{k-1}$), so the sum does indeed converge to 33/2, as claimed. □

> **Concept:** Many sums involving fractions can be rearranged so that they telescope.

Here's one more:

> **Problem 17.15:** Let $a_1, a_2, \ldots, a_n$ be an arithmetic sequence with nonzero common difference d. Find a closed form in terms of a_1, n, and a_n for
> $$\sum_{i=1}^{n-1} \frac{1}{a_i a_{i+1}}.$$

Solution for Problem 17.15: A summation of a fraction: let's take a closer look with our telescope! We'd like to break the fraction into a difference of two fractions that will allow us to telescope. Inspired by partial fraction decomposition, we look for constants B and C such that

$$\frac{1}{a_i a_{i+1}} = \frac{B}{a_i} + \frac{C}{a_{i+1}}.$$

Multiplying by $a_i a_{i+1}$ gives $1 = B a_{i+1} + C a_i$. We don't get to choose the a_i, but we do know that $a_{i+1} - a_i = d$, so we have $(d + a_i)B + a_i C = 1$. This must hold for all the a_i in the sequence, so we choose B and C to eliminate a_i. This means we want $B = -C$, which leaves us with $dB = 1$, so $B = 1/d$ and $C = -1/d$. Checking that these work, we have

$$\frac{1/d}{a_i} + \frac{-1/d}{a_{i+1}} = \frac{1}{d}\left(\frac{a_{i+1} - a_i}{a_i a_{i+1}}\right) = \frac{1}{d} \cdot \frac{d}{a_i a_{i+1}} = \frac{1}{a_i a_{i+1}},$$

as desired. Now, it's telescope time:

$$\sum_{i=1}^{n-1} \frac{1}{a_i a_{i+1}} = \sum_{i=1}^{n-1} \frac{1}{d}\left(\frac{1}{a_i} - \frac{1}{a_{i+1}}\right) = \frac{1}{d}\sum_{i=1}^{n-1}\left(\frac{1}{a_i} - \frac{1}{a_{i+1}}\right).$$

All the terms except $1/a_1$ and $-1/a_n$ cancel, leaving a sum of

$$\frac{1}{d}\left(\frac{1}{a_1} - \frac{1}{a_n}\right) = \frac{1}{d}\left(\frac{a_n - a_1}{a_1 a_n}\right) = \frac{1}{d}\left(\frac{(n-1)d}{a_1 a_n}\right) = \frac{n-1}{a_1 a_n}.$$

$\square$

> **Sidenote:** Partial fraction decomposition isn't only useful for evaluating telescoping series. You'll see it again when you study calculus.

Exercises

17.2.1 Calculate $\displaystyle\prod_{n=1}^{13} \frac{n(n+2)}{(n+4)^2}$, leaving your answer in simplest form. *(Source: Mandelbrot)*

17.2.2 Find $\dfrac{1}{1+\sqrt{2}} + \dfrac{1}{\sqrt{2}+\sqrt{3}} + \dfrac{1}{\sqrt{3}+\sqrt{4}} + \cdots + \dfrac{1}{\sqrt{99}+\sqrt{100}}$.

17.2.3 Consider the sequence $1, -2, 3, -4, 5, -6, \ldots$, whose n^{th} term is $(-1)^{n+1} \cdot n$. What is the average of the first 200 terms of the sequence? *(Source: AHSME)*

17.2.4 Let n be a positive integer. Find the sum $\dfrac{1}{2!} + \dfrac{2}{3!} + \dfrac{3}{4!} + \cdots + \dfrac{n-1}{n!}$.

17.2.5 Find the sum $\dfrac{1}{1\cdot3} + \dfrac{1}{3\cdot5} + \cdots + \dfrac{1}{(2n-1)(2n+1)} + \cdots + \dfrac{1}{255\cdot257}$. *(Source: AHSME)*

17.2.6 Find the product $\displaystyle\prod_{n=1}^{20}\left(1 + \dfrac{2n+1}{n^2}\right)$.

17.2.7★ Evaluate $\displaystyle\sum_{k=1}^{7} \dfrac{1}{\sqrt[3]{k^2} + \sqrt[3]{k(k+1)} + \sqrt[3]{(k+1)^2}}$. **Hints:** 312

17.2.8★ Let $H_n = 1 + \dfrac{1}{2} + \dfrac{1}{3} + \cdots + \dfrac{1}{n}$, and let $T_n = \dfrac{1}{(n+1)H_nH_{n+1}}$. Prove that $T_1 + T_2 + T_3 + \cdots = 1$.

17.3 Sums of Polynomial Series

We have seen a variety of ways to find a closed form for $\displaystyle\sum_{k=1}^{n} k$. We used induction to prove that

$$\sum_{k=1}^{n} k^2 = \frac{n(n+1)(2n+1)}{6} \qquad \text{and} \qquad \sum_{k=1}^{n} k^3 = \frac{n^2(n+1)^2}{4},$$

but these inductions didn't give any insight into how we could have found the expressions $\dfrac{n(n+1)(2n+1)}{6}$ and $\dfrac{n^2(n+1)^2}{4}$ in the first place. In this section, we'll find a way to derive them.

Problems

Problem 17.16: In this problem, we use telescoping series to find a closed form for $\displaystyle\sum_{k=1}^{n} k$.

(a) Simplify the expression $(k+1)^2 - k^2$.

(b) How are $\displaystyle\sum_{k=1}^{n}((k+1)^2 - k^2)$ and $\displaystyle\sum_{k=1}^{n} k$ related?

(c) Find a closed form for $\displaystyle\sum_{k=1}^{n} k$.

Problem 17.17: Let n be a positive integer. Express $\displaystyle\sum_{k=1}^{n} k^2$ as a polynomial. **Hints:** 12

Problem 17.18: Let $f(x) = 3x^2 - 7x + 2$. Evaluate $\displaystyle\sum_{n=1}^{20} f(n)$.

Problem 17.19: Find a closed form for the series $1 \cdot 4 + 3 \cdot 6 + \cdots + (2n - 1)(2n + 2)$.

Problem 17.20: Notice the pattern

$$3^2 + 4^2 = 5^2,$$
$$10^2 + 11^2 + 12^2 = 13^2 + 14^2,$$
$$21^2 + 22^2 + 23^2 + 24^2 = 25^2 + 26^2 + 27^2.$$

Prove that for any positive integer n, there exist $n + 1$ consecutive squares whose sum equals the sum of the next n larger consecutive squares. **Hints:** 220

Problem 17.21: Find a formula for $\displaystyle\sum_{k=1}^{n} k^3$.

Problem 17.16: Use telescoping series to find a closed form for $\displaystyle\sum_{k=1}^{n} k$.

Solution for Problem 17.16:   We need to find a telescoping series that we can relate to our desired sum. We try summing $(k + 1)^2 - k^2$ both because such a sum obviously telescopes, and because simplifying this expression gives us a linear expression in k:

$$(k + 1)^2 - k^2 = k^2 + 2k + 1 - k^2 = 2k + 1.$$

So we have

$$\sum_{k=1}^{n}(2k + 1) = \sum_{k=1}^{n}((k + 1)^2 - k^2) = (2^2 - 1^2) + (3^2 - 2^2) + \cdots + [(n + 1)^2 - n^2] = (n + 1)^2 - 1^2 = n^2 + 2n.$$

Now, we have our target in sight, because

$$\sum_{k=1}^{n}(2k + 1) = \sum_{k=1}^{n}(2k) + \sum_{k=1}^{n} 1 = 2\left(\sum_{k=1}^{n} k\right) + n.$$

So, we have $2\left(\displaystyle\sum_{k=1}^{n} k\right) + n = n^2 + 2n$, which means $\displaystyle\sum_{k=1}^{n} k = \dfrac{n^2 + 2n - n}{2} = \dfrac{n(n + 1)}{2}$. $\square$

Let's see if we can use a similar approach to find the sum of the first n squares.

Problem 17.17: Let n be a positive integer. Express $\displaystyle\sum_{k=1}^{n} k^2$ as a polynomial.

Solution for Problem 17.17: Inspired by our previous success, we start by summing $(k+1)^3 - k^3$, both because we know this sum will telescope, and because the k^3 terms will cancel when we expand $(k+1)^3$ in this expression, leaving $(k+1)^3 - k^3 = 3k^2 + 3k + 1$ for all positive integers k. Therefore, we have

$$\sum_{k=1}^{n}(3k^2 + 3k + 1) = \sum_{k=1}^{n}[(k+1)^3 - k^3] = (n+1)^3 - 1^3 = n^3 + 3n^2 + 3n.$$

We also have

$$\sum_{k=1}^{n}(3k^2 + 3k + 1) = \sum_{k=1}^{n} 3k^2 + \sum_{k=1}^{n} 3k + \sum_{k=1}^{n} 1 = 3\left(\sum_{k=1}^{n} k^2\right) + 3\left(\sum_{k=1}^{n} k\right) + n.$$

Therefore, we have $3\left(\displaystyle\sum_{k=1}^{n} k^2\right) + 3\left(\displaystyle\sum_{k=1}^{n} k\right) + n = n^3 + 3n^2 + 3n$, so

$$\sum_{k=1}^{n} k^2 = \frac{1}{3}\left(n^3 + 3n^2 + 2n - 3\sum_{k=1}^{n} k\right) = \frac{1}{3}\left(n^3 + 3n^2 + 2n - 3 \cdot \frac{n^2+n}{2}\right) = \frac{2n^3 + 3n^2 + n}{6} = \frac{n(n+1)(2n+1)}{6}.$$

We can test our formula by trying a few small values of n. If $n = 2$, our formula gives $2(3)(5)/6 = 5$, which is indeed the sum of the first 2 positive squares. If $n = 3$, our formula gives $3(4)(7)/6 = 14$, which matches the sum of the first 3 perfect squares.

Concept: Whenever you derive a general formula for a sequence or series, you can usually test it with a few simple cases.

□

Problem 17.18: Let $f(x) = 3x^2 - 7x + 2$. Evaluate $\displaystyle\sum_{n=1}^{20} f(n)$.

Solution for Problem 17.18: We simply break the sum into pieces and use the formulas we derived in the first two problems:

$$\sum_{n=1}^{20} f(n) = \sum_{n=1}^{20}(3n^2 - 7n + 2) = 3\sum_{n=1}^{20} n^2 - 7\sum_{n=1}^{20} n + 2\sum_{n=1}^{20} 1$$

$$= 3\left(\frac{20(21)(41)}{6}\right) - 7\left(\frac{20(21)}{2}\right) + 2(20) = 8610 - 1470 + 40 = 7180.$$

□

Problem 17.19: Find a closed form for the series $1 \cdot 4 + 3 \cdot 6 + \cdots + (2n-1)(2n+2)$.

Solution for Problem 17.19: We can use the general term to help us rewrite the series using sigma notation:

$$1 \cdot 4 + 3 \cdot 6 + \cdots + (2n-1)(2n+2) = \sum_{k=1}^{n}(2k-1)(2k+2).$$

Now we expand the summed expression, break it into parts, and evaluate:

$$\sum_{k=1}^{n}(2k-1)(2k+2) = \sum_{k=1}^{n}(4k^2 + 2k - 2)$$

$$=4\sum_{k=1}^{n}k^2 + 2\sum_{k=1}^{n}k - 2\sum_{k=1}^{n}1$$

$$=4\left(\frac{n(n+1)(2n+1)}{6}\right) + 2\left(\frac{n(n+1)}{2}\right) - 2n$$

$$=\frac{2n(n+1)(2n+1)}{3} + n(n+1) - 2n$$

$$=\frac{4n^3 + 9n^2 - n}{3}.$$

We went through some fairly complicated algebra to get this formula, so we test it on a few simple cases. When $n = 1$, the series is simply $1 \cdot 4 = 4$. Indeed, our formula gives $(4 + 9 - 1)/3 = 4$ for $n = 1$. When $n = 2$, the series is $1 \cdot 4 + 3 \cdot 6 = 22$, and our formula gives $(4 \cdot 8 + 9 \cdot 4 - 2)/3 = 22$. $\square$

While we could have solved Problem 17.19 in several ways, rewriting the series with sigma notation made it very clear how to relate the series to ones we already knew how to evaluate.

Concept: Notation itself can be a powerful problem solving tool. Rewriting some series with sigma notation can help evaluate them more easily.

Problem 17.20: Notice the pattern

$$3^2 + 4^2 = 5^2,$$
$$10^2 + 11^2 + 12^2 = 13^2 + 14^2,$$
$$21^2 + 22^2 + 23^2 + 24^2 = 25^2 + 26^2 + 27^2.$$

Prove that for any positive integer n, there exist $n+1$ consecutive squares whose sum equals the sum of the next n higher consecutive squares.

Solution for Problem 17.20: One way to prove that there are $n+1$ consecutive squares whose sum equals the sum of the next n consecutive squares is to find the squares in terms of n. We could write $2n+1$ consecutive squares as $m^2, (m+1)^2, (m+2)^2, \ldots, (m+2n)^2$, but if we instead let m be the middle square, then the $2n+1$ consecutive squares are $(m-n)^2, (m-n+1)^2, \ldots, (m-1)^2, m^2, (m+1)^2, \ldots, (m+n)^2$. Our second list has a nice symmetry.

Our problem now is to show that for every positive integer n, we can find an integer m with $m > n$ such that

$$(m - n)^2 + (m - n + 1)^2 + (m - n + 2)^2 + \cdots + m^2 = (m + 1)^2 + (m + 2)^2 + (m + 3)^3 + \cdots + (m + n)^2.$$

> **Concept:** When working with an odd number of consecutive integers, try assigning a variable to the middle integer and letting symmetry do some of your work for you.

To show that such an integer m always exists, we try to solve this equation for m in terms of n. We first write the equation more compactly with summations:

$$\sum_{k=0}^{n} (m - k)^2 = \sum_{k=1}^{n} (m + k)^2.$$

Expanding both squares gives $\sum_{k=0}^{n} (m^2 - 2mk + k^2) = \sum_{k=1}^{n} (m^2 + 2mk + k^2)$, and breaking both sides of this into several sums gives

$$\sum_{k=0}^{n} m^2 - \sum_{k=0}^{n} 2mk + \sum_{k=0}^{n} k^2 = \sum_{k=1}^{n} m^2 + \sum_{k=1}^{n} 2mk + \sum_{k=1}^{n} k^2.$$

Subtracting $\sum_{k=1}^{n} m^2 + \sum_{k=1}^{n} k^2$ from both sides, we get

$$m^2 - \sum_{k=0}^{n} 2mk = \sum_{k=1}^{n} 2mk.$$

(Make sure you see why we still have an m^2 on the left side; the summation of m^2 on the left side had $n + 1$ terms, while the one on the right side had n terms, so one m^2 is "left over" on the left. There's a "leftover" k^2 term, too, but that leftover term is for $k = 0$, which makes $k^2 = 0$.)

Since $2mk$ is 0 when $k = 0$, we have $\sum_{k=0}^{n} 2mk = \sum_{k=1}^{n} 2mk$, so our equation now is $m^2 - \sum_{k=1}^{n} 2mk = \sum_{k=1}^{n} 2mk$, which gives us $m^2 = 2\sum_{k=1}^{n} 2mk$. Factoring $2m$ out of the sum, we have $m^2 = 4m\sum_{k=1}^{n} k$, so

$$m = 4 \sum_{k=1}^{n} k = 4 \cdot \frac{n(n + 1)}{2} = 2n^2 + 2n.$$

Thus, we have shown that if $\sum_{k=0}^{n} (m - k)^2 = \sum_{k=1}^{n} (m + k)^2$, then $m = 2n^2 + 2n$. All of our steps are reversible, so we have

$$\sum_{k=0}^{n} (2n^2 + 2n - k)^2 = \sum_{k=1}^{n} (2n^2 + 2n + k)^2$$

for all positive integers n. Therefore, for any positive integer n, there exist $n + 1$ consecutive squares whose sum equals the sum of the next n higher consecutive squares. $\square$

Problem 17.21: Find a formula for $\displaystyle\sum_{k=1}^{n} k^3$.

Solution for Problem 17.21: Problems 17.16 and 17.17 have already paved the way for us. We note that $(k + 1)^4 - k^4 = 4k^3 + 6k^2 + 4k + 1$ for all positive integers k, and we have

$$\sum_{k=1}^{n}(4k^3 + 6k^2 + 4k + 1) = \sum_{k=1}^{n}[(k + 1)^4 - k^4] = (n + 1)^4 - 1 = n^4 + 4n^3 + 6n^2 + 4n.$$

We also have

$$\sum_{k=1}^{n}(4k^3 + 6k^2 + 4k + 1) = 4\sum_{k=1}^{n}k^3 + 6\sum_{k=1}^{n}k^2 + 4\sum_{k=1}^{n}k + \sum_{k=1}^{n}1$$

$$= 4\left(\sum_{k=1}^{n}k^3\right) + n(n + 1)(2n + 1) + 2n(n + 1) + n$$

$$= 4\left(\sum_{k=1}^{n}k^3\right) + 2n^3 + 5n^2 + 4n.$$

We therefore have $4\left(\displaystyle\sum_{k=1}^{n}k^3\right) + 2n^3 + 5n^2 + 4n = n^4 + 4n^3 + 6n^2 + 4n$, so

$$\sum_{k=1}^{n}k^3 = \frac{n^4 + 2n^3 + n^2}{4} = \left(\frac{n(n + 1)}{2}\right)^2.$$

$\square$

Exercises

17.3.1 Evaluate $\displaystyle\sum_{i=1}^{50}(i^3 - 2i^2)$.

17.3.2 Find the sum of the series $1 \cdot 2 + 3 \cdot 4 + \cdots + (2n - 1) \cdot 2n$.

17.3.3 Let $g(x) = x^2 - 11x - 28$. Find the positive integer n such that $\displaystyle\sum_{k=1}^{n} g(k) = 0$.

17.3.4 Find a closed form for $\displaystyle\sum_{k=1}^{n} k^4$.

17.3.5★ Compute the value of

$$\frac{1}{2} + \frac{1}{6}\left(1^2 + 2^2\right) + \frac{1}{12}\left(1^2 + 2^2 + 3^2\right) + \frac{1}{20}\left(1^2 + 2^2 + 3^2 + 4^2\right) + \cdots + \frac{1}{3660}\left(1^2 + 2^2 + \cdots + 60^2\right).$$

(Source: ARML) **Hints:** 339

17.4 Arithmetico-Geometric Series

We know how to handle arithmetic and geometric series. In this section, we see what happens when we combine an arithmetic series with a geometric series of the same length by multiplying each term in the arithmetic series by the corresponding term in the geometric series, then summing all these products to form what we call an **arithmetico-geometric** series.

Problems

Problem 17.22: Let $S = 1 \cdot 2^0 + 2 \cdot 2^1 + 3 \cdot 2^2 + \cdots + n \cdot 2^{n-1} + \cdots + 11 \cdot 2^{10}$.

  (a) Write out a similar series equal to $2S$.

  (b) Find the difference between your series from (a) and the given series.

  (c) Evaluate the given series.

Problem 17.23: Determine the value of the infinite product

$$2^{1/3} \cdot 4^{1/9} \cdot 8^{1/27} \cdot 16^{1/81} \cdots .$$

(Source: Mandelbrot)

Problem 17.24: Let r be a constant such that $|r| < 1$. Find a closed form for $\displaystyle\sum_{k=0}^{\infty} kr^k$.

Problem 17.25: Evaluate the sum $\dfrac{5}{4} + \dfrac{8}{4^2} + \dfrac{11}{4^3} + \dfrac{14}{4^4} + \cdots$.

Problem 17.26: Find a closed form for $1 + 4x + 9x^2 + \cdots + (n+1)^2 x^n + \cdots$, where $|x| < 1$.

Problem 17.22: Evaluate the series $1 \cdot 2^0 + 2 \cdot 2^1 + 3 \cdot 2^2 + \cdots + n \cdot 2^{n-1} + \cdots + 11 \cdot 2^{10}$.

Solution for Problem 17.22: This looks a lot like a geometric series, so we try to evaluate it the same way we tackled geometric series.

> **Concept:** You know what we're going to say here. Compare new problems to similar problems you already know how to do. The tactics that worked on the similar problems might solve the new problem, too.

We let

$$S = 1 \cdot 2^0 + 2 \cdot 2^1 + 3 \cdot 2^2 + \cdots + n \cdot 2^{n-1} + \cdots + 11 \cdot 2^{10},$$

then multiply this equation by 2 to get the series

$$2S = 1 \cdot 2^1 + 2 \cdot 2^2 + 3 \cdot 2^3 + \cdots + 11 \cdot 2^{11}.$$

Subtracting the first equation from the second one, we get

$$S = -2^0 - 2^1 - 2^2 - \cdots - 2^{10} + 11 \cdot 2^{11} = 11 \cdot 2^{11} - \sum_{k=0}^{10} 2^k.$$

The sum subtracted at the end is a geometric series, so we can now finish computing S:

$$S = 11 \cdot 2^{11} - \sum_{k=0}^{10} 2^k = 11 \cdot 2^{11} - \left(\frac{2^{11} - 1}{2 - 1}\right) = 11 \cdot 2^{11} - (2^{11} - 1) = 10 \cdot 2^{11} + 1 = 20481.$$

$\square$

When we multiply each term in an arithmetic series by the corresponding term in a geometric series, we form an **arithmetico-geometric series**. For example, the series in Problem 17.22 results from multiplying each term in the arithmetic series $1 + 2 + 3 + \cdots + 11$ by the corresponding term in the geometric series $2^0 + 2^1 + 2^2 + \cdots + 2^{10}$. The result is the arithmetico-geometric series

$$1 \cdot 2^0 + 2 \cdot 2^1 + 3 \cdot 2^2 + \cdots + 11 \cdot 2^{10}.$$

The common ratio of the geometric series is also called the common ratio of the resulting arithmetico-geometric series. For example, the common ratio of the series in Problem 17.22 is 2.

> **Concept:** Arithmetico-geometric series can be evaluated using the methods we developed for solving geometric series. We multiply or divide the arithmetico-geometric series by its common ratio to create a new series. We then evaluate the difference between the two series.

As we've mentioned many times, our methods are more important than our formulas. Indeed, way back in Section 10.4, we noted that the method for evaluating geometric series is more important than the formulas we derived for geometric series. Here's yet another problem on which we can use this method:

> **Problem 17.23:** Determine the value of the infinite product $2^{1/3} \cdot 4^{1/9} \cdot 8^{1/27} \cdot 16^{1/81} \cdots$. *(Source: Mandelbrot)*

Solution for Problem 17.23: We first write each term as a power of 2:

$$2^{1/3} \cdot 2^{2/9} \cdot 2^{3/27} \cdot 2^{4/81} \cdots.$$

The product then equals 2^S, where

$$S = \frac{1}{3} + \frac{2}{3^2} + \frac{3}{3^3} + \frac{4}{3^4} + \cdots.$$

That's an arithmetico-geometric series! We know how to tackle this. We multiply this series by $1/3$ to give

$$\frac{S}{3} = \frac{1}{3^2} + \frac{2}{3^3} + \frac{3}{3^4} + \cdots.$$

Subtracting this from our expression for S above gives

$$\frac{2S}{3} = \frac{1}{3} + \frac{1}{3^2} + \frac{1}{3^3} + \cdots = \frac{\frac{1}{3}}{1 - \frac{1}{3}} = \frac{1}{2},$$

so $S = 3/4$. Therefore, our product equals $2^{3/4} = \sqrt[4]{8}$.

Even if you didn't remember our method for deriving the formulas for geometric series, you still have another option for finding S above. We see that it resembles a geometric series with ratio $1/3$. So, we start by writing such a series:

$$\frac{1}{3} + \frac{1}{3^2} + \frac{1}{3^3} + \frac{1}{3^4} + \cdots .$$

To get S, we need one more $1/3^2$ term, 2 more $1/3^3$ terms, 3 more $1/3^4$ terms, and so on. This gives us an idea; we can write all these powers of $1/3$ in a grid:

$$\frac{1}{3} \; + \; \frac{1}{3^2} \; + \; \frac{1}{3^3} \; + \; \frac{1}{3^4} \; + \; \cdots$$
$$\frac{1}{3^2} \; + \; \frac{1}{3^3} \; + \; \frac{1}{3^4} \; + \; \cdots$$
$$\frac{1}{3^3} \; + \; \frac{1}{3^4} \; + \; \cdots$$
$$\frac{1}{3^4} \; + \; \cdots$$
$$\vdots$$

Adding the columns of our grid gives us S, as there are k copies of $1/3^k$ for each positive integer k. The rows of the grid form an infinite collection of infinite geometric series!

We have:

$$\frac{1}{3} \; + \; \frac{1}{3^2} \; + \; \frac{1}{3^3} \; + \; \frac{1}{3^4} \; + \; \cdots \; = \; \frac{\frac{1}{3}}{1 - \frac{1}{3}} \; = \; \frac{1}{2}$$
$$\frac{1}{3^2} \; + \; \frac{1}{3^3} \; + \; \frac{1}{3^4} \; + \; \cdots \; = \; \frac{\frac{1}{3^2}}{1 - \frac{1}{3}} \; = \; \frac{1}{2} \cdot \frac{1}{3}$$
$$\frac{1}{3^3} \; + \; \frac{1}{3^4} \; + \; \cdots \; = \; \frac{\frac{1}{3^3}}{1 - \frac{1}{3}} \; = \; \frac{1}{2} \cdot \frac{1}{3^2}$$
$$\frac{1}{3^4} \; + \; \cdots \; = \; \frac{\frac{1}{3^4}}{1 - \frac{1}{3}} \; = \; \frac{1}{2} \cdot \frac{1}{3^3}$$
$$\vdots$$

So, the sums of the rows form another geometric sequence. When we add these sums to form S, we get

$$S = \frac{1}{2} + \frac{1}{2} \cdot \frac{1}{3} + \frac{1}{2} \cdot \frac{1}{3^2} + \frac{1}{2} \cdot \frac{1}{3^3} + \cdots = \frac{1}{2} \cdot \frac{1}{1 - \frac{1}{3}} = \frac{3}{4}.$$

$\square$

Now, we're ready to tackle a general infinite arithmetico-geometric series.

Problem 17.24: Let r be a constant such that $|r| < 1$. Find a closed form for $\displaystyle\sum_{k=0}^{\infty} kr^k$.

Solution for Problem 17.24: We again use our tactic for summing a geometric series. Letting the sum be S, we have $rS = r \sum\limits_{k=0}^{\infty} kr^k = \sum\limits_{k=0}^{\infty} kr^{k+1}$. Subtracting this from our expression for S gives

$$S - rS = \sum_{k=0}^{\infty} kr^k - \sum_{k=0}^{\infty} kr^{k+1}.$$

We want to group the like powers of r, so we shift the index of the second summation by 1 to give

$$S - rS = \sum_{k=0}^{\infty} kr^k - \sum_{k=1}^{\infty} (k-1)r^k.$$

We also note that the $k = 0$ term of the first sum equals 0, so we can start that sum at $k = 1$ as well, and we have

$$S - rS = \sum_{k=0}^{\infty} kr^k - \sum_{k=1}^{\infty} (k-1)r^k = \sum_{k=1}^{\infty} kr^k - \sum_{k=1}^{\infty} (k-1)r^k = \sum_{k=1}^{\infty} (kr^k - (k-1)r^k).$$

Since $kr^k - (k-1)r^k = r^k$, our final summation is just an infinite geometric series with first term r and common ratio r:

$$S - rS = \sum_{k=1}^{\infty} (kr^k - (k-1)r^k) = \sum_{k=1}^{\infty} r^k = \frac{r}{1-r}.$$

Therefore, we have $S = \dfrac{r}{(1-r)^2}$. $\square$

What if the arithmetic series doesn't have common difference 1?

Problem 17.25: Evaluate the sum $\dfrac{5}{4} + \dfrac{8}{4^2} + \dfrac{11}{4^3} + \dfrac{14}{4^4} + \cdots$.

Solution for Problem 17.25: We'll be pretty surprised if our "multiply by r" tactic doesn't work. We let S be the sum, and we multiply by 1/4 to get

$$\frac{S}{4} = \frac{5}{4^2} + \frac{8}{4^3} + \frac{11}{4^4} + \frac{14}{4^5} + \cdots.$$

Subtracting this from the original series gives

$$S - \frac{S}{4} = \frac{5}{4} + \frac{3}{4^2} + \frac{3}{4^3} + \frac{3}{4^4} + \cdots = \frac{5}{4} + \frac{\frac{3}{4^2}}{1 - \frac{1}{4}} = \frac{5}{4} + \frac{1}{4} = \frac{3}{2}.$$

So, we have $3S/4 = 3/2$, which means $S = 2$. $\square$

We know how to sum $\sum\limits_{k=0}^{\infty} r^k$ and $\sum\limits_{k=0}^{\infty} kr^k$ for $|r| < 1$. You can guess what's next.

Problem 17.26: Find a closed form for $1 + 4x + 9x^2 + \cdots + (n+1)^2 x^n + \cdots$, where $|x| < 1$.

Solution for Problem 17.26: And you can guess how we'll attack it. We begin by setting a variable equal to the whole series:

$$S = 1 + 4x + 9x^2 + \cdots + (n+1)^2 x^n + \cdots.$$

Multiplying this series by x, we have

$$xS = x + 4x^2 + 9x^3 + \cdots + n^2 x^n + \cdots.$$

Subtracting the second equation from the first, we get

$$S(1-x) = 1 + 3x + 5x^2 + \cdots + (2n+1)x^n + \cdots,$$

which is an arithmetico-geometric series!

Continuing with more of the same, we multiply our latest equation by x:

$$Sx(1-x) = x + 3x^2 + 5x^3 + \cdots + (2n-1)x^n + \cdots.$$

Subtracting this equation from the previous one, we get

$$S(1-x)^2 = 1 + 2x + 2x^2 + 2x^3 + \cdots + 2x^n + \cdots,$$

which is 1 more than a geometric series. Summing the geometric series then gives us

$$S(1-x)^2 = 1 + \frac{2x}{1-x} = \frac{x+1}{1-x},$$

so $S = \dfrac{x+1}{(1-x)^3}$. $\square$

Another triumph of methods over formulas!

Exercises

17.4.1　Find the sum of each of the following arithmetico-geometric series:

(a)　$\dfrac{1}{5} + \dfrac{2}{5^2} + \dfrac{3}{5^3} + \dfrac{4}{5^4} + \cdots$　(b)　$\displaystyle\sum_{k=1}^{n} (2k-1)3^{k-1}$　(c)　$\displaystyle\sum_{n=1}^{\infty} \frac{5n-1}{2^n}$

17.4.2　Evaluate: $\log_2 \sqrt{2} + \log_2 \sqrt[4]{4} + \log_2 \sqrt[8]{8} + \cdots + \log_2 \sqrt[2^n]{2^n} + \cdots$.

17.4.3　Find a closed form for the series $1 + 2x + 3x^2 + \cdots + nx^{n-1}$ for $x \neq 1$.

17.5　Finite Differences

Given a sequence $a_0, a_1, a_2, a_3, \ldots$, we can construct a table as follows. First, write the terms of the sequence:

$$a_0 \quad a_1 \quad a_2 \quad a_3 \quad \cdots$$

Next, write the differences between consecutive terms under the original sequence:

$$a_0 \qquad a_1 \qquad a_2 \qquad a_3 \qquad \cdots$$
$$a_1 - a_0 \qquad a_2 - a_1 \qquad a_3 - a_2 \qquad \cdots$$

To make things easier, we introduce some notation. We let $\Delta a_n = a_{n+1} - a_n$ for all $n \geq 0$. (The symbol Δ is the uppercase Greek letter **delta**.) We can then rewrite the table as follows:

$$a_0 \qquad a_1 \qquad a_2 \qquad a_3 \qquad \cdots$$
$$\Delta a_0 \qquad \Delta a_1 \qquad \Delta a_2 \qquad \cdots$$

The next step, naturally, is to write the differences between consecutive terms of the sequence $\{\Delta a_n\}$:

$$a_0 \qquad a_1 \qquad a_2 \qquad a_3 \qquad \cdots$$
$$\Delta a_0 \qquad \Delta a_1 \qquad \Delta a_2 \qquad \cdots$$
$$\Delta a_1 - \Delta a_0 \qquad \Delta a_2 - \Delta a_1 \qquad \cdots$$

Again, to make things easier, we denote the difference $\Delta a_{n+1} - \Delta a_n$ by $\Delta^2 a_n$:

$$a_0 \qquad a_1 \qquad a_2 \qquad a_3 \qquad \cdots$$
$$\Delta a_0 \qquad \Delta a_1 \qquad \Delta a_2 \qquad \cdots$$
$$\Delta^2 a_0 \qquad \Delta^2 a_1 \qquad \cdots$$

In this way, we can produce as many rows as we please:

$$a_0 \qquad a_1 \qquad a_2 \qquad a_3 \qquad \cdots$$
$$\Delta a_0 \qquad \Delta a_1 \qquad \Delta a_2 \qquad \cdots$$
$$\Delta^2 a_0 \qquad \Delta^2 a_1 \qquad \cdots$$
$$\Delta^3 a_0 \qquad \cdots$$
$$\cdots$$

The sequence $\Delta a_0, \Delta a_1, \Delta a_2, \ldots$ is the "first difference sequence," the sequence $\Delta^2 a_0, \Delta^2 a_1, \Delta^2 a_2, \ldots$ is the "second difference sequence," and so on. The table is known as the **finite difference** table for the sequence $a_0, a_1, a_2, \ldots$. By convention, the rows are labeled by the power of Δ, so the row containing the sequence $\{a_n\}$ is known as the 0th row, the row containing the sequence $\{\Delta a_n\}$ is known as the first row, and so on.

In this section, we use finite difference tables to analyze polynomials. If $p(x)$ is a polynomial, we can construct a finite difference table for $p(x)$ by starting with the sequence $\{p(n)\}_{n=0}^{\infty}$. The first difference sequence is $\{\Delta p(n)\}_{n=0}^{\infty}$, where $\Delta p(n) = p(n+1) - p(n)$. The second difference sequence then is $\{\Delta^2 p(n)\}_{n=0}^{\infty}$, where $\Delta^2 p(n) = \Delta p(n+1) - \Delta p(n)$, and so on.

Problems

Problem 17.27: Let $a_n = n^3$. Construct the finite difference table for the sequence $a_0, a_1, a_2, \ldots$. Do you notice anything interesting?

Problem 17.28:

(a) Let $f(x) = ax^2 + bx + c$, where a, b, and c are constants and $a \neq 0$. Consider the sequence $\{f(n)\}_{n=0}^{\infty}$. Prove that $\Delta^2 f(n)$ is constant and nonzero, but $\Delta f(n)$ is not constant.

(b) Let $p(x)$ be a polynomial with degree m, where $m > 0$. Show that $\Delta^m p(n)$ is constant and nonzero, but $\Delta^{m-1} p(n)$ is not constant.

Problem 17.29: Find all polynomials $p(x)$ such that $p(x+1) - p(x) = 2x + 1$ for all x.

Problem 17.30: In order to draw a graph of $f(x) = ax^2 + bx + c$, a table of values was constructed. These values of the function for a set of equally spaced increasing values of x were 3844, 3969, 4096, 4227, 4356, 4489, 4624, and 4761. Unfortunately, one of these values was determined incorrectly. Find the incorrect value. *(Source: AHSME)*

Problem 17.31: Assume that $x_1, x_2, \ldots, x_7$ are real numbers such that

$$x_1 + 4x_2 + 9x_3 + 16x_4 + 25x_5 + 36x_6 + 49x_7 = 1,$$
$$4x_1 + 9x_2 + 16x_3 + 25x_4 + 36x_5 + 49x_6 + 64x_7 = 12,$$
$$9x_1 + 16x_2 + 25x_3 + 36x_4 + 49x_5 + 64x_6 + 81x_7 = 123.$$

Find the value of $16x_1 + 25x_2 + 36x_3 + 49x_4 + 64x_5 + 81x_6 + 100x_7$. *(Source: AIME)*

Problem 17.27: Let $a_n = n^3$. Construct the finite difference table for the sequence $a_0, a_1, a_2, \ldots$.

Solution for Problem 17.27: We write the first several terms of the sequence, and then take the finite differences:

a_n :	0		1		8		27		64		125		216		$\cdots$
Δa_n :		1		7		19		37		61		91		$\cdots$	
$\Delta^2 a_n$:			6		12		18		24		30		$\cdots$		
$\Delta^3 a_n$:				6		6		6		6		$\cdots$			
$\Delta^4 a_n$:					0		0		0		$\cdots$				

We notice that the third row seems to consist of all 6s, and that all further rows seem to consist of 0s. Since we have only written down a portion of the table, can we be certain that this pattern continues indefinitely?

We can, by explicitly computing the formula for $\Delta^k a_n$:

$$a_n = n^3,$$
$$\Delta a_n = a_{n+1} - a_n = (n+1)^3 - n^3 = 3n^2 + 3n + 1,$$
$$\Delta^2 a_n = \Delta a_{n+1} - \Delta a_n = 3(n+1)^2 + 3(n+1) + 1 - (3n^2 + 3n + 1) = 6n + 6,$$
$$\Delta^3 a_n = \Delta^2 a_{n+1} - \Delta^2 a_n = 6(n+1) + 6 - (6n + 6) = 6,$$
$$\Delta^4 a_n = \Delta^3 a_{n+1} - \Delta^3 a_n = 6 - 6 = 0.$$

We see that the third row does indeed consist entirely of 6s, and that the fourth row consists entirely of 0s. Furthermore, once a difference table hits a row of all 0s, then clearly all further rows are also all 0s. □

In Problem 17.27, we saw that if $a_n = n^3$, then the sequence $\{\Delta^2 a_n\}$ is not constant, but the sequence $\{\Delta^3 a_n\}$ is constant and nonzero. Let's generalize this observation by considering the difference table of a general polynomial.

Problem 17.28: Let $p(x)$ be a polynomial with degree m, where $m > 0$. Show that $\Delta^m p(n)$ is constant and nonzero, but $\Delta^{m-1} p(n)$ is not constant.

Solution for Problem 17.28: We let

$$p(x) = c_m x^m + c_{m-1} x^{m-1} + \cdots + c_1 x + c_0.$$

Building the whole difference table looks like a nightmare, but we can at least explore a little by finding $\Delta p(n)$. We find

$$\Delta p(n) = p(n+1) - p(n) = c_m(n+1)^m + c_{m-1}(n+1)^{m-1} + \cdots + c_1(n+1) + c_0 - (c_m n^m + c_{m-1} n^{m-1} + \cdots + c_1 n + c_0).$$

We're certainly not interested in multiplying that all out, so we look back to Problem 17.27 for guidance. There, we saw that for the cubic $a_n = n^3$, we can express Δa_n as a quadratic in n. So, we wonder if $\Delta p(n)$ similarly has degree 1 less than the degree of $p(n)$. Going back to our expression above for $\Delta p(n)$, we see that the only two terms of degree m come from $c_m(n+1)^m$ and $-c_m n^m$. The degree m term of $c_m(n+1)^m$ is $c_m n^m$, so the degree m terms in our expression for $\Delta p(n)$ cancel.

But what about the degree $m - 1$ terms? There are three:

$$c_m \binom{m}{1} n^{m-1} + c_{m-1} n^{m-1} - c_{m-1} n^{m-1}.$$

The last two cancel, and we're left with $c_m \binom{m}{1} n^{m-1}$. We know that c_m is nonzero because $p(n)$ has degree m. So, $\Delta p(n)$ is a degree $m - 1$ polynomial.

By the exact same argument, we can show that $\deg \Delta^2 p(n)$ is 1 less than $\deg \Delta p(n)$. Continuing in this vein, we can show with induction that if $k \le m$, then $\deg \Delta^k p(n) = m - k$. So, when $k = m$, we have $\deg \Delta^m p(n) = 0$, which means $\Delta^m p(n)$ is a nonzero constant. Because $\deg \Delta^{m-1} p(n) = 1$, we know that $\Delta^{m-1} p(n)$ is nonconstant. □

Important: If $p(n)$ is a degree m polynomial, then $\Delta^m p(n)$ is a nonzero constant, while $\Delta^{m-1} p(n)$ is linear.

Problem 17.29: Find all polynomials $p(x)$ such that $p(x+1) - p(x) = 2x + 1$ for all x.

Solution for Problem 17.29: From the given equation, we can deduce that $\Delta p(n) = 2n + 1$. Therefore, $\Delta^2 p(n) = \Delta p(n+1) - \Delta p(n) = 2(n+1) + 1 - (2n+1) = 2$. Because $\Delta^2 p(n)$ is constant and nonzero, we know that $p(x)$ is a quadratic. So, we let $p(x) = ax^2 + bx + c$, and we have

$$a(x+1)^2 + b(x+1) + c - ax^2 - bx - c = 2x + 1.$$

Simplifying the left side gives $2ax + a + b = 2x + 1$. Equating the coefficients of x gives $a = 1$, and equating the constants gives $b = 0$. So, the polynomials $p(x)$ such that $p(x + 1) - p(x) = 2x + 1$ are $p(x) = x^2 + c$, for any constant c. $\square$

Problem 17.30: In order to draw a graph of $f(x) = ax^2 + bx + c$, a table of values was constructed. These values of the function for a set of equally spaced increasing values of x were 3844, 3969, 4096, 4227, 4356, 4489, 4624, and 4761. Unfortunately, one of these values was determined incorrectly. Find the incorrect value. *(Source: AHSME)*

Solution for Problem 17.30: Because $f(x)$ is a quadratic, we know that $\{\Delta^2 f(n)\}$ is a nonzero constant sequence. So, we may be able to detect the incorrect entry by constructing the difference table for $f(x)$:

3844		3969		4096		4227		4356		4489		4624		4761
	125		127		131		129		133		135		137	
		2		4		-2		4		2		2		

Looking at the table, we suspect that the incorrect value is 4227. All the entries in the second row are 2, except the 4, -2, and 4, and the entry 4227 affects all three of these. Indeed, if we were to set all the entries in the second row to 2 and rebuild the table, then the 4227 would change to 4225:

3844		3969		4096		4225		4356		4489		4624		4761
	125		127		129		131		133		135		137	
		2		2		2		2		2		2		

Therefore, the incorrect value is 4227.

But wait a minute! The problem only said that the given outputs were the result of equally spaced inputs x. We weren't told that the outputs were $f(0)$, $f(1)$, $f(2)$, and so on. Can we still use the difference table as we did above?

Yes! We can transform the quadratic $f(x)$ into another quadratic $g(y)$ such that our outputs correspond to $g(0)$, $g(1)$, $g(2)$, etc.

Suppose our first input x is $x = k$, and that the x's are spaced d apart, where k and d are constants. So, our erroneous original table has $f(k) = 3844$, $f(k + d) = 3969$, $f(k + 2d) = 4096$, and so on. Now we see how to make $g(y)$. We let $g(y) = f(k + dy)$. Then, we have $g(0) = f(k) = 3844$, $g(1) = f(k + d) = 3969$, $g(2) = f(k + 2d) = 4096$, and so on. Because $f(x)$ is quadratic, $f(k + dy)$ is quadratic in y (remember, k and d are constants). Our difference tables above are therefore the difference tables for $g(n)$, which we can use as described to determine that 4227 was the incorrect value. $\square$

Problem 17.31: Assume that $x_1, x_2, \ldots, x_7$ are real numbers such that

$$x_1 + 4x_2 + 9x_3 + 16x_4 + 25x_5 + 36x_6 + 49x_7 = 1,$$
$$4x_1 + 9x_2 + 16x_3 + 25x_4 + 36x_5 + 49x_6 + 64x_7 = 12,$$
$$9x_1 + 16x_2 + 25x_3 + 36x_4 + 49x_5 + 64x_6 + 81x_7 = 123.$$

Find the value of $16x_1 + 25x_2 + 36x_3 + 49x_4 + 64x_5 + 81x_6 + 100x_7$. *(Source: AIME)*

Solution for Problem 17.31: We're not too thrilled with the idea of trying to solve the system of equations. Moreover, with 3 linear equations in 7 variables, we won't even have a unique solution. Rather than trying to solve the system and substitute into the desired equation, we look for a way to manipulate the given equations to give us the expression we seek.

We first notice that the coefficients of the left sides of the given equations are perfect squares. Moreover, we can model all three left sides at once with the expression

$$a_n = (n+1)^2 x_1 + (n+2)^2 x_2 + (n+3)^2 x_3 + \cdots + (n+7)^2 x_7.$$

Then a_0, a_1, and a_2 are the left sides of the three given equations. And a_3 is the desired expression! Furthermore, a_n is quadratic in n. We're given that $a_0 = 1$, $a_1 = 12$, and $a_2 = 123$. So, now we can build our difference table for the sequence of a_n.

We know that the $\Delta^2 a_n$ row is constant, so after using 1, 12, and 123 in the top row to find 100 in the bottom row, we can add another 100 in the $\Delta^2 a_n$ row and work backwards to the top, finding $a_3 = 334$. Therefore, we have

a_n :	1		12		123		334
Δa_n :		11		111		211	
$\Delta^2 a_n$:			100		100		

$$16x_1 + 25x_2 + 36x_3 + \cdots + 100x_7 = a_3 = 334.$$

$\Box$

Exercises

17.5.1 A quadratic polynomial $p(n)$ satisfies $p(0) = 1$, $p(1) = 2$, and $p(2) = 7$.

(a) Use a difference table to find $p(3)$.

(b) Use a difference table to find $p(n)$.

17.5.2 Part of the difference table of a sequence a_n is as shown:

a_n :	1		*		*		*
Δa_n :		2		*		*	
$\Delta^2 a_n$:			4		*		
$\Delta^3 a_n$:				8			

Find a_1, a_2, and a_3.

17.5.3 If $\Delta^3 p(n) = 3n + 5$, then must $\deg p = 4$?

17.5.4 For any sequence of real numbers $A = a_1, a_2, a_3, \ldots$, define ΔA as the sequence of numbers $a_2 - a_1, a_3 - a_2, a_4 - a_3, \ldots$, whose nth term is $a_{n+1} - a_n$. Suppose that all of the terms of the sequence $\Delta(\Delta A)$ are 1, and that $a_{19} = a_{92} = 0$. Find a_1. *(Source: AIME)*

17.6 Summary

A **recursive sequence** is a sequence in which, after one or more initial terms, each term is defined as a function of previous terms in the sequence. Such a sequence definition is called a **recursion**.

A series in which parts of each term cancel with parts of other terms in the series is called a **telescoping series**. Some products also telescope. One tool that is helpful in discovering and evaluating telescoping series is **partial fraction decomposition**, through which we write a rational function as a sum of rational functions with simpler denominators.

We can use telescoping series to find formulas for the sums of powers of consecutive integers, such as:

$$\sum_{k=1}^{n} k = \frac{n(n+1)}{2},$$

$$\sum_{k=1}^{n} k^2 = \frac{n(n+1)(2n+1)}{6},$$

$$\sum_{k=1}^{n} k^3 = \frac{n^2(n+1)^2}{4}.$$

We form an **arithmetico-geometric series** when we multiply each term of an arithmetic series by the corresponding term of a geometric series of the same length, then sum all these products. We can evaluate such series using similar tactics to those we used to derive formulas for geometric series.

Finite differences can be used to analyze the polynomial $p(x)$ by first constructing the sequence $\{p(n)\}_{n=0}^{\infty}$. We can then construct the first difference sequence, $\{\Delta p(n)\}$, where $\Delta p(n) = p(n+1) - p(n)$. From this, we can construct the second difference sequence, $\{\Delta^2 p(n)\}$, where $\Delta^2 p(n) = \Delta p(n+1) - \Delta p(n)$. And so on.

> **Important:** If $p(n)$ is a degree m polynomial, then $\Lambda^m p(n)$ is a nonzero constant, while $\Delta^{m-1} p(n)$ is linear.

Problem Solving Strategies

Concepts:

- We can often learn a lot about a sequence or series by comparing each term to neighboring terms.

- To learn more about a recursion, compute the first few terms and look for a pattern. Then, once you've found the pattern, use the recursion to figure out why the pattern exists.

- When looking for useful relationships to help solve sequence problems (recursion problems in particular), keep an eye out for periodicity.

- Shifting the index of a recursion can be an effective way to learn more about the recursion.

Continued on the next page. . .

Concepts: ... *continued from the previous page*

- When trying to prove that a pattern exists, write a general form that represents the pattern.

- There aren't many coincidences in mathematics. If two types of problems consistently give the same answer, look for a reason why.

- Many sums involving fractions can be rearranged so that they telescope.

- Whenever you derive a general formula for a sequence or series, you can usually test it with a few simple cases.

- Notation itself can be a powerful problem solving tool. Rewriting some series with sigma notation can help evaluate them more easily.

- When a strategy gives you information about a problem, but doesn't completely solve it, try using the strategy again. Maybe it has more to reveal.

REVIEW PROBLEMS

17.32 Let $a_1, a_2, \ldots$ be a sequence for which $a_1 = 2$, $a_2 = 3$, and $a_n = \frac{a_{n-1}}{a_{n-2}}$ for each positive integer $n \geq 3$. What is a_{2006}? *(Source: AMC 10)*

17.33 Let t_n be the nth term of an geometric sequence with first term a and common ratio r. Write a recursion for the sequence $\{t_n\}$.

17.34 Let a and b be nonzero constants such that $f(x) = ax + b$ and $a \neq 1$, and let n be a positive integer. Express $f^n(x)$ in terms of a, b, and n.

17.35 On Halloween, Mrs. Pi tells each of the neighborhood children that they can take a handful of candy. In order to decide how much candy to take, each kid takes the average of the amount of candy that the 3 kids immediately before her took. If this average is not an integer, she always rounds up. If the first kid took 3 pieces of candy, the second kid took 2, and the third kid took 5, how many pieces will the 2007th kid take?

17.36 For the sequence of numbers $n_1, n_2, n_3, \ldots$, the relation $n_i = 2n_{i-1} + a$, where a is a constant, holds for all $i > 1$. If $n_2 = 5$ and $n_8 = 257$, what is n_5? *(Source: HMMT)*

17.37 Find $\log_{10} \frac{1}{2} + \log_{10} \frac{2}{3} + \log_{10} \frac{3}{4} + \cdots + \log_{10} \frac{99}{100}$.

17.38 Determine the value of $1 \cdot 2 - 2 \cdot 3 + 3 \cdot 4 - 4 \cdot 5 + \cdots + 2001 \cdot 2002$. *(Source: HMMT)*

17.39 Simplify the product

$$\left(1 - \frac{1}{2^2}\right)\left(1 - \frac{1}{3^2}\right)\cdots\left(1 - \frac{1}{9^2}\right)\left(1 - \frac{1}{10^2}\right).$$

(Source: AHSME)

17.40 Compute each of the following:

(a) $1 + \dfrac{2}{7} + \dfrac{3}{49} + \cdots + \dfrac{n}{7^{n-1}} + \cdots$

(b) $\displaystyle\sum_{n=1}^{\infty} \dfrac{3n-2}{5^n}$

17.41 Compute the unique positive integer n such that $2 \cdot 2^2 + 3 \cdot 2^3 + 4 \cdot 2^4 + 5 \cdot 2^5 + \cdots + n \cdot 2^n = 2^{n+10}$. *(Source: ARML)*

17.42 Compute the product $\dfrac{(1998^2 - 1996^2)(1998^2 - 1995^2)\cdots(1998^2 - 0^2)}{(1997^2 - 1996^2)(1997^2 - 1995^2)\cdots(1997^2 - 0^2)}$. *(Source: Mandelbrot)*

17.43 Evaluate $\displaystyle\sum_{k=1}^{20}(2k^2 - k)$.

17.44 Define a sequence of numbers by $a_n = 3n^2 + 3n + 1$, so that $a_1 = 7$, $a_2 = 19$, $a_3 = 37$, and so on. Calculate $a_1 + a_2 + \cdots + a_{100}$. *(Source: Mandelbrot)*

17.45 If $1 \cdot 1987 + 2 \cdot 1986 + 3 \cdot 1985 + \cdots + 1986 \cdot 2 + 1987 \cdot 1 = 1987 \cdot 994 \cdot x$, compute the integer x. *(Source: ARML)*

17.46 From the table shown, find the polynomial of least degree that expresses y in terms of x:

x	1	2	3	4	5
y	3	7	13	21	31

(Source: AHSME)

17.47 A cubic polynomial $f(n)$ satisfies $f(0) = 5$, $f(1) = 4$, $f(2) = 17$, and $f(3) = 56$. Find $f(4)$.

Challenge Problems

17.48 For a finite sequence $a_1, a_2, \ldots, a_n$ of numbers, the *Cesàro sum* of the sequence is defined to be

$$\dfrac{S_1 + S_2 + \cdots + S_n}{n},$$

where

$$S_k = a_1 + a_2 + \cdots + a_k$$

for $1 \le k \le n$. If the Cesàro sum of the 99-term sequence $a_1, a_2, \ldots, a_{99}$ is 1000, what is the Cesàro sum of the 100-term sequence $1, a_1, a_2, \ldots, a_{99}$? *(Source: AHSME)*

17.49 Let $P_0(x) = x^3 + 313x^2 - 77x - 8$. For integers $n \ge 1$, define $P_n(x) = P_{n-1}(x - n)$. What is the coefficient of x in $P_{20}(x)$? *(Source: AIME)* **Hints:** 136

17.50 A sequence of integers $a_1, a_2, a_3, \ldots$ is chosen so that $a_n = a_{n-1} - a_{n-2}$ for each $n \geq 3$. What is the sum of the first 2001 terms of the sequence if the sum of the first 1492 terms is 1985, and the sum of the first 1985 terms is 1492? *(Source: AIME)* **Hints:** 82

17.51 Consider all sets of two distinct positive integers less than or equal to 21. Find the sum of the products of the elements in all such sets. *(Source: Mandelbrot)* **Hints:** 313, 155

17.52★ Evaluate $\displaystyle\sum_{k=1}^{\infty} \frac{k^3 + k}{2^k}$. **Hints:** 184

17.53 If $f(n + 1) = (-1)^{n+1} n - 2f(n)$ for integers $n \geq 1$, and $f(1) = f(1986)$, find the value of the sum $f(1) + f(2) + f(3) + \cdots + f(1985)$. *(Source: ARML)* **Hints:** 280

17.54 Find a closed form for $\displaystyle\sum_{k=1}^{n} \frac{1}{k(k + 1)(k + 2)}$.

17.55 Determine the value of

$$S = \sqrt{1 + \frac{1}{1^2} + \frac{1}{2^2}} + \sqrt{1 + \frac{1}{2^2} + \frac{1}{3^2}} + \cdots + \sqrt{1 + \frac{1}{1999^2} + \frac{1}{2000^2}}.$$

(Source: USAMTS) **Hints:** 268

17.56 Find the value of $\dfrac{1}{3^2 + 1} + \dfrac{1}{4^2 + 2} + \dfrac{1}{5^2 + 3} + \cdots$. *(Source: HMMT)*

17.57★ Find a closed form for $\displaystyle\sum_{k=1}^{n} \frac{k}{k^4 + k^2 + 1}$. **Hints:** 300

17.58 A sequence of numbers $a_1, a_2, a_3, \ldots$ satisfies $a_1 = 1/2$ and $a_1 + a_2 + \cdots + a_n = n^2 a_n$ for all $n \geq 1$. Determine the value of a_n for $n \geq 1$. *(Source: CMO)* **Hints:** 17

17.59 Let m be a positive integer, and let $a_0, a_1, \ldots, a_m$ be a sequence of real numbers such that $a_0 = 37$, $a_1 = 72$, $a_m = 0$, and $a_{k+1} = a_{k-1} - \frac{3}{a_k}$ for $k = 1, 2, \ldots, m - 1$. Find m. *(Source: AIME)*

17.60★ Evaluate the infinite product $\displaystyle\prod_{n=2}^{\infty} \frac{n^3 - 1}{n^3 + 1}$. *(Source: Putnam)* **Hints:** 139

17.61★ Let $x_0 = x_1 = 1$ and $x_n = 4x_{n-1} - x_{n-2}$ for $n \geq 2$. Prove that $x_{n-1} x_{n+1} - x_n^2 = 2$ for all $n \geq 1$. **Hints:** 135

17.62★ Find the sum $\dfrac{2^1}{4^1 - 1} + \dfrac{2^2}{4^2 - 1} + \dfrac{2^4}{4^4 - 1} + \dfrac{2^8}{4^8 - 1} + \cdots$. *(Source: HMMT)* **Hints:** 61

17.63★ Let $F_0 = 0$, $F_1 = 1$, and $F_n = F_{n-1} + F_{n-2}$. Find the value of the infinite sum

$$\frac{1}{3} + \frac{1}{9} + \frac{2}{27} + \cdots + \frac{F_n}{3^n} + \cdots.$$

(Source: Mandelbrot) **Hints:** 140

There is always inequality in life. – John F. Kennedy

CHAPTER 18

_____ More Inequalities

18.1 Mean Inequality Chain

We've already studied the Arithmetic Mean-Geometric Mean Inequality, which states that the arithmetic mean of a sequence of nonnegative numbers is greater than or equal to the geometric mean of those numbers. In this section, we introduce two more means, and find relationships between them and both the arithmetic mean and the geometric mean.

First, we'll define the two new means:

> **Definition:** The **quadratic mean** of a set of real numbers is the square root of the arithmetic mean of the squares of the numbers. In other words, the quadratic mean of the real numbers $a_1, a_2, \ldots, a_n$ is
>
> $$\sqrt{\frac{a_1^2 + a_2^2 + a_3^2 + \cdots + a_n^2}{n}}.$$
>
> The quadratic mean is also sometimes referred to as the **root mean square**.

> **Definition:** The **harmonic mean** of a set of nonzero real numbers is the reciprocal of the average of the reciprocals of the numbers. In other words, the harmonic mean of the nonzero numbers $a_1, a_2, a_3, \ldots, a_n$ is
>
> $$\frac{n}{\frac{1}{a_1} + \frac{1}{a_2} + \frac{1}{a_3} + \cdots + \frac{1}{a_n}}.$$

The quadratic mean, root mean square, and harmonic mean are sometimes abbreviated **QM**, **RMS**, and **HM**, respectively.

Problems

Problem 18.1: In this problem, we show that the quadratic mean of the nonnegative numbers $a_1, a_2,$ $a_3, \ldots, a_n$ is greater than or equal to the arithmetic mean of the numbers.

(a) Show that the desired inequality is equivalent to $n(a_1^2 + a_2^2 + \cdots + a_n^2) \geq (a_1 + a_2 + \cdots + a_n)^2$.

(b) Use an inequality from Chapter 12 to prove that $n(a_1^2 + a_2^2 + \cdots + a_n^2) \geq (a_1 + a_2 + \cdots + a_n)^2$.

(c) Prove that

$$\sqrt{\frac{a_1^2 + a_2^2 + \cdots + a_n^2}{n}} \geq \frac{a_1 + a_2 + \cdots + a_n}{n}.$$

Must this inequality still hold if some of the a_i are negative?

Problem 18.2: Let $x_1, x_2, \ldots, x_n > 0$. Prove the Geometric Mean-Harmonic Mean Inequality, which states that

$$\sqrt[n]{x_1 x_2 \cdots x_n} \geq \frac{n}{\frac{1}{x_1} + \frac{1}{x_2} + \cdots + \frac{1}{x_n}}.$$

What is the equality condition?

Problem 18.3: Let $a_1, a_2, \ldots, a_n$ be positive real numbers such that $a_1 + a_2 + \cdots + a_n = 1$. Prove that

$$\frac{1}{a_1} + \frac{1}{a_2} + \cdots + \frac{1}{a_n} \geq n^2.$$

Problem 18.4: Let a, b, c, and d be positive real numbers. Prove that

$$\frac{a^2 + b^2 + c^2}{a + b + c} + \frac{a^2 + b^2 + d^2}{a + b + d} + \frac{a^2 + c^2 + d^2}{a + c + d} + \frac{b^2 + c^2 + d^2}{b + c + d} \geq a + b + c + d.$$

Problem 18.5: Prove that $\sqrt{\dfrac{a^2 + b^2}{2}} - \dfrac{a + b}{2} \geq \sqrt{ab} - \dfrac{2}{\frac{1}{a} + \frac{1}{b}}$ for any positive numbers a and b. **Hints:** 342

Problem 18.6: Find the minimum possible value of

$$\left(1 + \frac{x}{2y}\right)\left(1 + \frac{y}{2z}\right)\left(1 + \frac{z}{2x}\right),$$

where x, y, and z are positive. **Hints:** 328, 320

Problem 18.1: Prove that the quadratic mean of the nonnegative numbers $a_1, a_2, a_3, \ldots, a_n$ is greater than or equal to the arithmetic mean of the numbers. Must this inequality still hold if some of the a_i are negative?

Solution for Problem 18.1: First, we write the inequality we wish to prove,

$$\sqrt{\frac{a_1^2 + a_2^2 + \cdots + a_n^2}{n}} \geq \frac{a_1 + a_2 + \cdots + a_n}{n}.$$

We avoid the square root by considering the squares of both sides. We wish to show that

$$\frac{a_1^2 + a_2^2 + \cdots + a_n^2}{n} \geq \left(\frac{a_1 + a_2 + \cdots + a_n}{n}\right)^2.$$

Both sides of this inequality are nonnegative. Therefore, if we prove this inequality, we can take the square root of both sides to establish the desired inequality.

Our new inequality has a sum of squares on the greater side and the square of a sum on the lesser side. That sounds like the Cauchy-Schwarz Inequality. We multiply both sides by n^2, both to get rid of the fractions and to isolate the square of a sum on the lesser side:

$$n(a_1^2 + a_2^2 + \cdots + a_n^2) \geq (a_1 + a_2 + \cdots + a_n)^2.$$

Now we can see the similarity between this and the Cauchy-Schwarz Inequality. But the Cauchy-Schwarz Inequality involves two sequences. It states that

$$(a_1^2 + a_2^2 + \cdots + a_n^2)(b_1^2 + b_2^2 + \cdots + b_n^2) \geq (a_1b_1 + a_2b_2 + \cdots + a_nb_n)^2.$$

Aha! We need an n on the greater side, and we want the b_i's to disappear on the lesser side. Letting $b_1 = b_2 = \cdots = b_n = 1$ in the Cauchy-Schwarz Inequality gives us

$$(a_1^2 + a_2^2 + \cdots + a_n^2)(n) \geq (a_1 + a_2 + \cdots + a_n)^2,$$

as desired. Once again, we've found our proof working backwards, but we write the proof forwards:

Proof: ☑ The Cauchy-Schwarz Inequality gives us $(a_1^2 + a_2^2 + \cdots + a_n^2)(1^2 + 1^2 + \cdots + 1^2) \geq (a_1 + a_2 + \cdots + a_n)^2$, so we have $n(a_1^2 + a_2^2 + \cdots + a_n^2) \geq (a_1 + a_2 + \cdots + a_n)^2$. Dividing both sides by n^2 gives

$$\frac{a_1^2 + a_2^2 + \cdots + a_n^2}{n} \geq \frac{(a_1 + a_2 + \cdots + a_n)^2}{n^2}.$$

Because both sides are nonnegative, we can take the square root of both sides to give

$$\sqrt{\frac{a_1^2 + a_2^2 + \cdots + a_n^2}{n}} \geq \frac{a_1 + a_2 + \cdots + a_n}{n}.$$

From the equality condition of the Cauchy-Schwarz Inequality, we see that equality holds if and only if $a_1/1 = a_2/1 = a_3/1 = \cdots = a_n/1$; that is, if all the a_i are equal.

If some of the a_i are negative, then the sum $a_1 + a_2 + \cdots + a_n$ might be negative. If $a_1 + a_2 + \cdots + a_n \geq 0$, then our proof above is valid, and shows that the arithmetic mean of the a_i is no greater than the quadratic mean of the a_i. If $a_1 + a_2 + \cdots + a_n < 0$, then the square root of $(a_1 + a_2 + \cdots + a_n)^2$ is $-(a_1 + a_2 + \cdots + a_n)$. This means our proof above doesn't work if $a_1 + a_2 + \cdots + a_n < 0$. However, if the sum of the a_i is negative, then their quadratic mean is positive, but the arithmetic mean is negative, which means the quadratic mean is greater than the arithmetic mean. So, the inequality still holds if some of the a_i are negative. $\square$

The inequality we proved in Problem 18.1 is sometimes referred to as the **Quadratic Mean-Arithmetic Mean Inequality**, or **QM-AM**.

One key step in our solution was recognizing that the sum of squares and the square of a sum suggest using the Cauchy-Schwarz Inequality.

> **Concept:** Much of success in proving inequalities is recognizing expressions that are part of well-known inequalities such as the Cauchy-Schwarz Inequality or the AM-GM Inequality.

This will be a common theme throughout this section, and you'll have plenty of practice with it when you tackle the problems at the end of the chapter. When stuck on an inequality, ask yourself, "Where have I seen an expression like this before?"

Recognizing the sum of squares as a clue to try Cauchy wasn't enough to solve Problem 18.1. We also had to figure out how to use the Cauchy-Schwarz Inequality, which is an inequality for two sequences, to prove an inequality involving only one sequence. The led us to a useful Cauchy-specific strategy:

> **Concept:** To derive some inequalities involving a single sequence of real numbers, we can set one of the sequences in the Cauchy-Schwarz Inequality equal to constants.

In fact, as we saw above, the QM-AM Inequality really isn't something new. It's just a specific application of the Cauchy-Schwarz Inequality in which we let one sequence be a sequence of 1's.

> **Problem 18.2:** Let $x_1, x_2, \ldots, x_n > 0$. Prove the Geometric Mean-Harmonic Mean Inequality, which states that
> $$\sqrt[n]{x_1 x_2 \cdots x_n} \geq \frac{n}{\frac{1}{x_1} + \frac{1}{x_2} + \cdots + \frac{1}{x_n}}.$$
> What is the equality condition?

Solution for Problem 18.2: We don't know much about the harmonic mean, but we do know how to relate the geometric mean to the arithmetic mean. The right side is the reciprocal of the arithmetic mean of the numbers $\frac{1}{x_1}, \frac{1}{x_2}, \ldots, \frac{1}{x_n}$. Applying AM-GM to these positive numbers gives

$$\frac{\frac{1}{x_1} + \frac{1}{x_2} + \cdots + \frac{1}{x_n}}{n} \geq \sqrt[n]{\frac{1}{x_1} \cdot \frac{1}{x_2} \cdots \frac{1}{x_n}} = \frac{1}{\sqrt[n]{x_1 x_2 \cdots x_n}}.$$

Both sides of the inequality $\dfrac{\frac{1}{x_1} + \frac{1}{x_2} + \cdots + \frac{1}{x_n}}{n} \geq \dfrac{1}{\sqrt[n]{x_1 x_2 \cdots x_n}}$ are positive, so taking the reciprocal of both sides reverses the inequality sign and gives us the desired

$$\frac{n}{\frac{1}{x_1} + \frac{1}{x_2} + \cdots + \frac{1}{x_n}} \leq \sqrt[n]{x_1 x_2 \cdots x_n}.$$

We have equality when equality holds in our application of AM-GM, which is when all the $1/x_i$ terms are equal. Therefore, equality holds if and only if all the x_i are equal. $\square$

Putting this together with the result of the previous problem, we have the **Mean Inequality Chain**, which is also known as the **QM-AM-GM-HM Inequality**:

> **Important:** Let $x_1, x_2, \ldots, x_n > 0$, and let QM, AM, GM, and HM denote the quadratic
> ⚠️ mean, arithmetic mean, geometric mean, and harmonic mean of these n
> numbers, respectively. Then, we have
>
> $$QM \geq AM \geq GM \geq HM.$$
>
> For each inequality sign, equality occurs if and only if all of the x_i are
> equal.

The AM-GM Inequality is the most commonly applied part of this inequality chain, but inequalities involving sums of reciprocals or sums of squares sometimes call for the harmonic mean or quadratic mean, respectively.

> **Problem 18.3:** Let $a_1, a_2, \ldots, a_n$ be positive real numbers, such that $a_1 + a_2 + \cdots + a_n = 1$. Prove that
>
> $$\frac{1}{a_1} + \frac{1}{a_2} + \cdots + \frac{1}{a_n} \geq n^2.$$

Solution for Problem 18.3: The sum of the reciprocals suggests comparing the harmonic mean of the numbers to some other mean. So, we manipulate the given inequality to isolate the HM:

$$\frac{1}{n} \geq \frac{n}{\frac{1}{a_1} + \frac{1}{a_2} + \cdots + \frac{1}{a_n}}.$$

> **Concept:** If you suspect a possible application of the QM-AM-GM-HM inequality
> 🔑 chain, try isolating one of those means in your desired inequality.

The right side is the harmonic mean of the a_i. But what about the left side? We know that the HM is no greater than any of the other means. Which other mean should we compare it to?

Both the $\frac{1}{n}$ on the greater side and the fact that $a_1 + a_2 + \cdots + a_n = 1$ suggest using the AM-HM Inequality, which gives us

$$\frac{a_1 + a_2 + \cdots + a_n}{n} \geq \frac{n}{\frac{1}{a_1} + \frac{1}{a_2} + \cdots + \frac{1}{a_n}}.$$

Because $a_1 + a_2 + \cdots + a_n = 1$, we have

$$\frac{1}{n} \geq \frac{n}{\frac{1}{a_1} + \frac{1}{a_2} + \cdots + \frac{1}{a_n}},$$

which we can rearrange to the desired inequality. $\square$

CHAPTER 18. MORE INEQUALITIES

Problem 18.4: Let $a, b, c,$ and d be positive real numbers. Prove that

$$\frac{a^2 + b^2 + c^2}{a+b+c} + \frac{a^2 + b^2 + d^2}{a+b+d} + \frac{a^2 + c^2 + d^2}{a+c+d} + \frac{b^2 + c^2 + d^2}{b+c+d} \geq a+b+c+d.$$

Solution for Problem 18.4: The sums of squares in the numerators on the left side of the inequality suggest comparing quadratic means to some other mean(s). In particular, the sums of variables on the right side and in the denominators of the left side make the QM-AM Inequality a prime candidate. So, we take a look at the QM-AM Inequality for three variables:

$$\sqrt{\frac{a^2 + b^2 + c^2}{3}} \geq \frac{a+b+c}{3}.$$

Both sides are positive, so we can square both sides to get an equivalent inequality without radicals:

$$\frac{a^2 + b^2 + c^2}{3} \geq \frac{(a+b+c)^2}{9}.$$

Rearranging this inequality to fit the fractions on the left side of the inequality we're trying to prove gives us

$$\frac{a^2 + b^2 + c^2}{a+b+c} \geq \frac{a+b+c}{3}.$$

Likewise, we have similar inequalities for all the terms on the left side of the desired inequality:

$$\frac{a^2 + b^2 + d^2}{a+b+d} \geq \frac{a+b+d}{3},$$
$$\frac{a^2 + c^2 + d^2}{a+c+d} \geq \frac{a+c+d}{3},$$
$$\frac{b^2 + c^2 + d^2}{b+c+d} \geq \frac{b+c+d}{3}.$$

Summing these four inequalities, we have the desired result. □

Notice that in our solutions to the last two problems, we aren't just blindly trying to apply different inequalities in a search for something that will work. We use the forms of the expressions in the inequalities we wish to prove as a guide. When we see a sum of reciprocals, we focus on inequalities involving the harmonic mean. When we see a sum of variables, we focus on the arithmetic mean. Products suggest the geometric mean, and sums of squares suggest the quadratic mean or Cauchy-Schwarz.

And if we're at a loss for anything to try, we can rearrange the desired inequality, hoping to find a familiar form:

Problem 18.5: Prove that $\sqrt{\frac{a^2 + b^2}{2}} - \frac{a+b}{2} \geq \sqrt{ab} - \frac{2}{\frac{1}{a} + \frac{1}{b}}$ for any positive numbers a and b.

Solution for Problem 18.5: Although all four of the means in the Mean Inequality Chain are present, it's not at all clear how we can use the Mean Inequality Chain to prove this inequality. So, we start by

586

rearranging the inequality. We put the radicals on one side and the other means on the other side. This gives us

$$\sqrt{\frac{a^2 + b^2}{2}} - \sqrt{ab} \geq \frac{a + b}{2} - \frac{2}{\frac{1}{a} + \frac{1}{b}}.$$

Aha! We see the QM on the greater side and the AM on the lesser side. Maybe if we start with QM ≥ AM, we can get to this desired inequality. Let's see what we need to show to accomplish this. The QM-AM Inequality gives us

$$\sqrt{\frac{a^2 + b^2}{2}} \geq \frac{a + b}{2}.$$

To get to the desired inequality, we must show that

$$-\sqrt{ab} \geq -\frac{2}{\frac{1}{a} + \frac{1}{b}},$$

since adding this to the QM-AM Inequality above gives us the desired inequality. But this inequality that we want to prove says that $-(\text{GM}) \geq -(\text{HM})$, and multiplying this by -1 gives $\text{GM} \leq \text{HM}$. Uh-oh. This last inequality isn't even true! So, if we start with QM ≥ AM, we'll be stuck, since it's impossible to show that $-(\text{GM}) \geq -(\text{HM})$, which is what we need to combine with QM-AM to get the desired inequality.

Back to the drawing board. We wish to show that

$$\sqrt{\frac{a^2 + b^2}{2}} - \sqrt{ab} \geq \frac{a + b}{2} - \frac{2}{\frac{1}{a} + \frac{1}{b}}.$$

Because QM ≥ GM and AM ≥ HM, both sides are nonnegative. So, we can square both sides of the inequality. Before doing so, we simplify the lesser side:

$$\frac{a + b}{2} - \frac{2}{\frac{1}{a} + \frac{1}{b}} = \frac{a + b}{2} - \frac{2ab}{a + b} = \frac{(a + b)^2 - 4ab}{2(a + b)} = \frac{a^2 - 2ab + b^2}{2(a + b)} = \frac{(a - b)^2}{2(a + b)}.$$

Now, our desired inequality is $\sqrt{\frac{a^2 + b^2}{2}} - \sqrt{ab} \geq \frac{(a - b)^2}{2(a + b)}$. Squaring both sides gives us

$$\frac{a^2 + b^2}{2} - 2\sqrt{ab}\sqrt{\frac{a^2 + b^2}{2}} + ab \geq \frac{(a - b)^4}{4(a + b)^2}.$$

This is a little scary at first, but when we combine the terms without radicals on the left, we have

$$\frac{a^2 + b^2}{2} + ab = \frac{a^2 + 2ab + b^2}{2} = \frac{(a + b)^2}{2}.$$

That's a lot simpler. We can also simplify the radical term and write our desired inequality as

$$\frac{(a + b)^2}{2} - \sqrt{2ab(a^2 + b^2)} \geq \frac{(a - b)^4}{4(a + b)^2}.$$

CHAPTER 18. MORE INEQUALITIES

The radical term is now the geometric mean of $2ab$ and $a^2 + b^2$. That's interesting, so we isolate it, and we have

$$\frac{(a+b)^2}{2} - \frac{(a-b)^4}{4(a+b)^2} \geq \sqrt{2ab(a^2+b^2)}.$$

Now, we combine the terms on the greater side, hoping to be able to simplify it. We find

$$\frac{(a+b)^2}{2} - \frac{(a-b)^4}{4(a+b)^2} = \frac{2(a+b)^4}{4(a+b)^2} - \frac{(a-b)^4}{4(a+b)^2} = \frac{2(a+b)^4 - (a-b)^4}{4(a+b)^2}.$$

Rather than multiplying out the numerator, we note that we can apply the difference of squares factorization to yield

$$\frac{2(a+b)^4 - (a-b)^4}{4(a+b)^2} = \frac{(a+b)^4 + (a+b)^4 - (a-b)^4}{4(a+b)^2}$$

$$= \frac{(a+b)^4 + ((a+b)^2 - (a-b)^2)((a+b)^2 + (a-b)^2)}{4(a+b)^2}$$

$$= \frac{(a+b)^4 + (4ab)(2a^2 + 2b^2)}{4(a+b)^2}.$$

Aha! This makes our desired inequality

$$\frac{(a+b)^4 + 8ab(a^2+b^2)}{4(a+b)^2} \geq \sqrt{2ab(a^2+b^2)}.$$

On the right, we have a geometric mean involving $ab(a^2 + b^2)$. On the left, we have the sum of two terms, one of which involves $ab(a^2 + b^2)$. Maybe we can connect the two sides using AM-GM. Applying AM-GM to $(a+b)^4$ and $8ab(a^2 + b^2)$ gives

$$\frac{(a+b)^4 + 8ab(a^2+b^2)}{2} \geq \sqrt{(a+b)^4(8ab)(a^2+b^2)} = 2(a+b)^2\sqrt{2ab(a^2+b^2)}.$$

Dividing both sides by $2(a+b)^2$ gives us

$$\frac{(a+b)^4 + 8ab(a^2+b^2)}{4(a+b)^2} \geq \sqrt{2ab(a^2+b^2)},$$

which is the last desired inequality from above! All of our steps are reversible, so we can reverse our steps above to complete our proof. $\square$

Problem 18.6: Find the minimum possible value of $\left(1 + \dfrac{x}{2y}\right)\left(1 + \dfrac{y}{2z}\right)\left(1 + \dfrac{z}{2x}\right)$, where x, y, and z are positive.

Solution for Problem 18.6: In order to find the minimum of the expression, we try to prove an inequality of the form

$$\left(1 + \frac{x}{2y}\right)\left(1 + \frac{y}{2z}\right)\left(1 + \frac{z}{2x}\right) \geq \text{some constant,}$$

and then show that the expression can, in fact, equal that "some constant" for some values of x, y, and z. A natural tool to try is AM-GM, since we have three sums on the larger side of the inequality. Applying AM-GM to each gives

$$\frac{1 + \frac{x}{2y}}{2} \geq \sqrt{\frac{x}{2y}},$$

$$\frac{1 + \frac{y}{2z}}{2} \geq \sqrt{\frac{y}{2z}},$$

$$\frac{1 + \frac{z}{2x}}{2} \geq \sqrt{\frac{z}{2x}}.$$

Taking the product of these three gives

$$\frac{\left(1 + \frac{x}{2y}\right)\left(1 + \frac{y}{2z}\right)\left(1 + \frac{z}{2x}\right)}{8} \geq \sqrt{\frac{x}{2y} \cdot \frac{y}{2z} \cdot \frac{z}{2x}} = \sqrt{\frac{1}{8}}.$$

Therefore, we have

$$\left(1 + \frac{x}{2y}\right)\left(1 + \frac{y}{2z}\right)\left(1 + \frac{z}{2x}\right) \geq \sqrt{8}.$$

> **WARNING!!** We're not finished. In a minimization or maximization problem, we must show that the claimed minimum or maximum value can be achieved.

It's not immediately obvious what choices of x, y, and z give us this minimal value. (This alone might make us suspicious about our answer.) In order to have equality in our final nonstrict inequality, we must have equality in each of the three applications of AM-GM. From the equality condition of AM-GM, we must therefore have $1 = \frac{x}{2y}$, $1 = \frac{y}{2z}$, and $1 = \frac{z}{2x}$. Rearranging these gives $x = 2y$, $y = 2z$, and $z = 2x$. Substituting for z in the second equation gives $y = 4x$, and substituting this into $x = 2y$ gives $x = 8x$. Uh-oh. That gives us $x = 0$, which is a big problem, since we can't divide by 0.

Back to the drawing board.

Solution 1: Focus on the equality condition. With a little experimentation, we might guess that the expression is minimized when $x = y = z$. Of course, a guess isn't good enough, but maybe we can find a solution starting from this guess.

If the expression is minimized when $x = y = z$, then we will have to build our inequality

$$\left(1 + \frac{x}{2y}\right)\left(1 + \frac{y}{2z}\right)\left(1 + \frac{z}{2x}\right) \geq \text{some constant}$$

from inequalities that give us an equality condition of $x = y = z$.

> **Concept:** We can sometimes use the equality conditions of inequalities as a guide to solving inequality problems.

We start with the first factor, $1 + \frac{x}{2y}$. We already know that directly applying AM-GM to 1 and $\frac{x}{2y}$ is a loser, so we look for ways to apply inequalities such that $x = y$ is the equality condition. Writing the sum with a common denominator gives us $\frac{2y+x}{2y}$. We don't want to apply AM-GM to $2y$ and x, since that will still give us an equality condition of $x = 2y$. But if we write $2y$ as $y + y$, our numerator then is a sum of 3 terms that we are happy to have equal: $\frac{y+y+x}{2y}$.

Applying AM-GM to y, y, and x gives $\frac{y+y+x}{3} \geq \sqrt[3]{y^2x}$, so $y + y + x \geq 3\sqrt[3]{y^2x}$. Therefore, we have $1 + \frac{x}{2y} = \frac{y+y+x}{2y} \geq \frac{3\sqrt[3]{y^2x}}{2y}$, where equality holds if $y = y = x$. Aha! We do the same to each of the other two factors in our original expression, and we have

$$1 + \frac{x}{2y} = \frac{y+y+x}{2y} \geq \frac{3\sqrt[3]{y^2x}}{2y},$$

$$1 + \frac{y}{2z} = \frac{z+z+y}{2z} \geq \frac{3\sqrt[3]{z^2y}}{2z},$$

$$1 + \frac{z}{2x} = \frac{x+x+z}{2x} \geq \frac{3\sqrt[3]{x^2z}}{2x}.$$

Multiplying these three, we have

$$\left(1 + \frac{x}{2y}\right)\left(1 + \frac{y}{2z}\right)\left(1 + \frac{z}{2x}\right) \geq \frac{27\sqrt[3]{y^2x \cdot z^2y \cdot x^2z}}{8xyz} = \frac{27}{8}.$$

Success! Equality is achieved in all three applications of AM-GM when $x = y = z$. A quick check reveals that the expression $\left(1 + \frac{x}{2y}\right)\left(1 + \frac{y}{2z}\right)\left(1 + \frac{z}{2x}\right)$ equals $\frac{27}{8}$ whenever $x = y = z$ (and all three are nonzero).

Solution 2: Substitution. The expression we must minimize is complicated. Worse yet, it's not even symmetric. If we exchange x and y, for example, the resulting expression is not the same as the original expression. Simple, symmetric expressions are a lot easier to work with than complicated, non-symmetric expressions. So, we make a substitution to simplify the expression we must minimize. We let $a = \frac{x}{2y}$, $b = \frac{y}{2z}$, and $c = \frac{z}{2x}$. This makes the expression we must minimize become

$$(1 + a)(1 + b)(1 + c).$$

That's much easier to deal with. We also have $abc = \left(\frac{x}{2y}\right)\left(\frac{y}{2z}\right)\left(\frac{z}{2x}\right) = \frac{1}{8}$, so when we expand the product $(1 + a)(1 + b)(1 + c)$, we have

$$(1 + a)(1 + b)(1 + c) = 1 + a + b + c + ab + bc + ca + abc = 1 + (a + b + c) + ab + bc + ca + \frac{1}{8}.$$

So, to minimize $(1 + a)(1 + b)(1 + c)$, we must minimize $a + b + c + ab + bc + ca$. Minimizing sums—that's a job for AM-GM. What's wrong with this approach:

> **Bogus Solution:** AM-GM gives us
>
> $$\frac{a + b + c + ab + bc + ca}{6} \geq \sqrt[6]{abc(ab)(bc)(ca)} = \sqrt[6]{(abc)^3} = \sqrt{abc} = \frac{1}{\sqrt{8}}.$$

The trouble here becomes apparent when we investigate the equality condition. For equality to hold, we must have $a = b = c = ab = bc = ca$. But this gives us $a = b = c = 1$, so $x = 2y$, $y = 2z$, and $z = 2x$, which we've already seen gives us the impossible $x = y = z = 0$.

Instead, we address $a + b + c$ and $ab + bc + ca$ separately. AM-GM gives us

$$\frac{a + b + c}{3} \geq \sqrt[3]{abc} = \frac{1}{2}, \text{ and}$$
$$\frac{ab + bc + ca}{3} \geq \sqrt[3]{(ab)(bc)(ca)} = \sqrt[3]{(abc)^2} = \frac{1}{4}.$$

Multiplying these inequalities by 3 and adding them gives us

$$a + b + c + ab + bc + ca \geq \frac{3}{2} + \frac{3}{4} = \frac{9}{4}.$$

Adding $1 + abc = \frac{9}{8}$ to this inequality then gives us

$$1 + a + b + c + ab + bc + ca + abc \geq \frac{27}{8}.$$

We must still show that this minimum can be achieved. Both of our applications of AM-GM have the same equality condition, $a = b = c$, which gives us $\frac{x}{2y} = \frac{y}{2z} = \frac{z}{2x}$. Applying one of the ratio manipulations from Section 11.2, this chain of equal ratios tells us that

$$\frac{x}{2y} = \frac{y}{2z} = \frac{z}{2x} = \frac{x + y + z}{2x + 2y + 2z} = \frac{1}{2}.$$

Therefore, we have $x = y = z$ as our equality condition. Indeed, when $x = y = z$, we have $a = b = c = \frac{1}{2}$, and $(1 + a)(1 + b)(1 + c) = \frac{27}{8}$, as expected. $\square$

Our second solution shows us a very powerful approach to difficult inequality problems:

> **Concept:** When faced with a problem involving unwieldy expressions, try making substitutions for those expressions to simplify the problem.

You might be wondering why we end a section about the Mean Inequality Chain with two complicated problems that we solve with plain old AM-GM. We do so because AM-GM is by far the most commonly used of the inequalities regarding means. In fact, a vast majority of the inequalities you will encounter can be proved with some combination of the Trivial Inequality, AM-GM, and the Cauchy-Schwarz Inequality. Other inequalities will have their moments, but if you master these three, you will be able to tackle most inequalities.

> **Extra!** Some of the inequalities in the Mean Inequality Chain are specific cases of the **Power Mean Inequality**, which tells us that if $x_1, x_2, \ldots, x_n$ are positive and r and s are nonzero numbers with $r > s$, then
>
> $$\left(\frac{x_1^r + x_2^r + \cdots + x_n^r}{n} \right)^{\frac{1}{r}} \geq \left(\frac{x_1^s + x_2^s + \cdots + x_n^s}{n} \right)^{\frac{1}{s}}.$$

Sidenote: Below is trapezoid $WXYZ$, with $\overline{WX} \parallel \overline{YZ}$.

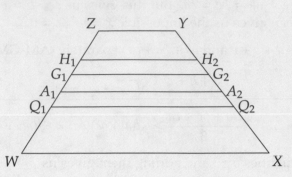

Inside the trapezoid, we have constructed four segments that are parallel to the bases:

- $\overline{Q_1 Q_2}$ is the segment that divides the trapezoid into two regions of equal area.

- $\overline{A_1 A_2}$ is the median of the trapezoid, connecting the midpoints of legs $\overline{XY}$ and $\overline{WZ}$.

- $\overline{G_1 G_2}$ is the segment that divides the trapezoid into two similar trapezoids.

- $\overline{H_1 H_2}$ passes through the intersection of the diagonals of the trapezoid.

Note that $Q_1 Q_2 \geq A_1 A_2 \geq G_1 G_2 \geq H_1 H_2$. Can you find a geometric explanation for the order of these lengths? Furthermore, suppose $WX = a$ and $YZ = b$. Find the four lengths above in terms of a and b. (As an additional challenge, find a geometric construction of each of these segments using a straightedge and compass.)

Exercises

18.1.1 Prove that if a, b, c are positive, then $\dfrac{1}{a+b} + \dfrac{1}{b+c} + \dfrac{1}{c+a} \leq \dfrac{1}{2}\left(\dfrac{1}{a} + \dfrac{1}{b} + \dfrac{1}{c}\right)$.

18.1.2 Prove that if a, b, c are positive, then $\dfrac{9}{a+b+c} \leq 2\left(\dfrac{1}{a+b} + \dfrac{1}{b+c} + \dfrac{1}{c+a}\right)$.

18.1.3 Which parts of the Mean Inequality Chain still hold even if some of the variables take on negative values?

18.1.4 Find all real numbers $0 \leq x_3 \leq x_2 \leq x_1 \leq 1$ such that $(1 - x_1)^2 + (x_1 - x_2)^2 + (x_2 - x_3)^2 + x_3^2 = \dfrac{1}{4}$.

18.1.5 Let a and b be positive. Show that $\dfrac{a+b}{2} - \sqrt{ab} \geq \sqrt{\dfrac{a^2 + b^2}{2}} - \dfrac{a+b}{2}$. *(Source: Mandelbrot)*

18.1.6★ Let m and n be positive integers. Find the minimum value of $x^m + \dfrac{1}{x^n}$ for $x > 0$. **Hints:** 137

18.2 The Rearrangement Inequality

Problems

> **Problem 18.7:** Suppose we have a game called *Who Has Common Sense?* The way the game works is that there are 4 giant bags full of money. The first bag contains \$100 bills. The second bag has \$10 bills. The third has \$5 bills, and the fourth has \$1 bills. You are allowed to take 50 bills from any one bag, and then 8 bills from another bag, and then 2 bills from a third bag, and then 1 bill from the last bag. How can you make the most money?

> **Problem 18.8:** Let $x_1 = -2$, $x_2 = 2$, and $x_3 = 3$. Let $y_1 = -1$, $y_2 = 3$, and $y_3 = 5$. Which of the following expressions has the greatest value? Which has the least value?
>
> $$x_1y_1 + x_2y_2 + x_3y_3 \qquad x_1y_1 + x_2y_3 + x_3y_2 \qquad x_1y_2 + x_2y_1 + x_3y_3$$
> $$x_1y_2 + x_2y_3 + x_3y_1 \qquad x_1y_3 + x_2y_1 + x_3y_2 \qquad x_1y_3 + x_2y_2 + x_3y_1$$
>
> Notice anything interesting? If so, create a few more examples of your own and test your ideas before continuing.

Problem 18.7: Suppose we have a game called *Who Has Common Sense?* The way the game works is that there are 4 giant bags full of money. The first bag contains \$100 bills. The second bag has \$10 bills. The third has \$5 bills, and the fourth has \$1 bills. You are allowed to take 50 bills from any one bag, and then 8 bills from another bag, and then 2 bills from a third bag, and then 1 bill from the last bag. How can you make the most money?

Solution for Problem 18.7: Clearly, you want as many \$100 bills as possible, so you start by taking 50 of them. Next, the best you can do is get as many \$10 bills as possible, so you grab 8 of them. Then, you take 2 of the \$5 bills and 1 of the \$1 bills, giving you a total of \$5091. □

This intuitively obvious way to maximize your winnings is an application of the **Rearrangement Inequality**. Before we get into the details of the Rearrangement Inequality, let's see what happens when we throw in some negative numbers, too.

Problem 18.8: Let $x_1 = -2$, $x_2 = 2$, and $x_3 = 3$. Let $y_1 = -1$, $y_2 = 3$, and $y_3 = 5$. Which of the following expressions has the greatest value? Which has the least value?
$$x_1y_1 + x_2y_2 + x_3y_3 \qquad x_1y_1 + x_2y_3 + x_3y_2 \qquad x_1y_2 + x_2y_1 + x_3y_3$$
$$x_1y_2 + x_2y_3 + x_3y_1 \qquad x_1y_3 + x_2y_1 + x_3y_2 \qquad x_1y_3 + x_2y_2 + x_3y_1$$

Solution for Problem 18.8: We find:

$$
\begin{aligned}
x_1y_1 + x_2y_2 + x_3y_3 &= (-2)\cdot(-1)+2\cdot3+3\cdot5 &= 23,\\
x_1y_1 + x_2y_3 + x_3y_2 &= (-2)\cdot(-1)+2\cdot5+3\cdot3 &= 21,\\
x_1y_2 + x_2y_1 + x_3y_3 &= (-2)\cdot3+2\cdot(-1)+3\cdot5 &= 7,\\
x_1y_2 + x_2y_3 + x_3y_1 &= (-2)\cdot3+2\cdot5+3\cdot(-1) &= 1,\\
x_1y_3 + x_2y_1 + x_3y_2 &= (-2)\cdot5+2\cdot(-1)+3\cdot3 &= -3,\\
x_1y_3 + x_2y_2 + x_3y_1 &= (-2)\cdot5+2\cdot3+3\cdot(-1) &= -7.
\end{aligned}
$$

We have the largest result when we pair the two sequences off in the same order, multiplying the least of the x_i by the least of the y_i, the middle of the x_i by the middle of the y_i, and the greatest of the x_i by the greatest of the y_i. We have the smallest result when we pair the sequences off in opposite orders, multiplying the least of the x_i by the greatest of the y_i, the middle of the x_i by the middle of the y_i, and the greatest of the x_i by the least of the y_i. □

After experimenting with a few more examples, we might think that this pattern holds with any two sequences of numbers. Let's see.

Before diving into the next problem, we'll introduce a little vocabulary to make the problem easier to discuss. We'll say that we take a "sequence-product" of two sequences of length n when we group the terms of the sequences into n pairs, one term from each sequence in each pair, take the product of each pair, and then sum all n products. For example, in Problem 18.8, we found all six sequence-products of the sequences $-2, 2, 3$ and $-1, 3, 5$.

The term "sequence-product" is not standard; we introduce it to make discussing the Rearrangement Inequality easier.

Problems

Problem 18.9: Consider all possible sequence-products of the two sequences $x_1, x_2, x_3, \ldots, x_n$ and $y_1, y_2, y_3, \ldots, y_n$, where $x_1 \le x_2 \le x_3 \le \cdots \le x_n$ and $y_1 \le y_2 \le y_3 \le \cdots \le y_n$. In this problem, we show that the sum $x_1y_1 + x_2y_2 + x_3y_3 + \cdots + x_ny_n$ is the maximum possible sequence-product of the two sequences.

(a) Show that $x_1y_1 + x_2y_2 \ge x_1y_2 + x_2y_1$.

(b) Show that $x_1y_2 + x_2y_1 + x_3y_3 \ge x_1y_2 + x_2y_3 + x_3y_1$.

(c) Show that if $n = 3$, then the maximum possible sequence-product is $x_1y_1 + x_2y_2 + x_3y_3$.

(d) Use the results of parts (a) and (c) to show that if $n = 4$, then the maximum possible sequence-product is $x_1y_1 + x_2y_2 + x_3y_3 + x_4y_4$.

(e) Show that $x_1y_1 + x_2y_2 + x_3y_3 + \cdots + x_ny_n$ is the maximum possible sequence-product of the two sequences for all n.

Problem 18.10: Let $a_1, a_2, a_3, \ldots, a_n$ be n different positive integers. Show that $\displaystyle\sum_{k=1}^{n} \frac{a_k}{k^2} \ge \sum_{k=1}^{n} \frac{1}{k}$.

Problem 18.11: Prove that $\dfrac{a^3}{bc} + \dfrac{b^3}{ac} + \dfrac{c^3}{ab} \ge a + b + c$ for all positive a, b, and c. *(Source: Canada)*

Problem 18.9: Consider all possible sequence-products of the two sequences $x_1, x_2, x_3, \ldots, x_n$ and $y_1, y_2, y_3, \ldots, y_n$, where $x_1 \le x_2 \le x_3 \le \cdots \le x_n$ and $y_1 \le y_2 \le y_3 \le \cdots \le y_n$. Show that the sum

$$x_1y_1 + x_2y_2 + x_3y_3 + \cdots + x_ny_n$$

is the maximum possible sequence-product of the two sequences.

Solution for Problem 18.9: We start by investigating the $n = 2$ case, hoping that this simpler problem will give us some insight into the general problem.

Concept: Tackling simple versions of a problem can often provide a guide to solving the problem.

When $n = 2$, we have x_1, x_2 and y_1, y_2 as our sequences, where $x_1 \le x_2$ and $y_1 \le y_2$. Then, we wish to prove that $x_1y_1 + x_2y_2 \ge x_1y_2 + x_2y_1$. Moving all the terms to the left gives us $x_1y_1 - x_1y_2 + x_2y_2 - x_2y_1 \ge 0$. Factoring the left side gives

$$x_1y_1 - x_1y_2 + x_2y_2 - x_2y_1 = (x_2 - x_1)(y_2 - y_1).$$

Because $x_2 \ge x_1$ and $y_2 \ge y_1$, we have $x_2 - x_1 \ge 0$ and $y_2 - y_1 \ge 0$, which means $(x_2 - x_1)(y_2 - y_1) \ge 0$. We can then reverse our steps above to show that $x_1y_1 + x_2y_2 \ge x_1y_2 + x_2y_1$, as desired.

Therefore, the maximum possible sequence-product of the sequences is $x_1y_1 + x_2y_2$ and the minimum possible sequence-product is $x_1y_2 + x_2y_1$.

Let's take a look at the $n = 3$ case, so our sequences are x_1, x_2, x_3 and y_1, y_2, y_3, where $x_1 \le x_2 \le x_3$ and $y_1 \le y_2 \le y_3$. We wish to show that the maximum possible sequence-product of these two sequences is $x_1y_1 + x_2y_2 + x_3y_3$.

We first consider what happens when we swap two of the y_i by comparing $x_1y_1 + x_2y_2 + x_3y_3$ to $x_1y_2 + x_2y_1 + x_3y_3$. But this is the same as the 2-term sequences case that we already solved! We showed that if $x_1 \le x_2$ and $y_1 \le y_2$, then $x_1y_1 + x_2y_2 \ge x_1y_2 + x_2y_1$. Adding x_3y_3 to both sides gives

$$x_1y_1 + x_2y_2 + x_3y_3 \ge x_1y_2 + x_2y_1 + x_3y_3.$$

Similarly, we know that exchanging any two of the y_i in $x_1y_1 + x_2y_2 + x_3y_3$ will produce a sequence-product that is no greater than $x_1y_1 + x_2y_2 + x_3y_3$.

But what about sequence-products like $x_1y_2 + x_2y_3 + x_3y_1$? We can't just swap a pair of y_i to compare this to $x_1y_1 + x_2y_2 + x_3y_3$. However, swapping y_1 and y_3 in $x_1y_2 + x_2y_3 + x_3y_1$ will give us $x_1y_2 + x_2y_1 + x_3y_3$, and we know that this latter sequence-product is no greater than $x_1y_1 + x_2y_2 + x_3y_3$. But which sequence-product is larger,

$$x_1y_2 + x_2y_3 + x_3y_1 \qquad \text{or} \qquad x_1y_2 + x_2y_1 + x_3y_3?$$

The first term in each is the same, so we must compare $x_2y_3 + x_3y_1$ to $x_2y_1 + x_3y_3$. But this is just our 2-term sequences case again! We have $x_2 \le x_3$ and $y_1 \le y_3$, so

$$x_2y_1 + x_3y_3 \ge x_2y_3 + x_3y_1.$$

Adding $x_1 y_2$ to both sides gives

$$x_1 y_2 + x_2 y_1 + x_3 y_3 \geq x_1 y_2 + x_2 y_3 + x_3 y_1.$$

We already know that $x_1 y_1 + x_2 y_2 + x_3 y_3 \geq x_1 y_2 + x_2 y_1 + x_3 y_3$, so we have $x_1 y_1 + x_2 y_2 + x_3 y_3 \geq x_1 y_2 + x_2 y_3 + x_3 y_1$. Similarly, we can show that $x_1 y_1 + x_2 y_2 + x_3 y_3 \geq x_1 y_3 + x_2 y_1 + x_3 y_2$, so we have shown that $x_1 y_1 + x_2 y_2 + x_3 y_3$ is indeed the largest sequence-product we can form.

We used the 2-term case to tackle the 3-term case. Maybe we can similarly use the 3-term case to prove the 4-term case, and then use the 4-term case to prove the 5-term case, and so on. In other words, maybe we can use induction.

We take the 2-term case as our base case. We then assume that if $x_1 \leq x_2 \leq \cdots \leq x_k$ and $y_1 \leq y_2 \leq \cdots \leq y_k$, then the largest sequence-product we can form of the k-term sequences $\{x_i\}$ and $\{y_i\}$ is $x_1 y_1 + x_2 y_2 + \cdots + x_k y_k$.

We then look at the $(k + 1)$-term case. Suppose $x_1 \leq x_2 \leq \cdots \leq x_{k+1}$ and $y_1 \leq y_2 \leq \cdots \leq y_{k+1}$. We wish to show that the largest sequence-product we can form of the sequences $\{x_i\}$ and $\{y_i\}$ is $x_1 y_1 + x_2 y_2 + \cdots + x_{k+1} y_{k+1}$. We do so by showing that every other possible sequence-product is no greater than this one. In our proof of the 3-term case, we tackled separately the sequence-products with $x_3 y_3$ and those without $x_3 y_3$. We do the same here.

Case 1: Sequence-products that contain $x_{k+1} y_{k+1}$. Each such sequence-product has the form $S + x_{k+1} y_{k+1}$, where S is a k-term sequence-product of the sequences $x_1, x_2, \ldots, x_k$ and $y_1, y_2, \ldots, y_k$. By our inductive assumption, each such sequence-product is no greater than $x_1 y_1 + x_2 y_2 + \cdots + x_k y_k$. Therefore, we have

$$x_1 y_1 + x_2 y_2 + \cdots + x_k y_k + x_{k+1} y_{k+1} \geq S + x_{k+1} y_{k+1}$$

for all possible S.

Case 2: Sequence-products that do not contain $x_{k+1} y_{k+1}$. Each such sequence-product has an $x_{k+1} y_i$ term and an $x_j y_{k+1}$ term, where $x_j \leq x_{k+1}$ and $y_i \leq y_{k+1}$. We can then write the sequence as $T + x_{k+1} y_i + x_j y_{k+1}$, where T consists of the other $k - 1$ terms in the sequence-product. Applying our 2-term case (that is, our base case), we have $x_j y_i + x_{k+1} y_{k+1} \geq x_j y_{k+1} + x_{k+1} y_i$. Adding T to both sides, we have

$$T + x_j y_i + x_{k+1} y_{k+1} \geq T + x_j y_{k+1} + x_{k+1} y_i.$$

The sequence-product on the greater side has an $x_{k+1} y_{k+1}$ term, so Case 1 tells us that it is no greater than $x_1 y_1 + x_2 y_2 + \cdots + x_k y_k + x_{k+1} y_{k+1}$, and we have

$$x_1 y_1 + x_2 y_2 + \cdots + x_k y_k + x_{k+1} y_{k+1} \geq T + x_j y_i + x_{k+1} y_{k+1} \geq T + x_j y_{k+1} + x_{k+1} y_i.$$

Combining our two cases, we have shown that $x_1 y_1 + x_2 y_2 + \cdots + x_{k+1} y_{k+1}$ is the largest possible sequence-product of the $(k + 1)$-term sequences $\{x_i\}$ and $\{y_i\}$, so our proof by induction is complete.

Therefore, the maximum possible sequence-product of the two sequences $x_1, x_2, x_3, \ldots, x_n$ and $y_1, y_2, y_3, \ldots, y_n$ is

$$x_1 y_1 + x_2 y_2 + \cdots + x_n y_n.$$

As an Exercise, you'll have the chance to prove that

$$x_1 y_n + x_2 y_{n-1} + x_3 y_{n-2} + \cdots + x_n y_1$$

is the minimum possible sequence-product of the two sequences. $\square$

> **Important:** Suppose we have sequences $x_1, x_2, \ldots, x_n$ and $y_1, y_2, \ldots, y_n$, where $x_1 \le x_2 \le x_3 \le \cdots \le x_n$ and $y_1 \le y_2 \le y_3 \le \cdots \le y_n$. If we group the terms of the sequences into n pairs with one term from each pair, take the product of each pair, then sum all n products, then the largest possible sum is
>
> $$x_1 y_1 + x_2 y_2 + \cdots + x_n y_n,$$
>
> and the smallest possible sum is
>
> $$x_1 y_n + x_2 y_{n-1} + \cdots + x_n y_1.$$
>
> This is called the **Rearrangement Inequality**.

Another way to look at the Rearrangement Inequality is that the maximum sum is formed when we pair off the terms of the sequences *in order*, and the smallest possible sum is formed when we pair off the terms of the sequences *in opposite order*.

Problem 18.10: Let $a_1, a_2, a_3, \ldots, a_n$ be n different positive integers. Show that $\displaystyle\sum_{k=1}^{n} \frac{a_k}{k^2} \ge \sum_{k=1}^{n} \frac{1}{k}$.

Solution for Problem 18.10: First, we note that we need only consider the sequences in which the a_i consist of the first n integers. If any $a_k = s$, where $s > n$, then there is some integer m with $1 \le m \le n$ such that none of the a_i equals m. We therefore have $s/k^2 > m/k^2$, so the sum on the left side of the desired inequality can be reduced by letting $a_k = m$ rather than $a_k = s$. Therefore, any sequence that contains numbers greater than n can be converted into a sequence with no terms greater than n such that the latter sequence produces a smaller sum $\displaystyle\sum_{k=1}^{n} \frac{a_k}{k^2}$. So, if we show that the inequality holds for all sequences that consist of the numbers 1 through n in some order, then we know it holds for all sequences.

If the a_k terms consist of the first n positive integers, we can view the left side of the inequality as a sequence-product of the sequence $1, 2, 3, \ldots, n$ and the sequence $\frac{1}{1^2}, \frac{1}{2^2}, \ldots, \frac{1}{n^2}$. We might be inspired to do this by noting that the right side of the inequality can be written as $\displaystyle\sum_{k=1}^{n} \frac{k}{k^2}$, which is also a sequence-product of these two sequences.

We're comparing two sequence-products of the same two sequences—this is a job for the Rearrangement Inequality! Because $1 \le 2 \le 3 \le \cdots \le n$ and $\frac{1}{n^2} \le \frac{1}{(n-1)^2} \le \frac{1}{(n-2)^2} \le \cdots \le \frac{1}{1^2}$, the Rearrangement Inequality tells us that the minimum possible sequence-product is $1 \cdot \frac{1}{1^2} + 2 \cdot \frac{1}{2^2} + \cdots + n \cdot \frac{1}{n^2}$. Therefore, we have

$$\sum_{k=1}^{n} \frac{a_k}{k^2} \ge \sum_{k=1}^{n} \frac{k}{k^2} = \sum_{k=1}^{n} \frac{1}{k},$$

as desired. $\square$

Problem 18.11: Prove that $\dfrac{a^3}{bc} + \dfrac{b^3}{ac} + \dfrac{c^3}{ab} \ge a + b + c$ for all positive a, b, and c. *(Source: Canada)*

Solution for Problem 18.11: We might first try the Mean Inequality Chain on this problem, but after trying AM-GM for the sum on the larger side, or QM-AM for the sum on the smaller, we feel a little stuck.

The Rearrangement Inequality seems like a possible candidate because if we could change the denominators on the left to a^2, b^2, c^2 instead of bc, ac, ab, then we would have the right side. In other words, if we could take a^3, b^3, c^3 as one of our sequences, then maybe we can choose another sequence as our second sequence and get somewhere with the Rearrangement Inequality.

However, we have a problem with using the sequence a^3, b^3, c^3. We don't know the order of a^3, b^3, and c^3. We could break the problem into cases. First, we'd assume $a \geq b \geq c$ (so $a^3 \geq b^3 \geq c^3$). Then, we'd assume $b \geq a \geq c$. And so on.

But all these cases are essentially the same! Both sides of the inequality are symmetric, so if we prove the inequality for the case $a \geq b \geq c$, then we know it's also true for $b \geq a \geq c$, because we know we can just exchange a and b.

We have a special term for when we work with one case that is equivalent to all possible cases. We say, "**Without loss of generality**, we let $a \geq b \geq c$." This means that we will consider the case $a \geq b \geq c$, and that this is the only case we must consider because the other cases are equivalent. The phrase "without loss of generality" is sometimes shortened to **WLOG**.

However, while "without loss of generality" can make writing proofs much easier, we do have to be careful that we don't go astray, as we do below:

> **Bogus Solution:** Without loss of generality, let $a \geq b \geq c$. Because $a \geq b \geq c$, we have $a^2 \geq bc$, so $\frac{1}{bc} \geq \frac{1}{a^2}$. Multiplying both sides of this inequality by a^3 gives us $\frac{a^3}{bc} \geq \frac{a^3}{a^2}$, so $\frac{a^3}{bc} \geq a$. Similarly, we have $\frac{b^3}{ac} \geq b$ and $\frac{c^3}{ab} \geq c$, and adding these three inequalities gives the desired result.

Our "Without loss of generality" start is OK. Exchanging any two of a, b, and c doesn't make the expression on either side of the desired inequality change, so if we prove the desired inequality for one order of the variables, the same proof will hold for any order of the variables.

The problem comes with the "Similarly" step.

> **WARNING!!** Be careful when you use "similarly" in a proof. You must be sure that the conclusions that you state are "similarly" true can, in fact, be proved by essentially the same steps. We particularly have to be careful about this when starting with "without loss of generality."

If we go through the same steps with $\frac{b^3}{ac}$ as we did for $\frac{a^3}{bc}$ in the Bogus Solution, we have a problem right away. Starting from $a \geq b \geq c$ (we can't change the order of the variables for each part of the proof, of course), we can't assume that $b^2 \geq ac$. So, we can't show that $\frac{1}{ac} \geq \frac{1}{b^2}$, or that $\frac{b^3}{ac} \geq b$.

However, the ordering of a, b, and c does allow us to break out the Rearrangement Inequality!

Solution 1: Use the Rearrangement Inequality. Without loss of generality, we let $a \geq b \geq c$. So, we have $a^3 \geq b^3 \geq c^3$. We need to multiply a^3, b^3, and c^3 by $\frac{1}{bc}$, $\frac{1}{ac}$, and $\frac{1}{ab}$, respectively. In other words, we need to form a sequence-product, which points us to the Rearrangement Inequality. But first, we have to order the terms $\frac{1}{bc}$, $\frac{1}{ac}$, and $\frac{1}{ab}$. Since $\frac{1}{bc} = a \cdot \frac{1}{abc}$, $\frac{1}{ac} = b \cdot \frac{1}{abc}$, and $\frac{1}{ab} = c \cdot \frac{1}{abc}$, and we have $a \geq b \geq c > 0$, we conclude that $\frac{1}{bc} \geq \frac{1}{ac} \geq \frac{1}{ab}$.

We can now apply the Rearrangement Inequality to $a^3 \geq b^3 \geq c^3$ and $\frac{1}{bc} \geq \frac{1}{ac} \geq \frac{1}{ab}$. The greatest possible sequence-product is the greater side of our desired inequality. We compare this sequence-product to one in which we pair each of the cubes with a denominator that allows us to do some cancellation:

$$\frac{a^3}{bc} + \frac{b^3}{ac} + \frac{c^3}{ab} \geq \frac{a^3}{ac} + \frac{b^3}{ab} + \frac{c^3}{bc} = \frac{a^2}{c} + \frac{b^2}{a} + \frac{c^2}{b}.$$

Well, that didn't finish the problem off, but it left us with something much simpler to deal with. We now must show that

$$\frac{a^2}{c} + \frac{b^2}{a} + \frac{c^2}{b} \geq a + b + c.$$

Concept: If a tactic simplifies a problem but doesn't solve it, look for ways to apply the tactic again.

This inequality looks a lot like the original one, so we try the Rearrangement Inequality again. Since $a \geq b \geq c > 0$, we have both $a^2 \geq b^2 \geq c^2 > 0$ and $\frac{1}{c} \geq \frac{1}{b} \geq \frac{1}{a} > 0$. So, the Rearrangement Inequality gives us

$$\frac{a^2}{c} + \frac{b^2}{a} + \frac{c^2}{b} \geq \frac{a^2}{a} + \frac{b^2}{b} + \frac{c^2}{c} = a + b + c,$$

as desired.

If we don't find the Rearrangement Inequality solution to this problem, we can still fall back on AM-GM.

Solution 2: Use AM-GM. In order to produce a from $\frac{a^3}{bc}$, we need to multiply by bc. AM-GM gives us a way to build an inequality that will leave $\frac{a^3}{bc}$ on the greater side, and to multiply $\frac{a^3}{bc}$ by b and c to leave a alone on the lesser side. Let's give it a try:

$$\frac{\frac{a^3}{bc} + b + c}{3} \geq \sqrt[3]{\frac{a^3}{bc} \cdot b \cdot c},$$

so we have

$$\frac{a^3}{bc} + b + c \geq 3a.$$

Well, we have some of what we wanted. We have a^3/bc on the greater side, and we have a on the lesser side. But we have b and c on the greater side, which we don't want, and we have $3a$, not just a, on the lesser side. Let's see what we happens if we repeat the process for $\frac{b^3}{ac}$ and $\frac{c^3}{ab}$. Maybe we'll get lucky and something good will happen. Applying AM-GM as above gives

$$\frac{b^3}{ac} + a + c \geq 3b,$$

$$\frac{c^3}{ab} + a + b \geq 3c.$$

Indeed, something good happened. Adding all three of our inequalities gives

$$\frac{a^3}{bc} + \frac{b^3}{ac} + \frac{c^3}{ab} + 2(a + b + c) \geq 3(a + b + c),$$

which then gives us the desired inequality. □

Not only does our second solution show yet again how versatile AM-GM is, but it also shows a general problem-solving approach we use with many inequalities. We've already seen how important it is to look for familiar forms when proving an inequality, but our second solution takes a slightly different approach:

> **Concept:** Don't expect to be able to prove an inequality in one step. Use your inequality tools to produce pieces of the inequality you want to prove. Look for ways to produce inequalities that have a term or two on the same side of the inequality sign as in the inequality you must prove. As you find these inequalities, you may then see ways to combine them to finish the problem.

Exercises

18.2.1   Let $a, b, c > 0$. Show that $\dfrac{a + b + c}{abc} \leq \dfrac{1}{a^2} + \dfrac{1}{b^2} + \dfrac{1}{c^2}$.

18.2.2   Suppose that $x_1 \leq x_2 \leq x_3 \leq \cdots \leq x_n$ and $y_1 \leq y_2 \leq y_3 \leq \cdots \leq y_n$. Show that the sum $x_1 y_n + x_2 y_{n-1} + x_3 y_{n-2} + \cdots + x_n y_1$ is the minimum possible sequence-product of the two sequences.

18.2.3   Let x and y be nonnegative real numbers. Show that $x^5 + y^5 \geq x^4 y + xy^4 \geq x^3 y^2 + x^2 y^3$.

18.2.4★   Prove that $\dfrac{a^3}{b} + \dfrac{b^3}{c} + \dfrac{c^3}{a} \geq ab + bc + ca$ for all positive a, b, c.

18.3   When Formulas Fail

After learning the Trivial Inequality, Cauchy-Schwarz, the Mean Inequality Chain, and the Rearrangement Inequality, it can become tempting to try to fit every inequality problem into one of these molds. This temptation becomes even stronger as you learn more and more of these "classical" inequalities—you attack every inequality by just trying to find which one of these tools solves the problem immediately. If you take this approach, you'll find that there are a great many inequality problems you simply can't solve.

In this section, we look at a variety of techniques we use to solve inequality problems when our formulas fail us.

> **WARNING!!** Don't rely too heavily on formulas and famous theorems. Many of the other problem solving strategies we've used to solve equations in this text can also be used to tackle inequality problems.

Problems

Problem 18.12: The mathematical constant e can be defined as $e = \dfrac{1}{0!} + \dfrac{1}{1!} + \dfrac{1}{2!} + \dfrac{1}{3!} + \cdots = \sum\limits_{n=0}^{\infty} \dfrac{1}{n!}$.

(a) Show that $\dfrac{1}{n!} \le \dfrac{1}{2^{n-1}}$ for all positive integers n.

(b) Show that $2 < e < 3$.

Problem 18.13: Prove that $2^n \ge n^2$ for all positive integers $n \ge 4$.

Problem 18.14: Show that if p and q are positive numbers less than 1, then $\dfrac{p}{q+1} + \dfrac{q}{p+1} < 1$.

Problem 18.15: Consider a square array of numbers consisting of m rows and m columns. For each i, let r_i denote the sum of the numbers in the i-th row, and c_j denote the sum of the numbers in the j-th column. Show that there are distinct numbers i and j for which $(r_i - c_i)(r_j - c_j) \le 0$.

Problem 18.12: The mathematical constant e can be defined as $e = \dfrac{1}{0!} + \dfrac{1}{1!} + \dfrac{1}{2!} + \dfrac{1}{3!} + \cdots = \sum\limits_{n=0}^{\infty} \dfrac{1}{n!}$.

Show that $2 < e < 3$.

Solution for Problem 18.12: We begin by investigating the series for e:

$$e = \frac{1}{0!} + \frac{1}{1!} + \frac{1}{2!} + \frac{1}{3!} + \cdots = 1 + 1 + \frac{1}{2} + \frac{1}{6} + \cdots .$$

The sum of the first two terms in the series is $1 + 1 = 2$. Since the other terms are all positive, we see that $e > 1 + 1 = 2$.

Proving that $e < 3$ is trickier. The series isn't arithmetic or geometric. It doesn't telescope. It seems that we are stuck—we can't evaluate the series at all.

But we don't have to evaluate it! All we have to do is show that it is less than 3. So, if we can show that the series is less than a series we know how to handle, and that this latter series has a sum less than or equal to 3, then we have a solution!

A natural candidate to consider is a geometric series, since we get each term of our series for e by multiplying the previous term by some factor. Suppose, instead of making that factor smaller at each step as we do for e, we keep that factor the same. Then, we get a geometric series, which we know how to handle. Let's try it.

The series for e starts $1 + 1 + \frac{1}{2} + \frac{1}{6}$. Clearly, we don't want to use 1 as our constant factor, since this will give us $1 + 1 + 1 + 1 + \cdots$. We might instead use $\frac{1}{2}$, which gives us $S = 1 + \frac{1}{2} + \frac{1}{4} + \cdots$. This doesn't quite work, because some of the terms of S are less than the corresponding terms of our series for e, so we cannot conclude that $e < S$. However, if we add 1 to S and call the result T, we have

$$T = 1 + 1 + \frac{1}{2} + \frac{1}{4} + \frac{1}{8} + \cdots .$$

Let's compare this to e:

$$e = 1 + 1 + \frac{1}{2} + \frac{1}{6} + \frac{1}{24} + \cdots .$$

Aha! It looks like each term of T is greater than or equal to the corresponding term of e. This is clearly true for the first 5 terms shown. Each subsequent term of T is half the preceding term, while each term of e after the fifth is found by multiplying the preceding term by a number that is smaller than $1/2$. So, each subsequent term of e is less than the corresponding term of T. We therefore have

$$e = 1 + 1 + \frac{1}{2} + \frac{1}{6} + \frac{1}{24} + \cdots < 1 + 1 + \frac{1}{2} + \frac{1}{4} + \frac{1}{8} + \cdots = 1 + \frac{1}{1 - \frac{1}{2}} = 3.$$

$\square$

The key step in our solution to Problem 18.12 was comparing the unwieldy series for e to a simple geometric series we know how to handle.

> **Concept:** To find an upper or lower bound of a complicated expression that we cannot evaluate directly, we can sometimes compare the complicated expression to an expression we do know how to evaluate.

Problem 18.13: Prove that $2^n \geq n^2$ for all positive integers $n \geq 4$.

Solution for Problem 18.13: It seems that the function 2^n "grows faster" than the function n^2 for $n \geq 4$, but this observation is not a proof.

Since we're proving a statement for all integers starting at a certain point, induction seems like a promising tool. For the base case, we note that when $n = 4$, we have $2^4 = 16 = 4^2$. For the inductive step, we assume that $2^k \geq k^2$ for some $k \geq 4$. We must use this to prove our inequality holds for $n = k + 1$; that is, we must show that $2^{k+1} \geq (k + 1)^2$.

Multiplying $2^k \geq k^2$ by 2 gives $2^{k+1} \geq 2k^2$, so we must now show that $2k^2 \geq (k + 1)^2$. Expanding the right side, we must show that $2k^2 \geq k^2 + 2k + 1$. Rearranging this leaves us $k^2 \geq 2k + 1$ to prove. We know that $k \geq 4$, so $k^2 \geq 4k = 2k + 2k \geq 2k + 1$.

We can now reverse our steps to complete our induction. Adding k^2 to both sides of $k^2 \geq 2k + 1$ gives $2k^2 \geq k^2 + 2k + 1$, so $2k^2 \geq (k + 1)^2$. From our inductive assumption, we have $2^k \geq k^2$. Multiplying this by 2 gives $2^{k+1} \geq 2k^2$, so $2^{k+1} \geq (k + 1)^2$, and our induction is complete. $\square$

> **Concept:** If you must prove an inequality for all integers greater than some value, then consider induction.

Problem 18.14: Show that if p and q are positive numbers less than 1, then $\dfrac{p}{q + 1} + \dfrac{q}{p + 1} < 1$.

Solution for Problem 18.14: We could try various pieces of the Mean Inequality Chain, but none seems to deal well with those fractions. We might also try our Cauchy-Schwarz trick for getting rid of denominators. Cauchy-Schwarz gives us

$$[(q+1)+(p+1)]\left(\frac{p}{q+1}+\frac{q}{p+1}\right) \geq \left(\sqrt{q+1}\sqrt{\frac{p}{q+1}}+\sqrt{p+1}\sqrt{\frac{q}{p+1}}\right)^2 = \left(\sqrt{p}+\sqrt{q}\right)^2.$$

We got rid of the fractions on the right side, but we still have square roots. Worse yet, the expression $\frac{p}{q+1}+\frac{q}{p+1}$ is on the lesser side in the desired inequality, but it is on the greater side in the inequality above.

> **WARNING!!** Pay attention to which side of the inequality your target expressions
> ☢ are on. If an expression is on the lesser side in your desired inequality,
> an inequality that has that expression on the greater side probably
> won't be helpful (and vice versa).

Another clue that Cauchy and the Mean Inequality Chain won't work is that our desired inequality is strict, whereas Cauchy and the Mean Inequality Chain are not. This doesn't completely eliminate the possibility that these inequalities can help, but it does suggest we'll need something else.

Back to the drawing board. We can get rid of the fractions by multiplying by $(q+1)(p+1)$. We don't have to worry about changing the direction of the inequality because p and q are positive. This gives us $p(p+1)+q(q+1) < (p+1)(q+1)$ to prove. Rearranging this gives us $p^2+q^2 < pq+1$ to prove. Now, we appear stuck again. The squares might make us think of the quadratic mean, but they are on the lesser side, so the QM portion of the Mean Inequality Chain probably won't help.

We appear stuck working backwards, so let's try working forwards. We only know that $0 < p < 1$ and $0 < q < 1$. We wish to build $p^2+q^2 < pq+1$ from these inequalities. We might try producing a pq term by multiplying the given inequalities, which gives $0 < pq < 1$. Unfortunately, $pq > 0$ doesn't appear helpful, and $pq < 1$ has pq on the lesser side, not the greater side. So, we try manipulating the given information in other ways to produce pq on the greater side.

We write $p < 1$ and $q < 1$ as $1-p > 0$ and $1-q > 0$. Aha! Multiplying these gives us $(1-p)(1-q) > 0$, which rearranges to $1 + pq > p + q$. This gives us both 1 and pq on the greater side. We're really close now! If only we could turn p and q on the lesser side into p^2 and q^2. Fortunately, we can do so easily. Because p is positive, we can multiply $1 > p$ by p to produce $p > p^2$. Similarly, we have $q > q^2$, so $1 + pq > p + q > p^2 + q^2$, as desired. Now, we're ready to write our proof:

> **Proof:** Because $1 > p$ and $1 > q$, we have $1-p > 0$ and $1-q > 0$. Multiplying these gives us
> ☑ $(1-p)(1-q) > 0$, and rearranging gives $1 + pq > p + q$. Because p and q are positive, we can
> multiply $1 > p$ and $1 > q$ by p and q, respectively, to produce $p > p^2$ and $q > q^2$. Therefore,
> we have $1 + pq > p + q > p^2 + q^2$. Adding $p + q$ to both sides of $1 + pq > p^2 + q^2$ gives
> $1 + p + q + pq > p^2 + p + q^2 + q$. Factoring then gives us $(1+p)(1+q) > p(p+1) + q(q+1)$. Since
> $p+1$ and $q+1$ are both positive, we can divide both sides of the inequality by $(p+1)(q+1)$
> without reversing the inequality to produce the desired
>
> $$1 > \frac{p}{q+1} + \frac{q}{p+1}.$$

□

Our solution to Problem 18.14 involves a lot of experimentation and reasoning both forwards and backwards. This is one key to not getting stuck on problems: don't get locked into one way of looking at a problem. When you feel stuck on an algebra problem, manipulate the expressions in the problem in a variety of ways, and look for familiar forms. If you've been working backwards for a while, stop and try going forwards, working from the given information to produce expressions like those you need to finish the problem. Conversely, if you're stuck working forwards, turn around and try working backwards.

Our previous two problems also exhibited a very common technique for tackling inequalities. In Problem 18.13, we showed that $2^{k+1} \geq (k+1)^2$ by proving $2^{k+1} \geq 2k^2$ and $2k^2 \geq (k+1)^2$, and then putting these together to get $2^{k+1} \geq 2k^2 \geq (k+1)^2$. In Problem 18.14, we combined $1+pq > p+q$ with $p+q > p^2+q^2$ to get $1 + pq > p^2 + q^2$. In both cases, we found a relationship between two expressions by finding a third expression that we could "sandwich" between them.

> **Concept:** To prove $A \geq B$ for some expressions A and B, we can sometimes find a third expression, C, such that $A \geq C$ and $C \geq B$.

We typically find expression C by focusing on either A or B alone, not both. For example, in Problem 18.14, we found $1 + pq > p + q$ by focusing on $1 + pq$ and ignoring $p^2 + q^2$. We then look for a way to relate this intermediary expression to the other expression in our desired inequality, as we did when we hunted for a way to show $p + q > p^2 + q^2$.

> **Problem 18.15:** Consider a square array of numbers consisting of m rows and m columns. For each i, let r_i denote the sum of the numbers in the i-th row, and c_j denote the sum of the numbers in the j-th column. Show that there are distinct numbers i and j for which $(r_i - c_i)(r_j - c_j) \leq 0$.

Solution for Problem 18.15: We must show that $(r_i - c_i)(r_j - c_j) \leq 0$ for some distinct i and j. This is a different problem from showing that we always have $(r_i - c_i)(r_j - c_j) \leq 0$. Indeed, it's not true that we must always have $(r_i - c_i)(r_j - c_j) \leq 0$. For example, in the array shown at right, we have $(r_1 - c_1)(r_2 - c_2) = (1)(1) = 1$, which is greater than 0.

$$\begin{array}{ccc} 1 & 3 & 2 \\ 1 & 2 & 4 \\ 3 & 1 & 2 \end{array}$$

So, this is a fundamentally different problem from most of the inequalities we have looked at so far. We must prove the existence of a pair of distinct numbers i and j such that $(r_i - c_i)(r_j - c_j) \leq 0$, but this statement doesn't need to be true for any specific i and j. Therefore, trying to prove the statement for a specific i and j will probably be fruitless. Proving the statement directly seems impossible, so what do we do?

We try a proof by contradiction. We think about what happens if there are no distinct i and j such that $(r_i - c_i)(r_j - c_j) \leq 0$. Then, we must have $(r_i - c_i)(r_j - c_j) > 0$ for all i and j. This has the advantage of giving us inequalities we can work with. If $(r_i - c_i)(r_j - c_j) > 0$ for all i and j, then the differences $r_k - c_k$ have the same sign for all k. Is this possible? Below, we compute all $r_k - c_k$ values for three example arrays:

$$\begin{array}{|ccc|l} 1 & 3 & 2 & r_1 - c_1 = 4 \\ 1 & 2 & 4 & r_2 - c_2 = 1 \\ 0 & 1 & 4 & r_3 - c_3 = -5 \end{array} \qquad \begin{array}{|ccc|l} -1 & -3 & 0 & r_1 - c_1 = -4 \\ 1 & 2 & 3 & r_2 - c_2 = 6 \\ 0 & 1 & 4 & r_3 - c_3 = -2 \end{array} \qquad \begin{array}{|ccc|l} 0 & 1 & 3 & r_1 - c_1 = 1 \\ 3 & 2 & -6 & r_2 - c_2 = -7 \\ 0 & 3 & 5 & r_3 - c_3 = 6 \end{array}$$

> **Concept:** Specific examples can be useful guides.

Above, one fact that stands out in all three examples is that, in each case, the sum of the $r_k - c_k$ equals 0. Is this always the case?

Yes! Adding all the $r_k - c_k$ is the same as adding all the r_k, then subtracting the sum of all the c_k. The sum of all the r_k and the sum of all the c_k are the same, because they both equal the sum of all the numbers in the array. So, the sum of all the $r_k - c_k$ must be 0. This gives us our contradiction. If the sum of all the $r_k - c_k$ is 0, then it is impossible for all the $r_k - c_k$ to be positive or for all of them to be negative. Either they are all 0 (in which case we have $(r_i - c_i)(r_j - c_j) = 0$ for all i and j), or there are two numbers i and j such that $r_i - c_i < 0$ and $r_j - c_j > 0$.

Now, we're ready to write our proof:

Proof: Because both $\displaystyle\sum_{k=1}^{m} r_k$ and $\displaystyle\sum_{k=1}^{m} c_k$ give the sum of all the entries in the array, we have $\displaystyle\sum_{k=1}^{m} r_k = \sum_{k=1}^{m} c_k$.

Rearranging this equation gives $\displaystyle\sum_{k=1}^{m}(r_k - c_k) = 0$. If any of the $r_k - c_k$ are zero, then we choose i such that $r_i - c_i = 0$, and let j be any other number from 1 to m, and we have $(r_i - c_i)(r_j - c_j) \le 0$, as desired. Otherwise, we proceed by contradiction.

We assume that the $r_k - c_k$ are either all positive or all negative for all k such that $1 \le k \le m$. Then, we must have

$$\sum_{k=1}^{m}(r_k - c_k) \ne 0,$$

which is a contradiction of the fact proved above. Therefore, the $r_k - c_k$ cannot all have the same sign, so there are two numbers i and j such that $r_i - c_i$ and $r_j - c_j$ have opposite signs. This means we have $(r_i - c_i)(r_j - c_j) \le 0$ for some distinct numbers i and j, as desired.

☐

> **Concept:** Existence problems, in which we must prove that something does or doesn't exist, can often be solved with contradiction.

Problem 18.15 is an example of using contradiction to prove existence. We showed that there exist numbers i and j such that $(r_i - c_i)(r_j - c_j) \le 0$ by assuming that no such i and j exist, and then showing that this leads to an impossible conclusion. Back on page 231, we used contradiction to show that $\sqrt{2}$ is irrational. This is an example of using contradiction to prove that something doesn't exist. Specifically, we assumed that there exists a rational number, p/q, that equals $\sqrt{2}$, and then showed that this leads to an impossible conclusion.

In both cases, the fact that the problem is about existence is a clue to try contradiction. Furthermore, in both cases, using contradiction gave us equations or inequalities to work with. In other words, it gave us a way to turn the words in the problem into equations or inequalities that we could manipulate to reach the desired conclusion.

Exercises

18.3.1 In the n-term series $S_n = 5.5 + 5.55 + 5.555 + \cdots$, the nth term has one five before the decimal point and n fives after it (e.g., the sixth term is 5.555555). Show that $550 < S_{100} < 556$. *(Source: CIMC)*

18.3.2 Show that $2 < 1.01^{100} < e$.

18.3.3 Prove that
$$2(x^3 + y^3 + z^3) \geq x^2y + xy^2 + y^2z + yz^2 + x^2z + xz^2$$
for all nonnegative x, y, and z.

18.3.4 Prove that $n! \geq 3^n$ for all integers $n \geq 7$.

18.3.5 Let $xyz = 1$. Find the minimum and maximum values of $S = \dfrac{1}{1 + x + xy} + \dfrac{1}{1 + y + yz} + \dfrac{1}{1 + z + zx}$.

18.3.6 Let a, b, and c be positive real numbers such that $a + b + c \leq 4$ and $ab + bc + ca \geq 4$. Prove that at least two of the inequalities $|a - b| \leq 2$, $|b - c| \leq 2$, and $|c - a| \leq 2$ are true. **Hints:** 171

18.3.7★ Let x, y, and z be real numbers such that $x + y + z = 2$ and $xy + xz + yz = 1$. Find the minimum and maximum value of x. **Hints:** 173

18.3.8★ Prove that if a, b, and c are positive real numbers, then $(a^a)(b^b)(c^c) \geq (abc)^{(a+b+c)/3}$. *(Source: USAMO)* **Hints:** 247

18.4 Summary

> **Definition:** The **quadratic mean** of a set of real numbers is the square root of the arithmetic mean of the squares of the numbers. In other words, the quadratic mean of the real numbers $a_1, a_2, \ldots, a_n$ is
> $$\sqrt{\frac{a_1^2 + a_2^2 + a_3^2 + \cdots + a_n^2}{n}}.$$
> The quadratic mean is also sometimes referred to as the **root mean square**.

> **Definition:** The **harmonic mean** of a set of nonzero real numbers is the reciprocal of the average of the reciprocals of the numbers. In other words, the harmonic mean of the nonzero numbers $a_1, a_2, a_3, \ldots, a_n$ is
> $$\frac{n}{\frac{1}{a_1} + \frac{1}{a_2} + \frac{1}{a_3} + \cdots + \frac{1}{a_n}}.$$

The quadratic mean, root mean square, and harmonic mean are sometimes abbreviated **QM**, **RMS**, and **HM**, respectively.

Important: Let $x_1, x_2, \ldots, x_n > 0$, and let QM, AM, GM, and HM denote the quadratic mean, arithmetic mean, geometric mean, and harmonic mean of these n numbers, respectively. Then, we have

$$QM \geq AM \geq GM \geq HM.$$

For each inequality sign, equality occurs if and only if all of the x_i are equal.

Important: Suppose we have sequences $x_1, x_2, \ldots, x_n$ and $y_1, y_2, \ldots, y_n$, where $x_1 \leq x_2 \leq x_3 \leq \cdots \leq x_n$ and $y_1 \leq y_2 \leq y_3 \leq \cdots \leq y_n$. If we group the terms of the sequences into n pairs with one term from each pair, take the product of each pair, then sum all n products, then the largest possible sum is

$$x_1y_1 + x_2y_2 + \cdots + x_ny_n,$$

and the smallest possible sum is

$$x_1y_n + x_2y_{n-1} + \cdots + x_ny_1.$$

This is called the **Rearrangement Inequality**.

Problem Solving Strategies

Concepts:

- Much of success in proving inequalities is recognizing expressions that are part of well-known inequalities such as the Cauchy-Schwarz Inequality or the AM-GM Inequality.

- To derive some inequalities involving a single sequence of real numbers, we can set one of the sequences in the Cauchy-Schwarz Inequality equal to constants.

- If you suspect a possible application of the QM-AM-GM-HM inequality chain, try isolating one of those means in your desired inequality.

- We can sometimes use the equality conditions of inequalities as a guide to solving inequality problems.

- When faced with a problem involving unwieldy expressions, try making substitutions for those expressions to simplify the problem.

Continued on the next page. . .

Concepts: . . . *continued from the previous page*

- If a tactic simplifies a problem but doesn't solve it, look for ways to apply the tactic again.

- To find an upper or lower bound of a complicated expression that we cannot evaluate directly, we can sometimes compare the complicated expression to an expression we do know how to evaluate.

- If you must prove an inequality for all integers greater than some value, consider induction.

- Existence problems, in which we must prove that something does or doesn't exist, can often be solved with contradiction.

REVIEW PROBLEMS

18.16 Show that $3a^2 + 3b^2 + 3c^2 \geq (a + b + c)^2$ for all real a, b, and c.

18.17

(a) What is the maximum perimeter of a rectangle inscribed in a circle of radius 1?

(b)★ What is the maximum perimeter of a rectangle inscribed in a semicircle of radius 1?

18.18 Let a, b, and c be real numbers such that $a^2 + b^2 + c^2 = 1$. Prove that $-\dfrac{1}{2} \leq ab + ac + bc \leq 1$.

18.19 Prove that $x^2y^2 + x^2z^2 + y^2z^2 \geq xyz(x + y + z)$ for all $x, y, z \geq 0$.

18.20 Show that if a, b, and c are positive numbers such that $a + b + c = 6$, then

$$\left(a + \frac{1}{b}\right)^2 + \left(b + \frac{1}{c}\right)^2 + \left(c + \frac{1}{a}\right)^2 \geq \frac{75}{4}.$$

(Source: ARML)

18.21 Find the maximum of x^2y if $x + y = 6$ and $x \geq 0$.

18.22 If a and b are positive integers such that $a < b$ and $ab \leq a + 3b$, how many possible values are there for a?

18.23 Let $a_1, a_2, \ldots, a_n, b_1, b_2, \ldots, b_n$ be positive real numbers. Show that either

$$\frac{a_1}{b_1} + \frac{a_2}{b_2} + \cdots + \frac{a_n}{b_n} \geq n \qquad \text{or} \qquad \frac{b_1}{a_1} + \frac{b_2}{a_2} + \cdots + \frac{b_n}{a_n} \geq n.$$

18.24 Let a, b, c, d, and e be positive integers such that

$$abcde = a + b + c + d + e.$$

Find the maximum possible value of e.

18.25 Let a, b, c be positive real numbers. Prove that $(a + b + c)\left(\dfrac{1}{a + b} + \dfrac{1}{a + c} + \dfrac{1}{b + c}\right) \geq \dfrac{9}{2}$.

18.26 Let a and b be positive real numbers such that $a + b = 1$. Prove that $\left(a + \frac{1}{a}\right)^2 + \left(b + \frac{1}{b}\right)^2 \geq \frac{25}{2}$.

18.27 Let $0 \leq x \leq 1$. Find the maximum value of $x^2(1 - x)$.

18.28 Prove that $x^7 + 1 \geq x^4 + x^3$ for all nonnegative x.

18.29 Show that $\dfrac{a^8}{b^2} + \dfrac{b^8}{a^2} \geq a^6 + b^6$ for any nonzero a and b.

18.30 Let a, b, and c be nonnegative real numbers. Prove that $abc \geq (a + b - c)(a - b + c)(-a + b + c)$.

18.31 Show that if $a, b, c > 0$, then

$$a(1 - b) > \frac{1}{4}, \quad b(1 - c) > \frac{1}{4}, \quad \text{and} \quad c(1 - a) > \frac{1}{4}$$

cannot all simultaneously be valid.

18.32 Show that if x, y, and z are positive numbers whose sum is 1, then $\dfrac{xy}{z} + \dfrac{yz}{x} + \dfrac{zx}{y} \geq 1$.

18.33 Prove that $n^n \geq 1 \cdot 3 \cdot 5 \cdots (2n - 1)$ for all positive integers n.

Challenge Problems

18.34 Show that $x^6 + 2 \geq x^3 + x^2 + x$ for all real values of x.

18.35 A walk consists of a sequence of steps of length 1 taken in directions north, south, east, or west. A walk is *self-avoiding* if it never passes through the same point twice. Let $f(n)$ denote the number of n-step self-avoiding walks that begin at the origin. Show that $2^n < f(n) \leq 4(3^{n-1})$. (*Source: Canada*)

18.36 Show that if a, b, and c are positive, then $\dfrac{1}{a^2} + \dfrac{1}{b^2} + \dfrac{1}{c^2} \geq \dfrac{1}{a\sqrt{bc}} + \dfrac{1}{b\sqrt{ac}} + \dfrac{1}{c\sqrt{ab}}$. **Hints:** 180

18.37 Let $S = x_1 + x_2 + \cdots + x_n$, where $x_i > 0$ for $i = 1, 2, \ldots, n$. Prove that

$$\frac{S}{S - x_1} + \frac{S}{S - x_2} + \cdots + \frac{S}{S - x_n} \geq \frac{n^2}{n - 1},$$

with equality if and only if $x_1 = x_2 = \cdots = x_n$.

18.38 Find the minimum value of $\dfrac{xy}{z} + \dfrac{xz}{y} + \dfrac{yz}{x}$, where $x^2 + y^2 + z^2 = 1$ and $x, y, z > 0$. **Hints:** 29

18.39 Let x, y, and z be nonnegative real numbers such that $x + y + z = 1$. Prove that

$$\left(1 + \frac{1}{x}\right)\left(1 + \frac{1}{y}\right)\left(1 + \frac{1}{z}\right) \geq 64,$$

and find when equality occurs. **Hints:** 321

18.40   Expanding $(1 + 0.2)^{1000}$ by the Binomial Theorem and doing no further manipulation gives

$$\binom{1000}{0}(0.2)^0 + \binom{1000}{1}(0.2)^1 + \binom{1000}{2}(0.2)^2 + \cdots + \binom{1000}{1000}(0.2)^{1000} = A_0 + A_1 + A_2 + \cdots + A_{1000},$$

where $A_k = \binom{1000}{k}(0.2)^k$ for $k = 0, 1, 2, \ldots, 1000$.   For which k is A_k the largest?   *(Source: AIME)*
Hints: 77, 178

18.41   Determine the minimum value of

$$(r-1)^2 + \left(\frac{s}{r} - 1\right)^2 + \left(\frac{t}{s} - 1\right)^2 + \left(\frac{4}{t} - 1\right)^2$$

for all real numbers r, s, t with $1 \le r \le s \le t \le 4$. *(Source: Putnam)*

18.42   Let a and b be positive real numbers. Prove that

$$\frac{x+a}{x} \cdot \frac{x-b}{x} \le \frac{(a+b)^2}{4ab}$$

for all nonzero x. **Hints:** 240

18.43   Let a, b, c be positive real numbers. Prove that $\dfrac{a}{b+c} + \dfrac{b}{a+c} + \dfrac{c}{a+b} \ge \dfrac{3}{2}$.   (This is **Nesbitt's Inequality**.) **Hints:** 166, 65

18.44★   Let a, b, and c be real numbers. Prove that the smallest of $(a-b)^2$, $(a-c)^2$, and $(b-c)^2$ is no greater than $(a^2 + b^2 + c^2)/2$. **Hints:** 37

18.45   If $a, b, c > 0$, what is the smallest possible value of $\left\lfloor \dfrac{a+b}{c} \right\rfloor + \left\lfloor \dfrac{b+c}{a} \right\rfloor + \left\lfloor \dfrac{c+a}{b} \right\rfloor$? *(Source: HMMT)*
Hints: 307

18.46★   Let $P(x)$ be a polynomial with positive coefficients. Prove that if $P\left(\frac{1}{x}\right) \ge \frac{1}{P(x)}$ holds for $x = 1$, then it holds for all $x > 0$. *(Source: Titu Andreescu)*

18.47★   Let a, b, and c be positive. Show that

$$\frac{a}{a + \sqrt{(a+b)(a+c)}} + \frac{b}{b + \sqrt{(b+c)(b+a)}} + \frac{c}{c + \sqrt{(c+a)(c+b)}} \le 1.$$

Hints: 124

18.48★   Determine the largest real number z such that there exist real numbers x, y, and z with $x + y + z = 5$ and $xy + xz + yz = 3$. *(Source: CMO)*

18.49★   Determine the value of $\lfloor x \rfloor$, where

$$x = 1 + \frac{1}{\sqrt{2}} + \frac{1}{\sqrt{3}} + \frac{1}{\sqrt{4}} + \frac{1}{\sqrt{5}} + \cdots + \frac{1}{\sqrt{1,000,000}}.$$

(Source: USAMTS) **Hints:** 272, 256

18.50★   Prove that $(a^5 - a^2 + 3)(b^5 - b^2 + 3)(c^5 - c^2 + 3) \ge (a + b + c)^3$ for all nonnegative numbers a, b, and c. *(Source: USAMO)* **Hints:** 50, 352

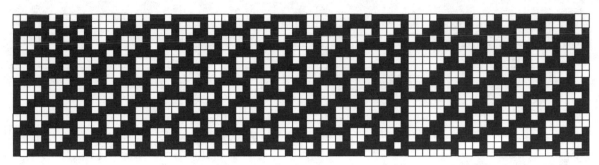

Equations are more important to me, because politics is for the present, but an equation is something for eternity.

– Albert Einstein

CHAPTER **19**

Functional Equations

> **Definition:** A **functional equation** is an equation that includes a function.

For example, $f(2x + 1) = 2f(x) + 3$ is a functional equation. So are each of the following:

$$f(x + y) = x + f(y), \qquad f\left(\frac{x + y}{2}\right) = \frac{f(x) + f(y)}{2}, \qquad f(x + y) = f(x)f(y).$$

19.1 Finding Values

Before we dive into solving functional equations, we'll first explore tactics for finding specific values of a function that satisfies a given functional equation

Problems

> **Problem 19.1:** Let $f(x)$ be a function such that $f(x + y) = x + f(y)$ for any two real numbers x and y, and $f(0) = 2$. What is the value of $f(1998)$? *(Source: AHSME)*

> **Problem 19.2:** The function f has the interval $[0, 1]$ as its domain and its range. It also satisfies $f(0) = 0$, $f(1) = 1$, and
>
> $$f\left(\frac{x + y}{2}\right) = \frac{f(x) + f(y)}{2}$$
>
> for all $x, y \in [0, 1]$.
>
> (a) Find all possible values of $f(1/2)$.   (b) Find all possible values of $f(1/3)$.

Problem 19.3: Let f be a real-valued function such that $f(m+n) = f(m)f(n)$ for all real numbers m and n. If $f(4) = 625$, then find a value of k such that $f(k) = 0.04$.

Problem 19.4: The sequence $a_0, a_1, a_2, \ldots$ satisfies

$$a_{m+n} + a_{m-n} = \frac{1}{2}(a_{2m} + a_{2n})$$

for all nonnegative integers m and n with $m \geq n$. If $a_1 = 1$, determine a_{1995}. *(Source: Russia)*

Problem 19.1: Let $f(x)$ be a function such that $f(x + y) = x + f(y)$ for any two real numbers x and y, and $f(0) = 2$. What is the value of $f(1998)$? *(Source: AHSME)*

Solution for Problem 19.1: We are given $f(0) = 2$, and we want $f(1998)$. By letting $y = 0$ in the functional equation, we get $f(x) = x + f(0) = x + 2$. Now, we just let $x = 1998$ to find $f(1998) = 1998 + 2 = 2000$.

Note that our work above tells us that the function $f(x) = x + 2$ is the only function that satisfies both properties of the problem. So, not only can we find $f(1998)$, but we can now evaluate $f(x)$ for any value of x. $\square$

The key step in Problem 19.1 is simply choosing appropriate values for the variables in the functional equation.

Concept: Choosing values for the variables in a functional equation is a great way to learn more about the function involved in the equation.

Typically we choose values for which we know the value of the function (as when we chose $y = 0$ above), or at which we want the value of the function (as when we chose $x = 1998$).

Problem 19.2: The function f has the interval $[0, 1]$ as its domain and its range. It also satisfies $f(0) = 0$, $f(1) = 1$, and

$$f\left(\frac{x + y}{2}\right) = \frac{f(x) + f(y)}{2}$$

for all $x, y \in [0, 1]$.

(a) Find all possible values of $f(1/2)$.   (b) Find all possible values of $f(1/3)$.

Solution for Problem 19.2:

(a) What's wrong with this solution:

Bogus Solution: The function $f(x) = x$ satisfies the constraints of the problem, since then we have $f(0) = 0$, $f(1) = 1$, and $f((x + y)/2) = (x + y)/2 = (f(x) + f(y))/2$. Therefore, we have $f(1/2) = 1/2$.

All the Bogus Solution tells us is that it is possible to have $f(1/2) = 1/2$. It does not tell us that this is the only possible value of $f(1/2)$.

We know $f(0)$ and $f(1)$, so letting $x = 0$ and $y = 1$ gives us $f(1/2)$ directly:

$$f\left(\frac{1}{2}\right) = \frac{f(0) + f(1)}{2} = \frac{1}{2}.$$

Notice that this solution does not assume that $f(x)$ equals some expression. We only use the given information in the problem, and show that any function $f(x)$ that satisfies the problem constraints must have $f(1/2) = 1/2$.

(b) To use the given functional equation to find $f(1/3)$, we seek x and y such that $(x + y)/2 = 1/3$. In other words, we want $1/3$ to be the average of x and y. Since we know $f(0) = 0$, we can try taking $x = 0$ and $y = 2/3$. Then we get

$$f\left(\frac{1}{3}\right) = \frac{f(0) + f(2/3)}{2} = \frac{f(2/3)}{2},$$

so $f(2/3) = 2f(1/3)$.

We need another equation relating $f(1/3)$ and $f(2/3)$. We note that the average of $1/3$ and $2/3$ is $1/2$, and we know $f(1/2)$. We have

$$f\left(\frac{1}{2}\right) = \frac{f(1/3) + f(2/3)}{2},$$

and we also have $f(1/2) = 1/2$, so $f(1/3) + f(2/3) = 1$. Substituting $f(2/3) = 2f(1/3)$ into this equation gives us $f(1/3) = 1/3$.

□

Problem 19.3: Let f be a real-valued function such that $f(m + n) = f(m)f(n)$ for all real numbers m and n. If $f(4) = 625$, then find a value of k such that $f(k) = 0.04$.

Solution for Problem 19.3: We start by experimenting. We find $f(m)$ for various values of m. Maybe we'll get lucky and find an m such that $f(m) = 0.04$. We know $f(4)$, so we start by letting $m = n = 4$, which gives us $f(4 + 4) = f(4)f(4)$, so $f(8) = 625^2$. Looks like we're moving in the wrong direction!

Another way we can use our value of $f(4)$ is to let $m = 4$ and $n = 0$. This gives us $f(4 + 0) = f(4)f(0)$, which means $625 = 625f(0)$. So, we have $f(0) = 1$. That's progress, but we still don't have $f(k) = 0.04$. Next, let's try $m = n = 2$, which gives us $f(2 + 2) = f(2)f(2)$, so $(f(2))^2 = 625$. Therefore, we have $f(2) = \pm 25$. But is $f(2)$ equal to 25 or −25? We can rule out $f(2) = -25$ by finding $f(1)$ the same way we just found $f(2)$. We let $m = n = 1$, and we have $f(2) = (f(1))^2$. Since $f(x)$ is real for all x and $f(2)$ is the square of $f(1)$, we know that $f(2)$ is nonnegative.

We have $f(2) = 25$, and we can similarly show that $f(1) = \sqrt{f(2)} = 5$. Hmmm.... We now have $f(0) = 1$, $f(1) = 5$, $f(2) = 25$, and $f(4) = 625$. We can easily fill in the gap to find $f(3) = f(2)f(1) = 5^3$. We could use induction to show that $f(n) = 5^n$ for all positive integers n, but that won't give us the value of k such that $f(k) = 0.04$. Writing 0.04 as a fraction, we have $f(k) = 1/25$, and now we suspect that $k = -2$. So, we let $n = -2$, and we have $f(m - 2) = f(m)f(-2)$. A natural choice for m now is $m = 2$, since we know both $f(0)$ and $f(2)$. We then find $f(0) = f(2)f(-2)$, so $1 = 25f(-2)$, which gives us the expected $f(-2) = 1/25$. Therefore, we have $k = -2$. □

> **Sidenote:** That the function in Problem 19.3 turns out to be an exponential function
> is no coincidence. In fact, the exponential function $f(x) = a^x$, where a is
> a positive constant, can be defined as the continuous function such that
> $f(1) = a$ and $f(x + y) = f(x)f(y)$ for all real numbers x and y.

Problem 19.3 exhibits one of our most useful strategies in dealing with new functional equations:

> **Concept:** If you can't find the answer to a functional equation question right away,
> explore the problem by trying various values for the variables. Evaluate
> the function at many different values, and hopefully a pattern will emerge.

Here's a harder problem on which to try this strategy.

> **Problem 19.4:** The sequence $a_0, a_1, a_2, \ldots$ satisfies
>
> $$a_{m+n} + a_{m-n} = \frac{1}{2}(a_{2m} + a_{2n})$$
>
> for all nonnegative integers m and n with $m \geq n$. If $a_1 = 1$, determine a_{1995}. (*Source: Russia*)

Solution for Problem 19.4: At first glance, this doesn't look like a functional equation problem. But the given recursion is essentially the same thing as a functional equation. Instead of being written in terms of $f(x)$, it is written in terms of a_i. For each nonnegative integer input i, there is an output a_i, just as we input x to a function f to get an output $f(x)$. So, we try the same substitution tactics we've used on the functional equations so far.

We can reduce our equation by one variable by letting $n = 0$, which gives us $2a_m = (a_{2m} + a_0)/2$. We'd like to evaluate a_m for some values of m to get a feel for our sequence. A natural substitution now is $m = 0$, since this gives us the equation $2a_0 = (a_0 + a_0)/2$, which we can solve to find $a_0 = 0$. Substituting this back into our expression for a_m, we have $2a_m = a_{2m}/2$, or $4a_m = a_{2m}$.

Now, we can let $m = 1$ to find $a_2 = 4a_1 = 4$. Letting $m = 2$ gives $a_4 = 4a_2 = 4^2$, and so on. Continuing in this vein, we can show that $a_{2^k} = 4^k$ for all k. Unfortunately, we still can't find a_m for any other specific values, so we turn back to our given equation. We note that a_{2m} and a_{2n} both appear, so we apply $a_{2m} = 4a_m$ and $a_{2n} = 4a_n$ to find

$$a_{m+n} + a_{m-n} = \frac{1}{2}(a_{2m} + a_{2n}) = 2a_m + 2a_n.$$

Now, letting $n = 1$ gives $a_{m+1} + a_{m-1} = 2a_m + 2a_1$. We solve this for a_{m+1} to get the recursion

$$a_{m+1} = 2a_m - a_{m-1} + 2a_1 = 2a_m - a_{m-1} + 2.$$

We know $a_2 = 4a_1 = 4$, and we can compute successive terms as follows:

$$a_3 = 2a_2 - a_1 + 2 = 8 - 1 + 2 = 9,$$
$$a_4 = 2a_3 - a_2 + 2 = 18 - 4 + 2 = 16,$$
$$a_5 = 2a_4 - a_3 + 2 = 32 - 9 + 2 = 25,$$

and so on. We've found our pattern! It looks like $a_m = m^2$. We'd like to show that this formula works for all positive integers—that sounds like a job for induction. We've already showed that $a_m = m^2$ for $m = 0$ and $m = 1$. Suppose that $k \geq 1$ and $a_m = m^2$ for all $m \leq k$, so that $a_k = k^2$ and $a_{k-1} = (k-1)^2$. Then, we have

$$a_{k+1} = 2a_k - a_{k-1} + 2 = 2k^2 - (k-1)^2 + 2 = k^2 + 2k + 1 = (k+1)^2.$$

So, by induction, we have $a_m = m^2$ for all positive integers m. In particular, we have $a_{1995} = 1995^2$. $\square$

Exercises

19.1.1 The function f is such that $f(0) = 0$ and $f(2x+1) = 2f(x) + 3$ for all real x. Find $f(1)$, $f(3)$, and $f(7)$.

19.1.2 $F(n)$ is a function such that $F(1) = F(2) = F(3) = 1$ and

$$F(n+1) = \frac{F(n)F(n-1) + 1}{F(n-2)}$$

for $n \geq 3$. Find $F(6)$. *(Source: AHSME)*

19.1.3 For all ordered pairs of positive integers (x, y), we define $f(x, y)$ as follows:

(a) $f(x, 1) = x$,

(b) $f(x, y) = 0$ if $y > x$,

(c) $f(x+1, y) = y[f(x, y) + f(x, y-1)]$ if $x + 1 \geq y > 1$.

Compute $f(5, 5)$. *(Source: ARML)*

19.1.4 If $f(x - y) = f(x) \cdot f(y)$ for all x and y, and $f(x)$ never equals zero, find all possible values of $f(1977)$.

19.1.5★ Assume that $f(1) = 0$, and that for all integers m and n, we have $f(m+n) = f(m) + f(n) + 3(4mn - 1)$. Determine $f(19)$. *(Source: USAMTS)*

19.2 Finding Functions with Substitution

In the previous section, we found specific values of functions that satisfy certain functional equations. In the next three sections, we explore strategies for finding the functions themselves.

We'll start by introducing some new notation. If we write, "Consider a function $f : \mathbb{R} \to [0, 1]$," we mean that function f only accepts real numbers as inputs and only outputs numbers in the interval $[0, 1]$. In other words, the domain of f is $\mathbb{R}$, and the range of f only has numbers in $[0, 1]$. Note that $f : \mathbb{R} \to [0, 1]$ does *not* indicate that the range is $[0, 1]$. It merely states that all values in the range must be in the interval $[0, 1]$. We say that $[0, 1]$ is the **codomain** of the function f, which means that $f(x) \in [0, 1]$ for all x.

To describe a function fully, we typically give its domain, its codomain, and a formula for the function. For example, the function $f : \mathbb{Z} \to [-1, 3]$, $f(x) = \frac{1}{2}$ is a function with domain $\mathbb{Z}$ (the set of all integers) and codomain $[-1, 3]$, but whose range is just the value $\frac{1}{2}$.

Finally, we have a special notation for excluding specific values from a set. For example, if a function is defined for all real numbers except 0, we can write its domain as $\mathbb{R}\backslash\{0\}$, where the "$\backslash\{0\}$" indicates that 0 is excluded from the domain. Similarly, $\mathbb{Z}\backslash\{2,5\}$ means "all integers except 2 and 5."

Don't worry too much about all this new notation. It will seem natural after you've seen it a little more.

Our goal in solving a functional equation is to find a definition for the function. Usually this means manipulating the equation until we have $f(x)$ equal to some expression in terms of x. In this section, we use substitutions to find these expressions.

Problems

Problem 19.5: Let $f : \mathbb{R} \to \mathbb{R}$ be a function that satisfies $f(xy) = x^2 f(y)$ for all x and y.

(a) Find a substitution for y that reduces the functional equation to a simpler one-variable equation that you can solve for $f(x)$.

(b) Find all functions f that satisfy the functional equation.

(c) Confirm that the functions you found in part (b) satisfy $f(xy) = x^2 f(y)$ for all x and y.

Problem 19.6: Find all functions $f : \mathbb{R} \to \mathbb{R}$ that satisfy the functional equation $f(x) + f(x + y) = y + 10$ for all real x and y.

Problem 19.7: Find $f(x)$ if
$$f(a + b) - f(a - b) = 4ab$$
for all $a, b \in \mathbb{R}$.

Problem 19.8: Find all functions $f : \mathbb{R} \to \mathbb{R}$ such that $f(x)f(y) - f(xy) = x + y$ for all x, y.

Problem 19.9: Determine all polynomials $P(x)$ such that $P(x^2 + 1) = [P(x)]^2 + 1$ and $P(0) = 0$. *(Source: Putnam)*

Problem 19.5: Find all functions $f : \mathbb{R} \to \mathbb{R}$ that satisfy $f(xy) = x^2 f(y)$ for all x and y.

Solution for Problem 19.5: We want to produce an equation of the form
$$f(x) = \text{some expression in terms of } x.$$

We might think to isolate $f(y)$ and write $f(y) = f(xy)/x^2$. But then we have that annoying $f(xy)$ on the right, and the x^2 as well. We can get rid of x by letting $x = 1$, but that just tells us $f(y) = f(y)/1$, which isn't very helpful.

Instead, we focus on the $f(xy)$ on the left side, and try substituting for y rather than for x. Letting $y = 1$ gives us $f(x) = x^2 f(1)$. We don't know what $f(1)$ is, but we do know that it is a constant. But what constant is it?

> **WARNING!!**
> ☢
> Don't get thrown off by $f(0)$ or $f(1)$, or f(some number) showing up in a functional equation after a substitution. Such an expression is just a constant. So, call that constant c, and then see if you can learn more about it.

We let $f(1) = c$, so we have $f(x) = x^2 f(1) = cx^2$. We now must check which values of c (if any) give us a valid solution to the functional equation. Letting $f(x) = cx^2$ in the original functional equation gives us $f(xy) = c(xy)^2 = cx^2y^2$ on the left and $x^2 f(y) = x^2(cy^2) = cx^2y^2$ on the right. The two sides of the equation are the same for all c, so $f(x) = cx^2$ is a solution for any real constant c. □

> **Concept:**
> ⚷
> When dealing with a functional equation that involves several variables, setting one or more of the variables equal to simple constants like 0, 1, or −1 often results in a simpler functional equation.

Our solution to Problem 19.5 essentially consisted of two parts:

1. We showed that every function that satisfies the functional equation must have the form $f(x) = cx^2$ for some constant c.

2. We showed that every function of the form $f(x) = cx^2$ satisfies the functional equation.

We must not forget the second step! Our next problem illustrates why this second step is necessary.

Problem 19.6: Find all functions $f : \mathbb{R} \to \mathbb{R}$ that satisfy the functional equation $f(x) + f(x+y) = y+10$ for all real x and y.

Solution for Problem 19.6: The complication here is the $f(x + y)$ term. We can reduce this term to $f(x)$ by letting $y = 0$. This makes our equation $2f(x) = 10$, so $f(x) = 5$ for all x.

We check our solution by substituting $f(x) = 5$ back into the functional equation. This gives us $5 + 5 = y + 10$. In order for our solution to be valid, this equation must be true for all possible values of x and y. This is clearly not the case; if y is anything but 0, then the equation is false. Did we make a mistake?

No! With our initial substitution, we showed that any function that satisfies the functional equation must also satisfy $f(x) = 5$. However, we showed that this function does not satisfy the functional equation. Therefore, there are no solutions to the functional equation. □

> **Important:**
> ⚠
> After you solve a functional equation, check that your solutions actually satisfy the given equation. This is not just to check your work; as seen in Problem 19.6, a solution that you derive from a functional equation does not automatically satisfy the functional equation.

Problem 19.7: Find $f(x)$ if $f(a + b) - f(a - b) = 4ab$ for all $a, b \in \mathbb{R}$.

Solution for Problem 19.7: With $f(a+b)$ and $f(a-b)$ as our main complications, we start off by letting $b = 0$, thus reducing both of these to $f(a)$. This gives us $f(a+b) - f(a-b) = f(a) - f(a) = 0$ on the left. Uh-oh. Over on the right side of the functional equation, we have $4ab = 4a(0) = 0$. So, we have $0 = 0$. Not very helpful.

Another way we can reduce the functional equation to a single variable is to let $b = a$, which makes the equation

$$f(2a) - f(0) = 4a^2.$$

Aha! This substitution also turned $f(a-b)$ into the constant $f(0)$. We can now let $2a = x$, so we have $f(x) - f(0) = x^2$, which gives us $f(x) = x^2 + f(0)$. Since $f(0)$ is a constant, we have the solution $f(x) = x^2 + c$, for some constant c.

We now let $f(x) = x^2 + c$ in the original equation. We find that

$$f(a+b) - f(a-b) = (a+b)^2 + c - (a-b)^2 - c = 4ab.$$

Therefore, any constant c works, so the solutions of the equation are all functions of the form $f(x) = x^2 + c$, where c is a constant. $\square$

Problem 19.8: Find all functions $f : \mathbb{R} \to \mathbb{R}$ such that $f(x)f(y) - f(xy) = x + y$ for all x, y.

Solution for Problem 19.8: Letting $y = 0$ turns two of our functional expressions into $f(0)$, which will leave us an equation we can solve for $f(x)$. This gives us $f(x)f(0) - f(0) = x$. Therefore, we have $f(x)f(0) = x + f(0)$. Notice that we don't divide by $f(0)$ right away; what if $f(0) = 0$? If we have $f(0) = 0$, then our equation $f(x)f(0) = x + f(0)$ becomes $0 = x + 0$. But this equation must be true for all x, which clearly isn't the case. So, we can't have $f(0) = 0$.

This means we can let $f(0) = c$, where $c \neq 0$, and divide $f(x)f(0) = x + f(0)$ by c to give $f(x) = \frac{x}{c} + 1$. Next, we substitute back into our functional equation. The left side becomes

$$\left(\frac{x}{c} + 1\right)\left(\frac{y}{c} + 1\right) - \left(\frac{xy}{c} + 1\right) = \frac{x}{c} + \frac{y}{c} + \frac{xy}{c^2} - \frac{xy}{c}.$$

Clearly, this doesn't equal $x + y$ for all values of c. It does equal $x + y$ for $c = 1$. To see that $c = 1$ is the only possible value of c, we let $x = 0$ in $f(x) = \frac{x}{c} + 1$, which gives us $f(0) = 1$. Since $c = f(0)$, we have $c = 1$, so the only solution to our functional equation is $f(x) = x + 1$, which does indeed satisfy the original equation. $\square$

WARNING!! Once you let $f(0)$ or $f(1)$ (or whatever) in your expression for $f(x)$ equal some constant c, you must remember to plug your expression for $f(x)$ back into the functional equation to check if you need any restrictions on c.

Problem 19.9: Determine all polynomials $P(x)$ such that $P(x^2 + 1) = [P(x)]^2 + 1$ and $P(0) = 0$. (*Source: Putnam*)

Solution for Problem 19.9: Let's start by substituting some values for x. Since we are given $P(0) = 0$, we should first try $x = 0$. This gives $P(1) = [P(0)]^2 + 1 = 1$.

We have $P(1)$ now, so we let $x = 1$. This gives us $P(2) = [P(1)]^2 + 1 = 2$.

We have $P(2)$ now, so we let $x = 2$. This gives us $P(5) = [P(2)]^2 + 1 = 5$.

Hmmm.... We think we know where this is headed. We let $x = 5$ and find $P(26) = 26$. Then we let $x = 26$ and find $P(677) = 677$.

We see that $P(x) = x$ for a lot of values of x. In fact, there are infinitely many values of x that satisfy this equation. We can prove this using induction. We have $P(1) = 1$, and if $P(n) = n$, then $P(n^2 + 1) = [P(n)]^2 + 1 = n^2 + 1$. Therefore, if we have $P(k) = k$ for any integer k, then there is a larger integer, namely $m = k^2 + 1$, for which $P(m) = m$. So, there are infinitely many solutions to $P(x) = x$.

But $P(x)$ is a polynomial! If there are infinitely many solutions to $P(x) = x$, then the polynomial $P(x) - x$ has infinitely many roots. The number of roots of a polynomial with finite degree must equal the degree of the polynomial. So, $P(x) - x$ must be the polynomial 0, which means $P(x) = x$ for all x. □

Exercises

19.2.1 Suppose that for all $x > 0$ we have $f(2x) = \frac{5}{2+x}$. What is $2f(x)$?

19.2.2 Find all functions $f(x)$ such that $f(x + 2y) + f(x - 2y) = 2x^2 + 8y^2$ for all real x and y.

19.2.3 The function $f(x)$ is defined for every positive integer x and satisfies the equation

$$f(x + y) = f(x)f(y) - f(xy) + 1$$

for all positive integers x and y. If $f(1) = 2$, find $f(x)$.

19.2.4 Find all functions $f : \mathbb{R} \to \mathbb{R}$ such that $1 + f(x + y) = 2f(x)f(y)$ for all real x, y.

19.2.5 Find all functions $f : \mathbb{Z} \to \mathbb{Z}$ such that $f(a + b) - f(a) + f(a)f(b) + f(b)$ for all a, b, and $f(1) = k - 1$, where k is some integer greater than 1.

19.3 Separation

Problems

Problem 19.10: Suppose that $xf(x) = yf(y)$ for all nonzero real numbers x and y.
  (a) Suppose that $3f(3) = 7$. What is $5f(5)$? What is $8f(8)$? What is $\pi f(\pi)$?
  (b) Find all functions that satisfy the equation.

Problem 19.11: Find all functions $f : \mathbb{R} \to \mathbb{R}$ such that $yf(x) = xf(y)$ for all real numbers x and y.

Problem 19.12: Find all functions $f : \mathbb{R}\backslash\{0\} \to \mathbb{R}$ such that $y^2 f(x) - x^2 f(y) = y^2 - x^2$ for all nonzero real numbers x, y.

Problem 19.10: Find all functions $f : \mathbb{R}\backslash\{0\} \rightarrow \mathbb{R}$ such that $xf(x) = yf(y)$ for all nonzero real numbers x and y.

Solution for Problem 19.10: If we let $y = 1$, then we have $xf(x) = f(1)$, so $f(x) = f(1)/x$. Letting $c = f(1)$, we have $f(x) = c/x$. Substituting this into the functional equation, we have $xf(x) = c$ and $yf(y) = c$, so the function $f(x) = c/x$ satisfies the functional equation for any value of the constant c.

We can also solve this functional equation by noticing that the two sides of the equation $xf(x) = yf(y)$ are essentially the same. This equation tells us that no matter what value we choose for x, the value of $xf(x)$ is always the same. For example, we have $xf(x) = 1f(1) = 2f(2) = 3f(3)$.

In other words, if we let $g(t) = tf(t)$, then we have $g(x) = g(y)$ for all x and y, which tells us that $g(t)$ is the same for all t. Therefore, $g(t)$ must be a constant. So, the expression $xf(x)$ is constant, and we have $xf(x) = c$ for some constant c. This gives us $f(x) = c/x$, which we can then substitute into the given functional equation to see that c can take on any value. $\square$

Problem 19.11: Find all functions $f : \mathbb{R} \rightarrow \mathbb{R}$ such that $yf(x) = xf(y)$ for all real numbers x and y.

Solution for Problem 19.11: Once again, we could use substitution. Letting $y = 1$ gives us $f(x) = xf(1) = cx$ for some constant c.

But let's see if we can also apply our second strategy from Problem 19.10. In that problem, we had an equation in which the two sides were essentially the same, but with different variables. We try to manipulate the present equation into a similar form by separating the variables, putting one variable on each side of the equation. Dividing by xy quickly separates both variables and gives us

$$\frac{f(x)}{x} = \frac{f(y)}{y}.$$

This equation must hold for all (nonzero) x and y, so the expression $f(x)/x$ must be constant. This means we have $f(x)/x = c$ for some constant c, which gives us $f(x) = cx$ for all $x \neq 0$. To confirm that $f(x) = cx$ for $x = 0$ as well, we let $x = 0$ in $yf(x) = xf(y)$, and we find $yf(0) = 0$ for all y. Therefore, $f(0) = 0$, so $f(x)$ must satisfy $f(x) = cx$ for all x.

Substituting this back into the original functional equation, we have $yf(x) = xf(y) = cxy$, so the function $f(x) = cx$ satisfies the equation for all constants c. $\square$

Concept: In some functional equations with two variables, we can separate the variables so that each variable appears only on one side of the equation. If the two sides are the same except for the different variables (in other words, swapping the variables results in the same equation), then the two expressions must equal a constant for all values of the variables.

Problem 19.12: Find all functions $f : \mathbb{R}\backslash\{0\} \rightarrow \mathbb{R}$ such that $y^2 f(x) - x^2 f(y) = y^2 - x^2$ for all nonzero real numbers x, y.

Solution for Problem 19.12: Inspired by our success with separation in the previous two problems, we try the same here. We'll have to separate the y^2 from $y^2 f(x)$ and the x^2 from $x^2 f(y)$, so we start by dividing both sides of the equation by $x^2 y^2$, which leaves

$$\frac{f(x)}{x^2} - \frac{f(y)}{y^2} = \frac{1}{x^2} - \frac{1}{y^2}.$$

Now, separating the variables is easy. We add $\frac{f(y)}{y^2} - \frac{1}{x^2}$ to both sides to get

$$\frac{f(x)}{x^2} - \frac{1}{x^2} = \frac{f(y)}{y^2} - \frac{1}{y^2}.$$

Aha! If we let $g(t) = \frac{f(t)}{t^2} - \frac{1}{t^2}$, our equation above tells us that $g(x) = g(y)$ for all x and y. In other words, $g(x)$ is constant, so we have $\frac{f(x)}{x^2} - \frac{1}{x^2} = c$ for some constant c. Solving for $f(x)$ gives $f(x) = cx^2 + 1$.

Substituting this back into the original functional equation gives us

$$y^2 f(x) - x^2 f(y) = cx^2 y^2 + y^2 - cx^2 y^2 - x^2 = y^2 - x^2,$$

so all functions of the form $f(x) = cx^2 + 1$ satisfy the equation. $\square$

Exercises

19.3.1 For every pair of numbers a and b, the function f satisfies $b^2 f(a) = a^2 f(b)$. If $f(2) \neq 0$, find the value of $\frac{f(5) - f(1)}{f(2)}$. *(Source: Canada)*

19.3.2 Determine all functions with domain $(0, +\infty)$ that satisfy the functional equation $y 2^{f(x)} = x 2^{f(y)}$.

19.3.3 Find all functions $f : \mathbb{R} \backslash \{-1\} \to \mathbb{R}$ such that $\dfrac{f(x)f(z)}{y + 1} = \dfrac{f(y)f(z)}{x + 1}$ whenever x, y, and z do not equal -1.

19.4 Cyclic Functions

Back in Section 2.3, we introduced a special notation for composing a function with itself. In this text, we write $f^2(x)$ to mean $f(f(x))$. Likewise, $f^n(x)$ refers to applying the function $f(x)$ exactly n times; for example, $f^5(x) = f(f(f(f(f(x)))))$.

> **WARNING!!** Some sources use $f^2(x)$ to mean $f(x) \cdot f(x)$ instead of $f(f(x))$. Likewise, $f^n(x)$ is sometimes used to denote $[f(x)]^n$. You'll often have to determine by context whether $f^n(x)$ refers to a function applied n times, or to a function raised to the nth power.

Problems

Problem 19.13: Let $u(x) = \dfrac{1}{1-x}$.

  (a) Find $u^2(x)$.

  (b) Find $u^{23}(x)$.

  (c) Find $u^{-1}(x)$.

Problem 19.14: Given $f_1(x) = \frac{2x-1}{x+1}$, let $f_{n+1}(x) = f_1(f_n(x))$ for $n = 1, 2, 3, \ldots$. It can be shown that $f_{35} = f_5$. Find a rational function that equals $f_{28}(x)$ for all values of x for which $f_{28}(x)$ is defined. *(Source: AHSME)*

Problem 19.15: The function $g : \mathbb{R}\backslash\{0\} \to \mathbb{R}$ satisfies the equation $g(x) + 2g\left(\frac{1}{x}\right) = -x + \frac{1}{x}$ for all $x \neq 0$.

  (a) Suppose $f(x) = 1/x$. What is $f^2(x)$? What is $f^3(x)$?

  (b) Find all functions $g(x)$ that satisfy the given functional equation.

Problem 19.16: Suppose that f satisfies the functional equation $2f(x) + 3f\left(\frac{2x+29}{x-2}\right) = 100x + 80$ for all $x \neq 2$. Find $f(3)$. *(Source: USAMTS)*

Problem 19.17: Let $F(x)$ be a real-valued function defined for all real x except for $x = 0$ and $x = 1$ such that $F(x)$ satisfies the functional equation

$$F(x) + F\left(\frac{x-1}{x}\right) = 1 + x$$

for all x besides 0 and 1. Find all functions $F(x)$ satisfying these properties. *(Source: Putnam)*

Problem 19.13: Let $u(x) = \dfrac{1}{1-x}$. Find $u^{23}(x)$ and $u^{-1}(x)$.

Solution for Problem 19.13: There's no obvious way to jump straight to $u^{23}(x)$, so we start by calculating $u^n(x)$ for smaller n. We find:

$$u(x) = \frac{1}{1-x},$$

$$u^2(x) = u(u(x)) = u\left(\frac{1}{1-x}\right) = \frac{1}{1-\frac{1}{1-x}} = \frac{\frac{1-x}{1-x}}{\frac{1-x}{1-x} - \frac{1}{1-x}} = \frac{1-x}{1-x-1} = -\frac{1-x}{x},$$

$$u^3(x) = u(u^2(x)) = u\left(-\frac{1-x}{x}\right) = \frac{1}{1+\frac{1-x}{x}} = \frac{x}{x+1-x} = x,$$

$$u^4(x) = u(u^3(x)) = u(x).$$

It repeats! Because $u^3(x) = x$, we have $u^5(x) = u^2(u^3(x)) = u^2(x)$, and $u^6(x) = u^3(u^3(x)) = u^3(x)$, and so on, so that $u^{n+3}(x) = u^n(x)$ for all positive integers n. Because 23 is one less than a multiple of 3, we have

$u^{23}(x) = u^2(x) = -(1-x)/x$.

To find $u^{-1}(x)$, we could go through the usual steps of solving $u(y) = x$ for y in terms of x, but instead, we note that we saw $u(u^2(x)) = x$ above. In other words, the function $f(x)$ that satisfies $u(f(x)) = x$ is $f(x) = u^2(x)$. This means that the inverse of $u(x)$ is $u^2(x)$, and we have $u^{-1}(x) = u^2(x) = -(1-x)/x$. $\square$

We say a function f is **cyclic** if $f^n(x) = x$ for some positive integer n. The smallest positive integer n for which $f^n(x) = x$ is the **order** of f, and $f^k(x) = f^{k+n}(x)$ for all k. For example, above we saw that $u(x) = 1/(1-x)$ has order 3, and we have $u(x) = u^4(x) = u^7(x) = \cdots$, $u^2(x) = u^5(x) = u^8(x) = \cdots$, and $u^3(x) = u^6(x) = u^9(x) = \cdots$.

Problem 19.14: Given $f_1(x) = \dfrac{2x-1}{x+1}$, let $f_{n+1}(x) = f_1(f_n(x))$ for $n = 1, 2, 3, \ldots$. It can be shown that $f_{35} = f_5$. Find a rational function that equals $f_{28}(x)$ for all values of x for which $f_{28}(x)$ is defined. *(Source: AHSME)*

Solution for Problem 19.14: Because $f_{35} = f_5$, we have $f_{30}(f_5(x)) = f_{35}(x) = f_5(x)$. Letting $y = f_5(x)$, we have $f_{30}(y) = y$. We might deduce from this that f has order 30, but it is possible that the order is some smaller n that evenly divides 30. So, we compute the first few $f_n(x)$ and hope we get lucky. We have

$$f_2(x) = f_1(f_1(x)) = \frac{2\left(\frac{2x-1}{x+1}\right) - 1}{\frac{2x-1}{x+1} + 1} = \frac{x-1}{x},$$

$$f_3(x) = f_1(f_2(x)) = \frac{2\left(\frac{x-1}{x}\right) - 1}{\frac{x-1}{x} + 1} = \frac{x-2}{2x-1}.$$

We could keep on plugging away, but what if the order is 10, or 15, or 30? Before chugging through all that algebra, let's stop and think a bit.

> **Concept:** Whenever it looks like you have a great deal of algebra ahead of you, take a few minutes to look for another approach to the problem.

We have $f_{30}(x) = x$, and we want to find $f_{28}(x)$. Since 28 isn't that far from 30, we look for a way to start with $f_{30}(x)$ and go back to $f_{28}(x)$. In other words, we want to undo f twice. But that's what an inverse function does! Since $f_{30}(x) = f_{29}(f_1(x)) = x$, we see that $f_{29}(x)$ is the inverse of $f_1(x)$. Solving $f_1(y) = x$ for y in terms of x gives us $f_1^{-1}(x) = y = (x+1)/(2-x)$. Since $f_{29}(x)$ is the inverse of $f_1(x)$, we have $f_{29}(x) = (x+1)/(2-x)$.

Now, we're just a step away from $f_{28}(x)$. We have $f_{29}(x) = f_1(f_{28}(x))$. Letting $y = f_{28}(x)$, we have $f_1(y) = f_{29}(x) = (x+1)/(2-x)$. So, to find $f_{28}(x)$, we solve $(2y-1)/(y+1) = (x+1)/(2-x)$ for y in terms of x. Multiplying both sides by $(2-x)(y+1)$ gives $(2y-1)(2-x) = (x+1)(y+1)$. Expanding and rearranging gives $3y - 3xy = 3$, so $f_{28}(x) = y = 1/(1-x)$. Note that we could have found $f_{28}(x)$ a little faster by noticing that because $f_{30}(x) = x$, we have $f_{29}(f_{29}(x)) = f_{58}(x) = f_{30}(f_{28}(x)) = f_{28}(x)$. Or, we could see that $f_1^{-1}(f_1^{-1}(f_{30}(x))) = f_{28}(x)$, so $f_{28}(x) = f_1^{-1}(f_1^{-1}(x))$. $\square$

> **Concept:** Just as we can go from $f^n(x)$ to $f^{n+1}(x)$ by performing the composition $f(f^n(x))$, we can also step backwards from $f^n(x)$ to $f^{n-1}(x)$ when f is invertible by performing the composition $f^{n-1}(x) = f^{-1}(f^n(x))$.

As we saw in Problem 19.14, this tactic can be particularly useful with cyclic functions, since these functions have values of n such that $f^n(x) = x$. This allows us to work both backwards and forwards to find $f^k(x)$ for high values of k.

> **Problem 19.15:** The function $g : \mathbb{R} \backslash \{0\} \to \mathbb{R}$ satisfies the equation $g(x) + 2g\left(\frac{1}{x}\right) = -x + \frac{1}{x}$ for all $x \neq 0$. Find $g(x)$.

Solution for Problem 19.15: We start with a few substitutions to get a feel for g. The most obvious is $x = 1$, which gives us $3g(1) = 0$, so $g(1) = 0$. Unfortunately, after that, there aren't any substitutions that give us immediate results for g.

We feel a bit stuck, so we examine the equation for any notable features that we might use as a starting point. We focus on $2g(1/x)$, since this unusual expression makes the equation difficult to handle. We let $t = 1/x$ to get the variable out of the denominator. This gives us $x = 1/t$ and makes our equation

$$g\left(\frac{1}{t}\right) + 2g(t) = -\frac{1}{t} + t.$$

That doesn't seem to have worked. We took the variable out of the denominator of $2g(1/x)$, but ended up with a variable in the denominator in the $g(1/t)$ term. But wait a minute—while our equations are similar, they aren't exactly the same. If we let $x = t$ in the original equation and write it next to our equation for $x = 1/t$, we get a system of equations:

$$g(t) + 2g\left(\frac{1}{t}\right) = -t + \frac{1}{t},$$
$$g\left(\frac{1}{t}\right) + 2g(t) = -\frac{1}{t} + t.$$

We can now get rid of the annoying $g(1/t)$ term. We subtract twice the second equation from the first to get $-3g(t) = -3t + \frac{3}{t}$, from which we find $g(t) = t - \frac{1}{t}$. We substitute this into the original equation to check, and we have

$$g(x) + 2g\left(\frac{1}{x}\right) = x - \frac{1}{x} + 2\left(\frac{1}{x} - x\right) = -x + \frac{1}{x},$$

so our solution is $g(x) = x - \frac{1}{x}$. $\square$

In hindsight, our clue that our approach in Problem 19.15 might work was the fact that $1/x$ is cyclic. So, by letting $x = t$ and $x = 1/t$ in our original equation, we produce two equations with $g(t)$ and $g(1/t)$ on the left side, which allows us to eliminate $g(1/t)$.

In general, any time we have a functional equation in which an argument of a function is itself a cyclic function, we should consider using appropriate substitutions to create a system of equations as we did in Problem 19.15.

Note that we used a second variable (namely t) in Problem 19.15 to generate our equations, because this makes the substitutions easier to follow and less prone to error. However, this substitution is not necessary. If you feel comfortable going straight from the equation

$$g(x) + 2g\left(\frac{1}{x}\right) = -x + \frac{1}{x}$$

to the equation

$$g\left(\frac{1}{x}\right) + 2g(x) = -\frac{1}{x} + x,$$

then feel free to do so. However, for more complicated problems, the extra variable will help avoid some confusion and errors.

Sometimes it's not so obvious that a function is cyclic.

Problem 19.16: Suppose that f satisfies the functional equation

$$2f(x) + 3f\left(\frac{2x+29}{x-2}\right) = 100x + 80$$

for all $x \neq 2$. Find $f(3)$. *(Source: USAMTS)*

Solution for Problem 19.16: Since we want to find $f(3)$, we start with substituting $x = 3$:

$$2f(3) + 3f(35) = 380.$$

The term $f(35)$ appears, so next we try substituting $x = 35$:

$$2f(35) + 3f(3) = 3580.$$

Luckily, we obtain another equation involving $f(3)$ and $f(35)$. Solving this system of equations gives $f(3) = 1996$.

On second thought, were we really so lucky to get that second equation?

Concept: There are few coincidences in mathematics. When you think you have found one, look more closely.

Let $r(x) = \dfrac{2x+29}{x-2}$. Then we have

$$r(r(x)) = r\left(\frac{2x+29}{x-2}\right) = \frac{2 \cdot \frac{2x+29}{x-2} + 29}{\frac{2x+29}{x-2} - 2} = \frac{2(2x+29) + 29(x-2)}{2x+29 - 2(x-2)} = \frac{33x}{33} = x.$$

And now the mysterious "coincidence" is solved—the function $r(x)$ is cyclic with order 2. In particular, $r(r(3)) = r(35) = 3$, which explains why when we substituted $x = 35$, the term $f(3)$ appeared again. $\square$

Problem 19.17: Let $F(x)$ be a real-valued function defined for all real x except for $x = 0$ and $x = 1$ such that $F(x)$ satisfies the functional equation

$$F(x) + F\left(\frac{x-1}{x}\right) = 1 + x$$

for all x besides 0 and 1. Find all functions $F(x)$ satisfying these properties. *(Source: Putnam)*

Solution for Problem 19.17: We have an unusual argument of a function in the functional equation, so we check if it's cyclic. We let $r(x) = (x-1)/x$. Then we have

$$r(r(x)) = \frac{\frac{x-1}{x} - 1}{\frac{x-1}{x}} = \frac{x-1-x}{x-1} = -\frac{1}{x-1},$$

and

$$r(r(r(x))) = -\frac{1}{r(x)-1} = -\frac{1}{\frac{x-1}{x}-1} = -\frac{x}{x-1-x} = x.$$

Therefore, the function r has order 3.

We've used functions with order 2 to solve functional equations. Let's see if we can do the same with functions with order 3. First, we let $x = t$, so

$$F(t) + F\left(\frac{t-1}{t}\right) = 1 + t. \tag{19.1}$$

Next, to get the term $F(\frac{t-1}{t})$, we let $x = \frac{t-1}{t}$, which gives us

$$F\left(\frac{t-1}{t}\right) + F\left(-\frac{1}{t-1}\right) = 1 + \frac{t-1}{t}. \tag{19.2}$$

Now we have 2 equations, but 3 functional terms: $F(t)$, $F(\frac{t-1}{t})$, and $F(-\frac{1}{t-1})$. Here's where the cyclic function comes in. To produce another equation with the term $F(-\frac{1}{t-1})$, we let $x = -\frac{1}{t-1}$:

$$F\left(-\frac{1}{t-1}\right) + F(t) = 1 - \frac{1}{t-1}. \tag{19.3}$$

Success! We now have three equations and three functional terms. We can eliminate $F(\frac{t-1}{t})$ and $F(-\frac{1}{t-1})$ and solve for $F(t)$ by adding equations (19.1) and (19.3), subtracting (19.2) from the result, and then dividing by 2 to get

$$F(t) = \frac{1}{2}\left(1 + t + 1 - \frac{1}{t-1} - \left(1 + \frac{t-1}{t}\right)\right) = \frac{1}{2}\left(t + \frac{1}{t} - \frac{1}{t-1}\right) = \frac{t^3 - t^2 - 1}{2t(t-1)}.$$

Our last step is to check that this function satisfies the given equation:

$$\begin{aligned}
F(x) + F\left(\frac{x-1}{x}\right) &= \frac{x^3 - x^2 - 1}{2x(x-1)} + \frac{\left(\frac{x-1}{x}\right)^3 - \left(\frac{x-1}{x}\right)^2 - 1}{2 \cdot \frac{x-1}{x} \cdot \left(\frac{x-1}{x} - 1\right)} \\
&= \frac{x^3 - x^2 - 1}{2x(x-1)} + \frac{(x-1)^3 - x(x-1)^2 - x^3}{2x(x-1)(x-1-x)} \\
&= \frac{x^3 - x^2 - 1}{2x(x-1)} + \frac{-x^3 - x^2 + 2x - 1}{-2x(x-1)} \\
&= \frac{2x^3 - 2x}{2x(x-1)} \\
&= x + 1,
\end{aligned}$$

for all real x not equal to 0 or 1, so the equation is satisfied. $\square$

Exercises

19.4.1 Let $f(t) = \dfrac{t + \sqrt{3}}{1 - t\sqrt{3}}$. Prove that f is cyclic, and find its order.

19.4.2 The function f has the property that for each real number x in its domain, $1/x$ is also in its domain, and $f(x) + f\left(\frac{1}{x}\right) = x$. What is the largest set of real numbers that can be in the domain of f? *(Source: AMC 12)*

19.4.3 If $2f(x) + 3f(1-x) = 5x^2 - 3x - 11$ for all real x, then find $f(x)$.

19.4.4 Find all functions $f : \mathbb{R}\backslash\{0, 1\} \to \mathbb{R}$ such that $xf(x) + f\left(\frac{x}{x-1}\right) = 2x$.

19.4.5 If $f(x)$ is a real-valued function such that $f(x) = [f(1/x)]^2$, find the value of $[f(1977)]^2 - f(1977)$.

19.5 Summary

> **Definition:** A **functional equation** is an equation that includes a function.

Perhaps the most common tactic for working with functional equations is **substitution**. We choose values for the variables in the functional equation to learn more about the function.

If a functional equation has two variables, sometimes we can manipulate the equation so that the variables are separated on opposite sides of the equation. If the two sides then have the same form, we can often use this information to solve for the function.

We say that a function is **cyclic** with **order** n if n is the smallest positive integer such that $f^n(x) = x$ for all x. If one of the arguments of a function in a functional equation is cyclic, we can sometimes use the cyclic nature of this argument to create a system of equations that we can then solve for the function.

> **Important:** After you solve a functional equation, check that your solutions actually satisfy the given equation.

Extra! The **Cauchy Functional Equation** is

$$f(x + y) = f(x) + f(y).$$

This simple-looking equation is one of the most researched functional equations. In 1821, Cauchy proved that the only continuous functions $f : \mathbb{R} \to \mathbb{R}$ that satisfy this equation are functions of the form $f(x) = cx$ for some real constant c. But what if we allow non-continuous functions, or restrict the domain to $\mathbb{Q}$ (rational numbers) or $\mathbb{Z}$ (integers)? Then, are there other functions that satisfy the Cauchy Functional Equation? These, and many other related questions, have been researched for nearly two centuries.

Problem Solving Strategies

Concepts:

- If you can't find the answer to a functional equation question right away, explore the problem by trying various values for the variables.

- Whenever it looks like you have a great deal of algebra ahead of you, take a few minutes to look for another approach to the problem.

- There are few coincidences in mathematics. When you think you have found one, look more closely.

REVIEW PROBLEMS

19.18 Suppose f is a function with domain $\mathbb{R}$, and suppose the equation

$$f(x_1 + x_2 + x_3 + x_4 + x_5) = f(x_1) + f(x_2) + f(x_3) + f(x_4) + f(x_5) - 8$$

holds for all real numbers x_1, x_2, x_3, x_4, x_5. What is $f(0)$? *(Source: HMMT)*

19.19 Let f be a function satisfying $f(xy) = f(x)/y$ for all positive real numbers x and y. If $f(500) = 3$, what is the value of $f(600)$? *(Source: AMC 12)*

19.20 Find all functions $f : \mathbb{R} \to \mathbb{R}$ such that $f(x) + f(x + y) + f(x + 2y) = 6x + 6y$ for all real numbers x, y.

19.21 The function f has the property that, for all nonzero real numbers x, we have

$$f(x) + 2f\left(\frac{1}{x}\right) = 3x.$$

How many nonzero solutions are there to the equation $f(x) = f(-x)$? *(Source: AHSME)*

19.22 A sequence $a_1, a_2, a_3, \ldots$ satisfies $a_1 = 2$ and $a_{m+n} = 4^{mn}a_m a_n$ for all m and n. Determine a_n.

19.23 Let $u(x) = \dfrac{1}{1-x}$. Show that $u(x)$ is cyclic, and find its order.

19.24 Determine all functions that satisfy the identity $f(xy) = (y^2 - y + 1)f(x)$.

19.25 If $f(x)$ is defined for all integers $x \geq 0$, $f(a + b) = f(a) + f(b) - 2f(ab)$, and $f(1) = 1$, compute $f(2007)$.

19.26 Find all functions $f : \mathbb{R} \to \mathbb{R}$ such that $f(z) + zf(1 - z) = 1 + z$ for all real z.

19.27 Let $f(x) = \dfrac{x+1}{x+c}$. If $f(f(x)) = x$ for all x when both sides are defined, then find c.

19.28 The function f satisfies the functional equation $f(x) + f(y) = f(x + y) - xy - 1$ for every pair x, y of real numbers. If $f(1) = 1$, then find the number of integers $n \neq 1$ for which $f(n) = n$. *(Source: AHSME)*

19.29 Let f be a function defined on the set of all integers, and assume that it satisfies the following properties:

(i) $f(0) \neq 0$,

(ii) $f(1) = 3$,

(iii) $f(x)f(y) = f(x + y) + f(x - y)$ for all integers x and y.

Determine $f(7)$. *(Source: USAMTS)*

19.30 The sequence of numbers $\ldots, a_{-3}, a_{-2}, a_{-1}, a_0, a_1, a_2, a_3, \ldots$ is defined by $a_n - (n+1)a_{2-n} = (n+3)^2$ for all integers n. Calculate a_0. *(Source: COMC)*

19.31 The function $f(x)$ is defined for all real numbers $x \neq 0$, and satisfies $xf(x) + 2xf(-x) = -1$ for all $x \neq 0$. Find $f(x)$.

19.32 Find all functions $f : \mathbb{R} \to \mathbb{R}$ such that for all real x and y, we have $yf(2x) - xf(2y) = 8xy(x^2 - y^2)$. *(Source: Saudi Arabia)*

Challenge Problems

19.33 Let $f(x) = 4x - x^2$. Consider the sequence $x_1, x_2, x_3, \ldots$ where $x_i = f(x_{i-1})$ for $i > 1$. Compute the number of values x_1 such that x_1, x_2, and x_3 are distinct, but $x_i = x_3$ for all $i > 3$. *(Source: ARML)* **Hints:** 304

19.34 The function f has the property that, for each real number x, we have $f(x) + f(x - 1) = x^2$. If $f(19) = 94$, find $f(94)$. *(Source: AIME)* **Hints:** 200

19.35 Let $g(x) = \dfrac{x}{x + 1}$. Find $g^{2000}(2)$. **Hints:** 6

19.36 Let $F(z) = \frac{z+i}{z-i}$ for all complex numbers $z \neq i$, and let $z_n = F(z_{n-1})$ for all positive integers n. Given that $z_0 = \frac{1}{137} + i$, find z_{2002}. *(Source: AIME)* **Hints:** 239

19.37 The function f satisfies $f(0) = 0$, $f(1) = 1$, and $f\left(\dfrac{x + y}{2}\right) = \dfrac{f(x) + f(y)}{2}$ for all $x, y \in \mathbb{R}$. Show that $f(x) = x$ for all rational numbers x. **Hints:** 44

19.38 Let $f(x) = \dfrac{1}{1 + x}$. Find $f^{12}(x)$.

19.39 The real function f has the property that, whenever a, b, n are positive integers such that $a + b = 2^n$, then $f(a) + f(b) = n^2$. What is $f(2002)$? *(Source: HMMT)* **Hints:** 323

19.40 Solve the functional equation $f(x + t) - f(x - t) = 24xt$.

19.41 Find all polynomials $p(x)$ such that $(x + 3)p(x) = xp(x + 1)$ for all real x. **Hints:** 263, 31

19.42 Find all functions $f : \mathbb{R}\backslash\{0\} \to \mathbb{R}$ such that $f(a + b) - f(b) = -\dfrac{a}{b(a + b)}$, for all real values of a and b for which both sides of the equation are defined.

19.43   Prove that $f(n) = 1 - n$ is the only integer-valued function defined on the integers that satisfies the following conditions:

(i) $f(f(n)) = n$ for all integers n,

(ii) $f(f(n+2)+2) = n$ for all integers n,

(iii) $f(0) = 1$.

(Source: Putnam) **Hints:** 108

19.44★   The function $f(x,y)$ satisfies

(1) $f(0,y) = y + 1$,

(2) $f(x+1,0) = f(x,1)$,

(3) $f(x+1,y+1) = f(x, f(x+1,y))$

for all integers x and y. Find $f(4,1981)$. *(Source: IMO)*

19.45★   Find all polynomials p such that $p(x)+p(y)+p(z)+p(x+y+z) = p(x+y)+p(x+z)+p(y+z)+p(0)$ for all real numbers x, y, and z. **Hints:** 138

19.46★   Find all solutions to the functional equation $f(1-x) = f(x)+1-2x$. *(Source: M&IQ)* **Hints:** 48, 274, 314

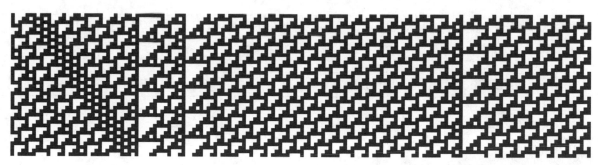

You sort of start thinking anything's possible if you've got enough nerve. – J. K. Rowling

CHAPTER **20**

_____Some Advanced Strategies

In this chapter, we discuss a variety of advanced problem solving strategies. Many of these are extensions of strategies that we have used several other times in the text.

20.1 Symmetry

We've worked a lot with symmetric sums already. In this section, we address symmetric systems and symmetric polynomials of one variable.

Problem 20.1: Determine $3x_4 + 2x_5$, if x_1, x_2, x_3, x_4, and x_5 satisfy the system of equations given below:

$$2x_1 + x_2 + x_3 + x_4 + x_5 = 6,$$
$$x_1 + 2x_2 + x_3 + x_4 + x_5 = 12,$$
$$x_1 + x_2 + 2x_3 + x_4 + x_5 = 24,$$
$$x_1 + x_2 + x_3 + 2x_4 + x_5 = 48,$$
$$x_1 + x_2 + x_3 + x_4 + 2x_5 = 96.$$

(Source: AIME)

Problem 20.2: Solve the following system of equations:

$$xy + xz = 8 - x^2,$$
$$xy + yz = 12 - y^2,$$
$$yz + zx = -4 - z^2.$$

Problem 20.3: Let x, y, and z be numbers satisfying $x + \frac{1}{y} = 4$, $y + \frac{1}{z} = 1$, and $z + \frac{1}{x} = \frac{7}{3}$. Find xyz. *(Source: AMC 12)*

Problem 20.4: Find all real solutions to $x^4 + (2-x)^4 = 34$. *(Source: HMMT)*

Problem 20.5: Suppose $x + \frac{1}{x} = 7$. Evaluate each of the following:

(a) $x^2 + \frac{1}{x^2}$.

(b) $x^3 + \frac{1}{x^3}$.

Problem 20.6: Let $f(x) = x^6 + x^4 - 115x^3 + x^2 + 1$. *(Source: Mandelbrot)*

(a) Show that if r is a root of $f(x)$, then $\frac{1}{r}$ is also a root.

(b) Consider the equation $f(x)/x^3 = 0$. What substitution would help you solve this equation?

(c) Find the sum of the real roots of the polynomial $x^6 + x^4 - 115x^3 + x^2 + 1$. *(Source: Mandelbrot)*

Problem 20.7: The polynomial function f satisfies $f(6+x) = f(6-x)$ for all real numbers x. Moreover, $f(x) = 0$ has exactly four distinct real roots. What is the sum of these roots?

Problem 20.1: Determine $3x_4 + 2x_5$, if x_1, x_2, x_3, x_4, and x_5 satisfy the system of equations given below:

$$
\begin{aligned}
2x_1 + x_2 + x_3 + x_4 + x_5 &= 6, \\
x_1 + 2x_2 + x_3 + x_4 + x_5 &= 12, \\
x_1 + x_2 + 2x_3 + x_4 + x_5 &= 24, \\
x_1 + x_2 + x_3 + 2x_4 + x_5 &= 48, \\
x_1 + x_2 + x_3 + x_4 + 2x_5 &= 96.
\end{aligned}
$$

(Source: AIME)

Solution for Problem 20.1: We can solve this system by eliminating or substituting for variables one-by-one. However, we notice that there is symmetry among the left sides of the equation, and exploiting this symmetry makes the solution much easier. If we don't see how to take advantage of the symmetry right away, we can let $s = x_1 + x_2 + x_3 + x_4 + x_5$, and our system becomes

$$
\begin{aligned}
x_1 + s &= 6, \\
x_2 + s &= 12, \\
x_3 + s &= 24, \\
x_4 + s &= 48, \\
x_5 + s &= 96.
\end{aligned}
$$

Now, our path is clear. We add all the equations to give $6s = 186$, so we have $s = 31$. We therefore have $x_4 = 48 - s = 17$ and $x_5 = 96 - s = 65$, so $3x_4 + 2x_5 = 3 \cdot 17 + 2 \cdot 65 = 181$. $\square$

Our solution to Problem 20.1 depended on our recognition of the pattern of the coefficients in the system of equations.

Concept: Look everywhere for patterns.

Problem 20.2: Solve the following system of equations:

$$xy + xz = 8 - x^2,$$
$$xy + yz = 12 - y^2,$$
$$yz + zx = -4 - z^2.$$

Solution for Problem 20.2: It's not immediately clear how we might make use of symmetry, but we can rewrite each equation with all the variables on the left sides in order to factor a variable out of each:

$$x^2 + xy + xz = x(x + y + z) = 8,$$
$$xy + y^2 + yz = y(x + y + z) = 12,$$
$$xz + yz + z^2 = z(x + y + z) = -4.$$

We let $S = x + y + z$, since we see this factor in all three equations, so we have

$$xS = 8,$$
$$yS = 12,$$
$$zS = -4.$$

Adding these equations gives us $(x + y + z)S = 16$. Aha—we have S again! Since $x + y + z = S$, our equation is now $S^2 = 16$, which means $S = \pm 4$.

If $S = x + y + z = 4$, then $x = 8/4 = 2$, $y = 12/4 = 3$, and $z = -4/4 = -1$. If $S = x + y + z = -4$, then $x = 8/(-4) = -2$, $y = 12/(-4) = -3$, and $z = -4/(-4) = 1$. So, our two solutions are $(x, y, z) = (2, 3, -1)$ and $(-2, -3, 1)$.

If you are very comfortable working with expressions with three variables, you might have added all the equations prior to any manipulation, giving

$$2(xy + yz + zx) = 16 - x^2 - y^2 - z^2.$$

Adding $x^2 + y^2 + z^2$ to both sides gives

$$x^2 + y^2 + z^2 + 2(xy + yz + zx) = 16.$$

Here's where familiarity with three-variable expressions comes in handy. The left side is just $(x+y+z)^2$. □

Note that we found a nice symmetry in the system of equations in Problem 20.2 after rearranging and factoring each equation.

> **Concept:** If you don't see how to work with an expression, try rewriting it. Many problems can be simplified by a different look at the same information.

Problem 20.3: Let x, y, and z be numbers satisfying

$$x + \frac{1}{y} = 4, \quad y + \frac{1}{z} = 1, \quad \text{and} \quad z + \frac{1}{x} = \frac{7}{3}.$$

Find xyz. *(Source: AMC 12)*

Solution for Problem 20.3: We could try isolating variables and substituting for them one-by-one, but that approach looks pretty messy. The similarity among the equations suggests we might be able to find a nicer solution.

We're looking for the value of xyz, which is an expression that will appear when we multiply the given equations together:

$$xyz + x + y + z + \frac{1}{x} + \frac{1}{y} + \frac{1}{z} + \frac{1}{xyz} = \frac{28}{3}.$$

In addition to xyz, we have terms we don't want. However, the unwanted terms all appear in the sum of the three given equations:

$$x + y + z + \frac{1}{x} + \frac{1}{y} + \frac{1}{z} = \left(x + \frac{1}{y}\right) + \left(y + \frac{1}{z}\right) + \left(z + \frac{1}{x}\right) = 4 + 1 + \frac{7}{3} = \frac{22}{3}.$$

Subtracting the sum of the given equations from their product, we have an equation involving only the product of the variables:

$$xyz + \frac{1}{xyz} = \frac{28}{3} - \frac{22}{3} = 2.$$

Letting $P = xyz$, we have $P + \frac{1}{P} = 2$. Rearranging this equation gives us $(P - 1)^2 = 0$, which means $P = 1$, so $xyz = 1$. $\square$

Not all problems have obvious symmetry to exploit. But in some problems, we can introduce a useful symmetry.

Problem 20.4: Find all real solutions to $x^4 + (2 - x)^4 = 34$. *(Source: HMMT)*

Solution for Problem 20.4: The obvious place to start with this equation is by expanding the left side. I'm guessing that's what you tried first, too. And you probably ran into a nasty fourth degree polynomial with no rational roots. Maybe you then recalled the Fundamental Theorem of Algebra, which tells us that any quartic with rational coefficients can be factored into the product of two quadratics with real coefficients. And maybe you even found those quadratics, and their roots.

Or maybe you saw how daunting all that would be, and looked for a slicker solution. We know that expanding the left side will be a chore, and the resulting quartic won't be much fun to work with. But maybe we can make a clever substitution to at least help us with the grindwork.

We see two terms raised to fourth powers. The sum of these terms is constant: $x + (2 - x) = 2$. This means x and $2 - x$ are symmetric around the number 1. Letting $y = 1 - x$ allows us to take advantage of this symmetry. We have $x = 1 - y$ and $2 - x = 1 + y$, so $x^4 + (2 - x)^4 = (1 - y)^4 + (1 + y)^4$. Doing this simplifies expanding the fourth powers considerably because the odd degree terms cancel out:

$$(1 - y)^4 + (1 + y)^4 = 1 - 4y + 6y^2 - 4y^3 + y^4 + 1 + 4y + 6y^2 + 4y^3 + y^4$$
$$= 2 + 12y^2 + 2y^4.$$

So, our equation now is $2 + 12y^2 + 2y^4 = 34$, so $y^4 + 6y^2 - 16 = 0$.

Our new equation has only terms with even exponents. This allows us to factor it as a quadratic equation in y^2, which gives us $(y^2 + 8)(y^2 - 2) = 0$. Since x is real, so is y. Therefore, we have $y = \pm\sqrt{2}$, which means $x = 1 \pm \sqrt{2}$ are the real solutions to the original equation. $\square$

Problem 20.5: Suppose $x + \frac{1}{x} = 7$. Evaluate each of the following:

(a) $x^2 + \frac{1}{x^2}$.

(b) $x^3 + \frac{1}{x^3}$.

Solution for Problem 20.5: We could start off by solving for x, but the possible values of x are irrational, which makes our arithmetic a bit messy. Instead, we try to manipulate the given equation to construct the expressions we intend to evaluate.

(a) We need squared terms, so we square the given equation, which gives us $x^2 + 2 + \frac{1}{x^2} = 49$. Subtracting 2 from both sides of the resulting equation, we get $x^2 + \frac{1}{x^2} = 47$.

(b) *Solution 1:* Proceeding as in (a), we cube the given equation, which gives $x^3 + 3x + \frac{3}{x} + \frac{1}{x^3} = 343$. We have cubed terms, but we also have other terms. Fortunately, we can evaluate $3x + \frac{3}{x}$ using the given equation:

$$x^3 + 3x + \frac{3}{x} + \frac{1}{x^3} = x^3 + \frac{1}{x^3} + 3\left(x + \frac{1}{x}\right) = x^3 + \frac{1}{x^3} + 3(7) = x^3 + \frac{1}{x^3} + 21.$$

So, we have $x^3 + \frac{1}{x^3} + 21 = 343$, which means $x^3 + \frac{1}{x^3} = 322$.

Solution 2: Building off our work from part (a), we note that $\left(x + \frac{1}{x}\right)\left(x^2 + \frac{1}{x^2}\right) = 7 \cdot 47$, so we have $x^3 + x + \frac{1}{x} + \frac{1}{x^3} = 329$. Subtracting $x + \frac{1}{x} = 7$ from this equation, we get $x^3 + \frac{1}{x^3} = 322$.

$\square$

We can also use manipulations similar to those in Problem 20.5 to simplify some problems involving one-variable polynomials.

Problem 20.6: Find the sum of the real roots of the polynomial $f(x) = x^6 + x^4 - 115x^3 + x^2 + 1$. *(Source: Mandelbrot)*

Solution for Problem 20.6: First, we try finding rational roots. By the Rational Root Theorem, the only possibilities are 1 and −1, and both fail. Guessing irrational roots seems impossible, so we'll have to look for some nice way to relate this sixth degree polynomial to something simpler.

> **Concept:** When stuck trying to deal with an algebraic expression, focus on any unusual aspects of the expression.

The most interesting aspect of the given polynomial is the symmetry of its coefficients; the coefficients read the same forwards as backwards. To investigate whether or not this is important, let's look at some simpler examples. We start by looking at quadratics whose coefficients read the same forwards as backwards. (Linear polynomials with this property are not very interesting; they all are of the form $rx + r$, which means -1 is the root of each such polynomial.)

So, we look for what's interesting about the roots of a quadratic of the form $ax^2 + bx + a$. We focus on what's important about the constant and quadratic coefficients being the same. This suggests considering Vieta's Formulas, which tell us that the product of the roots is $a/a = 1$. In other words, the roots are reciprocals of each other!

Hmmm.... Let's see if there's something similar true about our given sixth degree polynomial. Suppose r is a root of $f(x)$. We then have $f(r) = 0$, so

$$r^6 + r^4 - 115r^3 + r^2 + 1 = 0.$$

We want to see if $1/r$ is also a root. We have

$$f\left(\frac{1}{r}\right) = \left(\frac{1}{r}\right)^6 + \left(\frac{1}{r}\right)^4 - 115\left(\frac{1}{r}\right)^3 + \left(\frac{1}{r}\right)^2 + 1 = \frac{1}{r^6} + \frac{1}{r^4} - \frac{115}{r^3} + \frac{1}{r^2} + 1 = \frac{1 + r^2 - 115r^3 + r^4 + r^6}{r^6} = \frac{f(r)}{r^6} = 0.$$

Success! If r is a root of $f(x)$, then so is $1/r$.

But how can we use this information? We know that for each root, the reciprocal of that root is also a root of $f(x)$. So, we try rearranging the equation so that it contains both x and $\frac{1}{x}$. One way to do so is to divide $f(x) = 0$ by x^3, which gives us

$$x^3 + x - 115 + \frac{1}{x} + \frac{1}{x^3} = 0.$$

This introduces reciprocals, and highlights the fact that the coefficients read the same both forwards and backwards. We see that our new expression includes both $x + \frac{1}{x}$ and $x^3 + \frac{1}{x^3}$. We know how to relate these two expressions. We let $y = x + \frac{1}{x}$, and we cube y:

$$y^3 = x^3 + 3x + \frac{3}{x} + \frac{1}{x^3} = x^3 + \frac{1}{x^3} + 3\left(x + \frac{1}{x}\right) = x^3 + \frac{1}{x^3} + 3y.$$

So, $x^3 + \frac{1}{x^3} = y^3 - 3y$. Now, we can rewrite $x^3 + x - 115 + \frac{1}{x} + \frac{1}{x^3}$ in terms of y:

$$x^3 + x - 115 + \frac{1}{x} + \frac{1}{x^3} = \left(x^3 + \frac{1}{x^3}\right) + \left(x + \frac{1}{x}\right) - 115$$

$$= (y^3 - 3y) + y - 115 = y^3 - 2y - 115.$$

We therefore have $y^3 - 2y - 115 = 0$. Fortunately, there is a rational root, and we can factor the left side as $(y - 5)(y^2 + 5y + 23) = 0$. We need the sum of the real roots of x, and whenever x is real and nonzero, we know that y is also real. But the roots of $y^2 + 5y + 23$ are nonreal, so we are only concerned with the root $y = 5$, from which we have $x + \frac{1}{x} = 5$. Rearranging this equation gives $x^2 - 5x + 1 = 0$. Both roots of this quadratic equation are real, and by Vieta's Formulas, their sum is 5. $\square$

> **Concept:** The substitution $y = x + \frac{1}{x}$ can help us simplify problems involving polynomials in x in which the coefficients read the same forwards and backwards.

Problem 20.7: The polynomial function f satisfies $f(6+x) = f(6-x)$ for all real numbers x. Moreover, $f(x) = 0$ has exactly four distinct real roots. What is the sum of these roots?

Solution for Problem 20.7: Let r be a root of $f(x) = 0$. We can find another root in terms of r using $f(6+x) = f(6-x)$. Letting $r = 6 + x$, we have $x = r - 6$ and

$$f(r) = f(6+x) = f(6-x) = f(6-(r-6)) = f(12-r).$$

Since $f(r) = 0$, we have $f(12-r) = 0$ as well. Therefore, both r and $12-r$ are roots, which means we can group the roots of $f(x)$ into two pairs such that sum of each pair is 12. So, the sum of the roots of f is $2 \cdot 12 = 24$. $\square$

Exercises

20.1.1 Four positive integers are given. Select any three of these integers, find their average, and add this result to the fourth integer. In this fashion, the numbers 29, 23, 21, and 17 are obtained. Find the four original integers. *(Source: AHSME)*

20.1.2 Find the value of $r^3 + \dfrac{1}{r^3}$ given that $r + \dfrac{1}{r} = 3$. *(Source: AHSME)*

20.1.3 If $y = x + \frac{1}{x}$, then write the equation $x^4 + x^3 - 4x^2 + x + 1 = 0$ in terms of y. *(Source: AHSME)*

20.1.4 Solve the following system of equations:

$$ab = 1, \qquad bc = 2, \qquad cd = 3, \qquad de = 4, \qquad ea = 6.$$

(Source: Sweden)

20.1.5

(a) Show that the only solution to the system

$$x_1 + x_2 + x_3 = 0,$$
$$x_2 + x_3 + x_4 = 0,$$
$$\vdots$$
$$x_{98} + x_{99} + x_{100} = 0,$$
$$x_{99} + x_{100} + x_1 = 0,$$
$$x_{100} + x_1 + x_2 = 0.$$

is $x_1 = x_2 = x_3 = \cdots = x_{100} = 0$.

(b)★ In part (a), if there were 99 variables in 99 equations instead of 100 variables in 100 equations, would all of the variables still have to be equal to 0?

20.2 Substitution for Simplification

Way back in Chapter 1, we reviewed a most basic use of substitution: turning a system of two linear equations in two variables into a single one-variable linear equation. The substitution simplified the system of equations. In this section, we expand on this strategy, simplifying various equations and systems of equations with strategically chosen substitutions.

 Problems

Problem 20.8: Consider the following system of equations, where x and y are real numbers:

$$x + y + \sqrt{x + y} = 30,$$
$$x - y + \sqrt{x - y} = 12.$$

(a) Find *two* substitutions that simplify the system of equations.

(b) Solve each equation individually for your two new variables from (a).

(c) Solve the original system of equations.

Problem 20.9: In this problem, we find all solutions to the system of equations

$$2x^2 + 3y^2 - 4xy = 3,$$
$$2x^2 - y^2 = 7.$$

(a) Solve the second equation for $2x^2$ and substitute the result into the first equation.

(b) Can you solve the resulting equation in part (a) for x or y?

(c) Solve the problem.

(d) In this part, we find a second, more subtle, solution. Instead of eliminating a variable from the system, combine the equations in a way that eliminates the constants. In other words, combine the two equations into a new equation such that each nonzero term in the new equation has at least one variable in it. Use the new equation to solve the system.

Problem 20.10: In this problem, we find all ordered pairs of real numbers (x, y) that satisfy the system

$$x^3 - 4y^3 = 4,$$
$$3y^3 - x^2y + xy^2 = 1.$$

(a) Combine the equations to produce an equation that does not have a nonzero constant term.

(b) Suppose $(x, y) = (a, b)$ is a solution to your equation in part (a). Explain why $(x, y) = (ra, rb)$ is also a solution for any constant r.

(c) Suppose a and b are nonzero constants such that $(x, y) = (ra, rb)$ is a solution for any constant r. Notice that all of these solutions have the same ratio x/y. Use this fact to find a substitution that you can use to reduce the equation to a one-variable equation.

(d) Find all ordered pairs of real numbers (x, y) that satisfy the original system.

Problem 20.11: Consider the equation $\sqrt{3x^2 - 4x + 34} + \sqrt{3x^2 - 4x - 11} = 9$.

  (a) Set a and b equal to the two radical parts. Rewrite the original equation using a and b.

  (b) Find a second equation relating a^2 and b^2.

  (c) Solve the system of equations you came up with in parts (a) and (b).

  (d) Find all real values of x that satisfy the original equation.

Problem 20.12: Find the positive solution to

$$\frac{1}{x^2 - 10x - 29} + \frac{1}{x^2 - 10x - 45} - \frac{2}{x^2 - 10x - 69} = 0.$$

(Source: AIME)

Problem 20.13: Find all values of x such that $2x^2 + 3x + 3 = 7x\sqrt{x+1}$. *(Source: ARML)*

Problem 20.8: Solve the following system of equations, where x and y are real numbers:

$$x + y + \sqrt{x+y} = 30,$$
$$x - y + \sqrt{x-y} = 12.$$

Solution for Problem 20.8: There's no easy way to isolate x or y, and no easy way to eliminate the square roots. However, since $x + y$ is repeated in the first equation and $x - y$ repeated in the second, substitution might make the equations a lot easier to work with.

We could substitute for $x + y$ and $x - y$. However, this would still leave us with square roots in our equation. Instead, we let $u = \sqrt{x+y}$ and $v = \sqrt{x-y}$, which turns our equations into simple quadratics:

$$u^2 + u = 30,$$
$$v^2 + v = 12.$$

> **Concept:** Use substitutions to turn complicated equations into simple ones. Repeated expressions within an equation are a clue that substitution might be helpful.

The solutions to the first equation are $u = 5$ and $u = -6$. The solutions to the second equation are $v = 3$ and $v = -4$. However, u and v are equal to square roots of expressions, so they can't be negative. This means $u = 5$ and $v = 3$. Now, we have $\sqrt{x+y} = u = 5$ and $\sqrt{x-y} = v = 3$. Squaring both gives $x + y = 25$ and $x - y = 9$, from which we find that $(x, y) = (17, 8)$. $\square$

To solve Problem 20.8, we used two related substitutions. Though we introduced multiple variables, the result was still a significant simplification of the given problem.

> **Concept:** Multiple substitutions can make a complicated problem easier to solve.

Sometimes these substitutions aren't so obvious, though.

Problem 20.9: Find all solutions to the system of equations

$$2x^2 + 3y^2 - 4xy = 3,$$
$$2x^2 - y^2 = 7.$$

Solution for Problem 20.9: Solution 1: Substitute for x^2. We can't solve the first equation for x or y, but we can easily solve the second equation for either x^2 or y^2 in terms of the other. Since $2x^2$ appears in both equations, it's easiest to simply solve the second equation for $2x^2$, and substitute the result in the first equation.

Concept: Don't make problems harder than they have to be. When you have multiple options for substitution, choose the one that appears to offer the easiest path.

From the second equation, we have $2x^2 = 7 + y^2$, so substituting for $2x^2$ in the first equation gives us

$$4y^2 - 4xy = -4.$$

Dividing this equation by 4 gives $y^2 - xy = -1$. We still have two variables. We could solve $2x^2 = 7 + y^2$ for x and substitute for x, but this would both introduce radicals and fractions, not to mention a $\pm$ sign. So, back to the drawing board.

But not all the way back to the beginning. The equation $y^2 - xy = -1$ is linear in x, which means it is easy to solve this equation for x in terms of y, which gives

$$x = y + \frac{1}{y}.$$

Important: If you have an equation in a system that is linear in any variable, you can solve for that variable and substitute the result in the rest of the equations.

Substituting this expression for x into $2x^2 = 7 + y^2$ gives us

$$2y^2 + 4 + \frac{2}{y^2} = 7 + y^2.$$

Rearranging this equation gives us $y^4 - 3y^2 + 2 = 0$, which gives us $(y^2 - 1)(y^2 - 2) = 0$, so $y = \pm 1$ or $y = \pm \sqrt{2}$.

From these values of y, we find the four solutions $(x, y) = (2, 1), (-2, -1), \left(\frac{3}{2}\sqrt{2}, \sqrt{2}\right)$, and $\left(-\frac{3}{2}\sqrt{2}, -\sqrt{2}\right)$.

Solution 2: Turn the problem into one you know how to do. We have tackled equations like $6x^2 + xy - 15y^2 = 0$ by factoring the left side, or by letting $r = x/y$. In the present problem, both of our equations have the complication that there is a nonzero constant on the right side of the equation. So, we eliminate the constants on the right side in our system of equations by subtracting 3 times $2x^2 - y^2 = 7$ from 7 times $2x^2 + 3y^2 - 4xy = 3$ to give

$$8x^2 + 24y^2 - 28xy = 0.$$

Dividing by 4 gives $2x^2 + 6y^2 - 7xy = 0$. Now, we're in familiar territory. We factor this equation to find $(2x - 3y)(x - 2y) = 0$, from which we have $x = \frac{3}{2}y$ or $x = 2y$. Substituting $x = 2y$ into either of our original equations gives $(x, y) = (2, 1)$ and $(x, y) = (-2, -1)$ as solutions. Substituting $x = \frac{3}{2}y$ into either of our original equations gives $(x, y) = \left(\frac{3}{2}\sqrt{2}, \sqrt{2}\right)$, and $(x, y) = \left(-\frac{3}{2}\sqrt{2}, -\sqrt{2}\right)$ as solutions. $\square$

Our second solution is an example of a powerful problem solving strategy called **homogenization**. A **homogeneous equation** is one in which all nonzero terms have the same degree. So, $2x^2 + 6y^2 - 7xy = 0$ is homogeneous, but $2x^2 - y^2 = 7$ is not, because the 7 term has degree 0 while the other terms have degree 2. In our next problem, we explore why this strategy is so useful.

Problem 20.10: Find all ordered pairs of real numbers (x, y) that satisfy the system

$$x^3 - 4y^3 = 4,$$
$$3y^3 - x^2y + xy^2 = 1.$$

Solution for Problem 20.10: It's not too appealing to solve the first equation for x or y and substitute into the second equation, so we look for another approach. In the previous problem, eliminating the constants worked well, so we try the same here. Subtracting 4 times the second equation from the first equation gives

$$x^3 + 4x^2y - 4xy^2 - 16y^3 = 0.$$

Immediately, we see that $(0, 0)$ is a solution, and if x or y is nonzero, then the other must also be nonzero. We also see that all four nonzero terms have the same degree, so the equation is homogeneous. But how does that help?

Suppose we find one solution $(x, y) = (a, b)$ to this equation, so

$$a^3 + 4a^2b - 4ab^2 - 16b^3 = 0.$$

Because every term has degree 3, if we multiply x and y both by r, we multiply each term by r^3. So, if we let $(x, y) = (ra, rb)$, then

$$x^3 + 4x^2y - 4xy^2 - 16y^3 = (ra)^3 + 4(ra)^2(rb) - 4(ra)(rb)^2 - 16(rb)^3 = r^3(a^3 + 4a^2b - 4ab^2 - 16b^3) = 0.$$

Therefore, if $(x, y) = (a, b)$ is a solution of the equation, then so is $(x, y) = (ra, rb)$ for any constant r.

Concept: If we find a solution to a homogeneous equation, then multiplying every variable in this solution by the same constant produces another solution to the equation.

Of course, we still haven't found any solution besides $(0, 0)$ to our equation. But we now know that if $(x, y) = (a, b)$ is a solution, then so are all pairs of the form (ra, rb). Investigating what's special about all these pairs, we find that, as long as a, b, and r are nonzero, they all have the same ratio of x to y, namely $(ra)/(rb) = a/b$. Aha! Now, our problem is reduced to finding all possible values of this ratio. Letting $x/y = k$ gives us $x = yk$, and substituting this into our equation gives

$$k^3y^3 + 4k^2y^3 - 4ky^3 - 16y^3 = 0.$$

For $y \neq 0$, we can divide by y^3 to give $k^3 + 4k^2 - 4k - 16 = 0$. Finally, an equation we know how to solve!

> **Concept:** When working with a two-variable homogeneous equation, it's often help-ful to set the ratio of the variables equal to a new variable and write the equation in terms of the new variable.

Factoring this polynomial gives $(k-2)(k+2)(k+4) = 0$, so $k = 2$, $k = -2$, or $k = -4$.

We handle these three cases separately. When $k = 2$, we have $x = 2y$, and substituting this into the first equation gives $4y^3 = 4$. The only real solution to this equation is $y = 1$, which gives $(2,1)$ as a solution to the system. Similarly, $x = -2y$ gives us $-12y^3 = 4$, so $y = -\frac{1}{\sqrt[3]{3}}$, which gives us $\left(\frac{2}{\sqrt[3]{3}}, -\frac{1}{\sqrt[3]{3}}\right)$ as a solution. Finally, letting $x = -4y$ gives us $-68y^3 = 4$, so $y = -\frac{1}{\sqrt[3]{17}}$, which means $\left(\frac{4}{\sqrt[3]{17}}, -\frac{1}{\sqrt[3]{17}}\right)$ is a solution. Those last two solutions would have been very difficult to guess! $\square$

Problem 20.11: Solve $\sqrt{3x^2 - 4x + 34} + \sqrt{3x^2 - 4x - 11} = 9$.

Solution for Problem 20.11: We have a nasty equation with a repeated expression, so we look for a substitution to simplify the equation.

Solution 1: Let $y = 3x^2 - 4x - 11$, so our equation is

$$\sqrt{y + 45} + \sqrt{y} = 9.$$

We know how to tackle this equation! We isolate $\sqrt{y+45}$ to give $\sqrt{y+45} = 9 - \sqrt{y}$, and squaring this gives

$$y + 45 = y - 18\sqrt{y} + 81.$$

Simplifying this equation gives $18\sqrt{y} = 36$, from which we find $y = 4$. Since $y = 3x^2 - 4x - 11$, we have $4 = 3x^2 - 4x - 11$, which means $3x^2 - 4x - 15 = 0$. The roots of this quadratic are 3 and $-5/3$. Because we squared an equation to find these solutions, we must check if they are extraneous. Both solutions satisfy the original equation, so our solutions are $x = 3$ and $x = -5/3$.

Solution 2: Our first solution takes advantage of the fact that the expressions under the radicals differ only by a constant. Perhaps we can assign a variable to each radical expression and find another way to take advantage of this fact. We let $a = \sqrt{3x^2 - 4x + 34}$ and $b = \sqrt{3x^2 - 4x - 11}$, so that we are given the equation $a + b = 9$. We haven't yet used the similarity between the expressions under the radicals, so we focus on that.

> **Concept:** When stuck on a problem, focus on information you haven't used yet.

Because the expressions under the radicals differ by a constant, the difference of the squares $a^2 - b^2$ is a constant:

$$a^2 - b^2 = (3x^2 - 4x + 34) - (3x^2 - 4x - 11) = 45.$$

This makes us happy—now we can factor! We have $(a - b)(a + b) = 45$, and we know $a + b = 9$, so we have $a - b = 5$. Since $a + b = 9$ and $a - b = 5$, we have $(a, b) = (7, 2)$. Now, we have $\sqrt{3x^2 - 4x + 34} = a = 7$. Squaring this gives $3x^2 - 4x + 34 = 49$, which results in the same values of x as in our first solution. $\square$

Sometimes, more is better. While it's not always a good idea to bring lots of extra variables into a problem, our first solution to Problem 20.11 shows that a little substitution can go a long way. Our second solution shows that a little more substitution can sometimes go even further. The primary advantage of the second solution over the first is that the necessary algebraic manipulations after substitution are much simpler.

Here's another problem in which we have multiple options for substitution, and choosing the right one helps simplify the post-substitution algebra.

Problem 20.12: Find the positive solution to

$$\frac{1}{x^2 - 10x - 29} + \frac{1}{x^2 - 10x - 45} - \frac{2}{x^2 - 10x - 69} = 0.$$

(Source: AIME)

Solution for Problem 20.12: This problem seems intimidating at first glance. Not only does the given equation involve several quadratics, but those quadratics are in the denominators of fractions.

Fortunately, the quadratics are related in that they differ only by constants. This allows us to substitute for all three quadratics at once. We could choose many possible substitutions such as $a = x^2 - 10x$ or $b = x^2 - 10x - 29$. We choose to let $y = x^2 - 10x - 69$ with the goal of keeping the algebra as simple as possible. This makes one of the denominators equal y, and leaves no negative constants in the denominators after substitution:

$$\frac{1}{y + 40} + \frac{1}{y + 24} - \frac{2}{y} = 0.$$

Now we just get rid of the fractions by multiplying both sides by the product of the denominators, which gives $y(y + 24) + y(y + 40) - 2(y + 24)(y + 40) = 0$. Simplifying the left side gives $64y + 1920 = 0$, so $y = -30$.

From our original $y = x^2 - 10x - 69$, we have $x^2 - 10x - 69 = -30$. Rearranging and factoring gives $(x + 3)(x - 13) = 0$, so $x = 13$ is the positive solution. $\square$

Problem 20.13: Find all values of x such that $2x^2 + 3x + 3 = 7x\sqrt{x + 1}$. *(Source: ARML)*

Solution for Problem 20.13: We might start by squaring the equation, but that produces a huge mess on the left side. At a loss for anything else to do, we might try it anyway. What results is a quartic polynomial that has no rational roots.

Before we give up, we look for a helpful substitution. What makes this equation nasty is the square root expression, so we substitute $y = \sqrt{x + 1}$, which gives us $2x^2 + 3x + 3 = 7xy$. This, at least, is a

CHAPTER 20. SOME ADVANCED STRATEGIES

quadratic. We might then try to use the quadratic formula to find x in terms of y. But that looks messy, and besides, we already have a simple relationship between x and y, namely $y = \sqrt{x+1}$. So, what should we do?

We look for other ways to use our substitution, and see that $x + 1$ appears on the left side, too. We have $2x^2 + 3x + 3 = 2x^2 + 3(x+1)$. We can use our substitution again! Since $y = \sqrt{x+1}$, we have $y^2 = x+1$, so our equation is

$$2x^2 + 3y^2 = 7xy.$$

Now we're in familiar territory. Subtracting $7xy$ from both sides and factoring gives $(2x - y)(x - 3y) = 0$, so $2x = y$ or $x = 3y$.

If $2x = y$, then we have $2x = \sqrt{x+1}$. Squaring both sides gives $4x^2 = x + 1$, which gives us the solutions $x = (1 \pm \sqrt{17})/8$. Because $\sqrt{x+1} = 2x$, we can't have $x < 0$, so we discard the extraneous root $(1 - \sqrt{17})/8$.

If $x = 3y$, then we have $x = 3\sqrt{x+1}$. Squaring both sides gives $x^2 = 9x + 9$, which gives us the solutions $x = (9 \pm 3\sqrt{13})/2$. Once again, x must be positive, so we discard $(9 - 3\sqrt{13})/2$ as extraneous.

Finally, our two solutions are $x = (1 + \sqrt{17})/8$ and $x = (9 + 3\sqrt{13})/2$. $\square$

Exercises

20.2.1 Find all solutions to the equation $7^{2a} = 45 \cdot 7^a + 4 \cdot 7^2$.

20.2.2 Find all real numbers y such that $y^8 + 108 = 21y^4$.

20.2.3

(a) Find all values of r such that $(r^2 + 5r)(r^2 + 5r + 3) = 4$.

(b) If x is a positive integer and $x(x + 1)(x + 2)(x + 3) + 1 = 379^2$, compute x. (*Source: ARML*)

20.2.4 Find all solutions to the equation $(6\sqrt{4 - x} - 17)x = 36\sqrt{4 - x} - 68$.

20.2.5 Find all values of x such that $4x^2 - 6x - 41 + \dfrac{1}{2x^2 - 3x - 19} = 0$.

20.2.6 Find all values of x such that $\sqrt{x^2 + 17x + 59} - 2\sqrt{x^2 + 17x - 85} = 3$.

20.2.7 Find all ordered pairs (x, y) such that

$$2x^4 + 5x^3y + 45xy^3 = 34x^2y^2 + 18y^4 \qquad \text{and} \qquad 2x - 3y = 7.$$

20.3 Method of Undetermined Coefficients

We have used the **Method of Undetermined Coefficients** several times already in this book. Simply put, the Method of Undetermined Coefficients is the use of variables to represent coefficients whose values we don't yet know.

Problems

Problem 20.14: The polynomial $x^2 + 2x + 5$ is a factor of $x^4 + px^2 + q$. Find p and q. *(Source: AHSME)*

Problem 20.15: Let $f(x)$ and $g(x)$ be two quadratic polynomials such that $f(x) = 2x^2 - 3x + 5$ and $f(g(x)) = 18x^4 - 21x^2 + 10$. Find $g(x)$.

Problem 20.16: Suppose p is a polynomial such that $p(1) = 3$ and $p(x) = p(2x) - 6x^2 + 3x$.

  (a) Show that the degree of $p(x)$ is greater than 1.

  (b) Show that the degree of $p(x)$ is no greater than 2.

  (c) Let $p(x) = a_2 x^2 + a_1 x + a_0$. Find $p(x)$.

Problem 20.17: Real roots x_1 and x_2 of the equation $x^5 - 55x + 21 = 0$ satisfy $x_1 x_2 = 1$. In this problem, we find x_1 and x_2. *(Source: Bosnia and Herzegovnia)*

  (a) What must the constant term of a monic quadratic be if the product of its roots is 1?

  (b) Suppose $x^5 - 55x + 21$ is the product of a quadratic and a cubic. Suppose x_1 and x_2 are the roots of the quadratic. Why can we let the quadratic and cubic be monic? What are the constant terms of the quadratic and the cubic?

  (c) Find the rest of the coefficients of the quadratic, and find x_1 and x_2.

Problem 20.14: The polynomial $x^2 + 2x + 5$ is a factor of $x^4 + px^2 + q$. Find p and q. *(Source: AHSME)*

Solution for Problem 20.14: *Solution 1:* Regardless of the values of p and q, when we divide $x^4 + px^2 + q$ by $x^2 + 2x + 5$, the leading term of the quotient is x^2. So, we let the quotient be $x^2 + ax + b$, so that

$$x^4 + px^2 + q = (x^2 + 2x + 5)(x^2 + ax + b) = x^4 + (a + 2)x^3 + (2a + b + 5)x^2 + (5a + 2b)x + 5b.$$

Equating the coefficients on the far left side above to the corresponding coefficients on the far right, we have a system of equations:

$$a + 2 = 0,$$
$$2a + b + 5 = p,$$
$$5a + 2b = 0,$$
$$5b = q.$$

From the first of these equations, we have $a = -2$. The third equation then gives us $-10 + 2b = 0$, so $b = 5$. The second equation therefore gives $p = 2a + b + 5 = -4 + 5 + 5 = 6$, and the last equation yields $q = 5b = 25$. Now, we have the coefficients of the second quadratic factor, and the quartic itself:

$$(x^2 + 2x + 5)(x^2 - 2x + 5) = x^4 + 6x^2 + 25.$$

In practice, we can often use a number of shortcuts on a problem like this one. We start as before, with

$$x^4 + px^2 + q = (x^2 + 2x + 5)(x^2 + ax + b).$$

Rather than expanding the product on the right, we focus on important pieces of information that will help us find a, b, p, and q quickly. For example, the coefficient of x^3 on the left is 0. The only x^3 terms in the expansion of the right side come from $(2x)(x^2)$ and $(x^2)(ax)$, so we must have $2x^3 + ax^3 = 0$, which means $a = -2$. This means we have

$$x^4 + px^2 + q = (x^2 + 2x + 5)(x^2 - 2x + b).$$

Similarly, we can focus on the x term of the expansion on the right to deduce that $(2x)(b) + (5)(-2x) = 0$, which gives us $b = 5$, so $x^4 + px^2 + q = (x^2 + 2x + 5)(x^2 - 2x + 5)$. Now, looking at the constant terms gives $q = 25$, and looking at the x^2 terms (or simply letting $x = 1$) gives us $p = 6$.

Solution 2: While the Method of Undetermined Coefficients provides a nice solution to this problem, some students might cleverly sidestep its use by noting that $f(x) = x^4 + px^2 + q$ is an even polynomial function. So, $f(x) = f(-x)$, which means that if r and s are roots of $f(x)$, then $-r$ and $-s$ are also roots.

We are given that $f(x) = (x^2 + 2x + 5)g(x)$ for some quadratic polynomial $g(x)$. Since the roots r and s of $x^2 + 2x + 5$ don't sum to 0, they are not opposites. Therefore, the roots of $g(x)$ must be $-r$ and $-s$. Because r and s are roots of $x^2 + 2x + 5$, we have $r + s = -2$ and $rs = 5$. Therefore, we have $(-r) + (-s) = 2$ and $(-r)(-s) = 5$. So, the sum of the roots of $g(x)$ is 2 and the product of the roots is 5, which means $g(x) = x^2 - 2x + 5$. We can now compute $f(x) = (x^2 - 2x + 5)(x^2 + 2x + 5) = x^4 + 6x^2 + 25$. $\square$

While our second solution to Problem 20.14 is slicker than the first, the first solution is much more broadly applicable to problems. Specifically, it is an example of the Method of Undetermined Coefficients:

> **Concept:** When working with an unknown polynomial, it is often useful to assign variables to the coefficients of the polynomial, and then construct equations in terms of those variables.

> **Problem 20.15:** Let $f(x)$ and $g(x)$ be two quadratic polynomials such that $f(x) = 2x^2 - 3x + 5$ and $f(g(x)) = 18x^4 - 21x^2 + 10$. Find $g(x)$.

Solution for Problem 20.15: Let $g(x) = ax^2 + bx + c$, where a, b, and c are constants. Then

$$\begin{aligned}
f(g(x)) &= f(ax^2 + bx + c) \\
&= 2(ax^2 + bx + c)^2 - 3(ax^2 + bx + c) + 5 \\
&= 2a^2x^4 + 4abx^3 + (4ac + 2b^2 - 3a)x^2 + (4bc - 3b)x + 2c^2 - 3c + 5.
\end{aligned}$$

Equating the coefficients of this expression to the corresponding coefficients of the given $18x^4 - 21x^2 + 10$ gives us a system of equations:

$$2a^2 = 18,$$
$$4ab = 0,$$
$$4ac + 2b^2 - 3a = -21,$$
$$4bc - 3b = 0,$$
$$2c^2 - 3c + 5 = 10.$$

From the first equation, we have $a = \pm 3$. From the second equation, we have $ab = 0$, but $a \neq 0$, so $b = 0$. The last equation gives us $2c^2 - 3c - 5 = 0$, so $(c+1)(2c-5) = 0$, which means $c = -1$ or $c = 5/2$. Finally, we match the possible values of a and c using the third equation, $4ac - 3a = -21$, to find that $(a, c) = (3, -1)$ or $(-3, 5/2)$.

Each of the two possible ordered triples $(a, b, c) = (3, 0, -1)$ and $(a, b, c) = (-3, 0, 5/2)$ gives us a solution to the problem. The two possible polynomials g are $g(x) = 3x^2 - 1$ or $g(x) = -3x^2 + \frac{5}{2}$. $\square$

Now, we take a look at a functional equation that can be solved using the Method of Undetermined Coefficients.

Problem 20.16: Suppose p is a polynomial such that $p(1) = 3$ and $p(x) = p(2x) - 6x^2 + 3x$. Find $p(x)$.

Solution for Problem 20.16: By considering the leading terms of both sides of $p(x) = p(2x) - 6x^2 + 3x$, we suspect that the degree of p is 2. The degree of p cannot be less than 2, since this would make the degree of the left side less than that of the right. To show that $\deg p$ cannot be greater than 2, we let the leading term of $p(x)$ be $a_n x^n$, where $n > 2$. The leading term of $p(x)$ then is $a_n x^n$, while that of $p(2x) - 6x^2 + 3x$ is $2^n a_n x^n$. We must therefore have $a_n = 2^n a_n$, which means $a_n = 0$. However, this is a contradiction, since we can't have $a_n = 0$ if $a_n x^n$ is the leading term of $p(x)$. So, the degree of p cannot be greater than 2.

Since $\deg p$ cannot be greater than 2 or less than 2, it must equal 2, so we have $p(x) = a_2 x^2 + a_1 x + a_0$. Substituting this into $p(x) = p(2x) - 6x^2 + 3x$ gives

$$a_2 x^2 + a_1 x + a_0 = (4a_2 - 6)x^2 + (2a_1 + 3)x + a_0.$$

Equating the coefficients of x^2, we see that $4a_2 - 6 = a_2$, so $a_2 = 2$. From the coefficients of x, we have $a_1 = 2a_1 + 3$, so $a_1 = -3$. We now have $p(x) = 2x^2 - 3x + a_0$. How will we get a_0?

Concept: When stuck on a problem, try focusing on information in the problem that you haven't used yet.

We haven't used the fact that $p(1) = 3$, so we try applying that. We have $p(1) = 2 - 3 + a_0 = a_0 - 1$, so we must have $a_0 - 1 = 3$. Therefore, we have $a_0 = 4$, so $p(x) = 2x^2 - 3x + 4$. $\square$

Problem 20.17: Real roots x_1 and x_2 of the equation $x^5 - 55x + 21 = 0$ satisfy $x_1 x_2 = 1$. Find x_1 and x_2. *(Source: Bosnia and Herzegovnia)*

Solution for Problem 20.17: Let $p(x) = x^5 - 55x + 21$. A quick check reveals no rational roots, so it's not clear how we can factor $p(x)$. However, the fact that $x_1 x_2 = 1$ gives us some information about a quadratic factor of $p(x)$. Specifically, because $x_1 x_2 = 1$, we know that x_1 and x_2 are the roots of a quadratic of the form $x^2 + ax + 1$ for some constant a. This quadratic must be a factor of $p(x)$ because its roots are also roots of $p(x)$. So, we must have

$$x^5 - 55x + 21 = (x^2 + ax + 1)(bx^3 + cx^2 + dx + f).$$

From the leading term of both sides, we have $b = 1$, and from the constant term, we have $f = 21$. So, we have

$$x^5 - 55x + 21 = (x^2 + ax + 1)(x^3 + cx^2 + dx + 21).$$

There's no x^4 term on the left, so the two x^4 terms in the expansion of the right side, which are ax^4 and cx^4, must cancel each other. This gives us $a = -c$, and we have

$$x^5 - 55x + 21 = (x^2 - cx + 1)(x^3 + cx^2 + dx + 21).$$

We could continue this reasoning coefficient by coefficient, but it's not so obvious which one will give us immediately useful information next. So, we multiply out the right side, and we have

$$x^5 - 55x + 21 = x^5 + (d - c^2 + 1)x^3 + (21 - cd + c)x^2 + (-21c + d)x + 21.$$

Equating coefficients of like terms gives us the system

$$d - c^2 + 1 = 0,$$
$$21 - cd + c = 0,$$
$$-21c + d = -55.$$

The first equation gives us $d = c^2 - 1$. Substituting this into the third equation gives us $c^2 - 21c - 1 = -55$. Solving this quadratic gives us $c = 3$ or $c = 18$. If $c = 3$, then $a = -3$ and $d = 8$. The triple $(a, c, d) = (-3, 3, 8)$ does indeed satisfy all three equations. If $c = 18$, then $a = -18$ and $d = 323$. However, this solution does not satisfy the second equation in our system.

So, we have one solution for (a, c, d), and we have our factorization:

$$x^5 - 55x + 21 = (x^2 - 3x + 1)(x^3 + 3x^2 + 8x + 21).$$

The roots of $x^2 - 3x + 1$ have product 1. From the quadratic formula, we find that these roots are $(3 \pm \sqrt{5})/2$. As an Exercise, you'll show that the other factor, $x^3 + 3x^2 + 8x + 21$, does not have a pair of real roots with product 1. $\square$

> **Sidenote:** This probably isn't the last you'll see of the Method of Undetermined Coefficients. You'll see it again if you study differential equations in calculus.

Exercises

20.3.1 If a and b are integers such that $x^2 - x - 1$ is a factor of $ax^3 + bx^2 + 1$, then find b. (*Source: AHSME*)

20.3.2 If $3x^2 + kxy - 2y^2 - 7x + 7y - 6$ is the product of two linear factors with integral coefficients, find the value of k. (*Source: CMO*)

20.3.3 Let $p(x)$ be a fourth degree polynomial with leading coefficient 2 such that $p(-2) = 34$, $p(-1) = 10$, $p(1) = 10$, and $p(2) = 34$. Find $p(0)$.

20.3.4 Show that we have found the only possible solution to Problem 20.17 by showing that the polynomial $x^3 + 3x^2 + 8x + 21$ does not have a pair of real roots with product 1.

20.3.5★ Solve for all complex numbers z such that $z^4 + 4z^2 + 6 = z$. (*Source: HMMT*) **Hints:** 2

20.4 Constructing Polynomials From Roots

We've discussed several techniques and theorems that help us determine relationships between polynomials and their roots. However, we've focused mostly on finding those roots. In this section, we start with information about the roots of polynomials, and construct the polynomials themselves, or solve other related problems.

Problems

Problem 20.18: The roots of the polynomial $P(x)$ are $2, 3, 7$, and -5. The constant term of $P(x)$ is 105.

(a) What is the product of the roots of $P(x)$?

(b) What is the coefficient of x^4 of $P(x)$?

(c) What is $P(x)$?

(d) What is the sum of the coefficients of $P(x)$?

Problem 20.19: Let a, b, and c be the roots of $f(x) = x^3 + 8x + 5$.

(a) Construct the monic polynomial with roots $1 + a$, $1 + b$, and $1 + c$.

(b) Find the value of $(1 + a)(1 + b)(1 + c)$.

(c) Find the value of $\dfrac{1}{1+a} + \dfrac{1}{1+b} + \dfrac{1}{1+c}$.

Problem 20.20: Let the roots of the $ax^2 + bx + c = 0$ be r and s. Find the monic quadratic whose roots are $ar + b$ and $as + b$. *(Source: AHSME)*

Problem 20.21: Consider the following system of equations:

$$x + y + z = 2,$$
$$xy + xz + yz = -19,$$
$$xyz = -20.$$

(a) Find a monic cubic polynomial that has x, y, and z as its roots.

(b) Solve the system of equations.

Problem 20.22: If $P(x)$ denotes a polynomial of degree n such that $P(k) = \frac{k}{k+1}$ for $k = 0, 1, 2, \ldots, n$, determine $P(n + 1)$. *(Source: USAMO)*

Extra! Suppose we have a degree n polynomial whose graph passes through the $n + 1$ points $(x_0, y_0), (x_1, y_1), \ldots (x_n, y_n)$. We can find the polynomial as follows. First, we note that the graph of $f(x) = (x - x_0)g_1(x) + y_0$, where $g_1(x)$ is any polynomial, will pass through (x_0, y_0). Now, we need $g_1(x)$ to be a degree $n - 1$ polynomial whose graph passes through $(x_1, \frac{y_1-y_0}{x_1-x_0}), (x_2, \frac{y_2-y_0}{x_2-x_0}), \ldots, (x_n, \frac{y_n-y_0}{x_n-x_0})$. So, we just repeat the process, by letting $g_1(x) = (x - x_1)g_2(x) + \frac{y_1-y_0}{x_1-x_0}$. Continuing in this manner, we can find the polynomial $f(x)$ whose graph passes through all $n + 1$ given points.

Problem 20.23: Suppose $P(x)$ is a polynomial with finite degree such that $P(1) = 1$ and

$$\frac{P(2x)}{P(x+1)} = 8 - \frac{56}{x+7}$$

for all real x for which both sides are defined.

(a) Show that $P(x) = xQ(x)$ for some polynomial $Q(x)$.

(b) Express $Q(x)$ as the product of a linear factor and a polynomial $R(x)$.

(c) Express $R(x)$ as the product of a linear factor and a polynomial $S(x)$.

(d) Find $P(-1)$. *(Source: HMMT)*

Problem 20.18: The roots of the polynomial $P(x)$ are $2, 3, 7$, and -5. The constant term of $P(x)$ is 105. What is the sum of the coefficients of $P(x)$?

Solution for Problem 20.18: Let $P(x) = ax^4 + bx^3 + cx^2 + dx + 105$. By Vieta, the product of the roots is $105/a$. The product of the given roots is $2 \cdot 3 \cdot 7 \cdot (-5) = -210$, so we have $105/a = -210$, which means $a = -1/2$. We could plow ahead with Vieta's Formulas to find the rest of the coefficients, but we just want the sum of the coefficients. We know that the sum of the coefficients equals $P(1)$, so maybe we can find a faster way than Vieta's Formulas to construct an expression for $P(x)$ that we can use to find $P(1)$.

Looking for another way to construct $P(x)$, we note that we have all the roots, and thus know all the monic linear factors of $P(x)$. We also have the leading coefficient of $P(x)$, so we can use the Factor Theorem to write

$$P(x) = a(x-2)(x-3)(x-7)(x+5) = -\frac{1}{2}(x-2)(x-3)(x-7)(x+5).$$

We could expand this product to find the coefficients, but instead we note that $P(1)$ equals the sum of the coefficients of $P(x)$. So, the sum of the coefficients of $P(x)$ is $P(1) = -\frac{1}{2}(1-2)(1-3)(1-7)(1+5) = 36$. $\square$

> **Important:** Vieta's Formulas and the Factor Theorem can be quite helpful in constructing polynomials from information about their roots.

Problem 20.19: Let a, b, and c be the roots of $f(x) = x^3 + 8x + 5$.

(a) Find the value of $(1+a)(1+b)(1+c)$.

(b) Find the value of $\dfrac{1}{1+a} + \dfrac{1}{1+b} + \dfrac{1}{1+c}$.

Solution for Problem 20.19:

(a) We could expand the product $(1+a)(1+b)(1+c)$ and use Vieta's Formulas to find the values of the symmetric sums. However, if we could find a polynomial with roots $1+a$, $1+b$, and $1+c$, we could

---

use Vieta on that polynomial to quickly find the desired product. So, we seek the polynomial $g(y)$ with roots $1 + a$, $1 + b$, and $1 + c$. We relate the roots of $g(y)$ to those of $f(x)$ by letting $y = 1 + x$. This makes $y = 1 + a$ when $x = a$, and $y = 1 + b$ when $x = b$, and $y = 1 + c$ when $x = c$.

From $y = 1 + x$, we have $x = y - 1$. Because the roots of $f(x)$ are a, b, and c, we have $f(x) = f(y-1) = 0$ when y equals $a + 1$, $b + 1$, or $c + 1$. So, $f(y-1)$ produces our desired polynomial $g(y)$:

$$g(y) = f(y - 1) = (y - 1)^3 + 8(y - 1) + 5 = y^3 - 3y^2 + 11y - 4.$$

Since $(1+a)(1+b)(1+c)$ is the product of the roots of $g(y)$, Vieta gives us $(1+a)(1+b)(1+c) = -\frac{-4}{1} = 4$. Note that our answer equals $-f(-1)$. See if you can figure out why this is not a coincidence.

(b) *Solution 1: Find a polynomial with roots $\frac{1}{a+1}$, $\frac{1}{b+1}$, and $\frac{1}{c+1}$.* The polynomial $g(y) = y^3 - 3y^2 + 11y - 4$ from part (a) has roots $a+1$, $b+1$, and $c+1$. We seek the polynomial whose roots are the reciprocals of these roots, since we can then use Vieta on the new polynomial to find our desired sum. So, just as we let $y = x + 1$ in part (a), we let $z = \frac{1}{y}$ here. Since $g(y) = 0$ if $y = a + 1$, we have $g(\frac{1}{z}) = 0$ if $z = \frac{1}{a+1}$. Similarly, $\frac{1}{b+1}$ and $\frac{1}{c+1}$ are also solutions to the equation $g(\frac{1}{z}) = 0$. So, we find $g(\frac{1}{z})$:

$$g\left(\frac{1}{z}\right) = \frac{1}{z^3} - \frac{3}{z^2} + \frac{11}{z} - 4 = \frac{1 - 3z + 11z^2 - 4z^3}{z^3}.$$

Notice that the coefficients of the numerator are the same as those of $g(y)$, but *with their order reversed.* (You'll prove in the Exercises that this is not a coincidence.)

We have $g(\frac{1}{z}) = 0$ if and only if $1 - 3z + 11z^2 - 4z^3 = 0$, and $g(\frac{1}{z}) = 0$ for $z = \frac{1}{a+1}$, $z = \frac{1}{b+1}$, and $z = \frac{1}{c+1}$, so $1 - 3z + 11z^2 - 4z^3$ has roots $\frac{1}{a+1}$, $\frac{1}{b+1}$, and $\frac{1}{c+1}$. Finally, Vieta's Formulas tell us that the sum of these roots is $-11/(-4) = 11/4$.

Solution 2: Use Vieta's Formulas with $g(y)$ from part (a). The polynomial

$$g(y) = y^3 - 3y^2 + 11y - 4$$

from part (a) has roots $a + 1$, $b + 1$, and $c + 1$. If we let $a + 1 = r$, $b + 1 = s$, and $c + 1 = t$, we can write our desired expression in terms of symmetric sums of r, s, and t:

$$\frac{1}{a + 1} + \frac{1}{b + 1} + \frac{1}{c + 1} = \frac{1}{r} + \frac{1}{s} + \frac{1}{t} = \frac{st + rt + rs}{rst}.$$

By Vieta, we have $st + rt + rs = 11$ and $rst = -(-4) = 4$, so our desired sum is $11/4$.

□

Our key step in Problem 20.19 was transforming one polynomial into another that allows us to compute the desired expression directly with Vieta's Formulas. See if you can use the polynomial transformation tactics we developed in Problem 20.19 on the next problem.

Problem 20.20: Let the roots of the $ax^2 + bx + c = 0$ be r and s. Find the monic quadratic whose roots are $ar + b$ and $as + b$. *(Source: AHSME)*

Solution for Problem 20.20: We use our work in Problem 20.19 as guide. We seek a monic quadratic with roots $ar + b$ and $as + b$, so we let $y = ax + b$, which means that $y = ar + b$ when $x = r$ and $y = as + b$ when

$x = s$. Solving $y = ax + b$ for x gives $x = (y - b)/a$. Substituting this expression for x into $ax^2 + bx + c$ gives

$$ax^2 + bx + c = a\left(\frac{y-b}{a}\right)^2 + b\left(\frac{y-b}{a}\right) + c$$
$$= \frac{1}{a}y^2 - \frac{2b}{a}y + \frac{b^2}{a} + \frac{b}{a}y - \frac{b^2}{a} + c$$
$$= \frac{1}{a}y^2 - \frac{b}{a}y + c.$$

We have $ax^2 + bx + c = 0$ for $x = r$ and $x = s$, and we have $ax^2 + bx + c = \frac{1}{a}y^2 - \frac{b}{a}y + c$ when $y = ax + b$. So, when $x = r$, we have $y = ar + b$ and $\frac{1}{a}y^2 - \frac{b}{a}y + c = ax^2 + bx + c = 0$, which means that $ar + b$ is a root of $\frac{1}{a}y^2 - \frac{b}{a}y + c$. Similarly, $as + b$ is also a root of $\frac{1}{a}y^2 - \frac{b}{a}y + c$. Therefore, multiplying this quadratic by a gives us our desired monic quadratic, $y^2 - by + ca$. $\square$

Problem 20.21: Solve the following system of equations:

$$x + y + z = 2,$$
$$xy + xz + yz = -19,$$
$$xyz = -20.$$

Solution for Problem 20.21: This system of equations initially looks quite frightening. While we could grind out a series of substitutions to find the answer, we'd prefer to find a nicer solution.

The left sides of the system are the elementary symmetric sums of three variables, which makes us think of Vieta's Formulas. So, we let x, y, and z be roots of a cubic polynomial. We let the leading coefficient of the polynomial be 1, so that $-(x + y + z)$ gives us the quadratic coefficient, $xy + yz + zx$ gives us the linear coefficient, and $-xyz$ is the constant. Letting the variable be t, we therefore have

$$(t - x)(t - y)(t - z) = t^3 - 2t^2 - 19t + 20.$$

Factoring the polynomial on the right, we find that $t = -4$, $t = 1$, and $t = 5$ are its roots. We can assign these three to x, y, and z in any order. So, the six solutions to the system of equations are $(x, y, z) = (-4, 1, 5), (1, -4, 5), (-4, 5, 1), (1, 5, -4), (5, 1, -4)$, and $(5, -4, 1)$. More succinctly, we can say that these six solutions are the six **permutations** of $(-4, 1, 5)$. $\square$

Our fairly simple solution to Problem 20.21 comes from recognizing that the relationships between symmetric sums and roots of a polynomial work both ways. We can calculate symmetric sums of roots of a given polynomial, or we can construct a polynomial whose roots are the variables in a system of symmetric sum equations.

Concept: Just as we can find the values of symmetric sums given a polynomial, we can construct polynomials given symmetric sums.

Problem 20.22: If $P(x)$ denotes a polynomial of degree n such that $P(k) = \frac{k}{k+1}$ for $k = 0, 1, 2, \ldots, n$, determine $P(n + 1)$. *(Source: USAMO)*

Solution for Problem 20.22: It is tempting to set

$$Q(x) = P(x) - \frac{x}{x+1},$$

and then analyze $Q(x)$, since we are given $n + 1$ values of x for which $Q(x) = 0$. But $Q(x)$ is not a polynomial, so we cannot use all our polynomial tools on $Q(x)$. However, if we write the right side of the above equation as a rational function, we have

$$P(x) - \frac{x}{x+1} = \frac{(x+1)P(x) - x}{x+1}.$$

The numerator is a polynomial, so we instead let $R(x) = (x+1)P(x) - x$, and analyze $R(x)$.

Because $P(x) = \frac{x}{x+1}$ for $x = 0, 1, 2, \ldots, n$, but $x + 1 \neq 0$ for all of these values, we have

$$R(x) = (x+1)P(x) - x = (x+1) \cdot \frac{x}{x+1} - x = 0$$

for $x = 0, 1, 2, \ldots, n$. Furthermore, since $P(x)$ has degree n, the polynomial $R(x)$ has degree $n + 1$. We have already found all $n + 1$ roots of $R(x)$, so we have

$$R(x) = cx(x-1)(x-2)\cdots(x-n) \tag{20.1}$$

for some constant c. Now, we're in familiar territory. All we have to do is find c. We don't know $P(x)$ for any more values of x, but we can eliminate $P(x)$ from $R(x) = (x+1)P(x) - x$ by letting $x = -1$, which gives us $R(-1) = 1$. So, letting $x = -1$ in Equation (20.1), we get

$$R(-1) = c \cdot (-1) \cdot (-2) \cdot (-3) \cdots (-1-n) = c(-1)^{n+1}(n+1)!,$$

so $c = \dfrac{R(-1)}{(-1)^{n+1}(n+1)!} = \dfrac{(-1)^{n+1}}{(n+1)!}$, and

$$R(x) = \frac{(-1)^{n+1}}{(n+1)!} \cdot x(x-1)(x-2)\cdots(x-n).$$

Then, we have

$$R(n+1) = \frac{(-1)^{n+1}}{(n+1)!} \cdot (n+1) \cdot n \cdot (n-1) \cdots 1 = (-1)^{n+1}.$$

Solving $R(x) = (x+1)P(x) - x$ for $P(x)$ gives $P(x) = \dfrac{R(x) + x}{x+1}$ for $x \neq -1$, so

$$P(n+1) = \frac{(-1)^{n+1} + n + 1}{n+2} = \begin{cases} \frac{n}{n+2} & \text{if } n \text{ is even,} \\ 1 & \text{if } n \text{ is odd.} \end{cases}$$

$\square$

Problem 20.23: Suppose $P(x)$ is a polynomial with finite degree such that $P(1) = 1$ and

$$\frac{P(2x)}{P(x+1)} = 8 - \frac{56}{x+7}$$

for all real x for which both sides are defined. Find $P(-1)$. *(Source: HMMT)*

CHAPTER 20. SOME ADVANCED STRATEGIES

Solution for Problem 20.23: The left side is written as a ratio of polynomials. We begin by rewriting the right side that way so that we can compare apples to apples:

$$\frac{P(2x)}{P(x+1)} = \frac{8x}{x+7}.$$

We'd like to use the information that $P(1) = 1$. We could do so by letting $x = 1/2$ or by letting $x = 0$. We choose the latter since it makes the right side of our equation 0, and thus identifies a root of $P(x)$. Letting $x = 0$ gives us $P(0)/P(1) = 0$, so $P(0) = 0$.

Since $x = 0$ is a root of $P(x)$, we know that x is a factor of $P(x)$. So, we let $P(x) = xQ(x)$, and we have

$$\frac{8x}{x+7} = \frac{P(2x)}{P(x+1)} = \frac{2xQ(2x)}{(x+1)Q(x+1)}, \qquad \text{so} \qquad \frac{Q(2x)}{Q(x+1)} = \frac{4(x+1)}{x+7}.$$

This looks familiar. Similar to what we did in our first step, we let $x = -1$ to find a root of $Q(x)$.

> **Concept:** If a tactic gives some information in a problem, but doesn't completely solve it, try using the tactic again.

We find that $Q(-2)/Q(0) = 0$, so $Q(-2) = 0$. This means $Q(x) = (x+2)R(x)$ for some polynomial $R(x)$. Substituting for $Q(x)$, we get

$$\frac{4(x+1)}{x+7} = \frac{Q(2x)}{Q(x+1)} = \frac{(2x+2)R(2x)}{(x+3)R(x+1)}, \qquad \text{so} \qquad \frac{R(2x)}{R(x+1)} = \frac{2(x+3)}{x+7}.$$

Next, we let $x = -3$ to get $R(-6)/R(-2) = 0$, so $R(-6) = 0$. This means $R(x) = (x+6)S(x)$ for some polynomial $S(x)$. Substituting for $R(x)$, we get

$$\frac{2(x+3)}{x+7} = \frac{R(2x)}{R(x+1)} = \frac{(2x+6)S(2x)}{(x+7)S(x+1)}, \qquad \text{so} \qquad \frac{S(2x)}{S(x+1)} = 1,$$

which means that $S(2x) = S(x+1)$ for all x.

We've reduced the problem to a much simpler equation. Here are two ways to use $S(2x) = S(x+1)$ to show that $S(x)$ is constant for all x:

- *Method 1: Experiment.* Letting $x = 2$ in $S(2x) = S(x+1)$ gives $S(4) = S(3)$. Letting $x = 3$ gives $S(6) = S(4)$, and letting $x = 5$ gives $S(10) = S(6)$. Continuing in this way, we can find infinitely many values of x such that $S(x) = S(3)$. This means that the polynomial $S(x) - S(3)$ has infinitely many roots. We know that $S(x)$ is not the zero polynomial, since $S(2x)/S(x+1) = 1$, so $S(x)$ has finite degree. Therefore, the fact that $S(x) - S(3)$ has infinitely many roots tells us that $S(x) - S(3) = 0$ for all x. This means that $S(x) = S(3)$ for all x, so $S(x)$ is constant.

- *Method 2: Undetermined coefficients.* We know that $S(x)$ cannot be the 0 polynomial, since we would then have division by 0 at various points in the problem. Therefore, we can let $S(x) = a_n x^n + a_{n-1}x^{n-1} + \cdots + a_1 x + a_0$, where $a_n \neq 0$, so $\deg S = n$. The leading coefficient of $S(2x)$ then is $2^n a_n$ and the leading coefficient of $S(x+1)$ is a_n. If $S(2x) = S(x+1)$ for all x, we must then have $2^n a_n = a_n$. Since $a_n \neq 0$, we must have $2^n = 1$, so $n = 0$, which means $S(x)$ is a constant.

Now, we can find $P(x)$. We have

$$P(x) = xQ(x) = x(x+2)R(x) = x(x+2)(x+6)S(x) = cx(x+2)(x+6).$$

Because $P(1) = 1$, we have $c = 1/21$, so

$$P(-1) = \frac{(-1)(-1+2)(-1+6)}{21} = -\frac{5}{21}.$$

See if you can find another solution to this problem by using the Method of Undetermined Coefficients from the beginning, in the spirit of our solution to Problem 20.16. $\square$

Problem 20.23 is a great example of the importance of perseverance in problem solving.

Concept: Don't give up!

Exercises

20.4.1 Compute $\dfrac{1}{r-1} + \dfrac{1}{s-1} + \dfrac{1}{t-1}$ given that r, s, and t are the roots of $x^3 - 2x^2 + 3x - 4 = 0$.

20.4.2 Let p, q, r, s, t satisfy the three equations

$$p^3 + sp + t = 1,$$
$$q^3 + sq + t = 1,$$
$$r^3 + sr + t = 1.$$

If p, q, and r are distinct, evaluate $p^2 + q^2 + r^2$ in terms of s and t.

20.4.3 Let $f(x) = a_n x^n + a_{n-1} x^{n-1} + \cdots + a_1 x + a_0$ and $g(x) = a_0 x^n + a_1 x^{n-1} + \cdots + a_{n-1} x + a_n$, where $a_0 a_n \neq 0$. Show that each root of f is the reciprocal of a root of g.

20.4.4 In part (a) of Problem 20.19, we let a, b, and c be the roots of $f(x) = x^3 + 8x + 5$. We used a transformation and Vieta's Formulas to evaluate $(1+a)(1+b)(1+c)$. We then noticed that $(1+a)(1+b)(1+c) = -f(-1)$. Why is this not a coincidence?

20.4.5★ Let $p(x)$ be a degree 2 polynomial such that $p(1) = 1$, $p(2) = 3$, and $p(3) = 2$. Then $p(p(x)) = x$ has four real solutions. Find the only such solution that is not an integer. *(Source: Mandelbrot)* **Hints:** 114

20.4.6★ Let $P(x)$ be the polynomial of minimal degree that satisfies

$$P(k) = \frac{1}{k(k+1)}$$

for $k = 1, 2, 3, \ldots, 10$. Find $P(11)$. *(Source: Thomas Mildorf)* **Hints:** 76

20.5 Common Divisors of Polynomials

A common divisor of two integers is an integer that divides both integers. We have a similar definition for a common divisor of two polynomials.

> **Definition:** Let $f(x)$ and $g(x)$ be polynomials. A **common divisor** of $f(x)$ and $g(x)$ is a polynomial $p(x)$ such that $p(x)$ is a factor of both $f(x)$ and $g(x)$.

Polynomials with common polynomial divisors of degree greater than 0 share all the roots of those common divisors. For example, the quadratics $x^2 - 3x + 2$ and $x^2 - 4x + 3$ are both divisible by $x - 1$, and they share $x = 1$ as a root. This means that when we are looking for common roots of polynomials, we can find them by finding their common divisors. Similarly, we can use common roots to find common divisors.

> **Concept:** Finding common roots of polynomials is equivalent to finding common divisors of the polynomials.

Problems

> **Problem 20.24:** Let $f(x)$, $g(x)$, and $h(x)$ be polynomials such that $h(x)$ is a factor of both $f(x)$ and $g(x)$. Prove that $h(x)$ is a factor of $a(x)f(x) + b(x)g(x)$ for any polynomials $a(x)$ and $b(x)$.

> **Problem 20.25:** Let $p(x) = x^2 + bx + c$, where b and c are integers. If $p(x)$ is a factor of both $x^4 + 6x^2 + 25$ and $3x^4 + 4x^2 + 28x + 5$, then find b and c. (*Source: AHSME*)

> **Problem 20.26:** For what values of b do the equations $1988x^2 + bx + 8891 = 0$ and $8891x^2 + bx + 1988 = 0$ have a common root? (*Source: CMO*)

> **Problem 20.27:** The equations $x^3 + ax^2 + 7x - 6 = 0$ and $x^3 - x^2 + bx + 3 = 0$ have two common roots for some integers a and b. Find a and b.

> **Problem 20.24:** Let $f(x)$, $g(x)$, and $h(x)$ be polynomials such that $h(x)$ is a factor of both $f(x)$ and $g(x)$. Prove that $h(x)$ is a factor of $a(x)f(x) + b(x)g(x)$ for any polynomials $a(x)$ and $b(x)$.

Solution for Problem 20.24: Since $h(x)$ is a factor of both $f(x)$ and $g(x)$, we know that there are some polynomials $q_1(x)$ and $q_2(x)$ such that

$$f(x) = q_1(x)h(x) \qquad \text{and} \qquad g(x) = q_2(x)h(x).$$

Therefore, we have

$$a(x)f(x) + b(x)g(x) = a(x)q_1(x)h(x) + b(x)q_2(x)h(x) = [a(x)q_1(x) + b(x)q_2(x)]h(x),$$

so $h(x)$ is a factor of $a(x)f(x) + b(x)g(x)$. $\square$

> **Important:** If $f(x)$, $g(x)$, and $h(x)$ are polynomials such that $h(x)$ is a factor of both $f(x)$ and $g(x)$, then $h(x)$ is a factor of $a(x)f(x) + b(x)g(x)$ for any polynomials $a(x)$ and $b(x)$.

Let's put this fact to work on some problems.

Problem 20.25: Let $p(x) = x^2 + bx + c$, where b and c are integers. If $p(x)$ is a factor of both $x^4 + 6x^2 + 25$ and $3x^4 + 4x^2 + 28x + 5$, then find b and c. *(Source: AHSME)*

Solution for Problem 20.25: At first, we might approach this problem by trying to find rational roots of both polynomials, and use these to find a common quadratic factor. Unfortunately, this strategy fails, because neither $x^4 + 6x^2 + 25$ nor $3x^4 + 4x^2 + 28x + 5$ has any rational roots. We could try to factor each polynomial as a product of quadratics using the Method of Undetermined Coefficients, but instead, we note that the result of Problem 20.24 offers a more attractive strategy. Perhaps we can combine the two polynomials to produce a third polynomial that will allow us to find the desired quadratic quickly.

Specifically, we know that if r and s are constants, then any polynomial that divides both $x^4 + 6x^2 + 25$ and $3x^4 + 4x^2 + 28x + 5$ also divides

$$r(x^4 + 6x^2 + 25) + s(3x^4 + 4x^2 + 28x + 5).$$

Choosing $r = 3$ and $s = -1$ eliminates the x^4 term and leaves us

$$3(x^4 + 6x^2 + 25) - (3x^4 + 4x^2 + 28x + 5) = 14x^2 - 28x + 70 = 14(x^2 - 2x + 5).$$

Therefore, the polynomial $p(x) = x^2 - 2x + 5$ divides both $x^4 + 6x^2 + 25$ and $3x^4 + 4x^2 + 28x + 5$, so $b = -2$ and $c = 5$. We can check our answer by performing polynomial division to find

$$x^4 + 6x^2 + 25 = (x^2 - 2x + 5)(x^2 + 2x + 5),$$
$$3x^4 + 4x^2 + 28x + 5 = (x^2 - 2x + 5)(3x^2 + 6x + 1).$$

$\square$

Problem 20.26: For what values of b do the equations $1988x^2 + bx + 8891 = 0$ and $8891x^2 + bx + 1988 = 0$ have a common root? *(Source: CMO)*

Solution for Problem 20.26: We let $f(x) = 1988x^2 + bx + 8891$ and $g(x) = 8891x^2 + bx + 1988$. Finding a common root of $f(x)$ and $g(x)$ is equivalent to finding a common linear factor of the two polynomials. Any common factor of $f(x)$ and $g(x)$ is also a factor of the difference $g(x) - f(x)$, which conveniently eliminates b and leaves us

$$g(x) - f(x) = 6903x^2 - 6903 = 6903(x^2 - 1) = 6903(x - 1)(x + 1).$$

Therefore, if $f(x)$ and $g(x)$ have a common linear factor, it is either $x-1$ or $x+1$. So, the common root of $f(x)$ and $g(x)$ is either $x = 1$ or $x = -1$. When $x = 1$ is the common root, we have $b = -1988 - 8891 = -10879$. When $x = -1$ is the common root, we have $b = 1988 + 8891 = 10879$. These are the two values of b for which the given equations have a common root. $\square$

Problem 20.27: The polynomials $f(x) = x^3 + ax^2 + 7x - 6$ and $g(x) = x^3 - x^2 + bx + 3$ have two common roots for some integers a and b. Find a and b.

Solution for Problem 20.27: The given cubic polynomials have the same leading coefficient, so their difference is a quadratic whose roots are the two common roots of the cubics:

$$(x^3 + ax^2 + 7x - 6) - (x^3 - x^2 + bx + 3) = (a + 1)x^2 + (7 - b)x - 9.$$

Unfortunately, we still can't find a and b, so we need more information.

Constructing a second polynomial with the common roots of the given cubics might help. Just as we eliminated the cubic term in our first step, we could also eliminate the constant term. We have

$$f(x) + 2g(x) = x^3 + ax^2 + 7x - 6 + 2(x^3 - x^2 + bx + 3) = x[3x^2 + (a - 2)x + (2b + 7)].$$

The common roots of $f(x)$ and $g(x)$ must also be roots of $f(x) + 2g(x)$. Since x is clearly not a factor of either of the cubics, the roots of $3x^2 + (a - 2)x + (2b + 7)$ must be the common roots of the given cubics.

We now have two quadratics, $(a + 1)x^2 + (7 - b)x - 9$ and $3x^2 + (a - 2)x + (2b + 7)$, with the same roots. This means that one of them is a constant times the other. In other words, their corresponding coefficients are proportional:

$$\frac{a + 1}{3} = \frac{7 - b}{a - 2} = \frac{-9}{2b + 7}.$$

From $\frac{a+1}{3} = \frac{7-b}{a-2}$, we find that

$$b = 7 - \frac{1}{3}(a + 1)(a - 2) = \frac{-a^2 + a + 23}{3}.$$

From $\frac{a+1}{3} = \frac{-9}{2b+7}$, we have $(a+1)(2b+7) = -27$. Substituting for b and simplifying gives $2a^3 - 69a - 148 = 0$. Factoring gives $(a + 4)(2a^2 - 8a - 37) = 0$. The discriminant of the quadratic is $(-8)^2 + 4 \cdot 2 \cdot 37 = 360$, which is not a perfect square. Therefore, the only integer solution for a is $a = -4$. Substituting this into our expression above for b gives $b = 1$. We can check our work by factoring the given polynomials for these values of a and b:

$$x^3 - 4x^2 + 7x - 6 = (x - 2)(x^2 - 2x + 3),$$
$$x^3 - x^2 + x + 3 = (x + 1)(x^2 - 2x + 3).$$

□

Exercises

20.5.1 Determine all real numbers a such that the two polynomials $x^2 + ax + 1$ and $x^2 + x + a$ have at least one real root in common. *(Source: CMO)*

20.5.2 Find a and b so that the equations $x^3 + ax^2 + 11x + 6 = 0$ and $x^3 + bx^2 + 14x + 8 = 0$ have two roots in common.

20.6 Symmetric Sums Revisited

We finish our study of polynomials by taking a look at more complex problems involving symmetric sums. We've already seen with Vieta's Formulas that elementary symmetric sums of the roots of a polynomial are related to the coefficients of the polynomial. In this section, we'll frequently combine Vieta with the following concept:

> **Important:** We can think of elementary symmetric sums as building blocks because any symmetric polynomial can be built from them.

Problems

Problem 20.28: Let a and b be the roots of the equation $2x^2 - 10x + 23 = 0$.
(a) Find the value of $a^2 + b^2$.
(b) Find the value of $a^4 + b^4$.

Problem 20.29: If $a + b = 1$ and $a^2 + b^2 = 2$, then find $a^3 + b^3$. *(Source: AHSME)*

Problem 20.30: In this problem, we find all ordered pairs of solutions (x, y) that satisfy

$$2x + 3y + 6xy = 9,$$
$$2x^2 y + 3xy^2 = \frac{10}{3}.$$

(a) Factor the expression $2x^2 y + 3xy^2$.
(b) Make substitutions that turn the left sides of the two equations into symmetric polynomials.
(c) Solve the equations you formed in part (b), then solve the original system.

Problem 20.31: Suppose that the sum of the squares of two complex numbers x and y is 7 and the sum of their cubes is 10. What is the largest real value that $x + y$ can have? *(Source: AIME)*

Problem 20.32: In this problem, we solve the system of equations

$$x + y + z = 8,$$
$$x^2 + y^2 + z^2 = 62,$$
$$\frac{1}{x} + \frac{1}{y} + \frac{1}{z} = -\frac{1}{42}.$$

(a) Find the values of $xy + yz + zx$ and xyz.
(b) Find all solutions (x, y, z) of the system of equations above.

Problem 20.33: Let a, b, c, d be complex numbers satisfying

$$a + b + c + d = a^3 + b^3 + c^3 + d^3 = 0.$$

Prove that at least one pair of $a, b, c,$ and d must add up to 0. *(Source: Tournament of Towns)*

Problem 20.28: Let a and b be the roots of the equation $2x^2 - 10x + 23 = 0$. Find the value of $a^4 + b^4$.

Solution for Problem 20.28: The quadratic formula gives us $x = \frac{5}{2} \pm \frac{i\sqrt{21}}{2}$ as the roots. We could raise these to the fourth power using the Binomial Theorem and add the results. However, all that arithmetic (not to mention the radicals and imaginary numbers) leaves a lot of room for error. Let's see if we can find an approach that requires less complicated arithmetic.

By Vieta, the sum and product of the roots of $2x^2 - 10x + 23 = 0$ are $a + b = 5$ and $ab = \frac{23}{2}$.

Solution 1: We first find $a^2 + b^2$ using the given equations, and hope to use that as a stepping stone to $a^4 + b^4$. We need squares, so we square $a + b$ and find that $a^2 + b^2 = (a + b)^2 - 2ab = 5^2 - 2 \cdot \frac{23}{2} = 2$. Squaring $a^2 + b^2$ now gives us something very close to what we want:

$$2^2 = (a^2 + b^2)^2 = a^4 + 2a^2b^2 + b^4 = a^4 + b^4 + 2(ab)^2.$$

Substituting for ab, we have $2^2 = a^4 + b^4 + 2\left(\frac{23}{2}\right)^2 = a^4 + b^4 + \frac{529}{2}$, so $a^4 + b^4 = 2^2 - \frac{529}{2} = -\frac{521}{2}$.

Our first solution builds up to $a^4 + b^4$ by first finding $a^2 + b^2$. However, a straightforward attempt to expand $(a + b)^4$ to get $a^4 + b^4$ leads to a different set of manipulations.

Solution 2: We start by expanding $(a + b)^4$ in order to get an expression with a^4 and b^4 terms:

$$625 = (a + b)^4 = a^4 + 4a^3b + 6a^2b^2 + 4ab^3 + b^4.$$

This expansion includes terms we don't need. In particular, we have

$$625 = a^4 + b^4 + 2ab(2a^2 + 3ab + 2b^2) = a^4 + b^4 + 23(2a^2 + 3ab + 2b^2).$$

We need to determine the value of $2a^2 + 3ab + 2b^2$. The squares lead us to squaring $a + b$. We have $(a + b)^2 = a^2 + 2ab + b^2$, so $2a^2 + 3ab + 2b^2$ is ab less than $2(a + b)^2$:

$$2a^2 + 3ab + 2b^2 = 2(a + b)^2 - ab = 2 \cdot 5^2 - \frac{23}{2} = \frac{77}{2}.$$

We therefore have $625 = a^4 + b^4 + 23 \cdot \frac{77}{2}$, so $a^4 + b^4 = 625 - 23 \cdot \frac{77}{2} = -\frac{521}{2}$. $\square$

Problem 20.29: If $a + b = 1$ and $a^2 + b^2 = 2$, then find $a^3 + b^3$. *(Source: AHSME)*

Solution for Problem 20.29: We could find a and b, and then compute $a^3 + b^3$ directly. However, if we instead try to build $a^3 + b^3$ from $a + b$ and $a^2 + b^2$, we will reduce the amount of computation we must perform, and reduce the possibility of making an arithmetic error.

We need a sum of cubes, so the natural step is to multiply $a + b$ and $a^2 + b^2$ to produce the cubes. This gives us

$$1 \cdot 2 = (a + b)(a^2 + b^2) = a^3 + a^2b + ab^2 + b^3 = a^3 + b^3 + ab(a + b). \tag{20.2}$$

The value of ab would finish off the problem. We know we get an ab from squaring $a + b$. This also gives us $a^2 + b^2$, when we know equals 2. So, we have $(a + b)^2 = a^2 + b^2 + 2ab = 2 + 2ab$. Because $a + b = 1$, we have $1 = 2 + 2ab$, so $ab = -1/2$. Equation (20.2) now gives us

$$2 = a^3 + b^3 + ab(a + b) = a^3 + b^3 + \left(-\frac{1}{2}\right)(1) = a^3 + b^3 - \frac{1}{2},$$

so $a^3 + b^3 = \frac{5}{2}$. $\square$

Notice that in each of our first two problems, we squared expressions or multiplied expressions without knowing for sure that the result would be useful. We saw that these manipulations would produce some terms we needed, and hoped we'd be able to find a way to deal with the other terms that were produced.

> **Concept:** You don't have to see a path all the way to the end of a solution before taking your first step. Experiment! Use what you need as a guide to work with what you have.

> **Problem 20.30:** Find all ordered pairs of solutions (x, y) that satisfy
> $$2x + 3y + 6xy = 9,$$
> $$2x^2y + 3xy^2 = \frac{10}{3}.$$

Solution for Problem 20.30: What's wrong with this start:

> **Bogus Solution:** We apply Simon's Favorite Factoring Trick to the first equation to get $(2x + 1)(3y + 1) = 10$. So, we can use the factors of 10 to find possible values of x and y.

The problem here is that x and y are not necessarily integers, so the prime factorization of 10 doesn't help us find x and y.

Instead, we turn our factoring skills to the second equation, and find $2x^2y + 3xy^2 = xy(2x + 3y)$. The expressions xy, $2x$, and $3y$ are all present in the left side of the first equation in the problem. With this in mind, we let $t = 2x$ and $u = 3y$, which allows us to write the left sides of both equations as symmetric polynomials:

$$t + u + tu = 9,$$
$$\frac{tu(t + u)}{6} = \frac{10}{3}.$$

It's still not yet clear how to solve this system, but at least it involves symmetric expressions $t + u$ and tu, which we are a little more comfortable with than $2x + 3y + 6xy$ and $2xy^2 + 3xy^2$.

> **Concept:** A well-chosen substitution can turn an equation into a more familiar form.

Substitution worked once, so seeing $t + u$ and tu in both of our equations might inspire us to do it again. We let $s = t + u$ and $p = tu$, so we can now write our equations as

$$s + p = 9,$$
$$\frac{sp}{6} = \frac{10}{3}.$$

Therefore, we have $s + p = 9$ and $sp = 20$, which means that s and p are the roots of the quadratic $x^2 - 9x + 20 = 0$. So, we have $(s, p) = (4, 5)$ or $(s, p) = (5, 4)$.

If $(s, p) = (4, 5)$, we have $t + u = 4$ and $tu = 5$, which means that t and u are the roots of $z^2 - 4z + 5$. Applying the quadratic formula gives $2 \pm i$ as the roots, so $(t, u) = (2 + i, 2 - i)$ or $(t, u) = (2 - i, 2 + i)$. These values of t and u give us $(x, y) = (1 + \frac{1}{2}i, \frac{2}{3} - \frac{1}{3}i)$ or $(x, y) = (1 - \frac{1}{2}i, \frac{2}{3} + \frac{1}{3}i)$.

Similarly, if $(s, p) = (5, 4)$, then t and u are the roots of $z^2 - 5z + 4$, which are 1 and 4. These give us the solutions $(x, y) = (\frac{1}{2}, \frac{4}{3})$ and $(x, y) = (2, \frac{1}{3})$. $\square$

> **Concept:** Any symmetric expression with two variables can be written in terms of the sum and product of those two variables. Therefore, problems involving symmetric expressions in two variables can often be solved by substituting s for the sum of the variables and p for their product.

Let's see this strategy in action again:

Problem 20.31: Suppose that the sum of the squares of two complex numbers x and y is 7 and the sum of the cubes is 10. What is the largest real value that $x + y$ can have? *(Source: AIME)*

Solution for Problem 20.31: We're given two symmetric polynomial equations:

$$x^2 + y^2 = 7,$$
$$x^3 + y^3 = 10.$$

We seek the largest value of $x + y$, so we let $s = x + y$ and try to write the equations in terms of s. We have $s^2 = x^2 + y^2 + 2xy = 7 + 2xy$, which reminds us that we can write symmetric polynomials with two variables in terms of the sum and product of the variables. So, we let $p = xy$ and we have $s^2 = 7 + 2p$.

Turning to the other equation, we have $x^3 + y^3 = (x + y)(x^2 - xy + y^2) = s(s^2 - 3p)$, so $s(s^2 - 3p) = 10$.

Our equations in s and p aren't all that nice, but solving $s^2 = 7 + 2p$ for p gives us $p = \frac{s^2 - 7}{2}$. We can use this equation to substitute for p in the second equation, giving us a cubic equation in s:

$$s\left(s^2 - 3 \cdot \frac{s^2 - 7}{2}\right) = 10.$$

Simplifying this equation gives $s^3 - 21s + 20 = 0$. Using the Rational Root Theorem, we factor the left side to find $(s - 1)(s - 4)(s + 5) = 0$. The largest possible value of $x + y = s$ is therefore 4. $\square$

Now that we have some experience manipulating two-variable symmetric polynomials, let's try more variables.

Problem 20.32: Solve the following system of equations:

$$x + y + z = 8,$$
$$x^2 + y^2 + z^2 = 62,$$
$$\frac{1}{x} + \frac{1}{y} + \frac{1}{z} = -\frac{1}{42}.$$

Solution for Problem 20.32: The three symmetric polynomials on the left remind us of Problem 20.21, in which we let the variables be the roots of a polynomial, then found that polynomial using Vieta's Formulas. But in this problem, only one of our symmetric polynomials can be directly related to a polynomial via Vieta. We are given $x + y + z = 8$, but we don't know $xy + yz + zx$ or xyz. Can we find them?

> **Concept:** Start new problems by asking yourself if you have solved a similar problem before. If you have, then try to turn the new problem into the same problem as the old problem.

Squaring $x + y + z = 8$ will produce xy, yz, and zx terms. This gives us

$$8^2 = x^2 + y^2 + z^2 + 2xy + 2yz + 2zx = 62 + 2(xy + yz + zx),$$

so $xy + yz + zx = 1$. Success!

We'll probably have to use the third equation for something, so we start by writing $\frac{1}{x} + \frac{1}{y} + \frac{1}{z}$ with a common denominator:

$$-\frac{1}{42} = \frac{1}{x} + \frac{1}{y} + \frac{1}{z} = \frac{xy + yz + zx}{xyz}.$$

We know that $xy + yz + zx = 1$, so we have $xyz = -42$.

Now, by Vieta's Formulas, x, y, and z are the roots of the cubic

$$t^3 - (x + y + z)t^2 + (xy + xz + yz)t - xyz = t^3 - 8t^2 + t + 42.$$

This factors as $(t - 3)(t - 7)(t + 2) = 0$. Therefore, our six solutions are $(x, y, z) = (3, 7, -2)$, and each of the other five permutations (orders) of $(3, 7, -2)$. □

> **Problem 20.33:** Let a, b, c, d be complex numbers satisfying
>
> $$a + b + c + d = a^3 + b^3 + c^3 + d^3 = 0.$$
>
> Prove that a pair of the a, b, c, d must add up to 0. *(Source: Tournament of Towns)*

Solution for Problem 20.33: It's not at all clear even where to start, so we'll have to play around a little. We'll start by brainstorming a little bit, and see what we come up with.

> **Concept:** On a tough problem, it often pays to start off by listing a bunch of ideas about how to approach the problem. You may not start off with the right one, but this brainstorming process will help prevent you from getting stuck in one track of thought. It also might feed you ideas once you start working with the problem more.

Keeping in mind that we're looking for information about sums of pairs of variables, let's brainstorm:

Try 1: Rearrange and factor. We can rearrange $a + b + c + d = a^3 + b^3 + c^3 + d^3$, then factor, to get

$$a(a - 1)(a + 1) + b(b - 1)(b + 1) + c(c - 1)(c + 1) + d(d - 1)(d + 1) = 0.$$

No sums of variables. Probably not useful.

Try 2: Factor sums of cubes. We have $a^3 + b^3 + c^3 + d^3 = (a + b)(a^2 - ab + b^2) + (c + d)(c^2 - cd + d^2)$. We subtract $a^3 + b^3 + c^3 + d^3 = a + b + c + d$ from this equation to give

$$(a + b)(a^2 - ab + b^2 - 1) + (c + d)(c^2 - cd + d^2 - 1) = 0.$$

It's not clear where to go from here. We have sums of variables, which is good, but we have nasty terms we don't know what to do with, which is bad.

Try 3: Rearrange $a + b + c + d = 0$ and cube. We have $a + b = -(c + d)$, which is good because we have sums of pairs of variables. Then, we cube, so we can use our information about the cubes. This gives us

$$a^3 + 3a^2b + 3ab^2 + b^3 = -c^3 - 3c^2d - 3cd^2 - d^3.$$

Since we are given $a^3 + b^3 = -c^3 - d^3$, we have $3a^2b + 3ab^2 = -3c^2d - 3cd^2$. A little factoring gives

$$ab(a + b) = -cd(c + d).$$

Combining this with $a + b = -(c + d)$ gives us either $a + b = c + d = 0$, in which case we are done, or $ab = cd$. It's not too clear what we'll do with this. It tells us about products of the variables instead of sums, but it seems like it might be important. We might spend a few more minutes here exploring where $ab = cd$ will lead us, but if we don't get anywhere, it's back to brainstorming.

> **WARNING!!** ☢ If you find a promising method of attack on a problem, explore it, but don't get stuck to it. If it doesn't work out after a little while, go back to exploring the problem from different angles.

Try 4: Rearrange differently before cubing. Although $a + b = -(c + d)$ didn't give the answer right away, it did seem to give us progress. So, we feel close. We try rearranging a little differently, and we write $a = -(b + c + d)$. Cubing this is a bit tricky, but we find

$$a^3 = -b^3 - c^3 - d^3 - 3b^2c - 3c^2b - 3b^2d - 3d^2b - 3c^2d - 3d^2c - 6bcd.$$

We know that $a^3 = -b^3 - c^3 - d^3$, so we can get rid of those terms, then divide the equation by -3 to have

$$b^2c + c^2b + b^2d + d^2b + c^2d + d^2c + 2bcd = 0.$$

Before we get scared and quit, we remember what we're trying to prove. We're trying to prove that one of the sums of a pair of a, b, c, and d equals 0. Obviously, the messy equation we just hit has nothing to do with a, but it does have b, c, and d. We'd be happy if we could show that $b + c = 0$, which makes us wonder, "Is $b + c$ a factor of the messy expression?" We have a way to check! We let $b = -c$ in the expression, and we have

$$b^2c + c^2b + b^2d + d^2b + c^2d + d^2c + 2bcd = c^3 - c^3 + c^2d - d^2c + c^2d + d^2c - 2c^2d = 0.$$

Aha! We see that $b + c$ is a factor. The polynomial is symmetric, so we aren't surprised to find that $b + d$ and $c + d$ are factors, as well, and we have

$$b^2c + c^2b + b^2d + d^2b + c^2d + d^2c + 2bcd = (b + c)(c + d)(d + b).$$

So, if $a + b + c + d = a^3 + b^3 + c^3 + d^3 = 0$, then our manipulations above show that we must have $(b + c)(c + d)(d + b) = 0$, which means that one of the sums of a pair of b, c, and d equals 0. $\square$

Exercises

20.6.1 If $x + y + z = 1$ and $\frac{1}{x} + \frac{1}{y} + \frac{1}{z} = 0$, then find $x^2 + y^2 + z^2$.

20.6.2 If $xy = 5$ and $x^2 + y^2 = 21$, then compute $x^4 + y^4$. *(Source: HMMT)*

20.6.3 Find all pairs of real numbers (x, y) such that $x^2 + y^2 = 34$ and $x^3 + y^3 = 98$.

20.6.4 Show that *two* of the sums of pairs of a, b, c, and d must equal 0 in Problem 20.33.

20.6.5 Find all solutions of the equations

$$x + y + z = 2,$$
$$x^2 + y^2 + z^2 = 14,$$
$$xyz = -6.$$

20.6.6★ Find x, y, and z:

$$x + y + z = 6,$$
$$x^2 + y^2 + z^2 = 26,$$
$$x^3 + y^3 + z^3 = 90.$$

20.7 Summary

Concepts:

- Look everywhere for patterns.

- If you don't immediately see how to work with an expression, try rewriting it. Many problems can be simplified by a different look at the same information.

- The substitution $y = x + \frac{1}{x}$ can help us simplify problems involving polynomials in x with symmetric coefficients.

- When stuck trying to deal with an algebraic expression, focus on any unusual aspects of the expression.

- Use substitutions to turn complicated equations into simple ones. Repeated expressions within an equation are a clue that substitution might be helpful.

- Don't make problems harder than they have to be. When you have multiple options for substitution, choose the one that appears to offer the easiest path.

Continued on the next page. . .

Concepts: ... *continued from the previous page*

- A **homogeneous equation** is one in which all nonzero terms have the same degree. When working with a two-variable homogeneous equation, it's often helpful to set the ratio of the variables equal to a new variable and write the equation in terms of the new variable.

- When stuck on a problem, focus on information you haven't used yet.

- Try to turn new problems you don't know how to do into old problems you know how to do.

- Don't get intimidated by complicated notation. Take your time, read the problem closely, and play with simple examples. Often, complicated notation is just shorthand for a very simple idea.

- When working with an unknown polynomial, it is often useful to assign variables to the coefficients of the polynomial, and then construct equations in terms of those variables.

- Just as we can find the values of symmetric sums given a polynomial, we can construct polynomials given symmetric sums.

- If a tactic gives some information in a problem, but doesn't completely solve it, try using the tactic again.

- You don't have to see a path all the way to the end of a solution before taking your first step. Experiment! Use what you need as a guide to work with what you have.

- On a tough problem, it often pays to start off by listing a bunch of ideas about how to approach the problem. You may not start off with the right one, but this brainstorming process will help prevent you from getting stuck in one track of thought. It also might feed you ideas once you start working with the problem more.

REVIEW PROBLEMS

20.34 Solve the system of equations

$$(x + y)(x + y + z) = 15,$$
$$(y + z)(x + y + z) = 6,$$
$$(z + x)(x + y + z) = -3.$$

20.35 Find three constants p, q, and r such that $x^4 + 4x^3 - 2x^2 - 12x + 9 = (px^2 + qx + r)^2$ for all x.

20.36 Find all m for which $x^2 + 3xy + x + my - m$ has two linear factors with integer coefficients. *(Source: AHSME)*

20.37 A polynomial $P(x)$ has four roots, $\frac{1}{4}, \frac{1}{2}, 2, 4$. Find $P(0)$ if $P(1) = 1$. *(Source: HMMT)*

20.38 Find all ordered triples (a, b, c) that satisfy the system

$$a + b + c = 4,$$
$$ab + bc + ca = -4,$$
$$abc = -21.$$

20.39 Solve for x where $\sqrt{4x - 3} + \dfrac{10}{\sqrt{4x - 3}} = 7$.

20.40 Solve for a_{17} if

$$a_1 + a_2 + a_3 = 1,$$
$$a_2 + a_3 + a_4 = 2,$$
$$\vdots$$
$$a_{16} + a_{17} + a_1 = 16,$$
$$a_{17} + a_1 + a_2 = 17.$$

(Source: Mandelbrot)

20.41 Find all ordered triples (x, y, z) of real numbers that have the property that each number is the product of the other two.

20.42 What is the product of the real roots of the equation $x^2 + 18x + 30 = 2\sqrt{x^2 + 18x + 45}$? *(Source: AIME)*

20.43 The equation $x^4 - 16x^3 + 94x^2 + px + q = 0$ has two double roots. Find $p + q$.

20.44 Let $f(x) = 2x^3 - 3x^2 + 2x - 1$ have roots $r, s,$ and t.

(a) Evaluate $\dfrac{1}{r - 2} + \dfrac{1}{s - 2} + \dfrac{1}{t - 2}$.

(b)★ Evaluate $\dfrac{1}{2r + 1} + \dfrac{1}{2s + 1} + \dfrac{1}{2t + 1}$.

20.45 Let $P(x) = x^3 + x^2 - 1$ and $Q(x) = x^3 - x - 1$. Given that r and s are two distinct solutions of $P(x) = 0$, prove that rs is a solution of $Q(x) = 0$. *(Source: MMPC)*

20.46 Real numbers a, b, c satisfy the equations $a + b + c = 26$ and $\frac{1}{a} + \frac{1}{b} + \frac{1}{c} = 28$. Find the value of

$$\frac{a}{b} + \frac{b}{c} + \frac{c}{a} + \frac{a}{c} + \frac{c}{b} + \frac{b}{a}.$$

(Source: HMMT)

20.47   Find all ordered triples (x, y, z) that satisfy the system

$$\sqrt{x} + \sqrt{y} + \sqrt{z} = 10,$$
$$x + y + z = 38,$$
$$\sqrt{xy} + \sqrt{xz} + \sqrt{yz} = 30.$$

20.48   Find the four values of x that satisfy $(x - 3)^4 + (x - 5)^4 = -8$. *(Source: ARML)*

20.49   Given that $\sqrt{2x^2 - 3xy + 64} + \sqrt{2x^2 - 3xy} = 16$, find the value of $\sqrt{2x^2 - 3xy}$.

20.50   Find $x^2 + y^2$ given that x and y are natural numbers such that $xy + x + y = 71$ and $x^2y + xy^2 = 880$. *(Source: AIME)*

20.51   If a, b, c, d are the solutions of the equation $x^4 - bx - 3 = 0$, then find a polynomial whose roots are the numbers

$$\frac{a + b + c}{d^2}, \quad \frac{a + b + d}{c^2}, \quad \frac{a + c + d}{b^2}, \quad \frac{b + c + d}{a^2}.$$

(Source: AHSME)

20.52   The equations $x^3 - 7x^2 + px + q = 0$ and $x^3 - 9x^2 + px + r = 0$ have two roots in common. If the third root of each equation is represented by x_1 and x_2, respectively, compute the ordered pair (x_1, x_2).

20.53   What is the sum of the solutions to the equation $\sqrt[4]{x} = \dfrac{12}{7 - \sqrt[4]{x}}$? *(Source: AIME)*

20.54   Find all ordered pairs (x, y) such that $2x^2 + y^2 + 3xy = 12$ and $(x + y)^2 - \frac{y^2}{2} = 7$.

20.55   If $f(x)$ is a ninth degree polynomial such that $f(n) = \frac{n^2+1}{n}$ for $n = 1, 2, 3, \ldots, 10$, then find the value of $f(11)$.

Challenge Problems

Since this is the final group of Challenge Problems, we've mixed in problems that require strategies from some other chapters, as well.

20.56   Show that there exists a polynomial $f(x)$ with rational coefficients such that $f(\sqrt{2} + \sqrt{3}) = \sqrt{2}$. **Hints:** 38

20.57   Let $x, y,$ and z be nonzero real numbers such that

$$x + y + z = \frac{1}{x} + \frac{1}{y} + \frac{1}{z} = 1.$$

Prove that at least one of $x, y,$ and z is equal to 1.

20.58 If $x = 2007(a-b)$, $y = 2007(b-c)$, $z = 2007(c-a)$, and $xy + yz + zx \neq 0$, then compute the numerical value of

$$\frac{x^2 + y^2 + z^2}{xy + yz + zx}.$$

Hints: 123

20.59 Find all real solutions of $(2^x - 4)^3 + (4^x - 2)^3 = (4^x + 2^x - 6)^3$. *(Source: AHSME)* **Hints:** 116

20.60★ Let a, b, c, and d be distinct real numbers such that $a > b > c > d > 0$, $abcd = 1$, and

$$a + b + c + d = \frac{1}{a} + \frac{1}{b} + \frac{1}{c} + \frac{1}{d}.$$

Show that $ad = bc = 1$. **Hints:** 80, 337

20.61 Find the real values of x that satisfy the equation $(x + 1)(x^2 + 1)(x^3 + 1) = 30x^3$.

20.62 For each positive number x, let

$$f(x) = \frac{\left(x + \frac{1}{x}\right)^6 - \left(x^6 + \frac{1}{x^6}\right) - 2}{\left(x + \frac{1}{x}\right)^3 + \left(x^3 + \frac{1}{x^3}\right)}.$$

Find the minimum value of $f(x)$. *(Source: AHSME)* **Hints:** 225

20.63 Suppose that x, y, and z are three positive numbers that satisfy the equations

$$xyz = 1,$$
$$x + \frac{1}{z} = 5,$$
$$y + \frac{1}{x} = 29.$$

Find the value of $z + 1/y$. *(Source: AIME)* **Hints:** 343

20.64 Find all ordered pairs (x, y) such that $3xy - x^2 - y^2 = 5$ and $7x^2y^2 - x^4 - y^4 = 155$.

20.65 Find the value of

$$\frac{a}{a+1} + \frac{b}{b+1} + \frac{c}{c+1}$$

given that $a + b + c = -1$, $ab + bc + ca = -1$, and none of a, b, or c is equal to -1. **Hints:** 54, 348

20.66 Suppose that the roots of $x^3 + 3x^2 + 4x - 11 = 0$ are a, b, and c, and that the roots of $x^3 + rx^2 + sx + t = 0$ are $a + b$, $b + c$, and $c + a$. Find t. *(Source: AIME)* **Hints:** 196

20.67

(a) Let x, y, and z be real numbers such that $x + y + z$, $xy + xz + yz$, and xyz are all positive. Prove that x, y, and z are also positive.

(b) Is it possible for $x + y + z$, $xy + xz + yz$, and xyz to be positive real numbers if x, y, and z can be nonreal?

20.68 Let $P(x, y)$ be a polynomial in x and y such that

(i) $P(x, y)$ is symmetric, i.e., $P(x, y) = P(y, x)$ for all x and y,

(ii) $x - y$ is a factor of $P(x, y)$, i.e., $P(x, y) = (x - y)Q(x, y)$ for some polynomial $Q(x, y)$.

Prove that $(x - y)^2$ is a factor of $P(x, y)$. *(Source: Canada)*

20.69 If p, q, and r are the roots of $x^3 - 2x + 5 = 0$, find an equation with roots $\frac{1}{p} + \frac{1}{q}, \frac{1}{q} + \frac{1}{r}$, and $\frac{1}{r} + \frac{1}{p}$. **Hints:** 185

20.70★ A polynomial $P(x)$ of degree n satisfies $P(k) = 1/k$ for $k = 1, 2, 4, 8, \ldots, 2^n$. Find $P(0)$. *(Source: Poland)*

20.71 A sequence of numbers $x_1, x_2, x_3, \ldots, x_{100}$ has the property that, for every integer k between 1 and 100, inclusive, the number x_k is k less than the sum of the other 99 numbers. Find x_{50}. *(Source: AIME)* **Hints:** 168

20.72 Solve for x and y:

$$(x + y)^{2/3} + 2(x - y)^{2/3} = 3(x^2 - y^2)^{1/3},$$
$$3x - 2y = 13.$$

Hints: 255

20.73 Let $p(x, y)$ be a polynomial function whose domain and range are the real numbers. Prove or find a counterexample to the statement, "If $[p(x, y)]^2$ is a symmetric function, then $p(x, y)$ is also a symmetric function."

20.74 Show that every symmetric two-variable polynomial in x and y can be written as a polynomial in s and p, where $s = x + y$ and $p = xy$. **Hints:** 227, 203, 350

20.75 Find all solutions to the system

$$a + b + c = 1,$$
$$a^2 + b^2 + c^2 = 2,$$
$$a^4 + b^4 + c^4 = 3.$$

20.76 Let x, y, z be real numbers such that $x + y + z \neq 0$. Prove that if

$$\frac{x(y - z)}{y + z} + \frac{y(z - x)}{z + x} + \frac{z(x - y)}{x + y} = 0,$$

then two of x, y, and z must be equal. **Hints:** 83

20.77 The product of two of the four roots of the quartic equation

$$x^4 - 18x^3 + kx^2 + 200x - 1984 = 0$$

is -32. Determine the value of k. *(Source: USAMO)* **Hints:** 329, 21, 71

20.78★ Find all polynomials p such that $p(x + 2) = p(x) + 6x^2 + 12x + 8$.

20.79★ Let $f(x) = a_{2n}x^{2n} + a_{2n-1}x^{2n-1} + \cdots + a_1x + a_0$, where $a_{2n} \neq 0$ and $a_{2n-i} = a_i$ for all i such that $0 \leq i \leq 2n$. Show that there is a polynomial $g(x)$ such that the solutions of $g(x + \frac{1}{x}) = 0$ coincide with the solutions of $f(x) = 0$. **Hints:** 206, 9

20.80★ Find $ax^5 + by^5$ if the real numbers a, b, x, and y satisfy the equations

$$ax + by = 3,$$
$$ax^2 + by^2 = 7,$$
$$ax^3 + by^3 = 16,$$
$$ax^4 + by^4 = 42.$$

(Source: AIME) **Hints:** 201

20.81★ Let x_1, x_2, and x_3 be the roots of $x^3 - 2x^2 - x + 1 = 0$. Let $y_i = x_i^2 + x_i + 1$. Find the monic cubic whose roots are y_1, y_2, and y_3. **Hints:** 238, 103

20.82★ Find all solutions to the equation $(z^2 - 3z + 1)^2 - 3(z^2 - 3z + 1) + 1 = z$. **Hints:** 96

20.83★ Let $p(x) = x^5 + x^2 + 1$ have roots r_1, r_2, r_3, r_4, r_5. Let $q(x) = x^2 - 2$. Determine the product $q(r_1)q(r_2)q(r_3)q(r_4)q(r_5)$. *(Source: USAMTS)* **Hints:** 297, 223

20.84★ Curves A, B, C, and D are defined in the plane as follows:

$$A \text{ is the graph of } x^2 - y^2 = \frac{x}{x^2 + y^2},$$
$$B \text{ is the graph of } 2xy + \frac{y}{x^2 + y^2} = 3,$$
$$C \text{ is the graph of } x^3 - 3xy^2 + 3y = 1,$$
$$D \text{ is the graph of } 3x^2y - 3x - y^3 = 0.$$

Prove that the intersection of curves A and B is the same as the intersection of curves C and D. *(Source: Putnam)* **Hints:** 187

20.85★ Let a, b, c, d be distinct real numbers such that

$$a = \sqrt{4 + \sqrt{5 + a}},$$
$$b = \sqrt{4 - \sqrt{5 + b}},$$
$$c = \sqrt{4 + \sqrt{5 - c}},$$
$$d = \sqrt{4 - \sqrt{5 - d}}.$$

Find $abcd$. *(Source: Brazil)* **Hints:** 63

20.86★ Determine $x^2 + y^2 + z^2 + w^2$ if

$$\frac{x^2}{2^2 - 1^2} + \frac{y^2}{2^2 - 3^2} + \frac{z^2}{2^2 - 5^2} + \frac{w^2}{2^2 - 7^2} = 1,$$

$$\frac{x^2}{4^2 - 1^2} + \frac{y^2}{4^2 - 3^2} + \frac{z^2}{4^2 - 5^2} + \frac{w^2}{4^2 - 7^2} = 1,$$

$$\frac{x^2}{6^2 - 1^2} + \frac{y^2}{6^2 - 3^2} + \frac{z^2}{6^2 - 5^2} + \frac{w^2}{6^2 - 7^2} = 1,$$

$$\frac{x^2}{8^2 - 1^2} + \frac{y^2}{8^2 - 3^2} + \frac{z^2}{8^2 - 5^2} + \frac{w^2}{8^2 - 7^2} = 1.$$

(Source: AIME) **Hints:** 132

Extra! The Art of Problem Solving's Intermediate Series of textbooks continues with *Intermediate*
➡➡➡➡ *Counting & Probability* by David Patrick.

Intermediate Counting & Probability is the sequel to the acclaimed *Introduction to Counting & Probability* textbook that has been used by thousands of students. Some of the topics covered in *Intermediate Counting & Probability* are:

- The Principle of Inclusion and Exclusion

- 1-1 correspondences

- The Pigeonhole Principle

- Distributions

- Fibonacci and Catalan numbers

- Recursion

- Conditional probability

- Generating functions

- Graph theory

And many more! The book contains over 650 problems and exercises.

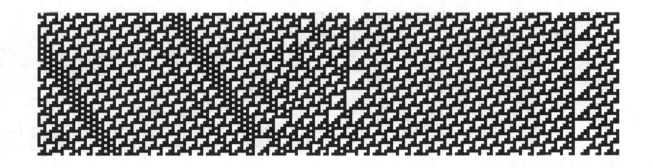

References

In addition to the books cited below, there are links to many useful websites on the book's links page at

`http://www.artofproblemsolving.com/BookLinks/IntermAlgebra/links.php`

1. J. Derbyshire, *Unknown Quantity*, Penguin Group, 2006.

2. A. Engel, *Problem-Solving Strategies*, Springer-Verlag, 1998.

3. D. Fomin, S. Genkin, and I. Itenberg, *Mathematical Circles (Russian Experience)*, American Mathematical Society, 1996.

4. L. Larson, *Problem-Solving Through Problems*, Springer-Verlag, 1983.

5. S. Lehoczky and R. Rusczyk, *the Art of Problem Solving, Volume 1: the Basics*, 7th edition, AoPS Incorporated, 2006.

6. S. Lehoczky and R. Rusczyk, *the Art of Problem Solving, Volume 2: and Beyond*, 7th edition, AoPS Incorporated, 2006.

7. D. Patrick, *Intermediate Counting & Probability*, AoPS Incorporated, 2007.

8. D. Patrick, *Introduction to Counting & Probability*, 2nd edition, AoPS Incorporated, 2007.

9. R. Rusczyk, *Introduction to Algebra*, AoPS Incorporated, 2007.

10. P. Zeitz, *The Art and Craft of Problem Solving*, 2nd edition, John Wiley & Sons, Inc., 2007.

Hints to Selected Problems

1. Note that 3 and -5 are both odd.

2. You've probably already moved all the terms to one side, and determined that there aren't any linear factors with rational coefficients. So, what kind of factors with rational coefficients must there be?

3. Let l and w be the length and width of the top of the table, and let h be the height of the table. Find h in terms of l and w. Express the volume in terms of h.

4. After rearranging due to the first hint, you probably have $n + 1$ terms on the greater side. One of them should stand out from the others. How can you split that term up, and join one piece with each of the other n terms?

5. Let $d = (a + b + c)/3$.

6. It is wise to generalize. Find $g^n(x)$.

7. Let r be the root you found in the first hint. Now, go backwards. Find the polynomial with roots α, $\alpha^2 + 2\alpha - 14$, and r if α is a root of the given quadratic. (You can't assume anything about the latter two roots!)

8. What happens if you let one sequence in Cauchy-Schwarz be $\sqrt{a}$, $\sqrt{b}$, $\sqrt{c}$, $\sqrt{d}$?

9. Manipulate the equation $f(x) = 0$ such that its highest power of x is the reciprocal of its lowest power of x. Pair off the terms, each power of x with its reciprocal. Can you express the sum of each pair in terms of $x + \frac{1}{x}$?

10. Compare the sum to $1 + \frac{1}{1\cdot 2} + \frac{1}{2\cdot 3} + \cdots + \frac{1}{(n-1)n}$. Can you find a simple expression that is equal to this sum with experimentation and induction?

11. Consider the product $(a + b + c)\left(\frac{1}{a} + \frac{1}{b} + \frac{1}{c}\right)$.

12. What is $(x + 1)^3 - x^3$?

13. How many times does $\frac{1}{2^3}$ appear? How about $\frac{1}{3^3}$? $\frac{1}{4^3}$?

14. Complete the square.

15. Such as $y = \sqrt{x}$.

16. Start with the 4-variable AM-GM applied to $a, b, c,$ and d. Choose an expression for d in terms of $a, b,$ and c that will give us the AM of $a, b,$ and c on the greater side of the 4-variable AM-GM Inequality.

17. Find the first few terms. See a pattern? Prove it!

18. Write the product with $\prod$ notation. You should have a polynomial. What do we usually do to learn more about a polynomial?

19. You used AM-GM for 4 variables to prove it for 3 variables. You should already have proved AM-GM for 2^3 variables, and for 2^4 variables, etc.

20. Can you find $f(a)$ in terms of $a, p,$ and q?

21. Can you factor the polynomial now using what you know about products of pairs of roots?

22. 3^{31} is way bigger than 2^{31}, and 3^{29} is way bigger than 2^{29}. So, the fraction is probably very close to $3^{31}/3^{29} = 9$. But is it a little larger, or a little smaller? (If a little larger, you must show it is less than 10; if a little smaller, you must show it is more than 8.)

23. Once you have an equation, simplify it as much as possible. Do you see any terms that can be grouped and factored?

24. Focus on the four key points: the two ends of the graph and the two points where the graph changes direction. Where are these points on the new graph?

25. For what values of x is f defined? Is f monotonically increasing or monotonically decreasing for these values of x? Is there a smallest value in the domain of f?

26. Multiply by $1 - x$.

27. What expression must $36x^2 + 25y^2$ be greater than if x and y have the same sign? What if they have opposite signs?

28. Can you express p^3 in terms of p^2 and p?

29. What can you do to the expression you must minimize in order to produce squared terms?

30. If $f(a) = f(b)$, where $1 \le a, b \le 5$, then what two possibilities are there for how a and b are related? How can we find solutions to $f(f(x))$ for each of these cases?

31. When a strategy works a little, try it again. Let $p(x) = d(x)q(x)$, where $d(x)$ is the factor you already found. Find a factor of $q(x)$.

32. Your setup should have plenty of sums, as well as square root signs to get rid of. Squaring gets rid of square root signs. Sums and squaring—what inequality should help with that?

33. It would probably be easier if all the logarithms had the same base. Notice that $2^3 = 8$.

34. Let $c = b - 1$.

35. Try some specific examples, such as $f(x) = x^3$ or $f(x) = x + 3$.

36. Suppose P and Q are on f and f^{-1}, respectively, such that PQ is the desired shortest distance. So, we have $P = (a, f(a))$ for some a. Can you express the coordinates of Q in terms of a?

37. Focus on just a pair of variables. Can you create an inequality that relates $(a - b)^2$ to $a^2 + b^2$?

38. It's a puzzle! Compute a few powers of $\sqrt{2} + \sqrt{3}$. Try to fit them together to get $\sqrt{2}$.

39. How are the graphs of f and f^{-1} related?

40. What is the smallest possible value of ab?

41. Because a is a root of the polynomial, we have $a^3 + 3a^2 - 24a + 1 = 0$. We'd like to find a simple expression for $\sqrt[3]{a}$. Reorganize $a^3 + 3a^2 - 24a + 1 = 0$ such that a is on one side, and the cube of a polynomial is on the other.

42. Think about the polynomial $P(x) - 10x$.

43. We know a lot more about finding the roots of polynomials whose coefficients are all real.

44. What is $f(x + y)$?

45. Express $f(x)$ as the product of two quadratics that are either monotonically increasing or monotonically decreasing.

46. Rearrange the equation, and remember that a and b are nonzero. You should get a quadratic. What must be true about a quadratic if it has integer roots?

47. Let $a = \log_{225} x$ and $b = \log_{64} y$. Rewrite the given equations in terms of a and b.

48. You probably already tried using the fact that $1 - x$ is cyclic. And that failed. Can you try another tactic from the chapter?

49. Compare $f(x)$ to $f(-x)$. For what values of x must these polynomials be equal?

50. Find a polynomial $f(a)$ such that $a^5 - a^2 + 3 \geq f(a)$, and $f(a)$ is easier to work with than $a^5 - a^2 + 3$. Note that $a^5 - a^2 = a^2(a^3 - 1)$.

51. Let a be the first term and r be the common ratio, and write the given information in terms of a and r. How can you factor $r^4 + r^2 + 1$? (If you don't remember, then try thinking of a notable polynomial that has $r^4 + r^2 + 1$ as a factor.)

52. Suppose you wrote an equation for each root. What can you do with these 20 equations to produce an equation that contains the sum of the 20th powers of the roots?

53. Each term in the series looks a lot like our formula for the sum of an infinite geometric series. Find infinite geometric series that correspond to each.

54. Can you find a substitution that simplifies the expression you must find, but doesn't greatly complicate the given equations?

55. Suppose you can find the length of wire needed to create the portion of the graph with $x \geq 0$ and $y \geq 0$. How can you then quickly find the answer?

56. What is true about the diagonals of a parallelogram?

57. How many different integers can appear in the sum on the right side?

58. Suppose the circle is the graph of $x^2 + y^2 = r^2$, and let P be $(p, 0)$.

59. Write the equation for the line in terms of w, then let $w = a + bi$.

60. Experiment! Compute a few of the z_i. Notice anything interesting?

61. Find pattern, prove pattern. Experiment!

62. Let $x = \sqrt[3]{49} + \sqrt[3]{7}$. How do you get rid of cube roots?

63. If we could find a polynomial with roots a, b, c, and d, we'd be done. But it's not so obvious how to do that. Can we at least find a polynomial with a as a root?

64. What expression can you add to $\left(15 + \sqrt{220}\right)^{19}$ that, when both are expanded, will eliminate all the $\sqrt{220}$ terms?

65. Forget about the numerators for a minute. Deal first with $\frac{1}{a+b} + \frac{1}{b+c} + \frac{1}{a+c}$.

66. Place the problem on the Cartesian plane. Let A be $(-2, 0)$ and B be $(2, 0)$. Where is C? What does $AX + XB = AC + CB$ tell us about point X?

67. How can you convert the problem to exponential form?

68. First, consider what inputs are allowed to f. Then, consider what inputs are allowed to g, then determine what inputs to f give outputs that are valid inputs to g.

69. Expand $(x + 1)^3$.

70. Write the expression as a sum of positive powers of z.

71. A product of a pair of roots won't tell you much about a linear factor of the polynomial. What type of factor will it tell you something about?

72. Combine the previous two problems in a way-cool induction. Start from AM-GM for 2^k variables, and use the same tactic you used to prove AM-GM for 3 variables to prove it for $2^k - 1$ variables. Can you then set up an induction to show that it holds for $2^k - 2$ variables, $2^k - 3$ variables, and so on?

73. An easy way to divide 23149 by 7 is to break into pieces that are divisible by 7. We have $23149 = 21000 + 2100 + 49$, so dividing by 7 gives $3000 + 300 + 7 = 3307$. Can you do something similar here?

74. If α and $\alpha^2 + 2\alpha - 14$ are roots of the cubic, what must the third root be?

75. For both inequalities, try completing the square in x and y. You should recognize one of the results. You may have to factor the other.

76. We have 10 solutions to $P(x) - \frac{1}{x(x+1)} = 0$. But that's not a polynomial on the left side.

77. What goes up must come down. But when?

78. Experiment. Try various small values of n.

79. A substitution makes the algebra easier. Let $a = \sqrt[3]{60-x}$ and $b = \sqrt[3]{x-11}$.

80. There are a whole lot of elementary symmetric polynomials (or parts thereof) in this problem.

81. Sum of squares; try Cauchy.

82. Experiment. Find several terms of the sequence in terms of a_1 and a_2. Can you find the two given sums in terms of a_1 and a_2?

83. What equation is equivalent to the statement, "Two of x, y, and z must be equal?" Use the answer to this, along with the restriction $x + y + z \neq 0$, as inspiration to find a factored form of the left side.

84. Can you pair the factors so that you can use the quadratic to evaluate the results of expanding the product of each pair?

85. Experiment. Choose simple values of p, q, and r, and look for a pattern in the factors of the resulting numbers.

86. Express both sides as a single base 10 logarithm.

87. It's not as hard as it looks. Think about it before cubing.

88. All the pieces are linear, so what's a good guess for the form of each of the polynomials?

89. Add 1 to each of the three ratios in the first given equation.

90. Let $c = a^2$.

91. Can you factor the cubic? Try factoring x^2 out of the first two terms. Can you then factor further?

92. What is $(x-1)(x-2)(x-3)$?

93. Let $x = 2^y$. What is 4^y in terms of x?

94. What happens if you multiply the product by $1 - \frac{1}{3}$?

95. We'd like to get rid of the cube roots. The polynomial we seek probably has a^3, b^3, and c^3 terms. Do some wishful thinking. Find $g(x,y,z)$ and $h(x,y,z)$ such that $x^3+y^3+z^3+g(x,y,z) = (x+y+z)h(x,y,z)$, and $g(a, b\sqrt[3]{2}, c\sqrt[3]{4})$ is a polynomial with integer coefficients.

96. Make the obvious substitution for simplification. You will then have two equations; the result of the substitution and the variable definition that you used to make the substitution. What's interesting about those equations? Can you combine them in any interesting ways?

97. In terms of n, how many integers satisfy the inequality chain from the first hint?

98. Experiment. Evaluate $\lfloor x \rfloor - 19 \cdot \lfloor \frac{x}{19} \rfloor$ for different positive integers x. Notice anything interesting?

99. Rearrange and factor the given equation.

100. Yikes, that's ugly. Let $a = x^2 - yz$, $b = y^2 - zx$, and $c = z^2 - xy$. (But don't forget what a, b, and c stand for as you work!)

101. All five choices are nonreal, so one of the factors of the polynomial is a quadratic with nonreal roots. Which of the given choices can be a root of a monic quadratic with integer coefficients?

102. Let $(y_1, y_2, y_3, y_4) = (1, 0, 0, 0)$. What does this suggest as a possible $f(a, b, c, d)$? Confirm it works.

103. Find x_i in terms of y_i. Don't forget what the x_i are!

104. What happens if you add all three equations?

105. Express a_1^3 in the form $ca_1 + d$, where c and d are constants.

106. What is $a^3 + b^3$? Use the given equation to find $a^3 b^3$.

107. You care most about e. Write Cauchy in a way that allows you to substitute expressions in terms of e, to produce an inequality that has e as the only variable.

108. Use the three rules to find $f(n)$ for various values of n. Can you use your experimentation to prove a pattern?

109. The polynomial $g(x)$ should look familiar. Find a linear expression $x - c$ such that $(x - c)g(x)$ is nice to work with.

110. Find each sum in terms of a and r. What happens to the desired sum? Do you find any more geometric series?

111. Graph S and g, then use transformations of the graph of S to build the graph of g.

112. Find two quadratics whose product is the given quartic.

113. Consider the portion of the graph with $x \geq 0$ and $y \geq 0$. Why does this simplify the equation? Can you repeat your first step? In other words, can you reduce the problem to considering a smaller portion of the graph that, if you could find the length of wire needed for that portion of the graph, then you could find the answer quickly?

114. Suppose r is a solution to $p(p(r)) = r$. What might $p(r)$ be?

115. Multiply both sides by the product of the denominators.

116. What is interesting about the three quantities in parentheses; how are they related?

117. Write the bases in terms of 2 and 3. Let $x = 2$ and $y = 3$, and factor.

118. Write the left side as the difference of two sums.

119. Express the original equation in standard form. Can you express the equations corresponding to the parabola's reflection and both translated parabolas in standard form?

120. Solve $a_1 x + b_1 y + c_1 z = 0$ for z.

121. What is the result of rotating the point (a, b) 90 degrees counterclockwise about the origin?

122. For what x and y does $(a^3b^2c)^x(a^2b^3c^5)^y = a^3b^7c^{14}$?

123. The expression we must evaluate is symmetric. What simple symmetric expression might be useful to form from the three given equations?

124. What icky denominators! How can you get the square roots out of them?

125. Let b be the number bordering 0 to the right, a be the number one box above 0, and let c be the number in the other box adjacent to both a and b. Find other boxes in terms of these three.

126. If we multiply both sides by $a + b + c + d$, we have a product of two sums. How will we then apply Cauchy?

127. Group the x^3 and x^2 terms, and group the terms that are linear in x and those without x. Factor. Then factor again. Then again.

128. Include an extra term in the numerator such that the denominator is a factor of the numerator.

129. How can you get the sum of all the coefficients? How can you eliminate those with odd subscripts?

130. Sum the first six rows and look for a pattern. Then, try to explain why the pattern exists.

131. Let $n = 2m + b_0$, where b_0 is the last digit of the binary representation of n. What happens to $f(n)$ if b_0 is 0? If it is 1?

132. Can we introduce a single variable, then express all four equations as a single equation in which letting the new variable take on 4 different values produces the four given equations?

133. Vieta doesn't appear to help with the difference of the roots. But what about the square of the difference of the roots?

134. Complete the square. (That means write the quadratic in the form $a(x - h)^2 + k$ for some constants a, h, and k. If you're not familiar with completing the square, don't worry about this part. We'll review completing the square in a Chapter 4.)

135. A proof of a statement for all positive integers n . . . .

136. Experiment. What is $P_1(x)$ in terms of P_0? What is $P_2(x)$ in terms of P_0? How about $P_3(x)$? Notice anything interesting?

137. How many copies of x^m do you need to cancel with how many $\frac{1}{x^n}$ in order to have a constant GM?

138. For any polynomial $p(x)$, how is the degree of $p(x + 1) - p(x)$ related to $\deg p$? What substitution does this suggest?

139. Factor. Write out a few terms. See anything interesting?

140. When stuck, go back to the beginning. Get back to your basic series-evaluation tools.

141. Shift that index.

142. Graph the piecewise function in the problem. The result should resemble other absolute value graphs you have seen. In those other graphs, how have the "corners" of the graphs been related to the functions the graphs represent?

143. The $a^2 + b^2$ suggests complex numbers. Let $z = a + bi$. Can you write an equation in terms of z that is equivalent to the given system?

144. Experiment. Find the first few x_i. Then find the first term of the series, the sum of the first two terms, the sum of the first three terms. Notice a pattern? Can you take a guess as to what the sum of the first k terms is in terms of one of the x_i?

145. Rearrange the inequality so that all terms on both sides are positive. Notice that you have a sum on the greater side. Looks like a job for AM-GM. How many terms does the $2^{1/n}$ suggest you need in your AM?

146. Use the described property of the function to write an equation in terms of a, b, and z. Remember, this equation must hold for all z.

147. Express $r + s$ in terms of t, and $s + t$ in terms of r, and $t + r$ in terms of s.

148. Expand $\left(x - \frac{n}{1}\right)\left(x - \frac{n}{2}\right)\cdots\left(x - \frac{n}{n}\right)$.

149. Let r be a root of $f(x)$. Produce an equation with r^k.

150. There are infinitely many solutions. Let $a = n - 2k$; your original equation should have an $(n - 2k)^2$ in it somewhere.

151. In many of the graphs of polynomials that appear in the text, there are portions of the graph that are nearly vertical lines. Where are these nearly vertical lines located on these graphs, and what causes them to be nearly vertical?

152. Suppose there are n numbers in the list, and m is their sum. Find n and m.

153. In the text, we found a polynomial that has $\sqrt{2} + \sqrt{3}$ as a root. How might this polynomial be helpful?

154. Do you see an expression that appears in more than one equation? We can substitute for whole expressions, not just variables!

155. If you have three variables, you are trying to evaluate a sum of the form $ab + bc + ca$. What algebraic expression will produce a sum like that?

156. Find an integer that is close to the given fraction.

157. Let $a = 3x - 5$, $b = 2x - 4$, and $c = 5x - 9$. Then cube.

158. We'd like to get rid of those fractional part symbols. Find expressions for $\{a^{-1}\}$ and $\{a^2\}$ that don't have any special functions in them.

159. Experiment. Forget about a_9 and a_{10}. Pick different values of a_2 and find the next few terms. Do you see a pattern?

160. Can you combine the equations in a way that allows you to factor the result?

161. We'd like to find an expression for the number of positive integers such that $f(k) = n$. If the closest integer to $\sqrt[4]{k}$ is n, then what inequality chain can we write with $\sqrt[4]{k}$ in the middle?

162. Note that dividing a number by 6 is the same as dividing it by 2, then dividing the result by 3. Can you factor the quadratic?

163. Let $z = a + bi$, and use the definition $|z| = \sqrt{a^2 + b^2}$. For another approach, what is the graph of the equation $|z - 1| = |z + 3|$ on the complex plane?

164. Write $(\alpha_1 + \beta_1)(\alpha_1 + \beta_2)(\alpha_1 + \beta_3)$ as a polynomial in α_1. You should be able to evaluate the coefficients of this polynomial.

165. Two useful tactics with quadratics are factoring and completing the square. Try them both!

166. Start by finding an inequality that will get $a + b$, $a + c$, and $b + c$ out of the denominators on one side of the inequality.

167. How can the first n terms of the sum be combined to produce the terms with n^2 as the numerator? To produce the terms with n^3 as the numerator? With n^n as the numerator?

168. If you were to write an equation for each x_k, you'd have a long list of very similar-looking equations. What strategy can you use with a list of very similar-looking equations?

169. Suppose $b_1 = \log_{64} y_1$ and $b_2 = \log_{64} y_2$. Write $y_1 y_2$ in terms of b_1 and b_2.

170. Suppose one root is $a + bi$. What is another root? What is the sum of the roots?

171. What's a tactic that will both get rid of absolute value symbols, and introduce terms like ab into the three inequalities that have absolute value symbols?

172. The cubic has no x^2 term. What does that tell us about m, n, and p? Does this help us find a root of the quadratic?

173. Back to basics. Get rid of a variable. When does the remaining equation have real solutions?

174. Factor $x^3 + y^3$. Can you use the given equations to write the factored expression in terms of z?

175. The right side is an integer power of a binomial. We have a tool for that.

176. After applying Cauchy, write the inequality you must prove to finish, then work backwards. Don't forget that $ab + bc + ca = 1/3$. And don't forget to factor, factor, factor.

177. It would be a lot easier if we could combine both instances of each variable in a single equation. What happens if we do so for all the variables?

178. Look at the ratio of consecutive terms. How do we know we've reached the maximum A_i?

179. Write the expression with a common denominator. Symmetric expressions suggest Vieta's Formulas.

180. When you arrange things twice, you rearrange them.

181. Suppose there were originally n numbers. In terms of n, what is the largest possible average of the numbers that remain after erasing one? What is the smallest possible such average?

182. Can you factor $xy - 3x + 7y - 21$?

183. Write the expression in terms of $\log_a b$ and $\log_b a$.

184. Back to basics. It looks like an arithmetico-geometric series; what general tactic did we use to tackle those?

185. Can you first find a polynomial with roots $\frac{1}{p}$, $\frac{1}{q}$, and $\frac{1}{r}$? We're interested in sums of pairs of these roots; what is the sum of all three?

186. Squaring what expression will give us x^2, $\frac{1}{y^2}$, and $2\frac{x}{y}$?

187. The expression $x^2 + y^2$ suggests complex numbers. What can we do to the complex number $x + yi$ to produce $x^2 - y^2$ and $2xy$?

188. Use $x^3 = 1$ and $x \neq 1$ to express x^2 as a linear function of x.

189. Look for squares of binomials. (And of other expressions!)

190. Can you find a polynomial in a, b, and c with integer coefficients such that the polynomial has $a + b\sqrt[3]{2} + c\sqrt[3]{4}$ as a factor?

191. Write the equation as a quadratic in a (or b). What must be true about the discriminant if a and b are real?

192. What kind of series is the polynomial in x?

193. Write the three quantities in terms of $\lfloor x \rfloor$ and $\{x\}$. Make an equation that includes both these quantities. Can you solve for one in terms of the other? (And don't forget what they mean!)

194. Start with $a - b = b - c$, rearrange, and factor relentlessly. You should be able to express the equation as the product of two factors (in terms of x, y, and z) that must equal 0.

195. Compare each term in the expansion of the right side to n.

196. Vieta asks, "What is $a + b$ in terms of c?"

197. First, try to figure out which polynomial of the type described in the problem has the largest root. Then, try to find which of the intervals the root is in.

198. Let x_i be the number picked by the person who announced the average i. What is x_8 in terms of x_6? Then, what is x_{10} in terms of x_6?

199. Let r equal the ratios that follow from the first hint. Find a useful way to combine the three equations you then have for r.

200. Experiment. Find $f(94)$ in terms of $f(92)$. In terms of $f(91)$. In terms of $f(90)$.

201. What can you do to one of the four equations to produce $ax^5 + by^5$? (You'll get some extra terms, but you might be able to evaluate some of those, too.)

202. If we write the equation in standard form, and use the vertex, what is c in terms of a?

203. We wish to show that we can express $cx^n y^m + cx^m y^n$, where $m \neq n$, in terms of $p = xy$ and $s = x + y$. Factoring leaves what?

204. Suppose k of the terms on the right equal a. Find both sides of the equality in terms of a, n, and k.

205. Can we write a single equation in the variable x such that letting x equal 1, then 2, then 3, then 4, produces the four equations?

206. Start simple. Let $f(x) = x^2 + 1$. How about $f(x) = x^4 + 1$? $f(x) = x^4 + 6x^2 + 1$?

207. The sum on the left consists of 73 integers. How many distinct integers are in this sum?

208. Can you find a root of the quadratic?

209. Before writing any equations, think about the problem. Let P be one of the endpoints of the common chord. What do you know about the distance between P and the x-axis?

210. We have a $2n$ on the right, and, if you look closely, there are $2n$ terms on the left.

211. What is $x - \lfloor x \rfloor$? What does the first equation look like?

212. Move everything to one side and factor, factor, factor. Be on the lookout for squares!

213. Suppose ki is one of the imaginary roots. Let $x = ki$ in the equation. What must both the resulting real and imaginary parts equal?

214. The numbers being squared on the right side are 3 more than multiples of 12. Express each in the form $12a + 3$. Do you recognize the values of a that give you the numbers?

215. Move all the variables to one side. Does that side suggest any of the inequalities we have studied?

216. $a + b + c + d = 8 - e$.

217. Let $g(x) = f(x) - 5$. Write $g(x)$ in factored form. What is $g(k)$?

218. When going from row k to row $k + 1$, how many times is each number in row k included in a sum?

219. The arguments of the logarithms are the same. It sure would be nice if the bases were the same instead. How can we make that happen?

220. Let m be the middle square.

221. Let $a = \frac{1}{x}$.

222. Let $a = \log_x y$. Find all possible values of a.

223. The first hint should inspire you to seek a polynomial whose roots are the squares of r_i. So, let $y = x^2$, and try to build the necessary polynomial in y from $p(x) = 0$.

224. Based on the previous hint, what are the possible values of $Q(1)$?

225. How are the terms in the numerator related to those in the denominator?

226. Ugh. Fractions with nasty denominators. Cauchy should help. What clever substitution could you use as a first step to move the cubes from the denominator to the numerator right away?

227. If the term cx^ny^m, where $m \neq n$, is in a symmetric two-variable polynomial, what other term must appear?

228. This looks a lot like the previous problem. In terms of f, what can you multiply $f(x)$ by to get the product in the previous problem?

229. Let $b = a^x$. Solve the equation for b in terms of y.

230. Expand, rearrange, factor. Summation notation will help a lot here.

231. What is $\deg Q$?

232. Use your equation from the first hint to express a^3 as a linear expression in a. Then, use this equation to express a^6 as a linear expression in a. And so on.

233. Let the number of people in the m groups be $x_1, x_2, \ldots, x_m$, and let the side lengths of the cakes be $s_1, s_2, \ldots, s_m$. Write inequalities and equations in terms of the x_i and s_i, as well as an expression for the total length of ribbon needed.

234. Take the base 10 logarithm of both sides to get the logarithm out of the exponent. Then, let $y = \log_{10} x$. Can you write your equation in terms of y?

235. Can you write the product in a more convenient form? What expressions in terms of a and b are easy to evaluate?

236. Let $a = (\sqrt{2} + 1)^x$ and $b = (\sqrt{2} - 1)^x$. Find two equations in terms of a and b.

237. Let a be the middle term of the arithmetic sequence.

238. The expression for y_i should look familiar.

239. Forget the nasty z_0. Experiment and learn more about $F(z)$. Notice that $z_2 = F(F(z_0))$.

240. This problem looks like a nail. Break out your hammer and bash away. Then, factor the pieces.

241. There is an n; induct! (Be careful—we need more than one base case.)

242. What is $x - \lfloor x \rfloor$?

243. Vieta says, "Let the roots be a, ar, ar^2, ar^3, and ar^4."

244. The desired inequality has squares. Square the given equation. Don't forget that we are given the order of a, b, and c.

245. Make a substitution that turns the equation into something you know how to handle.

246. Get rid of the fractions and move everything to one side. Factor, factor, factor.

247. Simplify the problem. Tackle a similar problem for 2 variables.

248. Think first. Are there values of y (or a whole range of values) that obviously cannot satisfy the inequality? Use the answer to this question to simplify the absolute value expression before you even start with the algebra.

249. Express every logarithm in terms of base a logarithms.

250. Can you find a point on the directrix? What is the slope of the axis? Of the directrix?

251. Think both about what happens when $x \geq 0$ and when $x < 0$, as well as about what happens when $f(x) \geq 0$ and when $f(x) < 0$.

252. Notice that $(3x - 5) + (2x - 4) = 5x - 9$.

253. What does $P(0) = 0$ tell us about $P(x)$? And what does this tell us about all possible *rational* solutions to the equation?

254. $2 + 2 = 4$. Try applying AM-GM to a and b, then to c and d.

255. That first equation is quite ugly. A substitution for simplification might help. What substitution would give you a two-variable equation that isn't so ugly?

256. Can you create a series with radicals that looks a lot like the given series, but telescopes?

257. We want the maximum of a product of three x's and a $4 - x$, so AM-GM is a natural candidate. But, we'll want our AM to have a constant sum, so that we can say our GM is at most some constant. Find four nonnegative expressions that have a constant sum and a GM that involves $x^3(4 - x)$.

258. Be careful! After simplifying the logarithm and finding what value of x minimizes the argument, check if $f(x)$ is defined for that value of x.

259. Play a bit. Find a_3 in terms of a_1 and a_0.

260. Start with your intuition. What number is the answer probably near?

261. "Smallest real number" means we probably have to build an inequality of some sort. How can we get an inequality from the equation?

262. Write out some terms.

263. Find a factor of $p(x)$.

264. Let $x = \frac{a+b}{a-b}$. We're given information about $a^2 + b^2$. What should we do with x?

265. Let the roots be a, $a + d$, $a + 2d$, and $a + 3d$. Use Vieta's Formulas to relate a and d. Vieta was helpful once; try again!

266. We know how to handle $x^3 + y^3$ and $x^3 - y^3$.

267. Find an equation in terms of a, b, and c that must hold if $x - 1$ divides $f(x) - 1$. What equation must be true if $x - 1$ divides the quotient of $f(x) - 1$ and $x - 1$?

268. What is the nth term in terms of n? Simplify it as much as possible.

269. The equation is a quadratic in y. If y is real, what must be true about the discriminant?

270. For what values of k does the graph of $y = kx$ intersect the ellipse?

271. In your experimenting, you should have done a lot of dividing by 2. Also, 1023 is very close to a power of 2. All those 2s might suggest thinking about the binary representation of n.

272. Compare the series to a very similar series that you can evaluate.

273. Let $p(x) = p_n x^n + p_{n-1}x^{n-1} + \cdots + p_1 x + p_0$ and $q(x) = q_n x^n + q_{n-1}x^{n-1} + \cdots + q_1 x + q_0$. What is the average of the roots of $p(q(x))$ in terms of the coefficients of these polynomials?

274. Can you separate x and $1 - x$, as in the separation section?

275. After writing $A + B + C = 1$ in terms of a, b, and c, you should get a cubic polynomial equal to 0. To find factors of this polynomial, experiment by choosing values for b and c and factoring the resulting polynomial in a. After trying a few, you should have a guess as to what the factors are. Find the product of these factors to confirm your guess.

276. Regroup the terms in a way that lets you factor piece by piece.

277. What is $f(t)$?

278. Don't forget that x can be negative. Handle $x < 0$ and $x \geq 0$ separately.

279. $1 + \frac{1}{n} = \frac{n+1}{n}$.

280. Imagine you wrote out the given recursion for every value of n from 2 to 1986 (remember, $f(1986) = f(1)$). What could you do with these equations to get an equation involving the desired sum?

281. We have a sum on each side. That's almost Cauchy! We need another sum on one side and a square of a sum on the other. What can we multiply the inequality by to achieve this?

282. How can you get $ab + bc + ca$ terms from the equation $a + b + c = 3$?

283. Write the expression as a difference of squares.

284. If a polynomial is divisible by $(x - c)g(x)$, is it divisible by $g(x)$?

285. What graph is produced when the graph of $y = f(x)$ is shifted 6 to the right, and then the result scaled horizontally by a factor of $1/2$?

286. Let $f(x) = a_5 x^5 + a_4 x^4 + a_3 x^3 + a_2 x^2 + a_1 x + a_0$. What does the fact that $f(x)$ is divisible by x^3 tell us about some of the coefficients of $f(x)$?

287. It's obviously divisible by $x - 1$. What is the quotient?

288. Factor those quadratics.

289. After factoring, you should have a bunch of powers of x. Don't forget that $0 \leq x \leq 1$.

290. Isolate the fractional part of x. What does the result tell us about $\lfloor x \rfloor$?

291. How can we get an equation with $x + y + z$ in it from the three given equations?

292. The second equation is quadratic in xy.

293. What do Vieta's Formulas tell us about $x_1 + x_2$ and $x_1 x_2$? Use these to find expressions for $x_1^3 + x_2^3$ and $x_1^3 x_2^3$.

294. Write the left sides as the sum or difference of two fractions. Does the result resemble a system of two-variable linear equations?

295. Let a be the first term and d be the common difference of the arithmetic sequence. Once you have an equation, solve for a. Remember that the sequence consists of *positive integers*.

296. How can you turn the y in $x + y + z$ into two y's?

297. We seek the product $(r_1^2 - 2)(r_2^2 - 2)(r_3^2 - 2)(r_4^2 - 2)(r_5^2 - 2)$. What does this look like if we replace the 2 with a variable?

298. Rewrite the equation in a form that allows you to use what you know about the monotonic nature of exponential functions to solve the problem.

299. Use Vieta's Formulas and the fact that γ is a root of $x^2 + Px + Q$ to express $(\alpha - \gamma)(\beta - \gamma)$ in the form $c\gamma + d$, where c and d are constants.

300. Can you factor the denominator? (Yes.)

301. Let $u = a + bi$. Find v and $\bar{u}$ in terms of a and b.

302. It sure would be nice if the logarithms had the same base instead of the same argument.

303. What does the given equation tell us about how the sum and the product of the would-be roots are related?

304. For what values of x_3 is $x_4 = x_3$?

305. For the second ratio, rearrange the desired ratio equation to have the x and y terms all in one ratio, and a and b in the other. Let k equal each of the given ratios.

306. We know how to handle rational function inequalities. Turn this inequality into one with a substitution.

307. Get rid of the floor function. In terms of x, what must $\lfloor x \rfloor$ be greater than?

308. We have $f(f(x)) = x$ for *all* x. Choose a convenient x.

309. After some algebra, you should have terms of the form $x_i^{19} x_j^{93} + x_j^{19} x_i^{93}$ on the greater side and terms of the form $x_i^{20} x_j^{92} + x_j^{20} x_i^{92}$ on the lesser side. Focus on a single i, j pair. Rearrange, factor.

310. Which terms in the expansion affect the first three digits after the decimal point?

311. Sums and nasty denominators. Try Cauchy. You'll get another nasty denominator, but one nasty denominator on the greater side might be better than three.

312. How can you get rid of that nasty denominator with those cube roots?

313. Start with fewer numbers; use 3 or 4 rather than 21. Write out the products (don't compute them), and look for a clever way to sum them.

314. Suppose we have a function such that $h(x) = h(1 - x)$. About what value of x is such a function symmetric? What other class of functions is symmetric about some specific value of x? How might h be related to one of these functions?

315. The graph of $xy = 1$ is the result of rotating the graph of what equation 45 degrees counterclockwise? What are the foci of the latter equation?

316. We took care of a difference of square roots by multiplying by a sum of square roots. What will we have to multiply a difference of cube roots by in order to get an expression without cube roots?

317. For what values of x does $x^3 + x = 0$?

318. Suppose (a, b) is on the graph of $y = \log_{10} x$. Express both coordinates of the corresponding point on G' in terms of a. Use these coordinates to express y in terms of x to produce an equation whose graph is G'.

319. Is it possible for $f(0)$ to be even?

320. Try letting $a = x/(2y)$, $b = y/(2z)$, and $c = z/(2x)$.

321. Multiply out the left side. Can you minimize groups of terms that result? Start with $\frac{1}{xyz}$.

322. After using a logarithm identity and the Pythagorean Theorem to write the logarithmic equation as an equation without logarithms in terms of just a and b, Simon would know how to finish the problem. Don't forget that $a \le b$.

323. Can you express $f(2002)$ in terms of $f(n)$, where $n < 2002$?

324. We can't speak of the roots of "polynomial $P(x)$ $\frac{1}{x}$," because this is not a polynomial. But $xP(x) - 1$ is.

325. The expression $\binom{n}{r}$ equals $\frac{n!}{r!(n-r)!}$, where $n! = n(n-1)(n-2)\cdots(1)$. Let the binomial coefficients be $\binom{n}{k-1}$, $\binom{n}{k}$, and $\binom{n}{k+1}$. Write some equations with the ratio information.

326. The sum consists of the coefficients of the even degree terms in the expansion of $(1 + x)^{99}$. How can we get rid of the odd degree terms?

327. What are the y-coordinates of the endpoints of the latus rectum?

328. What is likely to be the condition that will provide the desired minimum? Try using this to figure out what tools you need to build an inequality that you can use to prove the expression cannot be smaller than this minimum.

329. You have the product of two roots, so what can you immediately find by looking at the coefficients of the polynomial?

330. Let z be a complex number (point in the complex plane), and let $f(z)$ be its image. How can we write the distance between these two points as an algebraic expression?

331. That first equation resembles an equation whose graph is a circle. Experiment with values of x in different ranges. For example, what happens if $\lfloor x \rfloor = 0$?

332. Can you find any squares near $n^2 - 10n + 29$?

333. Confirm your pattern by letting $a_2 = n$. Use $a_9 + a_{10}$ to find n.

334. How is a real number related to its conjugate?

335. What can you sometimes do to get complicated expressions out of exponents?

336. Find $g^{-1}(x)$ in terms of a, b, c, and d.

337. Build a polynomial with roots a, b, c, and d. What is interesting about the coefficients of this polynomial?

338. Factor!

339. Write the whole series in summation notation. (And don't use $\cdots$ to express each term.)

340. In the previous problem, we used AM-GM for 2 variables to prove it for 4. How do you think we'll get to 8? To 16? To 2^k?

341. Work backwards; square and rearrange. Does a familiar inequality appear?

342. Rearrange the inequality into various different forms. Try squaring some of the radicals away. Don't give up! Keep manipulating it until you see some expressions you can work with.

343. Let $z + \frac{1}{y} = a$. Can you combine this with the final two given equations to produce xyz terms?

344. What inequality helps with finding the minimum value of a sum?

345. The $a - b$ in the denominator is annoying. How can you rewrite the expression so that there's an $a - b$ in a numerator, that you can then cancel with the denominator when you apply AM-GM?

346. Find an equation whose graph is the storm at time t. (Your equation should have t in it.) Where is the car at time t?

347. Find a perfect square or perfect cube near this number.

348. Those denominators are annoying.

349. Isolate $|y|$. How can you most easily examine the behavior of the other side of the equation after doing so?

350. Show that $x^2 + y^2$ can be expressed in terms of p and s. What about $x^3 + y^3$? $x^4 + y^4$? What, essentially, are you doing as you go from one power to the next?

351. Tackle it in two parts. Handle the left end by letting $m = x_1/x_2$.

352. Explain why $(a^3 - 1)(a^2 - 1) \geq 0$.

353. Simon would first substitute variables for $\log_{10} x$, $\log_{10} y$, and $\log_{10} z$.

354. Squared polynomials are often easier to work with. See if you can find a way to express the given polynomial as a linear combination of squared polynomials.

355. How many x's, y's, and z's do you need to produce a GM of xy^2z^3?

356. Find three roots of $P(Q(x))$.

357. How can you tell whether or not a quadratic equation has any real solutions?

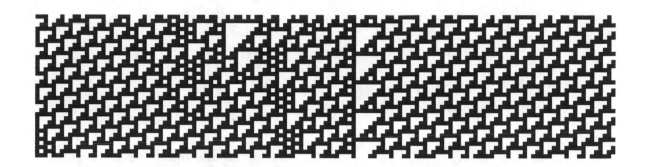

Index

WLOG, 598
Wolfram, Stephen, xviii

x-intercept, 32

y-intercept, 32

$\mathbb{Z}$, 230
$\bar{z}$, 58
zero (of a polynomial), 193
zero (of a quadratic), 79
zero polynomial, 162

www.artofproblemsolving.com

The Art of Problem Solving (AoPS) is:

• Books

For over 16 years, the classic *Art of Problem Solving* books have been used by students as a resource for the American Mathematics Competitions and other national and local math events.

> *Every school should have this in their math library.*
> – Paul Zeitz, past coach of the U.S. International Mathematical Olympiad team

Our Introduction series of textbooks—covering Algebra, Counting & Probability, Geometry, and Number Theory—constitutes a complete curriculum for outstanding math students in grades 6-10. Our Intermediate series of textbooks will form a complete curriculum for outstanding math students in grades 8-12. Intermediate books covering Counting & Probability and Algebra are available now.

> *The new book [Introduction to Counting & Probability] is great. I have started to use it in my classes on a regular basis. I can see the improvement in my kids over just a short period.*
> – Jeff Boyd, four time National MATHCOUNTS Championship coach from Texas

• Classes

The Art of Problem Solving offers online classes on topics such as number theory, counting, geometry, algebra, and more at beginning, intermediate, and Olympiad levels.

> *All the children were very engaged. It's the best use of technology I have ever seen.*
> – Mary Fay-Zenk, coach of National Champion California MATHCOUNTS teams

• Online Community

As of June 2010, the Art of Problem Solving Forum has over 70,000 members who have posted over 1,700,000 messages on our discussion board. Members can also participate in any of our free "Math Jams."

> *I'd just like to thank the coordinators of this site for taking the time to set it up... I think this is a great site, and I bet just about anyone else here would say the same...*
> – AoPS Community Member

• Resources

We have links to summer programs, book resources, problem sources, national and local competitions, scholarship listings, a math wiki, and a LATEX tutorial.

> *I'd like to commend you on your wonderful site. It's informative, welcoming, and supportive of the math community. I wish it had been around when I was growing up.*
> – AoPS Community Member

• . . . and more!

Membership is **FREE**! Come join the Art of Problem Solving community today!